Graduate Texts in Mathematics **185**

T0207476

Graduate Texts in Mathematics

(continued after index)

David A. Cox
John Little
Donal O'Shea

Using Algebraic Geometry

Second Edition

With 24 Illustrations

David Cox
Department of Mathematics
Amherst College
Amherst, MA 01002-5000
USA
dac@cs.amherst.edu

Donal O'Shea
Department of Mathematics
Mount Holyoke College
South Hadley, MA 01075
USA
doshea@mtholyoke.edu

John Little
Department of Mathematics
College of the Holy Cross
Worcester, MA 01610
USA
little@mathcs.holycross.edu

Mathematics Subject Classification (2000): 13Pxx, 13-01, 14-01, 14Qxx

Library of Congress Cataloging-in-Publication Data

Little, John B.
     Using algebraic geometry / John Little, David A. Cox, Donal O'Shea.
        p.   cm. — (Graduate texts in mathematics ; v. 185)
     Cox's name appears first on the earlier edition.
     Includes bibliographical references and index.
     ISBN 0-387-20706-6 (alk. paper) – ISBN 0-387-20733-3 (pbk. : alk. paper)
        1. Geometry, Algebraic.   I. Cox, David A.   II. O'Shea, Donal.   III. Title.   IV. Graduate
     texts in mathematics ; 185.
     QA564.C6883   2004
     516.3′5—dc22   2003070363

ISBN 0-387-20706-6 (hardcover)   Printed on acid-free paper.
ISBN 0-387-20733-3 (softcover)

Printed in the United States of America.   (EB/ING)

9 8 7 6 5 4 3 2 1   SPIN 10947098 (hardcover)   SPIN 10946961 (softcover)

springeronline.com

Preface to the Second Edition

Since the first edition of *Using Algebraic Geometry* was published in 1998, the field of computational algebraic geometry and its applications has developed rapidly. Many new results concerning topics we discussed have appeared. Moreover, a number of new introductory texts have been published. Our goals in this revision have been to update the references to reflect these additions to the literature, to add discussions of some new material, to improve some of the proofs, and to fix typographical errors. The major changes in this edition are the following:

- A unified discussion of how matrices can be used to specify monomial orders in §2 of Chapter 1.

- A rewritten presentation of the Mora normal form algorithm in §3 of Chapter 4 and the division of §4 into two sections.

- The addition of two sections in Chapter 8: §4 introduces the Gröbner fan of an ideal and §5 discusses the Gröbner Walk basis conversion algorithm.

- The replacement of §5 of Chapter 9 by a new Chapter 10 on the theory of order domains, associated codes, and the Berlekamp-Massey-Sakata decoding algorithm. The one-point geometric Goppa codes studied in the first edition are special cases of this construction.

- The Maple code has been updated and *Macaulay* has been replaced by *Macaulay 2*.

We would like to thank the many readers who helped us find typographical errors in the first edition. Special thanks go to Rainer Steinwandt for his heroic efforts. We also want to give particular thanks to Rex Agacy, Alicia Dickenstein, Dan Grayson, Serkan Hoşten, Christoph Kögl, Nick Loehr, Jim Madden, Mike O'Sullivan, Lyle Ramshaw, Hal Schenck, Hans Sterk, Mike Stillman, Bernd Sturmfels, and Irena Swanson for their help.

August, 2004

David Cox
John Little
Donal O'Shea

Preface to the First Edition

In recent years, the discovery of new algorithms for dealing with polynomial equations, coupled with their implementation on inexpensive yet fast computers, has sparked a minor revolution in the study and practice of algebraic geometry. These algorithmic methods and techniques have also given rise to some exciting new applications of algebraic geometry.

One of the goals of *Using Algebraic Geometry* is to illustrate the many uses of algebraic geometry and to highlight the more recent applications of Gröbner bases and resultants. In order to do this, we also provide an introduction to some algebraic objects and techniques more advanced than one typically encounters in a first course, but which are nonetheless of great utility. Finally, we wanted to write a book which would be accessible to nonspecialists and to readers with a diverse range of backgrounds.

To keep the book reasonably short, we often have to refer to basic results in algebraic geometry without proof, although complete references are given. For readers learning algebraic geometry and Gröbner bases for the first time, we would recommend that they read this book in conjunction with one of the following introductions to these subjects:

- *Introduction to Gröbner Bases*, by Adams and Loustaunau [AL]

- *Gröbner Bases*, by Becker and Weispfenning [BW]

- *Ideals, Varieties and Algorithms*, by Cox, Little and O'Shea [CLO]

We have tried, on the other hand, to keep the exposition self-contained outside of references to these introductory texts. We have made no effort at completeness, and have not hesitated to point the reader to the research literature for more information.

Later in the preface we will give a brief summary of what our book covers.

The Level of the Text

This book is written at the graduate level and hence assumes the reader knows the material covered in standard undergraduate courses, including abstract algebra.

But because the text is intended for beginning graduate students, it does not require graduate algebra, and in particular, the book does not assume that the reader is familiar with modules. Being a graduate text, *Using Algebraic Geometry* covers more sophisticated topics and has a denser exposition than most undergraduate texts, including our previous book [CLO].

However, it is possible to use this book at the undergraduate level, provided proper precautions are taken. With the exception of the first two chapters, we found that most undergraduates needed help reading preliminary versions of the text. That said, if one supplements the other chapters with simpler exercises and fuller explanations, many of the applications we cover make good topics for an upper-level undergraduate applied algebra course. Similarly, the book could also be used for reading courses or senior theses at this level. We hope that our book will encourage instructors to find creative ways for involving advanced undergraduates in this wonderful mathematics.

How to Use the Text

The book covers a variety of topics, which can be grouped roughly as follows:

- Chapters 1 and 2: Gröbner bases, including basic definitions, algorithms and theorems, together with solving equations, eigenvalue methods, and solutions over $\mathbb{R}$.

- Chapters 3 and 7: Resultants, including multipolynomial and sparse resultants as well as their relation to polytopes, mixed volumes, toric varieties, and solving equations.

- Chapters 4, 5 and 6: Commutative algebra, including local rings, standard bases, modules, syzygies, free resolutions, Hilbert functions and geometric applications.

- Chapters 8 and 9: Applications, including integer programming, combinatorics, polynomial splines, and algebraic coding theory.

One unusual feature of the book's organization is the early introduction of resultants in Chapter 3. This is because there are many applications where resultant methods are much more efficient than Gröbner basis methods. While Gröbner basis methods have had a greater theoretical impact on algebraic geometry, resultants appear to have an advantage when it comes to practical applications. There is also some lovely mathematics connected with resultants.

There is a large degree of independence among most chapters of the book. This implies that there are many ways the book can be used in teaching a course. Since there is more material than can be covered in one semester, some choices are necessary. Here are three examples of how to structure a course using our text.

- Solving Equations. This course would focus on the use of Gröbner bases and resultants to solve systems of polynomial equations. Chapters 1, 2, 3 and 7 would form the heart of the course. Special emphasis would be placed on §5 of Chapter 2, §5 and §6 of Chapter 3, and §6 of Chapter 7. Optional topics would include §1 and §2 of Chapter 4, which discuss multiplicities.

- Commutative Algebra. Here, the focus would be on topics from classical commutative algebra. The course would follow Chapters 1, 2, 4, 5 and 6, skipping only those parts of §2 of Chapter 4 which deal with resultants. The final section of Chapter 6 is a nice ending point for the course.

- Applications. A course concentrating on applications would cover integer programming, combinatorics, splines and coding theory. After a quick trip through Chapters 1 and 2, the main focus would be Chapters 8 and 9. Chapter 8 uses some ideas about polytopes from §1 of Chapter 7, and modules appear naturally in Chapters 8 and 9. Hence the first two sections of Chapter 5 would need to be covered. Also, Chapters 8 and 9 use Hilbert functions, which can be found in either Chapter 6 of this book or Chapter 9 of [CLO].

We want to emphasize that these are only three of many ways of using the text. We would be very interested in hearing from instructors who have found other paths through the book.

References

References to the bibliography at the end of the book are by the first three letters of the author's last name (e.g., [Hil] for Hilbert), with numbers for multiple papers by the same author (e.g., [Mac1] for the first paper by Macaulay). When there is more than one author, the first letters of the authors' last names are used (e.g., [AM] for Atiyah and Macdonald), and when several sets of authors have the same initials, other letters are used to distinguish them (e.g., [BoF] is by Bonnesen and Fenchel, while [BuF] is by Burden and Faires).

The bibliography lists books alphabetically by the full author's name, followed (if applicable) by any coauthors. This means, for instance, that [BS] by Billera and Sturmfels is listed before [Bla] by Blahut.

Comments and Corrections

We encourage comments, criticism, and corrections. Please send them to any of us:

David Cox	dac@cs.amherst.edu
John Little	little@math.holycross.edu
Don O'Shea	doshea@mhc.mtholyoke.edu

For each new typo or error, we will pay $1 to the first person who reports it to us. We also encourage readers to check out the web site for *Using Algebraic Geometry*, which is at

http://www.cs.amherst.edu/~dac/uag.html

This site includes updates and errata sheets, as well as links to other sites of interest.

Acknowledgments

We would like to thank everyone who sent us comments on initial drafts of the manuscript. We are especially grateful to thank Susan Colley, Alicia Dickenstein, Ioannis Emiris, Tom Garrity, Pat Fitzpatrick, Gert-Martin Greuel, Paul Pedersen, Maurice Rojas, Jerry Shurman, Michael Singer, Michael Stanfield, Bernd Sturmfels (and students), Moss Sweedler (and students), and Wiland Schmale for especially detailed comments and criticism.

We also gratefully acknowledge the support provided by National Science Foundation grant DUE-9666132, and the help and advice afforded by the members of our Advisory Board: Susan Colley, Keith Devlin, Arnie Ostebee, Bernd Sturmfels, and Jim White.

November, 1997   *David Cox*
                                                       John Little
                                                     Donal O'Shea

Contents

Chapter 1

Introduction

Algebraic geometry is the study of geometric objects defined by polynomial equations, using algebraic means. Its roots go back to Descartes' introduction of coordinates to describe points in Euclidean space and his idea of describing curves and surfaces by algebraic equations. Over the long history of the subject, both powerful general theories and detailed knowledge of many specific examples have been developed. Recently, with the development of computer algebra systems and the discovery (or rediscovery) of algorithmic approaches to many of the basic computations, the techniques of algebraic geometry have also found significant applications, for example in geometric design, combinatorics, integer programming, coding theory, and robotics. Our goal in *Using Algebraic Geometry* is to survey these algorithmic approaches and many of their applications.

For the convenience of the reader, in this introductory chapter we will first recall the basic algebraic structure of *ideals* in polynomial rings. In §2 and §3 we will present a rapid summary of the *Gröbner basis algorithms* developed by Buchberger for computations in polynomial rings, with several worked out examples. Finally, in §4 we will recall the geometric notion of an *affine algebraic variety*, the simplest type of geometric object defined by polynomial equations. The topics in §1, §2, and §3 are the common prerequisites for all of the following chapters. §4 gives the geometric context for the algebra from the earlier sections. We will make use of this language at many points. If these topics are familiar, you may wish to proceed directly to the later material and refer back to this introduction as needed.

§1 Polynomials and Ideals

To begin, we will recall some terminology. A *monomial* in a collection of variables $x_1, \ldots, x_n$ is a product

$$(1.1) \qquad x_1^{\alpha_1} x_2^{\alpha_2} \cdots x_n^{\alpha_n}$$

1

where the α_i are non-negative integers. To abbreviate, we will sometimes rewrite (1.1) as x^α where $\alpha = (\alpha_1, \ldots, \alpha_n)$ is the vector of exponents in the monomial. The *total degree* of a monomial x^α is the sum of the exponents: $\alpha_1 + \cdots + \alpha_n$. We will often denote the total degree of the monomial x^α by $|\alpha|$. For instance $x_1^3 x_2^2 x_4$ is a monomial of total degree 6 in the variables x_1, x_2, x_3, x_4, since $\alpha = (3, 2, 0, 1)$ and $|\alpha| = 6$.

If k is any field, we can form finite linear combinations of monomials with coefficients in k. The resulting objects are known as *polynomials* in $x_1, \ldots, x_n$. We will also use the word *term* on occasion to refer to a product of a nonzero element of k and a monomial appearing in a polynomial. Thus, a general polynomial in the variables $x_1, \ldots, x_n$ with coefficients in k has the form

$$f = \sum_\alpha c_\alpha x^\alpha,$$

where $c_\alpha \in k$ for each α, and there are only finitely many terms $c_\alpha x^\alpha$ in the sum. For example, taking k to be the field $\mathbb{Q}$ of rational numbers, and denoting the variables by x, y, z rather than using subscripts,

$$(1.2) \qquad p = x^2 + \tfrac{1}{2} y^2 z - z - 1$$

is a polynomial containing four terms.

In most of our examples, the field of coefficients will be either $\mathbb{Q}$, the field of real numbers, $\mathbb{R}$, or the field of complex numbers, $\mathbb{C}$. Polynomials over finite fields will also be introduced in Chapter 9. We will denote by $k[x_1, \ldots, x_n]$ the collection of all polynomials in $x_1, \ldots, x_n$ with coefficients in k. Polynomials in $k[x_1, \ldots, x_n]$ can be added and multiplied as usual, so $k[x_1, \ldots, x_n]$ has the structure of a *commutative ring* (with identity). However, only nonzero constant polynomials have multiplicative inverses in $k[x_1, \ldots, x_n]$, so $k[x_1, \ldots, x_n]$ is not a field. However, the set of *rational functions* $\{f/g : f, g \in k[x_1, \ldots, x_n], g \neq 0\}$ is a field, denoted $k(x_1, \ldots, x_n)$.

A polynomial f is said to be *homogeneous* if all the monomials appearing in it with nonzero coefficients have *the same* total degree. For instance, $f = 4x^3 + 5xy^2 - z^3$ is a homogeneous polynomial of total degree 3 in $\mathbb{Q}[x, y, z]$, while $g = 4x^3 + 5xy^2 - z^6$ is not homogeneous. When we study resultants in Chapter 3, homogeneous polynomials will play an important role.

Given a collection of polynomials, $f_1, \ldots, f_s \in k[x_1, \ldots, x_n]$, we can consider all polynomials which can be built up from these by multiplication by arbitrary polynomials and by taking sums.

(1.3) Definition. Let $f_1, \ldots, f_s \in k[x_1, \ldots, x_n]$. We let $\langle f_1, \ldots, f_s \rangle$ denote the collection

$$\langle f_1, \ldots, f_s \rangle = \{p_1 f_1 + \cdots + p_s f_s : p_i \in k[x_1, \ldots, x_n] \text{ for } i = 1, \ldots, s\}.$$

For example, consider the polynomial p from (1.2) above and the two polynomials

$$f_1 = x^2 + z^2 - 1$$
$$f_2 = x^2 + y^2 + (z - 1)^2 - 4.$$

We have

(1.4)
$$p = x^2 + \tfrac{1}{2}y^2 z - z - 1$$
$$= (-\tfrac{1}{2}z + 1)(x^2 + z^2 - 1) + (\tfrac{1}{2}z)(x^2 + y^2 + (z - 1)^2 - 4).$$

This shows $p \in \langle f_1, f_2 \rangle$.

Exercise 1.
a. Show that $x^2 \in \langle x - y^2, xy \rangle$ in $k[x, y]$ (k any field).
b. Show that $\langle x - y^2, xy, y^2 \rangle = \langle x, y^2 \rangle$.
c. Is $\langle x - y^2, xy \rangle = \langle x^2, xy \rangle$? Why or why not?

Exercise 2. Show that $\langle f_1, \ldots, f_s \rangle$ is closed under sums in $k[x_1, \ldots, x_n]$. Also show that if $f \in \langle f_1, \ldots, f_s \rangle$, and $p \in k[x_1, \ldots, x_n]$ is an arbitrary polynomial, then $p \cdot f \in \langle f_1, \ldots, f_s \rangle$.

The two properties in Exercise 2 are the defining properties of *ideals* in the ring $k[x_1, \ldots, x_n]$.

(1.5) Definition. Let $I \subset k[x_1, \ldots, x_n]$ be a non-empty subset. I is said to be an *ideal* if
a. $f + g \in I$ whenever $f \in I$ and $g \in I$, and
b. $pf \in I$ whenever $f \in I$, and $p \in k[x_1, \ldots, x_n]$ is an arbitrary polynomial.

Thus $\langle f_1, \ldots, f_s \rangle$ is an ideal by Exercise 2. We will call it the *ideal generated by* $f_1, \ldots, f_s$ because it has the following property.

Exercise 3. Show that $\langle f_1, \ldots, f_s \rangle$ is the *smallest* ideal in $k[x_1, \ldots, x_n]$ containing $f_1, \ldots, f_s$, in the sense that if J is any ideal containing $f_1, \ldots, f_s$, then $\langle f_1, \ldots, f_s \rangle \subset J$.

Exercise 4. Using Exercise 3, formulate and prove a general criterion for equality of ideals $I = \langle f_1, \ldots, f_s \rangle$ and $J = \langle g_1, \ldots, g_t \rangle$ in $k[x_1, \ldots, x_n]$. How does your statement relate to what you did in part b of Exercise 1?

Given an ideal, or several ideals, in $k[x_1, \ldots, x_n]$, there are a number of algebraic constructions that yield other ideals. One of the most important of these for geometry is the following.

(1.6) Definition. Let $I \subset k[x_1, \ldots, x_n]$ be an ideal. The *radical of I* is the set

$$\sqrt{I} = \{g \in k[x_1, \ldots, x_n] : g^m \in I \text{ for some } m \geq 1\}.$$

An ideal I is said to be a *radical ideal* if $\sqrt{I} = I$.

For instance,

$$x + y \in \sqrt{\langle x^2 + 3xy, 3xy + y^2 \rangle}$$

in $\mathbb{Q}[x, y]$ since

$$(x + y)^3 = x(x^2 + 3xy) + y(3xy + y^2) \in \langle x^2 + 3xy, 3xy + y^2 \rangle.$$

Since each of the generators of the ideal $\langle x^2 + 3xy, 3xy + y^2 \rangle$ is homogeneous of degree 2, it is clear that $x + y \notin \langle x^2 + 3xy, 3xy + y^2 \rangle$. It follows that $\langle x^2 + 3xy, 3xy + y^2 \rangle$ is *not* a radical ideal.

Although it is not obvious from the definition, we have the following property of the radical.

- (Radical Ideal Property) For every ideal $I \subset k[x_1, \ldots, x_n]$, $\sqrt{I}$ is an ideal containing I.

See [CLO], Chapter 4, §2, for example. We will consider a number of other operations on ideals in the exercises.

One of the most important general facts about ideals in $k[x_1, \ldots, x_n]$ is known as the Hilbert Basis Theorem. In this context, a *basis* is another name for a generating set for an ideal.

- (Hilbert Basis Theorem) Every ideal I in $k[x_1, \ldots, x_n]$ has a *finite* generating set. In other words, given an ideal I, there exists a finite collection of polynomials $\{f_1, \ldots, f_s\} \subset k[x_1, \ldots, x_n]$ such that $I = \langle f_1, \ldots, f_s \rangle$.

For polynomials in one variable, this is a standard consequence of the one-variable polynomial division algorithm.

- (Division Algorithm in $k[x]$) Given two polynomials $f, g \in k[x]$, we can divide f by g, producing a unique quotient q and remainder r such that

$$f = qg + r,$$

and either $r = 0$, or r has degree strictly smaller than the degree of g.

See, for instance, [CLO], Chapter 1, §5. The consequences of this result for ideals in $k[x]$ are discussed in Exercise 6 below. For polynomials in several variables, the Hilbert Basis Theorem can be proved either as a byproduct of the theory of Gröbner bases to be reviewed in the next section (see [CLO], Chapter 2, §5), or inductively by showing that if every ideal in a ring R is finitely generated, then the same is true in the ring $R[x]$ (see [AL], Chapter 1, §1, or [BW], Chapter 4, §1).

ADDITIONAL EXERCISES FOR §1

Exercise 5. Show that $\langle y - x^2, z - x^3 \rangle = \langle z - xy, y - x^2 \rangle$ in $\mathbb{Q}[x, y, z]$.

Exercise 6. Let k be any field, and consider the polynomial ring in one variable, $k[x]$. In this exercise, you will give one proof that every ideal in $k[x]$ is finitely generated. In fact, every ideal $I \subset k[x]$ is generated by a single polynomial: $I = \langle g \rangle$ for some g. We may assume $I \neq \{0\}$ for there is nothing to prove in that case. Let g be a nonzero element in I of minimal degree. Show using the division algorithm that every f in I is divisible by g. Deduce that $I = \langle g \rangle$.

Exercise 7.
a. Let k be any field, and let n be any positive integer. Show that in $k[x]$, $\sqrt{\langle x^n \rangle} = \langle x \rangle$.
b. More generally, suppose that
$$p(x) = (x - a_1)^{e_1} \cdots (x - a_m)^{e_m}.$$
   What is $\sqrt{\langle p(x) \rangle}$?
c. Let $k = \mathbb{C}$, so that *every* polynomial in one variable factors as in b. What are the radical ideals in $\mathbb{C}[x]$?

Exercise 8. An ideal $I \subset k[x_1, \ldots, x_n]$ is said to be *prime* if whenever a product fg belongs to I, either $f \in I$, or $g \in I$ (or both).
a. Show that a prime ideal is radical.
b. What are the prime ideals in $\mathbb{C}[x]$? What about the prime ideals in $\mathbb{R}[x]$ or $\mathbb{Q}[x]$?

Exercise 9. An ideal $I \subset k[x_1, \ldots, x_n]$ is said to be *maximal* if there are no ideals J satisfying $I \subset J \subset k[x_1, \ldots, x_n]$ other than $J = I$ and $J = k[x_1, \ldots, x_n]$.
a. Show that $\langle x_1, x_2, \ldots, x_n \rangle$ is a maximal ideal in $k[x_1, \ldots, x_n]$.
b. More generally show that if $(a_1, \ldots, a_n)$ is any point in k^n, then the ideal $\langle x_1 - a_1, \ldots, x_n - a_n \rangle \subset k[x_1, \ldots, x_n]$ is maximal.
c. Show that $I = \langle x^2 + 1 \rangle$ is a maximal ideal in $\mathbb{R}[x]$. Is I maximal considered as an ideal in $\mathbb{C}[x]$?

Exercise 10. Let I be an ideal in $k[x_1, \ldots, x_n]$, let $\ell \geq 1$ be an integer, and let I_ℓ consist of the elements in I that do not depend on the first ℓ variables:
$$I_\ell = I \cap k[x_{\ell+1}, \ldots, x_n].$$
I_ℓ is called the ℓth *elimination ideal* of I.
a. For $I = \langle x^2 + y^2, x^2 - z^3 \rangle \subset k[x, y, z]$, show that $y^2 + z^3$ is in the first elimination ideal I_1.

b. Prove that I_ℓ is an ideal in the ring $k[x_{\ell+1}, \ldots, x_n]$.

Exercise 11. Let I, J be ideals in $k[x_1, \ldots, x_n]$, and define

$$I + J = \{f + g : f \in I, g \in J\}.$$

a. Show that $I + J$ is an ideal in $k[x_1, \ldots, x_n]$.
b. Show that $I + J$ is the smallest ideal containing $I \cup J$.
c. If $I = \langle f_1, \ldots, f_s \rangle$ and $J = \langle g_1, \ldots, g_t \rangle$, what is a finite generating set for $I + J$?

Exercise 12. Let I, J be ideals in $k[x_1, \ldots, x_n]$.
a. Show that $I \cap J$ is also an ideal in $k[x_1, \ldots, x_n]$.
b. Define IJ to be the smallest ideal containing all the products fg where $f \in I$, and $g \in J$. Show that $IJ \subset I \cap J$. Give an example where $IJ \neq I \cap J$.

Exercise 13. Let I, J be ideals in $k[x_1, \ldots, x_n]$, and define $I : J$ (called the *quotient ideal* of I by J) by

$$I : J = \{f \in k[x_1, \ldots, x_n] : fg \in I \text{ for all } g \in J\}.$$

a. Show that $I : J$ is an ideal in $k[x_1, \ldots, x_n]$.
b. Show that if $I \cap \langle h \rangle = \langle g_1, \ldots, g_t \rangle$ (so each g_i is divisible by h), then a basis for $I : \langle h \rangle$ is obtained by cancelling the factor of h from each g_i:

$$I : \langle h \rangle = \langle g_1/h, \ldots, g_t/h \rangle.$$

§2 Monomial Orders and Polynomial Division

The examples of ideals that we considered in §1 were artificially simple. In general, it can be difficult to determine by inspection or by trial and error whether a given polynomial $f \in k[x_1, \ldots, x_n]$ is an element of a given ideal $I = \langle f_1, \ldots, f_s \rangle$, or whether two ideals $I = \langle f_1, \ldots, f_s \rangle$ and $J = \langle g_1, \ldots, g_t \rangle$ are equal. In this section and the next one, we will consider a collection of algorithms that can be used to solve problems such as deciding ideal membership, deciding ideal equality, computing ideal intersections and quotients, and computing elimination ideals. See the exercises at the end of §3 for some examples.

The starting point for these algorithms is, in a sense, the polynomial division algorithm in $k[x]$ introduced at the end of §1. In Exercise 6 of §1, we saw that the division algorithm implies that every ideal $I \subset k[x]$ has the form $I = \langle g \rangle$ for some g. Hence, if $f \in k[x]$, we can also use division to determine whether $f \in I$.

Exercise 1. Let $I = \langle g \rangle$ in $k[x]$ and let $f \in k[x]$ be any polynomial. Let q, r be the unique quotient and remainder in the expression $f = qg + r$ produced by polynomial division. Show that $f \in I$ if and only if $r = 0$.

Exercise 2. Formulate and prove a criterion for equality of ideals $I_1 = \langle g_1 \rangle$ and $I_2 = \langle g_2 \rangle$ in $k[x]$ based on division.

Given the usefulness of division for polynomials in one variable, we may ask: Is there a corresponding notion for polynomials in several variables? The answer is *yes*, and to describe it, we need to begin by considering different ways to *order* the monomials appearing within a polynomial.

(2.1) Definition. A *monomial order* on $k[x_1, \ldots, x_n]$ is any relation $>$ on the set of monomials x^α in $k[x_1, \ldots, x_n]$ (or equivalently on the exponent vectors $\alpha \in \mathbb{Z}_{\geq 0}^n$) satisfying:

a. $>$ is a *total (linear) ordering* relation;
b. $>$ is *compatible with multiplication* in $k[x_1, \ldots, x_n]$, in the sense that if $x^\alpha > x^\beta$ and x^γ is any monomial, then $x^\alpha x^\gamma = x^{\alpha+\gamma} > x^{\beta+\gamma} = x^\beta x^\gamma$;
c. $>$ is a *well-ordering*. That is, every nonempty collection of monomials has a smallest element under $>$.

Condition a implies that the terms appearing within any polynomial f can be uniquely listed in increasing or decreasing order under $>$. Then condition b shows that that ordering does not change if we multiply f by a monomial x^γ. Finally, condition c is used to ensure that processes that work on collections of monomials, e.g., the collection of all monomials less than some fixed monomial x^α, will terminate in a finite number of steps.

The division algorithm in $k[x]$ makes use of a monomial order *implicitly*: when we divide g into f by hand, we always compare the leading term (the term of highest degree) in g with the leading term of the intermediate dividend. In fact there is no choice in the matter in this case.

Exercise 3. Show that the *only* monomial order on $k[x]$ is the *degree order* on monomials, given by

$$\cdots > x^{n+1} > x^n > \cdots > x^3 > x^2 > x > 1.$$

For polynomial rings in several variables, there are many choices of monomial orders. In writing the exponent vectors α and β in monomials x^α and x^β as ordered n-tuples, we implicitly set up an ordering on the variables x_i in $k[x_1, \ldots, x_n]$:

$$x_1 > x_2 > \cdots > x_n.$$

With this choice, there are still many ways to define monomial orders. Some of the most important are given in the following definitions.

(2.2) Definition (Lexicographic Order). Let x^α and x^β be monomials in $k[x_1, \ldots, x_n]$. We say $x^\alpha >_{lex} x^\beta$ if in the difference $\alpha - \beta \in \mathbb{Z}^n$, the leftmost nonzero entry is positive.

Lexicographic order is analogous to the ordering of words used in dictionaries.

(2.3) Definition (Graded Lexicographic Order). Let x^α and x^β be monomials in $k[x_1, \ldots, x_n]$. We say $x^\alpha >_{grlex} x^\beta$ if $\sum_{i=1}^n \alpha_i > \sum_{i=1}^n \beta_i$, or if $\sum_{i=1}^n \alpha_i = \sum_{i=1}^n \beta_i$, and $x^\alpha >_{lex} x^\beta$.

(2.4) Definition (Graded Reverse Lexicographic Order). Let x^α and x^β be monomials in $k[x_1, \ldots, x_n]$. We say $x^\alpha >_{grevlex} x^\beta$ if $\sum_{i=1}^n \alpha_i > \sum_{i=1}^n \beta_i$, or if $\sum_{i=1}^n \alpha_i = \sum_{i=1}^n \beta_i$, and in the difference $\alpha - \beta \in \mathbb{Z}^n$, the rightmost nonzero entry is negative.

For instance, in $k[x, y, z]$, with $x > y > z$, we have

(2.5) $$x^3 y^2 z >_{lex} x^2 y^6 z^{12}$$

since when we compute the difference of the exponent vectors:

$$(3, 2, 1) - (2, 6, 12) = (1, -4, -11),$$

the leftmost nonzero entry is positive. Similarly,

$$x^3 y^6 >_{lex} x^3 y^4 z$$

since in $(3, 6, 0) - (3, 4, 1) = (0, 2, -1)$, the leftmost nonzero entry is positive. Comparing the *lex* and *grevlex* orders shows that the results can be quite different. For instance, it is true that

$$x^2 y^6 z^{12} >_{grevlex} x^3 y^2 z.$$

Compare this with (2.5), which contains the same monomials. Indeed, *lex* and *grevlex* are *different* orderings even on the monomials of the same total degree in three or more variables, as we can see by considering pairs of monomials such as $x^2 y^2 z^2$ and $xy^4 z$. Since $(2, 2, 2) - (1, 4, 1) = (1, -2, 1)$,

$$x^2 y^2 z^2 >_{lex} xy^4 z.$$

On the other hand by Definition (2.4),

$$xy^4 z >_{grevlex} x^2 y^2 z^2.$$

Exercise 4. Show that $>_{lex}$, $>_{grlex}$, and $>_{grevlex}$ are monomial orders in $k[x_1, \ldots, x_n]$ according to Definition (2.1).

Exercise 5. Show that the monomials of a *fixed* total degree d in *two* variables $x > y$ are ordered in the same sequence by $>_{lex}$ and $>_{grevlex}$. Are these orderings the same on all of $k[x, y]$ though? Why or why not?

For future reference, we next discuss a general method for specifying monomial orders on $k[x_1, \ldots, x_n]$. We start from any $m \times n$ real matrix M and write the rows of M as $\mathbf{w}_1, \ldots, \mathbf{w}_m$. Then we can compare monomials x^α and x^β by first comparing their $\mathbf{w}_1$-weights $\alpha \cdot \mathbf{w}_1$ and $\alpha \cdot \mathbf{w}_1$. If $\alpha \cdot \mathbf{w}_1 > \beta \cdot \mathbf{w}_1$ or $\beta \cdot \mathbf{w}_1 > \alpha \cdot \mathbf{w}_1$, then we order the monomials accordingly. If $\alpha \cdot \mathbf{w}_1 = \beta \cdot \mathbf{w}_1$, then we continue to the later rows, breaking ties successively with the $\mathbf{w}_2$-weights, the $\mathbf{w}_3$-weights, and so on through the $\mathbf{w}_m$-weights. This process defines an order relation $>_M$. In symbols: $x^\alpha >_M x^\beta$ if there is an $\ell \leq m$ such that $\alpha \cdot \mathbf{w}_i = \beta \cdot \mathbf{w}_i$ for $i = 1, \ldots, \ell - 1$, but $\alpha \cdot \mathbf{w}_\ell > \beta \cdot \mathbf{w}_\ell$.

To obtain a total order by this construction, it must be true that $\ker(M) \cap \mathbb{Z}^n = \{0\}$. If the entries of M are rational numbers, then this property implies that $m \geq n$, and M has full rank n. The same construction also works for M with irrational entries, but there is a small subtlety concerning what notion of rank is appropriate in that case. See Exercise 9 below. To guarantee the well-ordering property of monomial orders, it is sufficient (although not necessary) to require that M have all entries nonnegative.

Exercise 6. All the monomial orders we have seen can be specified as $>_M$ orders for appropriate matrices M.

a. Show that the *lex* order with $x > y > z$ is defined by the identity matrix

$$M = \begin{pmatrix} 1 & 0 & 0 \\ 0 & 1 & 0 \\ 0 & 0 & 1 \end{pmatrix},$$

and similarly in $k[x_1, \ldots, x_n]$ for all $n \geq 1$.

b. Show that the *grevlex* order with $x > y > z$ is defined by either the matrix

$$M = \begin{pmatrix} 1 & 1 & 1 \\ 1 & 1 & 0 \\ 1 & 0 & 0 \end{pmatrix}$$

or the matrix

$$M' = \begin{pmatrix} 1 & 1 & 1 \\ 0 & 0 & -1 \\ 0 & -1 & 0 \end{pmatrix}$$

and similarly in $k[x_1, \ldots, x_n]$ for all $n \geq 1$. This example shows that matrices with negative entries can also define monomial orders.

c. The *grlex* order compares monomials first by total degree (weight vector $\mathbf{w}_1 = (1, 1, 1)$), then breaks ties by the *lex order*. This, together with part a, shows $>_{grlex} => >_M$ for the matrix

$$M = \begin{pmatrix} 1 & 1 & 1 \\ 1 & 0 & 0 \\ 0 & 1 & 0 \\ 0 & 0 & 1 \end{pmatrix}.$$

Show that we could also use

$$M' = \begin{pmatrix} 1 & 1 & 1 \\ 1 & 0 & 0 \\ 0 & 1 & 0 \end{pmatrix}.$$

That is, show that the last row in M is actually superfluous. (Hint: Making comparisons, when would we ever need to use the last row?)

d. One very common way to define a monomial order is to compare weights with respect to one vector first, then break ties with another standard order such as *grevlex*. We denote such an order by $>_{\mathbf{w}, grevlex}$. These weight orders are studied, for instance, in [CLO], Chapter 2, §4, Exercise 12. Suppose $\mathbf{w} = (2, 4, 7)$ and ties are broken by *grevlex* with $x > y > z$. To define this order, it is most natural to use

$$M = \begin{pmatrix} 2 & 4 & 7 \\ 1 & 1 & 1 \\ 1 & 1 & 0 \\ 1 & 0 & 0 \end{pmatrix}.$$

However, some computer algebra systems (e.g., Maple V, Release 5 and later versions with the Groebner package) require square weight matrices. Consider the two matrices obtained from M by deleting a row:

$$M' = \begin{pmatrix} 2 & 4 & 7 \\ 1 & 1 & 1 \\ 1 & 1 & 0 \end{pmatrix} \qquad M'' = \begin{pmatrix} 2 & 4 & 7 \\ 1 & 1 & 1 \\ 1 & 0 & 0 \end{pmatrix}.$$

Both have rank 3 so the condition $\ker(M) \cap \mathbb{Z}^3 = \{0\}$ is satisfied. Which matrix defines the $>_{\mathbf{w}, grevlex}$ order?

e. Let $m > n$. Given an $m \times n$ matrix M defining a monomial order $>_M$, describe a general method for picking an $n \times n$ submatrix M' of M to define the same order.

In Exercise 8 below, you will prove that $>_M$ defines a monomial order for any suitable matrix M. In fact, by a result of Robbiano (see [Rob]), the $>_M$ construction gives all monomial orders on $k[x_1, \dots, x_n]$.

We will use monomial orders in the following way. The natural generalization of the leading term (term of highest degree) in a polynomial in $k[x]$ is defined as follows. Picking any particular monomial order $>$ on $k[x_1, \dots, x_n]$, we consider the terms in $f = \sum_\alpha c_\alpha x^\alpha$. Then the *leading term* of f (with respect to $>$) is the product $c_\alpha x^\alpha$ where x^α is the *largest* monomial appearing in f in the ordering $>$. We will use the notation $\mathrm{LT}_>(f)$ for the leading term, or just $\mathrm{LT}(f)$ if there is no chance of confusion about which monomial order is being used. Furthermore, if $\mathrm{LT}(f) = cx^\alpha$, then $\mathrm{LC}(f) = c$ is the *leading coefficient* of f and $\mathrm{LM}(f) = x^\alpha$ is the *leading monomial*. Note that $\mathrm{LT}(0)$, $\mathrm{LC}(0)$, and $\mathrm{LM}(0)$ are undefined.

For example, consider $f = 3x^3y^2 + x^2yz^3$ in $\mathbb{Q}[x, y, z]$ (with variables ordered $x > y > z$ as usual). We have

$$\mathrm{LT}_{>_{lex}}(f) = 3x^3y^2$$

since $x^3y^2 >_{lex} x^2yz^3$. On the other hand

$$\mathrm{LT}_{>_{grevlex}}(f) = x^2yz^3$$

since the total degree of the second term is 6 and the total degree of the first is 5.

Monomial orders are used in a generalized division algorithm.

- (Division Algorithm in $k[x_1, \ldots, x_n]$) Fix any monomial order $>$ in $k[x_1, \ldots, x_n]$, and let $F = (f_1, \ldots, f_s)$ be an ordered s-tuple of polynomials in $k[x_1, \ldots, x_n]$. Then every $f \in k[x_1, \ldots, x_n]$ can be written as

$$(2.6) \qquad f = a_1 f_1 + \cdots + a_s f_s + r,$$

  where $a_i, r \in k[x_1, \ldots, x_n]$, for each i, $a_i f_i = 0$ or $\mathrm{LT}_>(f) \geq \mathrm{LT}_>(a_i f_i)$, and either $r = 0$, or r is a linear combination of monomials, none of which is divisible by any of $\mathrm{LT}_>(f_1), \ldots, \mathrm{LT}_>(f_s)$. We will call r a *remainder* of f on division by F.

In the particular algorithmic form of the division process given in [CLO], Chapter 2, §3, and [AL], Chapter 1, §5, the intermediate dividend is reduced at each step using the divisor f_i with the *smallest possible i* such that $\mathrm{LT}(f_i)$ divides the leading term of the intermediate dividend. A characterization of the expression (2.6) that is produced by this version of division can be found in Exercise 11 of Chapter 2, §3 of [CLO]. More general forms of division or polynomial reduction procedures are considered in [AL] and [BW], Chapter 5, §1.

You should note two differences between this statement and the division algorithm in $k[x]$. First, we are allowing the possibility of dividing f by an s-tuple of polynomials with $s > 1$. The reason for this is that we will usually want to think of the divisors f_i as generators for some particular ideal I, and ideals in $k[x_1, \ldots, x_n]$ for $n \geq 2$ might not be generated by any single polynomial. Second, although any algorithmic version of division, such as the one presented in Chapter 2 of [CLO], produces one particular expression of the form (2.6) for each ordered s-tuple F and each f, there are always *different* expressions of this form for a given f as well. Reordering

F or changing the monomial order can produce different a_i and r in some cases. See Exercise 7 below for some examples.

We will sometimes use the notation

$$r = \overline{f}^F$$

for a remainder on division by F.

Most computer algebra systems that have Gröbner basis packages provide implementations of some form of the division algorithm. However, in most cases the output of the division command is just the remainder $\overline{f}^F$, the quotients a_i are not saved or displayed, and an algorithm different from the one described in [CLO], Chapter 2, §3 may be used. For instance, the Maple Groebner package contains a function **normalf** which computes a remainder on division of a polynomial by any collection of polynomials. To use it, one must start by loading the **Groebner** package (just once in a session) with

with(Groebner);

The format for the **normalf** command is

normalf(f, F, torder);

where **f** is the dividend polynomial, **F** is the ordered list of divisors (in square brackets, separated by commas), and **torder** specifies the monomial order. For instance, to use the $>_{lex}$ order, enter **plex**, then in parentheses, separated by commas, list the variables in descending order. Similarly, to use the $>_{grevlex}$ order, enter **tdeg**, then in parentheses, separated by commas, list the variables in descending order. Let us consider dividing $f_1 = x^2y^2 - x$ and $f_2 = xy^3 + y$ into $f = x^3y^2 + 2xy^4$ using the *lex* order on $\mathbb{Q}[x, y]$ with $x > y$. The Maple commands

(2.7)
```
f := x^3*y^2 + 2*x*y^4;
F := [x^2*y^2 - x, x*y^3 + y];
normalf(f,F,plex(x,y));
```

will produce as output

(2.8) $x^2 - 2y^2$.

Thus the remainder is $\overline{f}^F = x^2 - 2y^2$. The **normalf** procedure uses the algorithmic form of division presented, for instance, in [CLO], Chapter 2, §3.

The **Groebner** package contains several additional ways to specify monomial orders, including one to construct $>_M$ for a square matrix M with positive integer entries. Hence it can be used to work with general monomial orders on $k[x_1, \dots, x_n]$. We will present a number of examples in later chapters.

ADDITIONAL EXERCISES FOR §2

Exercise 7.
a. Verify by hand that the remainder from (2.8) occurs in an expression

$$f = a_1 f_1 + a_2 f_2 + x^2 - 2y^2,$$

where $a_1 = x$, $a_2 = 2y$, and f_i are as in the discussion before (2.7).
b. Show that reordering the variables and changing the monomial order to tdeg(x,y) has no effect in (2.8).
c. What happens if you change F in (2.7) to

$$F = [x^2 y^2 - x^4, xy^3 - y^4]$$

and take $f = x^2 y^6$? Does changing the order of the variables make a difference now?
d. Now change F to

$$F = [x^2 y^2 - z^4, xy^3 - y^4],$$

take $f = x^2 y^6 + z^5$, and change the monomial order to plex(x,y,z). Also try *lcx* orders with the variables permuted and other monomial orders.

Exercise 8. Let M be an $m \times n$ real matrix with nonnegative entries. Assume that $\ker(M) \cap \mathbb{Z}^n = \{0\}$. Show that $>_M$ is a monomial order on $k[x_1, \ldots, x_n]$.

Exercise 9. Given $\mathbf{w} \in (\mathbb{R}^n)^+$ define $x^\alpha >_{\mathbf{w}} x^\beta$ if $\alpha \cdot \mathbf{w} > \beta \cdot \mathbf{w}$.
a. Give an example to show that $>_{\mathbf{w}}$ is not necessarily a monomial order on $k[x_1, \ldots, x_n]$.
b. With $n = 2$, let $\mathbf{w} = (1, \sqrt{2})$. Show that $>_{\mathbf{w}}$ is a monomial order on $k[x_1, x_2]$ in this case.
c. What property of the components of the vector $\mathbf{w} \in (\mathbb{R}^n)^+$ guarantees that $>_{\mathbf{w}}$ *does* define a monomial order on $k[x_1, \ldots, x_n]$? Prove your assertion. (Hint: See Exercise 11 of Chapter 2, §4 of [CLO].)

§3 Gröbner Bases

Since we now have a division algorithm in $k[x_1, \ldots, x_n]$ that seems to have many of the same features as the one-variable version, it is natural to ask if deciding whether a given $f \in k[x_1, \ldots, x_n]$ is a member of a given ideal $I = \langle f_1, \ldots, f_s \rangle$ can be done along the lines of Exercise 1 in §2, by computing the remainder on division. One direction is easy. Namely, from (2.6) it follows that if $r = \overline{f}^F = 0$ on dividing by $F = (f_1, \ldots, f_s)$, then $f = a_1 f_1 + \cdots + a_s f_s$. By definition then, $f \in \langle f_1, \ldots, f_s \rangle$. On the other hand, the following exercise shows that we are not guaranteed to get $\overline{f}^F = 0$ for every $f \in \langle f_1, \ldots, f_s \rangle$ if we use an arbitrary basis F for I.

Exercise 1. Recall from (1.4) that $p = x^2 + \frac{1}{2}y^2 z - z - 1$ *is* an element of the ideal $I = \langle x^2 + z^2 - 1, x^2 + y^2 + (z - 1)^2 - 4 \rangle$. Show, however, that the remainder on division of p by this generating set F is not zero. For instance, using $>_{lex}$, we get a remainder

$$\overline{p}^F = \tfrac{1}{2}y^2 z - z - z^2.$$

What went wrong here? From (2.6) and the fact that $f \in I$ in this case, it follows that the remainder is *also an element of* I. However, $\overline{p}^F$ is not zero because it contains terms that cannot be removed by division by these particular generators for I. The leading terms of $f_1 = x^2 + z^2 - 1$ and $f_2 = x^2 + y^2 + (z - 1)^2 - 4$ do not divide the leading term of $\overline{p}^F$. In order for division to produce zero remainders for all elements of I, we need to be able to remove *all* leading terms of elements of I using the leading terms of the divisors. That is the motivation for the following definition.

(3.1) Definition. Fix a monomial order $>$ on $k[x_1, \ldots, x_n]$, and let $I \subset k[x_1, \ldots, x_n]$ be an ideal. A *Gröbner basis* for I (with respect to $>$) is a finite collection of polynomials $G = \{g_1, \ldots, g_t\} \subset I$ with the property that for every nonzero $f \in I$, $\mathrm{LT}(f)$ is divisible by $\mathrm{LT}(g_i)$ for some i.

We will see in a moment (Exercise 3) that a Gröbner basis for I is indeed a basis for I, i.e., $I = \langle g_1, \ldots, g_t \rangle$. Of course, it must be proved that Gröbner bases *exist* for all I in $k[x_1, \ldots, x_n]$. This can be done in a non-constructive way by considering the ideal $\langle \mathrm{LT}(I) \rangle$ generated by the leading terms of all the elements in I (a *monomial ideal*). By a direct argument (Dickson's Lemma: see [CLO], Chapter 2, §4, or [BW], Chapter 4, §3, or [AL], Chapter 1, §4), or by the Hilbert Basis Theorem, the ideal $\langle \mathrm{LT}(I) \rangle$ has a finite generating set consisting of monomials $x^{\alpha(i)}$ for $i = 1, \ldots, t$. By the definition of $\langle \mathrm{LT}(I) \rangle$, there is an element $g_i \in I$ such that $\mathrm{LT}(g_i) = x^{\alpha(i)}$ for each $i = 1, \ldots, t$.

Exercise 2. Show that if $\langle \mathrm{LT}(I) \rangle = \langle x^{\alpha(1)}, \ldots, x^{\alpha(t)} \rangle$, and if $g_i \in I$ are polynomials such that $\mathrm{LT}(g_i) = x^{\alpha(i)}$ for each $i = 1, \ldots, t$, then $G = \{g_1, \ldots, g_t\}$ is a Gröbner basis for I.

Remainders computed by division with respect to a Gröbner basis are much better behaved than those computed with respect to arbitrary sets of divisors. For instance, we have the following results.

Exercise 3.
a. Show that if G is a Gröbner basis for I, then for any $f \in I$, the remainder on division of f by G (listed in any order) is zero.

b. Deduce that $I = \langle g_1, \ldots, g_t \rangle$ if $G = \{g_1, \ldots, g_t\}$ is a Gröbner basis for I. (If $I = \langle 0 \rangle$, then $G = \emptyset$ and we make the convention that $\langle \emptyset \rangle = \{0\}$.)

Exercise 4. If G is a Gröbner basis for an ideal I, and f is an arbitrary polynomial, show that if the algorithm of [CLO], Chapter 2, §3 is used, the remainder on division of f by G is independent of the ordering of G. Hint: If two different orderings of G are used, producing remainders r_1 and r_2, consider the difference $r_1 - r_2$.

Generalizing the result of Exercise 4, we also have the following important statement.

- (Uniqueness of Remainders) Fix a monomial order $>$ and let $I \subset k[x_1, \ldots, x_n]$ be an ideal. Division of $f \in k[x_1, \ldots, x_n]$ by a Gröbner basis for I produces an expression $f = g + r$ where $g \in I$ and no term in r is divisible by any element of $\mathrm{LT}(I)$. If $f = g' + r'$ is any other such expression, then $r = r'$.

See [CLO], Chapter 2, §6, [AL], Chapter 1, §6, or [BW], Chapter 5, §2. In other words, the remainder on division of f by a Gröbner basis for I is a uniquely determined *normal form* for f modulo I depending only on the choice of monomial order and not on the way the division is performed. Indeed, uniqueness of remainders gives another characterization of Gröbner bases.

More useful for many purposes than the existence proof for Gröbner bases above is an *algorithm*, due to Buchberger, that takes an arbitrary generating set $\{f_1, \ldots, f_s\}$ for I and produces a Gröbner basis G for I from it. This algorithm works by forming new elements of I using expressions guaranteed to cancel leading terms and uncover other possible leading terms, according to the following recipe.

(3.2) Definition. Let $f, g \in k[x_1, \ldots, x_n]$ be nonzero. Fix a monomial order and let

$$\mathrm{LT}(f) = cx^\alpha \qquad \text{and} \qquad \mathrm{LT}(g) = dx^\beta,$$

where $c, d \in k$. Let x^γ be the least common multiple of x^α and x^β. The *S-polynomial* of f and g, denoted $S(f, g)$, is the polynomial

$$S(f, g) = \frac{x^\gamma}{\mathrm{LT}(f)} \cdot f - \frac{x^\gamma}{\mathrm{LT}(g)} \cdot g.$$

Note that by definition $S(f, g) \in \langle f, g \rangle$. For example, with $f = x^3 y - 2x^2 y^2 + x$ and $g = 3x^4 - y$ in $\mathbb{Q}[x, y]$, and using $>_{lex}$, we have $x^\gamma = x^4 y$, and

$$S(f, g) = xf - (y/3)g = -2x^3 y^2 + x^2 + y^2/3.$$

In this case, the leading term of the S-polynomial is divisible by the leading term of f. We might consider taking the remainder on division by $F = (f, g)$ to uncover possible new leading terms of elements in $\langle f, g \rangle$. And indeed in this case we find that the remainder is

(3.3) $$\overline{S(f,g)}^F = -4x^2y^3 + x^2 + 2xy + y^2/3$$

and $\mathrm{LT}(\overline{S(f,g)}^F) = -4x^2y^3$ is divisible by neither $\mathrm{LT}(f)$ nor $\mathrm{LT}(g)$. An important result about this process of forming S-polynomial remainders is the following statement.

- (Buchberger's Criterion) A finite set $G = \{g_1, \ldots, g_t\}$ is a Gröbner basis of $I = \langle g_1, \ldots, g_t \rangle$ if and only if $\overline{S(g_i, g_j)}^G = 0$ for all pairs $i \neq j$.

See [CLO], Chapter 2, §7, [BW], Chapter 5, §3, or [AL], Chapter 1, §7. Using this criterion above, we obtain a very rudimentary procedure for producing a Gröbner basis of a given ideal.

- (Buchberger's Algorithm)

    Input: $F = (f_1, \ldots, f_s)$

    Output: a Gröbner basis $G = \{g_1, \ldots, g_t\}$ for $I = \langle F \rangle$, with $F \subset G$

    $G := F$

    REPEAT

    $\quad G' := G$

    $\quad$ FOR each pair $p \neq q$ in G' DO

    $\qquad S := \overline{S(p,q)}^{G'}$

    $\qquad$ IF $S \neq 0$ THEN $G := G \cup \{S\}$

    $\quad$ UNTIL $G = G'$

See [CLO], Chapter 2, §6, [BW], Chapter 5, §3, or [AL], Chapter 1, §7. For instance, in the example above we would adjoin $h = \overline{S(f,g)}^F$ from (3.3) to our set of polynomials. There are two new S-polynomials to consider now: $S(f, h)$ and $S(g, h)$. Their remainders on division by (f, g, h) would be computed and adjoined to the collection if they are nonzero. Then we would continue, forming new S-polynomials and remainders to determine whether further polynomials must be included.

Exercise 5. Carry out Buchberger's Algorithm on the example above, continuing from (3.3). (You may want to use a computer algebra system for this.)

In Maple, there is an implementation of a more sophisticated version of Buchberger's algorithm in the **Groebner** package. The relevant command is called `gbasis`, and the format is

$$\texttt{gbasis(F,torder);}$$

Here `F` is a list of polynomials and `torder` specifies the monomial order. See the description of the `normalf` command in §2 for more details. For instance, the commands

$$\texttt{F := [x\textasciicircum 3*y - 2*x\textasciicircum 2*y\textasciicircum 2 + x,3*x\textasciicircum 4 - y];}$$

$$\texttt{gbasis(F,plex(x,y));}$$

will compute a *lex* Gröbner basis for the ideal from Exercise 4. The output is

$$(3.4) \qquad [-9y + 48y^{10} - 49y^7 + 6y^4, 252x - 624y^7 + 493y^4 - 3y]$$

(possibly up to the ordering of the terms, which can vary). This is not the same as the result of the rudimentary form of Buchberger's algorithm given before. For instance, notice that neither of the polynomials in F actually appears in the output. The reason is that the `gbasis` function actually computes what we will refer to as a *reduced* Gröbner basis for the ideal generated by the list F.

(3.5) Definition. A *reduced Gröbner basis* for an ideal $I \subset k[x_1, \ldots, x_n]$ is a Gröbner basis G for I such that for all distinct $p, q \in G$, no monomial appearing in p is a multiple of $\mathrm{LT}(q)$. A *monic Gröbner basis* is a reduced Gröbner basis in which the leading coefficient of every polynomial is 1, or $\emptyset$ if $I = \langle 0 \rangle$.

Exercise 6. Verify that (3.4) is a reduced Gröbner basis according to this definition.

Exercise 7. Compute a Gröbner basis G for the ideal I from Exercise 1 of this section. Verify that $\bar{p}^G = 0$ now, in agreement with the result of Exercise 3.

A comment is in order concerning (3.5). Many authors include the condition that the leading coefficient of each element in G is 1 in the definition of a reduced Gröbner basis. However, many computer algebra systems (including Maple, see (3.4)) do not perform that extra normalization because it often increases the amount of storage space needed for the Gröbner basis elements when the coefficient field is $\mathbb{Q}$. The reason that condition is often included, however, is the following statement.

- (Uniqueness of Monic Gröbner Bases) Fix a monomial order $>$ on $k[x_1, \ldots, x_n]$. Each ideal I in $k[x_1, \ldots, x_n]$ has a *unique* monic Gröbner basis with respect to $>$.

See [CLO], Chapter 2, §7, [AL], Chapter 1, §8, or [BW], Chapter 5, §2. Of course, varying the monomial order can change the reduced Gröbner basis guaranteed by this result, and one reason different monomial orders are considered is that the corresponding Gröbner bases can have different, useful properties. One interesting feature of (3.4), for instance, is that the second polynomial in the basis does not depend on x. In other words, it is an element of the elimination ideal $I \cap \mathbb{Q}[y]$. In fact, *lex* Gröbner bases systematically eliminate variables. This is the content of the Elimination Theorem from [CLO], Chapter 3, §1. Also see Chapter 2, §1 of this book for further discussion and applications of this remark. On the other hand, the *grevlex* order often minimizes the amount of computation needed to produce a Gröbner basis, so if no other special properties are required, it can be the best choice of monomial order. Other *product orders* and *weight orders* are used in many applications to produce Gröbner bases with special properties. See Chapter 8 for some examples.

ADDITIONAL EXERCISES FOR §3

Exercise 8. Consider the ideal $I = \langle x^2 y^2 - x, xy^3 + y \rangle$ from (2.7).
a. Using $>_{lex}$ in $\mathbb{Q}[x, y]$, compute a Gröbner basis G for I.
b. Verify that each basis element g you obtain is in I, by exhibiting equations $g = A(x^2 y^2 - x) + B(xy^3 + y)$ for suitable $A, B \in \mathbb{Q}[x, y]$.
c. Let $f = x^3 y^2 + 2xy^4$. What is $\overline{f}^G$? How does this compare with the result in (2.7)?

Exercise 9. What monomials can appear in *remainders* with respect to the Gröbner basis G in (3.4)? What monomials appear in leading terms of elements of the ideal generated by G?

Exercise 10. Let G be a Gröbner basis for an ideal $I \subset k[x_1, \dots, x_n]$ and suppose there exist distinct $p, q \in G$ such that $\mathrm{LT}(p)$ is divisible by $\mathrm{LT}(q)$. Show that $G \setminus \{p\}$ is also a Gröbner basis for I. Use this observation, together with division, to propose an algorithm for producing a reduced Gröbner basis for I given G as input.

Exercise 11. This exercise will sketch a Gröbner basis method for computing the intersection of two ideals. It relies on the Elimination Theorem for *lex* Gröbner bases, as stated in [CLO], Chapter 3, §1. Let $I = \langle f_1, \dots, f_s \rangle \subset k[x_1, \dots, x_n]$ be an ideal. Given $f(t)$, an arbitrary polynomial in $k[t]$, consider the ideal

$$f(t)I = \langle f(t)f_1, \dots, f(t)f_s \rangle \subset k[x_1, \dots, x_n, t].$$

a. Let I, J be ideals in $k[x_1, \dots, x_n]$. Show that

$$I \cap J = (tI + (1 - t)J) \cap k[x_1, \dots, x_n].$$

b. Using the Elimination Theorem, deduce that a Gröbner basis G for $I \cap J$ can be found by first computing a Gröbner basis H for $tI + (1 - t)J$ using a *lex* order on $k[x_1, \ldots, x_n, t]$ with the variables ordered $t > x_i$ for all i, and then letting $G = H \cap k[x_1, \ldots, x_n]$.

Exercise 12. Using the result of Exercise 11, derive a Gröbner basis method for computing the quotient ideal $I : \langle h \rangle$. Hint: Exercise 13 of §1 shows that if $I \cap \langle h \rangle$ is generated by $g_1, \ldots, g_t$, then $I : \langle h \rangle$ is generated by $g_1/h, \ldots, g_t/h$.

§4 Affine Varieties

We will call the set $k^n = \{(a_1, \ldots, a_n) : a_1, \ldots, a_n \in k\}$ the *affine n-dimensional space* over k. With $k = \mathbb{R}$, for example, we have the usual coordinatized Euclidean space $\mathbb{R}^n$. Each polynomial $f \in k[x_1, \ldots, x_n]$ defines a function $f : k^n \to k$. The value of f at $(a_1, \ldots, a_n) \in k^n$ is obtained by substituting $x_i = a_i$, and evaluating the resulting expression in k. More precisely, if we write $f = \sum_\alpha c_\alpha x^\alpha$ for $c_\alpha \in k$, then $f(a_1, \ldots, a_n) = \sum_\alpha c_\alpha a^\alpha \in k$, where

$$a^\alpha = a_1^{\alpha_1} \cdots a_n^{\alpha_n}.$$

We recall the following basic fact.

- (Zero Function) If k is an *infinite* field, then $f : k^n \to k$ is the zero function if and only if $f = 0 \in k[x_1, \ldots, x_n]$.

See, for example, [CLO], Chapter 1, §1. As a consequence, when k is infinite, two polynomials define the same function on k^n if and only if they are equal in $k[x_1, \ldots, x_n]$.

The simplest geometric objects studied in algebraic geometry are the subsets of affine space defined by one or more polynomial equations. For instance, in $\mathbb{R}^3$, consider the set of (x, y, z) satisfying the equation

$$x^2 + z^2 - 1 = 0,$$

a circular cylinder of radius 1 along the y-axis (see Fig. 1.1).

Note that any equation $p = q$, where $p, q \in k[x_1, \ldots, x_n]$, can be rewritten as $p - q = 0$, so it is customary to write all equations in the form $f = 0$ and we will always do this. More generally, we could consider the simultaneous solutions of a system of polynomial equations.

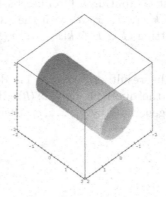

FIGURE 1.1. Circular cylinder

(4.1) Definition. The set of all simultaneous solutions $(a_1, \ldots, a_n) \in k^n$ of a system of equations

$$f_1(x_1, \ldots, x_n) = 0$$
$$f_2(x_1, \ldots, x_n) = 0$$
$$\vdots$$
$$f_s(x_1, \ldots, x_n) = 0$$

is known as the *affine variety* defined by $f_1, \ldots, f_s$, and is denoted by $\mathbf{V}(f_1, \ldots, f_s)$. A subset $V \subset k^n$ is said to be an *affine variety* if $V = \mathbf{V}(f_1, \ldots, f_s)$ for some collection of polynomials $f_i \in k[x_1, \ldots, x_n]$.

In later chapters we will also introduce projective varieties. For now, though, we will often say simply "variety" for "affine variety." For example, $\mathbf{V}(x^2 + z^2 - 1)$ in $\mathbb{R}^3$ is the cylinder pictured above. The picture was generated using the Maple command

```
implicitplot3d(x^2+z^2-1,x=-2..2,y=-2..2,z=-2..2,
    grid=[20,20,20]);
```

The variety $\mathbf{V}(x^2 + y^2 + (z - 1)^2 - 4)$ in $\mathbb{R}^3$ is the sphere of radius 2 centered at $(0, 0, 1)$ (see Fig. 1.2).

If there is more than one defining equation, the resulting variety can be considered as an *intersection* of other varieties. For example, the variety $\mathbf{V}(x^2 + z^2 - 1, x^2 + y^2 + (z - 1)^2 - 4)$ is the curve of intersection of the

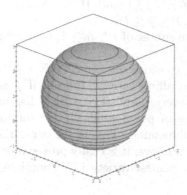

FIGURE 1.2. Sphere

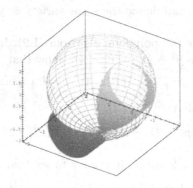

FIGURE 1.3. Cylinder-sphere intersection

cylinder and the sphere pictured above. This is shown, from a viewpoint below the xy-plane, in Fig. 1.3.

The *union* of the sphere and the cylinder is also a variety, namely $\mathbf{V}((x^2 + z^2 - 1)(x^2 + y^2 + (z - 1)^2 - 4))$. Generalizing examples like these, we have:

Exercise 1.

a. Show that any finite intersection of affine varieties is also an affine variety.

b. Show that any finite union of affine varieties is also an affine variety. Hint: If $V = \mathbf{V}(f_1, \ldots, f_s)$ and $W = \mathbf{V}(g_1, \ldots, g_t)$, then what is $\mathbf{V}(f_i g_j : 1 \le i \le s, 1 \le j \le t)$?

c. Show that any finite subset of k^n, $n \ge 1$, is an affine variety.

On the other hand, consider the set $S = \mathbb{R} \setminus \{0, 1, 2\}$, a subset of $\mathbb{R}$. We claim S is *not* an affine variety. Indeed, if f is any polynomial in $\mathbb{R}[x]$ that vanishes at every point of S, then f has infinitely many roots. By standard properties of polynomials in one variable, this implies that f must be the zero polynomial. (This is the one-variable case of the Zero Function property given above; it is easily proved in $k[x]$ using the division algorithm.) Hence the smallest variety in $\mathbb{R}$ containing S is the whole real line itself.

An affine variety $V \subset k^n$ can be described by many different systems of equations. Note that if $g = p_1 f_1 + p_2 f_2 + \cdots + p_s f_s$, where $p_i \in k[x_1, \ldots, x_n]$ are any polynomials, then $g(a_1, \ldots, a_n) = 0$ at each $(a_1, \ldots, a_n) \in \mathbf{V}(f_1, \ldots, f_s)$. So given any set of equations defining a variety, we can always produce infinitely many additional polynomials that also vanish on the variety. In the language of §1 of this chapter, the g as above are just the elements of the ideal $\langle f_1, \ldots, f_s \rangle$. Some collections of these new polynomials can define the same variety as the $f_1, \ldots, f_s$.

Exercise 2. Consider the polynomial p from (1.2). In (1.4) we saw that $p \in \langle x^2 + z^2 - 1, x^2 + y^2 + (z-1)^2 - 4 \rangle$. Show that

$$\langle x^2 + z^2 - 1, x^2 + y^2 + (z-1)^2 - 4 \rangle = \langle x^2 + z^2 - 1, y^2 - 2z - 2 \rangle$$

in $\mathbb{Q}[x, y, z]$. Deduce that

$$\mathbf{V}(x^2 + z^2 - 1, x^2 + y^2 + (z-1)^2 - 4) = \mathbf{V}(x^2 + z^2 - 1, y^2 - 2z - 2).$$

Generalizing Exercise 2 above, it is easy to see that

- (Equal Ideals Have Equal Varieties) If $\langle f_1, \ldots, f_s \rangle = \langle g_1, \ldots, g_t \rangle$ in $k[x_1, \ldots, x_n]$, then $\mathbf{V}(f_1, \ldots, f_s) = \mathbf{V}(g_1, \ldots, g_t)$.

See [CLO], Chapter 1, §4. By this result, together with the Hilbert Basis Theorem from §1, it also makes sense to think of a variety as being defined by an *ideal* in $k[x_1, \ldots, x_n]$, rather than by a specific system of equations. If we want to think of a variety in this way, we will write $V = \mathbf{V}(I)$ where $I \subset k[x_1, \ldots, x_n]$ is the ideal under consideration.

Now, given a variety $V \subset k^n$, we can also try to turn the construction of V from an ideal around, by considering the entire collection of polynomials that vanish at every point of V.

(4.2) Definition. Let $V \subset k^n$ be a variety. We denote by $\mathbf{I}(V)$ the set

$$\{f \in k[x_1, \ldots, x_n] : f(a_1, \ldots, a_n) = 0 \text{ for all } (a_1, \ldots, a_n) \in V\}.$$

We call $\mathbf{I}(V)$ the *ideal of V* for the following reason.

Exercise 3. Show that $\mathbf{I}(V)$ is an ideal in $k[x_1, \ldots, x_n]$ by verifying that the two properties in Definition (1.5) hold.

If $V = \mathbf{V}(I)$, is it always true that $\mathbf{I}(V) = I$? The answer is *no*, as the following simple example demonstrates. Consider $V = \mathbf{V}(x^2)$ in $\mathbb{R}^2$. The ideal $I = \langle x^2 \rangle$ in $\mathbb{R}[x, y]$ consists of all polynomials divisible by x^2. These polynomials are certainly contained in $\mathbf{I}(V)$, since the corresponding variety V consists of all points of the form $(0, b)$, $b \in \mathbb{R}$ (the y-axis). Note that $p(x, y) = x \in \mathbf{I}(V)$, but $x \notin I$. In this case, $\mathbf{I}(\mathbf{V}(I))$ is strictly larger than I.

Exercise 4. Show that the following inclusions are always valid:

$$I \subset \sqrt{I} \subset \mathbf{I}(\mathbf{V}(I)),$$

where $\sqrt{I}$ is the *radical* of I from Definition (1.6).

It is also true that the properties of the field k influence the relation between $\mathbf{I}(\mathbf{V}(I))$ and I. For instance, over $\mathbb{R}$, we have $\mathbf{V}(x^2 + 1) = \emptyset$ and $\mathbf{I}(\mathbf{V}(x^2 + 1)) = \mathbb{R}[x]$. On the other hand, if we take $k = \mathbb{C}$, then every polynomial in $\mathbb{C}[x]$ factors completely by the Fundamental Theorem of Algebra. We find that $\mathbf{V}(x^2 + 1)$ consists of the two points $\pm i \in \mathbb{C}$, and $\mathbf{I}(\mathbf{V}(x^2 + 1)) = \langle x^2 + 1 \rangle$.

Exercise 5. Verify the claims made in the preceding paragraph. You may want to start out by showing that if $a \in \mathbb{C}$, then $\mathbf{I}(\{a\}) = \langle x - a \rangle$.

The first key relationships between ideals and varieties are summarized in the following theorems.

- (Strong Nullstellensatz) If k is an *algebraically closed* field (such as $\mathbb{C}$) and I is an ideal in $k[x_1, \ldots, x_n]$, then

$$\mathbf{I}(\mathbf{V}(I)) = \sqrt{I}.$$

- (Ideal-Variety Correspondence) Let k be an arbitrary field. The maps

$$\text{affine varieties} \xrightarrow{\ \mathbf{I}\ } \text{ideals}$$

and

$$\text{ideals} \xrightarrow{\ \mathbf{V}\ } \text{affine varieties}$$

are inclusion-reversing, and $\mathbf{V}(\mathbf{I}(V)) = V$ for all affine varieties V. If k is algebraically closed, then

$$\text{affine varieties} \xrightarrow{\;\mathbf{I}\;} \text{radical ideals}$$

and

$$\text{radical ideals} \xrightarrow{\;\mathbf{V}\;} \text{affine varieties}$$

are inclusion-reversing bijections, and inverses of each other.

See, for instance [CLO], Chapter 4, §2, or [AL], Chapter 2, §2. We consider how the operations on ideals introduced in §1 relate to operations on varieties in the following exercises.

ADDITIONAL EXERCISES FOR §4

Exercise 6. In §1, we saw that the polynomial $p = x^2 + \frac{1}{2}y^2 z - z - 1$ is in the ideal $I = \langle x^2 + z^2 - 1, x^2 + y^2 + (z-1)^2 - 4 \rangle \subset \mathbb{R}[x, y, z]$.
a. What does this fact imply about the varieties $\mathbf{V}(p)$ and $\mathbf{V}(I)$ in $\mathbb{R}^3$? ($\mathbf{V}(I)$ is the curve of intersection of the cylinder and the sphere pictured in the text.)
b. Using a 3-dimensional graphing program (e.g. Maple's `implicitplot3d` function from the `plots` package) or otherwise, generate a picture of the variety $\mathbf{V}(p)$.
c. Show that $\mathbf{V}(p)$ contains the variety $W = \mathbf{V}(x^2 - 1, y^2 - 2)$. Describe W geometrically.
d. If we solve the equation

$$x^2 + \tfrac{1}{2}y^2 z - z - 1 = 0$$

for z, we obtain

(4.3)
$$z = \frac{x^2 - 1}{1 - \frac{1}{2}y^2}.$$

The right-hand side $r(x, y)$ of (4.3) is a quotient of polynomials or, in the terminology of §1, a rational function in x, y, and (4.3) is the equation of the *graph* of $r(x, y)$. Exactly how does this graph relate to the variety $\mathbf{V}(x^2 + \frac{1}{2}y^2 z - z - 1)$ in $\mathbb{R}^3$? (Are they the same? Is one a subset of the other? What is the domain of $r(x, y)$ as a function from $\mathbb{R}^2$ to $\mathbb{R}$?)

Exercise 7. Show that for any ideal $I \subset k[x_1, \ldots, x_n]$, $\sqrt{\sqrt{I}} = \sqrt{I}$. Hence $\sqrt{I}$ is automatically a radical ideal.

Exercise 8. Assume k is an algebraically closed field. Show that in the Ideal-Variety Correspondence, sums of ideals (see Exercise 11 of §1) correspond to intersections of the corresponding varieties:

$$\mathbf{V}(I + J) = \mathbf{V}(I) \cap \mathbf{V}(J).$$

Also show that if V and W are any varieties,

$$\mathbf{I}(V \cap W) = \sqrt{\mathbf{I}(V) + \mathbf{I}(W)}.$$

Exercise 9.
a. Show that the intersection of two radical ideals is also a radical ideal.
b. Show that in the Ideal-Variety Correspondence above, intersections of ideals (see Exercise 12 from §1) correspond to unions of the corresponding varieties:

$$\mathbf{V}(I \cap J) = \mathbf{V}(I) \cup \mathbf{V}(J).$$

Also show that if V and W are any varieties,

$$\mathbf{I}(V \cup W) = \mathbf{I}(V) \cap \mathbf{I}(W).$$

c. Show that products of ideals (see Exercise 12 from §1) also correspond to unions of varieties:

$$\mathbf{V}(IJ) = \mathbf{V}(I) \cup \mathbf{V}(J).$$

Assuming k is algebraically closed, how is the product $\mathbf{I}(V)\mathbf{I}(W)$ related to $\mathbf{I}(V \cup W)$?

Exercise 10. A variety V is said to be *irreducible* if in every expression of V as a union of other varieties, $V = V_1 \cup V_2$, either $V_1 = V$ or $V_2 = V$. Show that an affine variety V is irreducible if and only if $\mathbf{I}(V)$ is a prime ideal (see Exercise 8 from §1).

Exercise 11. Let k be algebraically closed.
a. Show by example that the set difference of two affine varieties:

$$V \setminus W = \{p \in V : p \notin W\}$$

need not be an affine variety. Hint: For instance, consider $k[x]$ and let $V = k = \mathbf{V}(0)$ and $W = \{0\} = \mathbf{V}(x)$.
b. Show that for any ideals I, J in $k[x_1, \ldots, x_n]$, $\mathbf{V}(I:J)$ contains $\mathbf{V}(I) \setminus \mathbf{V}(J)$, but that we may not have equality. (Here $I:J$ is the quotient ideal introduced in Exercise 13 from §1.)
c. If I is a radical ideal, show that $\mathbf{V}(I) \setminus \mathbf{V}(J) \subset \mathbf{V}(I : J)$ and that any variety containing $\mathbf{V}(I) \setminus \mathbf{V}(J)$ must contain $\mathbf{V}(I:J)$. Thus $\mathbf{V}(I:J)$ is the *smallest* variety containing the difference $\mathbf{V}(I) \setminus \mathbf{V}(J)$; it is called the *Zariski closure* of $\mathbf{V}(I) \setminus \mathbf{V}(J)$. See [CLO], Chapter 4, §4.
d. Show that if I is a radical ideal and J is any ideal, then $I:J$ is also a radical ideal. Deduce that $\mathbf{I}(V):\mathbf{I}(W)$ is the radical ideal corresponding to the Zariski closure of $V \setminus W$ in the Ideal-Variety Correspondence.

Chapter 2

Solving Polynomial Equations

In this chapter we will discuss several approaches to solving systems of polynomial equations. First, we will discuss a straightforward attack based on the elimination properties of lexicographic Gröbner bases. Combining elimination with numerical root-finding for one-variable polynomials we get a conceptually simple method that generalizes the usual techniques used to solve systems of linear equations. However, there are potentially severe difficulties when this approach is implemented on a computer using finite-precision arithmetic. To circumvent these problems, we will develop some additional algebraic tools for root-finding based on the algebraic structure of the quotient rings $k[x_1, \ldots, x_n]/I$. Using these tools, we will present alternative numerical methods for approximating solutions of polynomial systems and consider methods for real root-counting and root-isolation. In Chapters 3, 4 and 7, we will also discuss polynomial equation solving. Specifically, Chapter 3 will use resultants to solve polynomial equations, and Chapter 4 will show how to assign a well-behaved multiplicity to each solution of a system. Chapter 7 will consider other numerical techniques (homotopy continuation methods) based on bounds for the total number of solutions of a system, counting multiplicities.

§1 Solving Polynomial Systems by Elimination

The main tools we need are the Elimination and Extension Theorems. For the convenience of the reader, we recall the key ideas:

- (Elimination Ideals) If I is an ideal in $k[x_1, \ldots, x_n]$, then the ℓth *elimination ideal* is

$$I_\ell = I \cap k[x_{\ell+1}, \ldots, x_n].$$

  Intuitively, if $I = \langle f_1, \ldots, f_s \rangle$, then the elements of I_ℓ are the linear combinations of the $f_1, \ldots, f_s$, with polynomial coefficients, that eliminate $x_1, \ldots, x_\ell$ from the equations $f_1 = \cdots = f_s = 0$.

- (The Elimination Theorem) If G is a Gröbner basis for I with respect to the *lex* order $(x_1 > x_2 > \cdots > x_n)$ (or any order where monomials involving at least one of $x_1, \ldots, x_\ell$ are greater than all monomials involving only the remaining variables), then

$$G_\ell = G \cap k[x_{\ell+1}, \ldots, x_n]$$

is a Gröbner basis of the ℓth elimination ideal I_ℓ.
- (Partial Solutions) A point $(a_{\ell+1}, \ldots, a_n) \in \mathbf{V}(I_\ell) \subset k^{n-\ell}$ is called a *partial solution*. Any solution $(a_1, \ldots, a_n) \in \mathbf{V}(I) \subset k^n$ truncates to a partial solution, but the converse may fail—not all partial solutions extend to solutions. This is where the Extension Theorem comes in. To prepare for the statement, note that each f in $I_{\ell-1}$ can be written as a polynomial in x_ℓ, whose coefficients are polynomials in $x_{\ell+1}, \ldots, x_n$:

$$f = c_q(x_{\ell+1}, \ldots, x_n)x_\ell^q + \cdots + c_0(x_{\ell+1}, \ldots, x_n).$$

We call c_q the leading coefficient polynomial of f if x_ℓ^q is the highest power of x_ℓ appearing in f.
- (The Extension Theorem) If k is algebraically closed (e.g., $k = \mathbb{C}$), then a partial solution $(a_{\ell+1}, \ldots, a_n)$ in $\mathbf{V}(I_\ell)$ extends to $(a_\ell, a_{\ell+1}, \ldots, a_n)$ in $\mathbf{V}(I_{\ell-1})$ provided that the leading coefficient polynomials of the elements of a *lex* Gröbner basis for $I_{\ell-1}$ do not all vanish at $(a_{\ell+1}, \ldots, a_n)$.

For the proofs of these results and a discussion of their geometric meaning, see Chapter 3 of [CLO]. Also, the Elimination Theorem is discussed in §6.2 of [BW] and §2.3 of [AL], and [AL] discusses the geometry of elimination in §2.5.

The Elimination Theorem shows that a *lex* Gröbner basis G successively eliminates more and more variables. This gives the following strategy for finding all solutions of the system: start with the polynomials in G with the fewest variables, solve them, and then try to extend these partial solutions to solutions of the whole system, applying the Extension Theorem one variable at a time.

As the following example shows, this works especially nicely when $\mathbf{V}(I)$ is finite. Consider the system of equations

(1.1)
$$x^2 + y^2 + z^2 = 4$$
$$x^2 + 2y^2 = 5$$
$$xz = 1$$

from Exercise 4 of Chapter 3, §1 of [CLO]. To solve these equations, we first compute a *lex* Gröbner basis for the ideal they generate using Maple:

```
with(Groebner):
PList := [x^2+y^2+z^2-4, x^2+2*y^2-5, x*z-1];
G := gbasis(PList,plex(x,y,z));
```

This gives output

$$G := [1 + 2z^4 - 3z^2, y^2 - z^2 - 1, x + 2z^3 - 3z].$$

From the Gröbner basis it follows that the set of solutions of this system in $\mathbb{C}^3$ is finite (why?). To find all the solutions, note that the last polynomial depends only on z (it is a generator of the second elimination ideal $I_2 = I \cap \mathbb{C}[z]$) and factors nicely in $\mathbb{Q}[z]$. To see this, we may use

```
factor(2*z^4 - 3*z^2 + 1);
```

which generates the output

$$(z - 1)(z + 1)(2z^2 - 1).$$

Thus we have four possible z values to consider:

$$z = \pm 1, \pm 1/\sqrt{2}.$$

By the Elimination Theorem, the first elimination ideal $I_1 = I \cap \mathbb{C}[y, z]$ is generated by

$$y^2 - z^2 - 1$$
$$2z^4 - 3z^2 + 1.$$

Since the coefficient of y^2 in the first polynomial is a nonzero constant, *every* partial solution in $\mathbf{V}(I_2)$ extends to a solution in $\mathbf{V}(I_1)$. There are eight such points in all. To find them, we substitute a root of the last equation for z and solve the resulting equation for y. For instance,

```
subs(z=1,G);
```

will produce:

$$[-1 + x, y^2 - 2, 0],$$

so in particular, $y = \pm\sqrt{2}$. In addition, since the coefficient of x in the first polynomial in the Gröbner basis is a nonzero constant, we can extend each partial solution in $\mathbf{V}(I_1)$ (uniquely) to a point of $\mathbf{V}(I)$. For this value of z, we have $x = 1$.

Exercise 1. Carry out the same process for the other values of z as well. You should find that the eight points

$$(1, \pm\sqrt{2}, 1), \ (-1, \pm\sqrt{2}, -1), \ (\sqrt{2}, \pm\sqrt{6}/2, 1/\sqrt{2}), \ (-\sqrt{2}, \pm\sqrt{6}/2, -1/\sqrt{2})$$

form the set of solutions.

The system in (1.1) is relatively simple because the coordinates of the solutions can all be expressed in terms of square roots of rational numbers. Unfortunately, general systems of polynomial equations are rarely this nice. For instance it is known that there are *no general formulas* involving only the field operations in k and extraction of roots (i.e., radicals) for solving single variable polynomial equations of degree 5 and higher. This is a famous result of Ruffini, Abel, and Galois (see [Her]). Thus, if elimination leads to a one-variable equation of degree 5 or higher, then we may not be able to give radical formulas for the roots of that polynomial.

We take the system of equations given in (1.1) and change the first term in the first polynomial from x^2 to x^5. Then executing

```
PList2 := [x^5+y^2+z^2-4, x^2+2*y^2-5, x*z-1];
G2 := gbasis(PList2,plex(x,y,z));
```

produces the following *lex* Gröbner basis:

$$(1.2) \quad [2 + 2z^7 - 3z^5 - z^3, 4y^2 - 2z^5 + 3z^3 + z - 10, 2x + 2z^6 - 3z^4 - z^2].$$

In this case, the command

```
factor(2*z^7 - 3*z^5 - z^3 + 2);
```

gives the factorization

$$2z^7 - 3z^5 - z^3 + 2 = (z - 1)(2z^6 + 2z^5 - z^4 - z^3 - 2z^2 - 2z - 2),$$

and the second factor is irreducible in $\mathbb{Q}[z]$. In a situation like this, to go farther in equation solving, we need to decide what kind of answer is required.

If we want a purely algebraic, "structural" description of the solutions, then Maple can represent solutions of systems like this via the `solve` command. Let's see what this looks like. Entering

```
solve(convert(G2,set),{x,y,z});
```

you should generate the following output:

$\{\{y = \text{RootOf}(\_Z^2 - 2, \mathit{label} = \_L4), x = 1, z = 1\},$
$\quad \{y = 1/2\text{RootOf}(\_Z^2$
$\qquad - 2\text{RootOf}(2\_Z^6 + 2\_Z^5 - \_Z^4 - \_Z^3 - 2\_Z^2 - 2\_Z - 2)^5$
$\qquad + 3\text{RootOf}(2\_Z^6 + 2\_Z^5 - \_Z^4 - \_Z^3 - 2\_Z^2 - 2\_Z - 2)^3$
$\qquad + \text{RootOf}(2\_Z^6 + 2\_Z^5 - \_Z^4 - \_Z^3 - 2\_Z^2 - 2\_Z - 2)$
$\qquad - 10, \mathit{label} = \_L1),$
$\quad x = \text{RootOf}(2\_Z^6 + 2\_Z^5 - \_Z^4 - \_Z^3 - 2\_Z^2 - 2\_Z - 2)^4$
$\qquad - 1/2\text{RootOf}(2\_Z^6 + 2\_Z^5 - \_Z^4 - \_Z^3 - 2\_Z^2 - 2\_Z - 2)^2 - 1$
$\qquad + \text{RootOf}(2\_Z^6 + 2\_Z^5 - \_Z^4 - \_Z^3 - 2\_Z^2 - 2\_Z - 2)^5$
$\qquad - 1/2\text{RootOf}(2\_Z^6 + 2\_Z^5 - \_Z^4 - \_Z^3 - 2\_Z^2 - 2\_Z - 2)^3$
$\qquad - \text{RootOf}(2\_Z^6 + 2\_Z^5 - \_Z^4 - \_Z^3 - 2\_Z^2 - 2\_Z - 2),$
$\quad z = \text{RootOf}(2\_Z^6 + 2\_Z^5 - \_Z^4 - \_Z^3 - 2\_Z^2 - 2\_Z - 2)\}\}$

Here $\text{RootOf}(2\_Z^6 + 2\_Z^5 - \_Z^4 - \_Z^3 - 2\_Z^2 - 2\_Z - 2)$ stands for any one root of the polynomial equation $2\_Z^6 + 2\_Z^5 - \_Z^4 - \_Z^3 - 2\_Z^2 - 2\_Z - 2 = 0$. Similarly, the other RootOf expressions stand for any solution of the corresponding equation in the dummy variable $\_Z$.

Exercise 2. Verify that the expressions above are obtained if we solve for z from the Gröbner basis G_2 and then use the Extension Theorem. How many solutions are there of this system in $\mathbb{C}^3$?

On the other hand, in many practical situations where equations must be solved, knowing a *numerical approximation* to a real or complex solution is often more useful, and perfectly acceptable provided the results are sufficiently accurate. In our particular case, one possible approach would be to use a numerical root-finding method to find approximate solutions of the one-variable equation

$$(1.3) \qquad 2z^6 + 2z^5 - z^4 - z^3 - 2z^2 - 2z - 2 = 0,$$

and then proceed as before using the Extension Theorem, except that we now use floating point arithmetic in all calculations. In some examples, numerical methods will also be needed to solve for the other variables as we extend.

One well-known numerical method for solving one-variable polynomial equations in $\mathbb{R}$ or $\mathbb{C}$ is the *Newton-Raphson method* or, more simply but less accurately, *Newton's method*. This method may also be used for equations involving functions other than polynomials, although we will not discuss those here. For motivation and a discussion of the theory behind the method, see [BuF] or [Act].

The Newton-Raphson method works as follows. Choosing some initial approximation z_0 to a root of $p(z) = 0$, we construct a sequence of numbers by the rule

$$z_{k+1} = z_k - \frac{p(z_k)}{p'(z_k)} \qquad \text{for } k = 0, 1, 2, \ldots,$$

where $p'(z)$ is the usual derivative of p from calculus. In *most* situations, the sequence z_k will converge rapidly to a solution $\bar{z}$ of $p(z) = 0$, that is, $\bar{z} = \lim_{k \to \infty} z_k$ will be a root. Stopping this procedure after a finite number of steps (as we must!), we obtain an approximation to $\bar{z}$. For example, we might stop when z_{k+1} and z_k agree to some desired accuracy, or when a maximum allowed number of terms of the sequence have been computed. See [BuF], [Act], or the comments at the end of this section for additional information on the performance of this technique. When trying to find *all* roots of a polynomial, the trickiest part of the Newton-Raphson method is making appropriate choices of z_0. It is easy to find the same root repeatedly and to miss other ones if you don't know where to look!

Fortunately, there are elementary bounds on the absolute values of the roots (real or complex) of a polynomial $p(z)$. Here is one of the simpler bounds.

Exercise 3. Show that if $p(z) = z^n + a_{n-1}z^{n-1} + \cdots + a_0$ is a monic polynomial with complex coefficients, then all roots $\overline{z}$ of p satisfy $|\overline{z}| \leq B$, where

$$B = \max\{1, |a_{n-1}| + \cdots + |a_1| + |a_0|\}.$$

Hint: The triangle inequality implies that $|a + b| \geq |a| - |b|$.

See Exercise 10 below for another better bound on the roots. Given any bound of this sort, we can limit our attention to z_0 in this region of the complex plane to search for roots of the polynomial.

Instead of discussing searching strategies for finding roots, we will use a built-in Maple function to approximate the roots of the system from (1.2). The Maple function `fsolve` finds numerical approximations to all real (or complex) roots of a polynomial by a combination of root location and numerical techniques like Newton-Raphson. For instance, the command

```
fsolve(2*z^6+2*z^5-z^4-z^3-2*z^2-2*z-2);
```

will compute approximate values for the *real* roots of our polynomial (1.3). The output should be:

$$-1.395052015, \qquad 1.204042437.$$

(Note: In Maple, 10 digits are carried by default in decimal calculations; more digits can be used by changing the value of the Maple system variable `Digits`. Also, the actual digits in your output may vary slightly if you carry out this computation using another computer algebra system.) To get approximate values for the complex roots as well, try:

```
fsolve(2*z^6+2*z^5-z^4-z^3-2*z^2-2*z-2,complex);
```

We illustrate the Extension Step in this case using the approximate value

$$z = 1.204042437.$$

We substitute this value into the Gröbner basis polynomials using

```
subs(z=1.204042437,G2);
```

and obtain

$$[2x - 1.661071025, -8.620421528 + 4y^2, -.2 * 10^{-8}].$$

Note that the value of the last polynomial was *not exactly zero* at our approximate value of z. Nevertheless, as in Exercise 1, we can extend this approximate partial solution to two approximate solutions of the system:

$$(x, y, z) = (.8305355125, \pm 1.468027718, 1.204042437).$$

Checking one of these by substituting into the equations from (1.2), using

```
subs(z=1.204042437,y=1.468027718,x=.8305355125, G2);
```

we find

$$[0, -.4 * 10^{-8}, -.2 * 10^{-8}],$$

so we have a reasonably good approximate solution, in the sense that our computed solution gives values very close to zero in the polynomials of the system.

Exercise 4. Find approximate values for all other real solutions of this system by the same method.

In considering what we did here, one potential pitfall of this approach should be apparent. Namely, since our solutions of the one-variable equation are only approximate, when we substitute and try to extend, the remaining polynomials to be solved for x and y are themselves only approximate. Once we substitute approximate values for one of the variables, we are in effect solving a system of equations that is *different from the one we started with*, and there is little guarantee that the solutions of this new system are close to the solutions of the original one. Accumulated errors after several approximation and extension steps can build up quite rapidly in systems in larger numbers of variables, and the effect can be particularly severe if equations of high degree are present.

To illustrate how bad things can get, we consider a famous cautionary example due to Wilkinson, which shows how much the roots of a polynomial can be changed by very small changes in the coefficients.

Wilkinson's example involves the following polynomial of degree 20:

$$p(x) = (x + 1)(x + 2) \cdots (x + 20) = x^{20} + 210x^{19} + \cdots + 20!.$$

The roots are the 20 integers $x = -1, -2, \ldots, -20$. Suppose now that we "perturb" just the coefficient of x^{19}, adding a very small number. We carry 20 decimal digits in all calculations. First we construct $p(x)$ itself:

```
Digits := 20:
p := 1:
for k to 20 do p := p*(x+k) end do:
```

Printing **expand(p)** out at this point will show a polynomial with some large coefficients indeed! But the polynomial we want is actually this:

```
q := expand(p + .000000001*x^19):
fsolve(q,x,complex);
```

The approximate roots of $q = p + .000000001\, x^{19}$ (truncated for simplicity) are:

$$- 20.03899, -18.66983 - .35064\ I, \ -18.66983 + .35064\ I,$$
$$- 16.57173 - .88331\ I, \ -16.57173 + .88331\ I,$$
$$- 14.37367 - .77316\ I, \ -14.37367 + .77316\ I,$$
$$- 12.38349 - .10866\ I, \ -12.38349 + .10866\ I,$$
$$- 10.95660, \ -10.00771, \ -8.99916, \ -8.00005,$$
$$- 6.999997, \ -6.000000, \ -4.99999, \ -4.00000,$$
$$- 2.999999, \ -2.000000, \ -1.00000.$$

Instead of 20 real roots, the new polynomial has 12 real roots and 4 complex conjugate pairs of roots. Note that the imaginary parts are not even especially small!

While this example is admittedly pathological, it indicates that we should use care in finding roots of polynomials whose coefficients are only approximately determined. (The reason for the surprisingly bad behavior of this p is essentially the equal spacing of the roots! We refer the interested reader to Wilkinson's paper [Wil] for a full discussion.)

Along the same lines, even if nothing this spectacularly bad happens, when we take the approximate roots of a one-variable polynomial and try to extend to solutions of a system, the results of a numerical calculation can still be unreliable. Here is a simple example illustrating another situation that causes special problems.

Exercise 5. Verify that if $x > y$, then

$$G = [x^2 + 2x + 3 + y^5 - y,\ y^6 - y^2 + 2y]$$

is a *lex* Gröbner basis for the ideal that G generates in $\mathbb{R}[x, y]$.

We want to find all *real* points $(x, y) \in \mathbf{V}(G)$. Begin with the equation

$$y^6 - y^2 + 2y = 0,$$

which has exactly two real roots. One is $y = 0$, and the second is in the interval $[-2, -1]$ because the polynomial changes sign on that interval. Hence there must be a root there by the Intermediate Value Theorem from calculus. Using `fsolve` to find an approximate value, we find the nonzero root is

(1.4) -1.267168305

to 10 decimal digits. Substituting this approximate value for y into G yields

$$[x^2 + 2x + .999999995,\ .7 * 10^{-8}].$$

Then use

$$\texttt{fsolve(x\textasciicircum2 + 2*x + .999999995);}$$

to obtain

$$-1.000070711, \quad -.9999292893.$$

Clearly these are both close to $x = -1$, but they are different. Taken uncritically, this would seem to indicate two distinct real values of x when y is given by (1.4).

Now, suppose we used an approximate value for y with *fewer* decimal digits, say $y \doteq -1.2671683$. Substituting this value for y gives us the quadratic

$$x^2 + 2x + 1.000000054.$$

This polynomial has no real roots at all. Indeed, using the **complex** option in **fsolve**, we obtain two complex values for x:

$$-1. - .0002323790008 \ I, \qquad -1. + .0002323790008 \ I.$$

To see what is really happening, note that the nonzero real root of $y^6 - y^2 + 2y = 0$ satisfies $y^5 - y + 2 = 0$. When the exact root is substituted into G, we get

$$[x^2 + 2x + 1, 0]$$

and the resulting equation has a double root $x = -1$.

The conclusion to be drawn from this example is that equations with double roots, such as the *exact* equation

$$x^2 + 2x + 1 = 0$$

we got above, are *especially* vulnerable to the errors introduced by numerical root-finding. It can be very difficult to tell the difference between a pair of real roots that are close, a real double root, and a pair of complex conjugate roots.

From these examples, it should be clear that finding solutions of polynomial systems is a delicate task in general, especially if we ask for information about how many real solutions there are. For this reason, numerical methods, for all their undeniable usefulness, are not the whole story. And they should never be applied blindly. The more information we have about the structure of the set of solutions of a polynomial system, the better a chance we have to determine those solutions accurately. For this reason, in §2 and §3 we will go to the algebraic setting of the quotient ring $k[x_1, \ldots, x_n]/I$ to obtain some additional tools for this problem. We will apply those tools in §4 and §5 to give better methods for finding solutions.

For completeness, we conclude with a few additional words about the numerical methods for equation solving that we have used. First, if $\bar{z}$ is a multiple root of $p(z) = 0$, then the convergence of the Newton-Raphson sequence z_k can be quite slow, and a large number of steps and high precision may be required to get really close to a root (though we give a method for avoiding this difficulty in Exercise 8). Second, there are some choices of z_0 where the sequence z_k will *fail to converge* to a root of $p(z)$. See Exercise 9 below for some simple examples. Finally, the location of $\bar{z}$ in relation to z_0 can be somewhat unpredictable. There could be other roots lying closer to z_0. These last two problems are related to the fractal pictures associated to the Newton-Raphson method over $\mathbb{C}$—see, for example, [PR]. We should also mention that there are multivariable versions of Newton-Raphson for systems of equations and other iterative methods that do not depend on elimination. These have been much studied in numerical analysis. For more details on these and other numerical root-finding methods, see [BuF] and [Act]. Also, we will discuss homotopy continuation methods in Chapter 7, §5 of this book.

ADDITIONAL EXERCISES FOR §1

Exercise 6. Use elimination to solve the system

$$0 = x^2 + 2y^2 - y - 2z$$
$$0 = x^2 - 8y^2 + 10z - 1$$
$$0 = x^2 - 7yz.$$

How many solutions are there in $\mathbb{R}^3$; how many are there in $\mathbb{C}^3$?

Exercise 7. Use elimination to solve the system

$$0 = x^2 + y^2 + z^2 - 2x$$
$$0 = x^3 - yz - x$$
$$0 = x - y + 2z.$$

How many solutions are there in $\mathbb{R}^3$; how many are there in $\mathbb{C}^3$?

Exercise 8. In this exercise we will study exactly why the performance of the Newton-Raphson method is poor for multiple roots, and suggest a remedy. Newton-Raphson iteration for any equation $p(z) = 0$ is an example of *fixed point iteration*, in which a starting value z_0 is chosen and a sequence

$$(1.5) \qquad\qquad z_{k+1} = g(z_k) \qquad \text{for } k = 0, 1, 2, \dots$$

is constructed by iteration of a fixed function $g(z)$. For Newton-Raphson iteration, the function $g(z)$ is $g(z) = N_p(z) = z - p(z)/p'(z)$. If the sequence produced by (1.5) converges to some limit $\bar{z}$, then $\bar{z}$ is a *fixed point* of g (that is, a solution of $g(z) = z$). It is a standard result from analysis (a special case of the Contraction Mapping Theorem) that iteration as in

(1.5) will converge to a fixed point $\overline{z}$ of g provided that $|g'(\overline{z})| < 1$, and z_0 is chosen sufficiently close to $\overline{z}$. Moreover, the smaller $|g'(\overline{z})|$ is, the faster convergence will be. The case $g'(\overline{z}) = 0$ is especially favorable.

a. Show that each simple root of the polynomial equation $p(z) = 0$ is a *fixed point* of the rational function $N_p(z) = z - p(z)/p'(z)$.

b. Show that multiple roots of $p(z) = 0$ are *removable singularities* of $N_p(z)$ (that is, $|N_p(z)|$ is bounded in a neighborhood of each multiple root). How should N_p be defined at a multiple root of $p(z) = 0$ to make N_p continuous at those points?

c. Show that $N'_p(\overline{z}) = 0$ if $\overline{z}$ is a *simple* root of $p(z) = 0$ (that is, if $p(\overline{z}) = 0$, but $p'(\overline{z}) \neq 0$).

d. On the other hand, show that if $\overline{z}$ is a root of multiplicity k of $p(z)$ (that is, if $p(\overline{z}) = p'(\overline{z}) = \cdots = p^{(k-1)}(\overline{z}) = 0$ but $p^{(k)}(\overline{z}) \neq 0$), then

$$\lim_{z \to \overline{z}} N'_p(z) = 1 - \frac{1}{k}.$$

Thus Newton-Raphson iteration converges much faster to a simple root of $p(z) = 0$ than it does to a multiple root, and the larger the multiplicity, the slower the convergence.

e. Show that replacing $p(z)$ by

$$p_{red}(z) = \frac{p(z)}{\mathrm{GCD}(p(z), p'(z))}$$

(see [CLO], Chapter 1, §5, Exercises 14 and 15) eliminates this difficulty, in the sense that the roots of $p_{red}(z) = 0$ are all simple roots.

Exercise 9. There are cases when the Newton-Raphson method fails to find a root of a polynomial for *lots* of starting points z_0.

a. What happens if the Newton-Raphson method is applied to solve the equation $z^2 + 1 = 0$ starting from a *real* z_0? What happens if you take z_0 with nonzero imaginary parts? Note: It can be shown that Newton-Raphson iteration for the equation $p(z) = 0$ is *chaotic* if z_0 is chosen in the *Julia set* of the rational function $N_p(z) = z - p(z)/p'(z)$ (see [PR]), and exact arithmetic is employed.

b. Let $p(z) = z^4 - z^2 - 11/36$ and, as above, let $N_p(z) = z - p(z)/p'(z)$. Show that $\pm 1/\sqrt{6}$ satisfies $N_p(1/\sqrt{6}) = -1/\sqrt{6}$, $N_p(-1/\sqrt{6}) = 1/\sqrt{6}$, and $N'_p(1/\sqrt{6}) = 0$. In the language of dynamical systems, $\pm 1/\sqrt{6}$ is a *superattracting 2-cycle* for $N_p(z)$. One consequence is that for *any* z_0 close to $\pm 1/\sqrt{6}$, the Newton-Raphson method will *not* locate a root of p. This example is taken from Chapter 13 of [Dev].

Exercise 10. This exercise improves the bound on roots of a polynomial given in Exercise 3. Let $p(z) = z^n + a_{n-1}z^{n-1} + \cdots + a_1 z + a_0$ be a monic polynomial in $\mathbb{C}[z]$. Show that all roots $\overline{z}$ of p satisfy $|\overline{z}| \leq B$, where

$$B = 1 + \max\{|a_{n-1}|, \ldots, |a_1|, |a_0|\}.$$

This upper bound can be much smaller than the one given in Exercise 3. Hint: Use the Hint from Exercise 3, and consider the evaluation of $p(z)$ by nested multiplication:

$$p(z) = (\cdots((z + a_{n-1})z + a_{n-2})z + \cdots + a_1)z + a_0.$$

§2 Finite-Dimensional Algebras

This section will explore the "remainder arithmetic" associated to a Gröbner basis $G = \{g_1, \ldots, g_t\}$ of an ideal $I \subset k[x_1, \ldots, x_n]$. Recall from Chapter 1 that if we divide $f \in k[x_1, \ldots, x_n]$ by G, the division algorithm yields an expression

$$(2.1) \qquad f = h_1 g_1 + \cdots + h_t g_t + \overline{f}^G,$$

where the remainder $\overline{f}^G$ is a linear combination of the monomials $x^\alpha \notin \langle \mathrm{LT}(I) \rangle$. Furthermore, since G is a Gröbner basis, we know that $f \in I$ if and only if $\overline{f}^G = 0$, and the remainder is uniquely determined for all f. This implies

$$(2.2) \qquad \overline{f}^G = \overline{g}^G \iff f - g \in I.$$

Since polynomials can be added and multiplied, given $f, g \in k[x_1, \ldots, x_n]$ it is natural to ask how the remainders of $f + g$ and fg can be computed if we know the remainders of f, g themselves. The following observations show how this can be done.

- The sum of two remainders is again a remainder, and in fact one can easily show that $\overline{f}^G + \overline{g}^G = \overline{f + g}^G$.
- On the other hand, the product of remainders need not be a remainder. But it is also easy to see that $\overline{\overline{f}^G \cdot \overline{g}^G}^G = \overline{fg}^G$, and $\overline{\overline{f}^G \cdot \overline{g}^G}^G$ is a remainder.

We can also interpret these observations as saying that the set of remainders on division by G has naturally defined addition and multiplication operations which produce remainders as their results.

This "remainder arithmetic" is closely related to the quotient ring $k[x_1, \ldots, x_n]/I$. We will assume the reader is familiar with quotient rings, as described in Chapter 5 of [CLO] or in a course on abstract algebra. Recall how this works: given $f \in k[x_1, \ldots, x_n]$, we have the *coset*

$$[f] = f + I = \{f + h : h \in I\},$$

and the crucial property of cosets is

$$(2.3) \qquad [f] = [g] \iff f - g \in I.$$

The quotient ring $k[x_1, \ldots, x_n]/I$ consists of all cosets $[f]$ for $f \in k[x_1, \ldots, x_n]$.

From (2.1), we see that $\overline{f}^G \in [f]$, and then (2.2) and (2.3) show that we have a one-to-one correspondence

$$\text{remainders} \longleftrightarrow \text{cosets}$$

$$\overline{f}^G \longleftrightarrow [f].$$

Thus we can think of the remainder $\overline{f}^G$ as a standard representative of its coset $[f] \in k[x_1, \ldots, x_n]/I$. Furthermore, it follows easily that remainder arithmetic is *exactly* the arithmetic in $k[x_1, \ldots, x_n]/I$. That is, under the above correspondence we have

$$\overline{f}^G + \overline{g}^G \longleftrightarrow [f] + [g]$$

$$\overline{\overline{f}^G \cdot \overline{g}^G}^G \longleftrightarrow [f] \cdot [g].$$

Since we can add elements of $k[x_1, \ldots, x_n]/I$ and multiply by constants (the cosets $[c]$ for $c \in k$), $k[x_1, \ldots, x_n]/I$ also has the structure of a vector space over the field k. A ring that is also a vector space in this fashion is called an *algebra*. The algebra $k[x_1, \ldots, x_n]/I$ will be denoted by A throughout the rest of this section, which will focus on its vector space structure.

An important observation is that remainders are the linear combinations of the monomials $x^\alpha \notin \langle \text{LT}(I) \rangle$ in this vector space structure. (Strictly speaking, we should use cosets, but in much of this section we will identify a remainder with its coset in A.) Since this set of monomials is linearly independent in A (why?), it can be regarded as a basis of A. In other words, the monomials

$$B = \{x^\alpha : x^\alpha \notin \langle \text{LT}(I) \rangle\}$$

form a basis of A (more precisely, their cosets are a basis). We will refer to elements of B as *basis monomials*. In the literature, basis monomials are often called *standard monomials*.

The following example illustrates how to compute in A using basis monomials. Let

(2.4) $G = \{x^2 + 3xy/2 + y^2/2 - 3x/2 - 3y/2, xy^2 - x, y^3 - y\}.$

Using the *grevlex* order with $x > y$, it is easy to verify that G is a Gröbner basis for the ideal $I = \langle G \rangle \subset \mathbb{C}[x, y]$ generated by G. By examining the leading monomials of G, we see that $\langle \text{LT}(I) \rangle = \langle x^2, xy^2, y^3 \rangle$. The only monomials not lying in this ideal are those in

$$B = \{1, x, y, xy, y^2\}$$

so that by the above observation, these five monomials form a vector space basis for $A = \mathbb{C}[x, y]/I$ over $\mathbb{C}$.

We now turn to the structure of the quotient ring A. The addition operation in A can be viewed as an ordinary vector sum operation once we express elements of A in terms of the basis B in (2.4). Hence we will consider the addition operation to be completely understood.

Perhaps the most natural way to describe the multiplication operation in A is to give a table of the remainders of all products of pairs of elements from the basis B. Since multiplication in A distributes over addition, this information will suffice to determine the products of all pairs of elements of A.

For example, the remainder of the product $x \cdot xy$ may be computed as follows using Maple. Using the Gröbner basis G, we compute

```
normalf(x^2*y,G,tdeg(x,y));
```

and obtain

$$\frac{3}{2}xy - \frac{3}{2}x + \frac{3}{2}y^2 - \frac{1}{2}y.$$

Exercise 1. By computing all such products, verify that the multiplication table for the elements of the basis B is:

$\cdot$	1	x	y	xy	y^2
1	1	x	y	xy	y^2
x	x	α	xy	β	x
y	y	xy	y^2	x	y
xy	xy	β	x	α	xy
y^2	y^2	x	y	xy	y^2

(2.5)

where

$$\alpha = -3xy/2 - y^2/2 + 3x/2 + 3y/2$$
$$\beta = 3xy/2 + 3y^2/2 - 3x/2 - y/2.$$

This example was especially nice because A was finite-dimensional as a vector space over $\mathbb{C}$. In general, for any field $k \subset \mathbb{C}$, we have the following basic theorem which describes when $k[x_1, \ldots, x_n]/I$ is finite-dimensional.

- (Finiteness Theorem) Let $k \subset \mathbb{C}$ be a field, and let $I \subset k[x_1, \ldots, x_n]$ be an ideal. Then the following conditions are equivalent:
  a. The algebra $A = k[x_1, \ldots, x_n]/I$ is finite-dimensional over k.
  b. The variety $\mathbf{V}(I) \subset \mathbb{C}^n$ is a finite set.
  c. If G is a Gröbner basis for I, then for each i, $1 \leq i \leq n$, there is an $m_i \geq 0$ such that $x_i^{m_i} = \mathrm{LT}(g)$ for some $g \in G$.

For a proof of this result, see Theorem 6 of Chapter 5, §3 of [CLO], Theorem 2.2.7 of [AL], or Theorem 6.54 of [BW]. An ideal satisfying any of the above conditions is said to be *zero-dimensional*. Thus

A is a finite-dimensional algebra $\Longleftrightarrow$ I is a zero-dimensional ideal.

A nice consequence of this theorem is that I is zero-dimensional if and only if there is a nonzero polynomial in $I \cap k[x_i]$ for each $i = 1, \ldots, n$. To see why this is true, first suppose that I is zero-dimensional, and let G be a reduced Gröbner basis for any *lex* order with x_i as the "last" variable (i.e., $x_j > x_i$ for $j \neq i$). By item c above, there is some $g \in G$ with $\mathrm{LT}(g) = x_i^{m_i}$. Since we're using a *lex* order with x_i last, this implies $g \in k[x_i]$ and hence g is the desired nonzero polynomial. Note that g generates $I \cap k[x_i]$ by the Elimination Theorem.

Going the other way, suppose $I \cap k[x_i]$ is nonzero for each i, and let m_i be the degree of the unique monic generator of $I \cap k[x_i]$ (remember that $k[x_i]$ is a principal ideal domain—see Corollary 4 of Chapter 1, §5 of [CLO]). Then $x_i^{m_i} \in \langle \mathrm{LT}(I) \rangle$ for any monomial order, so that all monomials not in $\langle \mathrm{LT}(I) \rangle$ will contain x_i to a power strictly less than m_i. In other words, the exponents α of the monomials $x^\alpha \notin \langle \mathrm{LT}(I) \rangle$ will all lie in the "rectangular box"

$$R = \{\alpha \in \mathbb{Z}_{\geq 0}^n : \text{ for each } i, 0 \leq \alpha_i \leq m_i - 1\}.$$

This is a finite set of monomials, which proves that A is finite-dimensional over k.

Given a zero-dimensional ideal I, it is now easy to describe an algorithm for finding the set B of all monomials not in $\langle \mathrm{LT}(I) \rangle$. Namely, no matter what monomial order we are using, the exponents of the monomials in B will lie in the box R described above. For each $\alpha \in R$, we know that $x^\alpha \notin \langle \mathrm{LT}(I) \rangle$ if and only if $\overline{x^\alpha}^G = x^\alpha$. Thus we can list the $\alpha \in R$ in some systematic way and compute $\overline{x^\alpha}^G$ for each one. A vector space basis of A is given by the set of monomials

$$B = \{x^\alpha : \alpha \in R \text{ and } \overline{x^\alpha}^G = x^\alpha\}.$$

See Exercise 13 below for a Maple procedure implementing this method.

The vector space structure on $A = k[x_1, \ldots, x_n]/I$ for a zero-dimensional ideal I can be used in several important ways. To begin, let us consider the problem of finding the monic generators of the elimination ideals $I \cap k[x_i]$. As indicated above, we could find these polynomials by computing several different *lex* Gröbner bases, reordering the variables each time to place x_i last. This is an extremely inefficient method, however. Instead, let us consider the set of non-negative powers of $[x_i]$ in A:

$$S = \{1, [x_i], [x_i]^2, \ldots\}.$$

Since A is finite-dimensional as a vector space over the field k, S must be *linearly dependent* in A. Let m_i be the *smallest* positive integer for which $\{1, [x_i], [x_i]^2, \ldots, [x_i]^{m_i}\}$ is linearly dependent. Then there is a linear combination

$$\sum_{j=0}^{m_i} c_j [x_i]^j = [0]$$

in A in which the $c_j \in k$ are not all zero. In particular, $c_{m_i} \neq 0$ since m_i is minimal. By the definition of the quotient ring, this is equivalent to saying that

$$(2.6) \qquad\qquad p_i(x_i) = \sum_{j=0}^{m_i} c_j x_i^j \in I.$$

Exercise 2. Verify that $p_i(x_i)$ as in (2.6) is a generator of the ideal $I \cap k[x_i]$, and develop an algorithm based on this fact to find the monic generator of $I \cap k[x_i]$, given any Gröbner basis G for a zero-dimensional ideal I as input.

The algorithm suggested in Exercise 2 often requires far less computational effort than a *lex* Gröbner basis calculation. Any ordering (e.g. *grevlex*) can be used to determine G, then only standard linear algebra (matrix operations) are needed to determine whether the set $\{1, [x_i], [x_i]^2, \ldots, [x_i]^m\}$ is linearly dependent. We note that the `univpoly` function from Maple's `Groebner` package is an implementation of this method.

We will next discuss how to find the *radical* of a zero-dimensional ideal (see Chapter 1 for the definition of radical). To motivate what we will do, recall from §1 how multiple roots of a polynomial can cause problems when trying to find roots numerically. When dealing with a one-variable polynomial p with coefficients lying in a subfield of $\mathbb{C}$, it is easy to see that the polynomial

$$p_{red} = \frac{p}{\text{GCD}(p, p')}$$

has the same roots as p, but all with multiplicity one (for a proof of this, see Exercises 14 and 15 of Chapter 1, §5 of [CLO]). We call p_{red} the *square-free part* of p.

The radical $\sqrt{I}$ of an ideal I generalizes the idea of the square-free part of a polynomial. In fact, we have the following elementary exercise.

Exercise 3. If $p \in k[x]$ is a nonzero polynomial, show that $\sqrt{\langle p \rangle} = \langle p_{red} \rangle$.

Since $k[x]$ is a PID, this solves the problem of finding radicals for *all* ideals in $k[x]$. For a general ideal $I \subset k[x_1, \ldots, x_n]$, it is more difficult to find $\sqrt{I}$, though algorithms are known and have been implemented in *Macaulay 2*, REDUCE, and `Singular`. Fortunately, when I is zero-dimensional, computing the radical is much easier, as shown by the following proposition.

(2.7) Proposition. *Let $I \subset \mathbb{C}[x_1, \ldots, x_n]$ be a zero-dimensional ideal. For each $i = 1, \ldots, n$, let p_i be the unique monic generator of $I \cap \mathbb{C}[x_i]$, and let $p_{i,red}$ be the square-free part of p_i. Then*

$$\sqrt{I} = I + \langle p_{1,red}, \ldots, p_{n,red} \rangle.$$

PROOF. Write $J = I + \langle p_{1,red}, \ldots, p_{n,red} \rangle$. We first prove that J is a radical ideal, i.e., that $J = \sqrt{J}$. For each i, using the fact that $\mathbb{C}$ is algebraically closed, we can factor each $p_{i,red}$ to obtain $p_{i,red} = (x_i - a_{i1})(x_i - a_{i2}) \cdots (x_i - a_{id_i})$, where the a_{ij} are distinct. Then

$$J = J + \langle p_{1,red} \rangle = \bigcap_j (J + \langle x_1 - a_{1j} \rangle),$$

where the first equality holds since $p_{1,red} \in J$ and the second follows from Exercise 9 below since $p_{1,red}$ has distinct roots. Now use $p_{2,red}$ to decompose each $J + \langle x_1 - a_{1j} \rangle$ in the same way. This gives

$$J = \bigcap_{j,k} (J + \langle x_1 - a_{1j}, x_2 - a_{2k} \rangle).$$

If we do this for all $i = 1, 2, \ldots, n$, we get the expression

$$J = \bigcap_{j_1, \ldots, j_n} (J + \langle x_1 - a_{1j_1}, \ldots, x_n - a_{nj_n} \rangle).$$

Since $\langle x_1 - a_{1j_1}, \ldots, x_n - a_{nj_n} \rangle$ is a maximal ideal, the ideal $J + \langle x_1 - a_{1j_1}, \ldots, x_n - a_{nj_n} \rangle$ is either $\langle x_1 - a_{1j_1}, \ldots, x_n - a_{nj_n} \rangle$ or the whole ring $\mathbb{C}[x_1, \ldots, x_n]$. It follows that J is a finite intersection of maximal ideals. Since a maximal ideal is radical and an intersection of radical ideals is radical, we conclude that J is a radical ideal.

Now we can prove that $J = \sqrt{I}$. The inclusion $I \subset J$ is built into the definition of J, and the inclusion $J \subset \sqrt{I}$ follows from the Strong Nullstellensatz, since the square-free parts of the p_i vanish at all the points of $\mathbf{V}(I)$. Hence we have

$$I \subset J \subset \sqrt{I}.$$

Taking radicals in this chain of inclusions shows that $\sqrt{J} = \sqrt{I}$. But J is radical, so $\sqrt{J} = J$ and we are done. □

A Maple procedure that implements an algorithm for the radical of a zero-dimensional ideal based on Proposition (2.7) is discussed in Exercise 16 below. It is perhaps worth noting that even though we have proved Proposition (2.7) using the properties of $\mathbb{C}$, the actual computation of the polynomials $p_{i,red}$ will involve only rational arithmetic when the input polynomials are in $\mathbb{Q}[x_1, \ldots, x_n]$.

For example, consider the ideal

$$(2.8) \quad I = \langle y^4 x + 3x^3 - y^4 - 3x^2, x^2 y - 2x^2, 2y^4 x - x^3 - 2y^4 + x^2 \rangle$$

Exercise 4. Using Exercise 2 above, show that

$$I \cap \mathbb{Q}[x] = \langle x^3 - x^2 \rangle$$

and

$$I \cap \mathbb{Q}[y] = \langle y^5 - 2y^4 \rangle.$$

Writing $p_1(x) = x^3 - x^2$ and $p_2(y) = y^5 - 2y^4$, we can compute the square-free parts in Maple as follows. The command

```
p1red := simplify(p1/gcd(p1,diff(p1,x)));
```

will produce

$$p_{1,red}(x) = x(x-1).$$

Similarly,

$$p_{2,red}(y) = y(y-2).$$

Hence by Proposition (2.7), $\sqrt{I}$ is the ideal

$$\langle y^4 x + 3x^3 - y^4 - 3x^2, x^2 y - 2x^2, 2y^4 x - x^3 - 2y^4 + x^2, x(x-1), y(y-2) \rangle.$$

We note that Proposition (2.7) yields a basis, but usually *not a Gröbner basis*, for $\sqrt{I}$.

Exercise 5. How do the dimensions of the vector spaces $\mathbb{C}[x,y]/I$ and $\mathbb{C}[x,y]/\sqrt{I}$ compare in this example? How could you determine the number of distinct points in $\mathbf{V}(I)$? (There are *two*.)

We will conclude this section with a very important result relating the dimension of A and the number of points in the variety $\mathbf{V}(I)$, or what is the same, the number of solutions of the equations $f_1 = \cdots = f_s = 0$ in $\mathbb{C}^n$. To prepare for this we will need the following lemma.

(2.9) Lemma. *Let $S = \{p_1, \ldots, p_m\}$ be a finite subset of $\mathbb{C}^n$. There exist polynomials $g_i \in \mathbb{C}[x_1, \ldots, x_n]$, $i = 1, \ldots, m$, such that*

$$g_i(p_j) = \begin{cases} 0 & \text{if } i \neq j, \text{ and} \\ 1 & \text{if } i = j. \end{cases}$$

For instance, if $p_i = (a_{i1}, \ldots, a_{in})$ and the first coordinates a_{i1} are *distinct*, then we can take

$$g_i = g_i(x_1) = \frac{\prod_{j \neq i}(x_1 - a_{j1})}{\prod_{j \neq i}(a_{i1} - a_{j1})}$$

as in the *Lagrange interpolation formula*. In any case, a collection of polynomials g_i with the desired properties can be found in a similar fashion. We leave the proof to the reader as Exercise 11 below. The following theorem ties all of the results of this section together, showing how the dimension of the algebra A for a zero-dimensional ideal gives a bound on the number of points in $\mathbf{V}(I)$, and also how radical ideals are special in this regard.

(2.10) Theorem. *Let I be a zero-dimensional ideal in $\mathbb{C}[x_1, \ldots, x_n]$, and let $A = \mathbb{C}[x_1, \ldots, x_n]/I$. Then $\dim_{\mathbb{C}}(A)$ is greater than or equal to the* *number of points in* $\mathbf{V}(I)$. *Moreover, equality occurs if and only if* I *is a radical ideal.*

PROOF. Let I be a zero-dimensional ideal. By the Finiteness Theorem, $\mathbf{V}(I)$ is a finite set in $\mathbb{C}^n$, say $\mathbf{V}(I) = \{p_1, \ldots, p_m\}$. Consider the mapping

$$\varphi : \mathbb{C}[x_1, \ldots, x_n]/I \longrightarrow \mathbb{C}^m$$
$$[f] \mapsto (f(p_1), \ldots, f(p_m))$$

given by evaluating a coset at the points of $\mathbf{V}(I)$. In Exercise 12 below, you will show that φ is a well-defined linear map.

To prove the first statement in the theorem, it suffices to show that φ is onto. Let $g_1, \ldots, g_m$ be a collection of polynomials as in Lemma (2.9). Given an arbitrary $(\lambda_1, \ldots, \lambda_m) \in \mathbb{C}^m$, let $f = \sum_{i=1}^{m} \lambda_i g_i$. An easy computation shows that $\varphi([f]) = (\lambda_1, \ldots, \lambda_m)$. Thus φ is onto, and hence $\dim(A) \geq m$.

Next, suppose that I is radical. If $[f] \in \ker(\varphi)$, then $f(p_i) = 0$ for all i, so that by the Strong Nullstellensatz, $f \in \mathbf{I}(\mathbf{V}(I)) = \sqrt{I} = I$. Thus $[f] = [0]$, which shows that φ is one-to-one as well as onto. Then φ is an isomorphism, which proves that $\dim(A) = m$ if I is radical.

Conversely, if $\dim(A) = m$, then φ is an isomorphism since it is an onto linear map between vector spaces of the same dimension. Hence φ is one-to-one. We can use this to prove that I is radical as follows. Since the inclusion $I \subset \sqrt{I}$ always holds, it suffices to consider $f \in \sqrt{I} = \mathbf{I}(\mathbf{V}(I))$ and show that $f \in I$. If $f \in \sqrt{I}$, then $f(p_i) = 0$ for all i, which implies $\varphi([f]) = (0, \ldots, 0)$. Since φ is one-to-one, we conclude that $[f] = [0]$, or in other words that $f \in I$, as desired. $\qquad\square$

In Chapter 4, we will see that in the case I is *not* radical, there are well-defined multiplicities at each point in $\mathbf{V}(I)$ so that the sum of the multiplicities equals $\dim(A)$.

ADDITIONAL EXERCISES FOR §2

Exercise 6. Using the *grevlex* order, construct the monomial basis B for the quotient algebra $A = \mathbb{C}[x, y]/I$, where I is the ideal from (2.8) and construct the multiplication table for B in A.

Exercise 7. In this exercise, we will explain how the ideal $I = \langle x^2 + 3xy/2 + y^2/2 - 3x/2 - 3y/2, xy^2 - x, y^3 - y \rangle$ from (2.4) was constructed. The basic idea was to start from a finite set of points and construct a system of equations, rather than the reverse.

To begin, consider the maximal ideals

$$I_1 = \langle x, y \rangle, \qquad I_2 = \langle x - 1, y - 1 \rangle,$$
$$I_3 = \langle x + 1, y - 1 \rangle, \qquad I_4 = \langle x - 1, y + 1 \rangle,$$
$$I_5 = \langle x - 2, y + 1 \rangle$$

in $\mathbb{C}[x, y]$. Each variety $\mathbf{V}(I_j)$ is a single point in $\mathbb{C}^2$, indeed in $\mathbb{Q}^2 \subset \mathbb{C}^2$. The union of the five points forms an affine variety V, and by the algebra-geometry dictionary from Chapter 1, $V = \mathbf{V}(I_1 \cap I_2 \cap \cdots \cap I_5)$.

An algorithm for intersecting ideals is described in Chapter 1. Use it to compute the intersection $I = I_1 \cap I_2 \cap \cdots \cap I_5$ and find the reduced Gröbner basis for I with respect to the *grevlex* order $(x > y)$. Your result should be the Gröbner basis given in (2.4).

Exercise 8.
a. Use the method of Proposition (2.7) to show that the ideal I from (2.4) is a radical ideal.
b. Give a non-computational proof of the statement from part a using the following observation. By the form of the generators of each of the ideals I_j in Exercise 7, $\mathbf{V}(I_j)$ is a single point and I_j is the ideal $\mathbf{I}(\mathbf{V}(I_j))$. As a result, $I_j = \sqrt{I_j}$ by the Strong Nullstellensatz. Then use the general fact about intersections of radical ideals from part a Exercise 9 from §4 of Chapter 1.

Exercise 9. This exercise is used in the proof of Proposition (2.7). Suppose we have an ideal $I \subset k[x_1, \ldots, x_n]$, and let $p = (x_1 - a_1) \cdots (x_1 - a_d)$, where $a_1, \ldots, a_d$ are distinct. The goal of this exercise is to prove that

$$I + \langle p \rangle = \bigcap_j (I + \langle x_1 - a_j \rangle).$$

a. Prove that $I + \langle p \rangle \subset \bigcap_j (I + \langle x_1 - a_j \rangle)$.
b. Let $p_j = \prod_{i \neq j}(x_1 - a_i)$. Prove that $p_j \cdot (I + \langle x_1 - a_j \rangle) \subset I + \langle p \rangle$.
c. Show that $p_1, \ldots, p_n$ are relatively prime, and conclude that there are polynomials $h_1, \ldots, h_n$ such that $1 = \sum_j h_j p_j$.
d. Prove that $\bigcap_j (I + \langle x_1 - a_j \rangle) \subset I + \langle p \rangle$. Hint: Given h in the intersection, write $h = \sum_j h_j p_j h$ and use part b.

Exercise 10. (The Dual Space of $k[x_1, \ldots, x_n]/I$) Recall that if V is a vector space over a field k, then the *dual space* of V, denoted V^*, is the k-vector space of linear mappings $L : V \to k$. If V is finite-dimensional, then so is V^*, and $\dim V = \dim V^*$. Let I be a zero-dimensional ideal in $k[x_1, \ldots, x_n]$, and consider $A = k[x_1, \ldots, x_n]/I$ with its k-vector space structure. Let G be a Gröbner basis for I with respect to some monomial ordering, and let $B = \{x^{\alpha(1)}, \ldots, x^{\alpha(d)}\}$ be the corresponding monomial basis for A, so that for each $f \in k[x_1, \ldots, x_n]$,

$$\overline{f}^G = \sum_{j=1}^{d} c_j(f) x^{\alpha(j)}$$

for some $c_j(f) \in k$.

a. Show that each of the functions $c_j(f)$ is a linear function of $f \in k[x_1, \ldots, x_n]$. Moreover, show that $c_j(f) = 0$ for all j if and only if $f \in I$, or equivalently $[f] = 0$ in A.

b. Deduce that the collection B^* of mappings c_j given by $f \mapsto c_j(f)$, $j = 1, \ldots, d$ gives a *basis* of the dual space A^*.

c. Show that B^* is the *dual basis* corresponding to the basis B of A. That is, show that

$$c_j(x^{\alpha(i)}) = \begin{cases} 1 & \text{if } i = j \\ 0 & \text{otherwise.} \end{cases}$$

Exercise 11. Let $S = \{p_1, \ldots, p_m\}$ be a finite subset of $\mathbb{C}^n$.

a. Show that there exists a linear polynomial $\ell(x_1, \ldots, x_n)$ whose values at the points of S are *distinct*.

b. Using the linear polynomial ℓ from part a, show that there exist polynomials $g_i \in \mathbb{C}[x_1, \ldots, x_n]$, $i = 1, \ldots, m$, such that

$$g_i(p_j) = \begin{cases} 0 & \text{if } i \neq j, \text{ and} \\ 1 & \text{if } i = j. \end{cases}$$

Hint: Mimic the construction of the Lagrange interpolation polynomials in the discussion after the statement of Lemma (2.9).

Exercise 12. As in Theorem (2.10), suppose that $\mathbf{V}(I) = \{p_1, \ldots, p_m\}$.

a. Prove that the map $\varphi : \mathbb{C}[x_1, \ldots, x_n]/I \to \mathbb{C}^m$ given by evaluation at $p_1, \ldots, p_m$ is a well-defined linear map. Hint: $[f] = [g]$ implies $f - g \in I$.

b. We can regard $\mathbb{C}^m$ as a ring with coordinate-wise multiplication. Thus

$$(a_1, \ldots, a_m) \cdot (b_1, \ldots, b_m) = (a_1 b_1, \ldots, a_m b_m).$$

With this ring structure, $\mathbb{C}^m$ is a direct product of m copies of $\mathbb{C}$. Prove that the map φ of part a is a ring homomorphism.

c. Prove that φ is a ring isomorphism if and only if I is radical. This means that in the radical case, we can express A as a direct product of the simpler rings (namely, m copies of $\mathbb{C}$). In Chapter 4, we will generalize this result to the nonradical case.

Exercise 13. In Maple, the `SetBasis` command finds a monomial basis B for the quotient algebra $A = k[x_1, \ldots, x_n]/I$ for a zero-dimensional ideal I. However, it is instructive to have the following "home-grown" version called `kbasis` which makes it easier to see what is happening.

```
kbasis := proc(GB,VList,torder)

  # returns a list of monomials forming a basis of the quotient
  # ring, where GB is a Groebner basis for a zero-dimensional
  # ideal, and generates an error message if the ideal is not
  # 0-dimensional.

  local B,C,v,t,l,m,leadmons,i;

  if is_finite(GB,VList) then
     leadmons:={seq(leadterm(GB[i],torder),i=1..nops(GB))};
     B:=[1];
     for v in VList do
        m:=degree(univpoly(v,GB),v);
        C:=B;
        for t in C do
          for l to m-1 do
             t:=t*v;
             if evalb(not(1 in map(u->denom(t/u),leadmons))) then
                B:=[op(B),t];
             end if;
           end do;
         end do;
       end do;
     return B;
  else
     print('ideal is not zero-dimensional');
  end if
end proc:
```

a. Show that kbasis correctly computes $\{x^\alpha : x^\alpha \notin \langle \mathrm{LT}(I)\rangle\}$ if A is finite-dimensional over k and terminates for all inputs.
b. Use either kbasis or SetBasis to check the results for the ideal from (2.4).
c. Use either kbasis or SetBasis to check your work from Exercise 6 above.

Exercise 14. The algorithm used in the procedure from Exercise 13 can be improved considerably. The "box" R that kbasis searches for elements of the complement of $\langle \mathrm{LT}(I)\rangle$ is often much larger than necessary. This is because the call to univpoly, which finds a monic generator for $I \cap k[x_i]$ for each i, gives an m_i such that $x_i^{m_i} \in \langle \mathrm{LT}(I)\rangle$, but m_i might not be as small as possible. For instance, consider the ideal I from (2.4). The monic generator of $I \cap \mathbb{C}[x]$ has degree 4 (check this). Hence kbasis computes

$\overline{x^2}^G$, $\overline{x^3}^G$ and rejects these monomials since they are not remainders. But the Gröbner basis G given in (2.4) shows that $x^2 \in \langle \mathrm{LT}(I) \rangle$. Thus a smaller set of α containing the exponents of the monomial basis B can be determined directly by examining the leading terms of the Gröbner basis G, without using univpoly to get the monic generator for $I \cap k[x_i]$. Develop and implement an improved kbasis that takes this observation into account.

Exercise 15. Using either Setbasis or kbasis, develop and implement a procedure that computes the multiplication table for a finite-dimensional algebra A.

Exercise 16. Implement the following Maple procedure for finding the radical of a zero-dimensional ideal given by Proposition (2.7) and test it on the examples from this section.

```
zdimradical := proc(PList,VList)

   # constructs a set of generators for the radical of a
   # zero-dimensional ideal.

   local p,pred,v,RList;

   if is_finite(PList,VList) then
     RList := PList;
     for v in VList do
       p := univpoly(v,PList);
       pred := simplify(p/gcd(p,diff(p,v)));
       RList:=[op(RList),pred]
     end do;
     return RList
   else
     print('Ideal not zero-dimensional; method does not apply')
   end if
end proc:
```

Exercise 17. Let $I \subset \mathbb{C}[x_1, \ldots, x_n]$ be an ideal such that for every $1 \leq i \leq n$, there is a square-free polynomial p_i such that $p_i(x_i) \in I$. Use Proposition (2.7) to show that I is radical.

Exercise 18. For $1 \leq i \leq n$, let p_i be a square-free polynomial. Also let $d_i = \deg(p_i)$. The goal of this exercise is to prove that $\langle p_1(x_1), \ldots, p_n(x_n) \rangle$ is radical using only the division algorithm.

a. Let r be the remainder of $f \in \mathbb{C}[x_1, \ldots, x_n]$ on division by the $p_i(x_i)$. Prove that r has degree at most $d_i - 1$ in x_i.

b. Prove that r vanishes on $\mathbf{V}(p_1(x_1), \ldots, p_n(x_n))$ if and only if r is identically 0.

c. Conclude that $\langle p_1(x_1), \ldots, p_n(x_n) \rangle$ is radical without using Proposition (2.7).

Exercise 19. In this exercise, you will use Exercise 18 to give an elementary proof of the result of Exercise 17. Thus we assume that $I \subset \mathbb{C}[x_1, \ldots, x_n]$ is an ideal such that for every $1 \leq i \leq n$, there is a square-free polynomial p_i such that $p_i(x_i) \in I$. Take $f \in \mathbb{C}[x_1, \ldots, x_n]$ such that $f^N \in I$ for some $N > 0$. Let z be a new variable and set $J = \langle p_1(x_1), \ldots, p_n(x_n), z - f \rangle \subset \mathbb{C}[x_1, \ldots, x_n, z]$.

a. Prove that there is a ring isomorphism

$$\mathbb{C}[x_1, \ldots, x_n, z]/J \cong \mathbb{C}[x_1, \ldots, x_n]/\langle p_1(x_1), \ldots, p_n(x_n) \rangle$$

and conclude via Exercise 18 that J is zero-dimensional and radical.

b. Without using Proposition (2.7), show that there is a square-free polynomial g such that $g(z) \in J$.

c. Explain why $\mathrm{GCD}(g, z^N)$ is 1 or z, and conclude that $z = p(z)g(z) + q(z)z^N$ for some polynomials p, q.

d. Under the isomorphism of part a, show that $z = p(z)g(z) + q(z)z^N$ maps to $f = q(f)f^N + h$, where $h \in \langle p_1(x_1), \ldots, p_n(x_n) \rangle$. Conclude that $f \in I$.

This argument is due to M. Mereb.

§3 Gröbner Basis Conversion

In this section, we will use linear algebra in $A = k[x_1, \ldots, x_n]/I$ to show that a Gröbner basis G for a zero-dimensional ideal I with respect to one monomial order can be converted to a Gröbner basis G' for the same ideal with respect to *any* other monomial order. The process is sometimes called *Gröbner basis conversion*, and the idea comes from a paper of Faugère, Gianni, Lazard, and Mora [FGLM]. We will illustrate the method by converting from an arbitrary Gröbner basis G to a *lex* Gröbner basis G_{lex} (using any ordering on the variables). The Gröbner basis conversion method is often used in precisely this situation, so that a more favorable monomial order (such as *grevlex*) can be used in the application of Buchberger's algorithm, and the result can then be converted into a form more suited for equation solving via elimination. For another discussion of this topic, see [BW], §1 of Chapter 9.

The basic idea of the Faugère-Gianni-Lazard-Mora algorithm is quite simple. We start with a Gröbner basis G for a zero-dimensional ideal I, and we want to convert G to a *lex* Gröbner basis G_{lex} for some *lex* order. The algorithm steps through monomials in $k[x_1, \ldots, x_n]$ in increasing *lex* order. At each step of the algorithm, we have a list $G_{lex} = \{g_1, \ldots, g_k\}$ of elements in I (initially empty, and at each stage a subset of the eventual *lex* Gröbner basis), and a list B_{lex} of monomials (also initially empty, and at each stage a subset of the eventual *lex* monomial basis for A). For each input monomial x^α (initially 1), the algorithm consists of three steps:

(3.1) Main Loop. Given the input x^α, compute $\overline{x^\alpha}^G$. Then:

a. If $\overline{x^\alpha}^G$ is *linearly dependent* on the remainders (on division by G) of the monomials in B_{lex}, then we have a linear combination

$$\overline{x^\alpha}^G - \sum_j c_j \overline{x^{\alpha(j)}}^G = 0,$$

where $x^{\alpha(j)} \in B_{lex}$ and $c_j \in k$. This implies that

$$g = x^\alpha - \sum_j c_j x^{\alpha(j)} \in I.$$

We add g to the list G_{lex} as the last element. Because the x^α are considered in increasing *lex* order (see (3.3) below), whenever a polynomial g is added to G_{lex}, its leading term is $\mathrm{LT}(g) = x^\alpha$ with coefficient 1.

b. If $\overline{x^\alpha}^G$ is *linearly independent* from the remainders (on division by G) of the monomials in B_{lex}, then we add x^α to B_{lex} as the last element.

After the Main Loop acts on the monomial x^α, we test G_{lex} to see if we have the desired Gröbner basis. This test needs to be done only if we added a polynomial g to G_{lex} in part a of the Main Loop.

(3.2) Termination Test. If the Main Loop added a polynomial g to G_{lex}, then compute $\mathrm{LT}(g)$. If $\mathrm{LT}(g)$ is a power of x_1, where x_1 is the greatest variable in our *lex* order, then the algorithm terminates.

The proof of Theorem (3.4) below will explain why this is the correct way to terminate the algorithm. If the algorithm does not stop at this stage, we use the following procedure to find the next input monomial for the Main Loop:

(3.3) Next Monomial. Replace x^α with the next monomial in *lex* order which is not divisible by any of the monomials $\mathrm{LT}(g_i)$ for $g_i \in G_{lex}$.

Exercise 3 below will explain how the Next Monomial procedure works. Now repeat the above process by using the new x^α as input to the Main Loop, and continue until the Termination Test tells us to stop.

Before we prove the correctness of this algorithm, let's see how it works in an example.

Exercise 1. Consider the ideal

$$I = \langle xy + z - xz, x^2 - z, 2x^3 - x^2yz - 1 \rangle$$

in $\mathbb{Q}[x, y, z]$. For *grevlex* order with $x > y > z$, I has a Gröbner basis
$G = \{f_1, f_2, f_3, f_4\}$, where

$$f_1 = z^4 - 3z^3 - 4yz + 2z^2 - y + 2z - 2$$
$$f_2 = yz^2 + 2yz - 2z^2 + 1$$
$$f_3 = y^2 - 2yz + z^2 - z$$
$$f_4 = x + y - z.$$

Thus $\langle \mathrm{LT}(I) \rangle = \langle z^4, yz^2, y^2, x \rangle$, $B = \{1, y, z, z^2, z^3, yz\}$, and a remainder
$\overline{f}^G$ is a linear combination of elements of B. We will use basis conversion
to find a *lex* Gröbner basis for I, with $z > y > x$.

a. Carry out the Main Loop for $x^{\alpha} = 1, x, x^2, x^3, x^4, x^5, x^6$. At the end of
doing this, you should have

$$G_{lex} = \{x^6 - x^5 - 2x^3 + 1\}$$
$$B_{lex} = \{1, x, x^2, x^3, x^4, x^5\}.$$

Hint: The following computations will be useful:

$$\overline{1}^G = 1$$
$$\overline{x}^G = -y + z$$
$$\overline{x^2}^G = z$$
$$\overline{x^3}^G = -yz + z^2$$
$$\overline{x^4}^G = z^2$$
$$\overline{x^5}^G = z^3 + 2yz - 2z^2 + 1$$
$$\overline{x^6}^G = z^3.$$

Note that $\overline{1}^G, \ldots, \overline{x^5}^G$ are linearly independent while $\overline{x^6}^G$ is a linear
combination of $\overline{x^5}^G$, $\overline{x^3}^G$ and $\overline{1}^G$. This is similar to Exercise 2 of §2.

b. After we apply the Main Loop to x^6, show that the monomial provided
by the Next Monomial procedure is y, and after y passes through the
Main Loop, show that

$$G_{lex} = \{x^6 - x^5 - 2x^3 + 1, y - x^2 + x\}$$
$$B_{lex} = \{1, x, x^2, x^3, x^4, x^5\}.$$

c. Show that after y, Next Monomial produces z, and after z passes through
the Main Loop, show that

$$G_{lex} = \{x^6 - x^5 - 2x^3 + 1, y - x^2 + x, z - x^2\}$$
$$B_{lex} = \{1, x, x^2, x^3, x^4, x^5\}.$$

d. Check that the Termination Test (3.2) terminates the algorithm when
G_{lex} is as in part c. Hint: We're using *lex* order with $z > y > x$.

e. Verify that G_{lex} from part c is a *lex* Gröbner basis for I.

We will now show that the algorithm given by (3.1), (3.2) and (3.3) terminates and correctly computes a *lex* Gröbner basis for the ideal I.

(3.4) Theorem. *The algorithm described above terminates on every input Gröbner basis G generating a zero-dimensional ideal I, and correctly computes a lex Gröbner basis G_{lex} for I and the lex monomial basis B_{lex} for the quotient ring A.*

PROOF. We begin with the key observation that monomials are added to the list B_{lex} in strictly increasing *lex* order. Similarly, if $G_{lex} = \{g_1, \ldots, g_k\}$, then

$$\text{LT}(g_1) <_{lex} \cdots <_{lex} \text{LT}(g_k),$$

where $>_{lex}$ is the *lex* order we are using. We also note that when the Main Loop adds a new polynomial g_{k+1} to $G_{lex} = \{g_1, \ldots, g_k\}$, the leading term $\text{LT}(g_{k+1})$ is the input monomial in the Main Loop. Since the input monomials are provided by the Next Monomial procedure, it follows that for all k,

(3.5) $\text{LT}(g_{k+1})$ is divisible by none of $\text{LT}(g_1), \ldots, \text{LT}(g_k)$.

We can now prove that the algorithm terminates for all inputs G generating zero-dimensional ideals. If the algorithm did not terminate for some input G, then the Main Loop would be executed infinitely many times, so one of the two alternatives in (3.1) would be chosen infinitely often. If the first alternative were chosen infinitely often, G_{lex} would give an infinite list $\text{LT}(g_1), \text{LT}(g_2), \ldots$ of monomials. However, we have:

- (Dickson's Lemma) Given an infinite list $x^{\alpha(1)}, x^{\alpha(2)}, \ldots$ of monomials in $k[x_1, \ldots, x_n]$, there is an integer N such that every $x^{\alpha(i)}$ is divisible by one of $x^{\alpha(1)}, \ldots, x^{\alpha(N)}$.

(See, for example, Exercise 7 of [CLO], Chapter 2, §4). When applied to $\text{LT}(g_1), \text{LT}(g_2), \ldots$, Dickson's Lemma would contradict (3.5). On the other hand, if the second alternative were chosen infinitely often, then B_{lex} would give infinitely many monomials $x^{\alpha(j)}$ whose remainders on division by G were linearly independent in A. This would contradict the assumption that I is zero-dimensional. As a result, the algorithm always terminates for G generating a zero-dimensional ideal I.

Next, suppose that the algorithm terminates with $G_{lex} = \{g_1, \ldots, g_k\}$. By the Termination Test (3.2), $\text{LT}(g_k) = x_1^{a_1}$, where $x_1 >_{lex} \cdots >_{lex} x_n$. We will prove that G_{lex} is a *lex* Gröbner basis for I by contradiction. Suppose there were some $g \in I$ such that $\text{LT}(g)$ is not a multiple of any of the $\text{LT}(g_i)$, $i = 1, \ldots, k$. Without loss of generality, we may assume that g is *reduced* with respect to G_{lex} (replace g by $\overline{g}^{G_{lex}}$).

If $\mathrm{LT}(g)$ is greater than $\mathrm{LT}(g_k) = x_1^{a_1}$, then one easily sees that $\mathrm{LT}(g)$ is a multiple of $\mathrm{LT}(g_k)$ (see Exercise 2 below). Hence this case can't occur, which means that

$$\mathrm{LT}(g_i) < \mathrm{LT}(g) \le \mathrm{LT}(g_{i+1})$$

for some $i < k$. But recall that the algorithm places monomials into B_{lex} in strictly increasing order, and the same is true for the $\mathrm{LT}(g_i)$. All the non-leading monomials in g must be less than $\mathrm{LT}(g)$ in the *lex* order. They are not divisible by any of $\mathrm{LT}(g_j)$ for $j \le i$, since g is reduced. So, the non-leading monomials that appear in g would have been included in B_{lex} by the time $\mathrm{LT}(g)$ was reached by the Next Monomial procedure, and g would have been the next polynomial after g_i included in G_{lex} by the algorithm (i.e., g would equal g_{i+1}). This contradicts our assumption on g, which proves that G_{lex} is a *lex* Gröbner basis for I.

The final step in the proof is to show that when the algorithm terminates, B_{lex} consists of *all* basis monomials determined by the Gröbner basis G_{lex}. We leave this as an exercise for the reader. □

In the literature, the basis conversion algorithm discussed here is called the *FGLM algorithm* after the authors Faugère, Gianni, Lazard, and Mora of the paper [FGLM] in which the algorithm first appeared. We should also mention that while the FGLM algorithm assumes that I is zero-dimensional, there are methods which apply to the positive-dimensional case. For instance, if degree bounds on the elements of the Gröbner basis with respect to the desired order are known, then the approach described above can also be adapted to treat ideals that are not zero-dimensional. An interesting related "Hilbert function-driven" basis conversion method for homogeneous ideals has been proposed by Traverso (see [Trav]). However, general basis conversion methods that apply even when information such as degree bounds is not available are also desirable. Such a method is the *Gröbner Walk* to be described in Chapter 8.

The ideas used in Gröbner basis conversion can be applied in other contexts. In order to explain this, we need to recast the above discussion using linear maps. Recall that we began with a Gröbner basis G of a zero-dimensional ideal I and our goal was to find a *lex* Gröbner basis G_{lex} of I. However, for G, the main thing we used was the normal form $\overline{f}^G$ of a polynomial $f \in k[x_1, \dots, x_n]$.

Let's write this out carefully. Let B be the monomial basis of $A = k[x_1, \dots, x_n]/I$ determined by G. Denote $\overline{f}^G$ by $L(f)$ and $\mathrm{Span}(B)$ by V, so that $L(f) = \overline{f}^G \in V = \mathrm{Span}(B)$. Thus we have a map

$$(3.6) \qquad\qquad L : k[x_1, \dots, x_n] \longrightarrow V.$$

In Exercise 10 of §2, you showed that L is linear with kernel equal to I. Using this, the Main Loop (3.1) can be written as follows.

(3.7) Main Loop, Restated. Given the input x^α, compute $L(x^\alpha)$. Then:

a. If $L(x^\alpha)$ is *linearly dependent* on the images under L of the monomials in B_{lex}, then we have a linear combination

$$L(x^\alpha) - \sum_j c_j L(x^{\alpha(j)}) = 0,$$

where $x^{\alpha(j)} \in B_{lex}$ and $c_j \in k$. This implies that $L\big(x - \sum_j c_j x^{\alpha(j)}\big) = 0$. Since I is the kernel of L, we have

$$g = x^\alpha - \sum_j c_j x^{\alpha(j)} \in I.$$

We add g to G_{lex} as the last element.

b. If $L(x^\alpha)$ is *linearly independent* from the images under L of the monomials in B_{lex}, then we add x^α to B_{lex} as the last element.

If we combine (3.7) with the Termination Test (3.2) and Next Monomial (3.3), then we get the same algorithm as before. But even more is true, for this algorithm computes a *lex* Gröbner basis of the kernel for *any* linear map (3.6), provided that V has finite dimension and the kernel is an ideal of $k[x_1, \ldots, x_n]$. You will prove this in Exercise 9 below.

As an example of how this works, pick distinct points $p_1, \ldots, p_m \in k^n$ and consider the evaluation map

$$L : k[x_1, \ldots, x_n] \longrightarrow k^m, \quad L(f) = (f(p_1), \ldots, f(p_m)).$$

The kernel is the ideal $\mathbf{I}(p_1, \ldots, p_m)$ of polynomials vanishing at the given points. It follows that we now have an algorithm for computing a *lex* Gröbner basis of this ideal! This is closely related to the Buchberger-Möller algorithm described in [BuM]. You will work out an explicit example in Exercise 10.

For another example, consider

(3.8) $I = \{f \in \mathbb{C}[x, y] : f(0, 0) = f_x(0, 0) = f_y(0, 0) - f_{xx}(0, 0) = 0\}.$

In Exercise 11, you will show that I is an ideal of $\mathbb{C}[x, y]$. Since I is the kernel of the linear map

$$L : \mathbb{C}[x, y] \longrightarrow \mathbb{C}^3, \quad L(f) = (f(0, 0), f_x(0, 0), f_y(0, 0) - f_{xx}(0, 0)),$$

the above algorithm can be used to show that $\{y^2, xy, x^2 + 2y\}$ is a *lex* Gröbner basis with $x > y$ for the ideal I. See Exercise 11 for the details.

There are some very interesting ideas related to these examples. Differential conditions like those in (3.8), when combined with primary decomposition, can be used to describe any zero-dimensional ideal in $k[x_1, \ldots, x_n]$. This is explained in [MMM1] and [MöS] (and is where we got (3.8)). The paper [MMM1] also describes other situations where these ideas are useful, and [MMM2] makes a systematic study of the different representations of a zero-dimensional ideal and how one can pass from one representation to another.

ADDITIONAL EXERCISES FOR §3

Exercise 2. Consider the *lex* order with $x_1 > \cdots > x_n$ and fix a power x_1^a of x_1. Then, for any monomial x^α in $k[x_1, \ldots, x_n]$, prove that $x^\alpha > x_1^a$ if and only if x^α is divisible by x_1^a.

Exercise 3. Suppose $G_{lex} = \{g_1, \ldots, g_k\}$, where $\mathrm{LT}(g_1) < \cdots < \mathrm{LT}(g_k)$, and let x^α be a monomial. This exercise will show how the Next Monomial (3.3) procedure works, assuming that our *lex* order satisfies $x_1 > \cdots > x_n$. Since this procedure is only used when the Termination Test fails, we can assume that $\mathrm{LT}(g_k)$ is *not* a power of x_1.
a. Use Exercise 2 to show that none of the $\mathrm{LT}(g_i)$ divide $x_1^{a_1+1}$.
b. Now consider the *largest* $1 \le k \le n$ such that none of the $\mathrm{LT}(g_i)$ divide the monomial

$$x_1^{a_1} \cdots x_{k-1}^{a_{k-1}} x_k^{a_k+1}.$$

By part a, $k = 1$ has this property, so there must be a largest such k. If x^β is the monomial corresponding to the largest k, prove that $x^\beta > x^\alpha$ is the smallest monomial (relative to our *lex* order) greater than x^α which is not divisible by any of the $\mathrm{LT}(g_i)$.

Exercise 4. Complete the proof of Theorem (3.4) by showing that when the basis conversion algorithm terminates, the set B_{lex} gives a monomial basis for the quotient ring A.

Exercise 5. Use Gröbner basis conversion to find *lex* Gröbner bases for the ideals in Exercises 6 and 7 from §1. Compare with your previous results.

Exercise 6. What happens if you try to apply the basis conversion algorithm to an ideal that is *not* zero-dimensional? Can this method be used for general Gröbner basis conversion? What if you have more information about the *lex* basis elements, such as their total degrees, or bounds on those degrees?

Exercise 7. Show that the output of the basis conversion algorithm is actually a monic *reduced lex* Gröbner basis for $I = \langle G \rangle$.

Exercise 8. Implement the basis conversion algorithm outlined in (3.1), (3.2) and (3.3) in a computer algebra system. Hint: Exercise 3 will be useful. For a more complete description of the algorithm, see pages 428–433 of [BW].

Exercise 9. Consider a linear map $L : k[x_1, \ldots, x_n] \to V$, where V has finite dimension and the kernel of L is an ideal. State and prove a version of Theorem (3.4) which uses (3.7), (3.2), and (3.3).

Exercise 10. Use the method described at the end of the section to find a *lex* Gröbner basis with $x > y$ for the ideal of all polynomials vanishing at $(0,0), (1,0), (0,1) \in k^2$.

Exercise 11. Prove that (3.8) is an ideal of $\mathbb{C}[x, y]$ and use the method described at the end of the section to find a *lex* Gröbner basis with $x > y$ for this ideal.

§4 Solving Equations via Eigenvalues and Eigenvectors

The central problem of this chapter, finding the solutions of a system of polynomial equations $f_1 = f_2 = \cdots = f_s = 0$ over $\mathbb{C}$, rephrases in fancier language to finding the points of the variety $\mathbf{V}(I)$, where I is the ideal generated by $f_1, \ldots, f_s$. When the system has only finitely many solutions, i.e., when $\mathbf{V}(I)$ is a finite set, the Finiteness Theorem from §2 says that I is a zero-dimensional ideal and the algebra $A = \mathbb{C}[x_1, \ldots, x_n]/I$ is a finite-dimensional vector space over $\mathbb{C}$. The first half of this section exploits the structure of A in this case to evaluate an arbitrary polynomial f at the points of $\mathbf{V}(I)$; in particular, evaluating the polynomials $f = x_i$ gives the coordinates of the points (Corollary (4.6) below). The values of f on $\mathbf{V}(I)$ turn out to be *eigenvalues* of certain linear mappings on A. We will discuss techniques for computing these eigenvalues and show that the corresponding *eigenvectors* contain useful information about the solutions.

We begin with the easy observation that given a polynomial $f \in \mathbb{C}[x_1, \ldots, x_n]$, we can use multiplication to define a linear map m_f from $A = \mathbb{C}[x_1, \ldots, x_n]/I$ to itself. More precisely, f gives the coset $[f] \in A$, and we define $m_f : A \to A$ by the rule: if $[g] \in A$, then

$$m_f([g]) = [f] \cdot [g] = [fg] \in A.$$

Then m_f has the following basic properties.

(4.1) Proposition. *Let $f \in \mathbb{C}[x_1, \ldots, x_n]$. Then*
a. *The map m_f is a linear mapping from A to A.*
b. *We have $m_f = m_g$ exactly when $f - g \in I$. Thus two polynomials give the same linear map if and only if they differ by an element of I. In particular, m_f is the zero map exactly when $f \in I$.*

PROOF. The proof of part a is just the distributive law for multiplication over addition in the ring A. If $[g], [h] \in A$ and $c \in k$, then

$$m_f(c[g] + [h]) = [f] \cdot (c[g] + [h]) = c[f] \cdot [g] + [f] \cdot [h] = cm_f([g]) + m_f([h]).$$

Part b is equally easy. Since $[1] \in A$ is a multiplicative identity, if $m_f = m_g$, then

$$[f] = [f] \cdot [1] = m_f([1]) = m_g([1]) = [g] \cdot [1] = [g],$$

so $f - g \in I$. Conversely, if $f - g \in I$, then $[f] = [g]$ in A, so $m_f = m_g$. □

Since A is a finite-dimensional vector space over $\mathbb{C}$, we can represent m_f by its matrix with respect to a basis. For our purposes, a monomial basis B such as the ones we considered in §2 will be the most useful, because once we have the multiplication table for the elements in B, the matrices of the multiplication operators m_f can be read off immediately from the table. We will denote this matrix also by m_f, and whether m_f refers to the matrix or the linear operator will be clear from the context. Proposition (4.1) implies that $m_f = m_{\overline{f}^G}$, so that we may assume that f is a remainder.

For example, for the ideal I from (2.4) of this chapter, the matrix for the multiplication operator by f may be obtained from the table (2.5) in the usual way. Ordering the basis monomials as before,

$$B = \{1, x, y, xy, y^2\},$$

we make a 5×5 matrix whose jth column is the vector of coefficients in the expansion in terms of B of the image under m_f of the jth basis monomial. With $f = x$, for instance, we obtain

$$m_x = \begin{pmatrix} 0 & 0 & 0 & 0 & 0 \\ 1 & 3/2 & 0 & -3/2 & 1 \\ 0 & 3/2 & 0 & -1/2 & 0 \\ 0 & -3/2 & 1 & 3/2 & 0 \\ 0 & -1/2 & 0 & 3/2 & 0 \end{pmatrix}.$$

Exercise 1. Find the matrices m_1, m_y, m_{xy-y^2} with respect to B in this example. How do m_{y^2} and $(m_y)^2$ compare? Why?

We note the following useful general properties of the matrices m_f (the proof is left as an exercise).

(4.2) Proposition. *Let f, g be elements of the algebra A. Then*
a. $m_{f+g} = m_f + m_g$.
b. $m_{f \cdot g} = m_f \cdot m_g$ *(where the product on the right means composition of linear operators or matrix multiplication).*

This proposition says that the map sending $f \in \mathbb{C}[x_1, \ldots, x_n]$ to the matrix m_f defines a *ring homomorphism* from $\mathbb{C}[x_1, \ldots, x_n]$ to the ring $M_{d \times d}(\mathbb{C})$ of $d \times d$ matrices, where d is the dimension of A as a $\mathbb{C}$-vector space. Furthermore, part b of Proposition (4.1) and the Fundamental Theorem of Homomorphisms show that $[f] \mapsto m_f$ induces a one-to-one homomorphism $A \to M_{d \times d}(\mathbb{C})$. A discussion of ring homomorphisms and the

Fundamental Theorem of Homomorphisms may be found in Chapter 5, §2 of [CLO], especially Exercise 16. But the reader should note that $M_{d\times d}(\mathbb{C})$ is not a commutative ring, so we have here a slightly more general situation than the one discussed there.

For use later, we also point out a corollary of Proposition (4.2). Let $h(t) = \sum_{i=0}^{m} c_i t^i \in \mathbb{C}[t]$ be a polynomial. The expression $h(f) = \sum_{i=0}^{m} c_i f^i$ makes sense as an element of $\mathbb{C}[x_1, \ldots, x_n]$. Similarly $h(m_f) = \sum_{i=0}^{m} c_i (m_f)^i$ is a well-defined matrix (the term c_0 should be interpreted as $c_0 I$, where I is the $d \times d$ identity matrix).

(4.3) Corollary. *In the situation of Proposition (4.2), let $h \in \mathbb{C}[t]$ and $f \in \mathbb{C}[x_1, \ldots, x_n]$. Then*

$$m_{h(f)} = h(m_f).$$

Recall that a polynomial $f \in \mathbb{C}[x_1, \ldots, x_n]$ gives the coset $[f] \in A$. Since A is finite-dimensional, as we noted in §2 for $f = x_i$, the set $\{1, [f], [f]^2, \ldots\}$ must be *linearly dependent* in the vector space structure of A. In other words, there is a linear combination

$$\sum_{i=0}^{m} c_i [f]^i = [0]$$

in A, where $c_i \in \mathbb{C}$ are not all zero. By the definition of the quotient ring, this is equivalent to saying that

(4.4)
$$\sum_{i=0}^{m} c_i f^i \in I.$$

Hence $\sum_{i=0}^{m} c_i f^i$ vanishes at every point of $\mathbf{V}(I)$.

Now we come to the most important part of this discussion, culminating in Theorem (4.5) and Corollary (4.6) below. We are looking for the points in $\mathbf{V}(I)$, I a zero-dimensional ideal. Let $h(t) \in \mathbb{C}[t]$, and let $f \in \mathbb{C}[x_1, \ldots, x_n]$. By Corollary (4.3),

$$h(m_f) = 0 \qquad \Longleftrightarrow \qquad h([f]) = [0] \text{ in } A.$$

The polynomials h such that $h(m_f) = 0$ form an ideal in $\mathbb{C}[t]$ by the following exercise.

Exercise 2. Given a $d \times d$ matrix M with entries in a field k, consider the collection I_M of polynomials $h(t)$ in $k[t]$ such that $h(M) = 0$, the $d \times d$ zero matrix. Show that I_M is an ideal in $k[t]$.

The nonzero monic generator h_M of the ideal I_M is called the *minimal polynomial* of M. By the basic properties of ideals in $k[t]$, if h is any polynomial with $h(M) = 0$, then the minimal polynomial h_M divides h. In particular, the Cayley-Hamilton Theorem from linear algebra tells us that

h_M divides the characteristic polynomial of M. As a consequence, if $k = \mathbb{C}$, the roots of h_M are eigenvalues of M. Furthermore, all eigenvalues of M occur as roots of the minimal polynomial. See [Her] for a more complete discussion of the Cayley-Hamilton Theorem and the minimal polynomial of a matrix.

Let h_f denote the minimal polynomial of the multiplication operator m_f on A. We then have three interesting sets of numbers:

- the *roots* of the equation $h_f(t) = 0$,
- the *eigenvalues* of the matrix m_f, and
- the *values* of the function f on $\mathbf{V}(I)$, the set of points we are looking for.

The amazing fact is that all three sets are equal.

(4.5) Theorem. *Let* $I \subset \mathbb{C}[x_1, \ldots, x_n]$ *be zero-dimensional, let* $f \in \mathbb{C}[x_1, \ldots, x_n]$, *and let* h_f *be the minimal polynomial of* m_f *on* $A = \mathbb{C}[x_1, \ldots, x_n]/I$. *Then, for* $\lambda \in \mathbb{C}$, *the following are equivalent:*
a. λ *is a root of the equation* $h_f(t) = 0$,
b. λ *is an eigenvalue of the matrix* m_f, *and*
c. λ *is a value of the function* f *on* $\mathbf{V}(I)$.

PROOF. a $\Leftrightarrow$ b follows from standard results in linear algebra.

b $\Rightarrow$ c: Let λ be an eigenvalue of m_f. Then there is a corresponding eigenvector $[z] \neq [0] \in A$ such that $[f - \lambda][z] = [0]$. Aiming for a contradiction, suppose that λ is not a value of f on $\mathbf{V}(I)$. That is, letting $\mathbf{V}(I) = \{p_1, \ldots, p_m\}$, suppose that $f(p_i) \neq \lambda$ for all $i = 1, \ldots, m$.

Let $g = f - \lambda$, so that $g(p_i) \neq 0$ for all i. By Lemma (2.9) of this chapter, there exist polynomials g_i such that $g_i(p_j) = 0$ if $i \neq j$, and $g_i(p_i) = 1$. Consider the polynomial $g' = \sum_{i=1}^{m} 1/g(p_i)g_i$. It follows that $g'(p_i)g(p_i) = 1$ for all i, and hence $1 - g'g \in \mathbf{I}(\mathbf{V}(I))$. By the Nullstellensatz, $(1 - g'g)^\ell \in I$ for some $\ell \geq 1$. Expanding by the binomial theorem and collecting the terms that contain g as a factor, we get $1 - \tilde{g}g \in I$ for some $\tilde{g} \in \mathbb{C}[x_1, \ldots, x_n]$. In A, this last inclusion implies that $[1] = [\tilde{g}][g]$, hence g has a multiplicative inverse $[\tilde{g}]$ in A.

But from the above we have $[g][z] = [f - \lambda][z] = [0]$ in A. Multiplying both sides by $[\tilde{g}]$, we obtain $[z] = [0]$, which is a contradiction. Therefore λ must be a value of f on $\mathbf{V}(I)$.

c $\Rightarrow$ a: Let $\lambda = f(p)$ for $p \in \mathbf{V}(I)$. Since $h_f(m_f) = 0$, Corollary (4.3) shows $h_f([f]) = [0]$, and then (4.4) implies $h_f(f) \in I$. This means $h_f(f)$ vanishes at every point of $\mathbf{V}(I)$, so that $h_f(\lambda) = h_f(f(p)) = 0$. $\square$

Exercise 3. We saw earlier that the matrix of multiplication by x in the 5-dimensional algebra $A = \mathbb{C}[x, y]/I$ from (2.4) of this chapter is given by the matrix displayed before Exercise 1 in this section.

a. Using the `minpoly` command in Maple (part of the `linalg` package) or otherwise, show that the minimal polynomial of this matrix is

$$h_x(t) = t^4 - 2t^3 - t^2 + 2t.$$

The roots of $h_x(t) = 0$ are thus $t = 0, -1, 1, 2$.

b. Now find all points of $\mathbf{V}(I)$ using the methods of §1 and show that the roots of h_x are exactly the distinct values of the function $f(x, y) = x$ at the points of $\mathbf{V}(I)$. (Two of the points have the same x-coordinate, which explains why the degree and the number of roots are 4 instead of 5!) Also see Exercise 7 from §2 to see how the ideal I was constructed.

c. Finally, find the minimal polynomial of the matrix m_y, determine its roots, and explain the degree you get.

When we apply Theorem (4.5) with $f = x_i$, we get a general result exactly parallel to this example.

(4.6) Corollary. *Let $I \subset \mathbb{C}[x_1, \ldots, x_n]$ be zero-dimensional. Then the eigenvalues of the multiplication operator m_{x_i} on A coincide with the x_i-coordinates of the points of $\mathbf{V}(I)$. Moreover, substituting $t = x_i$ in the minimal polynomial h_{x_i} yields the unique monic generator of the elimination ideal $I \cap \mathbb{C}[x_i]$.*

Corollary (4.6) indicates that it is possible to solve equations by computing eigenvalues of the multiplication operators m_{x_i}. This has been studied in papers such as [Laz], [Möl], and [MöS], among others. As a result a whole array of numerical methods for approximating eigenvalues can be brought to bear on the root-finding problem, at least in favorable cases. We include a brief discussion of some of these methods for the convenience of some readers; the following two paragraphs may be safely ignored if you are familiar with numerical eigenvalue techniques. For more details, we suggest [BuF] or [Act].

In elementary linear algebra, eigenvalues of a matrix M are usually determined by solving the *characteristic polynomial equation*:

$$\det(M - tI) = 0.$$

The degree of the polynomial on the left hand side is the size of the matrix M. But computing $\det(M - tI)$ for large matrices is a large job itself, and as we have seen in §1, exact solutions (and even accurate approximations to solutions) of polynomial equations of high degree over $\mathbb{R}$ or $\mathbb{C}$ can be hard to come by, so the characteristic polynomial is almost never used in practice. So other methods are needed.

The most basic numerical eigenvalue method is known as the *power method*. It is based on the fact that if a matrix M has a unique *dominant eigenvalue* (i.e., an eigenvalue λ satisfying $|\lambda| > |\mu|$ for all other eigenvalues μ of M), then starting from a randomly chosen vector x_0, and forming the sequence

$$x_{k+1} = \text{unit vector in direction of } Mx_k,$$

we almost always approach an eigenvector for λ as $k \to \infty$. An approximate value for the dominant eigenvalue λ may be obtained by computing the norm $\|Mx_k\|$ at each step. If there is no unique dominant eigenvalue, then the iteration may not converge, but the power method can also be modified to eliminate that problem and to find other eigenvalues of M. In particular, we can find the eigenvalue of M *closest to* some fixed s by applying the power method to the matrix $M' = (M - sI)^{-1}$. For almost all choices of s, there will be a unique dominant eigenvalue of M'. Moreover, if λ' is that dominant eigenvalue of M', then $1/\lambda' + s$ is the eigenvalue of M closest to s. This observation makes it possible to search for *all* the eigenvalues of a matrix as we would do in using the Newton-Raphson method to find all the roots of a polynomial. Some of the same difficulties arise, too. There are also much more sophisticated iterative methods, such as the LR and QR algorithms, that can be used to determine *all* the (real or complex) eigenvalues of a matrix except in some very uncommon degenerate situations. It is known that the QR algorithm, for instance, converges for all matrices having no more than two eigenvalues of any given magnitude in $\mathbb{C}$. Some computer algebra systems (e.g., Maple and Mathematica) provide built-in procedures that implement these methods.

A legitimate question at this point is this: *Why* might one consider applying these eigenvalue techniques for root finding instead of using elimination? There are two reasons.

The first concerns the amount of calculation necessary to carry out this approach. The direct attack—solving systems via elimination as in §1—imposes a *choice of monomial order* in the Gröbner basis we use. Pure *lex* Gröbner bases frequently require a large amount of computation. As we saw in §3, it is possible to compute a *grevlex* Gröbner basis first, then convert it to a *lex* basis using the FGLM basis conversion algorithm, with some savings in total effort. But basis conversion is unnecessary if we use Corollary (4.6), because the algebraic structure of $\mathbb{C}[x_1, \ldots, x_n]/I$ is *independent of the monomial order* used for the Gröbner basis and remainder calculations. Hence any monomial order can be used to determine the matrices of the multiplication operators m_{x_i}.

The second reason concerns the amount of numerical versus symbolic computation involved, and the potential for numerical instability. In the frequently-encountered case that the generators for I have rational coefficients, the entries of the matrices m_{x_i} will also be rational, and hence can be determined *exactly* by symbolic computation. Thus the numerical component of the calculation is restricted to the eigenvalue calculations.

There is also a significant difference even between a naive first idea for implementing this approach and the elimination method discussed in §1. Namely, we could begin by computing all the m_{x_i} and their eigenvalues separately. Then with some additional computation we could determine exactly which vectors $(x_1, \ldots, x_n)$ formed using values of the coordinate functions actually give approximate solutions. The difference here is that the computed values of x_i are *not used* in the determination of the x_j, $j \neq i$. In §1, we saw that a major source of error in approximate solutions was the fact that small errors in one variable could produce larger errors in the other variables when we substitute them and use the Extension Theorem. Separating the computations of the values x_i from one another, we can avoid those *accumulated error* phenomena (and also the numerical stability problems encountered in other non-elimination methods).

We will see shortly that it is possible to reduce the computational effort involved even further. Indeed, it suffices to consider the eigenvalues of only one suitably-chosen multiplication operator $m_{c_1 x_1 + \cdots + c_n x_n}$. Before developing this result, however, we present an example using the more naive approach.

Exercise 4. We will apply the ideas sketched above to find approximations to the complex solutions of the system:

$$0 = x^2 - 2xz + 5$$
$$0 = xy^2 + yz + 1$$
$$0 = 3y^2 - 8xz.$$

a. First, compute a Gröbner basis to determine the monomial basis for the quotient algebra. We can use the *grevlex* (Maple `tdeg`) monomial order:

```
PList := [x^2 - 2*x*z + 5, x*y^2 + y*z + 1, 3*y^2 - 8*x*z];
G := gbasis(PList,tdeg(x,y,z));
B := SetBasis(G,tdeg(x,y,z))[1];
```

(this can also be done using the `kbasis` procedure from Exercise 13 in §2) and obtain the eight monomials:

$$[1, x, y, xy, z, z^2, xz, yz].$$

(You should compare this with the output of `SetBasis` or `kbasis` for *lex* order. Also print out the *lex* Gröbner basis for this ideal if you have a taste for complicated polynomials.)

b. Using the monomial basis B, check that the matrix of the full multiplication operator m_x is

$$\begin{pmatrix} 0 & -5 & 0 & 0 & 0 & -3/16 & -3/8 & 0 \\ 1 & 0 & 0 & 0 & 0 & 0 & 0 & 0 \\ 0 & 0 & 0 & -5 & 0 & 0 & 0 & 0 \\ 0 & 0 & 1 & 3/20 & 0 & 0 & 0 & 3/40 \\ 0 & 0 & 0 & 0 & 0 & 5/2 & 0 & 0 \\ 0 & 0 & 0 & -2 & 0 & 0 & 0 & -1 \\ 0 & 2 & 0 & 0 & 1 & 0 & 0 & 0 \\ 0 & 0 & 0 & -3/10 & 0 & -3/16 & -3/8 & -3/20 \end{pmatrix}.$$

This matrix can also be computed using the `MulMatrix` command in Maple.

c. Now, applying the numerical eigenvalue routine `eigenvals` from Maple, check that there are two approximate real eigenvalues:

$$-1.100987715, \qquad .9657124563,$$

and 3 complex conjugate pairs. (This computation can be done in several different ways and, due to roundoff effects, the results can be slightly different depending on the method used. The values above were found by expressing the entries of the matrix of m_x as floating point numbers, and applying Maple's `eigenvals` routine to that matrix.)

d. Complete the calculation by finding the multiplication operators m_y, m_z, computing their real eigenvalues, and determining which triples (x, y, z) give solutions. (There are exactly two real points.) Also see Exercises 9 and 10 below for a second way to compute the eigenvalues of m_x, m_y, and m_z.

In addition to eigenvalues, there are also eigenvectors to consider. In fact, every matrix M has two sorts of eigenvectors. The *right eigenvectors* of M are the usual ones, which are column vectors $v \neq 0$ such that

$$M v = \lambda v$$

for some $\lambda \in \mathbb{C}$. Since the transpose M^T has the same eigenvalues λ as M, we can find a column vector $v' \neq 0$ such that

$$M^T v' = \lambda v'.$$

Taking transposes, we can write this equation as

$$w M = \lambda w,$$

where $w = v'^T$ is a row vector. We call w a *left eigenvector* of M.

The right and left eigenvectors for a matrix are connected in the following way. For simplicity, suppose that M is a *diagonalizable* $n \times n$ matrix, so that there is a basis for $\mathbb{C}^n$ consisting of right eigenvectors for M. In Exercise 7 below, you will show that there is a matrix equation $MQ = QD$, where Q is the matrix whose columns are the right eigenvectors in a basis for $\mathbb{C}^n$, and D is a diagonal matrix whose diagonal entries are the eigenvalues of M. Rearranging the last equation, we have $Q^{-1}M = DQ^{-1}$. By the second part of Exercise 7 below, the rows of Q^{-1} are a collection of left eigenvectors of M that also form a basis for $\mathbb{C}^n$.

For a zero-dimensional ideal I, there is also a strong connection between the points of $\mathbf{V}(I)$ and the left eigenvectors of the matrix m_f relative to the monomial basis B coming from a Gröbner basis. We will assume that I is radical. In this case, Theorem (2.10) implies that A has dimension m, where m is the number of points in $\mathbf{V}(I)$. Hence, we can write the monomial basis B as the cosets

$$B = \{[x^{\alpha(1)}], \ldots, [x^{\alpha(m)}]\}.$$

Using this basis, let m_f be the matrix of multiplication by f. We can relate the left eigenvectors of m_f to points of $\mathbf{V}(I)$ as follows.

(4.7) Proposition. *Suppose $f \in \mathbb{C}[x_1, \ldots, x_n]$ is chosen such that the values $f(p)$ are distinct for $p \in \mathbf{V}(I)$, where I is a radical ideal not containing 1. Then the left eigenspaces of the matrix m_f are 1-dimensional and are spanned by the row vectors $(p^{\alpha(1)}, \ldots, p^{\alpha(m)})$ for $p \in \mathbf{V}(I)$.*

PROOF. If we write $m_f = (m_{ij})$, then for each j between 1 and m,

$$[x^{\alpha(j)}f] = m_f([x^{\alpha(j)}]) = m_{1j}[x^{\alpha(1)}] + \cdots + m_{mj}[x^{\alpha(m)}].$$

Now fix $p \in \mathbf{V}(f_1, \ldots, f_n)$ and evaluate this equation at p to obtain

$$p^{\alpha(j)}f(p) = m_{1j}p^{\alpha(1)} + \cdots + m_{mj}p^{\alpha(m)}$$

(this makes sense by Exercise 12 of §2). Doing this for $j = 1, \ldots, m$ gives

$$f(p)(p^{\alpha(1)}, \ldots, p^{\alpha(m)}) = (p^{\alpha(1)}, \ldots, p^{\alpha(m)})\, m_f.$$

Exercise 14 at the end of the section asks you to check this computation carefully. Note that one of the basis monomials in B is the coset $[1]$ (do you see why this follows from $1 \notin I$?), which shows that $(p^{\alpha(1)}, \ldots, p^{\alpha(m)})$ is nonzero and hence is a left eigenvector for m_f, with $f(p)$ as the corresponding eigenvalue.

By hypothesis, the $f(p)$ are distinct for $p \in \mathbf{V}(I)$, which means that the $m \times m$ matrix m_f has m distinct eigenvalues. Linear algebra then implies that the corresponding eigenspaces (right and left) are 1-dimensional. $\square$

This proposition can be used to find the points in $\mathbf{V}(I)$ for *any* zero-dimensional ideal I. The basic idea is as follows. First, we can assume that I is radical by replacing I with $\sqrt{I}$ as computed by Proposition (2.7). Then compute a Gröbner basis G and monomial basis B as usual. Now consider the function

$$f = c_1 x_1 + \cdots + c_n x_n,$$

where $c_1, \ldots, c_n$ are randomly chosen integers. This will ensure (with small probability of failure) that the values $f(p)$ are distinct for $p \in \mathbf{V}(I)$. Relative to the monomial basis B, we get the matrix m_f, so that we can use standard numerical methods to find an eigenvalue λ and corresponding left eigenvector v of m_f. This eigenvector, when combined with the Gröbner basis G, makes it *trivial* to find a solution $p \in \mathbf{V}(I)$.

To see how this is done, first note that Proposition (4.7) implies

$$(4.8) \qquad\qquad v = c(p^{\alpha(1)}, \ldots, p^{\alpha(m)})$$

for some nonzero constant c and some $p \in \mathbf{V}(I)$. Write $p = (a_1, \ldots, a_n)$. Our goal is to compute the coordinates a_i of p in terms of the coordinates of v. Equation (4.8) implies that each coordinate of v is of the form $cp^{\alpha(j)}$.

The Finiteness Theorem implies that for each i between 1 and n, there is $m_i \geq 1$ such that $x_i^{m_i}$ is the leading term of some element of G. If $m_i > 1$, it follows that $[x_i] \in B$ (do you see why?), so that ca_i is a coordinate of v. As noted above, we have $[1] \in B$, so that c is also a coordinate of v. Consequently,

$$a_i = \frac{ca_i}{c}$$

is a ratio of coordinates of v. This way, we get the x_i-coordinate of p for all i satisfying $m_i > 1$.

It remains to study the coordinates with $m_i = 1$. These variables appear in *none* of the basis monomials in B (do you see why?), so that we turn instead to the Gröbner basis G for guidance. Suppose the variables with $m_i = 1$ are $x_{i_1}, \ldots, x_{i_\ell}$. We will assume that the variables are labeled so that $x_1 > \cdots > x_n$ and $i_1 > \cdots > i_\ell$. In Exercise 15 below, you will show that for $j = 1, \ldots, \ell$, there are elements $g_j \in G$ such that

$$g_j = x_{i_j} + \text{terms involving } x_i \text{ for } i > i_j.$$

If we evaluate this at $p = (a_1, \ldots, a_n)$, we obtain

$$(4.9) \qquad\qquad 0 = a_{i_j} + \text{terms involving } a_i \text{ for } i > i_j.$$

Since we already know a_i for $i \notin \{i_1, \ldots, i_\ell\}$, these equations make it a simple matter to find $a_{i_1}, \ldots, a_{i_\ell}$. We start with a_{i_ℓ}. For $j = \ell$, (4.9) implies that a_{i_ℓ} is a polynomial in the coordinates of p we already know. Hence we get a_{i_ℓ}. But once we know a_{i_ℓ}, (4.9) shows that $a_{i_{\ell-1}}$ is also a polynomial in known coordinates. Continuing in this way, we get *all* of the coordinates of p.

Exercise 5. Apply this method to find the solutions of the equations given in Exercise 4. The x-coordinates of the solutions are distinct, so you can assume $f = x$. Thus it suffices to compute the left eigenvectors of the matrix m_x of Exercise 4.

The idea of using eigenvectors to find solutions first appears in the pioneering work of Auzinger and Stetter [AS] in 1988 and was further developed in [MöS], [MT], and [Ste]. Our treatment focused on the radical case since our first step was to replace I with $\sqrt{I}$. In general, whenever a multiplication map m_f is *nonderogatory* (meaning that all eigenspaces have dimension one), one can use Proposition (4.7) to find the solutions. Unfortunately, when I is not radical, it can happen that m_f is derogatory for *all* $f \in k[x_1, \ldots, x_n]$. Rather than replacing I with $\sqrt{I}$ as we did above, another approach is to realize that the *family* of operators $\{m_f : f \in k[x_1, \ldots, x_n]\}$ is nonderogatory, meaning that its joint left eigenspaces are one-dimensional and hence are spanned by the eigenvectors described in Proposition (4.7). This result and its consequences are discussed in [MT] and [Mou1]. We will say more about multiplication maps in §2 of Chapter 4.

Since the left eigenvectors of m_f help us find solutions in $\mathbf{V}(I)$, it is natural to ask about the right eigenvectors. In Exercise 17 below, you will show that these eigenvectors solve the *interpolation problem*, which asks for a polynomial that takes preassigned values at the points of $\mathbf{V}(I)$.

This section has discussed several ideas for solving polynomial equations using linear algebra. We certainly do not claim that these ideas are a computational panacea for all polynomial systems, but they do give interesting alternatives to other, more traditional methods in numerical analysis, and they are currently an object of study in connection with the implementation of the next generation of computer algebra systems. We will continue this discussion in §5 (where we study real solutions) and Chapter 3 (where we use resultants to solve polynomial systems).

ADDITIONAL EXERCISES FOR §4

Exercise 6. Prove Proposition (4.2).

Exercise 7. Let M, Q, P, D be $n \times n$ complex matrices, and assume D is a diagonal matrix.
a. Show that the equation $MQ = QD$ holds if and only if each nonzero column of Q is a right eigenvector of M and the corresponding diagonal entry of D is the corresponding eigenvalue.
b. Show that the equation $PM = DP$ holds if and only if each nonzero row of P is a left eigenvector of M and the corresponding diagonal entry of D is the corresponding eigenvalue.
c. If $MQ = QD$ and Q is invertible, deduce that the rows of Q^{-1} are left eigenvectors of M.

Exercise 8.
a. Apply the eigenvalue method from Corollary (4.6) to solve the system from Exercise 6 of §1. Compare your results.

b. Apply the eigenvalue method from Corollary (4.6) to solve the system from Exercise 7 from §1. Compare your results.

Exercise 9. Let V_i be the subspace of A spanned by the non-negative powers of $[x_i]$, and consider the *restriction* of the multiplication operator $m_{x_i} : A \to A$ to V_i. Assume $\{1, [x_i], \ldots, [x_i]^{m_i-1}\}$ is a basis for V_i.
a. What is the matrix of the restriction $m_{x_i}|_{V_i}$ with respect to this basis? Show that it can be computed by the same calculations used in Exercise 4 of §2 to find the monic generator of $I \cap \mathbb{C}[x_i]$, without computing a *lex* Gröbner basis. Hint: See also Exercise 11 of §1 of Chapter 3.
b. What is the characteristic polynomial of $m_{x_i}|_{V_i}$ and what are its roots?

Exercise 10. Use part b of Exercise 9 and Corollary (4.6) to give another determination of the roots of the system from Exercise 4.

Exercise 11. Let I be a zero-dimensional ideal in $\mathbb{C}[x_1, \ldots, x_n]$, and let $f \in \mathbb{C}[x_1, \ldots, x_n]$. Show that $[f]$ has a multiplicative inverse in $\mathbb{C}[x_1, \ldots, x_n]/I$ if and only if $f(p) \neq 0$ for all $p \in \mathbf{V}(I)$. Hint: See the proof of Theorem (4.5).

Exercise 12. Prove that a zero-dimensional ideal is radical if and only if the matrices m_{x_i} are diagonalizable for each i. Hint: Linear algebra tells us that a matrix is diagonalizable if and only if its minimal polynomial is square-free. Proposition (2.7) and Corollary (4.6) of this chapter will be useful.

Exercise 13. Let $A = \mathbb{C}[x_1, \ldots, x_n]/I$ for a zero-dimensional ideal I, and let $f \in \mathbb{C}[x_1, \ldots, x_n]$. If $p \in \mathbf{V}(I)$, we can find $g \in \mathbb{C}[x_1, \ldots, x_n]$ with $g(p) = 1$, and $g(p') = 0$ for all $p' \in \mathbf{V}(I)$, $p' \neq p$ (see Lemma (2.9)). Prove that there is an $\ell \geq 1$ such that the coset $[g^\ell] \in A$ is a *generalized eigenvector* for m_f with eigenvalue $f(p)$. (A generalized eigenvector of a matrix M is a nonzero vector v such that $(M - \lambda I)^m v = 0$ for some $m \geq 1$.) Hint: Apply the Nullstellensatz to $(f - f(p))g$. In Chapter 4, we will study the generalized eigenvectors of m_f in more detail.

Exercise 14. Verify carefully the formula $f(p)(p^{\alpha(1)}, \ldots, p^{\alpha(m)}) = (p^{\alpha(1)}, \ldots, p^{\alpha(m)}) m_f$ used in the proof of Proposition (4.7).

Exercise 15. Let $>$ be some monomial order, and assume $x_1 > \cdots > x_n$. If $g \in k[x_1, \ldots, x_n]$ satisfies $\mathrm{LT}(g) = x_j$, then prove that

$$g = x_j + \text{terms involving } x_i \text{ for } i > j.$$

Exercise 16. (The Shape Lemma) Let I be a zero-dimensional radical ideal such that the x_n-coordinates of the points in $\mathbf{V}(I)$ are distinct. Let

G be a reduced Gröbner basis for I relative to a *lex* monomial order with x_n as the *last* variable.

a. If $\mathbf{V}(I)$ has m points, prove that the cosets $1, [x_n], \ldots, [x_n^{m-1}]$ are linearly independent and hence are a basis of $A = k[x_1, \ldots, x_n]/I$.

b. Prove that G consists of n polynomials

$$g_1 = x_1 + h_1(x_n)$$

$$\vdots$$

$$g_{n-1} = x_{n-1} + h_{n-1}(x_n)$$
$$g_n = x_n^m + h_n(x_n),$$

where $h_1, \ldots, h_n$ are polynomials in x_n of degree at most $m - 1$. Hint: Start by expressing $[x_1], \ldots, [x_{n-1}], [x_n^m]$ in terms of the basis of part a.

c. Explain how you can find *all* points of $\mathbf{V}(I)$ once you know their x_n-coordinates. Hint: Adapt the discussion following (4.9).

Exercise 17. This exercise will study the right eigenvectors of the matrix m_f and their relation to interpolation. Assume that I is a zero-dimensional radical ideal and that the values $f(p)$ are distinct for $p \in \mathbf{V}(I)$. We write the monomial basis B as $\{[x^{\alpha(1)}], \ldots, [x^{\alpha(m)}]\}$.

a. If $p \in \mathbf{V}(I)$, Lemma (2.9) of this chapter gives us g such that $g(p) = 1$ and $g(p') = 0$ for all $p' \neq p$ in $\mathbf{V}(I)$. Prove that the coset $[g] \in A$ is a right eigenvector of m_f and that the corresponding eigenspace has dimension 1. Conclude that *all* eigenspaces of m_f are of this form.

b. If $v = (v_1, \ldots, v_m)^t$ is a right eigenvector of m_f corresponding to the eigenvalue $f(p)$ for p as in part a, then prove that the polynomial

$$\tilde{g} = v_1 x^{\alpha(1)} + \cdots + v_m x^{\alpha(m)}$$

satisfies $\tilde{g}(p) \neq 0$ and $\tilde{g}(p') = 0$ for $p' \neq p$ in $\mathbf{V}(I)$.

c. Show that we can take the polynomial g of part a to be

$$g = \frac{1}{\tilde{g}(p)} \tilde{g}.$$

Thus, once we know the solution p and the corresponding right eigenvector of m_f, we get an *explicit formula* for the polynomial g.

d. Given $\mathbf{V}(I) = \{p_1, \ldots, p_m\}$ and the corresponding right eigenvectors of m_f, we get polynomials $g_1, \ldots, g_m$ such that $g_i(p_j) = 1$ if $i = j$ and 0 otherwise. Each g_i is given explicitly by the formula in part c. The *interpolation problem* asks to find a polynomial h which takes preassigned values $\lambda_1, \ldots, \lambda_m$ at the points $p_1, \ldots, p_m$. This means $h(p_i) = \lambda_i$ for all i. Prove that one choice for h is given by

$$h = \lambda_1 g_1 + \cdots + \lambda_m g_m.$$

Exercise 18. Let $A = k[x_1, \ldots, x_n]/I$, where I is zero-dimensional. In Maple, `MulMatrix` computes the matrix of the multiplication map m_{x_i} relative to a monomial basis computed by `SetBasis`. However, in §5, we will need to compute the matrix of m_f, where $f \in k[x_1, \ldots, x_n]$ is an arbitrary polynomial. Develop and code a Maple procedure `getmatrix` which, given a polynomial f, a monomial basis B, a Gröbner basis G, and a term order, produces the matrix of m_f relative to B. You will use `getmatrix` in Exercise 6 of §5.

§5 Real Root Location and Isolation

The eigenvalue techniques for solving equations from §4 are only a first way that we can use the results of §2 for finding roots of systems of polynomial equations. In this section we will discuss a second application that is more sophisticated. We follow a recent paper of Pedersen, Roy, and Szpirglas [PRS] and consider the problem of determining the *real* roots of a system of polynomial equations with coefficients in a field $k \subset \mathbb{R}$ (usually $k = \mathbb{Q}$ or a finite extension field of $\mathbb{Q}$). The underlying principle here is that for many purposes, explicitly determined, bounded regions $R \subset \mathbb{R}^n$, each guaranteed to contain *exactly one* solution of the system can be just as useful as a collection of numerical approximations. Note also that if we wanted numerical approximations, once we had such an R, the job of finding that one root would generally be much simpler than a search for *all* of the roots! (Think of the choice of the initial approximation for an iterative method such as Newton-Raphson.) For one-variable equations, this is also the key idea of the *interval arithmetic* approach to computation with real algebraic numbers (see [Mis]). We note that there are also other methods known for locating and isolating the real roots of a polynomial system (see §8.8 of [BW] for a different type of algorithm).

To define our regions R in $\mathbb{R}^n$, we will use polynomial functions in the following way. Let $h \in k[x_1, \ldots, x_n]$ be a nonzero polynomial. The real points where h takes the value 0 form the variety $\mathbf{V}(h) \cap \mathbb{R}^n$. We will denote this by $\mathbf{V}_\mathbb{R}(h)$ in the discussion that follows. In typical cases, $\mathbf{V}_\mathbb{R}(h)$ will be a *hypersurface*—an $(n-1)$-dimensional variety in $\mathbb{R}^n$. The complement of $\mathbf{V}_\mathbb{R}(h)$ in $\mathbb{R}^n$ is the union of connected open subsets on which h takes either all positive values or all negative values. We obtain in this way a decomposition of $\mathbb{R}^n$ as a disjoint union

$$(5.1) \qquad \mathbb{R}^n = H^+ \cup H^- \cup \mathbf{V}_\mathbb{R}(h),$$

where $H^+ = \{a \in \mathbb{R}^n : h(a) > 0\}$, and similarly for H^-. Here are some concrete examples.

Exercise 1.

a. Let $h = (x^2 + y^2 - 1)(x^2 + y^2 - 2)$ in $\mathbb{R}[x, y]$. Identify the regions H^+ and H^- for this polynomial. How many connected components does each of them have?

b. In this part of the exercise, we will see how regions like rectangular "boxes" in $\mathbb{R}^n$ may be obtained by intersecting several regions H^+ or H^-. For instance, consider the box

$$R = \{(x, y) \in \mathbb{R}^2 : a < x < b, \ c < y < d\}.$$

If $h_1(x, y) = (x - a)(x - b)$ and $h_2(x, y) = (y - c)(y - d)$, show that

$$R = H_1^- \cap H_2^- = \{(x, y) \in \mathbb{R}^2 : h_i(x, y) < 0, \ i = 1, 2\}.$$

What do H_1^+, H_2^+ and $H_1^+ \cap H_2^+$ look like in this example?

Given a region R like the box from part b of the above exercise, and a system of equations, we can ask whether there are roots of the system in R. The results of [PRS] give a way to answer questions like this, using an extension of the results of §2 and §4. Let I be a zero-dimensional ideal and let B be the monomial basis of $A = k[x_1, \ldots, x_n]/I$ for any monomial order. Recall that the *trace* of a square matrix is just the sum of its diagonal entries. This gives a mapping Tr from $d \times d$ matrices to k. Using the trace, we define a *symmetric bilinear form* S by the rule:

$$S(f, g) = \mathrm{Tr}(m_f \cdot m_g) = \mathrm{Tr}(m_{fg})$$

(the last equality follows from part b of Proposition (4.2)).

Exercise 2.

a. Prove that S defined as above is a symmetric bilinear form on A, as claimed. That is, show that S is symmetric, meaning $S(f, g) = S(g, f)$ for all $f, g \in A$, and linear in the first variable, meaning

$$S(cf_1 + f_2, g) = cS(f_1, g) + S(f_2, g)$$

for all $f_1, f_2, g \in A$ and all $c \in k$. It follows that S is linear in the second variable as well.

b. Given a symmetric bilinear form S on a vector space V with basis $\{v_1, \ldots, v_d\}$, the matrix of S is the $d \times d$ matrix $M = (S(v_i, v_j))$. Show that the matrix of S with respect to the monomial basis $B = \{x^{\alpha(i)}\}$ for A is given by:

$$M = (\mathrm{Tr}(m_{x^{\alpha(i)} x^{\alpha(j)}})) = (\mathrm{Tr}(m_{x^{\alpha(i) + \alpha(j)}})).$$

Similarly, given the polynomial $h \in k[x_1, \ldots, x_n]$ used in the decomposition (5.1), we can construct a bilinear form

$$S_h(f, g) = \mathrm{Tr}(m_{hf} \cdot m_g) = \mathrm{Tr}(m_{hfg}).$$

Let M_h be the matrix of S_h with respect to B.

Exercise 3. Show that S_h is also a symmetric bilinear form on A. What is the i,j entry of M_h?

Since we assume $k \subset \mathbb{R}$, the matrices M and M_h are symmetric matrices with *real* entries. It follows from the real spectral theorem (or principal axis theorem) of linear algebra that all of the eigenvalues of M and M_h will be *real*. For our purposes the exact values of these eigenvalues are much less important than their *signs*.

Under a change of basis defined by an invertible matrix Q, the matrix M of a symmetric bilinear form S is taken to $Q^t M Q$. There are two fundamental invariants of S under such changes of basis—the *signature* $\sigma(S)$, which equals the difference between the number of positive eigenvalues and the number of negative eigenvalues of M, and the *rank* $\rho(S)$, which equals the rank of the matrix M. (See, for instance, Chapter 6 of [Her] for more information on the signature and rank of bilinear forms.)

We are now ready to state the main result of this section.

(5.2) Theorem. *Let I be a zero-dimensional ideal generated by polynomials in $k[x_1, \ldots, x_n]$ ($k \subset \mathbb{R}$), so that $\mathbf{V}(I) \subset \mathbb{C}^n$ is finite. Then, for $h \in k[x_1, \ldots, x_n]$, the signature and rank of the bilinear form S_h satisfy:*

$$\sigma(S_h) = \#\{a \in \mathbf{V}(I) \cap \mathbb{R}^n : h(a) > 0\} - \#\{a \in \mathbf{V}(I) \cap \mathbb{R}^n : h(a) < 0\}$$
$$\rho(S_h) = \#\{a \in \mathbf{V}(I) : h(a) \neq 0\}.$$

PROOF. This result is essentially a direct consequence of the reasoning leading up to Theorem (4.5) of this chapter. However, to give a full proof it is necessary to take into account the *multiplicities* of the points in $\mathbf{V}(I)$ as defined in Chapter 4. Hence we will only sketch the proof in the special case when I is radical. By Theorem (2.10), this means that $\mathbf{V}(I) = \{p_1, \ldots, p_m\}$, where m is the dimension of the algebra A. Given the basis $B = \{[x^{\alpha(i)}]\}$ of A, Proposition (4.7) implies that $(p_j^{\alpha(i)})$ is an invertible matrix.

By Theorem (4.5), for any f, we know that the set of eigenvalues of m_f coincides with the set of values of the f at the points in $\mathbf{V}(I)$. The key new fact we will need is that using the structure of the algebra A, for each point p in $\mathbf{V}(I)$ it is possible to define a positive integer $m(p)$ (the multiplicity) so that $\sum_p m(p) = d = \dim(A)$, and so that $(t - f(p))^{m(p)}$ is a factor of the characteristic polynomial of m_f. (See §2 of Chapter 4 for the details.)

By definition, the i,j entry of the matrix M_h is equal to

$$\mathrm{Tr}(m_{h \cdot x^{\alpha(i)} \cdot x^{\alpha(j)}}).$$

The trace of the multiplication operator equals the sum of its eigenvalues. By the previous paragraph, the sum of these eigenvalues is

$$(5.3) \qquad \sum_{p \in \mathbf{V}(I)} m(p) h(p) p^{\alpha(i)} p^{\alpha(j)},$$

where $p^{\alpha(i)}$ denotes the value of the monomial $x^{\alpha(i)}$ at the point p. List the points in $\mathbf{V}(I)$ as $p_1, \ldots, p_d$, where each point p in $\mathbf{V}(I)$ is repeated $m(p)$ times consecutively. Let U be the $d \times d$ matrix whose jth column consists of the values $p_j^{\alpha(i)}$ for $i = 1, \ldots, d$. From (5.3), we obtain a matrix factorization $M_h = UDU^t$, where D is the diagonal matrix with entries $h(p_1), \ldots, h(p_d)$. The equation for the rank follows since U is invertible. Both U and D may have nonreal entries. However, the equation for the signature follows from this factorization as well, using the facts that M_h has real entries and that the nonreal points in $\mathbf{V}(I)$ occur in complex conjugate pairs. We refer the reader to Theorem 2.1 of [PRS] for the details. $\square$

The theorem may be used to determine how the real points in $\mathbf{V}(I)$ are distributed among the sets H^+, H^- and $\mathbf{V}_{\mathbb{R}}(h)$ determined by h in (5.1). Theorem (5.2) implies that we can count the number of real points of $\mathbf{V}(I)$ in H^+ and in H^- as follows. The signature of S_h gives the *difference* between the number of solutions in H^+ and the number in H^-. By the same reasoning, computing the signature of S_{h^2} we get the number of solutions in $H^+ \cup H^-$, since $h^2 > 0$ at every point of $H^+ \cup H^-$. From this we can recover $\#\mathbf{V}(I) \cap H^+$ and $\#\mathbf{V}(I) \cap H^-$ by simple arithmetic. Finally, we need to find $\#\mathbf{V}(I) \cap \mathbf{V}_{\mathbb{R}}(h)$, which is done in the following exercise.

Exercise 4. Using the form S_1 in addition to S_h and S_{h^2}, show that the three signatures $\sigma(S), \sigma(S_h), \sigma(S_{h^2})$ give all the information needed to determine $\#\mathbf{V}(I) \cap H^+$, $\#\mathbf{V}(I) \cap H^-$ and $\#\mathbf{V}(I) \cap \mathbf{V}_{\mathbb{R}}(h)$.

From the discussion above, it might appear that we need to compute the eigenvalues of the forms S_h to count the numbers of solutions of the equations in H^+ and H^-, but the situation is actually *much better than that*. Namely, the entire calculation can be done symbolically, so no recourse to numerical methods is needed. The reason is the following consequence of the classical Descartes Rule of Signs.

(5.4) Proposition. *Let M_h be the matrix of S_h, and let*

$$p_h(t) = \det(M_h - tI)$$

be its characteristic polynomial. Then the number of positive eigenvalues of S_h is equal to the number of sign changes in the sequence of coefficients of $p_h(t)$. (In counting sign changes, any zero coefficients are ignored.)

PROOF. See Proposition 2.8 of [PRS], or Exercise 5 below for a proof. $\square$

For instance, consider the real symmetric matrix

$$M = \begin{pmatrix} 3 & 1 & 5 & 4 \\ 1 & 2 & 6 & 9 \\ 5 & 6 & 7 & -1 \\ 4 & 9 & -1 & 0 \end{pmatrix}.$$

The characteristic polynomial of M is $t^4 - 12t^3 - 119t^2 + 1098t - 1251$, giving *three* sign changes in the sequence of coefficients. Thus M has three positive eigenvalues, as one can check.

Exercise 5. The usual version of Descartes' Rule of Signs asserts that the number of positive roots of a polynomial $p(t)$ in $\mathbb{R}[t]$ equals the number of sign changes in its coefficient sequence minus a non-negative even integer.
a. Using this, show that the number of negative roots equals the number of sign changes in the coefficient sequence of $p(-t)$ minus another non-negative even integer.
b. Deduce (5.4) from Descartes' Rule of Signs, part a, and the fact that all eigenvalues of M_h are real.

Using these ideas to find and isolate roots requires a good searching strategy. We will not consider such questions here. For an example showing how to certify the presence of exactly one root of a system in a given region, see Exercise 6 below.

The problem of counting real solutions of polynomial systems in regions $R \subset \mathbb{R}^n$ defined by several polynomial inequalities and/or equalities has been considered in general by Ben-Or, Kozen, and Reif (see, for instance, [BKR]). Using the signature calculations as above gives an approach which is very well suited to *parallel* computation, and whose complexity is relatively manageable. We refer the interested reader to [PRS] once again for a discussion of these issues.

For a recent exposition of the material in this section, we refer the reader to Chapter 6 of [GRRT]. One topic not mentioned in our treatment is *semidefinite programming*. As explained in Chapter 7 of [Stu5], this has interesting relations to real solutions and sums of squares.

ADDITIONAL EXERCISES FOR §5

Exercise 6. In this exercise, you will verify that the equations
$$0 = x^2 - 2xz + 5$$
$$0 = xy^2 + yz + 1$$
$$0 = 3y^2 - 8xz$$
have exactly one real solution in the rectangular box
$$R = \{(x, y, z) \in \mathbb{R}^3 : 0 < x < 1, \ -3 < y < -2, \ 3 < z < 4\}.$$

a. Using *grevlex* monomial order with $x > y > z$, compute a Gröbner basis G for the ideal I generated by the above equations. Also find the corresponding monomial basis B for $\mathbb{C}[x, y, z]/I$.

b. Implement the following Maple procedure `getform` which computes the matrix of the symmetric bilinear form S_h.

```
getform := proc(h,B,G,torder)


    # computes the matrix of the symmetric bilinear form S_h,
    # with respect to the monomial basis B for the quotient
    # ring. G should be a Groebner basis with respect to
    # torder.

    local d,M,i,j,p,q;

    d:=nops(B);
    M := array(symmetric,1..d,1..d);
    for i to d do
      for j from i to d do
        p := normalf(h*B[i]*B[j],G,torder);
        M[i,j]:=trace(getmatrix(p,B,G,torder));
        end do;
      end do;
    return eval(M)
    end proc:
```

The call to `getmatrix` computes the matrix $m_{hx^{\alpha(i)}x^{\alpha(j)}}$ with respect to the monomial basis $B = \{x^{\alpha(i)}\}$ for A. Coding `getmatrix` was Exercise 18 in §4 of this chapter.

c. Then, using

```
h := x*(x-1);

S := getform(h,B,G,tdeg(x,y,z));
```

compute the matrix of the bilinear form S_h for $h = x(x - 1)$.

d. The actual entries of this 8×8 rational matrix are rather complicated and not very informative; we will omit reproducing them. Instead, use

$$\texttt{charpoly(S,t);}$$

to compute the characteristic polynomial of the matrix. Your result should be a polynomial of the form:

$$t^8 - a_1 t^7 + a_2 t^6 + a_3 t^5 - a_4 t^4 - a_5 t^3 - a_6 t^2 + a_7 t + a_8,$$

where each a_i is a positive rational number.

e. Use Proposition (5.4) to show that S_h has 4 positive eigenvalues. Since $a_8 \neq 0$, $t = 0$ is not an eigenvalue. Explain why the other 4 eigenvalues are strictly negative, and conclude that S_h has signature

$$\sigma(S_h) = 4 - 4 = 0.$$

f. Use the second equation in Theorem (5.2) to show that h is nonvanishing
   on the real or complex points of $\mathbf{V}(I)$. Hint: Show that S_h has rank 8.
g. Repeat the computation for h^2:

$$\texttt{T := getform(h*h,B,G,tdeg(x,y,z));}$$

and show that in this case, we get a second symmetric matrix with exactly 5 positive and 3 negative eigenvalues. Conclude that the signature
of S_{h^2} (which counts the total number of real solutions in this case) is

$$\sigma(S_{h^2}) = 5 - 3 = 2.$$

h. Using Theorem (5.2) and combining these two calculations, show that

$$\#\mathbf{V}(I) \cap H^+ = \#\mathbf{V}(I) \cap H^- = 1,$$

and conclude that there is exactly one real root between the two planes
$x = 0$ and $x = 1$ in $\mathbb{R}^3$. Our desired region R is contained in this infinite
slab in $\mathbb{R}^3$. What can you say about the other real solution?
i. Complete the exercise by applying Theorem (5.2) to polynomials in y
   and z chosen according to the definition of R.

Exercise 7. Use the techniques of this section to determine the number
of real solutions of

$$0 = x^2 + 2y^2 - y - 2z$$
$$0 = x^2 - 8y^2 + 10z - 1$$
$$0 = x^2 - 7yz$$

in the box $R = \{(x, y, z) \in \mathbb{R}^3 : 0 < x < 1, 0 < y < 1, 0 < z < 1\}$. (This
is the same system as in Exercise 6 of §1. Check your results using your
previous work.)

Exercise 8. The alternative real root isolation methods discussed in §8.8
of [BW] are based on a result for real one-variable polynomials known as
Sturm's Theorem. Suppose $p(t) \in \mathbb{Q}[t]$ is a polynomial with no multiple
roots in $\mathbb{C}$. Then $\mathrm{GCD}(p(t), p'(t)) = 1$, and the sequence of polynomials
produced by

$$p_0(t) = p(t)$$
$$p_1(t) = p'(t)$$
$$p_i(t) = -\mathrm{rem}(p_{i-1}(t), p_{i-2}(t), t), i \geq 2$$

(so $p_i(t)$ is the *negative* of the remainder on division of $p_{i-1}(t)$ by $p_{i-2}(t)$ in
$\mathbb{Q}[t]$) will eventually reach a nonzero constant, and all subsequent terms will

be zero. Let $p_m(t)$ be the last nonzero term in the sequence. This sequence of polynomials is called the *Sturm sequence* associated to $p(t)$.

a. (Sturm's Theorem) If $a < b$ in $\mathbb{R}$, and neither is a root of $p(t) = 0$, then show that the number of real roots of $p(t) = 0$ in the interval $[a, b]$ is the difference between the number of sign changes in the sequence of real numbers $p_0(a), p_1(a), \ldots, p_m(a)$ and the number of sign changes in the sequence $p_0(b), p_1(b), \ldots, p_m(b)$. (Sign changes are counted in the same way as for Descartes' Rule of Signs.)

b. Give an algorithm based on part a that takes as input a polynomial $p(t) \in \mathbb{Q}[t]$ with no multiple roots in $\mathbb{C}$, and produces as output a collection of intervals $[a_i, b_i]$ in $\mathbb{R}$, each of which contains exactly one root of p. Hint: Start with an interval guaranteed to contain all the real roots of $p(t) = 0$ (see Exercise 3 of §1, for instance) and bisect repeatedly, using Sturm's Theorem on each subinterval.

Chapter 3

Resultants

In Chapter 2, we saw how Gröbner bases can be used in Elimination Theory. An alternate approach to the problem of elimination is given by *resultants*. The resultant of two polynomials is well known and is implemented in many computer algebra systems. In this chapter, we will review the properties of the resultant and explore its generalization to several polynomials in several variables. This *multipolynomial resultant* can be used to eliminate variables from three or more equations and, as we will see at the end of the chapter, it is a surprisingly powerful tool for finding solutions of equations.

§1 The Resultant of Two Polynomials

Given two polynomials $f, g \in k[x]$ of positive degree, say

$$(1.1) \qquad \begin{aligned} f &= a_0 x^l + \cdots + a_l, \quad a_0 \neq 0, \quad l > 0 \\ g &= b_0 x^m + \cdots + b_m, \quad b_0 \neq 0, \quad m > 0. \end{aligned}$$

Then the *resultant* of f and g, denoted $\mathrm{Res}(f, g)$, is the $(l + m) \times (l + m)$ determinant

$$(1.2) \qquad \mathrm{Res}(f, g) = \det \begin{pmatrix} a_0 & & & & b_0 & & & \\ a_1 & a_0 & & & b_1 & b_0 & & \\ a_2 & a_1 & \ddots & & b_2 & b_1 & \ddots & \\ \vdots & a_2 & \ddots & a_0 & \vdots & b_2 & \ddots & b_0 \\ a_l & \vdots & \ddots & a_1 & b_m & \vdots & \ddots & b_1 \\ & a_l & & a_2 & & b_m & & b_2 \\ & & \ddots & \vdots & & & \ddots & \vdots \\ & & & a_l & & & & b_m \end{pmatrix}$$

$$\underbrace{}_{m \text{ columns}} \quad \underbrace{}_{l \text{ columns}}$$

where the blank spaces are filled with zeros. When we want to emphasize the dependence on x, we will write $\mathrm{Res}(f, g, x)$ instead of $\mathrm{Res}(f, g)$. As a simple example, we have

$$(1.3) \quad \mathrm{Res}(x^3 + x - 1, 2x^2 + 3x + 7) = \det \begin{pmatrix} 1 & 0 & 2 & 0 & 0 \\ 0 & 1 & 3 & 2 & 0 \\ 1 & 0 & 7 & 3 & 2 \\ -1 & 1 & 0 & 7 & 3 \\ 0 & -1 & 0 & 0 & 7 \end{pmatrix} = 159.$$

Exercise 1. Show that $\mathrm{Res}(f, g) = (-1)^{lm} \mathrm{Res}(g, f)$. Hint: What happens when you interchange two columns of a determinant?

Three basic properties of the resultant are:

- (Integer Polynomial) $\mathrm{Res}(f, g)$ is an integer polynomial in the coefficients of f and g.
- (Common Factor) $\mathrm{Res}(f, g) = 0$ if and only if f and g have a nontrivial common factor in $k[x]$.
- (Elimination) There are polynomials $A, B \in k[x]$ such that $A f + B g = \mathrm{Res}(f, g)$. The coefficients of A and B are integer polynomials in the coefficients of f and g.

Proofs of these properties can be found in [CLO], Chapter 3, §5. The Integer Polynomial property says that there is a polynomial

$$\mathrm{Res}_{l,m} \in \mathbb{Z}[u_0, \dots, u_l, v_0, \dots, v_m]$$

such that if f, g are as in (1.1), then

$$\mathrm{Res}(f, g) = \mathrm{Res}_{l,m}(a_0, \dots, a_l, b_0, \dots, b_m).$$

Over the complex numbers, the Common Factor property tells us that $f, g \in \mathbb{C}[x]$ have a common root if and only if their resultant is zero. Thus (1.3) shows that $x^3 + x - 1$ and $2x^2 + 3x + 7$ have no common roots in $\mathbb{C}$ since $159 \neq 0$, even though we don't know the roots themselves.

To understand the Elimination property, we need to explain how resultants can be used to eliminate variables from systems of equations. As an example, consider the equations

$$f = xy - 1 = 0$$
$$g = x^2 + y^2 - 4 = 0.$$

Here, we have two variables to work with, but if we regard f and g as polynomials in x whose coefficients are polynomials in y, we can compute the resultant with respect to x to obtain

$$\mathrm{Res}(f, g, x) = \det \begin{pmatrix} y & 0 & 1 \\ -1 & y & 0 \\ 0 & -1 & y^2 - 4 \end{pmatrix} = y^4 - 4y^2 + 1.$$

By the Elimination property, there are polynomials $A, B \in k[x, y]$ with $A \cdot (xy - 1) + B \cdot (x^2 + y^2 - 4) = y^4 - 4y^2 + 1$. This means $\mathrm{Res}(f, g, x)$ is in the elimination ideal $\langle f, g \rangle \cap k[y]$ as defined in §1 of Chapter 2, and it follows that $y^4 - 4y^2 + 1$ vanishes at any common solution of $f = g = 0$. Hence, by solving $y^4 - 4y^2 + 1 = 0$, we can find the y-coordinates of the solutions. Thus resultants relate nicely to what we did in Chapter 2.

Exercise 2. Use resultants to find all solutions of the above equations $f = g = 0$. Also find the solutions using $\mathrm{Res}(f, g, y)$. In Maple, the command for resultant is **resultant**.

More generally, if f and g are *any* polynomials in $k[x, y]$ in which x appears to a positive power, then we can compute $\mathrm{Res}(f, g, x)$ in the same way. Since the coefficients are polynomials in y, the Integer Polynomial property guarantees that $\mathrm{Res}(f, g, x)$ is again a polynomial in y. Thus, we can use the resultant to eliminate x, and as above, $\mathrm{Res}(f, g, x)$ is in the elimination ideal $\langle f, g \rangle \cap k[y]$ by the Elimination property. For a further discussion of the connection between resultants and elimination theory, the reader should consult Chapter 3 of [CLO] or Chapter XI of [vdW].

One interesting aspect of the resultant is that it can be expressed in many different ways. For example, given $f, g \in k[x]$ as in (1.1), suppose their roots are $\xi_1, \ldots, \xi_l$ and $\eta_1, \ldots, \eta_m$ respectively (note that these roots might lie in some bigger field). Then one can show that the resultant is given by

$$\mathrm{Res}(f, g) = a_0^m b_0^l \prod_{i=1}^{l} \prod_{j=1}^{m} (\xi_i - \eta_j)$$

(1.4)
$$= a_0^m \prod_{i=1}^{l} g(\xi_i)$$

$$= (-1)^{lm} b_0^l \prod_{j=1}^{m} f(\eta_j).$$

A proof of this is given in the exercises at the end of the section.

Exercise 3.
a. Show that the three products on the right hand side of (1.4) are all equal. Hint: $g = b_0(x - \eta_1) \cdots (x - \eta_m)$.
b. Use (1.4) to show that $\mathrm{Res}(f_1 f_2, g) = \mathrm{Res}(f_1, g)\mathrm{Res}(f_2, g)$.

The formulas given in (1.4) may seem hard to use since they involve the roots of f or g. But in fact there is a relatively simple way to compute the above products. For example, to understand the formula $\mathrm{Res}(f, g) = a_0^m \prod_{i=1}^{l} g(\xi_i)$, we will use the techniques of §2 of Chapter 2. Thus, consider the quotient ring $A_f = k[x]/\langle f \rangle$, and let the multiplication map m_g be defined by

$$m_g([h]) = [g] \cdot [h] = [gh] \in A_f,$$

where $[h] \in A_f$ is the coset of $h \in k[x]$. If we think in terms of remainders on division by f, then we can regard A_f as consisting of all polynomials h of degree $< l$, and under this interpretation, $m_g(h)$ is the remainder of gh on division by f. Then we can compute the resultant $\text{Res}(f, g)$ in terms of m_g as follows.

(1.5) Proposition. $\text{Res}(f, g) = a_0^m \det(m_g : A_f \to A_f)$.

PROOF. Note that A_f is a vector space over k of dimension l (this is clear from the remainder interpretation of A_f). Further, as explained in §2 of Chapter 2, $m_g : A_f \to A_f$ is a linear map. Recall from linear algebra that the determinant $\det(m_g)$ is defined to be the determinant of any matrix M representing the linear map m_g. Since M and m_g have the same eigenvalues, it follows that $\det(m_g)$ is the product of the eigenvalues of m_g, counted with multiplicity.

In the special case when $g(\xi_1), \ldots, g(\xi_l)$ are distinct, we can prove our result using the theory of Chapter 2. Namely, since $\{\xi_1, \ldots, \xi_l\} = \mathbf{V}(f)$, it follows from Theorem (4.5) of Chapter 2 that the numbers $g(\xi_1), \ldots, g(\xi_l)$ are the eigenvalues of m_g. Since these are distinct and A_f has dimension l, it follows that the eigenvalues have multiplicity one, so that $\det(m_g) = g(\xi_1) \cdots g(\xi_l)$, as desired. The general case will be covered in the exercises at the end of the section. □

Exercise 4. For $f = x^3 + x - 1$ and $g = 2x^2 + 3x + 7$ as in (1.3), use the basis $\{1, x, x^2\}$ of A_f (thinking of A_f in terms of remainders) to show

$$\text{Res}(f, g) = 1^2 \det(m_g) = \det \begin{pmatrix} 7 & 2 & 3 \\ 3 & 5 & -1 \\ 2 & 3 & 5 \end{pmatrix} = 159.$$

Note that the 3×3 determinant in this example is smaller than the 5×5 determinant required by the definition (1.2). In general, Proposition (1.5) tells us that $\text{Res}(f, g)$ can be represented as an $l \times l$ determinant, while the definition of resultant uses an $(l + m) \times (l + m)$ matrix. The getmatrix procedure from Exercise 18 of Chapter 2, §4 can be used to construct the smaller matrix. Also, by interchanging f and g, we can represent the resultant using an $m \times m$ determinant.

For the final topic of this section, we will discuss a variation on $\text{Res}(f, g)$ which will be important for §2. Namely, instead of using polynomials in the single variable x, we could instead work with *homogeneous* polynomials in variables x, y. Recall that a polynomial is homogeneous if every term has the same total degree. Thus, if $F, G \in k[x, y]$ are homogeneous polynomials of total degrees l, m respectively, then we can write

(1.6)
$$F = a_0 x^l + a_1 x^{l-1} y + \cdots + a_l y^l$$
$$G = b_0 x^m + b_1 x^{m-1} y + \cdots + b_m y^m.$$

Note that a_0 or b_0 (or both) might be zero. Then we define $\operatorname{Res}(F, G) \in k$ using the same determinant as in (1.2).

Exercise 5. Show that $\operatorname{Res}(x^l, y^m) = 1$.

If we homogenize the polynomials f and g of (1.1) using appropriate powers of y, then we get F and G as in (1.6). In this case, it is obvious that $\operatorname{Res}(f, g) = \operatorname{Res}(F, G)$. However, going the other way is a bit more subtle, for if F and G are given by (1.6), then we can dehomogenize by setting $y = 1$, but we might fail to get polynomials of the proper degrees since a_0 or b_0 might be zero. Nevertheless, the resultant $\operatorname{Res}(F, G)$ still satisfies the following basic properties.

(1.7) Proposition. *Fix positive integers l and m.*
a. *There is a polynomial $\operatorname{Res}_{l,m} \in \mathbb{Z}[a_0, \ldots, a_l, b_0, \ldots, b_m]$ such that*

$$\operatorname{Res}(F, G) = \operatorname{Res}_{l,m}(a_0, \ldots, a_l, b_0, \ldots, b_m)$$

for all F, G as in (1.6).
b. *Over the field of complex numbers, $\operatorname{Res}(F, G) = 0$ if and only if the equations $F = G = 0$ have a solution $(x, y) \neq (0, 0)$ in $\mathbb{C}^2$ (this is called a **nontrivial solution**).*

PROOF. The first statement is an obvious consequence of the determinant formula for the resultant. As for the second, first observe that if $(u, v) \in \mathbb{C}^2$ is a nontrivial solution, then so is $(\lambda u, \lambda v)$ for any nonzero complex number λ. We now break up the proof into three cases.

First, if $a_0 = b_0 = 0$, then note that the resultant vanishes and that we have the nontrivial solution $(x, y) = (1, 0)$. Next, suppose that $a_0 \neq 0$ and $b_0 \neq 0$. If $\operatorname{Res}(F, G) = 0$, then, when we dehomogenize by setting $y = 1$, we get polynomials $f, g \in \mathbb{C}[x]$ with $\operatorname{Res}(f, g) = 0$. Since we're working over the complex numbers, the Common Factor property implies f and g must have a common root $x = u$, and then $(x, y) = (u, 1)$ is the desired nontrivial solution. Going the other way, if we have a nontrival solution (u, v), then our assumption $a_0 b_0 \neq 0$ implies that $v \neq 0$. Then $(u/v, 1)$ is also a solution, which means that u/v is a common root of the dehomogenized polynomials. From here, it follows easily that $\operatorname{Res}(F, G) = 0$.

The final case is when exactly one of a_0, b_0 is zero. The argument is a bit more complicated and will be covered in the exercises at the end of the section.   □

We should also mention that many other properties of the resultant, along with proofs, are contained in Chapter 12 of [GKZ].

ADDITIONAL EXERCISES FOR §1

Exercise 6. As an example of how resultants can be used to eliminate variables from equations, consider the parametric equations

$$x = 1 + s + t + st$$
$$y = 2 + s + st + t^2$$
$$z = s + t + s^2.$$

Our goal is to eliminate s, t from these equations to find an equation involving only x, y, z.

a. Use Gröbner basis methods to find the desired equation in x, y, z.

b. Use resultants to find the desired equations. Hint: Let $f = 1 + s + t + st - x$, $g = 2 + s + st + t^2 - y$ and $h = s + t + s^2 - z$. Then eliminate t by computing $\mathrm{Res}(f, g, t)$ and $\mathrm{Res}(f, h, t)$. Now what resultant do you use to get rid of s?

c. How are the answers to parts a and b related?

Exercise 7. Let f, g be as in (1.1). If we divide g by f, we get $g = q f + r$, where $\deg(r) < \deg(g) = m$. Then, assuming that r is nonconstant, show that

$$\mathrm{Res}(f, g) = a_0^{m - \deg(r)} \mathrm{Res}(f, r).$$

Hint: Let $g_1 = g - (b_0/a_0)x^{m-l}f$ and use column operations to subtract b_0/a_0 times the first l columns in the f part of the matrix from the columns in the g part. Expanding repeatedly along the first row gives $\mathrm{Res}(f, g) = a_0^{m - \deg(g_1)} \mathrm{Res}(f, g_1)$. Continue this process to obtain the desired formula.

Exercise 8. Our definition of $\mathrm{Res}(f, g)$ requires that f, g have positive degrees. Here is what to do when f or g is constant.

a. If $\deg(f) > 0$ but g is a nonzero constant b_0, show that the determinant (1.2) still makes sense and gives $\mathrm{Res}(f, b_0) = b_0^l$.

b. If $\deg(g) > 0$ and $a_0 \neq 0$, what is $\mathrm{Res}(a_0, g)$? Also, what is $\mathrm{Res}(a_0, b_0)$? What about $\mathrm{Res}(f, 0)$ or $\mathrm{Res}(0, g)$?

c. Exercise 7 assumes that the remainder r has positive degree. Show that the formula of Exercise 7 remains true even if r is constant.

Exercise 9. By Exercises 1, 7 and 8, resultants have the following three properties: $\mathrm{Res}(f, g) = (-1)^{lm} \mathrm{Res}(g, f)$; $\mathrm{Res}(f, b_0) = b_0^l$; and $\mathrm{Res}(f, g) = a_0^{m - \deg(r)} \mathrm{Res}(f, r)$ when $g = q f + r$. Use these properties to describe an algorithm for computing resultants. Hint: Your answer should be similar to the Euclidean algorithm.

Exercise 10. This exercise will give a proof of (1.4).

a. Given f, g as usual, define $\text{res}(f, g) = a_0^m \prod_{i=1}^{l} g(\xi_i)$, where $\xi_1, \ldots, \xi_l$ are the roots of f. Then show that $\text{res}(f, g)$ has the three properties of resultants mentioned in Exercise 9.

b. Show that the algorithm for computing $\text{res}(f, g)$ is the same as the algorithm for computing $\text{Res}(f, g)$, and conclude that the two are equal for all f, g.

Exercise 11. Let $f = a_0 x^l + a_1 x^{l-1} + \cdots + a_l \in k[x]$ be a polynomial with $a_0 \neq 0$, and let $A_f = k[x]/\langle f \rangle$. Given $g \in k[x]$, let $m_g : A_f \to A_f$ be multiplication by g.

a. Use the basis $\{1, x, \ldots, x^{l-1}\}$ of A_f (so we are thinking of A_f as consisting of remainders) to show that the matrix of m_x is

$$C_f = \begin{pmatrix} 0 & 0 & \cdots & 0 & -a_l/a_0 \\ 1 & 0 & \cdots & 0 & -a_{l-1}/a_0 \\ 0 & 1 & \cdots & 0 & -a_{l-2}/a_0 \\ \vdots & \vdots & \ddots & \vdots & \vdots \\ 0 & 0 & \cdots & 1 & -a_1/a_0 \end{pmatrix}.$$

This matrix (or more commonly, its transpose) is called the *companion matrix* of f.

b. If $g = b_0 x^m + \cdots + b_m$, then explain why the matrix of m_g is given by

$$g(C_f) = b_0 C_f^m + b_1 C_f^{m-1} + \cdots + b_m I,$$

where I is the $l \times l$ identity matrix. Hint: By Proposition (4.2) of Chapter 2, the map sending $g \in k[x]$ to $m_g \in M_{l \times l}(k)$ is a ring homomorphism.

c. Conclude that $\text{Res}(f, g) = a_0^m \det(g(C_f))$.

Exercise 12. In Proposition (1.5), we interpreted $\text{Res}(f, g)$ as the determinant of a linear map. It turns out that the original definition (1.2) of resultant has a similar interpretation. Let P_n denote the vector space of polynomials of degree $\leq n$. Since such a polynomial can be written $a_0 x^n + \cdots + a_n$, it follows that $\{x^n, \ldots, 1\}$ is a basis of P_n.

a. Given f, g as in (1.1), show that if $(A, B) \in P_{m-1} \oplus P_{l-1}$, then $A f + B g$ is in P_{l+m-1}. Conclude that we get a linear map $\Phi_{f,g} : P_{m-1} \oplus P_{l-1} \to P_{l+m-1}$.

b. If we use the bases $\{x^{m-1}, \ldots, 1\}$ of P_{m-1}, $\{x^{l-1}, \ldots, 1\}$ of P_{l-1} and $\{x^{l+m-1}, \ldots, 1\}$ of P_{l+m-1}, show that the matrix of the linear map $\Phi_{f,g}$ from part a is exactly the matrix used in (1.2). Thus, $\text{Res}(f, g) = \det(\Phi_{f,g})$, provided we use the above bases.

c. If $\text{Res}(f, g) \neq 0$, conclude that every polynomial of degree $\leq l + m - 1$ can be written uniquely as $A f + B g$ where $\deg(A) < m$ and $\deg(B) < l$.

Exercise 13. In the text, we only proved Proposition (1.5) in the special case when $g(\xi_1), \ldots, g(\xi_l)$ are distinct. For the general case, suppose $f = a_0(x - \xi_1)^{a_1} \cdots (x - \xi_r)^{a_r}$, where $\xi_1, \ldots, \xi_r$ are distinct. Then we want to prove that $\det(m_g) = \prod_{i=1}^{r} g(\xi_i)^{a_i}$.

a. First, suppose that $f = (x - \xi)^a$. In this case, we can use the basis of A_f given by $\{(x - \xi)^{a-1}, \ldots, x - \xi, 1\}$ (as usual, we think of A_f as consisting of remainders). Then show that the matrix of m_g with respect to the above basis is upper triangular with diagonal entries all equal to $g(\xi)$. Conclude that $\det(m_g) = g(\xi)^a$. Hint: Write $g = b_0 x^m + \cdots + b_m$ in the form $g = c_0(x - \xi)^m + \cdots + c_{m-1}(x - \xi) + c_m$ by replacing x with $(x - \xi) + \xi$ and using the binomial theorem. Then let $x = \xi$ to get $c_m = g(\xi)$.

b. In general, when $f = a_0(x - \xi_1)^{a_1} \cdots (x - \xi_r)^{a_r}$, show that there is a well-defined map

$$A_f \longrightarrow (k[x]/\langle (x - \xi_1)^{a_1} \rangle) \oplus \cdots \oplus (k[x]/\langle (x - \xi_r)^{a_r} \rangle)$$

which preserves sums and products. Hint: This is where working with cosets is a help. It is easy to show that the map sending $[h] \in A_f$ to $[h] \in k[x]/\langle (x - \xi_i)^{a_i} \rangle$ is well-defined since $(x - \xi_i)^{a_i}$ divides f.

c. Show that the map of part b is a ring isomorphism. Hint: First show that the map is one-to-one, and then use linear algebra and a dimension count to show it is onto.

d. By considering multiplication by g on

$$(k[x]/\langle (x - \xi_1)^{a_1} \rangle) \oplus \cdots \oplus (k[x]/\langle (x - \xi_r)^{a_r} \rangle)$$

and using part a, conclude that $\det(m_g) = \prod_{i=1}^{r} g(\xi_i)^{a_i}$ as desired.

Exercise 14. This exercise will complete the proof of Proposition (1.7). Suppose that F, G are given by (1.6) and assume $a_0 \neq 0$ and $b_0 = \cdots = b_{r-1} = 0$ but $b_r \neq 0$. If we dehomogenize by setting $y = 1$, we get polynomials f, g of degree $l, m - r$ respectively.

a. Show that $\mathrm{Res}(F, G) = a_0^r \mathrm{Res}(f, g)$.

b. Show that $\mathrm{Res}(F, G) = 0$ if and only $F = G = 0$ has a nontrivial solution. Hint: Modify the argument given in the text for the case when a_0 and b_0 were both nonzero.

§2 Multipolynomial Resultants

In §1, we studied the resultant of two homogeneous polynomials F, G in variables x, y. Generalizing this, suppose we are given $n + 1$ homogeneous polynomials $F_0, \ldots, F_n$ in variables $x_0, \ldots, x_n$, and assume that each F_i has positive total degree. Then we get $n + 1$ equations in $n + 1$ unknowns:

(2.1) $F_0(x_0, \ldots, x_n) = \cdots = F_n(x_0, \ldots, x_n) = 0.$

Because the F_i are homogeneous of positive total degree, these equations always have the solution $x_0 = \cdots = x_n = 0$, which we call the *trivial* solution. Hence, the crucial question is whether there is a *nontrivial* solution. For the rest of this chapter, we will work over the complex numbers, so that a nontrivial solution will be a point in $\mathbb{C}^{n+1} \setminus \{(0, \ldots, 0)\}$.

In general, the existence of a nontrivial solution depends on the coefficients of the polynomials $F_0, \ldots, F_n$: for most values of the coefficients, there are no nontrivial solutions, while for certain special values, they exist.

One example where this is easy to see is when the polynomials F_i are all linear, i.e., have total degree 1. Since they are homogeneous, the equations (2.1) can be written in the form:

$$F_0 = c_{00}x_0 + \cdots + c_{0n}x_n = 0$$

(2.2)
$$\vdots$$

$$F_n = c_{n0}x_0 + \cdots + c_{nn}x_n = 0.$$

This is an $(n + 1) \times (n + 1)$ system of linear equations, so that by linear algebra, there is a nontrivial solution if and only if the determinant of the coefficient matrix vanishes. Thus we get the *single* condition $\det(c_{ij}) = 0$ for the existence of a nontrivial solution. Note that this determinant is a polynomial in the coefficients c_{ij}.

Exercise 1. There was a single condition for a nontrivial solution of (2.2) because the number of equations $(n + 1)$ equaled the number of unknowns (also $n + 1$). When these numbers are different, here is what can happen.
a. If we have $r < n + 1$ linear equations in $n + 1$ unknowns, explain why there is *always* a nontrivial solution, no matter what the coefficients are.
b. When we have $r > n + 1$ linear equations in $n + 1$ unknowns, things are more complicated. For example, show that the equations

$$F_0 = c_{00}x + c_{01}y = 0$$
$$F_1 = c_{10}x + c_{11}y = 0$$
$$F_2 = c_{20}x + c_{21}y = 0$$

have a nontrivial solution if and only if the *three* conditions

$$\det \begin{pmatrix} c_{00} & c_{01} \\ c_{10} & c_{11} \end{pmatrix} = \det \begin{pmatrix} c_{00} & c_{01} \\ c_{20} & c_{21} \end{pmatrix} = \det \begin{pmatrix} c_{10} & c_{11} \\ c_{20} & c_{21} \end{pmatrix} = 0$$

are satisfied.

In general, when we have $n + 1$ homogeneous polynomials $F_0, \ldots, F_n \in \mathbb{C}[x_0, \ldots, x_n]$, we get the following Basic Question: *What conditions must the coefficients of $F_0, \ldots, F_n$ satisfy in order that $F_0 = \cdots = F_n = 0$ has a nontrivial solution?* To state the answer precisely, we need to introduce some notation. Suppose that d_i is the total degree of F_i, so that F_i can be written

$$F_i = \sum_{|\alpha|=d_i} c_{i,\alpha} x^\alpha.$$

For each possible pair of indices i, α, we introduce a variable $u_{i,\alpha}$. Then, given a polynomial $P \in \mathbb{C}[u_{i,\alpha}]$, we let $P(F_0, \ldots, F_n)$ denote the number obtained by replacing each variable $u_{i,\alpha}$ in P with the corresponding coefficient $c_{i,\alpha}$. This is what we mean by a *polynomial in the coefficients of the F_i*. We can now answer our Basic Question.

(2.3) Theorem. *If we fix positive degrees $d_0, \ldots, d_n$, then there is a unique polynomial* Res $\in \mathbb{Z}[u_{i,\alpha}]$ *which has the following properties:*
a. *If $F_0, \ldots, F_n \in \mathbb{C}[x_0, \ldots, x_n]$ are homogeneous of degrees $d_0, \ldots, d_n$, then the equations (2.1) have a nontrivial solution over $\mathbb{C}$ if and only if* Res$(F_0, \ldots, F_n) = 0$.
b. Res$(x_0^{d_0}, \ldots, x_n^{d_n}) = 1$.
c. Res *is irreducible, even when regarded as a polynomial in $\mathbb{C}[u_{i,\alpha}]$.*

PROOF. A complete proof of the existence of the resultant is beyond the scope of this book. See Chapter 13 of [GKZ] or §78 of [vdW] for proofs. At the end of this section, we will indicate some of the intuition behind the proof when we discuss the geometry of the resultant. The question of uniqueness will be considered in Exercise 5.   □

We call Res$(F_0, \ldots, F_n)$ the *resultant* of $F_0, \ldots, F_n$. Sometimes we write Res$_{d_0, \ldots, d_n}$ instead of Res if we want to make the dependence on the degrees more explicit. In this notation, if each $F_i = \sum_{j=0}^n c_{ij} x_j$ is linear, then the discussion following (2.2) shows that

$$\text{Res}_{1,\ldots,1}(F_0, \ldots, F_n) = \det(c_{ij}).$$

Another example is the resultant of two polynomials, which was discussed in §1. In this case, we know that Res(F_0, F_1) is given by the determinant (1.2). Theorem (2.3) tells us that this determinant is an irreducible polynomial in the coefficients of F_0, F_1.

Before giving further examples of multipolynomial resultants, we want to indicate their usefulness in applications. Let's consider the *implicitization problem*, which asks for the equation of a parametric curve or surface. For concreteness, suppose a surface is given parametrically by the equations

$$x = f(s, t)$$
(2.4)
$$y = g(s, t)$$
$$z = h(s, t),$$

where $f(s, t), g(s, t), h(s, t)$ are polynomials (not necessarily homogeneous) of total degrees d_0, d_1, d_2. There are several methods to find the equation $p(x, y, z) = 0$ of the surface described by (2.4). For example, Chapter 3 of

[CLO] uses Gröbner bases for this purpose. We claim that in many cases, multipolynomial resultants can be used to find the equation of the surface.

To use our methods, we need homogeneous polynomials, and hence we will homogenize the above equations with respect to a third variable u. For example, if we write $f(s,t)$ in the form

$$f(s,t) = f_{d_0}(s,t) + f_{d_0-1}(s,t) + \cdots + f_0(s,t),$$

where f_j is homogeneous of total degree j in s, t, then we get

$$F(s,t,u) = f_{d_0}(s,t) + f_{d_0-1}(s,t)u + \cdots + f_0(s,t)u^{d_0},$$

which is now homogeneous in s, t, u of total degree d_0. Similarly, $g(s,t)$ and $h(s,t)$ homogenize to $G(s,t,u)$ and $H(s,t,u)$, and the equations (2.4) become

$$(2.5) \quad F(s,t,u) - xu^{d_0} = G(s,t,u) - yu^{d_1} = H(s,t,u) - zu^{d_2} = 0.$$

Note that x, y, z are regarded as coefficients in these equations.

We can now solve the implicitization problem for (2.4) as follows.

(2.6) Proposition. *With the above notation, assume that the system of homogeneous equations*

$$f_{d_0}(s,t) = g_{d_1}(s,t) = h_{d_2}(s,t) = 0$$

has only the trivial solution. Then, for a given triple $(x,y,z) \in \mathbb{C}^3$, the equations (2.4) have a solution $(s,t) \in \mathbb{C}^2$ if and only if

$$\mathrm{Res}_{d_0,d_1,d_2}(F - xu^{d_0}, G - yu^{d_1}, H - zu^{d_2}) = 0.$$

PROOF. By Theorem (2.3), the resultant vanishes if and only if (2.5) has a nontrivial solution (s,t,u). If $u \neq 0$, then $(s/u, t/u)$ is a solution to (2.4). However, if $u = 0$, then (s,t) is a nontrivial solution of $f_{d_0}(s,t) = g_{d_1}(s,t) = h_{d_2}(s,t) = 0$, which contradicts our hypothesis. Hence, $u = 0$ can't occur. Going the other way, note that a solution (s,t) of (2.4) gives the nontrivial solution $(s,t,1)$ of (2.5). □

Since the resultant is a polynomial in the coefficients, it follows that

$$(2.7) \quad p(x,y,z) = \mathrm{Res}_{d_0,d_1,d_2}(F - xu^{d_0}, G - yu^{d_1}, H - zu^{d_2})$$

is a polynomial in x, y, z which, by Proposition (2.6), vanishes *precisely* on the image of the parametrization. In particular, this means that the parametrization covers *all* of the surface $p(x,y,z) = 0$, which is not true for all polynomial parametrizations—the hypothesis that $f_{d_0}(s,t) = g_{d_1}(s,t) = h_{d_2}(s,t) = 0$ has only the trivial solution is important here.

Exercise 2.

a. If $f_{d_0}(s,t) = g_{d_1}(s,t) = h_{d_2}(s,t) = 0$ has a nontrivial solution, show that the resultant (2.7) vanishes identically. Hint: Show that (2.5) always has a nontrivial solution, no matter what x, y, z are.

b. Show that the parametric equations $(x, y, z) = (st, s^2t, st^2)$ define the surface $x^3 = yz$. By part a, we know that the resultant (2.7) can't be used to find this equation. Show that in this case, it is also true that the parametrization is not onto—there are points on the surface which don't come from any s, t.

We should point out that for some systems of equations, such as

$$x = 1 + s + t + st$$
$$y = 2 + s + 3t + st$$
$$z = s - t + st,$$

the resultant (2.7) vanishes identically by Exercise 2, yet a resultant can still be defined—this is one of the *sparse resultants* which we will consider in Chapter 7.

One difficulty with multipolynomial resultants is that they tend to be *very* large expressions. For example, consider the system of equations given by 3 quadratic forms in 3 variables:

$$F_0 = c_{01}x^2 + c_{02}y^2 + c_{03}z^2 + c_{04}xy + c_{05}xz + c_{06}yz = 0$$
$$F_1 = c_{11}x^2 + c_{12}y^2 + c_{13}z^2 + c_{14}xy + c_{15}xz + c_{16}yz = 0$$
$$F_2 = c_{21}x^2 + c_{22}y^2 + c_{23}z^2 + c_{24}xy + c_{25}xz + c_{26}yz = 0.$$

Classically, this is a system of "three ternary quadrics". By Theorem (2.3), the resultant $\mathrm{Res}_{2,2,2}(F_0, F_1, F_2)$ vanishes exactly when this system has a nontrivial solution in x, y, z.

The polynomial $\mathrm{Res}_{2,2,2}$ is very large: it has 18 variables (one for each coefficient c_{ij}), and the theory of §3 will tell us that it has total degree 12. Written out in its full glory, $\mathrm{Res}_{2,2,2}$ has 21,894 terms (we are grateful to Bernd Sturmfels for this computation). Hence, to work effectively with this resultant, we need to learn some more compact ways of representing it. We will study this topic in more detail in §3 and §4, but to whet the reader's appetite, we will now give one of the many interesting formulas for $\mathrm{Res}_{2,2,2}$.

First, let J denote the Jacobian determinant of F_0, F_1, F_2:

$$J = \det \begin{pmatrix} \dfrac{\partial F_0}{\partial x} & \dfrac{\partial F_0}{\partial y} & \dfrac{\partial F_0}{\partial z} \\[2mm] \dfrac{\partial F_1}{\partial x} & \dfrac{\partial F_1}{\partial y} & \dfrac{\partial F_1}{\partial z} \\[2mm] \dfrac{\partial F_2}{\partial x} & \dfrac{\partial F_2}{\partial y} & \dfrac{\partial F_2}{\partial z} \end{pmatrix},$$

which is a cubic homogeneous polynomial in x, y, z. This means that the partial derivatives of J are quadratic and hence can be written in the following form:

$$\frac{\partial J}{\partial x} = b_{01}x^2 + b_{02}y^2 + b_{03}z^2 + b_{04}xy + b_{05}xz + b_{06}yz$$

$$\frac{\partial J}{\partial y} = b_{11}x^2 + b_{12}y^2 + b_{13}z^2 + b_{14}xy + b_{15}xz + b_{16}yz$$

$$\frac{\partial J}{\partial z} = b_{21}x^2 + b_{22}y^2 + b_{23}z^2 + b_{24}xy + b_{25}xz + b_{26}yz.$$

Note that each b_{ij} is a cubic polynomial in the c_{ij}. Then, by a classical formula of Salmon (see [Sal], Art. 90), the resultant of three ternary quadrics is given by the 6×6 determinant

$$(2.8) \quad \text{Res}_{2,2,2}(F_0, F_1, F_2) = \frac{-1}{512} \det \begin{pmatrix} c_{01} & c_{02} & c_{03} & c_{04} & c_{05} & c_{06} \\ c_{11} & c_{12} & c_{13} & c_{14} & c_{15} & c_{16} \\ c_{21} & c_{22} & c_{23} & c_{24} & c_{25} & c_{26} \\ b_{01} & b_{02} & b_{03} & b_{04} & b_{05} & b_{06} \\ b_{11} & b_{12} & b_{13} & b_{14} & b_{15} & b_{16} \\ b_{21} & b_{22} & b_{23} & b_{24} & b_{25} & b_{26} \end{pmatrix}.$$

Exercise 3.
a. Use (2.8) to explain why $\text{Res}_{2,2,2}$ has total degree 12 in the variables $c_{01}, \ldots, c_{26}$.
b. Why is the fraction $-1/512$ needed in (2.8)? Hint: Compute the resultant $\text{Res}_{2,2,2}(x^2, y^2, z^2)$.
c. Use (2.7) and (2.8) to find the equation of the surface defined by the equations

$$x = 1 + s + t + st$$
$$y = 2 + s + st + t^2$$
$$z = s + t + s^2.$$

Note that $st = st + t^2 = s^2 = 0$ has only the trivial solution, so that Proposition (2.6) applies. You should compare your answer to Exercise 6 of §1.

In §4 we will study the general question of how to find a formula for a given resultant. Here is an example which illustrates one of the methods we will use. Consider the following system of three homogeneous equations in three variables:

$$(2.9) \quad \begin{aligned} F_0 &= a_1 x + a_2 y + a_3 z = 0 \\ F_1 &= b_1 x + b_2 y + b_3 z = 0 \\ F_2 &= c_1 x^2 + c_2 y^2 + c_3 z^2 + c_4 xy + c_5 xz + c_6 yz = 0. \end{aligned}$$

Since F_0 and F_1 are linear and F_2 is quadratic, the resultant involved is $\text{Res}_{1,1,2}(F_0, F_1, F_2)$. We get the following formula for this resultant.

(2.10) Proposition. $\mathrm{Res}_{1,1,2}(F_0, F_1, F_2)$ *is given by the polynomial*

$$a_1^2 b_2^2 c_3 - a_1^2 b_2 b_3 c_6 + a_1^2 b_3^2 c_2 - 2a_1 a_2 b_1 b_2 c_3 + a_1 a_2 b_1 b_3 c_6$$
$$+ a_1 a_2 b_2 b_3 c_5 - a_1 a_2 b_3^2 c_4 + a_1 a_3 b_1 b_2 c_6 - 2a_1 a_3 b_1 b_3 c_2 - a_1 a_3 b_2^2 c_5$$
$$+ a_1 a_3 b_2 b_3 c_4 + a_2^2 b_1^2 c_3 - a_2^2 b_1 b_3 c_5 + a_2^2 b_3^2 c_1 - a_2 a_3 b_1^2 c_6$$
$$+ a_2 a_3 b_1 b_2 c_5 + a_2 a_3 b_1 b_3 c_4 - 2a_2 a_3 b_2 b_3 c_1 + a_3^2 b_1^2 c_2 - a_3^2 b_1 b_2 c_4 + a_3^2 b_2^2 c_1.$$

PROOF. Let R denote the above polynomial, and suppose we have a non-trivial solution (x, y, z) of (2.9). We will first show that this forces a slight variant of R to vanish. Namely, consider the six equations

$$(2.11) \qquad x \cdot F_0 = y \cdot F_0 = z \cdot F_0 = y \cdot F_1 = z \cdot F_1 = 1 \cdot F_2 = 0,$$

which we can write as

$$
\begin{array}{ccccccccccccc}
a_1 x^2 & + & 0 & + & 0 & + & a_2 xy & + & a_3 xz & + & 0 & = & 0 \\
0 & + & a_2 y^2 & + & 0 & + & a_1 xy & + & 0 & + & a_3 yz & = & 0 \\
0 & + & 0 & + & a_3 z^2 & + & 0 & + & a_1 xz & + & a_2 yz & = & 0 \\
0 & + & b_2 y^2 & + & 0 & + & b_1 xy & + & 0 & + & b_3 yz & = & 0 \\
0 & + & 0 & + & b_3 z^3 & + & 0 & + & b_1 xz & + & b_2 yz & = & 0 \\
c_1 x^2 & + & c_2 y^2 & + & c_3 z^2 & + & c_4 xy & + & c_5 xz & + & c_6 yz & = & 0.
\end{array}
$$

If we regard $x^2, y^2, z^2, xy, xz, yz$ as "unknowns", then this system of six linear equations has a nontrivial solution, which implies that the determinant D of its coefficient matrix is zero. Using a computer, one easily checks that the determinant is $D = -a_1 R$.

Thinking geometrically, we have proved that in the 12 dimensional space $\mathbb{C}^{12}$ with $a_1, \ldots, c_6$ as coordinates, the polynomial D vanishes on the set

$$(2.12) \qquad \{(a_1, \ldots, c_6) : (2.9) \text{ has a nontrivial solution}\} \subset \mathbb{C}^{12}.$$

However, by Theorem (2.3), having a nontrivial solution is equivalent to the vanishing of the resultant, so that D vanishes on the set

$$\mathbf{V}(\mathrm{Res}_{1,1,2}) \subset \mathbb{C}^{12}.$$

This means that $D \in \mathbf{I}(\mathbf{V}(\mathrm{Res}_{1,1,2})) = \sqrt{\langle \mathrm{Res}_{1,1,2} \rangle}$, where the last equality is by the Nullstellensatz (see §4 of Chapter 1). But $\mathrm{Res}_{1,1,2}$ is irreducible, which easily implies that $\sqrt{\langle \mathrm{Res}_{1,1,2} \rangle} = \langle \mathrm{Res}_{1,1,2} \rangle$. This proves that $D \in \langle \mathrm{Res}_{1,1,2} \rangle$, so that $D = -a_1 R$ is a multiple of $\mathrm{Res}_{1,1,2}$. Irreducibility then implies that $\mathrm{Res}_{1,1,2}$ divides either a_1 or R. The results of §3 will tell us that $\mathrm{Res}_{1,1,2}$ has total degree 5. It follows that $\mathrm{Res}_{1,1,2}$ divides R, and since R also has total degree 5, it must be a constant multiple of $\mathrm{Res}_{1,1,2}$. By computing the value of each when $(F_0, F_1, F_2) = (x, y, z^2)$, we see that the constant must be 1, which proves that $R = \mathrm{Res}_{1,1,2}$, as desired. $\qquad \square$

Exercise 4. Verify that $R = 1$ when $(F_0, F_1, F_2) = (x, y, z^2)$.

The equations (2.11) may seem somewhat unmotivated. In §4 we will see that there is a systematic reason for choosing these equations.

The final topic of this section is the geometric interpretation of the resultant. We will use the same framework as in Theorem (2.3). This means that we consider homogeneous polynomials of degree $d_0, \ldots, d_n$, and for each monomial x^α of degree d_i, we introduce a variable $u_{i,\alpha}$. Let M be the total number of these variables, so that $\mathbb{C}^M$ is an affine space with coordinates $u_{i,\alpha}$ for all $0 \leq i \leq n$ and $|\alpha| = d_i$. A point of $\mathbb{C}^M$ will be written $(c_{i,\alpha})$. Then consider the "universal" polynomials

$$\mathbf{F}_i = \sum_{|\alpha| = d_i} u_{i,\alpha} x^\alpha, \quad i = 0, \ldots, n.$$

Note that the coefficients of the x^α are the variables $u_{i,\alpha}$. If we evaluate $\mathbf{F}_0, \ldots, \mathbf{F}_n$ at $(c_{i,\alpha}) \in \mathbb{C}^M$, we get the polynomials $F_0, \ldots, F_n$, where $F_i = \sum_{|\alpha| = d_i} c_{i,\alpha} x^\alpha$. Thus, we can think of points of $\mathbb{C}^M$ as parametrizing all possible $(n + 1)$-tuples of homogeneous polynomials of degrees $d_0, \ldots, d_n$.

To keep track of nontrivial solutions of these polynomials, we will use projective space $\mathbb{P}^n(\mathbb{C})$, which we write as $\mathbb{P}^n$ for short. Recall the following:

- A point in $\mathbb{P}^n$ has homogeneous coordinates $(a_0, \ldots, a_n)$, where $a_i \in \mathbb{C}$ are not all zero, and another set of coordinates $(b_0, \ldots, b_n)$ gives the same point in $\mathbb{P}^n$ if and only if there is a complex number $\lambda \neq 0$ such that $(b_0, \ldots, b_n) = \lambda(a_0, \ldots, a_n)$.
- If $F(x_0, \ldots, x_n)$ is homogeneous of degree d and $(b_0, \ldots, b_n) = \lambda(a_0, \ldots, a_n)$ are two sets of homogeneous coordinates for some point $p \in \mathbb{P}^n$, then

$$F(b_0, \ldots, b_n) = \lambda^d F(a_0, \ldots, a_n).$$

Thus, we can't define the value of F at p, but the equation $F(p) = 0$ makes perfect sense. Hence we get the *projective variety* $\mathbf{V}(F) \subset \mathbb{P}^n$, which is the set of points of $\mathbb{P}^n$ where F vanishes.

For a homogeneous polynomial F, notice that $\mathbf{V}(F) \subset \mathbb{P}^n$ is determined by the *nontrivial* solutions of $F = 0$. For more on projective space, see Chapter 8 of [CLO].

Now consider the product $\mathbb{C}^M \times \mathbb{P}^n$. A point $(c_{i,\alpha}, a_0, \ldots, a_n) \in \mathbb{C}^M \times \mathbb{P}^n$ can be regarded as $n + 1$ homogeneous polynomials and a point of $\mathbb{P}^n$. The "universal" polynomials $\mathbf{F}_i$ are actually polynomials on $\mathbb{C}^M \times \mathbb{P}^n$, which gives the subset $W = \mathbf{V}(\mathbf{F}_0, \ldots, \mathbf{F}_n)$. Concretely, this set is given by

(2.13)
$$\begin{aligned} W = \{&(c_{i,\alpha}, a_0, \ldots, a_n) \in \mathbb{C}^M \times \mathbb{P}^n : (a_0, \ldots, a_n) \text{ is a} \\ &\text{nontrivial solution of } F_0 = \cdots = F_n = 0, \text{ where} \\ &F_0, \ldots, F_n \text{ are determined by } (c_{i,\alpha})\} \\ = \{&\text{all possible pairs consisting of a set of equations} \\ &F_0 = \cdots = F_n = 0 \text{ of degrees } d_0, \ldots, d_n \text{ and} \\ &\text{a nontrivial solution of the equations}\}. \end{aligned}$$

Now comes the interesting part: there is a natural projection map

$$\pi : \mathbb{C}^M \times \mathbb{P}^n \longrightarrow \mathbb{C}^M$$

defined by $\pi(c_{i,\alpha}, a_0, \ldots, a_n) = (c_{i,\alpha})$, and under this projection, the variety $W \subset \mathbb{C}^M \times \mathbb{P}^n$ maps to

$$\pi(W) = \{(c_{i,\alpha}) \in \mathbb{C}^M : \text{there is } (a_0, \ldots, a_n) \in \mathbb{P}^n$$
$$\text{such that } (c_{i,\alpha}, a_0, \ldots, a_n) \in W\}$$
$$= \{\text{all possible sets of equations } F_0 = \cdots = F_n = 0 \text{ of}$$
$$\text{degrees } d_1, \ldots, d_n \text{ which have a nontrivial solution}\}.$$

Note that when the degrees are $(d_0, d_1, d_2) = (1, 1, 2)$, $\pi(W)$ is as in (2.12).

The essential content of Theorem (2.3) is that the set $\pi(W)$ is defined by the *single irreducible* equation $\text{Res}_{d_0,\ldots,d_n} = 0$. To prove this, first note that $\pi(W)$ is a variety in $\mathbb{C}^M$ by the following result of elimination theory.

- (Projective Extension Theorem) Given a variety $W \subset \mathbb{C}^M \times \mathbb{P}^n$ and the projection map $\pi : \mathbb{C}^M \times \mathbb{P}^n \to \mathbb{C}^M$, the image $\pi(W)$ is a variety in $\mathbb{C}^M$.

(See, for example, §5 of Chapter 8 of [CLO].) This is one of the key reasons we work with projective space (the corresponding assertion for affine space is false in general). Hence $\pi(W)$ is defined by the vanishing of certain polynomials on $\mathbb{C}^M$. In other words, the existence of a nontrivial solution of $F_0 = \cdots = F_n = 0$ is determined by polynomial conditions on the coefficients of $F_0, \ldots, F_n$.

The second step in the proof is to show that we need only one polynomial and that this polynomial is irreducible. Here, a rigorous proof requires knowing certain facts about the dimension and irreducible components of a variety (see, for example, [Sha], §6 of Chapter I). If we accept an intuitive idea of dimension, then the basic idea is to show that the variety $\pi(W) \subset \mathbb{C}^M$ is irreducible (can't be decomposed into smaller pieces which are still varieties) of dimension $M - 1$. In this case, the theory will tell us that $\pi(W)$ must be defined by exactly one irreducible equation, which is the resultant $\text{Res}_{d_0,\ldots,d_n} = 0$.

To prove this, first note that $\mathbb{C}^M \times \mathbb{P}^n$ has dimension $M + n$. Then observe that $W \subset \mathbb{C}^M \times \mathbb{P}^n$ is defined by the $n + 1$ equations $\mathbf{F}_0 = \cdots = \mathbf{F}_n = 0$. Intuitively, each equation drops the dimension by one, though strictly speaking, this requires that the equations be "independent" in an appropriate sense. In our particular case, this is true because each equation involves a disjoint set of coefficient variables $u_{i,\alpha}$. Thus the dimension of W is $(M + n) - (n + 1) = M - 1$. One can also show that W is irreducible (see Exercise 9 below). From here, standard arguments imply that $\pi(W)$ is irreducible. The final part of the argument is to show that the map $W \to \pi(W)$ is one-to-one "most of the time". Here, the idea is that if $F_0 = \cdots = F_n = 0$ do happen to have a nontrivial solution, then this solution is usually unique (up to a scalar multiple). For the special case when all of the F_i are linear, we will prove this in Exercise 10 below. For the general case, see Proposition 3.1 of Chapter 3 of [GKZ]. Since $W \to \pi(W)$ is onto and one-to-one most of the time, $\pi(W)$ also has dimension $M - 1$.

ADDITIONAL EXERCISES FOR §2

Exercise 5. To prove the uniqueness of the resultant, suppose there are two polynomials Res and Res' satisfying the conditions of Theorem (2.3).
a. Adapt the argument used in the proof of Proposition (2.10) to show that Res divides Res' and Res' divides Res. Note that this uses conditions a and c of the theorem.
b. Now use condition b of Theorem (2.3) to conclude that Res = Res'.

Exercise 6. A homogeneous polynomial in $\mathbb{C}[x]$ is written in the form ax^d. Show that $\mathrm{Res}_d(ax^d) = a$. Hint: Use Exercise 5.

Exercise 7. When the hypotheses of Proposition (2.6) are satisfied, the resultant (2.7) gives a polynomial $p(x, y, z)$ which vanishes precisely on the parametrized surface. However, p need not have the smallest possible total degree: it can happen that $p = q^d$ for some polynomial q of smaller total degree. For example, consider the (fairly silly) parametrization given by $(x, y, z) = (s, s, t^2)$. Use the formula of Proposition (2.10) to show that in this case, p is the square of another polynomial.

Exercise 8. The method used in the proof of Proposition (2.10) can be used to explain how the determinant (1.2) arises from nontrivial solutions $F = G = 0$, where F, G are as in (1.6). Namely, if (x, y) is a nontrivial solution of (1.6), then consider the $l + m$ equations

$$x^{m-1} \cdot F = 0$$
$$x^{m-2}y \cdot F = 0$$
$$\vdots$$
$$y^{m-1} \cdot F = 0$$
$$x^{l-1} \cdot G = 0$$
$$x^{l-2}y \cdot G = 0$$
$$\vdots$$
$$y^{l-1} \cdot G = 0.$$

Regarding this as a system of linear equations in unknowns x^{l+m-1}, $x^{l+m-2}y, \ldots, y^{l+m-1}$, show that the coefficient matrix is exactly the transpose of (1.2), and conclude that the determinant of this matrix must vanish whenever (1.6) has a nontrivial solution.

Exercise 9. In this exercise, we will give a rigorous proof that the set W from (2.13) is irreducible of dimension $M - 1$. For convenience, we will write a point of $\mathbb{C}^M$ as $(F_0, \ldots, F_n)$.

a. If $p = (a_0, \ldots, a_n)$ are fixed homogeneous coordinates for a point $p \in \mathbb{P}^n$, show that the map $\mathbb{C}^M \to \mathbb{C}^{n+1}$ defined by $(F_0, \ldots, F_n) \mapsto (F_0(p), \ldots, F_n(p))$ is linear and onto. Conclude that the kernel of this map has dimension $M - n - 1$. Denote this kernel by $K(p)$.

b. Besides the projection $\pi : \mathbb{C}^M \times \mathbb{P}^n \to \mathbb{C}^M$ used in the text, we also have a projection map $\mathbb{C}^M \times \mathbb{P}^n \to \mathbb{P}^n$, which is projection on the second factor. If we restrict this map to W, we get a map $\tilde{\pi} : W \to \mathbb{P}^n$ defined by $\tilde{\pi}(F_0, \ldots, F_n, p) = p$. Then show that

$$\tilde{\pi}^{-1}(p) = K(p) \times \{p\},$$

where as usual $\tilde{\pi}^{-1}(p)$ is the inverse image of $p \in \mathbb{P}^n$ under $\tilde{\pi}$, i.e., the set of all points of W which map to p under $\tilde{\pi}$. In particular, this shows that $\tilde{\pi} : W \to \mathbb{P}^n$ is onto and that all inverse images of points are irreducible (being linear subspaces) of the same dimension.

c. Use Theorem 8 of [Sha], §6 of Chapter 1, to conclude that W is irreducible.

d. Use Theorem 7 of [Sha], §6 of Chapter 1, to conclude that W has dimension $M - 1 = n$ (dimension of $\mathbb{P}^n$) + $M - n - 1$ (dimension of the inverse images).

Exercise 10. In this exercise, we will show that the map $W \to \pi(W)$ is usually one-to-one in the special case when $F_0, \ldots, F_n$ have degree 1. Here, we know that if $F_i = \sum_{j=0}^n c_{ij} x_j$, then $\mathrm{Res}(F_0, \ldots, F_n) = \det(A)$, where $A = (c_{ij})$. Note that A is an $(n+1) \times (n+1)$ matrix.

a. Show that $F_0 = \cdots = F_n = 0$ has a nontrivial solution if and only if A has rank $< n + 1$.

b. If A has rank n, prove that there is a unique nontrivial solution (up to a scalar multiple).

c. Given $0 \le i, j \le n$, let $A^{i,j}$ be the $n \times n$ matrix obtained from A by deleting row i and column j. Prove that A has rank $< n$ if and only if $\det(A^{i,j}) = 0$ for all i, j. Hint: To have rank $\ge n$, it must be possible to find n columns which are linearly independent. Then, looking at the submatrix formed by these columns, it must be possible to find n rows which are linearly independent. This leads to one of the matrices $A^{i,j}$.

d. Let $Y = \mathbf{V}(\det(A^{i,j}) : 0 \le i, j \le n)$. Show that $Y \subset \pi(W)$ and that $Y \ne \pi(W)$. Since $\pi(W)$ is irreducible, standard arguments show that Y has dimension strictly smaller than $\pi(W)$ (see, for example, Corollary 2 to Theorem 4 of [Sha], §6 of Chapter I).

e. Show that if $a, b \in W$ and $\pi(a) = \pi(b) \in \pi(W) \setminus Y$, then $a = b$. Since Y has strictly smaller dimension than $\pi(W)$, this is a precise version of what we mean by saying the map $W \to \pi(W)$ is "usually one-to-one". Hint: Use parts b and c.

§3 Properties of Resultants

In Theorem (2.3), we saw that the resultant $\mathrm{Res}(F_0, \ldots, F_n)$ vanishes if and only if $F_0 = \cdots = F_n = 0$ has a nontrivial solution, and is irreducible over $\mathbb{C}$ when regarded as a polynomial in the coefficients of the F_i. These conditions characterize the resultant up to a constant, but they in no way exhaust the many properties of this remarkable polynomial. This section will contain a summary of the other main properties of the resultant. No proofs will be given, but complete references will be provided.

Throughout this section, we will fix total degrees $d_0, \ldots, d_n > 0$ and let $\mathrm{Res} = \mathrm{Res}_{d_0, \ldots, d_n} \in \mathbb{Z}[u_{i,\alpha}]$ be the resultant polynomial from §2.

We begin by studying the degree of the resultant.

(3.1) Theorem. *For a fixed j between 0 and n, Res is homogeneous in the variables $u_{j,\alpha}$, $|\alpha| = d_j$, of degree $d_0 \cdots d_{j-1} d_{j+1} \cdots d_n$. This means that*

$$\mathrm{Res}(F_0, \ldots, \lambda F_j, \ldots, F_n) = \lambda^{d_0 \cdots d_{j-1} d_{j+1} \cdots d_n} \mathrm{Res}(F_0, \ldots, F_n).$$

Furthermore, the total degree of Res is $\sum_{j=0}^{n} d_0 \cdots d_{j-1} d_{j+1} \cdots d_n$.

PROOF. A proof can be found in §2 of [Jou1] or Chapter 13 of [GKZ]. $\square$

Exercise 1. Show that the final assertion of Theorem (3.1) is an immediate consequence of the formula for $\mathrm{Res}(F_0, \ldots, \lambda F_j, \ldots, F_n)$. Hint: What is $\mathrm{Res}(\lambda F_0, \ldots, \lambda F_n)$?

Exercise 2. Show that formulas (1.2) and (2.8) for $\mathrm{Res}_{l,m}$ and $\mathrm{Res}_{2,2,2}$ satisfy Theorem (3.1).

We next study the symmetry and multiplicativity of the resultant.

(3.2) Theorem.
a. *If $i < j$, then*

$$\mathrm{Res}(F_0, \ldots, F_i, \ldots, F_j, \ldots, F_n) =$$
$$(-1)^{d_0 \cdots d_n} \mathrm{Res}(F_0, \ldots, F_j, \ldots, F_i, \ldots, F_n),$$

where the bottom resultant is for degrees $d_0, \ldots, d_j, \ldots, d_i, \ldots, d_n$.
b. *If $F_j = F_j' F_j''$ is a product of homogeneous polynomials of degrees d_j' and d_j'', then*

$$\mathrm{Res}(F_0, \ldots, F_j, \ldots, F_n) =$$
$$\mathrm{Res}(F_0, \ldots, F_j', \ldots, F_n) \cdot \mathrm{Res}(F_0, \ldots, F_j'', \ldots, F_n),$$

where the resultants on the bottom are for degrees $d_0, \ldots, d_j', \ldots, d_n$ and $d_0, \ldots, d_j'', \ldots, d_n$.

PROOF. A proof of the first assertion of the theorem can be found in §5 of [Jou1]. As for the second, we can assume $j = n$ by part a. This case will be covered in Exercise 9 at the end of the section.   □

Exercise 3. Prove that formulas (1.2) and (2.8) for $\text{Res}_{l,m}$ and $\text{Res}_{2,2,2}$ satisfy part a of Theorem (3.2).

Our next task is to show that the analog of Proposition (1.5) holds for general resultants. We begin with some notation. Given homogeneous polynomials $F_0, \ldots, F_n \in \mathbb{C}[x_0, \ldots, x_n]$ of degrees $d_0, \ldots, d_n$, let

$$(3.3) \quad \begin{aligned} f_i(x_0, \ldots, x_{n-1}) &= F_i(x_0, \ldots, x_{n-1}, 1) \\ \overline{F}_i(x_0, \ldots, x_{n-1}) &= F_i(x_0, \ldots, x_{n-1}, 0). \end{aligned}$$

Note that $\overline{F}_0, \ldots, \overline{F}_{n-1}$ are homogeneous in $\mathbb{C}[x_0, \ldots, x_{n-1}]$ of degrees $d_0, \ldots, d_{n-1}$.

(3.4) Theorem. *If* $\text{Res}(\overline{F}_0, \ldots, \overline{F}_{n-1}) \neq 0$, *then the quotient ring* $A = \mathbb{C}[x_0, \ldots, x_{n-1}]/\langle f_0, \ldots, f_{n-1} \rangle$ *has dimension* $d_0 \cdots d_{n-1}$ *as a vector space over* $\mathbb{C}$, *and*

$$\text{Res}(F_0, \ldots, F_n) = \text{Res}(\overline{F}_0, \ldots, \overline{F}_{n-1})^{d_n} \det(m_{f_n} : A \to A),$$

where $m_{f_n} : A \to A$ *is the linear map given by multiplication by* f_n.

PROOF. Although we will not prove this result (see [Jou1], §§2, 3 and 4 for a complete proof), we will explain (non-rigorously) why the above formula is reasonable. The first step is to show that the ring A is a finite-dimensional vector space over $\mathbb{C}$ when $\text{Res}(\overline{F}_0, \ldots, \overline{F}_{n-1}) \neq 0$. The crucial idea is to think in terms of the projective space $\mathbb{P}^n$. We can decompose $\mathbb{P}^n$ into two pieces using x_n: the affine space $\mathbb{C}^n \subset \mathbb{P}^n$ defined by $x_n = 1$, and the "hyperplane at infinity" $\mathbb{P}^{n-1} \subset \mathbb{P}^n$ defined by $x_n = 0$. Note that the other variables $x_0, \ldots, x_{n-1}$ play two roles: they are ordinary coordinates for $\mathbb{C}^n \subset \mathbb{P}^n$, and they are homogeneous coordinates for the hyperplane at infinity.

The equations $F_0 = \cdots = F_{n-1} = 0$ determine a projective variety $V \subset \mathbb{P}^n$. By (3.3), $f_0 = \cdots = f_{n-1} = 0$ defines the "affine part" $\mathbb{C}^n \cap V \subset V$, while $\overline{F}_0 = \cdots = \overline{F}_{n-1} = 0$ defines the "part at infinity" $\mathbb{P}^{n-1} \cap V \subset V$. Hence, the hypothesis $\text{Res}(\overline{F}_0, \ldots, \overline{F}_{n-1}) \neq 0$ implies that there are no solutions at infinity. In other words, the projective variety V is contained in $\mathbb{C}^n \subset \mathbb{P}^n$. Now we can apply the following result from algebraic geometry:

- (Projective Varieties in Affine Space) If a projective variety in $\mathbb{P}^n$ is contained in an affine space $\mathbb{C}^n \subset \mathbb{P}^n$, then the projective variety must consist of a finite set of points.

(See, for example, [Sha], §5 of Chapter I.) Applied to V, this tells us that V must be a finite set of points. Since $\mathbb{C}$ is algebraically closed and $V \subset \mathbb{C}^n$

is defined by $f_0 = \cdots = f_{n-1} = 0$, the Finiteness Theorem from §2 of Chapter 2 implies that $A = \mathbb{C}[x_0, \ldots, x_{n-1}]/\langle f_0, \ldots, f_{n-1}\rangle$ is finite dimensional over $\mathbb{C}$. Hence $\det(m_{f_n} : A \to A)$ is defined, so that the formula of the theorem makes sense.

We also need to know the dimension of the ring A. The answer is provided by Bézout's Theorem:

- (Bézout's Theorem) If the equations $F_0 = \cdots = F_{n-1} = 0$ have degrees $d_0, \ldots, d_{n-1}$ and finitely many solutions in $\mathbb{P}^n$, then the number of solutions (counted with multiplicity) is $d_0 \cdots d_{n-1}$.

(See [Sha], §2 of Chapter II.) This tells us that V has $d_0 \cdots d_{n-1}$ points, counted with multiplicity. Because $V \subset \mathbb{C}^n$ is defined by $f_0 = \cdots = f_{n-1} = 0$, Theorem (2.2) from Chapter 4 implies that the number of points in V, counted with multiplicity, is the dimension of $A = \mathbb{C}[x_0, \ldots, x_{n-1}]/\langle f_0, \ldots, f_{n-1}\rangle$. Thus, Bézout's Theorem shows that $\dim A = d_0 \cdots d_{n-1}$.

We can now explain why $\mathrm{Res}(\overline{F}_0, \ldots, \overline{F}_{n-1})^{d_n} \det(m_{f_n})$ behaves like a resultant. The first step is to prove that $\det(m_{f_n})$ vanishes if and only if $F_0 = \cdots = F_n = 0$ has a solution in $\mathbb{P}^n$. If we have a solution p, then $p \in V$ since $F_0(p) = \cdots = F_{n-1}(p) = 0$. But $V \subset \mathbb{C}^n$, so we can write $p = (a_0, \ldots, a_{n-1}, 1)$, and $f_n(a_0, \ldots, a_{n-1}) = 0$ since $F_n(p) = 0$. Then Theorem (2.6) of Chapter 2 tells us that $f_n(a_0, \ldots, a_{n-1}) = 0$ is an eigenvalue of m_{f_n}, which proves that $\det(m_{f_n}) = 0$. Conversely, if $\det(m_{f_n}) = 0$, then one of its eigenvalues must be zero. Since the eigenvalues are $f_n(p)$ for $p \in V$ (Theorem (2.6) of Chapter 2 again), we have $f_n(p) = 0$ for some p. Writing p in the form $(a_0, \ldots, a_{n-1}, 1)$, we get a nontrivial solution of $F_0 = \cdots = F_n = 0$, as desired.

Finally, we will show that $\mathrm{Res}(\overline{F}_0, \ldots, \overline{F}_{n-1})^{d_n} \det(m_{f_n})$ has the homogeneity properties predicted by Theorem (3.1). If we replace F_j by λF_j for some $j < n$ and $\lambda \in \mathbb{C} \setminus \{0\}$, then $\overline{\lambda F_j} = \lambda \overline{F}_j$, and neither A nor m_{f_n} are affected. Since

$$\mathrm{Res}(\overline{F}_0, \ldots, \lambda\overline{F}_j, \ldots, \overline{F}_{n-1}) =$$
$$\lambda^{d_0 \cdots d_{j-1} d_{j+1} \cdots d_{n-1}} \mathrm{Res}(\overline{F}_0, \ldots, \overline{F}_j, \ldots, \overline{F}_{n-1}),$$

we get the desired power of λ because of the exponent d_n in the formula of the theorem. On the other hand, if we replace F_n with λF_n, then $\mathrm{Res}(\overline{F}_0, \ldots, \overline{F}_{n-1})$ and A are unchanged, but m_{f_n} becomes $m_{\lambda f_n} = \lambda m_{f_n}$. Since

$$\det(\lambda m_{f_n}) = \lambda^{\dim A} \det(m_{f_n})$$

it follows that we get the correct power of λ because, as we showed above, A has dimension $d_0 \cdots d_{n-1}$.

This discussion shows that the formula $\mathrm{Res}(\overline{F}_0, \ldots, \overline{F}_{n-1})^{d_n} \det(m_{f_n})$ has many of the properties of the resultant, although some important points were left out (for example, we didn't prove that it is a polynomial in the coefficients of the F_i). We also know what this formula means geometrically: it asserts that the resultant is a product of two terms, one coming from the behavior of $F_0, \ldots, F_{n-1}$ at infinity and the other coming from the behavior of $f_n = F_n(x_0, \ldots, x_{n-1}, 1)$ on the affine variety determined by vanishing of $f_0, \ldots, f_{n-1}$.   □

Exercise 4. When $n = 2$, show that Proposition (1.5) is a special case of Theorem (3.4). Hint: Start with f, g as in (1.1) and homogenize to get (1.6). Use Exercise 6 of §2 to compute $\mathrm{Res}(\overline{F})$.

Exercise 5. Use Theorem (3.4) and `getmatrix` to compute the resultant of the polynomials $x^2 + y^2 + z^2, xy + xz + yz, xyz$.

The formula given in Theorem (3.4) is sometimes called the *Poisson Formula*. Some further applications of this formula will be given in the exercises at the end of the section.

In the special case when $F_0, \ldots, F_n$ all have the same total degree $d > 0$, the resultant $\mathrm{Res}_{d,\ldots,d}$ has degree d^n in the coefficients of each F_i, and its total degree is $(n + 1)d^n$. Besides all of the properties listed so far, the resultant has some other interesting properties in this case:

(3.5) Theorem. $\mathrm{Res} = \mathrm{Res}_{d,\ldots,d}$ *has the following properties:*
a. *If F_j are homogeneous of total degree d and $G_i = \sum_{j=0}^{n} a_{ij} F_j$, where (a_{ij}) is an invertible matrix with entries in $\mathbb{C}$, then*

$$\mathrm{Res}(G_0, \ldots, G_n) = \det(a_{ij})^{d^n} \mathrm{Res}(F_0, \ldots, F_n).$$

b. *If we list all monomials of total degree d as $x^{\alpha(1)}, \ldots, x^{\alpha(N)}$ and pick $n + 1$ distinct indices $1 \leq i_0 < \cdots < i_n \leq N$, the bracket $[i_0 \ldots i_n]$ is defined to be the determinant*

$$[i_0 \ldots i_n] = \det(u_{i,\alpha(i_j)}) \in \mathbb{Z}[u_{i,\alpha(j)}].$$

Then Res is a polynomial in the brackets $[i_0 \ldots i_n]$.

PROOF. See Proposition 5.11.2 of [Jou1] for a proof of part a. For part b, note that if (a_{ij}) has determinant 1, then part a implies $\mathrm{Res}(G_0, \ldots, G_n) = \mathrm{Res}(F_0, \ldots, F_n)$, so Res is invariant under the action of $\mathrm{SL}(n + 1, \mathbb{C}) = \{A \in M_{(n+1) \times (n+1)}(\mathbb{C}) : \det(A) = 1\}$ on $(n + 1)$-tuples of homogeneous polynomials of degree d. If we regard the coefficients of the universal polynomials $\mathbf{F}_i$ as an $(n + 1) \times N$ matrix $(u_{i,\alpha(j)})$, then this action is matrix multiplication by elements of $\mathrm{SL}(n+1, \mathbb{C})$. Since Res is invariant under this action, the *First Fundamental Theorem of Invariant Theory* (see [Stu1], Section 3.2) asserts that Res is a polynomial in the $(n + 1) \times (n + 1)$ minors of $(u_{i,\alpha(j)})$, which are exactly the brackets $[i_0 \ldots i_n]$.   □

Exercise 6. Show that each bracket $[i_0 \ldots i_n] = \det(u_{i,\alpha(i_j)})$ is invariant under the action of $SL(n+1, \mathbb{C})$.

We should mention that the expression of Res in terms of the brackets $[i_0 \ldots i_n]$ is not unique. The different ways of doing this are determined by the algebraic relations among the brackets, which are described by the *Second Fundamental Theorem of Invariant Theory* (see Section 3.2 of [Stu1]).

As an example of Theorem (3.5), consider the resultant of three ternary quadrics

$$
\begin{aligned}
F_0 &= c_{01}x^2 + c_{02}y^2 + c_{03}z^2 + c_{04}xy + c_{05}xz + c_{06}yz = 0 \\
F_1 &= c_{11}x^2 + c_{12}y^2 + c_{13}z^2 + c_{14}xy + c_{15}xz + c_{16}yz = 0 \\
F_2 &= c_{21}x^2 + c_{22}y^2 + c_{23}z^2 + c_{24}xy + c_{25}xz + c_{26}yz = 0.
\end{aligned}
$$

In §2, we gave a formula for $\mathrm{Res}_{2,2,2}(F_0, F_1, F_2)$ as a certain 6×6 determinant. Using Theorem (3.5), we get quite a different formula. If we list the six monomials of total degree 2 as $x^2, y^2, z^2, xy, xz, yz$, then the bracket $[i_0 i_1 i_2]$ is given by

$$
[i_0 i_1 i_2] = \det \begin{pmatrix} c_{0i_0} & c_{0i_1} & c_{0i_2} \\ c_{1i_0} & c_{1i_1} & c_{1i_2} \\ c_{2i_0} & c_{2i_1} & c_{2i_2} \end{pmatrix}.
$$

By [KSZ], the resultant $\mathrm{Res}_{2,2,2}(F_0, F_1, F_2)$ is the following polynomial in the brackets $[i_0 i_1 i_2]$:

$[145][246][356][456] - [146][156][246][356] - [145][245][256][356]$

$\quad - [145][246][346][345] + [125][126][356][456] - 2[124][156][256][356]$

$\quad - [134][136][246][456] - 2[135][146][346][246] + [235][234][145][456]$

$\quad - 2[236][345][245][145] - [126]^2[156][356] - [125]^2[256][356]$

$\quad - [134]^2[246][346] - [136]^2[146][246] - [145][245][235]^2$

$\quad - [145][345][234]^2 + 2[123][124][356][456] - [123][125][346][456]$

$\quad - [123][134][256][456] + 2[123][135][246][456] - 2[123][145][246][356]$

$\quad - [124]^2[356]^2 + 2[124][125][346][356] - 2[124][134][256][356]$

$\quad - 3[124][135][236][456] - 4[124][135][246][356] - [125]^2[346]^2$

$\quad + 2[125][135][246][346] - [134]^2[256]^2 + 2[134][135][246][256]$

$\quad - 2[135]^2[246]^2 - [123][126][136][456] + 2[123][126][146][356]$

$\quad - 2[124][136]^2[256] - 2[125][126][136][346] + [123][125][235][456]$

$\quad - 2[123][125][245][356] - 2[124][235]^2[156] - 2[126][125][235][345]$

$\quad - [123][234][134][456] + 2[123][234][346][145] - 2[236][134]^2[245]$

$\quad - 2[235][234][134][146] + 3[136][125][235][126] - 3[126][135][236][125]$

$$- [136][125]^2[236] - [126]^2[135][235] - 3[134][136][126][234]$$
$$+ 3[124][134][136][236] + [134]^2[126][236] + [124][136]^2[234]$$
$$- 3[124][135][234][235] + 3[134][234][235][125] - [135][234]^2[125]$$
$$- [124][235]^2[134] - [136]^2[126]^2 - [125]^2[235]^2$$
$$- [134]^2[234]^2 + 3[123][124][135][236] + [123][134][235][126]$$
$$+ [123][135][126][234] + [123][134][236][125] + [123][136][125][234]$$
$$+ [123][124][235][136] - 2[123]^2[126][136] + 2[123]^2[125][235]$$
$$- 2[123]^2[134][234] - [123]^4.$$

This expression for $\text{Res}_{2,2,2}$ has total degree 4 in the brackets since the resultant has total degree 12 and each bracket has total degree 3 in the c_{ij}. Although this formula is rather complicated, its 68 terms are a lot simpler than the 21,894 terms we get when we express $\text{Res}_{2,2,2}$ as a polynomial in the c_{ij}!

Exercise 7. When $F_0 = a_0 x^2 + a_1 xy + a_2 y^2$ and $F_1 = b_0 x^2 + b_1 xy + b_2 y^2$, the only brackets to consider are $[01] = a_0 b_1 - a_1 b_0$, $[02] = a_0 b_2 - a_2 b_0$ and $[12] = a_1 b_2 - a_2 b_1$ (why?). Express $\text{Res}_{2,2}$ as a polynomial in these three brackets. Hint: In the determinant (1.2), expand along the first row and then expand along the column containing the zero.

Theorem (3.5) also shows that the resultant of two homogeneous polynomials $F_0(x, y), F_1(x, y)$ of degree d can be written in terms of the brackets $[ij]$. The resulting formula is closely related to the *Bézout Formula* described in Chapter 12 of [GKZ].

For further properties of resultants, the reader should consult Chapter 13 of [GKZ] or Section 5 of [Jou1].

ADDITIONAL EXERCISES FOR §3

Exercise 8. The product formula (1.4) can be generalized to arbitrary resultants. With the same hypotheses as Theorem (3.4), let $V = \mathbf{V}(f_0, \ldots, f_{n-1})$ be as in the proof of the theorem. Then

$$\text{Res}(F_0, \ldots, F_n) = \text{Res}(\overline{F}_0, \ldots, \overline{F}_{n-1})^{d_n} \prod_{p \in V} f_n(p)^{m(p)},$$

where $m(p)$ is the multiplicity of p in V. This concept is defined in [Sha], §2 of Chapter II, and §2 of Chapter 4. For this exercise, assume that V consists of $d_0 \cdots d_{n-1}$ distinct points (which means that all of the multiplicities $m(p)$ are equal to 1) and that f_n takes distinct values on these points. Then use Theorem (2.6) of Chapter 2, together with Theorem (3.4), to show that the above formula for the resultant holds in this case.

Exercise 9. In Theorem (3.4), we assumed that the field was $\mathbb{C}$. It turns out that the result is true over any field k. In this exercise, we will use this version of the theorem to prove part b of Theorem (3.2) when $F_n = F'_n F''_n$. The trick is to choose k appropriately: we will let k be the field of rational functions in the coefficients of $F_0, \ldots, F_{n-1}, F'_n, F''_n$. This means we regard each coefficient as a separate variable and then k is the field of rational functions in these variables with coefficients in $\mathbb{Q}$.

a. Explain why $\overline{F}_0, \ldots, \overline{F}_{n-1}$ are the "universal" polynomials of degrees $d_0, \ldots, d_{n-1}$ in $x_0, \ldots, x_{n-1}$, and conclude that $\text{Res}(\overline{F}_0, \ldots, \overline{F}_{n-1})$ is nonzero.

b. Use Theorem (3.4) (over the field k) to show that

$$\text{Res}(F_0, \ldots, F_n) = \text{Res}(F_0, \ldots, F'_n) \cdot \text{Res}(F_0, \ldots, F''_n).$$

Notice that you need to use the theorem three times. Hint: $m_{f_n} = m_{f'_n} \circ m_{f''_n}$.

Exercise 10. The goal of this exercise is to generalize Proposition (2.10) by giving a formula for $\text{Res}_{1,1,d}$ for any $d > 0$. The idea is to apply Theorem (3.4) when the field k consists of rational functions in the coefficients of F_0, F_1, F_2 (so we are using the version of the theorem from Exercise 9). For concreteness, suppose that

$$F_0 = a_1 x + a_2 y + a_3 z = 0$$
$$F_1 = b_1 x + b_2 y + b_3 z = 0.$$

a. Show that $\text{Res}(\overline{F}_0, \overline{F}_1) = a_1 b_2 - a_2 b_1$ and that the only solution of $f_0 = f_1 = 0$ is

$$x_0 = \frac{a_2 b_3 - a_3 b_2}{a_1 b_2 - a_2 b_1} \qquad y_0 = -\frac{a_1 b_3 - a_3 b_1}{a_1 b_2 - a_2 b_1}.$$

b. By Theorem (3.4), $k[x, y]/\langle f_0, f_1 \rangle$ has dimension one over $\mathbb{C}$. Use Theorem (2.6) of Chapter 2 to show that

$$\det(m_{f_2}) = f_2(x_0, y_0).$$

c. Since $f_2(x, y) = F_2(x, y, 1)$, use Theorem (3.4) to conclude that

$$\text{Res}_{1,1,d}(F_0, F_1, F_2) = F_2(a_2 b_3 - a_3 b_2, -(a_1 b_3 - a_3 b_1), a_1 b_2 - a_2 b_1).$$

Note that $a_2 b_3 - a_3 b_2, a_1 b_3 - a_3 b_1, a_1 b_2 - a_2 b_1$ are the 2×2 minors of the matrix

$$\begin{pmatrix} a_1 & a_2 & a_3 \\ b_1 & b_2 & b_3 \end{pmatrix}.$$

d. Use part c to verify the formula for $\text{Res}_{1,1,2}$ given in Proposition (2.10).

e. Formulate and prove a formula similar to part c for the resultant $\text{Res}_{1,\ldots,1,d}$. Hint: Use Cramer's Rule. The formula (with proof) can be found in Proposition 5.4.4 of [Jou1].

Exercise 11. Consider the elementary symmetric functions $\sigma_1, \ldots, \sigma_n \in$ $\mathbb{C}[x_1, \ldots, x_n]$. These are defined by

$$\sigma_1 = x_1 + \cdots + x_n$$

$$\vdots$$

$$\sigma_r = \sum_{i_1 < i_2 < \cdots < i_r} x_{i_1} x_{i_2} \cdots x_{i_r}$$

$$\vdots$$

$$\sigma_n = x_1 x_2 \cdots x_n.$$

Since σ_i is homogeneous of total degree i, the resultant $\operatorname{Res}(\sigma_1, \ldots, \sigma_n)$ is defined. The goal of this exercise is to prove that this resultant equals -1 for all $n > 1$. Note that this exercise deals with n polynomials and n variables rather than $n + 1$.

a. Show that $\operatorname{Res}(x + y, xy) = -1$.

b. To prove the result for $n > 2$, we will use induction and Theorem (3.4). Thus, let

$$\overline{\sigma}_i = \sigma_i(x_1, \ldots, x_{n-1}, 0)$$

$$\tilde{\sigma}_i = \sigma_i(x_1, \ldots, x_{n-1}, 1)$$

as in (3.3). Prove that $\overline{\sigma}_i$ is the ith elementary symmetric function in $x_1, \ldots, x_{n-1}$ and that $\tilde{\sigma}_i = \overline{\sigma}_i + \overline{\sigma}_{i-1}$ (where $\overline{\sigma}_0 = 1$).

c. If $A = \mathbb{C}[x_1, \ldots, x_{n-1}]/\langle \tilde{\sigma}_1, \ldots, \tilde{\sigma}_{n-1} \rangle$, then use part b to prove that the multiplication map $m_{\tilde{\sigma}_n} : A \to A$ is multiplication by $(-1)^n$. Hint: Observe that $\tilde{\sigma}_n = \overline{\sigma}_{n-1}$.

d. Use induction and Theorem (3.5) to show that $\operatorname{Res}(\sigma_1, \ldots, \sigma_n) = -1$ for all $n > 1$.

Exercise 12. Using the notation of Theorem (3.4), show that

$$\operatorname{Res}(F_0, \ldots, F_{n-1}, x_n^d) = \operatorname{Res}(\overline{F}_0, \ldots, \overline{F}_{n-1})^d.$$

§4 Computing Resultants

Our next task is to discuss methods for computing resultants. While Theorem (3.4) allows one to compute resultants inductively (see Exercise 5 of §3 for an example), it is useful to have other tools for working with resultants. In this section, we will give some further formulas for the resultant and then discuss the practical aspects of computing $\operatorname{Res}_{d_0, \ldots, d_n}$. We will begin by generalizing the method used in Proposition (2.10) to find a formula for $\operatorname{Res}_{1,1,2}$. Recall that the essence of what we did in (2.11) was to multiply each equation by appropriate monomials so that we got a square matrix whose determinant we could take.

To do this in general, suppose we have $F_0, \ldots, F_n \in \mathbb{C}[x_0, \ldots, x_n]$ of total degrees $d_0, \ldots, d_n$. Then set

$$d = \sum_{i=0}^{n}(d_i - 1) + 1 = \sum_{i=0}^{n} d_i - n.$$

For instance, when $(d_0, d_1, d_2) = (1, 1, 2)$ as in the example in Section 2, one computes that $d = 2$, which is precisely the degree of the monomials on the left hand side of the equations following (2.11).

Exercise 1. Monomials of total degree d have the following special property which will be very important below: each such monomial is divisible by $x_i^{d_i}$ for at least one i between 0 and n. Prove this. Hint: Argue by contradiction.

Now take the monomials $x^\alpha = x_0^{a_0} \cdots x_n^{a_n}$ of total degree d and divide them into $n + 1$ sets as follows:

$$S_0 = \{x^\alpha : |\alpha| = d, \ x_0^{d_0} \text{ divides } x^\alpha\}$$
$$S_1 = \{x^\alpha : |\alpha| = d, \ x_0^{d_0} \text{ doesn't divide } x^\alpha \text{ but } x_1^{d_1} \text{ does}\}$$

$$\vdots$$

$$S_n = \{x^\alpha : |\alpha| = d, \ x_0^{d_0}, \ldots, x_{n-1}^{d_{n-1}} \text{ don't divide } x^\alpha \text{ but } x_n^{d_n} \text{ does}\}.$$

By Exercise 1, every monomial of total degree d lies in one of $S_0, \ldots, S_n$. Note also that these sets are mutually disjoint. One observation we will need is the following:

$$\text{if } x^\alpha \in S_i, \text{ then we can write } x^\alpha = x_i^{d_i} \cdot x^\alpha / x_i^{d_i}.$$

Notice that $x^\alpha / x_i^{d_i}$ is a monomial of total degree $d - d_i$ since $x^\alpha \in S_i$.

Exercise 2. When $(d_0, d_1, d_2) = (1, 1, 2)$, show that $S_0 = \{x^2, xy, xz\}$, $S_1 = \{y^2, yz\}$, and $S_2 = \{z^2\}$, where we are using x, y, z as variables. Write down *all* of the $x^\alpha / x_i^{d_i}$ in this case and see if you can find these monomials in the equations (2.11).

Exercise 3. Prove that the number of monomials in S_n is exactly $d_0 \cdots d_{n-1}$. This fact will play an extremely important role in what follows. Hint: Given integers $a_0, \ldots, a_{n-1}$ with $0 \leq a_i \leq d_i - 1$, prove that there is a unique a_n such that $x_0^{a_0} \cdots x_n^{a_n} \in S_n$. Exercise 1 will also be useful.

Now we can write down a system of equations that generalizes (2.11). Namely, consider the equations

$$x^\alpha / x_0^{d_0} \cdot F_0 = 0 \quad \text{for all } x^\alpha \in S_0$$

(4.1) $$\vdots$$

$$x^\alpha / x_n^{d_n} \cdot F_n = 0 \quad \text{for all } x^\alpha \in S_n.$$

Exercise 4. When $(d_0, d_1, d_2) = (1, 1, 2)$, check that the system of equations given by (4.1) is *exactly* what we wrote down in (2.11).

Since F_i has total degree d_i, it follows that $x^\alpha / x_i^{d_i} \cdot F_i$ has total degree d. Thus each polynomial on the left side of (4.1) can be written as a linear combination of monomials of total degree d. Suppose that there are N such monomials. (In the exercises at the end of the section, you will show that N equals the binomial coefficient $\binom{d+n}{n}$.) Then observe that the total number of equations is the number of elements in $S_0 \cup \cdots \cup S_n$, which is also N. Thus, regarding the monomials of total degree d as unknowns, we get a system of N linear equations in N unknowns.

(4.2) Definition. The determinant of the coefficient matrix of the $N \times N$ system of equations given by (4.1) is denoted D_n.

For example, if we have

(4.3)
$$F_0 = a_1 x + a_2 y + a_3 z = 0$$
$$F_1 = b_1 x + b_2 y + b_3 z = 0$$
$$F_2 = c_1 x^2 + c_2 y^2 + c_3 z^2 + c_4 xy + c_5 xz + c_6 yz = 0,$$

then the equations following (2.11) imply that

(4.4) $$D_2 = \det \begin{pmatrix} a_1 & 0 & 0 & a_2 & a_3 & 0 \\ 0 & a_2 & 0 & a_1 & 0 & a_3 \\ 0 & 0 & a_3 & 0 & a_1 & a_2 \\ 0 & b_2 & 0 & b_1 & 0 & b_3 \\ 0 & 0 & b_3 & 0 & b_1 & b_2 \\ c_1 & c_2 & c_3 & c_4 & c_5 & c_6 \end{pmatrix}.$$

Exercise 5. When we have polynomials $F_0, F_1 \in \mathbb{C}[x, y]$ as in (1.6), show that the coefficient matrix of (4.1) is exactly the transpose of the matrix (1.2). Thus, $D_1 = \mathrm{Res}(F_0, F_1)$ in this case.

Here are some general properties of D_n:

Exercise 6. Since D_n is the determinant of the coefficient matrix of (4.1), it is clearly a polynomial in the coefficients of the F_i.

a. For a fixed i between 0 and n, show that D_n is homogeneous in the coefficients of F_i of degree equal to the number μ_i of elements in S_i. Hint: Show that replacing F_i by λF_i has the effect of multiplying a certain number (how many?) equations of (4.1) by λ. How does this affect the determinant of the coefficient matrix?

b. Use Exercise 3 to show that D_n has degree $d_0 \cdots d_{n-1}$ as a polynomial in the coefficients of F_n. Hint: If you multiply each coefficient of F_n by $\lambda \in \mathbb{C}$, show that D_n gets multiplied by $\lambda^{d_0 \cdots d_{n-1}}$.

c. What is the total degree of D_n? Hint: Exercise 19 will be useful.

Exercise 7. In this exercise, you will prove that D_n is divisible by the resultant.

a. Prove that D_n vanishes whenever $F_0 = \cdots = F_n = 0$ has a nontrivial solution. Hint: If the F_i all vanish at $(c_0, \ldots, c_n) \neq (0, \ldots, 0)$, then show that the monomials of total degree d in $c_0, \ldots, c_n$ give a nontrivial solution of (4.1).

b. Using the notation from the end of §2, we have $\mathbf{V}(\mathrm{Res}) \subset \mathbb{C}^N$, where $\mathbb{C}^N$ is the affine space whose variables are the coefficients $u_{i,\alpha}$ of $F_0, \ldots, F_n$. Explain why part a implies that D_n vanishes on $\mathbf{V}(\mathrm{Res})$.

c. Adapt the argument of Proposition (2.10) to prove that $D_n \in \langle \mathrm{Res} \rangle$, so that Res divides D_n.

Exercise 7 shows that we are getting close to the resultant, for it enables us to write

(4.5) $D_n = \mathrm{Res} \cdot \text{extraneous factor}.$

We next show that the extraneous factor doesn't involve the coefficients of F_n and in fact uses only some of the coefficients of $F_0, \ldots, F_{n-1}$.

(4.6) Proposition. *The extraneous factor in (4.5) is an integer polynomial in the coefficients of $\overline{F}_0, \ldots, \overline{F}_{n-1}$, where $\overline{F}_i = F_i(x_0, \ldots, x_{n-1}, 0)$.*

PROOF. Since D_n is a determinant, it is a polynomial in $\mathbb{Z}[u_{i,\alpha}]$, and we also know that Res $\in \mathbb{Z}[u_{i,\alpha}]$. Exercise 7 took place in $\mathbb{C}[u_{i,\alpha}]$ (because of the Nullstellensatz), but in fact, the extraneous factor (let's call it E_n) must lie in $\mathbb{Q}[u_{i,\alpha}]$ since dividing D_n by Res produces at worst rational coefficients. Since Res is irreducible in $\mathbb{Z}[u_{i,\alpha}]$, standard results about polynomial rings over $\mathbb{Z}$ imply that $E_n \in \mathbb{Z}[u_{i,\alpha}]$ (see Exercise 20 for details).

Since $D_n = \mathrm{Res} \cdot E_n$ is homogeneous in the coefficients of F_n, Exercise 20 at the end of the section implies that Res and E_n are also homogeneous in these coefficients. But by Theorem (3.1) and Exercise 6, both Res and D_n have degree $d_0 \cdots d_{n-1}$ in the coefficients of F_n. It follows immediately that E_n has degree zero in the coefficients of F_n, so that it depends only on the coefficients of $F_0, \ldots, F_{n-1}$.

To complete the proof, we must show that E_n depends only on the coefficients of the $\overline{F}_i$. This means that coefficients of $F_0, \ldots, F_{n-1}$ with x_n to a positive power don't appear in E_n. To prove this, we use the following clever argument of Macaulay (see [Mac1]). As above, we think of Res, D_n and E_n as polynomials in the $u_{i,\alpha}$, and we define the *weight* of $u_{i,\alpha}$ to be the exponent a_n of x_n (where $\alpha = (a_0, \ldots, a_n)$). Then, the weight of a monomial in the $u_{i,\alpha}$, say $u_{i_1,\alpha_1}^{m_1} \cdots u_{i_l,\alpha_l}^{m_l}$, is defined to be the sum of the weights of each u_{i_j,α_j} multiplied by the corresponding exponents. Finally, a polynomial in the $u_{i,\alpha}$ is said to be *isobaric* if every term in the polynomial has the same weight.

In Exercise 23 at the end of the section, you will prove that every term in D_n has weight $d_0 \cdots d_n$, so that D_n is isobaric. The same exercise will show that $D_n = \mathrm{Res} \cdot E_n$ implies that Res and E_n are isobaric and that the weight of D_n is the sum of the weights of Res and E_n. Hence, it suffices to prove that E_n has weight zero (be sure you understand this). To simplify notation, let u_i be the variable representing the coefficient of $x_i^{d_i}$ in F_i. Note that $u_0, \ldots, u_{n-1}$ have weight zero while u_n has weight d_n. Then Theorems (2.3) and (3.1) imply that one of the terms of Res is

$$\pm u_0^{d_1 \cdots d_n} u_1^{d_0 d_2 \cdots d_n} \cdots u_n^{d_0 \cdots d_{n-1}}$$

(see Exercise 23). This term has weight $d_0 \cdots d_n$, which shows that the weight of Res is $d_0 \cdots d_n$. We saw above that D_n has the same weight, and it follows that E_n has weight zero, as desired.   □

Although the extraneous factor in (4.5) involves fewer coefficients than the resultant, it can have a very large degree, as shown by the following example.

Exercise 8. When $d_i = 2$ for $0 \leq i \leq 4$, show that the resultant has total degree 80 while D_4 has total degree 420. What happens when $d_i = 3$ for $0 \leq i \leq 4$? Hint: Use Exercises 6 and 19.

Notice that Proposition (4.6) also gives a method for computing the resultant: just factor D_n into irreducibles, and the only irreducible factor in which all variables appear is the resultant! Unfortunately, this method is wildly impractical owing to the slowness of multivariable factorization (especially for polynomials as large as D_n).

In the above discussion, the sets $S_0, \ldots, S_n$ and the determinant D_n depended on how the variables $x_0, \ldots, x_n$ were ordered. In fact, the notation D_n was chosen to emphasize that the variable x_n came last. If we fix i between 0 and $n - 1$ and order the variables so that x_i comes last, then we get slightly different sets $S_0, \ldots, S_n$ and a slightly different system of equations (4.1). We will let D_i denote the determinant of this system of equations. (Note that there are many different orderings of the variables for which x_i is the last. We pick just one when computing D_i.)

Exercise 9. Show that D_i is homogeneous in the coefficients of each F_j and in particular, is homogeneous of degree $d_0 \cdots d_{i-1} d_{i+1} \cdots d_n$ in the coefficients of F_i.

We can now prove the following classical formula for Res.

(4.7) Proposition. *When* $\mathbf{F}_0, \dots, \mathbf{F}_n$ *are universal polynomials as at the end of* §2, *the resultant is the greatest common divisor of the polynomials* $D_0, \dots, D_n$ *in the ring* $\mathbb{Z}[u_{i,\alpha}]$, *i.e.*,

$$\mathrm{Res} = \pm \mathrm{GCD}(D_0, \dots, D_n).$$

PROOF. For each i, there are many choices for D_i (corresponding to the $(n-1)!$ ways of ordering the variables with x_i last). We need to prove that no matter which of the various D_i we pick for each i, the greatest common divisor of $D_0, \dots, D_n$ is the resultant (up to a sign).

By Exercise 7, we know that Res divides D_n, and the same is clearly true for $D_0, \dots, D_{n-1}$. Furthermore, the argument used in the proof of Proposition (4.6) shows that $D_i = \mathrm{Res} \cdot E_i$, where $E_i \in \mathbb{Z}[u_{i,\alpha}]$ doesn't involve the coefficients of F_i. It follows that

$$\mathrm{GCD}(D_0, \dots, D_n) = \mathrm{Res} \cdot \mathrm{GCD}(E_0, \dots, E_n).$$

Since each E_i doesn't involve the variables $u_{i,\alpha}$, the GCD on the right must be constant, i.e., an integer. However, since the coefficients of D_n are relatively prime (see Exercise 10 below), this integer must be ± 1, and we are done. Note that GCD's are only determined up to invertible elements, and in $\mathbb{Z}[u_{i,\alpha}]$, the only invertible elements are ± 1. $\square$

Exercise 10. Show that $D_n(x_0^{d_0}, \dots, x_n^{d_n}) = \pm 1$, and conclude that as a polynomial in $\mathbb{Z}[u_{i,\alpha}]$, the coefficients of D_n are relatively prime. Hint: If you order the monomials of total degree d appropriately, the matrix of (4.1) will be the identity matrix when $F_i = x_i^{d_i}$.

While the formula of Proposition (4.7) is very pretty, it is not particularly useful in practice. This brings us to our final resultant formula, which will tell us exactly how to find the extraneous factor in (4.5). The key idea, due to Macaulay, is that the extraneous factor is in fact a minor (i.e., the determinant of a submatrix) of the $N \times N$ matrix from (4.1). To describe this minor, we need to know which rows and columns of the matrix to delete. Recall also that we can label the rows and columns of the matrix of (4.1) using all monomials of total degree $d = \sum_{i=0}^{n} d_i - n$. Given such a monomial x^α, Exercise 1 implies that $x_i^{d_i}$ divides x^α for at least one i.

(4.8) Definition. Let $d_0, \dots, d_n$ and d be as usual.

a. A monomial x^α of total degree d is *reduced* if $x_i^{d_i}$ divides x^α for *exactly* one i.

b. D'_n is the determinant of the submatrix of the coefficient matrix of (4.1) obtained by deleting all rows and columns corresponding to reduced monomials x^α.

Exercise 11. When $(d_0, d_1, d_2) = (1, 1, 2)$, we have $d = 2$. Show that all monomials of degree 2 are reduced except for xy. Then show that the $D'_3 = a_1$ corresponding to the submatrix (4.4) obtained by deleting everything but row 2 and column 4.

Exercise 12. Here are some properties of reduced monomials and D'_n.
a. Show that the number of reduced monomials is equal to

$$\sum_{j=0}^{n} d_0 \cdots d_{j-1} d_{j+1} \cdots d_n.$$

Hint: Adapt the argument used in Exercise 3.
b. Show that D'_n has the same total degree as the extraneous factor in (4.5) and that it doesn't depend on the coefficients of F_n. Hint: Use part a and note that all monomials in S_n are reduced.

Macaulay's observation is that the extraneous factor in (4.5) is exactly D'_n up to a sign. This gives the following formula for the resultant as a quotient of two determinants.

(4.9) Theorem. *When $F_0, \ldots, F_n$ are universal polynomials, the resultant is given by*

$$\mathrm{Res} = \pm \frac{D_n}{D'_n}.$$

Further, if k is any field and $F_0, \ldots, F_n \in k[x_0, \ldots, x_n]$, then the above formula for Res holds whenever $D'_n \neq 0$.

PROOF. This is proved in Macaulay's paper [Mac2]. For a modern proof, see [Jou2]. □

Exercise 13. Using x_0, x_1, x_2 as variables with x_0 regarded as last, write $\mathrm{Res}_{1,2,2}$ as a quotient D_0/D'_0 of two determinants and write down the matrices involved (of sizes 10×10 and 2×2 respectively). The reason for using D_0/D'_0 instead of D_2/D'_2 will become clear in Exercise 2 of §5. A similar example is worked out in detail in [BGW].

While Theorem (4.9) applies to all resultants, it has some disadvantages. In the universal case, it requires dividing two very large polynomials, which can be very time consuming, and in the numerical case, we have the awkward situation where both D'_n and D_n vanish, as shown by the following exercise.

Exercise 14. Give an example of polynomials of degrees $1, 1, 2$ for which the resultant is nonzero yet the determinants D_2 and D_2' both vanish. Hint: See Exercise 10.

Because of this phenomenon, it would be nice if the resultant could be expressed as a single determinant, as happens with $\mathrm{Res}_{l,m}$. It is not known if this is possible in general, though many special cases have been found. We saw one example in the formula (2.8) for $\mathrm{Res}_{2,2,2}$. This can be generalized (in several ways) to give formulas for $\mathrm{Res}_{l,l,l}$ and $\mathrm{Res}_{l,l,l,l}$ when $l \geq 2$ (see [GKZ], Chapter 3, §4 and Chapter 13, §1, and [Sal], Arts. 90 and 91). As an example of these formulas, the following exercise will show how to express $\mathrm{Res}_{l,l,l}$ as a single determinant of size $2l^2 - l$ when $l \geq 2$.

Exercise 15. Suppose that $F_0, F_1, F_2 \in \mathbb{C}[x, y, z]$ have total degree $l \geq 2$. Before we can state our formula, we need to create some auxiliary equations. Given nonnegative integers a, b, c with $a + b + c = l - 1$, show that every monomial of total degree l in x, y, z is divisible by either x^{a+1}, y^{b+1}, or z^{c+1}, and conclude that we can write F_0, F_1, F_2 in the form

$$
\begin{aligned}
F_0 &= x^{a+1} P_0 + y^{b+1} Q_0 + z^{c+1} R_0 \\
F_1 &= x^{a+1} P_1 + y^{b+1} Q_1 + z^{c+1} R_1 \\
F_2 &= x^{a+1} P_2 + y^{b+1} Q_2 + z^{c+1} R_2.
\end{aligned}
$$

(4.10)

There may be many ways of doing this. We will regard F_0, F_1, F_2 as universal polynomials and pick one particular choice for (4.10). Then set

$$
F_{a,b,c} = \det \begin{pmatrix} P_0 & Q_0 & R_0 \\ P_1 & Q_1 & R_1 \\ P_2 & Q_2 & R_2 \end{pmatrix}.
$$

You should check that $F_{a,b,c}$ has total degree $2l - 2$.
Then consider the equations

(4.11)
$$
\begin{aligned}
x^\alpha \cdot F_0 &= 0, & x^\alpha \text{ of total degree } l - 2 \\
x^\alpha \cdot F_1 &= 0, & x^\alpha \text{ of total degree } l - 2 \\
x^\alpha \cdot F_2 &= 0, & x^\alpha \text{ of total degree } l - 2 \\
F_{a,b,c} &= 0, & x^a y^b z^c \text{ of total degree } l - 1.
\end{aligned}
$$

Each polynomial on the left hand side has total degree $2l - 2$, and you should prove that there are $2l^2 - l$ monomials of this total degree. Thus we can regard the equations in (4.11) as having $2l^2 - l$ unknowns. You should also prove that the number of equations is $2l^2 - l$. Thus the coefficient matrix of (4.11), which we will denote C_l, is a $(2l^2 - l) \times (2l^2 - l)$ matrix.
In the following steps, you will prove that the resultant is given by

$$
\mathrm{Res}_{l,l,l}(F_0, F_1, F_2) = \pm \det(C_l).
$$

a. If $(u, v, w) \neq (0, 0, 0)$ is a solution of $F_0 = F_1 = F_2 = 0$, show that $F_{a,b,c}$ vanishes at (u, v, w). Hint: Regard (4.10) as a system of equations in unknowns $x^{a+1}, y^{b+1}, z^{c+1}$.

b. Use standard arguments to show that $\text{Res}_{l,l,l}$ divides $\det(C_l)$.

c. Show that $\det(C_l)$ has degree l^2 in the coefficients of F_0. Show that the same is true for F_1 and F_2.

d. Conclude that $\text{Res}_{l,l,l}$ is a multiple of $\det(C_l)$.

e. When $(F_0, F_1, F_2) = (x^l, y^l, z^l)$, show that $\det(C_l) = \pm 1$. Hint: Show that $F_{a,b,c} = x^{l-1-a} y^{l-1-b} z^{l-1-c}$ and that all monomials of total degree $2l - 2$ not divisible by x^l, y^l, z^l can be written uniquely in this form. Then show that C_l is the identity matrix when the equations and monomials in (4.11) are ordered appropriately.

f. Conclude that $\text{Res}_{l,l,l}(F_0, F_1, F_2) = \pm \det(C_l)$.

Exercise 16. Use Exercise 15 to compute the following resultants.

a. $\text{Res}(x^2 + y^2 + z^2, xy + xz + yz, x^2 + 2xz + 3y^2)$.

b. $\text{Res}(st + su + tu + u^2(1 - x), st + su + t^2 + u^2(2 - y), s^2 + su + tu - u^2 z)$, where the variables are s, t, u, and x, y, z are part of the coefficients. Note that your answer should agree with what you found in Exercise 3 of §2.

Other determinantal formulas for resultants can be found in [DD], [SZ], and [WZ]. We should also mention that besides the quotient formula given in Theorem (4.9), there are other ways to represent resultants as quotients. These go under a variety of names, including *Morley forms* [Jou1], *Bezoutians* [ElM1], and *Dixon matrices* [KSY]. See [EmM] for a survey. Computer implementations of resultants are available in [Lew] (for the Dixon formulation of [KSY]) and [WEM] (for the Macaulay formulation of Theorem (4.9)). Also, the Maple package MR implementing Theorem (4.9) can be found at http://minimair.org/MR.mpl.

We will end this section with a brief discussion of some of the practical aspects of computing resultants. All of the methods we've seen involve computing determinants or ratios of determinants. Since the usual formula for an $N \times N$ determinant involves $N!$ terms, we will need some clever methods for computing large determinants.

As Exercise 16 illustrates, the determinants can be either *numerical*, with purely numerical coefficients (as in part a of the exercise), or *symbolic*, with coefficients involving other variables (as in part b). Let's begin with numerical determinants. In most cases, this means determinants whose entries are rational numbers, which can be reduced to integer entries by clearing denominators. The key idea here is to reduce modulo a prime p and do arithmetic over the finite field $\mathbb{F}_p$ of the integers mod p. Computing the determinant here is easier since we are working over a field, which allows us to use standard algorithms from linear algebra (using row and column operations) to find the determinant. Another benefit is that we don't have to worry how big the numbers are getting (since we always reduce mod p). Hence we can compute the determinant mod p fairly easily. Then we do this for several primes $p_1, \ldots, p_r$ and use the Chinese Remainder Theorem to recover the original determinant. Strategies for how to choose the size and number of primes p_i are discussed in [CM] and [Man2], and the sparseness properties of the matrices in Theorem (4.9) are exploited in [CKL].

This method works fine provided that the resultant is given as a single determinant or a quotient where the denominator is nonzero. But when we have a situation like Exercise 14, where the denominator of the quotient is zero, something else is needed. One way to avoid this problem, due to Canny [Can1], is to prevent determinants from vanishing by making some coefficients symbolic. Suppose we have $F_0, \ldots, F_n \in \mathbb{Z}[x_0, \ldots, x_n]$. The determinants D_n and D'_n from Theorem (4.9) come from matrices we will denote M_n and M'_n. Thus the formula of the theorem becomes

$$\mathrm{Res}(F_0, \ldots, F_n) = \pm \frac{\det(M_n)}{\det(M'_n)}$$

provided $\det(M'_n) \neq 0$. When $\det(M'_n) = 0$, Canny's method is to introduce a new variable u and consider the resultant

(4.12) $$\mathrm{Res}(F_0 - u\, x_0^{d_0}, \ldots, F_n - u\, x_n^{d_n}).$$

Exercise 17. Fix an ordering of the monomials of total degree d. Since each equation in (4.1) corresponds to such a monomial, we can order the equations in the same way. The ordering of the monomials and equations determines the matrices M_n and M'_n. Then consider the new system of equations we get by replacing F_i by $F_i - u\, x_i^{d_i}$ in (4.1) for $0 \leq i \leq n$.
a. Show that the matrix of the new system of equations is $M_n - u\, I$, where I is the identity matrix of the same size as M_n.
b. Show that the matrix we get by deleting all rows and columns corresponding to reduced monomials, show that the matrix we get is $M'_n - u\, I$ where I is the appropriate identity matrix.

This exercise shows that the resultant (4.12) is given by

$$\mathrm{Res}(F_0 - u\, x_0^{d_0}, \ldots, F_n - u\, x_n^{d_n}) = \pm \frac{\det(M_n - u\, I)}{\det(M'_n - u\, I)}$$

since $\det(M'_n - u\, I) \neq 0$ (it is the characteristic polynomial of M'_n). It follows that the resultant $\mathrm{Res}(F_0, \ldots, F_n)$ is the constant term of the polynomial obtained by dividing $\det(M_n - u\, I)$ by $\det(M'_n - u\, I)$. In fact, as the following exercise shows, we can find the constant term directly from these polynomials:

Exercise 18. Let F and G be polynomials in u such that F is a multiple of G. Let $G = b_r u^r + $ higher order terms, where $b_r \neq 0$. Then $F = a_r u^r + $ higher order terms. Prove that the constant term of F/G is a_r/b_r.

It follows that the problem of finding the resultant is reduced to computing the determinants $\det(M_n - u\,I)$ and $\det(M'_n - u\,I)$. These are called *generalized characteristic polynomials* in [Can1].

This brings us to the second part of our discussion, the computation of symbolic determinants. The methods described above for the numerical case don't apply here, so something new is needed. One of the most interesting methods involves interpolation, as described in [CM]. The basic idea is that one can reconstruct a polynomial from its values at a sufficiently large number of points. More precisely, suppose we have a symbolic determinant, say involving variables $u_0, \ldots, u_n$. The determinant is then a polynomial $D(u_0, \ldots, u_n)$. Substituting $u_i = a_i$, where $a_i \in \mathbb{Z}$ for $0 \leq i \leq n$, we get a numerical determinant, which we can evaluate using the above method. Then, once we determine $D(a_0, \ldots, a_n)$ for sufficiently many points $(a_0, \ldots, a_n)$, we can reconstruct $D(u_0, \ldots, u_n)$. Roughly speaking, the number of points chosen depends on the degree of D in the variables $u_0, \ldots, u_n$. There are several methods for choosing points $(a_0, \ldots, a_n)$, leading to various interpolation schemes (Vandermonde, dense, sparse, probabilistic) which are discussed in [CM]. We should also mention that in the case of a single variable, there is a method of Manocha [Man2] for finding the determinant without interpolation.

Now that we know how to compute resultants, it's time to put them to work. In the next section, we will explain how resultants can be used to solve systems of polynomial equations. We should also mention that a more general notion of resultant, called the *sparse resultant*, will be discussed in Chapter 7.

ADDITIONAL EXERCISES FOR §4

Exercise 19. Show that the number of monomials of total degree d in $n + 1$ variables is the binomial coefficient $\binom{d+n}{n}$.

Exercise 20. This exercise is concerned with the proof of Proposition (4.6).
a. Suppose that $E \in \mathbb{Z}[u_{i,\alpha}]$ is irreducible and nonconstant. If $F \in \mathbb{Q}[u_{i,\alpha}]$ is such that $D = EF \in \mathbb{Z}[u_{i,\alpha}]$, then prove that $F \in \mathbb{Z}[u_{i,\alpha}]$. Hint: We can find a positive integer m such that $mF \in \mathbb{Z}[u_{i,\alpha}]$. Then apply unique factorization to $m \cdot D = E \cdot mF$.
b. Let $D = EF$ in $\mathbb{Z}[u_{i,\alpha}]$, and assume that for some j, D is homogeneous in the $u_{j,\alpha}$, $|\alpha| = d_j$. Then prove that E and F are also homogeneous in the $u_{j,\alpha}$, $|\alpha| = d_j$.

Exercise 21. In this exercise and the next we will prove the formula for $\mathrm{Res}_{2,2,2}$ given in equation (2.8). Here we prove two facts we will need.

a. Prove Euler's formula, which states that if $F \in k[x_0, \ldots, x_n]$ is homogeneous of total degree d, then

$$d F = \sum_{i=0}^{n} x_i \frac{\partial F}{\partial x_i}.$$

Hint: First prove it for a monomial of total degree d and then use linearity.

b. Suppose that

$$M = \det \begin{pmatrix} A_1 & A_2 & A_3 \\ B_1 & B_2 & B_3 \\ C_1 & C_2 & C_3 \end{pmatrix},$$

where $A_1, \ldots, C_3$ are in $k[x_0, \ldots, x_n]$. Then prove that

$$\frac{\partial M}{\partial x_i} = \det \begin{pmatrix} \partial A_1/\partial x_i & A_2 & A_3 \\ \partial B_1/\partial x_i & B_2 & B_3 \\ \partial C_1/\partial x_i & C_2 & C_3 \end{pmatrix} + \det \begin{pmatrix} A_1 & \partial A_2/\partial x_i & A_3 \\ B_1 & \partial B_2/\partial x_i & B_3 \\ C_1 & \partial C_2/\partial x_i & C_3 \end{pmatrix}$$

$$+ \det \begin{pmatrix} A_1 & A_2 & \partial A_3/\partial x_i \\ B_1 & B_2 & \partial B_3/\partial x_i \\ C_1 & C_2 & \partial C_3/\partial x_i \end{pmatrix}.$$

Exercise 22. We can now prove formula (2.8) for $\text{Res}_{2,2,2}$. Fix $F_0, F_1, F_2 \in \mathbb{C}[x, y, z]$ of total degree 2. As in §2, let J be the Jacobian determinant

$$J = \det \begin{pmatrix} \partial F_0/\partial x & \partial F_0/\partial y & \partial F_0/\partial z \\ \partial F_1/\partial x & \partial F_1/\partial y & \partial F_1/\partial z \\ \partial F_2/\partial x & \partial F_2/\partial y & \partial F_2/\partial z \end{pmatrix}.$$

a. Prove that J vanishes at every nontrivial solution of $F_0 = F_1 = F_2 = 0$. Hint: Apply Euler's formula (part a of Exercise 21) to F_0, F_1, F_2.

b. Show that

$$x \cdot J = 2 \det \begin{pmatrix} F_0 & \partial F_0/\partial y & \partial F_0/\partial z \\ F_1 & \partial F_1/\partial y & \partial F_1/\partial z \\ F_2 & \partial F_2/\partial y & \partial F_2/\partial z \end{pmatrix},$$

and derive similar formulas for $y \cdot J$ and $z \cdot J$. Hint: Use column operations and Euler's formula.

c. By differentiating the formulas from part b for $x \cdot J$, $y \cdot J$ and $z \cdot J$ with respect to x, y, z, show that the partial derivatives of J vanish at all nontrivial solutions of $F_0 = F_1 = F_2 = 0$. Hint: Part b of Exercise 21 and part a of this exercise will be useful.

d. Use part c to show that the determinant in (2.8) vanishes at all nontrivial solutions of $F_0 = F_1 = F_2 = 0$.

e. Now prove (2.8). Hint: The proof is similar to what we did in parts b–f of Exercise 15.

Exercise 23. This exercise will give more details needed in the proof of Proposition (4.6). We will use the same terminology as in the proof. Let the weight of the variable $u_{i,\alpha}$ be $w(u_{i,\alpha})$.

a. Prove that a polynomial $P(u_{i,\alpha})$ is isobaric of weight m if and only if $P(\lambda^{w(u_{i,\alpha})}u_{i,\alpha}) = \lambda^m P(u_{i,\alpha})$ for all nonzero $\lambda \in \mathbb{C}$.

b. Prove that if $P = QR$ is isobaric, then so are Q and R. Also show that the weight of P is the sum of the weights of Q and R. Hint: Use part a.

c. Prove that D_n is isobaric of weight $d_0 \cdots d_n$. Hint: Assign the variables $x_0, \ldots, x_{n-1}, x_n$ respective weights $0, \ldots, 0, 1$. Let x^γ be a monomial with $|\gamma| = d$ (which indexes a column of D_n), and let $\alpha \in S_i$ (which indexes a row in D_n). If the corresponding entry in D_n is $c_{\gamma,\alpha,i}$, then show that

$$w(c_{\gamma,\alpha,i}) = w(x^\gamma) - w(x^\alpha/x_i^{d_i})$$

$$= w(x^\gamma) - w(x^\alpha) + \begin{cases} 0 & i < n \\ d_n & i = n. \end{cases}$$

Note that x^γ and x^α range over *all* monomials of total degree d.

d. Use Theorems (2.3) and (3.1) to prove that if u_i represents the coefficient of $x_i^{d_i}$ in F_i, then $\pm u_0^{d_1 \cdots d_n} \cdots u_n^{d_0 \cdots d_{n-1}}$ is in Res.

§5 Solving Equations via Resultants

In this section, we will show how resultants can be used to solve polynomial systems. To start, suppose we have n homogeneous polynomials $F_1, \ldots, F_n$ of degree $d_1, \ldots, d_n$ in variables $x_0, \ldots, x_n$. We want to find the nontrivial solutions of the system of equations

$$(5.1) \qquad\qquad F_1 = \cdots = F_n = 0.$$

But before we begin our discussion of finding solutions, we first need to review Bézout's Theorem and introduce the important idea of *genericity*.

As we saw in §3, Bézout's Theorem tells us that when (5.1) has finitely many solutions in $\mathbb{P}^n$, the number of solutions is $d_1 \cdots d_n$, counting multiplicities. In practice, it is often convenient to find solutions in affine space. In §3, we dehomogenized by setting $x_n = 1$, but in order to be compatible with Chapter 7, we now dehomogenize using $x_0 = 1$. Hence, we define:

$$(5.2) \qquad \begin{aligned} f_i(x_1, \ldots, x_n) &= F_i(1, x_1, \ldots, x_n) \\ \overline{F}_i(x_1, \ldots, x_n) &= F_i(0, x_1, \ldots, x_n). \end{aligned}$$

Note that f_i has total degree at most d_i. Inside $\mathbb{P}^n$, we have the affine space $\mathbb{C}^n \subset \mathbb{P}^n$ defined by $x_0 = 1$, and the solutions of the affine equations

(5.3) $f_1 = \cdots = f_n = 0$

are precisely the solutions of (5.1) which lie in $\mathbb{C}^n \subset \mathbb{P}^n$. Similarly, the nontrivial solutions of the homogeneous equations

$$\overline{F}_1 = \cdots = \overline{F}_n = 0$$

may be regarded as the solutions which lie "at ∞". We say that (5.3) has *no solutions at* ∞ if $\overline{F}_1 = \cdots = \overline{F}_n = 0$ has no nontrivial solutions. By Theorem (2.3), this is equivalent to the condition

(5.4) $\mathrm{Res}_{d_1,\ldots,d_n}(\overline{F}_1,\ldots,\overline{F}_n) \neq 0.$

The proof of Theorem (3.4) implies the following version of Bézout's Theorem.

(5.5) Theorem (Bézout's Theorem). *Assume that $f_1,\ldots,f_n$ are defined as in (5.2) and that the affine equations (5.3) have no solutions at ∞. Then these equations have $d_1 \cdots d_n$ solutions (counted with multiplicity), and the ring*

$$A = \mathbb{C}[x_1,\ldots,x_n]/\langle f_1,\ldots,f_n\rangle$$

has dimension $d_1 \cdots d_n$ as a vector space over $\mathbb{C}$.

Note that this result does not hold for all systems of equations (5.3). In general, we need a language which allows us to talk about properties which are true for most but not necessarily all polynomials $f_1,\ldots,f_n$. This brings us to the idea of genericity.

(5.6) Definition. A property is said to *hold generically* for polynomials $f_1,\ldots,f_n$ of degree at most $d_1,\ldots,d_n$ if there is a nonzero polynomial in the coefficients of the f_i such that the property holds for all $f_1,\ldots,f_n$ for which the polynomial is nonvanishing.

Intuitively, a property of polynomials is *generic* if it holds for "most" polynomials $f_1,\ldots,f_n$. Our definition makes this precise by defining "most" to mean that some polynomial in the coefficients of the f_i is nonvanishing. As a simple example, consider a single polynomial $ax^2 + bx + c$. We claim that the property "$ax^2 + bx + c = 0$ has two solutions, counting multiplicity" holds generically. To prove this, we must find a polynomial in the coefficients a, b, c whose nonvanishing implies the desired property. Here, the condition is easily seen to be $a \neq 0$ since we are working over the complex numbers.

Exercise 1. Show that the property "$ax^2 + bx + c = 0$ has two distinct solutions" is generic. Hint: By the quadratic formula, $a(b^2 - 4ac) \neq 0$ implies the desired property.

A more relevant example is given by Theorem (5.5). Having no solutions at ∞ is equivalent to the nonvanishing of the resultant (5.4), and since $\mathrm{Res}_{d_1,\ldots,d_n}(\overline{F}_1,\ldots,\overline{F}_n)$ is a nonzero polynomial in the coefficients of the f_i, it follows that this version of Bézout's Theorem holds generically. Thus, for most choices of the coefficients, the equations $f_1 = \cdots = f_n = 0$ have $d_1 \cdots d_n$ solutions, counting multiplicity. In particular, if we choose polynomials $f_1,\ldots,f_n$ with random coefficients (say given by some random number generator), then, with a very high probability, Bézout's Theorem will hold for the corresponding system of equations.

In general, genericity comes in different "flavors". For instance, consider solutions of the equation $ax^2 + bx + c = 0$:

- Generically, $ax^2 + bx + c = 0$ has two solutions, counting multiplicity. This happens when $a \neq 0$.
- Generically, $ax^2 + bx + c = 0$ has two distinct solutions. By Exercise 1, this happens when $a(b^2 - 4ac) \neq 0$.

Similarly, there are different versions of Bézout's Theorem. In particular, one can strengthen Theorem (5.5) to prove that generically, the equations $f_1 = \cdots = f_n = 0$ have $d_1 \cdots d_n$ distinct solutions. This means that generically, (5.3) has no solutions at ∞ and all solutions have multiplicity one. A proof of this result will be sketched in Exercise 6 at the end of the section.

With this genericity assumption on $f_1,\ldots,f_n$, we know the number of distinct solutions of (5.3), and our next task is to find them. We could use the methods of Chapter 2, but it is also possible to find the solutions using resultants. This section will describe two closely related methods, u-resultants and hidden variables, for solving equations. The next section will discuss further methods which use eigenvalues and eigenvectors.

The u-Resultant

The basic idea of van der Waerden's *u-resultant* (see [vdW]) is to start with the homogeneous equations $F_1 = \cdots = F_n = 0$ of (5.1) and add another equation $F_0 = 0$ to (5.1), so that we have $n + 1$ homogeneous equations in $n + 1$ variables. We will use

$$F_0 = u_0 x_0 + \cdots + u_n x_n,$$

where $u_0,\ldots,u_n$ are independent variables. Because the number of equations equals the number of variables, we can form the resultant

$$\mathrm{Res}_{1,d_1,\ldots,d_n}(F_0, F_1, \ldots, F_n),$$

which is called the *u-resultant*. Note that the u-resultant is a polynomial in $u_0,\ldots,u_n$.

As already mentioned, we will sometimes work in the affine situation, where we dehomogenize $F_0, \ldots, F_n$ to obtain $f_0, \ldots, f_n$. This is the notation of (5.2), and in particular, observe that

$$(5.7) \qquad f_0 = u_0 + u_1 x_1 + \cdots + u_n x_n.$$

Because $f_0, \ldots, f_n$ and $F_0, \ldots, F_n$ have the same coefficients, we write the u-resultant as $\mathrm{Res}(f_0, \ldots, f_n)$ instead of $\mathrm{Res}(F_0, \ldots, F_n)$ in this case.

Before we work out the general theory of the u-resultant, let's do an example. The following exercise will seem like a lot of work at first, but its surprising result will be worth the effort.

Exercise 2. Let

$$F_1 = x_1^2 + x_2^2 - 10x_0^2 = 0$$
$$F_2 = x_1^2 + x_1 x_2 + 2x_2^2 - 16x_0^2 = 0$$

be the intersection of a circle and an ellipse in $\mathbb{P}^2$. By Bézout's Theorem, there are four solutions. To find the solutions, we add the equation

$$F_0 = u_0 x_0 + u_1 x_1 + u_2 x_2 = 0.$$

a. The theory of §4 computes the resultant using 10×10 determinants D_0, D_1 and D_2. Using D_0, Theorem (4.9) implies

$$\mathrm{Res}_{1,2,2}(F_0, F_1, F_2) = \pm \frac{D_0}{D_0'}.$$

If the variables are ordered x_2, x_1, x_0, show that $D_0 = \det(M_0)$, where M_0 is the matrix

$$M_0 = \begin{pmatrix} u_0 & u_1 & u_2 & 0 & 0 & 0 & 0 & 0 & 0 & 0 \\ 0 & u_0 & 0 & u_2 & u_1 & 0 & 0 & 0 & 0 & 0 \\ 0 & 0 & u_0 & u_1 & 0 & u_2 & 0 & 0 & 0 & 0 \\ 0 & 0 & 0 & u_0 & 0 & 0 & 0 & u_1 & u_2 & 0 \\ -10 & 0 & 0 & 0 & 1 & 1 & 0 & 0 & 0 & 0 \\ 0 & -10 & 0 & 0 & 0 & 0 & 1 & 0 & 1 & 0 \\ 0 & 0 & -10 & 0 & 0 & 0 & 0 & 1 & 0 & 1 \\ -16 & 0 & 0 & 1 & 1 & 2 & 0 & 0 & 0 & 0 \\ 0 & -16 & 0 & 0 & 0 & 0 & 1 & 1 & 2 & 0 \\ 0 & 0 & -16 & 0 & 0 & 0 & 0 & 1 & 1 & 2 \end{pmatrix}.$$

Also show that $D_0' = \det(M_0')$, where M_0' is given by

$$M_0' = \begin{pmatrix} 1 & 1 \\ 1 & 2 \end{pmatrix}.$$

Hint: Using the order x_2, x_1, x_0 gives $S_0 = \{x_0^3, x_0^2 x_1, x_0^2 x_2, x_0 x_1 x_2\}$, $S_1 = \{x_0 x_1^2, x_1^3, x_1^2 x_2\}$ and $S_2 = \{x_0 x_2^2, x_1 x_2^2, x_2^3\}$. The columns in M_0 correspond to the monomials $x_0^3, x_0^2 x_1, x_0^2 x_2, x_0 x_1 x_2, x_0 x_1^2, x_0 x_2^2, x_1^3, x_1^2 x_2, x_1 x_2^2, x_2^3$. Exercise 13 of §4 will be useful.

b. Conclude that

$$\text{Res}_{1,2,2}(F_0, F_1, F_2) = \pm \, (2u_0^4 + 16u_1^4 + 36u_2^4 - 80u_1^3u_2 + 120u_1u_2^3$$
$$- 18u_0^2u_1^2 - 22u_0^2u_2^2 + 52u_1^2u_2^2 - 4u_0^2u_1u_2).$$

c. Using a computer to factor this, show that $\text{Res}_{1,2,2}(F_0, F_1, F_2)$ equals

$$(u_0 + u_1 - 3u_2)(u_0 - u_1 + 3u_2)(u_0^2 - 8u_1^2 - 2u_2^2 - 8u_1u_2)$$

up to a constant. By writing the quadratic factor as $u_0^2 - 2(2u_1 + u_2)^2$, conclude that $\text{Res}_{1,2,2}(F_0, F_1, F_2)$ equals

$$(u_0 + u_1 - 3u_2)(u_0 - u_1 + 3u_2)(u_0 + 2\sqrt{2}u_1 + \sqrt{2}u_2)(u_0 - 2\sqrt{2}u_1 - \sqrt{2}u_2)$$

times a nonzero constant. Hint: If you are using Maple, let the resultant be **res** and use the command **factor(res)**. Also, the command **factor(res,RootOf(x^2-2))** will do the complete factorization.

d. The coefficients of the linear factors of $\text{Res}_{1,2,2}(F_0, F_1, F_2)$ give four points

$$(1, 1, -3), \; (1, -1, 3), \; (1, 2\sqrt{2}, \sqrt{2}), \; (1, -2\sqrt{2}, -\sqrt{2})$$

in $\mathbb{P}^2$. Show that these points are the four solutions of the equations $F_1 = F_2 = 0$. Thus the solutions in $\mathbb{P}^2$ are *precisely* the coefficients of the linear factors of $\text{Res}_{1,2,2}(F_0, F_1, F_2)$!

In this exercise, all of the solutions lay in the affine space $\mathbb{C}^2 \subset \mathbb{P}^2$ defined by $x_0 = 1$. In general, we will study the u-resultant from the affine point of view. The key fact is that when all of the multiplicities are one, the solutions of (5.3) can be found using $\text{Res}_{1,d_1,\dots,d_n}(f_0, \dots, f_n)$.

(5.8) Proposition. *Assume that $f_1 = \dots = f_n = 0$ have total degrees bounded by $d_1, \dots, d_n$, no solutions at ∞, and all solutions of multiplicity one. If $f_0 = u_0 + u_1x_1 + \dots + u_nx_n$, where $u_0, \dots, u_n$ are independent variables, then there is a nonzero constant C such that*

$$\text{Res}_{1,d_1,\dots,d_n}(f_0, \dots, f_n) = C \prod_{p \in \mathbf{V}(f_1,\dots,f_n)} f_0(p).$$

PROOF. Let $C = \text{Res}_{d_1,\dots,d_n}(\overline{F}_1, \dots, \overline{F}_n)$, which is nonzero by hypothesis. Since the coefficients of f_0 are the variables $u_0, \dots, u_n$, we need to work over the field $K = \mathbb{C}(u_0, \dots, u_n)$ of rational functions in $u_0, \dots, u_n$. Hence, in this proof, we will work over K rather than over $\mathbb{C}$. Fortunately, the results we need are true over K, even though we proved them only over $\mathbb{C}$.

Adapting Theorem (3.4) to the situation of (5.2) (see Exercise 8) yields

$$\text{Res}_{1,d_1,\dots,d_n}(f_0, \dots, f_n) = C \, \det(m_{f_0}),$$

where $m_{f_0} : A \to A$ is the linear map given by multiplication by f_0 on the quotient ring

$$A = K[x_1, \ldots, x_n]/\langle f_1, \ldots, f_n \rangle.$$

By Theorem (5.5), A is a vector space over K of dimension $d_1 \cdots d_n$, and Theorem (4.5) of Chapter 2 implies that the eigenvalues of m_{f_0} are the values $f_0(p)$ for $p \in \mathbf{V}(f_1, \ldots, f_n)$. Since all multiplicities are one, there are $d_1 \cdots d_n$ such points p, and the corresponding values $f_0(p)$ are distinct since $f_0 = u_0 + u_1 x_1 + \cdots + u_n x_n$ and $u_0, \ldots, u_n$ are independent variables. Thus m_{f_0} has $d_1 \cdots d_n$ distinct eigenvalues $f_0(p)$, so that

$$\det(m_{f_0}) = \prod_{p \in \mathbf{V}(f_1, \ldots, f_n)} f_0(p).$$

This proves the proposition.   □

To see more clearly what the proposition says, let the points of $\mathbf{V}(f_1, \ldots, f_n)$ be p_i for $1 \le i \le d_1 \cdots d_n$. If we write each point as $p_i = (a_{i1}, \ldots, a_{in}) \in \mathbb{C}^n$, then (5.7) implies

$$f_0(p_i) = u_0 + a_{i1} u_1 + \cdots + a_{in} u_n,$$

so that by Proposition (5.8), the u-resultant is given by

$$(5.9) \quad \mathrm{Res}_{1, d_1, \ldots, d_n}(f_0, \ldots, f_n) = C \prod_{i=1}^{d_1 \cdots d_n} (u_0 + a_{i1} u_1 + \cdots + a_{in} u_n).$$

We see clearly that the u-resultant is a polynomial in $u_0, \ldots, u_n$. Furthermore, we get the following method for finding solutions of (5.3): compute $\mathrm{Res}_{1, d_1, \ldots, d_n}(f_0, \ldots, f_n)$, factor it into linear factors, and then read off the solutions! Hence, once we have the u-resultant, solving (5.3) is reduced to a problem in multivariable factorization.

To compute the u-resultant, we use Theorem (4.9). Because of our emphasis on f_0, we represent the resultant as the quotient

$$(5.10) \quad \mathrm{Res}_{1, d_1, \ldots, d_n}(f_0, \ldots, f_n) = \pm \frac{D_0}{D_0'}.$$

This is the formula we used in Exercise 2. In §4, we got the determinant D_0 by working with the homogenizations F_i of the f_i, regarding x_0 as the last variable, and decomposing monomials of degree $d = 1 + d_1 + \cdots + d_n - n$ into disjoint subsets $S_0, \ldots, S_n$. Taking x_0 last means that S_0 consists of the $d_1 \cdots d_n$ monomials

$$(5.11) \quad S_0 = \{x_0^{a_0} x_1^{a_1} \cdots x_n^{a_n} : 0 \le a_i \le d_i - 1 \text{ for } i > 0, \ \textstyle\sum_{i=0}^{n} a_i = d\}.$$

Then D_0 is the determinant of the matrix M_0 representing the system of equations (4.1). We saw an example of this in Exercise 2.

The following exercise simplifies the task of computing u-resultants.

Exercise 3. Assuming that $D_0' \ne 0$ in (5.10), prove that D_0' does not involve $u_0, \ldots, u_n$ and conclude that $\mathrm{Res}_{1, d_1, \ldots, d_n}(f_0, \ldots, f_n)$ and D_0 differ by a constant factor when regarded as polynomials in $\mathbb{C}[u_0, \ldots, u_n]$.

We will write D_0 as $D_0(u_0, \ldots, u_n)$ to emphasize the dependence on $u_0, \ldots, u_n$. We can use $D_0(u_0, \ldots, u_n)$ only when $D_0' \neq 0$, but since D_0' is a polynomial in the coefficients of the f_i, Exercise 3 means that generically, the linear factors of the determinant $D_0(u_0, \ldots, u_n)$ give the solutions of our equations (5.3). In this situation, we will apply the term *u-resultant* to both $\mathrm{Res}_{1,d_1,\ldots,d_n}(f_0, \ldots, f_n)$ and $D_0(u_0, \ldots, u_n)$.

Unfortunately, the u-resultant has some serious limitations. First, it is not easy to compute symbolic determinants of large size (see the discussion at the end of §4). And even if we can find the determinant, multivariable factorization as in (5.9) is very hard, especially since in most cases, floating point numbers will be involved.

There are several methods for dealing with this situation. We will describe one, as presented in [CM]. The basic idea is to specialize some of the coefficients in $f_0 = u_0 + u_1 x_1 + \cdots + u_n x_n$. For example, the argument of Proposition (5.8) shows that when the x_n-coordinates of the solution points are distinct, the specialization $u_1 = \cdots = u_{n-1} = 0, u_n = -1$ transforms (5.9) into the formula

$$(5.12) \qquad \mathrm{Res}_{1,d_1,\ldots,d_n}(u_0 - x_n, f_1, \ldots, f_n) = C \prod_{i=1}^{d_1 \cdots d_n} (u_0 - a_{in}),$$

where a_{in} is the x_n-coordinate of $p_i = (a_{i1}, \ldots, a_{in}) \in \mathbf{V}(f_1, \ldots, f_n)$. This resultant is a univariate polynomial in u_0 whose roots are precisely the x_n-coordinates of solutions of (5.3). There are similar formulas for the other coordinates of the solutions.

If we use the numerator $D_0(u_0, \ldots, u_n)$ of (5.10) as the u-resultant, then setting $u_1 = \cdots = u_n = 0, u_n = -1$ gives $D_0(u_0, 0, \ldots, 0, -1)$, which is a polynomial in u_0. The argument of Exercise 3 shows that generically, $D_0(u_0, 0, \ldots, 0, -1)$ is a constant multiple $\mathrm{Res}(u_0 - x_n, f_1, \ldots, f_n)$, so that its roots are also the x_n-coordinates. Since $D_0(u_0, 0, \ldots, 0, -1)$ is given by a symbolic determinant depending on the single variable u_0, it is *much* easier to compute than in the multivariate case. Using standard techniques (discussed in Chapter 2) for finding the roots of univariate polynomials such as $D_0(u_0, 0, \ldots, 0, -1)$, we get a computationally efficient method for finding the x_n-coordinates of our solutions. Similarly, we can find the other coordinates of the solutions by this method.

Exercise 4. Let $D_0(u_0, u_1, u_2)$ be the determinant in Exercise 2.
a. Compute $D_0(u_0, -1, 0)$ and $D_0(u_0, 0, -1)$.
b. Find the roots of these polynomials numerically. Hint: Try the Maple command `fsolve`. In general, `fsolve` should be used with the `complex` option, though in this case it's not necessary since the roots are real.
c. What does this say about the coordinates of the solutions of the equations $x_1^2 + x_2^2 = 10$, $x_1^2 + x_1 x_2 + 2x_2^2 = 16$? Can you figure out what the solutions are?

As this exercise illustrates, the univariate polynomials we get from the u-resultant enable us to find the individual coordinates of the solutions, but they don't tell us how to match them up. One method for doing this (based on [CM]) will be explained in Exercise 7 at the end of the section. We should also mention that a different u-resultant method for computing solutions is given in [Can2].

All of the u-resultant methods make strong genericity assumptions on the polynomials $f_0, \ldots, f_n$. In practice, one doesn't know in advance if a given system of equations is generic. Here are some of the things that can go wrong when trying to apply the above methods to non-generic equations:

- There might be solutions at infinity. This problem can be avoided by making a generic linear change of coordinates.
- If too many coefficients are zero, it might be necessary to use the sparse resultants of Chapter 7.
- The equations (5.1) might have infinitely many solutions. In the language of algebraic geometry, the projective variety $\mathbf{V}(F_1, \ldots, F_n)$ might have components of positive dimension, together with some isolated solutions. One is still interested in the isolated solutions, and techniques for finding them are described in Section 4 of [Can1].
- The denominator D_0' in the resultant formula (5.10) might vanish. When this happens, one can use the *generalized characteristic polynomials* described in §4 to avoid this difficulty. See Section 4.1 of [CM] for details.
- Distinct solutions might have the same x_i-coordinate for some i. The polynomial giving the x_i-coordinates would have multiple roots, which are computationally unstable. This problem can be avoided with a generic change of coordinates. See Section 4.2 of [CM] for an example.

Also, Chapter 4 will give versions of (5.12) and Proposition (5.8) for the case when $f_1 = \cdots = f_n = 0$ has solutions of multiplicity > 1.

Hidden Variables

One of the better known resultant techniques for solving equations is the *hidden variable* method. The basic idea is to regard one of variables as a constant and then take a resultant. To illustrate how this works, consider the affine equations we get from Exercise 2 by setting $x_0 = 1$:

(5.13)
$$f_1 = x_1^2 + x_2^2 - 10 = 0$$
$$f_2 = x_1^2 + x_1 x_2 + 2x_2^2 - 16 = 0.$$

If we regard x_2 as a constant, we can use the resultant of §1 to obtain

$$\mathrm{Res}(f_1, f_2) = 2x_2^4 - 22x_2^2 + 36 = 2(x_2 - 3)(x_2 + 3)(x_2 - \sqrt{2})(x_2 + \sqrt{2}).$$

The resultant is a polynomial in x_2, and its roots are *precisely* the x_2-coordinates of the solutions of the equations (as we found in Exercise 2).

To generalize this example, we first review the affine form of the resultant. Given $n+1$ homogeneous polynomials $G_0, \ldots, G_n$ of degrees $d_0, \ldots, d_n$ in $n+1$ variables $x_0, \ldots, x_n$, we get $\mathrm{Res}_{d_0, \ldots, d_n}(G_0, \ldots, G_n)$. Setting $x_0 = 1$ gives

$$g_i(x_1, \ldots, x_n) = G_i(1, x_1, \ldots, x_n),$$

and since the g_i and G_i have the same coefficients, we can write the resultant as $\mathrm{Res}_{d_0, \ldots, d_1}(g_0, \ldots, g_n)$. Thus, $n+1$ polynomials $g_0, \ldots, g_n$ in n variables $x_1, \ldots, x_n$ have a resultant. It follows that from the affine point of view, forming a resultant requires that *the number of polynomials be one more than the number of variables*.

Now, suppose we have n polynomials $f_1, \ldots, f_n$ of degrees $d_1, \ldots, d_n$ in n variables $x_1, \ldots, x_n$. In terms of resultants, we have the wrong numbers of equations and variables. One solution is to add a new polynomial, which leads to the u-resultant. Here, we will pursue the other alternative, which is to get rid of one of the variables. The basic idea is what we did above: we *hide* a variable, say x_n, by regarding it as a constant. This gives n polynomials $f_1, \ldots, f_n$ in $n-1$ variables $x_1, \ldots, x_{n-1}$, which allows us to form their resultant. We will write this resultant as

$$(5.14) \qquad \mathrm{Res}^{x_n}_{d_1, \ldots, d_n}(f_1, \ldots, f_n).$$

The superscript x_n reminds us that we are regarding x_n as constant. Since the resultant is a polynomial in the coefficients of the f_i, (5.14) is a polynomial in x_n.

We can now state the main result of the hidden variable technique.

(5.15) Proposition. *Generically, $\mathrm{Res}^{x_n}_{d_1, \ldots, d_n}(f_1, \ldots, f_n)$ is a polynomial in x_n whose roots are the x_n-coordinates of the solutions of (5.3).*

PROOF. The basic strategy of the proof is that by (5.12), we already know a polynomial whose roots are the x_n-coordinates of the solutions, namely

$$\mathrm{Res}_{1, d_1, \ldots, d_n}(u_0 - x_n, f_1, \ldots, f_n).$$

We will prove the theorem by showing that this polynomial is the same as the hidden variable resultant (5.14). However, (5.14) is a polynomial in x_n, while $\mathrm{Res}(u_0 - x_n, f_1, \ldots, f_n)$ is a polynomial in u_0. To compare these two polynomials, we will write

$$\mathrm{Res}^{x_n = u_0}_{d_1, \ldots, d_n}(f_1, \ldots, f_n)$$

to mean the polynomial obtained from (5.14) by the substitution $x_n = u_0$. Using this notation, the theorem will follow once we show that

$$\mathrm{Res}^{x_n = u_0}_{d_1, \ldots, d_n}(f_1, \ldots, f_n) = \pm \mathrm{Res}_{1, d_1, \ldots, d_n}(u_0 - x_n, f_1, \ldots, f_n).$$

We will prove this equality by applying Theorem (3.4) separately to the two resultants in this equation.

Beginning with $\mathrm{Res}(u_0 - x_n, f_1, \ldots, f_n)$, first recall that it equals the homogeneous resultant $\mathrm{Res}(u_0 x_0 - x_n, F_1, \ldots, F_n)$ via (5.2). Since u_0 is a coefficient, we will work over the field $\mathbb{C}(u_0)$ of rational functions in u_0. Then, adapting Theorem (3.4) to the situation of (5.2) (see Exercise 8), we see that $\mathrm{Res}(u_0 x_0 - x_n, F_1, \ldots, F_n)$ equals

$$(5.16) \qquad \mathrm{Res}_{1,d_1,\ldots,d_{n-1}}(-x_n, \overline{F}_1, \ldots, \overline{F}_{n-1})^{d_n} \det(m_{f_n}),$$

where $-x_n, \overline{F}_1, \ldots, \overline{F}_{n-1}$ are obtained from $u_0 x_0 - x_n, F_1, \ldots, F_{n-1}$ by setting $x_0 = 0$, and $m_{f_n} : A \to A$ is multiplication by f_n in the ring

$$A = \mathbb{C}(u)[x_1, \ldots, x_n]/\langle u - x_n, f_1, \ldots, f_n \rangle.$$

Next, consider $\mathrm{Res}^{x_n = u_0}(f_1, \ldots, f_n)$, and observe that if we define

$$\hat{f}_i(x_1, \ldots, x_{n-1}) = f_i(x_1, \ldots, x_{n-1}, u_0),$$

then $\mathrm{Res}^{x_n = u_0}(f_1, \ldots, f_n) = \mathrm{Res}(\hat{f}_1, \ldots, \hat{f}_n)$. If we apply Theorem (3.4) to the latter resultant, we see that it equals

$$(5.17) \qquad \mathrm{Res}_{d_1,\ldots,d_{n-1}}(\widetilde{F}_1, \ldots, \widetilde{F}_{n-1})^{d_n} \det(m_{\hat{f}_n}),$$

where $\widetilde{F}_i$ is obtained from $\hat{f}_i$ by first homogenizing with respect to x_0 and then setting $x_0 = 0$, and $m_{\hat{f}_n} : \widehat{A} \to \widehat{A}$ is multiplication by $\hat{f}_n$ in

$$\widehat{A} = \mathbb{C}(u_0)[x_1, \ldots, x_{n-1}]/\langle \hat{f}_1, \ldots, \hat{f}_n \rangle.$$

To show that (5.16) and (5.17) are equal, we first examine (5.17). We claim that if f_i homogenizes to F_i, then $\widetilde{F}_i$ in (5.17) is given by

$$(5.18) \qquad \widetilde{F}_i(x_1, \ldots, x_{n-1}) = F_i(0, x_1, \ldots, x_{n-1}, 0).$$

To prove this, take a term of F_i, say

$$c\, x_0^{a_0} \cdots x_n^{a_n}, \qquad a_0 + \cdots + a_n = d_i.$$

Since $x_0 = 1$ gives f_i and $x_n = u_0$ then gives $\hat{f}_i$, the corresponding term in $\hat{f}_i$ is

$$c\, 1^{a_0} x_1^{a_1} \cdots x_{n-1}^{a_{n-1}} u_0^{a_n} = c u_0^{a_n} \cdot x_1^{a_1} \cdots x_{n-1}^{a_{n-1}}.$$

When homogenizing $\hat{f}_i$ with respect to x_0, we want a term of total degree d_i in $x_0, \ldots, x_{n-1}$. Since $c u_0^{a_n}$ is a constant, we get

$$c u_0^{a_n} \cdot x_0^{a_0 + a_n} x_1^{a_1} \cdots x_{n-1}^{a_{n-1}} = c \cdot x_0^{a_0} \cdots x_{n-1}^{a_{n-1}} (u_0 x_0)^{a_n}.$$

It follows that the homogenization of $\hat{f}_i$ is $F_i(x_0, \ldots, x_{n-1}, u_0 x_0)$, and since $\widetilde{F}_i$ is obtained by setting $x_0 = 0$ in this polynomial, we get (5.18).

Once we know (5.18), Exercise 12 of §3 shows that

$$\mathrm{Res}_{1,d_1,\ldots,d_{n-1}}(-x_n, \overline{F}_1, \ldots, \overline{F}_{n-1}) = \pm \mathrm{Res}_{d_1,\ldots,d_{n-1}}(\widetilde{F}_1, \ldots, \widetilde{F}_{n-1})$$

since $\overline{F}_i(x_1, \ldots, x_n) = F_i(0, x_1, \ldots, x_n)$. Also, the ring homomorphism

$$\mathbb{C}(u_0)[x_1, \ldots, x_n] \to \mathbb{C}(u_0)[x_1, \ldots, x_{n-1}]$$

defined by $x_n \mapsto u_0$ carries f_i to $\hat{f}_i$. It follows that this homomorphism induces a ring isomorphism $A \cong \hat{A}$ (you will check the details of this in Exercise 8). Moreover, multiplication by f_n and $\hat{f}_n$ give a diagram

$$
\begin{array}{ccc}
A & \cong & \hat{A} \\
m_{f_n} \downarrow & & \downarrow m_{\hat{f}_n} \\
A & \cong & \hat{A}
\end{array}
$$

(5.19)

In Exercise 8, you will show that going across and down gives the same map $A \to \hat{A}$ as going down and across (we say that (5.19) is a *commutative diagram*). From here, it is easy to show that $\det(m_{f_n}) = \det(m_{\hat{f}_n})$, and it follows that (5.16) and (5.17) are equal. $\square$

The advantage of the hidden variable method is that it involves resultants with fewer equations and variables than the u-resultant. For example, when dealing with the equations $f_1 = f_2 = 0$ from (5.13), the u-resultant $\mathrm{Res}_{1,2,2}(f_0, f_1, f_2)$ uses the 10×10 matrix from Exercise 2, while $\mathrm{Res}_{2,2}^{x_2}(f_1, f_2)$ only requires a 4×4 matrix.

In general, we can compute $\mathrm{Res}^{x_n}(f_1, \ldots, f_n)$ by Theorem (4.9), and as with the u-resultant, we can again ignore the denominator. More precisely, if we write

$$
\mathrm{Res}_{d_1, \ldots, d_n}^{x_n}(f_1, \ldots, f_n) = \pm \frac{\widehat{D}_0}{\widehat{D}_0'},
$$

(5.20)

then $\widehat{D}_0'$ doesn't involve x_n. The proof of this result is a nice application of Proposition (4.6), and the details can be found in Exercise 10 at the end of the section. Thus, when using the hidden variable method, it suffices to use the numerator $\widehat{D}_0$—when $f_1, \ldots, f_n$ are generic, its roots give the x_n-coordinates of the affine equations (5.3).

Of course, there is nothing special about hiding x_n—we can hide any of the variables in the same way, so that the hidden variable method can be used to find the x_i-coordinates of the solutions for any i. One limitation of this method is that it only gives the individual coordinates of the solution points and doesn't tell us how they match up.

Exercise 5. Consider the affine equations

$$
\begin{aligned}
f_1 &= x_1^2 + x_2^2 + x_3^2 - 3 \\
f_2 &= x_1^2 + x_3^2 - 2 \\
f_3 &= x_1^2 + x_2^2 - 2x_3.
\end{aligned}
$$

a. If we compute the u-resultant with $f_0 = u_0 + u_1 x_1 + u_2 x_2 + u_3 x_3$, show that Theorem (4.9) expresses $\mathrm{Res}_{1,2,2,2}(f_0, f_1, f_2, f_3)$ as a quotient of determinants of sizes 35×35 and 15×15 respectively.

b. If we hide x_3, show that $\mathrm{Res}_{2,2,2}^{x_3}(f_1, f_2, f_3)$ is a quotient of determinants of sizes 15×15 and 3×3 respectively.

c. Hiding x_3 as in part b, use (2.8) to express $\mathrm{Res}_{2,2,2}^{x_3}(f_1, f_2, f_3)$ as the determinant of a 6×6 matrix, and show that up to a constant, the resultant is $(x_3^2 + 2x_3 - 3)^4$. Explain the significance of the exponent 4. Hint: You will need to regard x_3 as a constant and homogenize the f_i with respect to x_0. Then (2.8) will be easy to apply.

The last part of Exercise 5 illustrates how formulas such as (2.8) allow us, in special cases, to represent a resultant as a *single* determinant of relatively small size. This can reduce dramatically the amount of computation involved and explains the continuing interest in finding determinant formulas for resultants (see, for example, [DD], [SZ], and [WZ]).

ADDITIONAL EXERCISES FOR §5

Exercise 6. In the text, we claimed that generically, the solutions of n affine equations $f_1 = \cdots = f_n = 0$ have multiplicity one. This exercise will prove this result. Assume as usual that the f_i come from homogeneous polynomials F_i of degree d_i by setting $x_0 = 1$. We will also use the following fact from multiplicity theory: if $F_1 = \cdots = F_n = 0$ has finitely many solutions and p is a solution such that the gradient vectors

$$\nabla F_i(p) = \left(\frac{\partial F_i}{\partial x_0}(p), \ldots, \frac{\partial F_i}{\partial x_n}(p) \right), \quad 1 \leq i \leq n$$

are linearly independent, then p is a solution of multiplicity one.

a. Consider the affine space $\mathbb{C}^M$ consisting of all possible coefficients of the F_i. As in the discussion at the end of §2, the coordinates of $\mathbb{C}^M$ are $c_{i,\alpha}$, where for fixed i, the $c_{i,\alpha}$ are the coefficients of F_i. Now consider the set $W \subset \mathbb{C}^M \times \mathbb{P}^n \times \mathbb{P}^{n-1}$ defined by

$$W = \{(c_{i,\alpha}, p, a_1, \ldots, a_n) \in \mathbb{C}^M \times \mathbb{P}^n \times \mathbb{P}^{n-1} : p \text{ is a}$$
$$\text{nontrivial solution of } F_0 = \cdots = F_n = 0 \text{ and}$$
$$a_1 \nabla F_1(p) + \cdots + a_n \nabla F_n(p) = 0\}.$$

Under the projection map $\pi : \mathbb{C}^M \times \mathbb{P}^n \times \mathbb{P}^{n-1} \to \mathbb{C}^M$, explain why a generalization of the Projective Extension Theorem from §2 would imply that $\pi(W) \subset \mathbb{C}^M$ is a variety.

b. Show that $\pi(W) \subset \mathbb{C}^M$ is a proper variety, i.e., find $F_1, \ldots, F_n$ such that $(F_1, \ldots, F_n) \in \mathbb{C}^M \setminus \pi(W)$. Hint: Let $F_i = \Pi_{j=1}^{d_i}(x_i - jx_0)$ for $1 \leq i \leq n$.

c. By parts a and b, we can find a nonzero polynomial G in the coefficients of the F_i such that G vanishes on $\pi(W)$. Then consider $G \cdot \mathrm{Res}(\overline{F}_1, \ldots, \overline{F}_n)$. We can regard this as a polynomial in the coefficients of the f_i. Prove that if this polynomial is nonvanishing at $f_1, \ldots, f_n$, then the equations $f_0 = \cdots = f_n = 0$ have $d_1 \cdots d_n$ many solutions in $\mathbb{C}^n$, all of which have multiplicity one. Hint: Use Theorem (5.5).

Exercise 7. As we saw in (5.12), we can find the x_n-coordinates of the solutions using $\mathrm{Res}(u - x_n, f_1, \ldots, f_n)$, and in general, the x_i-coordinates can be found by replacing $u - x_n$ by $u - x_i$ in the resultant. In this exercise, we will describe the method given in [CM] for matching up coordinates to get the solutions. We begin by assuming that we've found the x_1- and x_2-coordinates of the solutions. To match up these two coordinates, let α and β be randomly chosen numbers, and consider the resultant

$$R_{1,2}(u) = \mathrm{Res}_{1,d_1,\ldots,d_n}(u - (\alpha x_1 + \beta x_2), f_1, \ldots, f_n).$$

a. Use (5.9) to show that

$$R_{1,2}(u) = C' \prod_{i=1}^{d_1 \cdots d_n} \left(u - (\alpha a_{i1} + \beta a_{i2}) \right),$$

where C' is a nonzero constant and, as in (5.9), the solutions are $p_i = (a_{i1}, \ldots, a_{in})$.

b. A random choice of α and β will ensure that for solutions p_i, p_j, p_k, we have $\alpha a_{i1} + \beta a_{j2} \neq \alpha a_{k1} + \beta a_{k2}$ except when $p_i = p_j = p_k$. Conclude that the only way the condition

$$\alpha \cdot (\text{an } x_1\text{-coordinate}) + \beta \cdot (\text{an } x_2\text{-coordinate}) = \text{root of } R_{1,2}(u)$$

can hold is when the x_1-coordinate and x_2-coordinate come from the same solution.

c. Explain how we can now find the first two coordinates of the solutions.

d. Explain how a random choice of α, β, γ will enable us to construct a polynomial $R_{1,2,3}(u)$ which will tell us how to match up the x_3-coordinates with the two coordinates already found.

e. In the affine equations $f_1 = f_2 = 0$ coming from (5.13), compute $\mathrm{Res}(u - x_1, f_1, f_2)$, $\mathrm{Res}(u - x_2, f_1, f_2)$ and (in the notation of part a) $R_{1,2}(u)$, using $\alpha = 1$ and $\beta = 2$. Find the roots of these polynomials numerically and explain how this gives the solutions of our equations. Hint: Try the Maple command `fsolve`. In general, `fsolve` should be used with the `complex` option, though in this case it's not necessary since the roots are real.

Exercise 8. This exercise is concerned with Proposition (5.15).
a. Explain what Theorem (3.4) looks like if we use (5.2) instead of (3.3), and apply this to (5.16), (5.17) and Proposition (5.8).
b. Show carefully that the the ring homomorphism

$$\mathbb{C}(u)[x_1, \ldots, x_n] \longrightarrow \mathbb{C}(u)[x_1, \ldots, x_{n-1}]$$

defined by $x_n \mapsto u$ carries f_i to $\hat{f}_i$ and induces a ring isomorphism $A \cong \hat{A}$.
c. Show that the diagram (5.19) is commutative and use it to prove that $\det(m_{f_n}) = \det(m_{\hat{f}_n})$.

Exercise 9. In this exercise, you will develop a homogeneous version of the hidden variable method. Suppose that we have homogeneous polynomials $F_1, \ldots, F_n$ in $x_0, \ldots, x_n$ such that

$$f_i(x_1, \ldots, x_n) = F_i(1, x_1, \ldots, x_n).$$

We assume that F_i has degree d_i, so that f_i has degree at most d_i. Also define

$$\hat{f}_i(x_1, \ldots, x_{n-1}) = f_i(x_1, \ldots, x_{n-1}, u).$$

As we saw in the proof of Proposition (5.15), the hidden variable resultant can be regarded as the affine resultant $\mathrm{Res}_{d_1, \ldots, d_n}(\hat{f}_1, \ldots, \hat{f}_n)$. To get a homogeneous resultant, we homogenize $\hat{f}_i$ with respect to x_0 to get a homogeneous polynomial $\widehat{F}_i(x_0, \ldots, x_{n-1})$ of degree d_i. Then

$$\mathrm{Res}_{d_1, \ldots, d_n}(\hat{f}_1, \ldots, \hat{f}_n) = \mathrm{Res}_{d_1, \ldots, d_n}(\widehat{F}_1, \ldots, \widehat{F}_n).$$

a. Prove that

$$\widehat{F}_i(x_0, \ldots, x_{n-1}) = F_i(x_0, x_1, \ldots, x_0 u).$$

Hint: This is done in the proof of Proposition (5.15).
b. Explain how part a leads to a purely homogeneous construction of the hidden variable resultant. This resultant is a polynomial in u.
c. State a purely homogeneous version of Proposition (5.15) and explain how it follows from the affine version stated in the text. Also explain why the roots of the hidden variable resultant are a_n/a_0 as $p = (a_0, \ldots, a_n)$ varies over all homogeneous solutions of $F_1 = \cdots = F_n = 0$ in $\mathbb{P}^n$.

Exercise 10. In (5.20), we expressed the hidden variable resultant as a quotient of two determinants $\pm \widehat{D}_0 / \widehat{D}_0'$. If we think of this resultant as a polynomial in u, then use Proposition (4.6) to prove that the denominator $\widehat{D}_0'$ does *not* involve u. This will imply that the numerator $\widehat{D}_0$ can be regarded as the hidden variable resultant. Hint: By the previous exercise, we can write the hidden variable resultant as $\mathrm{Res}(\widehat{F}_1, \ldots, \widehat{F}_n)$. Also note that Proposition (4.6) assumed that x_n is last, while here $\widehat{D}_0$ and $\widehat{D}_0'$ mean that x_0 is taken last. Thus, applying Proposition (4.6) to the $\widehat{F}_i$ means setting $x_0 = 0$ in $\widehat{F}_i$. Then use part a of Exercise 9 to explain why u disappears from the scene.

Exercise 11. Suppose that $f_1, \ldots, f_n$ are polynomials of total degrees $d_1, \ldots, d_n$ in $k[x_1, \ldots, x_n]$.
a. Use Theorem (2.10) of Chapter 2 to prove that the ideal $\langle f_1, \ldots, f_n \rangle$ is radical for $f_1, \ldots, f_n$ generic. Hint: Use the notion of generic discussed in Exercise 6.
b. Explain why Exercise 16 of Chapter 2, §4, describes a *lex* Gröbner basis (assuming x_n is the last variable) for the ideal $\langle f_1, \ldots, f_n \rangle$ when the f_i are generic.

§6 Solving Equations via Eigenvalues and Eigenvectors

In Chapter 2, we learned that solving the equations $f_1 = \cdots = f_n = 0$ can be reduced to an eigenvalue problem. We did this as follows. The monomials not divisible by the leading terms of a Gröbner basis G for $\langle f_1, \ldots, f_n \rangle$ give a basis for the quotient ring

$$(6.1) \qquad A = \mathbb{C}[x_1, \ldots, x_n]/\langle f_1, \ldots, f_n \rangle.$$

(see §2 of Chapter 2). Using this basis, we find the matrix of a multiplication map m_{f_0} by taking a basis element x^α and computing the remainder of $x^\alpha f_0$ on division by G (see §4 of Chapter 2). Once we have this matrix, its eigenvalues are the values $f_0(p)$ for $p \in \mathbf{V}(f_1, \ldots, f_n)$ by Theorem (4.5) of Chapter 2. In particular, the eigenvalues of the matrix for m_{x_i} are the x_i-coordinates of the solution points.

The amazing fact is that we can do all of this using resultants! We first show how to find a basis for the quotient ring.

(6.2) Theorem. *If $f_1, \ldots, f_n$ are generic polynomials of total degree $d_1, \ldots, d_n$, then the cosets of the monomials*

$$x_1^{a_1} \cdots x_n^{a_n}, \text{ where } 0 \le a_i \le d_i - 1 \text{ for } i = 1, \ldots, n$$

form a basis of the ring A of (6.1).

PROOF. Note that these monomials are *precisely* the monomials obtained from S_0 in (5.11) by setting $x_0 = 1$. As we will see, this is no accident.

By $f_1, \ldots, f_n$ generic, we mean that there are no solutions at ∞, that all solutions have multiplicity one, and that the matrix M_{11} which appears below is invertible.

Our proof will follow [ER] (see [PS1] for a different proof). There are $d_1 \cdots d_n$ monomials $x_1^{a_1} \cdots x_n^{a_n}$ with $0 \le a_i \le d_i - 1$. Since this is the dimension of A in the generic case by Theorem (5.5), it suffices to show that the cosets of these polynomials are linearly independent.

To prove this, we will use resultants. However, we have the wrong number of polynomials: since $f_1, \ldots, f_n$ are not homogeneous, we need $n + 1$ polynomials in order to form a resultant. Hence we will add the polynomial $f_0 = u_0 + u_1 x_1 + \cdots + u_n x_n$, where $u_0, \ldots, u_n$ are independent variables. This gives the resultant $\mathrm{Res}_{1,d_1,\ldots,d_n}(f_0, \ldots, f_n)$, which we recognize as the u-resultant. By (5.10), this resultant is the quotient D_0/D_0', where $D_0 = \det(M_0)$ and M_0 is the matrix coming from the equations (4.1).

We first need to review in detail how the matrix M_0 is constructed. Although we did this in (4.1), our present situation is different in two ways: first, (4.1) ordered the variables so that x_n was last, while here, we want x_0 to be last, and second, (4.1) dealt with homogeneous polynomials, while here we have dehomogenized by setting $x_0 = 1$. Let's see what changes this makes.

As before, we begin in the homogeneous situation and consider monomials $x^\gamma = x_0^{a_0} \cdots x_n^{a_n}$ of total degree $d = 1 + d_1 + \cdots + d_n - n$ (remember that the resultant is $\mathrm{Res}_{1,d_1,\ldots,d_n}$). Since we want to think of x_0 as last, we divide these monomials into $n + 1$ disjoint sets as follows:

$$S_n = \{x^\gamma : |\gamma| = d, \ x_n^{d_n} \text{ divides } x^\gamma\}$$

$$S_{n-1} = \{x^\gamma : |\gamma| = d, \ x_n^{d_n} \text{ doesn't divide } x^\gamma \text{ but } x_{n-1}^{d_{n-1}} \text{ does}\}$$

$$\vdots$$

$$S_0 = \{x^\gamma : |\gamma| = d, \ x_n^{d_n}, \ldots, x_1^{d_1} \text{ don't divide } x^\gamma \text{ but } x_0 \text{ does}\}$$

(remember that $d_0 = 1$ in this case). You should check that S_0 is precisely as described in (5.11). The next step is to dehomogenize the elements of S_i by setting $x_0 = 1$. If we denote the resulting set of monomials as S_i', then $S_0' \cup S_1' \cup \cdots \cup S_n'$ consists of all monomials of total degree $\le d$ in $x_1, \ldots, x_n$. Furthermore, we see that S_0' consists of the $d_1 \cdots d_n$ monomials in the statement of the theorem.

Because of our emphasis on S_0', we will use x^α to denote elements of S_0' and x^β to denote elements of $S_1' \cup \cdots \cup S_n'$. Then observe that

$$\text{if } x^\alpha \in S_0', \text{ then } x^\alpha \text{ has degree } \le d - 1,$$
$$\text{if } x^\beta \in S_i', \ i > 0, \text{ then } x^\alpha/x_i^{d_i} \text{ has degree } \le d - d_i.$$

Then consider the equations:

$$x^\alpha\, f_0 = 0 \quad \text{for all } x^\alpha \in S_0'$$
$$(x^\beta/x_1^{d_1})\, f_1 = 0 \quad \text{for all } x^\beta \in S_1'$$
$$\vdots$$
$$(x^\beta/x_n^{d_n})\, f_n = 0 \quad \text{for all } x^\beta \in S_n'.$$

Since the $x^\alpha\, f_0$ and $x^\beta/x_i^{d_i}\, f_i$ have total degree $\le d$, we can write these polynomials as linear combinations of the x^α and x^β. We will order these monomials so that the elements $x^\alpha \in S_0'$ come first, followed by the elements $x^\beta \in S_1' \cup \cdots \cup S_n'$. This gives a square matrix M_0 such that

$$M_0 \begin{pmatrix} x^{\alpha_1} \\ x^{\alpha_2} \\ \vdots \\ x^{\beta_1} \\ x^{\beta_2} \\ \vdots \end{pmatrix} = \begin{pmatrix} x^{\alpha_1}\, f_0 \\ x^{\alpha_2}\, f_0 \\ \vdots \\ x^{\beta_1}/x_1^{d_1}\, f_1 \\ x^{\beta_2}/x_1^{d_1}\, f_1 \\ \vdots \end{pmatrix},$$

where, in the column on the left, the first two elements of S_0' and the first two elements of S_1' are listed explicitly. This should make it clear what the whole column looks like. The situation is similar for the column on the right.

For $p \in \mathbf{V}(f_1, \ldots, f_n)$, we have $f_1(p) = \cdots = f_n(p) = 0$. Thus, evaluating the above equation at p yields

$$M_0 \begin{pmatrix} p^{\alpha_1} \\ p^{\alpha_2} \\ \vdots \\ p^{\beta_1} \\ p^{\beta_2} \\ \vdots \end{pmatrix} = \begin{pmatrix} p^{\alpha_1}\, f_0(p) \\ p^{\alpha_2}\, f_0(p) \\ \vdots \\ 0 \\ 0 \\ \vdots \end{pmatrix}.$$

To simplify notation, we let $\mathbf{p}^\alpha$ be the column vector $(p^{\alpha_1}, p^{\alpha_2}, \ldots)^T$ given by evaluating all monomials in S_0' at p (and T means transpose). Similarly, we let $\mathbf{p}^\beta$ be the column vector $(p^{\beta_1}, p^{\beta_2}, \ldots)^T$ given by evaluating all monomials in $S_1' \cup \cdots \cup S_n'$ at p. With this notation, we can rewrite the above equation more compactly as

$$(6.3) \qquad M_0 \begin{pmatrix} \mathbf{p}^\alpha \\ \mathbf{p}^\beta \end{pmatrix} = \begin{pmatrix} f_0(p)\, \mathbf{p}^\alpha \\ \mathbf{0} \end{pmatrix}.$$

The next step is to partition M_0 so that the rows and columns of M_0 corresponding to elements of S_0' lie in the upper left hand corner. This means writing M_0 in the form

$$M_0 = \begin{pmatrix} M_{00} & M_{01} \\ M_{10} & M_{11} \end{pmatrix},$$

where M_{00} is a $\mu \times \mu$ matrix for $\mu = d_1 \cdots d_n$, and M_{11} is also a square matrix. With this notation, (6.3) can be written

(6.4)
$$\begin{pmatrix} M_{00} & M_{01} \\ M_{10} & M_{11} \end{pmatrix} \begin{pmatrix} \mathbf{p}^\alpha \\ \mathbf{p}^\beta \end{pmatrix} = \begin{pmatrix} f_0(p)\,\mathbf{p}^\alpha \\ \mathbf{0} \end{pmatrix}.$$

By Lemma 4.4 of [Emi1], M_{11} is invertible for most choices of $f_1, \ldots, f_n$. Note that this condition is generic since it is given by $\det(M_{11}) \neq 0$ and $\det(M_{11})$ is a polynomial in the coefficients of the f_i. Hence, for generic $f_1, \ldots, f_n$, we can define the $\mu \times \mu$ matrix

(6.5)
$$\widetilde{M} = M_{00} - M_{01}M_{11}^{-1}M_{10}.$$

Note that the entries of $\widetilde{M}$ are polynomials in $u_0, \ldots, u_n$ since these variables only appear in M_{00} and M_{01}. If we multiply each side of (6.4) on the left by the matrix

$$\begin{pmatrix} I & -M_{01}M_{11}^{-1} \\ 0 & I \end{pmatrix},$$

then an easy computation gives

$$\begin{pmatrix} \widetilde{M} & 0 \\ M_{10} & M_{11} \end{pmatrix} \begin{pmatrix} \mathbf{p}^\alpha \\ \mathbf{p}^\beta \end{pmatrix} = \begin{pmatrix} f_0(p)\,\mathbf{p}^\alpha \\ \mathbf{0} \end{pmatrix}.$$

This implies

(6.6)
$$\widetilde{M}\,\mathbf{p}^\alpha = f_0(p)\,\mathbf{p}^\alpha,$$

so that for each $p \in \mathbf{V}(f_1, \ldots, f_n)$, $f_0(p)$ is an eigenvalue of $\widetilde{M}$ with $\mathbf{p}^\alpha$ as the corresponding eigenvector. Since $f_0 = u_0 + u_1x_1 + \cdots + u_nx_n$, the eigenvalues $f_0(p)$ are distinct for $p \in \mathbf{V}(f_1, \ldots, f_n)$. Standard linear algebra implies that the corresponding eigenvectors $\mathbf{p}^\alpha$ are linearly independent.

We can now prove the theorem. Write the elements of S_0' as $x^{\alpha_1}, \ldots, x^{\alpha_\mu}$, where as usual $\mu = d_1 \cdots d_n$, and recall that we need only show that the cosets $[x^{\alpha_1}], \ldots, [x^{\alpha_\mu}]$ are linearly independent in the quotient ring A. So suppose we have a linear relation among these cosets, say

$$c_1[x^{\alpha_1}] + \cdots + c_\mu[x^{\alpha_\mu}] = 0.$$

Evaluating this equation at $p \in \mathbf{V}(f_1, \ldots, f_n)$ makes sense by Exercise 12 of Chapter 2, §4 and implies that $c_1 p^{\alpha_1} + \cdots + c_\mu p^{\alpha_\mu} = 0$. In the generic case, $\mathbf{V}(f_1, \ldots, f_n)$ has $\mu = d_1 \cdots d_n$ points $p_1, \ldots, p_\mu$, which gives μ equations

$$c_1 p_1^{\alpha_1} + \cdots + c_\mu p_1^{\alpha_\mu} = 0$$

$$\vdots$$

$$c_1 p_\mu^{\alpha_1} + \cdots + c_\mu p_\mu^{\alpha_\mu} = 0.$$

In the matrix of these equations, the ith row is $(p_i^{\alpha_1}, \ldots, p_i^{\alpha_\mu})$, which in the notation used above, is the transpose of the column vector $\mathbf{p}_i^\alpha$ obtained by evaluating the monomials in S_0' at p_i. The discussion following (6.6) showed that the vectors $\mathbf{p}_i^\alpha$ are linearly independent. Thus the rows are linearly independent, so $c_1 = \cdots = c_\mu = 0$. We conclude that the cosets $[x^{\alpha_1}], \ldots, [x^{\alpha_\mu}]$ are linearly independent. $\qquad\qquad\qquad\qquad\square$

Now that we know a basis for the quotient ring A, our next task it to find the matrix of the multiplication map m_{f_0} relative to this basis. Fortunately, this is easy since we already know the matrix!

(6.7) Theorem. *Let $f_1, \ldots, f_n$ be generic polynomials of total degrees $d_1, \ldots, d_n$, and let $f_0 = u_0 + u_1 x_1 + \cdots + u_n x_n$. Using the basis of $A = \mathbb{C}[x_1, \ldots, x_n]/\langle f_1, \ldots, f_n \rangle$ from Theorem (6.2), the matrix of the multiplication map m_{f_0} is the* **transpose** *of the matrix*

$$\widetilde{M} = M_{00} - M_{01} M_{11}^{-1} M_{10}$$

from (6.5).

PROOF. Let $M_{f_0} = (m_{ij})$ be the matrix of m_{f_0} relative to the basis $[x^{\alpha_1}], \ldots, [x^{\alpha_\mu}]$ of A from Theorem (6.2), where $\mu = d_1 \cdots d_n$. The proof of Proposition (4.7) of Chapter 2 shows that for $p \in \mathbf{V}(f_1, \ldots, f_n)$, we have

$$f_0(p)(p^{\alpha_1}, \ldots, p^{\alpha_\mu}) = (p^{\alpha_1}, \ldots, p^{\alpha_\mu}) M_{f_0}.$$

Letting $\mathbf{p}^\alpha$ denote the column vector $(p^{\alpha_1}, \ldots, p^{\alpha_\mu})^T$ as in the previous proof, we can take the transpose of each side of this equation to obtain

$$f_0(p)\, \mathbf{p}^\alpha = \left(f_0(p)(p^{\alpha_1}, \ldots, p^{\alpha_\mu})\right)^T$$
$$= \left((p^{\alpha_1}, \ldots, p^{\alpha_\mu}) M_{f_0}\right)^T$$
$$= (M_{f_0})^T\, \mathbf{p}^\alpha,$$

where $(M_{f_0})^T$ is the transpose of M_{f_0}. Comparing this to (6.6), we get

$$(M_{f_0})^T\, \mathbf{p}^\alpha = \widetilde{M}\, \mathbf{p}^\alpha$$

for all $p \in \mathbf{V}(f_1, \ldots, f_n)$. Since $f_1, \ldots, f_n$ are generic, we have μ points $p \in \mathbf{V}(f_1, \ldots, f_n)$, and the proof of Theorem (6.2) shows that the corresponding eigenvectors $\mathbf{p}^\alpha$ are linearly independent. This implies $(M_{f_0})^T = \widetilde{M}$, and then $M_{f_0} = \widetilde{M}^T$ follows easily. $\qquad\qquad\square$

Since $f_0 = u_0 + u_1 x_1 + \cdots + u_n x_n$, Corollary (4.3) of Chapter 2 implies

$$M_{f_0} = u_0\, I + u_1\, M_{x_1} + \cdots + u_n\, M_{x_n},$$

where M_{x_i} is the matrix of m_{x_i} relative to the basis of Theorem (6.2). By Theorem (6.7), it follows that if we write

(6.8) $$\widetilde{M} = u_0\, I + u_1\, \widetilde{M}_1 + \cdots + u_n\, \widetilde{M}_n,$$

where each $\widetilde{M_i}$ has constant entries, then $M_{f_0} = \widetilde{M}^T$ implies that $M_{x_i} = (\widetilde{M_i})^T$ for all i. Thus $\widetilde{M}$ *simultaneously* computes the matrices of the n multiplication maps $m_{x_1}, \ldots, m_{x_n}$.

Exercise 1. For the equations

$$f_1 = x_1^2 + x_2^2 - 10 = 0$$
$$f_2 = x_1^2 + x_1 x_2 + 2x_2^2 - 16 = 0$$

(this is the affine version of Exercise 2 of §5), show that $\widetilde{M}$ is the matrix

$$\widetilde{M} = \begin{pmatrix} u_0 & u_1 & u_2 & 0 \\ 4u_1 & u_0 & 0 & u_1 + u_2 \\ 6u_2 & 0 & u_0 & u_1 - u_2 \\ 0 & 3u_1 + 3u_2 & 2u_1 - 2u_2 & u_0 \end{pmatrix}.$$

Use this to determine the matrices M_{x_1} and M_{x_2}. What is the basis of $\mathbb{C}[x_1, x_2]/\langle f_1, f_2 \rangle$ in this case? Hint: The matrix M_0 of Exercise 2 of §5 is already partitioned into the appropriate submatrices.

Now that we have the matrices M_{x_i}, we can find the x_i-coordinates of the solutions of (5.3) using the eigenvalue methods mentioned in Chapter 2 (see especially the discussion following Corollary (4.6)). This still leaves the problem of finding how the coordinates match up. We will follow Chapter 2 and show how the left eigenvectors of M_{f_0}, or equivalently, the right eigenvectors of $\widetilde{M} = (M_{f_0})^T$, give the solutions of our equations.

Since $\widetilde{M}$ involves the variables $u_0, \ldots, u_n$, we need to specialize them before we can use numerical methods for finding eigenvectors. Let

$$f_0' = c_0 + c_1 x_1 + \cdots + c_n x_n,$$

where $c_0, \ldots, c_n$ are constants chosen so that the values $f_0'(p)$ are distinct for $p \in \mathbf{V}(f_1, \ldots, f_n)$. In practice, this can be achieved by making a random choice of $c_0, \ldots, c_n$. If we let $\widetilde{M}'$ be the matrix obtained from $\widetilde{M}$ by letting $u_i = c_i$, then (6.6) shows that $\mathbf{p}^\alpha$ is a right eigenvector for $\widetilde{M}'$ with eigenvalue $f_0'(p)$. Since we have $\mu = d_1 \cdots d_n$ distinct eigenvalues in a vector space of the same dimension, the corresponding eigenspaces all have dimension 1.

To find the solutions, suppose that we've used a standard numerical method to find an eigenvector v of $\widetilde{M}'$. Since the eigenspaces all have dimension 1, it follows that $v = c\,\mathbf{p}^\alpha$ for some solution $p \in \mathbf{V}(f_1, \ldots, f_n)$ and nonzero constant c. This means that whenever x^α is a monomial in S_0', the corresponding coordinate of v is cp^α. The following exercise shows how to reconstruct p from the coordinates of the eigenvector v.

Exercise 2. As above, let $p = (a_1, \ldots, a_n) \in \mathbf{V}(f_1, \ldots, f_n)$ and let v be an eigenvector of $\widetilde{M}'$ with eigenvalue $f_0'(p)$. This exercise will explain how to recover p from v when $d_1, \ldots, d_n$ are all > 1, and Exercise 5 at the end of the section will explore what happens when some of the degrees equal 1.

a. Show that $1, x_1, \ldots, x_n \in S_0'$, and conclude that for some $c \neq 0$, the numbers $c, ca_1, \ldots, ca_n$ are among the coordinates of v.

b. Prove that a_j can be computed from the coordinates of v by the formula

$$a_j = \frac{ca_j}{c} \quad \text{for } j = 1, \ldots, n.$$

This shows that the solution p can be easily found using ratios of certain coordinates of the eigenvector v.

Exercise 3. For the equations $f_1 = f_2 = 0$ of Exercise 1, consider the matrix $\widetilde{M}'$ coming from $(u_0, u_1, u_2, u_3) = (0, 1, 0, 0)$. In the notation of (6.8), this means $\widetilde{M}' = \widetilde{M}_1 = (M_{x_1})^T$. Compute the eigenvectors of this matrix and use Exercise 2 to determine the solutions of $f_1 = f_2 = 0$.

While the right eigenvectors of $\widetilde{M}$ relate to the solutions of $f_1 = \cdots = f_n = 0$, the left eigenvectors give a nice answer to the *interpolation problem*. This was worked out in detail in Exercise 17 of Chapter 2, §4, which applies without change to the case at hand. See Exercise 6 at the end of this section for an example.

Eigenvalue methods can also be applied to the hidden variable resultants discussed earlier in this section. We will discuss this very briefly. In Proposition (5.15), we showed that the x_n-coordinates of the solutions of the equations $f_1 = \cdots = f_n = 0$ could be found using the resultant $\mathrm{Res}^{x_n}_{d_1, \ldots, d_n}(f_1, \ldots, f_n)$ obtained by regarding x_n as a constant. As we learned in (5.20),

$$\mathrm{Res}^{x_n}_{d_1, \ldots, d_n}(f_1, \ldots, f_n) = \pm \frac{\widehat{D}_0}{\widehat{D}_0'},$$

and if $\widehat{M}_0$ is the corresponding matrix (so that $\widehat{D}_0 = \det(\widehat{M}_0)$), one could ask about the eigenvalues and eigenvectors of $\widehat{M}_0$. It turns out that this is not quite the right question to ask. Rather, since $\widehat{M}_0$ depends on the variable x_n, we write the matrix as

(6.9) $$\widehat{M}_0 = A_0 + x_n A_1 + \cdots + x_n^l A_l,$$

where each A_i has constant entries and $A_l \neq 0$. Suppose that $\widehat{M}_0$ and the A_i are $m \times m$ matrices. If A_l is invertible, then we can define the *generalized companion matrix*

$$C = \begin{pmatrix} 0 & I_m & 0 & \cdots & 0 \\ 0 & 0 & I_m & \cdots & 0 \\ \vdots & \vdots & \vdots & \ddots & \vdots \\ 0 & 0 & 0 & \cdots & I_m \\ -A_l^{-1}A_0 & -A_l^{-1}A_1 & -A_l^{-1}A_2 & \cdots & -A_l^{-1}A_{l-1} \end{pmatrix},$$

where I_m is the $m \times m$ identity matrix. Then the correct question to pose concerns the eigenvalues and eigenvectors of C. One can show that the eigenvalues of the generalized companion matrix are precisely the roots of the polynomial $\widehat{D}_0 = \det(\widehat{M}_0)$, and the corresponding eigenvectors have a nice interpretation as well. Further details of this technique can be found in [Man2] and [Man3].

Finally, we should say a few words about how eigenvalue and eigenvector methods behave in the non-generic case. As in the discussion of u-resultants in §5, there are many things which can go wrong. All of the problems listed earlier are still present when dealing with eigenvalues and eigenvectors, and there are two new difficulties which can occur:

- In working with the matrix M_0 as in the proof of Theorem (6.2), it can happen that M_{11} is not invertible, so that $\widetilde{M} = M_{00} - M_{01}M_{11}^{-1}M_{10}$ doesn't make sense.
- In working with the matrix $\widehat{M}_0$ as in (6.9), it can happen that the leading term A_l is not invertible, so that the generalized companion matrix C doesn't make sense.

Techniques for avoiding both of these problems are described in [Emi2], [Man1], [Man2], and [Man3].

Exercise 4. Express the 6×6 matrix of part c of Exercise 5 of §5 in the form $A_0 + x_3 A_1 + x_3^2 A_2$ and show that A_2 is *not* invertible.

The idea of solving equations by a combination of eigenvalue/eigenvector methods and resultants goes back to the work of Auzinger and Stetter [AS]. This has now become an active area of research, not only for the resultants discussed here (see [BMP], [Man3], [Mou1] and [Ste], for example) but also for the *sparse resultants* to be introduced in Chapter 7. Also, we will say more about multiplication maps in §2 of Chapter 4.

ADDITIONAL EXERCISES FOR §6

Exercise 5. This exercise will explain how to recover the solution $p = (a_1, \ldots, a_n)$ from an eigenvector v of the matrix $\widetilde{M}'$ in the case when some of the degrees $d_1, \ldots, d_n$ are equal to 1. Suppose for instance that $d_i = 1$. This means that $x_i \notin S_0'$, so that the ith coordinate a_i of the solution p doesn't appear in the eigenvector $\mathbf{p}^\alpha$. The idea is that the matrix M_{x_i} (which we know by Theorem (6.7)) has all of the information we need. Let $c_1, \ldots, c_\mu$ be the entries of the column of M_{x_i} corresponding to $1 \in S_0'$.
a. Prove that $[x_i] = c_1[x^{\alpha_1}] + \cdots + c_\mu[x^{\alpha_\mu}]$ in A, where $S_0' = \{x^{\alpha_1}, \ldots, x^{\alpha_\mu}\}$.
b. Prove that $a_i = c_1 p^{\alpha_1} + \cdots + c_\mu p^{\alpha_\mu}$.

It follows that if we have an eigenvector v as in the discussion preceding Exercise 2, it is now straightforward to recover *all* coordinates of the solution p.

Exercise 6. The equations $f_1 = f_2 = 0$ from Exercise 1 have solutions p_1, p_2, p_3, p_4 (they are listed in projective form in Exercise 2 of §5). Apply Exercise 17 of Chapter 2, §4, to find the polynomials g_1, g_2, g_3, g_4 such that $g_i(p_j) = 1$ if $i = j$ and 0 otherwise. Then use this to write down explicitly a polynomial h which takes preassigned values $\lambda_1, \lambda_2, \lambda_3, \lambda_4$ at the points p_1, p_2, p_3, p_4. Hint: Since the x_1-coordinates are distinct, it suffices to find the eigenvectors of M_{x_1}. Exercise 1 will be useful.

Chapter 4

Computation in Local Rings

Many questions in algebraic geometry involve a study of *local properties* of varieties, that is, properties of a single point, or of a suitably small neighborhood of a point. For example, in analyzing $\mathbf{V}(I)$ for a zero-dimensional ideal $I \subset k[x_1, \ldots, x_n]$, even when k is algebraically closed, it sometimes happens that $\mathbf{V}(I)$ contains fewer distinct points than the dimension $d = \dim k[x_1, \ldots, x_n]/I$. In this situation, thinking back to the consequences of unique factorization for polynomials in one variable, it is natural to ask whether there is an algebraic *multiplicity* that can be computed locally at each point in $\mathbf{V}(I)$, with the property that the sum of the multiplicities is equal to d. Similarly in the study of *singularities* of varieties, one major object of study is local invariants of singular points. These are used to distinguish different types of singularities and study their local structure. In §1 of this chapter, we will introduce the algebra of *local rings* which is useful for both these types of questions. Multiplicities and some invariants of singularities will be introduced in §2. In §3 and §4, we will develop algorithmic techniques for computation in local rings parallel to the theory of Gröbner bases in polynomial rings. Applications of these techniques are given in §5.

In this chapter, we will often assume that k is an algebraically closed field containing $\mathbb{Q}$. The results of Chapters 2 and 3 are valid for such fields.

§1 Local Rings

One way to study properties of a variety V is to study functions on the variety. The elements of the ring $k[x_1, \ldots, x_n]/\mathbf{I}(V)$ can be thought of as the polynomial functions on V. Near a particular point $p \in V$ we can also consider rational functions defined at the point, power series convergent at the point, or even formal series centered at the point. Considering the collections of each of these types of functions in turn leads us to new rings that strictly contain the ring of polynomials. In a sense which we shall make precise as we go along, consideration of these larger rings corresponds to looking at smaller neighborhoods of points. We will begin with the following example. Let $V = k^n$, and let $p = (0, \dots, 0)$ be the origin. The single point set $\{p\}$ is a variety, and $\mathbf{I}(\{p\}) = \langle x_1, \dots, x_n \rangle \subset k[x_1, \dots, x_n]$. Furthermore, a rational function f/g has a well-defined value at p provided $g(p) \neq 0$.

(1.1) Definition. We denote by $k[x_1, \dots, x_n]_{\langle x_1, \dots, x_n \rangle}$ the collection of all rational functions f/g of $x_1, \dots, x_n$ with $g(p) \neq 0$, where $p = (0, \dots, 0)$.

The main properties of $k[x_1, \dots, x_n]_{\langle x_1, \dots, x_n \rangle}$ are as follows.

(1.2) Proposition. *Let $R = k[x_1, \dots, x_n]_{\langle x_1, \dots, x_n \rangle}$. Then*
a. *R is a subring of the field of rational functions $k(x_1, \dots, x_n)$ containing $k[x_1, \dots, x_n]$.*
b. *Let $M = \langle x_1, \dots, x_n \rangle \subset R$ (the ideal generated by $x_1, \dots, x_n$ in R). Then every element in $R \setminus M$ is a unit in R (i.e., has a multiplicative inverse in R).*
c. *M is a maximal ideal in R, and R has no other maximal ideals.*

PROOF. As above, let $p = (0, \dots, 0)$. Part a follows easily since R is closed under sums and products in $k(x_1, \dots, x_n)$. For instance, if f_1/g_1 and f_2/g_2 are two rational functions with $g_1(p), g_2(p) \neq 0$, then

$$f_1/g_1 + f_2/g_2 = (f_1 g_2 + f_2 g_1)/(g_1 g_2).$$

Since $g_1(p) \neq 0$ and $g_2(p) \neq 0$, $g_1(p) \cdot g_2(p) \neq 0$. Hence the sum is an element of R. A similar argument shows that the product $(f_1/g_1) \cdot (f_2/g_2)$ is in R. Finally, since $f = f/1$ is in R for all $f \in k[x_1, \dots, x_n]$, the polynomial ring is contained in R.

For part b, we will use the fact that the elements in $M = \langle x_1, \dots, x_n \rangle$ are exactly the rational functions $f/g \in R$ such that $f(p) = 0$. Hence if $f/g \notin M$, then $f(p) \neq 0$ and $g(p) \neq 0$, and g/f is a multiplicative inverse for f/g in R.

Finally, for part c, if $N \neq M$ is an ideal in R with $M \subset N \subset R$, then N must contain an element f/g in the complement of M. By part b, f/g is a unit in R, so $1 = (f/g)(g/f) \in N$, and hence $N = R$. Therefore M is maximal. M is the only maximal ideal in R, because it also follows from part b that every proper ideal $I \subset R$ is contained in M. □

Exercise 1. In this exercise you will show that if $p = (a_1, \dots, a_n) \in k^n$ is any point and

$$R = \{f/g : f, g \in k[x_1, \dots, x_n], g(p) \neq 0\},$$

then we have the following statements parallel to Proposition (1.2).
a. R is a subring of the field of rational functions $k(x_1, \dots, x_n)$.
b. Let M be the ideal generated by $x_1 - a_1, \dots, x_n - a_n$ in R. Then every element in $R \setminus M$ is a unit in R (i.e., has a multiplicative inverse in R).

c. M is a maximal ideal in R, and R has no other maximal ideals.

An alternative notation for the ring R in Exercise 1 is

$$R = k[x_1, \ldots, x_n]_{\langle x_1 - a_1, \ldots, x_n - a_n \rangle},$$

where $\langle x_1 - a_1, \ldots, x_n - a_n \rangle$ is the ideal $\mathbf{I}(\{p\})$ in $k[x_1, \ldots, x_n]$, and in R we allow denominators that are *not* elements of this ideal.

In the following discussion, the term *ring* will always mean a commutative ring with identity. Every ring has maximal ideals. As we will see, the rings that give *local* information are the ones with the property given by part c of Proposition (1.2) above.

(1.3) Definition. A *local ring* is a ring that has exactly one maximal ideal.

The idea of the argument used in the proof of part c of the proposition also gives one general criterion for a ring to be a local ring.

(1.4) Proposition. *A ring R with a proper ideal $M \subset R$ is a local ring if every element of $R \setminus M$ is a unit in R.*

PROOF. If every element of $R \setminus M$ is a unit in R, the unique maximal ideal is M. Exercise 5 below asks you to finish the proof.   □

Definition (1.1) above is actually a special case of a general procedure called *localization* that can be used to construct many additional examples of local rings. See Exercise 8 below. An even more general construction of rings of fractions is given in Exercise 9. We will need to use that construction in §3 and §4.

We also obtain important examples of local rings by considering functions more general than rational functions. One way such functions arise is as follows. When studying a curve or, more generally, a variety near a point, one often tries to *parametrize* the variety near the point. For example, the curve

$$x^2 + 2x + y^2 = 0$$

is a circle of radius 1 centered at the point $(-1, 0)$. To study this curve near the origin, we might use parametrizations of several different types.

Exercise 2. Show that one parametrization of the circle near the origin is given by

$$x = \frac{-2t^2}{1 + t^2}, \qquad y = \frac{2t}{1 + t^2}.$$

Note that both components are elements of the local ring $k[t]_{\langle t \rangle}$.

In this case, we might also use the parametrization in terms of trigonometric functions:

$$x = -1 + \cos t, \qquad y = \sin t.$$

The functions $\sin t$ and $\cos t$ are not polynomials or rational functions, but recall from elementary calculus that they can be expressed as convergent power series in t:

$$\sin t = \sum_{k=0}^{\infty}(-1)^k t^{2k+1}/(2k+1)!$$

$$\cos t = \sum_{k=0}^{\infty}(-1)^k t^{2k}/(2k)!\ .$$

In this case parametrizing leads us to consider functions more general than polynomials or rational functions.

If $k = \mathbb{C}$ or $k = \mathbb{R}$, then we can consider the set of convergent power series in n variables (expanding about the origin)

$$(1.5) \qquad k\{x_1, \ldots, x_n\} = \{\sum_{\alpha \in \mathbb{Z}_{\geq 0}^n} c_\alpha x^\alpha : c_\alpha \in k \text{ and the series}$$

$$\text{converges in some neighborhood of } 0 \in k^n\}.$$

With the usual notion of addition and multiplication, this set is a ring (we leave the verification to the reader; see Exercise 3). In fact, it is not difficult to see that $k\{x_1, \ldots, x_n\}$ is also a local ring with maximal ideal generated by $x_1, \ldots, x_n$.

No matter what field k is, we can also consider the set $k[[x_1, \ldots, x_n]]$ of formal power series

$$(1.6) \qquad k[[x_1, \ldots, x_n]] = \{\sum_{\alpha \in \mathbb{Z}_{\geq 0}^n} c_\alpha x^\alpha : c_\alpha \in k\},$$

where, now, we waive the condition that the series need converge. Algebraically, a formal power series is a perfectly well defined object and can easily be manipulated—one must, however, give up the notion of evaluating it at any point of k^n other than the origin. As a result, a formal power series defines a *function* only in a rather limited sense. But in any case we can define addition and multiplication of formal series in the obvious way and this makes $k[[x_1, \ldots, x_n]]$ into a ring (see Exercise 3). Formal power series are also useful in constructing parametrizations of varieties over arbitrary fields (see Exercise 7 below).

At the beginning of the section, we commented that the three rings $k[x_1, \ldots, x_n]_{\langle x_1, \ldots, x_n \rangle}$, $k\{x_1, \ldots, x_n\}$, and $k[[x_1, \ldots, x_n]]$ correspond to looking at smaller and smaller neighborhoods of the origin. Let us make this more precise. An element $f/g \in k[x_1, \ldots, x_n]_{\langle x_1, \ldots, x_n \rangle}$ is defined not just at the origin but at every point in the complement of $\mathbf{V}(g)$. The domain of convergence of a power series can be a much smaller set than the complement of a variety. For instance, the geometric series $1 + x + x^2 + \cdots$

converges to the sum $1/(1 - x) \in k[x]_{\langle x \rangle}$ only on the set of x with $|x| < 1$ in $k = \mathbb{R}$ or $\mathbb{C}$. A formal series in $k[[x_1, \ldots, x_n]]$ is only guaranteed to converge at the origin. Nevertheless, both $k\{x_1, \ldots, x_n\}$ and $k[[x_1, \ldots, x_n]]$ share the key algebraic property of $k[x_1, \ldots, x_n]_{\langle x_1, \ldots, x_n \rangle}$.

(1.7) Proposition. $k[[x_1, \ldots, x_n]]$ *is a local ring. If* $k = \mathbb{R}$ *or* $k = \mathbb{C}$ *then* $k\{x_1, \ldots, x_n\}$ *is also a local ring.*

PROOF. To show that $k[[x_1, \ldots, x_n]]$ is a local ring, consider the ideal $M = \langle x_1, \ldots, x_n \rangle \subset k[[x_1, \ldots, x_n]]$ generated by $x_1, \ldots, x_n$. If $f \notin M$, then $f = c_0 + g$ with $c_0 \neq 0$, and $g \in M$. Using the formal geometric series expansion

$$\frac{1}{1 + t} = 1 - t + t^2 + \cdots + (-1)^n t^n + \cdots,$$

we see that

$$\frac{1}{c_0 + g} = \frac{1}{c_0(1 + g/c_0)}$$
$$= (1/c_0)\left(1 - g/c_0 + (g/c_0)^2 + \cdots\right).$$

In Exercise 4 below, you will show that this expansion makes sense as an element of $k[[x_1, \ldots, x_n]]$. Hence f has a multiplicative inverse in $k[[x_1, \ldots, x_n]]$. Since this is true for every $f \notin M$, Proposition (1.4) implies that $k[[x_1, \ldots, x_n]]$ is a local ring.

To show that $k\{x_1, \ldots, x_n\}$ is also a local ring, we only need to show that the formal series expansion for $1/(c_0 + g)$ gives a convergent series. See Exercise 4.   ◻

All three types of local rings share other key algebraic properties with rings of polynomials. See the exercises in §4. By considering the power series expansion of a rational function defined at the origin, as in the proof above, we have $k[x_1, \ldots, x_n]_{\langle x_1, \ldots, x_n \rangle} \subset k[[x_1, \ldots, x_n]]$. In the case $k = \mathbb{R}$ or $\mathbb{C}$, we also have inclusions:

$$k[x_1, \ldots, x_n]_{\langle x_1, \ldots, x_n \rangle} \subset k\{x_1, \ldots, x_n\} \subset k[[x_1, \ldots, x_n]].$$

In general, we would like to be able to do operations on ideals in these rings in much the same way that we can carry out operations on ideals in a polynomial ring. For instance, we would like to be able to settle the ideal membership question, to form intersections of ideals, compute quotients, compute syzygies on a collection of elements, and the like. We will return to these questions in §3 and §4.

ADDITIONAL EXERCISES FOR §1

Exercise 3. The product operations in $k[[x_1, \ldots, x_n]]$ and $k\{x_1, \ldots, x_n\}$ can be described in the following fashion. Grouping terms by total degree, rewrite each power series

$$f(x) = \sum_{\alpha \in \mathbb{Z}_{\geq 0}^n} c_\alpha x^\alpha$$

as $\sum_{n \geq 0} f_n(x)$, where

$$f_n(x) = \sum_{\substack{\alpha \in \mathbb{Z}_{\geq 0}^n \\ |\alpha|=n}} c_\alpha x^\alpha$$

is a homogeneous polynomial of degree n. The product of two series $f(x)$ and $g(x)$ is the series $h(x)$ for which

$$h_n = f_n g_0 + f_{n-1} g_1 + \cdots + f_0 g_n.$$

a. Show that with this product and the obvious sum, $k[[x_1, \ldots, x_n]]$ is a (commutative) ring (with identity).
b. Now assume $k = \mathbb{R}$ or $k = \mathbb{C}$, and suppose $f, g \in k\{x_1, \ldots, x_n\}$. From part a, we know that sums and products of power series give other formal series. Show that if f and g are both convergent on some neighborhood U of $(0, \ldots, 0)$, then $f + g$ and $f \cdot g$ are also convergent on U.

Exercise 4. Let $h \in \langle x_1, \ldots, x_n \rangle \subset k[[x_1, \ldots, x_n]]$.
a. Show that the formal geometric series expansion

$$\frac{1}{1+h} = 1 - h + h^2 - h^3 + \cdots$$

gives a well-defined element of $k[[x_1, \ldots, x_n]]$. (What are the homogeneous components of the series on the right?)
b. Show that if h is convergent on some neighborhood of the origin, then the expansion in part a is also convergent on some (generally smaller) neighborhood of the origin. (Recall that

$$\frac{1}{1+t} = 1 - t + t^2 - t^3 + \cdots$$

is convergent only for t satisfying $|t| < 1$.)

Exercise 5. Give a complete proof for Proposition (1.4).

Exercise 6. Let F be a field. A *discrete valuation* of F is an onto mapping $v : F \setminus \{0\} \to \mathbb{Z}$ with the properties that
1. $v(x + y) \geq \min\{v(x), v(y)\}$, and
2. $v(xy) = v(x) + v(y)$.

The subset of F consisting of all elements x satisfying $v(x) \geq 0$, together with 0, is called the *valuation ring* of v.

a. Show that the valuation ring of a discrete valuation is a local ring. Hint: Use Proposition (1.4).

b. For example, let $F = k(x)$ (the rational function field in one variable), and let f be an irreducible polynomial in $k[x] \subset F$. If $g \in k(x)$, then by unique factorization in $k[x]$, there is a unique expression for g of the form $g = f^a \cdot n/d$, where $a \in \mathbb{Z}$, and $n, d \in k[x]$ are not divisible by f. Let $v(g) = a \in \mathbb{Z}$. Show that v defines a discrete valuation on $k(x)$. Identify the maximal ideal of the valuation ring.

c. Let $F = \mathbb{Q}$, and let p be a prime integer. Show that if $g \in \mathbb{Q}$, then by unique factorization in $\mathbb{Z}$, there is a unique expression for g of the form $g = p^a \cdot n/d$, where $a \in \mathbb{Z}$, and $n, d \in \mathbb{Z}$ are not divisible by p. Let $v(g) = a \in \mathbb{Z}$. Show that v defines a discrete valuation on $\mathbb{Q}$. Identify the maximal ideal of this valuation ring.

Exercise 7. (A Formal Implicit Function Theorem) Let $f(x, y) \in k[x, y]$ be a polynomial of the form

$$f(x, y) = y^n + c_1(x)y^{n-1} + \cdots + c_{n-1}(x)y + c_n(x),$$

where $c_i(x) \in k[x]$. Assume that $f(0, y) = 0$ has n *distinct* roots $a_i \in k$.
a. Starting from $y_i^{(0)}(x) = a_i$, show that there is a unique $a_{i1} \in k$ such that $y_i^{(1)}(x) = a_i + a_{i1}x$ satisfies

$$f(x, y_i^{(1)}(x)) \equiv 0 \bmod \langle x^2 \rangle.$$

b. Show that if we have a polynomial $y_i^{(\ell)}(x) = a_i + a_{i1}x + \cdots + a_{i\ell}x^\ell$, that satisfies

$$f(x, y_i^{(\ell)}(x)) \equiv 0 \bmod \langle x^{\ell+1} \rangle,$$

then there exists a unique $a_{i,\ell+1} \in k$ such that

$$y_i^{(\ell+1)}(x) = y_i^{(\ell)}(x) + a_{i,\ell+1}x^{\ell+1}$$

satisfies

$$f(x, y_i^{(\ell+1)}(x)) \equiv 0 \bmod \langle x^{\ell+2} \rangle.$$

c. From parts a and b, deduce that there is a unique power series $y_i(x) \in k[[x]]$ that satisfies $f(x, y_i(x)) = 0$ and $y_i(0) = a_i$.

Geometrically, this gives a formal series parametrization of the branch of the curve $f(x, y)$ passing through $(0, a_i)$: $(x, y_i(x))$. It also follows that $f(x, y)$ *factors* in the ring $k[[x]][y]$:

$$f(x, y) = \prod_{i=1}^{n}(y - y_i(x)).$$

Exercise 8. Let R be an integral domain (that is, a ring with no zero-divisors), and let $P \subset R$ be a prime ideal (see Exercise 8 of Chapter 1, §1 for the definition, which is the same in any ring R). The *localization of* R

with respect to P, denoted R_P, is a new ring containing R, in which every element in R not in the specified prime ideal P becomes a unit. We define

$$R_P = \{r/s : r, s \in R, s \notin P\},$$

so that R_P is a subset of the field of fractions of R.

a. Using Proposition (1.4), show that R_P is a local ring, with maximal ideal $M = \{p/s : p \in P, s \notin P\}$.

b. Show that every ideal in R_P has the form $I_P = \{a/s : a \in I, s \notin P\}$, where I is an ideal of R contained in P.

Exercise 9. The construction of R_P in Exercise 8 can be generalized in the following way. If R is any ring, and $S \subset R$ is a set which is closed under multiplication (that is, $s_1, s_2 \in S$ implies $s_1 \cdot s_2 \in S$), then we can form "fractions" a/s, with $a \in R$, $s \in S$. We will say two fractions a/s and b/t are *equivalent* if there is some $u \in S$ such that $u(at - bs) = 0$ in R. We call the collection of equivalence classes for this relation $S^{-1}R$.

a. Show that forming sums and products as with ordinary fractions gives *well-defined* operations on $S^{-1}R$.

b. Show that $S^{-1}R$ is a ring under these sum and product operations.

c. If R is any ring (not necessarily an integral domain) and $P \subset R$ is a prime ideal, show that $S = R \setminus P$ is closed under multiplication. The resulting ring of fractions $S^{-1}R$ is also denoted R_P (as in Exercise 8).

Exercise 10. Let $R = k[x_1, \ldots, x_n]$ and $I = \langle f_1, \ldots, f_m \rangle$ be an ideal in R. Let $M = \langle x_1, \ldots, x_n \rangle$ be the maximal ideal of polynomials vanishing at the origin and suppose that $I \subset M$.

a. Show that the ideal M/I generated by the cosets of $x_1, \ldots, x_n$ in R/I is a prime ideal.

b. Let IR_M denote the ideal generated by the f_i in the ring R_M, and let $(R/I)_{M/I}$ be constructed as in Exercise 8. Let $r/s \in R_M$, let $[r], [s]$ denote the cosets of the numerator and denominator in R/I, and let $[r/s]$ denote the coset of the fraction in R_M/IR_M. Show that the mapping

$$\varphi : R_M/IR_M \to (R/I)_{M/I}$$
$$[r/s] \mapsto [r]/[s]$$

is well defined and gives an isomorphism of rings.

Exercise 11. Let $R = k[x_1, \ldots, x_n]_{\langle x_1, \ldots, x_n \rangle}$. Show that every ideal $I \subset R$ has a generating set consisting of polynomials $f_1, \ldots, f_s \in k[x_1, \vdots \ldots, x_n]$.

Exercise 12. (Another interpretation of $k\{x_1, \ldots, x_n\}$) Let $k = \mathbb{R}$ or $\mathbb{C}$ and let $U \subset k^n$ be open. A function $f : U \to k$ is *analytic* if it can be represented by a power series with coefficients in k at each point of U. One can prove that every element of $k\{x_1, \ldots, x_n\}$ defines an analytic function on some neighborhood of the origin. We can describe $k\{x_1, \ldots, x_n\}$ in terms of analytic functions as follows. Two analytic functions, each defined on some neighborhood of the origin, are *equivalent* if there is some (smaller) neighborhood of the origin on which they are equal. An equivalence class of analytic functions with respect to this relation is called a *germ* of an analytic function (at the origin).

a. Show that the set of germs of analytic functions at the origin is a ring under the usual sum and product of functions.

b. Show that this ring can be identified with $k\{x_1, \ldots, x_n\}$ and that the maximal ideal is precisely the set of germs of analytic functions which vanish at the origin.

c. Consider the function $f : \mathbb{R} \to \mathbb{R}$ defined by

$$f(x) = \begin{cases} e^{-1/x^2} & \text{if } x > 0 \\ 0 & \text{if } x \le 0. \end{cases}$$

Show that f is C^∞ on $\mathbb{R}$, and construct its Taylor series, expanding at $a = 0$. Does the Taylor series converge to $f(x)$ for all x in some neighborhood of $0 \in \mathbb{R}$?

If $k = \mathbb{R}$, the example given in part c shows that the ring of germs of infinitely differentiable real functions is not equal to $k\{x_1, \ldots, x_n\}$. On the other hand, it is a basic theorem of complex analysis that a complex differentiable function is analytic.

§2 Multiplicities and Milnor Numbers

In this section we will see how local rings can be used to assign local multiplicities at the points in $\mathbf{V}(I)$ for a zero-dimensional ideal I. We will also use local rings to define the Milnor and Tjurina numbers of an isolated singular point of a hypersurface.

To see what the issues are, let us turn to one of the most frequent computations that one is called to do in a local ring, that of computing the dimension of the quotient ring by a zero-dimensional ideal. In Chapter 2, we learned how to compute the dimension of $k[x_1, \ldots, x_n]/I$ when I is a zero-dimensional polynomial ideal. Recall how this works. For any monomial order, we have

$$\dim k[x_1, \ldots, x_n]/I = \dim k[x_1, \ldots, x_n]/\langle \mathrm{LT}(I) \rangle,$$

and the latter is just the number of monomials x^α such that $x^\alpha \notin \langle \mathrm{LT}(I) \rangle$. For example, if

$$I = \langle x^2 + x^3, y^2 \rangle \subset k[x, y],$$

then using the *lex* order with $y > x$ for instance, the given generators form a Gröbner basis for I. So

$$\dim k[x, y]/I = \dim k[x, y]/\langle \mathrm{LT}(I) \rangle = \dim k[x, y]/\langle x^3, y^2 \rangle = 6.$$

The rightmost equality follows because the cosets of $1, x, x^2, y, xy, x^2y$ form a vector space basis of $k[x, y]/\langle x^3, y^2 \rangle$. The results of Chapter 2 show that there are at most six common zeros of $x^2 + x^3$ and y^2 in k^2. In fact, from the simple form of the generators of I we see there are precisely *two* distinct points in $\mathbf{V}(I)$: $(-1, 0)$ and $(0, 0)$.

To define the local multiplicity of a solution of a system of equations, we use a local ring instead of the polynomial ring, but the idea is much the same as above. We will need the following notation. If I is an ideal in $k[x_1, \ldots, x_n]$, then we sometimes denote by $Ik[x_1, \ldots, x_n]_{\langle x_1, \ldots, x_n \rangle}$ the ideal generated by I in the larger ring $k[x_1, \ldots, x_n]_{\langle x_1, \ldots, x_n \rangle}$.

(2.1) Definition. Let I be a zero-dimensional ideal in $k[x_1, \ldots, x_n]$, so that $\mathbf{V}(I)$ consists of finitely many points in k^n, and assume that $(0, 0, \ldots, 0)$ is one of them. Then the *multiplicity* of $(0, 0, \ldots, 0)$ as a point in $\mathbf{V}(I)$ is

$$\dim_k k[x_1, \ldots, x_n]_{\langle x_1, \ldots, x_n \rangle}/Ik[x_1, \ldots, x_n]_{\langle x_1, \ldots, x_n \rangle}.$$

More generally, if $p = (a_1, \ldots, a_n) \in \mathbf{V}(I)$, then the *multiplicity* of p, denoted $m(p)$, is the dimension of the ring obtained by localizing $k[x_1, \ldots, x_n]$ at the maximal ideal $M = \mathbf{I}(\{p\}) = \langle x_1 - a_1, \ldots, x_n - a_n \rangle$ corresponding to p, and taking the quotient:

$$\dim k[x_1, \ldots, x_n]_M/Ik[x_1, \ldots, x_n]_M.$$

Since $k[x_1, \ldots, x_n]_M$ is a local ring, it is easy to show that the quotient $k[x_1, \ldots, x_n]_M/Ik[x_1, \ldots, x_n]_M$ is also local (see Exercise 6 below). The intuition is that since M is the maximal ideal of $p \in \mathbf{V}(I)$, the ring $k[x_1, \ldots, x_n]_M/Ik[x_1, \ldots, x_n]_M$ should reflect the local behavior of I at p. Hence the multiplicity $m(p)$, which is the dimension of this ring, is a measure of how complicated I is at p. Theorem (2.2) below will guarantee that $m(p)$ is finite.

We can also define the multiplicity of a solution p of a specific system $f_1 = \cdots = f_s = 0$, provided that p is an *isolated solution* (that is, there exists a neighborhood of p in which the system has no other solutions). From a more sophisticated point of view, this multiplicity is sometimes called the *local intersection multiplicity* of the variety $\mathbf{V}(f_1, \ldots, f_s)$ at p. However, we caution the reader that there is a more sophisticated notion of multiplicity called the *Hilbert-Samuel multiplicity* of I at p. This is denoted $e(p)$ and is discussed in [BH], Section 4.6.

Let us check Definition (2.1) in our example. Let $R = k[x, y]_{\langle x, y \rangle}$ be the local ring of k^2 at $(0, 0)$ and consider the ideal J generated by the polynomials $x^2 + x^3$ and y^2 in R. The multiplicity of their common zero $(0, 0)$ is $\dim R/J$.

Exercise 1. Notice that $x^2 + x^3 = x^2(1 + x)$.
a. Show that $1 + x$ is a unit in R, so $1/(1 + x) \in R$.

b. Show that x^2 and y^2 generate the same ideal in R as $x^2 + x^3$ and y^2.

c. Show that every element $f \in R$ can be written uniquely as $f = g/(1 + h)$, where $g \in k[x, y]$ and $h \in \langle x, y \rangle \subset k[x, y]$.

d. Show that for each $f \in R$, the coset $[f] \in R/\langle x^2, y^2 \rangle R$ is equal to the coset $[g(1 - h + h^2)]$, where g, h are as in part c.

e. Deduce that every coset in $R/\langle x^2, y^2 \rangle R$ can be written as $[a + bx + cy + dxy]$ for some unique $a, b, c, d \in k$.

By the result of Exercise 1,

$$\dim R/J = \dim R/\langle x^2, y^2 \rangle R = 4.$$

Thus the multiplicity of $(0, 0)$ as a solution of $x^2 + x^3 = y^2 = 0$ is 4.

Similarly, let us compute the multiplicity of $(-1, 0)$ as a solution of this system. Rather than localizing at the prime ideal $\langle x + 1, y \rangle$, we change coordinates to translate the point $(-1, 0)$ to the origin and compute the multiplicity there. (This often simplifies the calculations; we leave the fact that these two procedures give the same results to the exercises.) So, set $X = x + 1, Y = y$ (we want X and Y to be 0 when $x = -1$ and $y = 0$) and let $S = k[X, Y]_{\langle X, Y \rangle}$. Then $x^2 + x^3 = (X - 1)^2 + (X - 1)^3 = X^3 - 2X^2 + X$ and $y^2 = Y^2$ and we want to compute the multiplicity of $(0, 0)$ as a solution of $X^3 - 2X^2 + X = Y^2 = 0$. Now we note that $X^3 - 2X^2 + X = X(1 - 2X + X^2)$ and $1/(1 - 2X + X^2) \in S$. Thus, the ideal generated by X and Y^2 in S is the same as that generated by $X^3 - 2X + X$ and Y^2 and, therefore,

$$\dim S/\langle X^3 - 2X^2 + X, Y^2 \rangle S = \dim S/\langle X, Y^2 \rangle S = 2.$$

Again, the equality on the right follows because the cosets of $1, Y$ are a basis of $S/\langle X, Y^2 \rangle$. We conclude that the multiplicity of $(-1, 0)$ as a solution of $x^3 + x^2 = y^2 = 0$ is 2.

Thus, we have shown that the polynomials $x^3 + x^2$ and y^2 have two common zeros, one of multiplicity 4 and the other of multiplicity 2. When the total number of zeros is counted with multiplicity, we obtain 6, in agreement with the fact that the dimension of the quotient ring of $k[x, y]$ by the ideal generated by these polynomials is 6.

Exercise 2.

a. Find all points in $\mathbf{V}(x^2 - 2x + y^2, x^2 - 4x + 4y^4) \subset \mathbb{C}^2$ and compute the multiplicity of each as above.

b. Verify that the sum of the multiplicities is equal to

$$\dim \mathbb{C}[x, y]/\langle x^2 - 2x + y^2, x^2 - 4x + 4y^4 \rangle.$$

c. What is the geometric explanation for the solution of multiplicity > 1 in this example?

Before turning to the question of computing the dimension of a quotient of a local ring in more complicated examples, we will verify that the total number of solutions of a system $f_1 = \cdots = f_s = 0$, counted with multiplicity, is the dimension of $k[x_1, \ldots, x_n]/I$ when k is algebraically closed and $I = \langle f_1, \ldots, f_s \rangle$ is zero-dimensional. In a sense, this is confirmation that our definition of multiplicity behaves as we would wish. In the following discussion, if $\{p_1, \ldots, p_m\}$ is a finite subset of k^n, and $M_i = \mathbf{I}(\{p_i\})$ is the maximal ideal of $k[x_1, \ldots, x_n]$ corresponding to p_i, we will write

$$k[x_1, \ldots, x_n]_{M_i} = \{f/g : g(p_i) \neq 0\} = \mathcal{O}_i$$

for simplicity of notation.

(2.2) Theorem. *Let I be a zero-dimensional ideal in $k[x_1, \ldots, x_n]$ (k algebraically closed) and let $\mathbf{V}(I) = \{p_1, \ldots, p_m\}$. Then, there is an isomorphism between $k[x_1, \ldots, x_n]/I$ and the direct product of the rings $A_i = \mathcal{O}_i/I\mathcal{O}_i$, for $i = 1, \ldots, m$.*

PROOF. For each i, $i = 1, \ldots, m$, there are ring homomorphisms

$$\varphi_i : k[x_1, \ldots, x_n] \to A_i$$
$$f \mapsto [f]_i,$$

where $[f]_i$ is the coset of f in the quotient ring $\mathcal{O}_i/I\mathcal{O}_i$. Hence we get a ring homomorphism

$$\varphi : k[x_1, \ldots, x_n] \to A_1 \times \cdots \times A_m$$
$$f \mapsto ([f]_1, \ldots, [f]_m).$$

Since $f \in I$ implies $[f]_i = 0 \in A_i$ for all i, we have $I \subset \ker(\varphi)$. So to prove the theorem, we need to show first that $I = \ker(\varphi)$ (by the fundamental theorem on ring homomorphisms, this will imply that $\mathrm{im}(\varphi) \cong k[x_1, \ldots, x_n]/I$), and second that φ is onto.

To prepare for this, we need to establish three basic facts. We use the notation $f \equiv g \bmod I$ to mean $f - g \in I$.

(2.3) Lemma. *Let $M_i = \mathbf{I}(\{p_i\})$ in $k[x_1, \ldots, x_n]$.*
a. *There exists an integer $d \geq 1$ such that $(\cap_{i=1}^m M_i)^d \subset I$.*
b. *There are polynomials $e_i \in k[x_1, \ldots, x_n]$, $i = 1, \ldots, m$, such that $\sum_{i=1}^m e_i \equiv 1 \bmod I$, $e_i e_j \equiv 0 \bmod I$ if $i \neq j$, and $e_i^2 \equiv e_i \bmod I$. Furthermore, $e_i \in I\mathcal{O}_j$ if $i \neq j$ and $e_i - 1 \in I\mathcal{O}_i$ for all i.*
c. *If $g \in k[x_1, \ldots, x_n] \setminus M_i$, then there exists $h \in k[x_1, \ldots, x_n]$ such that $hg \equiv e_i \bmod I$.*

PROOF OF THE LEMMA.   Part a is an easy consequence of the Nullstellensatz. We leave the details to the reader as Exercise 7 below.

Turning to part b, Lemma (2.9) of Chapter 2 implies the existence of polynomials $g_i \in k[x_1, \ldots, x_n]$ such that $g_i(p_j) = 0$ if $i \neq j$, and $g_i(p_i) = 1$

for each i. Let

$$(2.4) \qquad e_i = 1 - (1 - g_i^d)^d,$$

where d is as in part a. Expanding the right-hand side of (2.4) with the binomial theorem and canceling the 1s, we see that $e_j \in M_i^d$ for $j \neq i$. On the other hand, (2.4) implies $e_i - 1 \in M_i^d$ for all i. Hence for each i,

$$\sum_j e_j - 1 = e_i - 1 + \sum_{j \neq i} e_j$$

is an element of M_i^d. Since this is true for all i, $\sum_j e_j - 1 \in \cap_{i=1}^m M_i^d$. Because the M_i are distinct maximal ideals, $M_i + M_j = k[x_1, \ldots, x_n]$ whenever $i \neq j$. It follows that $\cap_{i=1}^m M_i^d = (\cap_{i=1}^m M_i)^d$ (see Exercise 8 below). Hence $\sum_j e_j - 1 \in (\cap_{i=1}^m M_i)^d \subset I$. Similarly, $e_i e_j \in \cap_{i=1}^m M_i^d = (\cap_{i=1}^m M_i)^d \subset I$ whenever $i \neq j$, and the congruence $e_i^2 \equiv e_i \bmod I$ now follows easily (see Exercise 9 below). This implies $e_i(e_i - 1) \in I\mathcal{O}_j$ for all i, j. If $i \neq j$, then $e_i - 1$ is a unit in $\mathcal{O}_j$ since $e_i(p_j) = 0$. Thus $e_i \in I\mathcal{O}_j$. The proof that $e_i - 1 \in I\mathcal{O}_i$ follows similarly using $e_i(p_i) = 1$.

For part c, by multiplying by a constant, we may assume $g(p_i) = 1$. Then $1 - g \in M_i$, and hence taking $h = (1 + (1 - g) + \cdots + (1 - g)^{d-1})e_i$,

$$hg = h(1 - (1 - g)) = (1 - (1 - g)^d)e_i = e_i - (1 - g)^d e_i.$$

Since $(1 - g)^d \in M_i^d$ and $e_i \in M_j^d$ for all $j \neq i$, as shown above, we have $(1 - g)^d e_i \in I$ by part a, and the lemma is established. □

We can now complete the proof of Theorem (2.2). Let $f \in \ker(\varphi)$, and note that that kernel is characterized as follows:

$$\ker(\varphi) = \{f \in k[x_1, \ldots, x_n] : [f]_i = 0 \text{ for all } i\}$$
$$= \{f : f \in I\mathcal{O}_i \text{ for all } i\}$$
$$= \{f : \text{ there exists } g_i \notin M_i \text{ with } g_i f \in I\}.$$

For each of the g_i, by part c of the lemma, there exists some h_i such that $h_i g_i \equiv e_i \bmod I$. As a result, $f \cdot \sum_{i=1}^m h_i g_i = \sum_{i=1}^m h_i(g_i f)$ is an element of I, since each $g_i f \in I$. But on the other hand, $f \cdot \sum_{i=1}^m h_i g_i \equiv f \cdot \sum_i e_i \equiv f \bmod I$ by part b of the lemma. Combining these two observations, we see that $f \in I$. Hence $\ker(\varphi) \subset I$. Since we proved earlier that $I \subset \ker(\varphi)$, we have $I = \ker(\varphi)$.

To conclude the proof, we need to show that φ is onto. So let $([n_1/d_1], \ldots, [n_m/d_m])$ be an arbitrary element of $A_1 \times \cdots \times A_m$, where $n_i, d_i \in k[x_1, \ldots, x_n]$, $d_i \notin M_i$, and the brackets denote the coset in A_i. By part c of the lemma again, there are $h_i \in k[x_1, \ldots, x_n]$ such that $h_i d_i \equiv e_i \bmod I$. Now let $F = \sum_{i=1}^m h_i n_i e_i \in k[x_1, \ldots, x_n]$. It is easy to see that $\varphi_i(F) = [n_i/d_i]$ for each i since $e_i \in I\mathcal{O}_j$ for $i \neq j$ and $e_i - 1 \in I\mathcal{O}_i$ by part b of the lemma. Hence φ is onto. □

An immediate corollary of this theorem is the result we want.

(2.5) Corollary. *Let k be algebraically closed, and let I be a zero-dimensional ideal in $k[x_1, \ldots, x_n]$. Then $\dim k[x_1, \ldots, x_n]/I$ is the number of points of $\mathbf{V}(I)$ counted with multiplicity. Explicitly, if $p_1, \ldots, p_m$ are the distinct points of $\mathbf{V}(I)$ and $\mathcal{O}_i$ is the ring of rational functions defined at p_i, then*

$$\dim k[x_1, \ldots, x_n]/I = \textstyle\sum_{i=1}^m \dim \mathcal{O}_i/I\mathcal{O}_i = \textstyle\sum_{i=1}^m m(p_i).$$

PROOF. The corollary follows immediately from the theorem by taking dimensions as vector spaces over k. $\qquad\qquad\square$

A second corollary tells us when a zero-dimensional ideal is radical.

(2.6) Corollary. *Let k be algebraically closed, and let I be a zero-dimensional ideal in $k[x_1, \ldots, x_n]$. Then I is radical if and only if every $p \in \mathbf{V}(I)$ has multiplicity $m(p) = 1$.*

PROOF. If $\mathbf{V}(I) = \{p_1, \ldots, p_m\}$, then Theorem (2.10) of Chapter 2 shows that $\dim k[x_1, \ldots, x_n]/I \geq m$, with equality if and only if I is radical. By Corollary (2.5), this inequality can be written $\sum_{i=1}^m m(p_i) \geq m$. Since $m(p_i)$ is always ≥ 1, it follows that $\sum_{i=1}^m m(p_i) \geq m$ is an equality if and only if all $m(p_i) = 1$. $\qquad\qquad\square$

We next discuss how to compute multiplicities. Given a zero-dimensional ideal $I \subset k[x_1, \ldots, x_n]$ and a polynomial $f \in k[x_1, \ldots, x_n]$, let m_f be multiplication by f on $k[x_1, \ldots, x_n]/I$. Then the characteristic polynomial $\det(m_f - uI)$ is determined by the points in $\mathbf{V}(I)$ *and* their multiplicities. More precisely, we have the following result.

(2.7) Proposition. *Let k be an algebraically closed field and let I be a zero-dimensional ideal in $k[x_1, \ldots, x_n]$. If $f \in k[x_1, \ldots, x_n]$, then*

$$\det(m_f - uI) = (-1)^d \prod_{p \in \mathbf{V}(I)} (u - f(p))^{m(p)},$$

where $d = \dim k[x_1, \ldots, x_n]/I$ and m_f is the map given by multiplication by f on $k[x_1, \ldots, x_n]/I$.

PROOF. Let $\mathbf{V}(I) = \{p_1, \ldots, p_m\}$. Using Theorem (2.2), we get a diagram:

$$
\begin{array}{ccc}
k[x_1, \ldots, x_n]/I & \cong & A_1 \times \cdots \times A_m \\[2pt]
m_f \downarrow & & \downarrow m_f \\[2pt]
k[x_1, \ldots, x_n]/I & \cong & A_1 \times \cdots \times A_m
\end{array}
$$

where $m_f : A_1 \times \cdots \times A_m \to A_1 \times \cdots \times A_m$ is multiplication by f on each factor. This diagram commutes in the same sense as the diagram (5.19) of Chapter 3.

Hence we can work with $m_f : A_1 \times \cdots \times A_m \to A_1 \times \cdots \times A_m$. If we restrict to $m_f : A_i \to A_i$, it suffices to show that $\det(m_f - uI) = (-1)^{m(p_i)}(u - f(p_i))^{m(p_i)}$. Equivalently, we must show that $f(p_i)$ is the only eigenvalue of m_f on A_i.

To prove this, consider the map $\varphi_i : k[x_1, \ldots, x_n] \to A_i$ defined in the proof of Theorem (2.2), and let $Q_i = \ker(\varphi_i)$. In Exercise 11 below, you will study the ideal Q_i, which is part of the *primary decomposition* of I. In particular, you will show that $\mathbf{V}(Q_i) = \{p_i\}$ and that $k[x_1, \ldots, x_n]/Q_i \cong A_i$. Consequently, the eigenvalues of m_f on A_i equal the eigenvalues of m_f on $k[x_1, \ldots, x_n]/Q_i$, which by Theorem (4.5) of Chapter 2 are the values of f on $\mathbf{V}(Q_i) = \{p_i\}$. It follows that $f(p_i)$ is the only eigenvalue, as desired. $\qquad\square$

The ideas used in the proof of Proposition (2.7) make it easy to determine the *generalized eigenvectors* of m_f. See Exercise 12 below for the details.

If we know the points $p_1, \ldots, p_m$ of $\mathbf{V}(I)$ (for example, we could find them using the methods of Chapters 2 or 3), then it is a simple matter to compute their multiplicities using Proposition (2.7). First pick f so that $f(p_1), \ldots, f(p_m)$ are distinct, and then compute the matrix of m_f relative to a monomial basis of $k[x_1, \ldots, x_n]/I$ as in Chapters 2 or 3. In typical cases, the polynomials generating I have coefficients in $\mathbb{Q}$, which means that the characteristic polynomial $\det(m_f - uI)$ is in $\mathbb{Q}[u]$. Then factor $\det(m_f - uI)$ over $\mathbb{Q}$, which can easily be done by computer (the Maple command is `factor`). This gives

$$\det(m_f - uI) = h_1^{m_1} \cdots h_r^{m_r},$$

where $h_1, \ldots, h_r$ are distinct irreducible polynomials over $\mathbb{Q}$. For each $p_i \in \mathbf{V}(I)$, $f(p_i)$ is a root of a unique h_j, and the corresponding exponent m_j is the multiplicity $m(p_i)$. This follows from Proposition (2.7) and the properties of irreducible polynomials (see Exercise 13). One consequence is that those points of $\mathbf{V}(I)$ corresponding to the same irreducible factor of $\det(m_f - uI)$ all have the same multiplicity.

We can also extend some of the results proved in Chapter 3 about resultants. For example, the techniques used to prove Theorem (2.2) give the following generalization of Proposition (5.8) of Chapter 3 (see Exercise 14 below for the details).

(2.8) Proposition. *Let $f_1, \ldots, f_n \in k[x_1, \ldots, x_n]$ (k algebraically closed) have total degrees at most $d_1, \ldots, d_n$ and no solutions at ∞. If $f_0 = u_0 + u_1 x_1 + \cdots + u_n x_n$, where $u_0, \ldots, u_n$ are independent variables, then there is a nonzero constant C such that*

$$\mathrm{Res}_{1,d_1,\ldots,d_n}(f_0, \ldots, f_n) = C \prod_{p \in \mathbf{V}(f_1,\ldots,f_n)} (u_0 + u_1 a_1 + \cdots + u_n a_n)^{m(p)},$$

where a point $p \in \mathbf{V}(f_1, \ldots, f_n)$ is written $p = (a_1, \ldots, a_n)$.

This tells us that the *u-resultant* of Chapter 3, §5, computes not only the points of $\mathbf{V}(f_1, \ldots, f_n)$ but also their multiplicities. In Chapter 3, we also studied the *hidden variable method*, where we set $x_n = u$ in the equations $f_1 = \cdots = f_n = 0$ and regard u as a constant. After homogenizing with respect to x_0, we get the resultant $\mathrm{Res}_{x_0, \ldots, x_{n-1}}(\widehat{F}_1, \ldots, \widehat{F}_n)$ from Proposition (5.9) in Chapter 3, which tells us about the x_n-coordinates of the solutions. In Chapter 3, we needed to assume that the x_n-coordinates were distinct. Now, using Proposition (2.8), it is easy to show that when $f_1, \ldots, f_n$ have no solutions at ∞,

$$
\mathrm{Res}_{1, d_1, \ldots, d_n}(u - x_n, f_1, \ldots, f_n) = \mathrm{Res}_{x_0, \ldots, x_{n-1}}(\widehat{F}_1, \ldots, \widehat{F}_n)
$$
(2.9)
$$
= C \prod_{p \in \mathbf{V}(f_1, \ldots, f_n)} (u - a_n)^{m(p)}
$$

where $p \in \mathbf{V}(f_1, \ldots, f_n)$ is written $p = (a_1, \ldots, a_n)$. See Exercise 14 for the proof.

The formulas given in (2.9) and Proposition (2.8) indicate a deep relation between multiplicities using resultants. In fact, in the case of two equations in two unknowns, one can use resultants to *define* multiplicities. This is done, for example, in Chapter 8 of [CLO] and Chapter 3 of [Kir].

Exercise 3. Consider the equations

$$
f_1 = y^2 - 3 = 0
$$
$$
f_2 = 6y - x^3 + 9x,
$$

and let $I = \langle f_1, f_2 \rangle \subset k[x, y]$.
a. Show that these equations have four solutions with distinct x coordinates.
b. Draw the graphs of $f_1 = 0$ and $f_2 = 0$. Use your picture to explain geometrically why two of the points should have multiplicity > 1.
c. Show that the characteristic polynomial of m_x on $\mathbb{C}[x, y]/I$ is $u^6 - 18u^4 + 81u^2 - 108 = (u^2 - 3)^2(u^2 - 12)$.
d. Use part c and Proposition (2.7) to compute the multiplicities of the four solution points.
e. Explain how you would compute the multiplicities using $\mathrm{Res}(f_1, f_2, y)$ and Proposition (2.8). This is the hidden variable method for computing multiplicities. Also explain the meaning of the exponent 3 in $\mathrm{Res}(f_1, f_2, x) = (y^2 - 3)^3$.

Besides resultants and multiplicities, Theorem (2.2) has other interesting consequences. For instance, suppose that a collection of n polynomials $f_1, \ldots, f_n$ has a *single* zero in k^n, which we may take to be the origin. Let $I = \langle f_1, \ldots, f_n \rangle$. Then the theorem implies

$$(2.10) \quad k[x_1, \ldots, x_n]/I \cong k[x_1, \ldots, x_n]_{\langle x_1, \ldots, x_n \rangle}/Ik[x_1, \ldots, x_n]_{\langle x_1, \ldots, x_n \rangle}.$$

This is very satisfying, but there is more to the story. With the above hypotheses on $f_1, \ldots, f_n$, one can show that most small perturbations of $f_1, \ldots, f_n$ result in a system of equations with distinct zeroes, each of which has multiplicity one, and that the number of such zeroes is precisely equal to the multiplicity of the origin as a solution of $f_1 = \cdots = f_n = 0$. Moreover, the ring $k[x_1, \ldots, x_n]/I$ turns out to be a limit, in a rather precise sense, of the set of functions on these distinct zeroes. Here is a simple example.

Exercise 4. Let $k = \mathbb{C}$ so that we can take limits in an elementary sense. Consider the ideals $I_t = \langle y - x^2, x^3 - t \rangle$ where $t \in \mathbb{C}$ is a parameter.
a. What are the points in $\mathbf{V}(I_t)$ for $t \neq 0$? Show that each point has multiplicity 1, so $A_i \cong k$ for each i.
b. Now let $t \to 0$. What is $\mathbf{V}(I_0)$ and its multiplicity?
c. Using the proof of Theorem (2.2), work out an explicit isomorphism between $\mathbb{C}[x, y]/I_t$, and the product of the A_i for $t \neq 0$.
d. What happens as $t \to 0$? Identify the image of a general f in $\mathbb{C}[x, y]/I_0$, and relate to the image of f in the product of A_i for $t \neq 0$.

Local rings give us the ability to discuss what's happening near a particular solution of a zero-dimensional ideal. This leads to some rich mathematics, including the following.

- As explained in Exercise 11, the isomorphism $A \cong A_1 \times \cdots \times A_m$ of Theorem (2.2) is related to primary decomposition. A method for computing this decomposition using the characteristic polynomial of a multiplication map is discussed in [Mon] and [YNT].
- The local ring A_i can be described in terms of the vanishing of certain linear combinations of partial derivatives. This is explained in [MMM1], [MMM2], [Möl], and [MöS], among others.
- When the number of equations equals the number of unknowns as in Chapter 3, the ring A is a *complete intersection*. Some of the very deep algebra related to this situation, including *Gorenstein duality*, is discussed in [ElM2].

The book [Stu5] gives a nice introduction to the first two bullets. The reader should also consult [YNT] for many other aspects of the ring A and [Rou] for an interesting method of representing the solutions and their multiplicities.

We also remark that we can compute multiplicities by passing to the formal power series ring or, in the cases $k = \mathbb{R}$ or $\mathbb{C}$, to the ring of convergent power series. More precisely, the following holds.

(2.11) Proposition. *Let $I \subset k[x_1, \ldots, x_n]$ be a zero-dimensional ideal such that the origin is a point of $\mathbf{V}(I)$ of multiplicity m. Then*

$$m = \dim k[x_1, \ldots, x_n]_{\langle x_1, \ldots, x_n \rangle} / I k[x_1, \ldots, x_n]_{\langle x_1, \ldots, x_n \rangle}$$
$$= \dim k[[x_1, \ldots, x_n]] / I k[[x_1, \ldots, x_n]].$$

If, moreover, $k = \mathbb{R}$ or $\mathbb{C}$, so that we can talk about whether a power series converges, then

$$m = \dim k\{x_1, \ldots, x_n\} / I k\{x_1, \ldots, x_n\}$$

as well.

To see the idea behind why this is so, consider the example we looked at in Exercise 1 above. We showed that $\dim k[x, y]_{\langle x, y \rangle} / \langle x^2 + x^3, y^2 \rangle = 4$ by noting that in $k[x, y]_{\langle x, y \rangle}$, we have

$$\langle x^2 + x^3, y^2 \rangle = \langle x^2, y^2 \rangle$$

because $1/(1 + x) \in k[x, y]_{\langle x, y \rangle}$. As in §1, we can represent $1/(1 + x)$ as the formal power series $1 - x + x^2 - x^3 + x^4 - \cdots \in k[[x, y]]$ and then

$$(x^2 + x^3)(1 - x + x^2 - x^3 + x^4 - \cdots) = x^2$$

in $k[[x, y]]$. This shows that, in $k[[x, y]]$, $\langle x^2 + x^3, y^2 \rangle = \langle x^2, y^2 \rangle$. It follows that

$$\dim k[[x, y]] / \langle x^2, y^2 \rangle = 4$$

(as before, the four monomials $1, x, y, xy$ form a vector space basis of $k[[x, y]] / \langle x^2, y^2 \rangle$). If $k = \mathbb{C}$, the power series $1 - x + x^2 - x^3 + x^4 - \cdots$ is convergent for x with $|x| < 1$, and precisely the same reasoning shows that $\langle x^2 + x^3, y^2 \rangle = \langle x^2, y^2 \rangle$ in $k\{x, y\}$ as well. Therefore,

$$\dim k\{x, y\} / \langle x^2, y^2 \rangle k\{x, y\} = 4.$$

It is possible to prove the proposition by generalizing these observations, but it will be more convenient to defer it to §5, so that we can make use of some additional computational tools for local rings.

We will conclude this section by introducing an important invariant in singularity theory—the *Milnor number* of a singularity. See [Mil] for the topological meaning of this integer. One says that an analytic function $f(x_1, \ldots, x_n)$ on an open set $U \subset \mathbb{C}^n$ has a *singularity* at a point $p \in U$ if the n first-order partial derivatives of f have a common zero at p. We say that the singular point p is *isolated* if there is some neighborhood of p containing no other singular points of f. As usual, when considering a given singular point p, one translates p to the origin. If we do this, then the assertion that the origin is isolated is enough to guarantee that

$$\dim \mathbb{C}\{x_1, \ldots, x_n\} / \langle \partial f / \partial x_1, \ldots, \partial f / \partial x_n \rangle < \infty.$$

Here, we are using the fact that in a neighborhood of the origin, any analytic function can be represented by a convergent power series. Thus f and its partial derivatives can be regarded as elements of $\mathbb{C}\{x_1, \dots, x_n\}$.

(2.12) Definition. Let $f \in \mathbb{C}\{x_1, \dots, x_n\}$ have an isolated singularity at the origin. The *Milnor number* of the singular point, denoted μ, is given by

$$\mu = \dim \mathbb{C}\{x_1, \dots, x_n\}/\langle \partial f/\partial x_1, \dots, \partial f/\partial x_n \rangle.$$

In view of Proposition (2.11), if the function f is a polynomial, the Milnor number of a singular point p of f is just the multiplicity of the common zero p of the partials of f.

Exercise 5. Each of the following $f(x, y) \in \mathbb{C}[x, y]$ has an isolated singular point at $(0, 0)$. For each, determine the Milnor number by computing

$$\mu = \dim \mathbb{C}[[x, y]]/\langle \partial f/\partial x, \partial f/\partial y \rangle.$$

a. $f(x, y) = y^2 - x^2 - x^3$.
b. $f(x, y) = y^2 - x^3$.
c. $f(x, y) = y^2 - x^5$.

In intuitive terms, the larger the Milnor number is, the more complicated the structure of the singular point is. To conclude this section, we mention that there is a closely related invariant of singularities called the Tjurina number, which is defined by

$$\tau = \dim k[[x_1, \dots, x_n]]/\langle f, \partial f/\partial x_1, \dots, \partial f/\partial x_n \rangle.$$

Over any field k, the Tjurina number is finite precisely when f has an isolated singular point.

ADDITIONAL EXERCISES FOR §2

Exercise 6. If $p \in \mathbf{V}(I)$ and $M = \mathbf{I}(\{p\})$ is the maximal ideal of p, then prove that $k[x_1, \dots, x_n]_M/Ik[x_1, \dots, x_n]_M$ is a local ring. Also show that the dimension of this ring, which is the multiplicity $m(p)$, is ≥ 1. Hint: Show that the map $k[x_1, \dots, x_n]_M/Ik[x_1, \dots, x_n]_M \to k$ given by evaluating a coset at p is a well-defined linear map which is onto.

Exercise 7. Using the Nullstellensatz, prove part a of Lemma (2.3).

Exercise 8. Let I and J be any two ideals in a ring R such that $I + J = R$ (we sometimes say I and J are *comaximal*).
a. Show that $IJ = I \cap J$.

b. From part a, deduce that if $d \geq 1$, then $I^d \cap J^d = (I \cap J)^d$.

c. Generalize part b to any number of ideals $I_1, \ldots, I_r$ if $I_i + I_j = R$ whenever $i \neq j$.

Exercise 9. Show that if e_i are the polynomials constructed in (2.4) for part b of Lemma (2.3), then $e_i^2 \equiv e_i \mod I$. Hint: Use the other two statements in part b.

Exercise 10. In this exercise, we will use Theorem (2.2) to give a new proof of Theorem (4.5) of Chapter 2. Let A_i be the local ring $\mathcal{O}_i / I\mathcal{O}_i$ as in the proof of Theorem (2.2). For $f \in k[x_1, \ldots, x_n]$, let $m_f : A_i \to A_i$ be multiplication by f. Also, the coset of f in A_i will be denoted $[f]_i$.

a. Prove that m_f is a vector space isomorphism if and only if $[f]_i \in A_i$ is invertible; i.e., there is $[g]_i \in A_i$ such that $[f]_i [g]_i = [1]_i$.

b. Explain why $[f]_i$ is in the maximal ideal of A_i if and only if $f(p_i) = 0$.

c. Explain why each of the following equivalences is true for a polynomial $f \in k[x_1, \ldots, x_n]$ and $\lambda \in \mathbb{C}$: λ is an eigenvalue of $m_f \Leftrightarrow m_{f-\lambda}$ is not invertible $\Leftrightarrow [f - \lambda]_i \in A_i$ is not invertible $\Leftrightarrow [f - \lambda]_i$ is in the maximal ideal of $A_i \Leftrightarrow f(p) = \lambda$. Hint: Use parts a and b of this exercise and part b of Exercise 1 from §1.

d. Combine part c with the isomorphism $k[x_1, \ldots, x_n]/I \cong A_1 \times \cdots \times A_m$ and the commutative diagram from Proposition (2.7) to give a new proof of Theorem (4.5) of Chapter 2.

Exercise 11. (Primary Decomposition) Let I be a zero-dimensional ideal with $\mathbf{V}(I) = \{p_1, \ldots, p_m\}$. This exercise will explore the relation between the isomorphism $A = k[x_1, \ldots, x_n]/I \cong A_1 \times \cdots \times A_m$ and the *primary decomposition* of I. More details on primary decomposition can be found in [CLO], Chapter 4, §7. We begin with the homomorphism $\varphi_i : k[x_1, \ldots, x_n] \to A_i$ defined by $\varphi(f) = [f]_i \in A_i$ (this is the notation used in the proof of Theorem (2.2)). Consider the ideal Q_i defined by

$$Q_i = \ker(\varphi_i) = \{f \in k[x_1, \ldots, x_n] : [f]_i = [0]_i \text{ in } A_i\}.$$

We will show that the ideals $Q_1, \ldots, Q_m$ give the primary decomposition of I. Let $M_i = \mathbf{I}(\{p_i\})$.

a. Show that $I \subset Q_i$ and that $Q_i = \{f \in k[x_1, \ldots, x_n] : \text{there exists } u \text{ in } k[x_1, \ldots, x_n] \setminus M_i \text{ such that } u \cdot f \in I\}$.

b. If $g_1, \ldots, g_m$ are as in the proof of Theorem (2.2), show that for $j \neq i$, some power of g_j lies in Q_i. Hint: Use part a and the Nullstellensatz.

c. Show that $\mathbf{V}(Q_i) = \{p_i\}$ and conclude that $\sqrt{Q_i} = M_i$. Hint: Use part b and the Nullstellensatz.

d. Show that Q_i is a *primary ideal*, which means that if $fg \in Q_i$, then either $f \in Q_i$ or some power of g is in Q_i. Hint: Use part c. Also, A_i is a local ring.

e. Prove that $I = Q_1 \cap \cdots \cap Q_m$. This is the primary decomposition of I (see Theorem 7 of [CLO], Chapter 4, §7).

f. Show that $k[x_1, \ldots, x_n]/Q_i \cong A_i$. Hint: Show that φ_i is onto using the proof of Theorem (2.2).

Exercise 12. (Generalized Eigenspaces) Given a linear map $T : V \to V$, where V is a finite-dimensional vector space, a *generalized eigenvector of* $\lambda \in k$ is a nonzero vector $v \in V$ such that $(T - \lambda I)^m(v) = 0$ for some $m \geq 1$. The *generalized eigenspace of* λ is the space of the generalized eigenvectors for λ. When k is algebraically closed, V is the direct sum of its generalized eigenspaces (see Section 7.1 of [FIS]). We will apply this theory to the linear map $m_f : A \to A$ to see how the generalized eigenspaces of m_f relate to the isomorphism $A \cong A_1 \times \cdots \times A_m$ of Theorem (2.2).

a. In the proof of Proposition (2.7), we proved that $f(p_i)$ is the only eigenvalue of $m_f : A_i \to A_i$. Use this to show that the generalized eigenspace of m_f is all of A_i.

b. If $f(p_1), \ldots, f(p_m)$ are distinct, prove that the decomposition of $A = k[x_1, \ldots, x_n]/I$ into a direct sum of generalized eigenspaces for m_f is *precisely* the isomorphism $A \cong A_1 \times \cdots \times A_m$ of Theorem (2.2).

Exercise 13.

a. If $h \in \mathbb{Q}[u]$ is irreducible, prove that all roots of h have multiplicity one. Hint: Compute h_{red}.

b. Let $h \in \mathbb{Q}[u]$ be irreducible and let $\lambda \in \mathbb{C}$ be a root of h. If $g \in \mathbb{Q}[u]$ and $g(\lambda) = 0$, prove that h divides g. Hint: If $\gcd(h, g) = 1$, there are polynomials $A, B \in \mathbb{Q}[u]$ such that $Ah + Bg = 1$.

c. If h_1 and h_2 are distinct irreducible polynomials in $\mathbb{Q}[u]$, prove that h_1 and h_2 have no common roots.

d. Use parts a and c to justify the method for computing multiplicities given in the discussion following Proposition (2.7).

Exercise 14. Prove Proposition (2.8) and the formulas given in (2.9). Hint: Use Exercise 12 and Proposition (5.8) of Chapter 3.

Exercise 15.

a. Let $\ell_1, \ldots, \ell_n$ be homogeneous linear polynomials in $k[x_1, \ldots, x_n]$ with $\mathbf{V}(\ell_1, \ldots, \ell_n) = \{(0, \ldots, 0)\}$. Compute the multiplicity of the origin as a solution of $\ell_1 = \cdots = \ell_n = 0$.

b. Now let $f_1, \ldots, f_n$ generate a zero-dimensional ideal in $k[x_1, \ldots, x_n]$, and suppose that the origin is in $\mathbf{V}(f_1, \ldots, f_n)$ and the Jacobian matrix

$$J = (\partial f_i / \partial x_j)$$

has nonzero determinant at the origin. Compute the multiplicity of the origin as a solution of $f_1 = \cdots = f_n = 0$. Hint: Use part a.

Exercise 16. We say $f \in \mathbb{C}[x_1, \ldots, x_n]$ has an *ordinary double point* at the origin 0 in $\mathbb{C}^n$ if $f(0) = \partial f/\partial x_i(0) = 0$ for all i, but the matrix of second-order partial derivatives is invertible at 0:

$$\det(\partial^2 f/\partial x_i \partial x_j)\big|_{(x_1, \ldots, x_n)=(0, \ldots, 0)} \neq 0.$$

Find the Milnor number of an ordinary double point. Hint: Use Exercise 15.

Exercise 17. Let I be a zero-dimensional ideal in $k[x_1, \ldots, x_n]$ and let $p = (a_1, \ldots, a_n) \in \mathbf{V}(I)$. Let $X_1, \ldots, X_n$ be a new set of variables, and consider the set $\bar{I} \subset k[X_1, \ldots, X_n]$ consisting of all $f(X_1 + a_1, \ldots, X_n + a_n)$ where $f \in I$.
a. Show that $\bar{I}$ is an ideal in $k[X_1, \ldots, X_n]$, and that the origin is a point in $\mathbf{V}(\bar{I})$.
b. Show that the multiplicity of p as a point in $\mathbf{V}(I)$ is the same as the multiplicity of the origin as a point in $\mathbf{V}(\bar{I})$. Hint: One approach is to show that

$$\varphi : k[x_1, \ldots, x_n] \to k[X_1, \ldots, X_n]$$
$$f(x_1, \ldots, x_n) \mapsto f(X_1 + a_1, \ldots, X_n + a_n)$$

defines an isomorphism of rings.

§3 Term Orders and Division in Local Rings

When working with an ideal $I \subset k[x_1, \ldots, x_n]$, for some purposes we can replace I with its ideal of leading terms $\langle \mathrm{LT}(I) \rangle$. For example, if I is zero-dimensional, we can compute the dimension of the quotient ring $k[x_1, \ldots, x_n]/I$ by using the fact that $\dim k[x_1, \ldots, x_n]/I = \dim k[x_1, \ldots, x_n]/\langle \mathrm{LT}(I) \rangle$. The latter dimension is easy to compute since $\langle \mathrm{LT}(I) \rangle$ is a *monomial* ideal—the dimension is just the number of monomials not in the ideal). The heart of the matter is to compute $\langle \mathrm{LT}(I) \rangle$, which is done by computing a Gröbner basis of I.

A natural question to ask is whether something similar might work in a local ring. An instructive example occurred in the last section, where we considered the ideal $I = \langle x^2 + x^3, y^2 \rangle$. For $R = k[x, y]_{\langle x, y \rangle}$ or $k[[x, y]]$ or $k\{x, y\}$, we computed $\dim R/IR$ by replacing I by the monomial ideal

$$\tilde{I} = \langle x^2, y^2 \rangle.$$

Note that $\tilde{I}$ is generated by the *lowest degree terms* in the generators of I. This is in contrast to the situation in the polynomial ring, where $\dim k[x, y]/I$ was computed from $\langle \mathrm{LT}(I) \rangle = \langle x^3, y^2 \rangle$ using the *lex leading terms*.

To be able to pick out terms of lowest degree in polynomials as leading terms, it will be necessary to extend the class of orders on monomials we can use. For instance, to make the leading term of a polynomial or a power series be one of the terms of minimal total degree, we could consider what are known as *degree-anticompatible* (or anti-graded) orders. By definition these are orders that satisfy

(3.1) $$|\alpha| < |\beta| \implies x^\alpha > x^\beta.$$

We still insist that our orders be total orderings and be compatible with multiplication. As in Definition (2.1) of Chapter 1, being a total ordering means that for any $\alpha, \beta \in \mathbb{Z}_{\geq 0}^n$, exactly one of the following is true:

$$x^\alpha > x^\beta, \quad x^\alpha = x^\beta, \text{ or } x^\alpha < x^\beta.$$

Compatibility with multiplication means that for any $\gamma \in \mathbb{Z}_{\geq 0}^n$, if $x^\alpha > x^\beta$, then $x^{\alpha+\gamma} > x^{\beta+\gamma}$. Notice that property (3.1) implies that $1 > x_i$ for all $i, 1 \leq i \leq n$. Here is a first example.

Exercise 1. Consider terms in $k[x]$.
a. Show that the only degree-anticompatible order is the antidegree order:

$$1 > x > x^2 > x^3 > \cdots.$$

b. Explain why the antidegree order is not a well-ordering.

Any total ordering that is compatible with multiplication and that satisfies $1 > x_i$ for all $i, 1 \leq i \leq n$ is called a *local order*. A degree-anticompatible order is a local order (but not conversely—see Exercise 2 below).

Perhaps the simplest example of a local order in n variables is degree-anticompatible lexicographic order, abbreviated *alex*, which first sorts by total degree, lower degree terms preceding higher degree terms, and which sorts monomials of the same total degree lexicographically.

(3.2) Definition (Antigraded Lex Order). Let $\alpha, \beta \in \mathbb{Z}_{\geq 0}^n$. We say $x^\alpha >_{alex} x^\beta$ if

$$|\alpha| = \sum_{i=1}^n \alpha_i < |\beta| = \sum_{i=1}^n \beta_i,$$

or if

$$|\alpha| = |\beta| \text{ and } x^\alpha >_{lex} x^\beta.$$

Thus, for example, in $k[x, y]$, with $x > y$, we have

$$1 >_{alex} x >_{alex} y >_{alex} x^2 >_{alex} xy >_{alex} y^2 >_{alex} x^3 >_{alex} \cdots.$$

Similarly one defines degree-anticompatible reverse lexicographic, or *arevlex*, order as follows.

(3.3) Definition (Antigraded Revlex Order). Let $\alpha, \beta \in \mathbb{Z}_{\geq 0}^n$. We say $x^\alpha >_{arevlex} x^\beta$ if

$$|\alpha| < |\beta|, \text{ or } |\alpha| = |\beta| \text{ and } x^\alpha >_{revlex} x^\beta.$$

So, for example, we have

$$1 >_{arevlex} x >_{arevlex} y >_{arevlex} z >_{arevlex} x^2 >_{arevlex}$$
$$xy >_{arevlex} y^2 >_{arevlex} xz >_{arevlex} yz >_{arevlex} z^2 >_{arevlex} \cdots.$$

Degree-anticompatible and local orders lack one of the key properties of the monomial orders that we have used up to this point. Namely, the third property in Definition (2.1) from Chapter 1, which requires that a monomial order be a *well-ordering* relation, does not hold. Local orders are not well-orderings. This can be seen even in the one-variable case in Exercise 1 above.

In §4 of this chapter, we will need to make use of even more general orders than degree-anticompatible or local orders. Moreover, and somewhat surprisingly, the whole theory can be simplified somewhat by generalizing at once to consider the whole class of *semigroup orders* as in the following definition.

(3.4) Definition. An order $>$ on $\mathbb{Z}_{\geq 0}^n$ or, equivalently, on the set of monomials $x^\alpha, \alpha \in \mathbb{Z}_{\geq 0}^n$ in $k[x_1, \ldots, x_n]$ or any of the local rings $k[x_1, \ldots, x_n]_{\langle x_1, \ldots, x_n \rangle}$, $k\{x_1, \ldots, x_n\}$, or $k[[x_1, \ldots, x_n]]$, is said to be a *semigroup order* if it satisfies:
a. $>$ is a total ordering on $\mathbb{Z}_{\geq 0}^n$;
b. $>$ is compatible with multiplication of monomials.

Semigroup orders include the monomial orders, which have the additional well-ordering property, as well as local orders and other orders which do not. Since the property of being a well-ordering is often used to assert that algorithms terminate, we will need to be especially careful in checking that procedures using semigroup orders terminate.

Recall that in §2 of Chapter 1 we discussed how monomial orders can be specified by matrices. If M is an $m \times n$ real matrix with rows $\mathbf{w}_1, \ldots, \mathbf{w}_m$, then we define $x^\alpha >_M x^\beta$ if there is an $\ell \leq m$ such that $\alpha \cdot \mathbf{w}_i = \beta \cdot \mathbf{w}_i$ for $i = 1, \ldots, \ell - 1$, but $\alpha \cdot \mathbf{w}_\ell > \beta \cdot \mathbf{w}_\ell$. Every semigroup order can be described by giving a suitable matrix M. The following exercise describes the necessary properties of M and gives some examples.

Exercise 2.
a. Show that $>_M$ is compatible with multiplication for every matrix M as above.
b. Show that $>_M$ is a total ordering if and only if $\ker(M) \cap \mathbb{Z}_{\geq 0}^n = \{(0, \ldots, 0)\}$.

c. Show that the *lex* monomial order with $x_1 > x_2 > \cdots > x_n$ is the order $>_I$, where I is the $n \times n$ identity matrix.

d. Show that the *alex* order is the order $>_M$ defined by the matrix

$$M = \begin{pmatrix} -1 & -1 & \cdots & -1 \\ 0 & -1 & \cdots & -1 \\ \vdots & \vdots & \ddots & \vdots \\ 0 & 0 & \cdots & -1 \end{pmatrix}.$$

e. Show that the *arevlex* order is the order $>_M$ for

$$M = \begin{pmatrix} -1 & -1 & \cdots & -1 & -1 \\ 0 & 0 & \cdots & 0 & -1 \\ 0 & 0 & \cdots & -1 & 0 \\ \vdots & \vdots & \ddots & \vdots & \vdots \\ 0 & -1 & \cdots & 0 & 0 \end{pmatrix}.$$

f. Find a local order that is not degree-anticompatible. Hint: What is it about the corresponding matrices that makes *alex* and *arevlex* degree-anticompatible, resp. local?

If $f = \sum_\alpha c_\alpha x^\alpha \in k[x_1, \ldots, x_n]$ is a polynomial and $>$ is a semigroup order, we define the multidegree, the leading coefficient, the leading monomial, and the leading term of f exactly as we did for a monomial order:

$$\text{multideg}(f) = \max\{\alpha \in \mathbb{Z}^n_{\geq 0} : c_\alpha \neq 0\}$$
$$\text{LC}(f) = c_{\text{multideg}(f)}$$
$$\text{LM}(f) = x^{\text{multideg}(f)}$$
$$\text{LT}(f) = \text{LC}(f) \cdot \text{LM}(f).$$

In addition, each semigroup order $>$ defines a particular ring of fractions in $k(x_1, \ldots, x_n)$ as in Exercise 9 of §1 of this chapter. Namely, given $>$, we consider the set

$$S = \{1 + g \in k[x_1, \ldots, x_n] : g = 0, \text{ or } \text{LT}_>(g) < 1\}.$$

S is closed under multiplication since if $\text{LT}_>(g) < 1$ and $\text{LT}_>(g') < 1$, then $(1 + g)(1 + g') = 1 + g + g' + gg'$, and $\text{LT}(g + g' + gg') < 1$ as well by the definition of a semigroup order.

(3.5) Definition. Let $>$ be a semigroup order on monomials in the ring $k[x_1, \ldots, x_n]$ and let $S = \{1 + g : \text{LT}(g) < 1\}$. The *localization* of $k[x_1, \ldots, x_n]$ with respect to $>$ is the ring

$$\text{Loc}_>(k[x_1, \ldots, x_n]) = S^{-1} k[x_1, \ldots, x_n] = \{f/(1 + g) : 1 + g \in S\}.$$

For example, if $>$ is a monomial order, then there are no nonzero monomials smaller than 1 so $S = \{1\}$ and $\mathrm{Loc}_>(k[x_1, \ldots, x_n]) = k[x_1, \ldots, x_n]$. On the other hand, if $>$ is a local order, then since $1 > x_i$ for all i,

$$\{g : g = 0, \text{ or } \mathrm{LT}_>(g) < 1\} = \langle x_1, \ldots, x_n \rangle.$$

Hence, for a local order, we have that S is contained in the set of units in $k[x_1, \ldots, x_n]_{\langle x_1, \ldots, x_n \rangle}$ so $\mathrm{Loc}_>(k[x_1, \ldots, x_n]) \subset k[x_1, \ldots, x_n]_{\langle x_1, \ldots, x_n \rangle}$. But in fact, by adjusting constants between the numerator and the denominator in a general $f/h \in k[x_1, \ldots, x_n]_{\langle x_1, \ldots, x_n \rangle}$, it is easy to see that $f/h = f'/(1 + g)$ for some $1 + g \in S$. Hence if $>$ is a local order, then

$$\mathrm{Loc}_>(k[x_1, \ldots, x_n]) = k[x_1, \ldots, x_n]_{\langle x_1, \ldots, x_n \rangle}.$$

The next two exercises give some additional, more general, and also quite suggestive examples of semigroup orders and their associated rings of fractions.

Exercise 3. Using $>_{alex}$ on the x-terms, and $>_{lex}$ on the y-terms, define a *mixed order* $>_{mixed}$ by $x^\alpha y^\beta >_{mixed} x^{\alpha'} y^{\beta'}$ if either $y^\beta >_{lex} y^{\beta'}$, or $y^\beta = y^{\beta'}$ and $x^\alpha >_{alex} x^{\alpha'}$.
a. Show that $>_{mixed}$ is a semigroup order and find a matrix M such that $>_{mixed} = >_M$.
b. Show that $>_{mixed}$ is *neither* a well-ordering, *nor* degree-anticompatible.
c. Let $g \in k[x_1, \ldots, x_n, y_1, \ldots, y_m]$. Show that $1 >_{mixed} \mathrm{LT}_{>_{mixed}}(g)$ if and only if g depends only on $x_1, \ldots, x_n$, and is in $\langle x_1, \ldots, x_n \rangle \subset k[x_1, \ldots, x_n]$.
d. Let $R = k[x_1, \ldots, x_n, y_1, \ldots, y_m]$. Deduce that $\mathrm{Loc}_{>_{mixed}}(R)$ is the ring $k[x_1, \ldots, x_n]_{\langle x_1, \ldots, x_n \rangle}[y_1, \ldots, y_m]$, whose elements can be written as polynomials in the y_j, with coefficients that are rational functions of the x_i in $k[x_1, \ldots, x_n]_{\langle x_1, \ldots, x_n \rangle}$.

Exercise 4. If we proceed as in Exercise 3 but compare the x-terms first, we get a new order defined by $>_{mixed'}$ by $x^\alpha y^\beta >_{mixed'} x^{\alpha'} y^{\beta'}$ if either $x^\alpha >_{alex} x^{\alpha'}$, or $x^\alpha = x^{\alpha'}$ and $y^\beta >_{lex} y^{\beta'}$.
a. Show that $>_{mixed'}$ is a semigroup order and find a matrix U such that $>_{mixed'} = >_U$.
b. Show that $>_{mixed'}$ is *neither* a well-ordering, *nor* degree-anticompatible.
c. Which elements $f \in k[x_1, \ldots, x_n, y_1, \ldots, y_n]$ satisfy $1 >_{mixed'} \mathrm{LT}_{>_{mixed'}}(f)$?
d. What is $\mathrm{Loc}_{>_{mixed'}}(k[x_1, \ldots, x_n, y_1, \ldots, y_m])$?

Note that the order $>_{mixed}$ from Exercise 3 has the following *elimination property*: if $x^\alpha >_{mixed} x^{\alpha'} y^{\beta'}$, then $\beta' = 0$. Equivalently, any monomial containing one of the y_j is greater than all monomials containing only the x_i. It follows that if the $>_{mixed}$ leading term of a polynomial depends only on the x_i, then the polynomial does not depend on any of the y_j.

We will return to this comment in §4 after developing analogs of the division algorithm and Gröbner bases for general term orders, because this is precisely the property we need for elimination theory.

Given any semigroup order $>$ on monomials in $k[x_1, \ldots, x_n]$, there is a natural extension of $>$ to $\mathrm{Loc}_>(k[x_1, \ldots, x_n])$, which we will also denote by $>$. Namely, if $1 + g \in S$ as in Definition (3.5), the rational function $1/(1 + g)$ is a unit in $\mathrm{Loc}_>(k[x_1, \ldots, x_n])$, so it shouldn't matter in defining the leading term of $f/(1 + g)$. For any $h \in \mathrm{Loc}_>(k[x_1, \ldots, x_n])$, we write $h = f/(1 + g)$ and define

$$\mathrm{multideg}(h) = \mathrm{multideg}(f)$$
$$\mathrm{LC}(h) = \mathrm{LC}(f)$$
$$\mathrm{LM}(h) = \mathrm{LM}(f)$$
$$\mathrm{LT}(h) = \mathrm{LT}(f).$$

Exercise 5. Write $A = k[x_1, \ldots, x_n]$ and let $h \in A$.
a. Show that $\mathrm{multideg}(h)$, $\mathrm{LC}(h)$, $\mathrm{LM}(h)$, $\mathrm{LT}(h)$ are well-defined in $\mathrm{Loc}_>(A)$ in the sense that if $h = f/(1+g) = f'/(1+g')$, then $\mathrm{multideg}(h)$, $\mathrm{LC}(h)$, $\mathrm{LM}(h)$, $\mathrm{LT}(h)$ will be the same whether f or f' is used to compute them.
b. Let $r \in R$ be defined by the equation

$$h = \mathrm{LT}(h) + r.$$

Show that either $r = 0$ or $\mathrm{LT}(r) < \mathrm{LT}(h)$.

In Exercise 8, you will show that if $>$ is a local order, then every nonempty subset has a maximal element. This allows us to define $\mathrm{multideg}(h)$, $\mathrm{LC}(h)$, $\mathrm{LM}(h)$, $\mathrm{LT}(h)$ when $h \in k[[x_1, \ldots, x_n]]$ (or $h \in k\{x_1, \ldots, x_n\}$ if $k = \mathbb{R}$ or $\mathbb{C}$). Moreover, in this case, the multidegree and leading term of $h = f/(1 + g) \in k[x_1, \ldots, x_n]_{\langle x_1, \ldots, x_n \rangle}$ agree with what one obtains upon viewing h as a power series (via the series expansion of $1/(1 + g)$).

The goal of this section is to use general semigroup orders to develop an extension of the division algorithm in $k[x_1, \ldots, x_n]$ which will yield information about ideals in $R = \mathrm{Loc}_>(k[x_1, \ldots, x_n])$. The key step in the division algorithm for polynomials is the reduction of a polynomial f by a polynomial g. If $\mathrm{LT}(f) = m \cdot \mathrm{LT}(g)$, for some term $m = cx^\alpha$, we define

$$\mathrm{Red}\,(f, g) = f - m\,g,$$

and say that we have *reduced* f by g. The polynomial $\mathrm{Red}\,(f, g)$ is just what is left after the first step in dividing f by g—it is the first partial dividend. In general, the division algorithm divides a polynomial by a set of other polynomials by repeatedly reducing the polynomial by members of the set and adding leading terms to the remainder when no reductions are possible. This terminates in the case of polynomials because successive leading terms form a strictly decreasing sequence, and such sequences always terminate because a monomial order is always a well-ordering.

In the case of a local order on a power series ring, one can define Red (f, g) exactly as above. However, a sequence of successive reductions need no longer terminate. For example, suppose $f = x$ and we decide to divide f by $g = x - x^2$, so that we successively reduce by $x - x^2$. This gives the reductions:

$$f_1 = \text{Red}\,(f, g) = x^2$$
$$f_2 = \text{Red}\,(f_1, g) = x^3$$
$$\vdots$$
$$f_n = \text{Red}\,(f_{n-1}, g) = x^{n+1},$$

and so on, which clearly does not terminate. The difficulty, of course, is that under the antidegree order in $k[x]_{\langle x \rangle}$ or $k[[x]]$, we have the infinite strictly decreasing sequence of terms $x > x^2 > x^3 > \cdots$.

We can evade this difficulty with a splendid idea of Mora's. When dividing f_i by g, for instance, we allow ourselves to reduce not just by g, but also *by the result of any previous reduction*. That is, we allow reductions by f itself (which we can regard as the "zeroth" reduction), or by any of $f_1, \ldots, f_{i-1}$. More generally, when dividing a set of polynomials or power series, we allow ourselves to reduce by the original set together with the results of any previous reduction. So, in our example, where we are dividing $f = x$ by $g = x - x^2$, the first reduction is $f_1 = \text{Red}\,(f, g) = x^2$. For the next reduction, we allow ourselves to reduce f_1 by f as well as g. One checks that

$$\text{Red}\,(f_1, f) = \text{Red}\,(x^2, x) = 0,$$

so that we halt. Moreover, this reduction being zero implies $x^2 = xf$. If we combine this with the equation $f = 1 \cdot g + x^2$ which gives $f_1 = \text{Red}\,(f, g) = x^2$, we obtain the relation $f = g + xf$, or $(1 - x)f = g$. This last equation tells us that in $k[x]_{\langle x \rangle}$, we have

$$f = \frac{1}{1 - x}\,g.$$

In other words, the remainder on division of f by g is zero since x and $x - x^2 = x(1 - x)$ generate the same ideal in $k[x]_{\langle x \rangle}$ or $k[[x]]$.

Looking at the above example, one might ask whether it would always suffice to first reduce by g, then subsequently reduce by f. Sadly, this is not the case: it is easy to construct examples where the sequence of reductions does not terminate. Suppose, for example, that we wish to divide $f = x + x^2$ by $g = x + x^3 + x^5$.

Exercise 6. Show that in this case too, f and g generate the same ideal in $k[[x]]$ or $k[x]_{\langle x \rangle}$.

Reducing f by g and then subsequently reducing the results by $f_0 = f$ gives the sequence

$$f_1 = \text{Red}\,(f, g) = x^2 - x^3 - x^5$$
$$f_2 = \text{Red}\,(f_1, f) = -2x^3 - x^5$$

$$f_3 = \text{Red}\,(f_2, f) = 2x^4 - x^5$$
$$f_4 = \text{Red}\,(f_3, f) = -3x^5$$
$$f_5 = \text{Red}\,(f_4, f) = 3x^6$$
$$f_6 = \text{Red}\,(f_5, f) = -3x^7,$$

and so on, which again clearly does not terminate. However, we get something which does terminate by reducing f_5 by f_4:

$$f_5 = \text{Red}\,(f_4, f) = 3x^6$$
$$\tilde{f}_6 = \text{Red}\,(f_5, f_4) = 0.$$

From this, we can easily give an expression for f:

$$f = 1 \cdot g + (x - 2x^2 + 2x^3 - 3x^4) \cdot f + f_5.$$

However, we also have

$$f_5 = 3x^6 = 3x^5 \cdot x = 3x^5 \cdot \frac{x + x^2}{1 + x} = \frac{3x^5}{1 + x} f.$$

Backsubstituting this into the previous equation for f and multiplying by $1 + x$, we obtain

$$(1 + x)f = (1 + x)g + (1 + x)(x - 2x^2 + 2x^3 - 3x^4)f + 3x^5 f.$$

Then moving xf to the right-hand side gives an equation of the form

$$f = (\text{unit}) \cdot g + (\text{polynomial vanishing at } 0) \cdot f.$$

This, of course, is what we want according to Exercise 6; upon transposing and solving for f, we have $f = (\text{unit}) \cdot g$.

Our presentation will now follow the recent book [GrP], which describes the algorithms underlying the latest version of the computer algebra system **Singular**. We will introduce this system in the next section. Since we deal with orders that are not well-orderings, the difficult part is to give a division process that is guaranteed to terminate. The algorithm and termination proof from [GrP] use a clever synthesis of ideas due to Lazard and Mora, but the proof is (rather amazingly) both simpler and more general than Mora's original one. Using reductions by results of previous reductions as above, Mora developed a division process for polynomials based on a *local* order. His proof used a notion called the *écart* of a polynomial, a measurement of the failure of the polynomial to be homogeneous, and the strategy in the division process was to perform reductions that decrease the écart. This is described, for instance, in [MPT]. Also see Exercise 11 below for the basics of this approach. Lazard had shown how to do the same sort of division by homogenizing the polynomials and using an appropriate monomial order defined using the local order. In implementing **Singular**, the authors of [GrP] found that Mora's algorithm could be made to work for any semigroup order. The same result was found independently by Gräbe (see [Grä]). Theorem (3.10) below gives the precise statement.

To prepare, we need to describe Lazard's idea mentioned above. We will specify the algorithm by using the homogenizations of f and the f_i with respect to a new variable t. If $g \in k[x_1, \ldots, x_n]$ is any polynomial, we will write g^h for the homogenization of g with respect to t. That is, if $g = \sum_\alpha c_\alpha x^\alpha$ and d is the total degree of g, then

$$g^h = \sum_\alpha c_\alpha t^{d-|\alpha|} x^\alpha.$$

(3.6) Definition. Each semigroup order $>$ on monomials in the x_i extends to a semigroup order $>'$ on monomials in $t, x_1, \ldots, x_n$ in the following way. We define $t^a x^\alpha >' t^b x^\beta$ if either $a + |\alpha| > b + |\beta|$, or $a + |\alpha| = b + |\beta|$ and $x^\alpha > x^\beta$.

In Exercise 12 below, you will show that $>'$ is actually a *monomial order* on $k[t, x_1, \ldots, x_n]$.

By the definition of $>'$, it follows that if $t^a > t^{a'} x^\beta$ for some a, a', β with $a = a' + |\beta|$, then $1 > x^\beta$. Hence, writing $R = \mathrm{Loc}_>(k[x_1, \ldots, x_n])$,

(3.7) $t^a > t^{a'} x^\beta$ and $a = a' + |\beta| \Rightarrow 1 + x^\beta$ is a unit in R.

It is also easy to see from the definition that if $g \in k[x_1, \ldots, x_n]$, then homogenization takes the $>$-leading term of g to the $>'$-leading term of g^h— that is, $\mathrm{LT}_{>'}(g^h) = t^a \mathrm{LT}_>(g)$, where $a = d - |\mathrm{multideg}_>(g)|$. Conversely, if G is homogeneous in $k[t, x_1, \ldots, x_n]$, then dehomogenizing (setting $t = 1$) takes the leading term $\mathrm{LT}_{>'}(G)$ to $\mathrm{LT}_>(g)$, where $g = G|_{t=1}$.

Given polynomials $f, f_1, \ldots, f_s$ and a semigroup order $>$, we want to show that there is an algorithm (called Mora's normal form algorithm) for producing polynomials $h, u, a_1, \ldots, a_s \in k[x_1, \ldots, x_n]$, where $u = 1 + g$ and $\mathrm{LT}(g) < 1$ (so u is a unit in $\mathrm{Loc}_>(k[x_1, \ldots, x_n])$), such that

(3.8) $u \cdot f = a_1 f_1 + \cdots + a_s f_s + h,$

where $\mathrm{LT}(a_i)\mathrm{LT}(f_i) \leq \mathrm{LT}(f)$ for all i, and either $h = 0$, or $\mathrm{LT}(h) \leq \mathrm{LT}(f)$ and $\mathrm{LT}(h)$ is not divisible by any of $\mathrm{LT}(f_1), \ldots, \mathrm{LT}(f_s)$.

Several comments are in order here. First, note that the inputs $f, f_1, \ldots, f_s$, the remainder h, the unit u, and the quotients $a_1, \ldots, a_s$ in (3.8) are all *polynomials*. The equation (3.8) holds in $k[x_1, \ldots, x_n]$, and as we will see, all the computations necessary to produce it also take place in a polynomial ring. We get a corresponding statement in $\mathrm{Loc}_>(k[x_1, \ldots, x_n])$

by multiplying both sides by $1/u$:

$$f = (a_1/u)f_1 + \cdots + (a_s/u)f_s + (h/u).$$

By Exercise 11 of §1, restricting to ideals generated by polynomials entails no loss of generality when we are studying ideals in $k[x_1, \ldots, x_n]_{\langle x_1, \ldots, x_n \rangle} = \text{Loc}_>(k[x_1, \ldots, x_n])$ for a local order $>$. But the major reason for restricting the inputs to be polynomials is that that allows us to specify a completely *algorithmic* (i.e., finite) division process. In $k[[x_1, \ldots, x_n]]$ or $k\{x_1, \ldots, x_n\}$, even a single reduction—computing Red (f, g)—would take infinitely many computational steps if f or g were power series with infinitely many non-zero terms.

Second, when dividing f by $f_1, \ldots, f_s$ as in (3.8), we get a "remainder" h whose *leading term* is not divisible by any of the $\text{LT}(f_i)$. In contrast, if we divide using the division algorithm of Chapter 1, §2, we get a remainder containing *no terms* divisible by any of the $\text{LT}(f_i)$. Conceptually, there would be no problem with removing a term not divisible by any of the $\text{LT}(f_i)$ and continuing to divide. But as in the first comment, this process may not be finite.

On the surface, these differences make the results of the Mora normal form algorithm seem weaker than those of the division algorithm. Even so, we will see in the next section that the Mora algorithm is strong enough for many purposes, including local versions of Buchberger's criterion and Buchberger's algorithm.

Instead of working with the f, f_i, h, a_i, and u directly, our statement of the algorithm will work with their homogenizations, and with the order $>'$ from Definition (3.6). Let $F = f^h$ and $F_i = f_i^h$ for $i = 1, \ldots, s$. We first show that there are homogeneous polynomials $U, A_1, \ldots, A_n$ such that

$$(3.9) \qquad U \cdot F = A_1 F_1 + \cdots + A_s F_s + H,$$

where $\text{LT}(U) = t^a$ for some a,

$$a + \deg(F) = \deg(A_i) + \deg(F_i) = \deg(H)$$

whenever $A_i, H \neq 0$. Note that since U is homogeneous, if $\text{LT}(U) = t^a$, then by (3.7) when we set $t = 1$, the dehomogenization u is a unit in $\text{Loc}_>(k[x_1, \ldots, x_n])$. The other conditions satisfied by $U, A_1, \ldots, A_s, H$ are described in the following theorem.

(3.10) Theorem (Homogeneous Mora Normal Form Algorithm).
Given nonzero homogeneous polynomials $F, F_1, \ldots, F_s$ in $k[t, x_1, \ldots, x_n]$ and the monomial order $>'$ extending the semigroup order $>$ on monomials in the x_i, there is an algorithm for producing homogeneous polynomials $U, A_1, \ldots, A_s, H \in k[t, x_1, \ldots, x_n]$ satisfying

$$U \cdot F = A_1 F_1 + \cdots + A_s F_s + H,$$

where $\text{LT}(U) = t^a$ *for some a,*

$$a + \deg(F) = \deg(A_i) + \deg(F_i) = \deg(H)$$

whenever $A_i, H \neq 0$, $t^a\text{LT}(F) \geq' \text{LT}(A_i)\text{LT}(F_i)$, *and no* $\text{LT}(F_i)$ *divides* $t^b\text{LT}(H)$ *for any* $b \geq 0$.

PROOF. We give below the algorithm for computing the remainder H. (The computation of the A_i and U is described in the correctness argument below.) An important component of the algorithm is a set L consisting of possible divisors for reduction steps. As the algorithm proceeds, this set records the results of previous reductions for later use, according to Mora's idea.

> Input: $F, F_1, \ldots, F_s \in k[t, x_1, \ldots, x_n]$ homogeneous and nonzero
> Output: H as in the statement of Theorem (3.10)
>
> $H := F; L := \{F_1, \ldots, F_s\}; M := \{G \in L : \text{LT}(G)|\text{LT}(t^a H) \text{ for some } a\}$
> WHILE ($H \neq 0$ AND $M \neq \emptyset$) DO
>   SELECT $G \in M$ with a minimal
>   IF $a > 0$ THEN
> $L := L \cup \{H\}$
> $H := \text{Red}(t^a H, G)$
>   IF $H \neq 0$ THEN
> $M := \{G \in L : \text{LT}(G)|\text{LT}(t^a H) \text{ for some } a\}$

We claim that the algorithm terminates on all inputs and correctly computes H as described in the statement of the theorem.

To prove termination, let $\mathcal{M}_j$ denote the monomial ideal

$$\langle \text{LT}(L) \rangle = \langle \text{LT}(G) : G \in L \rangle \subset k[t, x_1, \ldots, x_n]$$

after the jth pass through the WHILE loop ($j \geq 0$). The loop either leaves L unchanged or adds the polynomial H. Thus

$$\mathcal{M}_j \subset \mathcal{M}_{j+1}.$$

Notice that when H is added to L, $\text{LT}(H)$ does not lie in $\mathcal{M}_j$, for if it did, then we would have

$$\text{LT}(G)|\text{LT}(H)$$

for some $G \in L$. Thus $\text{LT}(G)|\text{LT}(t^0 H)$, which would contradict our choice of H since a was chosen to be minimal, yet adding H to L requires $a > 0$. It follows that $\mathcal{M}_j \subset \mathcal{M}_{j+1}$ is a strict inclusion when a new element is added to L during the jth pass.

Since the polynomial ring $k[t, x_1, \ldots, x_n]$ satisfies the ascending chain condition on ideals, there is some N such that $\mathcal{M}_N = \mathcal{M}_{N+1} = \cdots$. By what we just proved, it follows that no new elements are added to L after the Nth pass through the WHILE loop. Thus, from this point on, the algorithm continues with a fixed set of divisors L, and at each step a reduction takes place decreasing the $>'$-leading term of H. Since $>'$ is a monomial order on $k[t, x_1, \ldots, x_n]$, the process must terminate as in the proof of the usual division algorithm.

To prove correctness, observe that the algorithm terminates when $H = 0$ or $M = \emptyset$. In the latter case, $\{F_1, \ldots, F_s\} \subset L$ tells us that $\mathrm{LT}(F_i)$ doesn't divide $\mathrm{LT}(t^b H) = t^b \mathrm{LT}(H)$ for any $1 \leq i \leq s$ and $b \geq 0$. Thus H has the correct divisibility properties when it is nonzero.

It remains to show that H satisfies an identity of the form (3.9) with $\mathrm{LT}(U) = t^a$. We will count passes through the WHILE loop starting at $j = 0$ and let H_j be the value of H at the beginning of the jth pass through the loop (so $H_0 = F$ at the start of the 0th pass). We will prove by induction on $j \geq 0$ that we have identities of the form

$$(3.11) \qquad U_k F = A_{1,k} F_1 + \cdots + A_{s,k} F_s + H_k, \quad 0 \leq k \leq j,$$

where U_k and $A_{i,k}$ are homogeneous with

$$\mathrm{LT}(U_k) = t^{a_k}$$

such that $a_k + \deg(F) = \deg(A_{i,k}) + \deg(F_i) = \deg(H_k)$ and, for $0 < k \leq j$,

$$(3.12) \qquad a_{k-1} \leq a_k \quad \text{and} \quad t^{a_k} \mathrm{LT}(H_{k-1}) >' t^{a_{k-1}} \mathrm{LT}(H_k).$$

Since $H_0 = F$, setting $U_0 = 1$ and $A_{l,0} = 0$ for all l shows that everything works for $j = 0$. Now assume $j > 0$. We need to prove that the polynomial H_{j+1} produced by the jth pass through the loop satisfies the above conditions.

If no $\mathrm{LT}(G)$ divides $t^b \mathrm{LT}(H_j)$ for any $b \geq 0$ and $G \in L$, then the algorithm terminates with H_j and we are done. Otherwise some $G \in L$ satisfies $\mathrm{LT}(G) | \mathrm{LT}(t^a H_j)$ with a minimal. Hence there is a term M such that

$$\mathrm{LT}(t^a H_j) = \mathrm{M}\, \mathrm{LT}(G).$$

There are two possibilities to consider: either $G = F_i$ for some i, or $G = H_\ell$ for some $\ell < j$.

If $G = F_i$ for some i, and a is chosen as above, then $H_{j+1} = \mathrm{Red}(t^a H_j, F_i)$ means that

$$t^a H_j = \mathrm{M}\, F_i + H_{j+1}.$$

If we multiply the equation (3.11) with $k = j$ by t^a and substitute, then we obtain

$$t^a U_j F = t^a A_{1,j} F_1 + \cdots + t^a A_{s,j} F_s + t^a H_j$$
$$= t^a A_{1,j} F_1 + \cdots + t^a A_{s,j} F_s + \mathrm{M}\, F_i + H_{j+1}.$$

Taking $U_{j+1} = t^a U_j$ and

$$A_{l,j+1} = \begin{cases} t^a A_{l,j} & \text{if } l \neq i \\ t^a A_{l,j} + \mathrm{M} & \text{if } l = i, \end{cases}$$

we get an expression of the form (3.11) with $k = j + 1$. Also note that $\text{LT}(U_{j+1}) = t^{a+a_j}$.

On the other hand, if G is a result H_ℓ of a previous reduction, then $H_{j+1} = \text{Red}(t^a H_j, H_\ell)$ means that

$$t^a H_j = \text{M} H_\ell + H_{j+1}.$$

Now take (3.11) with $k = j$ (resp. $k = \ell$) and multiply by t^a (resp. M). Subtracting gives the equation

$$(t^a U_j - \text{M} U_\ell)F = (t^a A_{1,j} - \text{M} A_{1,\ell})F_1 + \cdots + (t^a A_{s,j} - \text{M} A_{s,\ell})F_s + H_{j+1}.$$

Setting $U_{j+1} = t^a U_j - \text{M} U_\ell$ and $A_{l,j+1} = t^a A_{l,j} - \text{M} A_{l,\ell}$, we see that (3.11) holds for $k = j + 1$. As for $\text{LT}(U_{j+1})$, note that (3.12) implies $t^{a_j}\text{LT}(H_\ell) >'$ $t^{a_\ell}\text{LT}(H_j)$ since $\ell < j$. Thus

$$t^{a+a_j}\text{LT}(H_\ell) = t^a t^{a_j}\text{LT}(H_\ell) >' t^a t^{a_\ell}\text{LT}(H_j) = t^{a_\ell}\text{LT}(t^a H_j) = t^{a_\ell}\text{M}\,\text{LT}(H_\ell),$$

which gives $t^{a+a_j} >' t^{a_\ell}\text{M}$. Using $\text{LT}(U_j) = t^{a_j}$ and $\text{LT}(U_\ell) = t^{a_\ell}$, we obtain

$$\text{LT}(U_{j+1}) = \text{LT}(t^a U_j - \text{M} U_\ell) = t^{a+a_j}.$$

Finally, note that $\text{LT}(U_{j+1}) = t^{a+a_j}$ in both cases, so that $a_{j+1} = a + a_j \geq a_j$. Also

$$\text{LT}(t^a H_j) >' \text{LT}(H_{j+1})$$

since H_{j+1} is a reduction of $t^a H_j$. From here, it is straightforward to show that (3.12) holds for $k = j + 1$. This completes the induction and shows that H has the required properties.

To finish the proof, we need to show that

$$a + \deg(F) = \deg(A_i) + \deg(F_i) \quad \text{and} \quad t^a \text{LT}(F) \geq' \text{LT}(A_i)\text{LT}(F_i)$$

when $A_i \neq 0$. You will do this in Exercise 13.   □

Next, we claim that after homogenizing, applying the homogeneous Mora normal form algorithm, and dehomogenizing, we obtain an expression (3.8) satisfying the required conditions. Here is the precise result.

(3.13) Corollary (Mora Normal Form Algorithm). *Suppose that $f, f_1, \ldots, f_s \in k[x_1, \ldots, x_n]$ are nonzero and $>$ is a semigroup order on monomials in the x_i. Then there is an algorithm for producing polynomials $u, a_1, \ldots, a_s, h \in k[x_1, \ldots, x_n]$ such that*

$$uf = a_1 f_1 + \cdots + a_s f_s + h,$$

where $\text{LT}(u) = 1$ (so u is a unit in $\text{Loc}_>(k[x_1, \ldots, x_n]))$, $\text{LT}(a_i)\text{LT}(f_i) \leq \text{LT}(f)$ for all i with $a_i \neq 0$, and either $h = 0$, or $\text{LT}(h)$ is not divisible by any $\text{LT}(f_i)$.

PROOF. See Exercise 14.   □

Exercise 7. Carry out the Mora normal form algorithm dividing $f = x^2 + y^2$ by $f_1 = x - xy$, $f_2 = y^2 + x^3$ using the *alex* order in $k[x, y]$.

In $\mathrm{Loc}_>(k[x_1, \ldots, x_n])$, we get a version of the Mora algorithm that doesn't require f to be a polynomial. Recall from Exercise 5 that $\mathrm{LT}(f)$ makes sense for any nonzero $f \in \mathrm{Loc}_>(k[x_1, \ldots, x_n])$.

(3.14) Corollary. *Let $>$ be a semigroup order on monomials in the ring $k[x_1, \ldots, x_n]$ and let $R = \mathrm{Loc}_>(k[x_1, \ldots, x_n])$. Let $f \in R$ and $f_1, \ldots, f_s \in k[x_1, \ldots, x_n]$ be nonzero. Then there is an algorithm for computing $h, a_1, \ldots, a_s \in R$ such that*

$$f = a_1 f_1 + \cdots + a_s f_s + h,$$

where $\mathrm{LT}(a_i)\mathrm{LT}(f_i) \leq \mathrm{LT}(f)$ for all i with $a_i \neq 0$, and either $h = 0$, or $\mathrm{LT}(h) \leq \mathrm{LT}(f)$ and $\mathrm{LT}(h)$ is not divisible by any of $\mathrm{LT}(f_1), \ldots, \mathrm{LT}(f_s)$.

PROOF. If we write f in the form f'/u' where $f', u' \in k[x_1, \ldots, x_n]$ and u' is a unit in R, then dividing f' by $f_1, \ldots, f_s$ via Corollary (3.13) gives

$$u \cdot f' = a_1' f_1 + \cdots + a_s' f_s + h',$$

where $u, h', a_1', \ldots, a_s'$ are as in the corollary. Also observe that $\mathrm{LT}(h') \leq \mathrm{LT}(h)$ follows from $\mathrm{LT}(a_i')\mathrm{LT}(f_i) \leq \mathrm{LT}(f')$. Since the leading term of a unit is a nonzero constant (see Exercise 2), dividing a polynomial by a unit doesn't affect the leading term (up to multiplication by a nonzero constant). Thus, dividing the above equation by the unit $u\,u'$ gives

$$f = a_1 f_1 + \cdots + a_s f_s + h,$$

where $a_i = a_i'/(uu')$, $h = h'/(uu')$ clearly have the required properties. $\square$

In the next section, we will use the Mora normal form algorithm to extend Buchberger's algorithm for Gröbner bases to ideals in local rings.

ADDITIONAL EXERCISES FOR §3

Exercise 8. Let $>$ be a local order on monomials in $k[x_1, \ldots, x_n]_{\langle x_1, \ldots, x_n \rangle}$ and $k[[x_1, \ldots, x_n]]$.
a. Show that every nonempty set of monomials has a maximal element under $>$. Hint: Define $>_r$ by $x^\alpha >_r x^\beta$ if and only if $x^\alpha < x^\beta$. Use Corollary 6 of Chapter 2, §4 of [CLO] to prove that $>_r$ is a well-ordering.
b. Use part a to define multideg(h) and $\mathrm{LT}(h)$ for $h \in k[[x_1, \ldots, x_n]]$.
c. Let $i : k[x_1, \ldots, x_n]_{\langle x_1, \ldots, x_n \rangle} \hookrightarrow k[[x_1, \ldots, x_n]]$ denote the inclusion obtained by writing each $h \in k[x_1, \ldots, x_n]_{\langle x_1, \ldots, x_n \rangle}$ in the form $f/(1+g)$ and then expanding $1/(1 + h)$ in a formal geometric series. Show that multideg$(h) =$ multideg$(i(h))$.

d. Deduce that

$$\text{LM}_>(h) = \text{LM}_>(i(h)), \quad \text{LC}_>(h) = \text{LC}_>(i(h)), \quad \text{and} \quad \text{LT}_>(h) = \text{LT}_>(i(h)).$$

Exercise 9. In the homogeneous Mora normal form algorithm (3.10), suppose that $h = 0$ after dehomogenizing. Show that f belongs to the ideal generated by $f_1, \ldots, f_s$ in the ring $R = \text{Loc}_>(k[x_1, \ldots, x_n])$. Is the converse always true?

Exercise 10. How should the homogeneous Mora normal form algorithm (3.10) be extended to return the quotients A_i and the unit U as well as the polynomial H? Hint: Use the proof of correctness.

Exercise 11. This exercise describes the way Mora based the original version of the normal form algorithm (for local orders) on the écart of a polynomial. Let $g \neq 0 \in k[x_1, \ldots, x_n]$, and write g as a finite sum of homogeneous nonzero polynomials of distinct total degrees:

$$g = \sum_{i=1}^{k} g_i, \quad g_i \text{ homogeneous},$$

with $\deg(g_1) < \cdots < \deg(g_k)$. The *order* of g, denoted $\text{ord}(g)$, is the total degree of g_1. The *total degree* of g, denoted $\deg(g)$ is the total degree of g_k. The *écart* of g, denoted $E(g)$, is the difference of the degree of g and the order of g:

$$E(g) = \deg(g) - \text{ord}(g).$$

By convention, we set $E(0) = -1$. Thus $E(g) \geq -1$ for all g. (The word *écart* is French for "difference" or "separation"—clearly a good description of the meaning of $E(g)$!)

a. Let $>$ be a local order and let f and g be two nonzero polynomials such that $\text{LT}(g)$ divides $\text{LT}(f)$. Then show that

$$E(\text{Red}\,(f, g)) \leq \max(E(f), E(g)).$$

b. In the one-variable case, part a gives a strategy that guarantees termination of division. Namely, at each stage, among all the polynomials by which we can reduce, we reduce by the polynomial whose écart is least. Show that this will ensure that the écarts of the sequence of partial dividends decreases to zero, at which point we have a monomial which can be used to reduce any subsequent partial dividend to 0.

c. Apply this strategy, reducing by the polynomial with the smallest possible écart at each step, to show that g divides f in $k[x]_{\langle x \rangle}$ in each of the following cases.

1. $g = x + x^2 + x^3$, $f = x^2 + 2x^7$. Note that there is no way to produce a sequence of partial dividends with *strictly* decreasing écarts in this case.

2. $g = x + x^2 + x^3$, $f = x + x^2 + x^3 + x^4$. Note that after producing a monomial with the first reduction, the écart must increase.

Exercise 12. Let $>$ be a semigroup order on monomials in $k[x_1, \ldots, x_n]$ and extend to $>'$ on monomials in $t, x_1, \ldots, x_n$ as in the text: define $t^a x^\alpha >' t^b x^\beta$ if either $a + |\alpha| > b + |\beta|$ or $a + |\alpha| = b + |\beta|$, but $x^\alpha > x^\beta$.
a. Show that $>'$ is actually a *monomial order* on $k[t, x_1, \ldots, x_n]$.
b. Show that if $> \, = \, >_M$ for an $m \times n$ matrix M, then $>'$ is the order $>_{M'}$ where M' is the $(m + 1) \times (n + 1)$ matrix

$$
\begin{pmatrix}
1 & 1 & \cdots & 1 \\
0 & & & \\
\vdots & & M & \\
0 & & &
\end{pmatrix}.
$$

Exercise 13. Prove that at every stage of the homogeneous Mora normal form algorithm from Theorem (3.10), the polynomials $U, A_1, \ldots, A_s, H$ are homogeneous and satisfy the conditions

$$
a + \deg(F) = \deg(A_i) + \deg(F_i) = \deg(H)
$$
$$
t^a \mathrm{LT}(F) \geq' \mathrm{LT}(A_i)\mathrm{LT}(F_i)
$$

whenever $A_i, H \neq 0$.

Exercise 14. Prove Corollary (3.13) using the homogeneous polynomials produced by the homogeneous Mora normal form algorithm described in the proof of Theorem (3.10). Hint: See the paragraph following (3.7).

Exercise 15. In [GrP], Mora's original notion of écart (described in Exercise 11) is modified to create a version of the Mora normal form algorithm which works directly with the polynomial ring $k[x_1, \ldots, x_n]$ and the semigroup order $>$. Define the écart of $f \in k[x_1, \ldots, x_n]$ to be

$$
\mathrm{ecart}(f) = \deg(f) - \deg(\mathrm{LT}(f)).
$$

Given nonzero polynomials $f, f_1, \ldots, f_s \in k[x_1, \ldots, x_n]$, prove that the remainder h from Corollary (3.13) is produced by the following algorithm.

```
h := f; L := {f_1, ..., f_s}; M := {g ∈ L : LT(g)|LT(h)}
WHILE (h ≠ 0 AND M ≠ ∅) DO
      SELECT g ∈ M with ecart(g) minimal
      IF ecart(g) > ecart(h) THEN
          L := L ∪ {h}
      h := Red(h, g)
      IF h ≠ 0 THEN
          M := {g ∈ L : LT(g)|LT(h)}
```

§4 Standard Bases in Local Rings

In this section, we want to develop analogs of Gröbner bases for ideals in any one of our local rings $R = k[x_1, \ldots, x_n]_{\langle x_1, \ldots, x_n \rangle}$, $R = k\{x_1, \ldots, x_n\}$, or $R = k[[x_1, \ldots, x_n]]$. Just as for well-orderings, given an ideal I in R, we define the *set of leading terms* of I, denoted $\mathrm{LT}(I)$, to be the set of all leading terms of elements of I with respect to $>$. Also, we define the ideal of leading terms of I, denoted $\langle \mathrm{LT}(I) \rangle$, to be the ideal generated by the set $\mathrm{LT}(I)$ in R. Also just as for ideals in polynomial rings, it can happen that $I = \langle f_1, \ldots, f_s \rangle$ but $\langle \mathrm{LT}(I) \rangle \neq \langle \mathrm{LT}(f_1), \ldots, \mathrm{LT}(f_s) \rangle$ for an ideal $I \subset R$. By analogy with the notion of a Gröbner basis, we make the following definition.

(4.1) Definition. Let $>$ be a semigroup order and let R be the ring of fractions $\mathrm{Loc}_> (k[x_1, \ldots, x_n])$ as in Definition (3.5), or let $>$ be a local order and let $R = k[[x_1, \ldots, x_n]]$ or $k\{x_1, \ldots, x_n\}$. Let $I \subset R$ be an ideal. A *standard basis* of I is a set $\{g_1, \ldots, g_t\} \subset I$ such that $\langle \mathrm{LT}(I) \rangle = \langle \mathrm{LT}(g_1), \ldots, \mathrm{LT}(g_t) \rangle$.

In the literature, the term "standard basis" is more common than "Gröbner basis" when working with local orders and the local rings $R = k[x_1, \ldots, x_n]_{\langle x_1, \ldots, x_n \rangle}$, $k[[x_1, \ldots, x_n]]$, or $k\{x_1, \ldots, x_n\}$ so we use that terminology here.

Every nonzero ideal in these local rings has standard bases. As a result, there is an analog of the Hilbert Basis Theorem for these rings: every ideal has a finite generating set. The proof is the same as for polynomials (see Exercise 2 of Chapter 1, §3 and Exercise 2 below). Moreover, the Mora normal form algorithm—Corollary (3.13)—is well behaved when dividing by a standard basis. In particular, we obtain a zero remainder if and only if f is in the ideal generated by the standard basis (see Exercise 2).

However, in order to construct algorithms for *computing* standard bases, we will restrict our attention once more to ideals that are generated in these rings by collections of *polynomials*. Most of the ideals of interest in questions from algebraic geometry have this form. This will give us algorithmic control over such ideals. For example, we obtain a solution of the *ideal membership problem* for ideals generated by polynomials in the local rings under consideration.

Given polynomial generators for an ideal, how can we compute a standard basis for the ideal? For the polynomial ring $k[x_1, \ldots, x_n]$ and Gröbner bases, the key elements were the division algorithm and Buchberger's algorithm. Since we have the Mora algorithm, we now need to see if we can carry Buchberger's algorithm over to the case of local or other semigroup orders. That is, given a collection $f_1, \ldots, f_s$ of polynomials, we would like to find a standard basis with respect to some local order of the ideal $\langle f_1, \ldots, f_s \rangle$ they generate in a local ring R. More generally, one could also look for algorithms for computing standard bases of ideals in $\mathrm{Loc}_>(k[x_1, \ldots, x_n])$ for any semigroup order.

It is a pleasant surprise that the ingredients fall into place with no difficulty. First, the definition of S-polynomials in this new setting is exactly the same as in $k[x_1, \ldots, x_n]$ (see Definition (3.2) of Chapter 1), but here we use the leading terms with respect to our chosen semigroup order.

Next, recall that Buchberger's algorithm consists essentially of forming S-polynomials of all elements in the input set $F = \{f_1, \ldots, f_s\}$ of polynomials, finding remainders upon division by F, adding to F any nonzero remainders, and iterating this process (see §3 of Chapter 1). Since we have the Mora normal form algorithm, whose output is a sort of remainder on division, we can certainly carry out the same steps as in Buchberger's algorithm. As with any algorithm, though, we have to establish its correctness (that is, that it gives us what we want) and that it terminates.

In the case of well-orders, correctness of Buchberger's algorithm is guaranteed by Buchberger's criterion, which states that a finite set G is a Gröbner basis if and only if the remainder upon division by G of every S-polynomial formed from pairs of elements of G is 0 (see Chapter 1, §3).

The following theorem gives analogs of Buchberger's criterion and Buchberger's algorithm for the ring of a semigroup order.

(4.2) Theorem. *Let $S \subset k[x_1, \ldots, x_n]$ be finite, let $>$ be any semigroup order, and let I be the ideal in $R = \mathrm{Loc}_>(k[x_1, \ldots, x_n])$ generated by S.*

a. *(Analog of Buchberger's Criterion) $S = \{g_1, \ldots, g_t\}$ is a standard basis for I if and only if applying the Mora normal form algorithm given in Corollary (3.13) to every S-polynomial formed from elements of the set S yields a zero remainder.*

b. *(Analog of Buchberger's Algorithm) Buchberger's algorithm, using the Mora normal form algorithm in place of the usual polynomial division algorithm, computes a polynomial standard basis for the ideal generated by S, and terminates after finitely many steps.*

PROOF. Let $\overline{f}^{\,S,\mathrm{Mora}}$ be the remainder h computed by Corollary (3.13) on division of f by S. If S is a standard basis of I, then since $S(g_i, g_j) \in I$ for all i, j, Exercise 2 implies that $\overline{S(g_i, g_j)}^{\,S,\mathrm{Mora}} = 0$ for all i, j.

Conversely, we need to show that $\overline{S(g_i, g_j)}^{\,S,\mathrm{Mora}} = 0$ for all i, j implies that S is a standard basis, or equivalently that $\langle \mathrm{LT}(I) \rangle = \langle \mathrm{LT}(g_1), \ldots, \mathrm{LT}(g_t) \rangle$, using the order $>$. We will give the proof in the special case when $>$ is degree-anticompatible, meaning that $|\alpha| > |\beta| \Rightarrow x^\alpha < x^\beta$. Examples are the orders $>_{alex}$ or $>_{arevlex}$ from Definitions (3.2) and (3.3). Given $f \in I = \langle g_1, \ldots, g_t \rangle$, we prove that $\mathrm{LT}(f) \in \langle \mathrm{LT}(g_1), \ldots, \mathrm{LT}(g_t) \rangle$ as follows. Consider the nonempty set

$$\mathcal{S}_f = \{\max\{\mathrm{LT}(a_i g_i)\} : a_1, \ldots, a_s \in R \text{ satisfy } f = \textstyle\sum_{i=1}^t a_i g_i\}.$$

For a general semigroup order, we can't claim that $\mathcal{S}_f$ has a minimal element, even though $\mathcal{S}_f$ is bounded below by $\mathrm{LT}(f)$. However, in Exercise 3, you will show that this is true for degree-anticompatible orders. Hence we can let $\delta = \min \mathcal{S}_f$. From here, the rest of the argument that $\mathrm{LT}(f) \in \langle \mathrm{LT}(g_1), \ldots, \mathrm{LT}(g_t) \rangle$ is a straightforward adaptation of the proof of Theorem 6 of Chapter 2, §6 of [CLO] (you will verify this in Exercise 4). This proves Buchberger's criterion for degree-anticompatible orders. The general case requires an analysis of the syzygy module of $g_1, \ldots, g_s$ (see Theorem 2.5.9 of [GrP] for the details).

For part b, observe that the usual proof that Buchberger's algorithm terminates and yields a Gröbner basis depends only on the ascending chain condition for polynomial ideals (applied to the chain of monomial ideals generated by the leading terms of the "partial bases" constructed as the algorithm proceeds—see the proof of Theorem 2 of [CLO], Chapter 2, §2). It does not require that the order used for the division process be a well-order. It follows that, replacing each ordinary remainder computation by a computation of the remainder from Mora's algorithm, we get an algorithm that terminates after a finite number of steps. Moreover, on termination, the result gives a standard basis for I by part a.   □

The Mora normal form algorithm and standard basis algorithms using local orders or more general semigroup orders > are not implemented directly in the Gröbner basis packages in Maple or *Mathematica*. They could be programmed directly in those systems, however, using the homogenization process and the order >′ from Definition (3.6). Alternatively, according to Lazard's original idea, the standard Buchberger algorithm could be applied to the homogenizations of a generating set for I. This approach is sketched in Exercise 5 below and can be carried out in any Gröbner basis implementation. Experience seems to indicate that standard basis computation with Mora's normal form algorithm is more efficient than computation using Lazard's approach, however. The CALI package for REDUCE does contain an implementation of Buchberger's algorithm using semigroup orders including local orders.

There is also a powerful package called **Singular** described in [GrP] and available via the World Wide Web from the University of Kaiserslautern (see the **Singular** homepage at http://www.singular.uni-kl.de/) that carries out these and many other calculations. In particular, **Singular** is set up so that local orders, monomial orders (well-orderings), and mixed orders can be specified in a unified way as $>_M$ orders for integer matrices M. This means that it can be used for both Gröbner and standard basis computations. Here is a very simple **Singular** session computing a standard basis of the ideal generated by

$$x^5 - xy^6 + z^7, \ xy + y^3 + z^3, \ x^2 + y^2 - z^2$$

in $R = k[x, y, z]_{\langle x,y,z \rangle}$ using the *alex* order, and computing the multiplicity
of the origin as a solution of the corresponding system of equations.

```
> ring r = 32003, (x,y,z), Ds;
> ideal i = x5-xy6+z7, xy+y3+z3, x2+y2-z2;
> ideal j=std(i);
4(2)s5.8-s(2)s9..s(3).10.---sH(11)
product criterion:8 chain criterion:7
> j;
j[1]=x2+y2-1z2
j[2]=xy+y3+z3
j[3]=y3-1yz2-1xy3-1xz3
j[4]=xz4-1y6+2y4z2-1y3z3+2yz5-1xy6+z7
j[5]=y2z4-1z6+xy6-2xy4z2+xy3z3-2xyz5+x2y6-1xz7
j[6]=yz7
j[7]=z9
> vdim(j);
24
```

Singular can work either with a finite field of coefficients or with $k = \mathbb{Q}$
or a finite extension of $\mathbb{Q}$. The first line here defines the characteristic of the
field, the ring variables, and the monomial order. The Ds is an abbreviation
for the *alex* order, which could also be specified by a matrix as follows

```
> ring r = 32003, (x,y,z), ((-1,-1,-1),(0,-1,-1),(0,0,-1));
```

as in Exercise 2 of §3. The ideal I is defined by the three polynomials above,
J contains the standard basis (seven polynomials in all), and the vdim
command computes the dimension of dim $R/\langle \text{LT}(J) \rangle$. For more information
about this very flexible package, we refer the interested reader to [GrP].

We've already commented on how standard bases enable one to solve
the ideal membership problem in local rings, just as Gröbner bases solve
the corresponding problem in polynomial rings. Another important use of
Gröbner bases is the computation of dim $k[x_1, \ldots, x_n]/I$ when this dimension is finite. For the local version of this result, we will use the following
terminology: given a local order $>$ and an ideal I in one of the local
rings $k[x_1, \ldots, x_n]_{\langle x_1,\ldots,x_n \rangle}$, $k[[x_1, \ldots, x_n]]$ or $k\{x_1, \ldots, x_n\}$, we say that
a monomial x^α is *standard* if

$$x^\alpha \notin \langle \text{LT}(I) \rangle.$$

Then we have the following result about standard monomials.

(4.3) Theorem. *Let R be one of the local rings $k[x_1, \ldots, x_n]_{\langle x_1,\ldots,x_n \rangle}$,
$k[[x_1, \ldots, x_n]]$ or $k\{x_1, \ldots, x_n\}$. If $I \subset R$ is an ideal and $>$ is a local
order, then the following are equivalent.*
a. *dim R/I is finite.*

b. $\dim R/\langle \mathrm{LT}(I)\rangle$ *is finite.*
c. *There are only finitely many standard monomials.*

Furthermore, when any of these conditions is satisfied, we have

$$\dim R/I = \dim R/\langle \mathrm{LT}(I)\rangle = \text{number of standard monomials}$$

and every $f \in R$ *can be written uniquely as a sum*

$$f = g + r,$$

where $g \in I$ *and* r *is a linear combination of standard monomials. In addition, this decomposition can be computed algorithmically when* $R = k[x_1, \ldots, x_n]_{\langle x_1, \ldots, x_n\rangle}$.

PROOF. We first prove a $\Rightarrow$ c. Suppose that $x^{\alpha(1)}, \ldots, x^{\alpha(m)}$ are standard monomials with $m > \dim R/I$. It follows easily that there is a nontrivial linear combination

$$f = \sum_{i=1}^{\ell} c_i x^{\alpha(i)} \in I, \quad c_i \in k.$$

Then $\mathrm{LT}(f) \in \langle \mathrm{LT}(I)\rangle$ implies that some $x^{\alpha(i)} \in \langle \mathrm{LT}(I)\rangle$, which is impossible since $x^{\alpha(i)}$ is standard. This shows that the number of standard monomials is bounded above by $\dim R/I$.

For c $\Rightarrow$ a, suppose that $R = k[x_1, \ldots, x_n]_{\langle x_1, \ldots, x_n\rangle}$. Then Exercise 11 of §1 implies that I is generated by polynomials, which means that we can compute a polynomial standard basis G of I. Now take $f \in R$ and divide f by G using Corollary (3.14) to obtain

$$f = g_1 + h_1,$$

where $g_1 \in I$ and either $h_1 = 0$ or $\mathrm{LT}(h_1) \notin \langle \mathrm{LT}(G)\rangle = \langle \mathrm{LT}(I)\rangle$ (since G is a standard basis) and $\mathrm{LT}(f) \geq \mathrm{LT}(h_1)$. Note that we are using the extension of LT to R studied in Exercise 5 of §3.

If $h_1 \neq 0$, let $\mathrm{LT}(h_1) = c_1 x^{\alpha(1)}$, $c_1 \in k$, $c_1 \neq 0$. Thus $x^{\alpha(1)}$ is standard and, by Exercise 5 of §3, $h_1 = c_1 x^{\alpha(1)} + r_1$, where $r_1 = 0$ or $x^{\alpha(1)} > \mathrm{LT}(r_1)$. If $r_1 \neq 0$, then applying the above process gives

$$r_1 = g_2 + h_2 = g_2 + c_2 x^{\alpha(2)} + r_2$$

with $g_2 \in I$, $x^{\alpha(2)}$ standard, and $r_2 = 0$ or $x^{\alpha(2)} > \mathrm{LT}(r_2)$. If we combine this with the formula for f, we obtain

$$f = g_1 + h_1 = g_1 + c_1 x^{\alpha(1)} + r_1 = (g_1 + g_2) + c_1 x^{\alpha(1)} + c_2 x^{\alpha(2)} + r_2,$$

where $g_1 + g_2 \in I$, $x^{\alpha(1)}, x^{\alpha(2)}$ standard, and $x^{\alpha(1)} > x^{\alpha(2)} > \mathrm{LT}(r_2)$ if $r_2 \neq 0$. We can continue this process as long as we have nonzero terms to work with. However, since there are only finitely many standard monomials, this process must eventually terminate, which shows that f has the form $g + r$ described in the statement of the theorem. We will leave it for the reader to prove uniqueness and describe an algorithm that carries out this process (see Exercise 6 below). It follows that the cosets of the standard monomials give a basis of R/I, proving

$$\dim R/I = \text{number of standard monomials}$$

when $R = k[x_1, \ldots, x_n]_{\langle x_1, \ldots, x_n \rangle}$.

When $R = k\{x_1, \ldots, x_n\}$ or $R = k[[x_1, \ldots, x_n]]$, if we assume that we can perform the Mora Normal Form Algorithm on inputs from R, then the above argument applies for any $f \in R$. The details of how this works will be discussed in Exercise 2 below. This completes the proof of c $\Rightarrow$ a and the final assertions of the theorem.

It remains to prove b $\Leftrightarrow$ c. This follows immediately from what we have already proved since I and $\langle \text{LT}(I) \rangle$ have the same standard monomials. $\square$

When $R = k[[x_1, \ldots, x_n]]$ or $R = k\{x_1, \ldots, x_n\}$, there are more powerful versions of Theorem (4.3) that don't assume that $\dim R/\langle \text{LT}(I) \rangle$ is finite. In these situations, the remainder r is an infinite series, none of whose terms are in $\langle \text{LT}(I) \rangle$. See, for example, [Hir] or [MPT]. However, for $R = k[x_1, \ldots, x_n]_{\langle x_1, \ldots, x_n \rangle}$, it is possible to find ideals $I \subset R$ where nice remainders don't exist (see [AMR], Example 2).

ADDITIONAL EXERCISES FOR §4

Exercise 1. In this exercise and the next, we will show that every ideal I in one of our local rings R has standard bases, and derive consequences about the structure of R. Let $>$ be any local order on R.
a. Explain why $\langle \text{LT}(I) \rangle$ has a finite set of generators.
b. For each $x^{\alpha(i)}$, $i = 1, \ldots, t$, in a finite set of generators of $\langle \text{LT}(I) \rangle$, let $g_i \in I$ be an element with $\text{LT}(g_i) = x^{\alpha(i)}$. Deduce that $G = \{g_1, \ldots, g_t\}$ is a standard basis for I.

Exercise 2. If we ignore the fact that infinitely many computational steps are needed to perform reductions on power series in $k[[x_1, \ldots, x_n]]$ or $k\{x_1, \ldots, x_n\}$, then the Mora Normal Form Algorithm can be performed with inputs that are not polynomials. Hence we can assume that the Mora algorithm works for R, where R is either $k[[x_1, \ldots, x_n]]$ or $k\{x_1, \ldots, x_n\}$.
a. Let G be a standard basis for an ideal $I \subset R$. Show that we obtain a zero remainder on division of f by G if and only if $f \in I$.
b. Using part a, deduce that every ideal $I \subset R$ has a finite basis. (This is the analog of the Hilbert Basis Theorem for $k[x_1, \ldots, x_n]$.)
c. Deduce that the ascending chain condition holds for ideals in R. Hint: See Exercise 13 of §2 of Chapter 5.

Exercise 3. Let $>$ be a degree-anticompatible order on one of our local rings R. Show that any nonempty set of monomials S that is bounded below (meaning that there exists a monomial x^α such that $x^\beta \geq x^\alpha$ for all $x^\beta \in S$) has a smallest element.

Exercise 4. Carry out the proof of the analog of Buchberger's Criterion for degree-anticompatible orders, using Exercise 3 and the discussion before the statement of Theorem (4.2).

Exercise 5. This exercise discusses an alternative method due to Lazard for computing in local rings. Let $>'$ be the order in $k[t, x_1, \ldots, x_n]$ from Definition (3.6). Given polynomials $f_1, \ldots, f_s$, let $f_1^h, \ldots, f_s^h$ be their homogenizations in $k[t, x_1, \ldots, x_n]$, and let G be a Gröbner basis for $\langle f_1^h, \ldots, f_s^h \rangle$ with respect to the $>'$ consisting of *homogeneous* polynomials (such Gröbner bases always exist—see Theorem 2 in Chapter 8, §3 of [CLO], for instance). Show that the *dehomogenizations* of the elements of G (that is, the polynomials in $k[x_1, \ldots, x_n]$ obtained from the elements of G by setting $t = 1$) are a standard basis for the ideal generated by F in the local ring R with respect to the semigroup order $>$.

Exercise 6. Let $I \subset R = k[x_1, \ldots, x_n]_{\langle x_1, \ldots, x_n \rangle}$ be an ideal such that $\dim R/\langle \mathrm{LT}(I) \rangle$ is finite for some local order on R. Describe an algorithm which for the input $f \in R$ computes the remainder r from Theorem (4.3).

§5 Applications of Standard Bases

We will consider some applications of standard bases in this section. The multiplicity, and Milnor and Tjurina number computations we introduced in §2 can be carried out in an algorithmic fashion using standard bases. We begin by using Theorem (4.3) to prove Proposition (2.11), which asserts that if I is a zero-dimensional ideal of $k[x_1, \ldots, x_n]$ such that $0 \in \mathbf{V}(I)$, then the multiplicity of 0 is

$$\dim k[x_1, \ldots, x_n]_{\langle x_1, \ldots, x_n \rangle}/Ik[x_1, \ldots, x_n]_{\langle x_1, \ldots, x_n \rangle}$$

(5.1) $$= \dim k[[x_1, \ldots, x_n]]/Ik[[x_1, \ldots, x_n]]$$

$$= \dim k\{x_1, \ldots, x_n\}/Ik\{x_1, \ldots, x_n\},$$

where the last equality assumes $k = \mathbb{R}$ or $\mathbb{C}$. The proof begins with the observation that by Theorem (2.2), we know that

$$\dim k[x_1, \ldots, x_n]_{\langle x_1, \ldots, x_n \rangle}/Ik[x_1, \ldots, x_n]_{\langle x_1, \ldots, x_n \rangle} < \infty.$$

By Theorem (4.3), it follows that this dimension is the number of standard monomials for a standard basis S for $I \subset k[x_1, \ldots, x_n]_{\langle x_1, \ldots, x_n \rangle}$. However, S is also a standard basis for $Ik[[x_1, \ldots, x_n]]$ and $Ik\{x_1, \ldots, x_n\}$ by Buchberger's criterion. Thus, for a fixed local order, the standard monomials are *the same* no matter which of the local rings R we are considering. Then (5.1) follows immediately from Theorem (4.3).

This gives an algorithm for computing multiplicities. Exercises 2 and 3 below give some nice examples. In the same way, we can compute the Milnor and Tjurina numbers defined in §2 (see Exercise 4).

Standard bases in local rings have other geometric applications as well. For instance, suppose that $V \subset k^n$ is a variety and that $p = (a_1, \ldots, a_n)$ is a point of V. Then the *tangent cone* to V at p, denoted $C_p(V)$, is defined to be the variety

$$C_p(V) = \mathbf{V}(f_{p,min} : f \in \mathbf{I}(V)),$$

where $f_{p,min}$ is the homogeneous component of lowest degree in the polynomial $f(x_1 + a_1, \ldots, x_n + a_n)$ obtained by translating p to the origin (see part b of Exercise 17 of §2). A careful discussion of tangent cones, including a Gröbner basis method for computing them, can be found in Chapter 9, §7 of [CLO]. However, standard bases give a more direct way to compute tangent cones than the Gröbner basis method. See Exercise 5 below for an outline of the main ideas.

Here is another sort of application, where localization is used to concentrate attention on one irreducible component of a reducible variety. To illustrate the idea, we will use an example from Chapter 6, §4 of [CLO]. In that section, we showed that the hypotheses and the conclusions of a large class of theorems in Euclidean plane geometry can be expressed as polynomial equations on the coordinates of points specified in the construction of the geometric figures involved in their statements. For instance, consider the theorem which states that the diagonals of a parallelogram $ABCD$ in the plane intersect at a point that bisects both diagonals (Example 1 of [CLO], Chapter 6, §4). We place the vertices A, B, C, D of the parallelogram as follows:

$$A = (0,0), \quad B = (u,0), \quad C = (v,w), \quad D = (a,b),$$

and write the intersection point of the diagonals $\overline{AD}$ and $\overline{BC}$ as $N = (c,d)$. We think of the coordinates u, v, w as arbitrary; their values determine the values of a, b, c, d. The conditions that $ABCD$ is a parallelogram and N is the intersection of the diagonals can be written as the following polynomial equations:

$$h_1 = b - w = 0$$
$$h_2 = (a - u)w - bv = 0$$
$$h_3 = ad - cw = 0$$
$$h_4 = d(v - u) - (c - u)w = 0,$$

as can the conclusions of the theorem (the equalities between the lengths $AN = DN$ and $BN = CN$)

$$g_1 = a^2 - 2ac - 2bd + b^2 = 0$$
$$g_2 = 2cu - 2cv - 2dw - u^2 + v^2 + w^2 = 0.$$

Since the geometric theorem is true, we might naively expect that the conclusions $g_1 = g_2 = 0$ are satisfied whenever the hypothesis equations $h_1 = h_2 = h_3 = h_4 = 0$ are satisfied. If we work over the algebraically closed field $\mathbb{C}$, then the Strong Nullstellensatz shows that our naive hope is equivalent to

$$g_i \in \mathbf{I}(\mathbf{V}(h_1, h_2, h_3, h_4)) = \sqrt{\langle h_1, h_2, h_3, h_4 \rangle}.$$

However, as the following exercise illustrates, this is unfortunately not true.

Exercise 1. Use the radical membership test from [CLO], Chapter 4, §2 to show that

$$g_1, g_2 \notin \sqrt{\langle h_1, h_2, h_3, h_4 \rangle} \subset \mathbb{C}[u, v, w, a, b, c, d].$$

Thus neither conclusion g_1, g_2 follows directly from the hypothesis equations.

In fact, in [CLO], Chapter 6, §4 we saw that the reason for this was that the variety $\mathbf{V}(h_1, h_2, h_3, h_4) \subset \mathbb{C}^7$ defined by the hypotheses is actually *reducible*, and the conclusion equations $g_i = 0$ are not identically satisfied on several of the irreducible components of H. The points on the "bad" components correspond to *degenerate special cases* of the configuration A, B, C, D, N such as "parallelograms" in which two of the vertices A, B, C, D coincide. In [CLO], Chapter 6, §4 we analyzed this situation very carefully and found the "good" component of H, on which the conclusions $g_1 = g_2 = 0$ do hold. Our purpose here is to point out that what we did in [CLO] can also be accomplished more easily by *localizing* appropriately.

Note that taking $(u, v, w) = (1, 1, 1)$ gives an "honest" parallelogram. If we now translate $(1, 1, 1)$ to the origin as in Exercise 17 of §2, and write the translated coordinates as (U, V, W, a, b, c, d), the hypotheses and conclusions become

$$h_1 = b - W - 1 = 0$$
$$h_2 = (a - U - 1)(W + 1) - b(V + 1) = 0$$
$$h_3 = ad - c(W + 1) = 0$$
$$h_4 = d(V - U) - (c - U - 1)(W + 1)$$
$$g_1 = a^2 - 2ac - 2cd + b^2 = 0$$
$$g_2 = 2c(U + 1) - 2c(V + 1) - 2d(W + 1) - (U + 1)^2$$
$$+ (V + 1)^2 + (W + 1)^2 = 0.$$

Using **Singular**, we can compute a standard basis for the ideal generated by the h_i in the localization $R = \mathbb{Q}[U, V, W]_{\langle U, V, W \rangle}[a, b, c, d]$ as follows.

```
> ring r = 0, (a,b,c,d,U,V,W), (Dp(4),Ds(3));
> ideal i = b-W-1, (a-U-1)*(W+1)-b*(V+1), ad-c*(W+1), d*(V-U)-
(c-U-1)*(W+1);
> ideal j = std(i);
> j;
j[1]=a+aW-1b-1bV-1-1U-1W-1UW
j[2]=b-1-1W
j[3]=c+cW+dU-1dV-1-1U-1W-1UW
j[4]=2d+2dU+2dW+2dUW-1-1U-2W-2UW-1W2-1UW2
```

The first line sets up the ring R by specifying the coefficient field $k = \mathbb{Q}$ and a mixed order on the variables as in Exercise 3 of §3 of this chapter, with *alex* on the variables U, V, W, ordinary *lex* on a, b, c, d, and all monomials containing a, b, c, d greater than any monomial in U, V, W alone. If we now apply the Mora algorithm from Corollary (3.13), which is provided in the Singular command reduce, we find that both conclusions are actually in the ideal generated by h_1, h_2, h_3, h_4 in R.

```
> poly g=a2-2ac-2bd+b2;
> poly h=reduce(g,j);
> h;
0
> poly m = 2c*(U+1)-2c*(V+1)-2d*(W+1)-(U+1)^2+(V+1)^2+(W+1)^2;
> poly n = reduce(m,j);
> n;
0
```

This shows that *locally* near the point with $(u, v, w) = (1, 1, 1)$ on the variety $\mathbf{V}(h_1, h_2, h_3, h_4)$, the conclusions do follow from the hypotheses. Using the mixed order in the Mora algorithm, we have an equation

$$u \cdot g_1 = a_1 h_1 + \cdots + a_4 h_4,$$

where $u \in \mathbb{Q}[U, V, W]$ is a unit in $\mathbb{Q}[U, V, W]_{\langle U, V, W \rangle}$, and a similar equation for g_2. In particular, this shows that Proposition 8 of Chapter 6, §4 of [CLO] applies and the conclusions g_1, g_2 *follow generically* from the hypotheses h_i, as defined there.

Along the same lines we have the following general statement, showing that localizing at a point p in a variety V implies that we ignore components of V that do not contain p.

(5.2) Proposition. *Let $I \subset k[x_1, \ldots, x_n]$ and suppose that the origin in k^n is contained in an irreducible component W of $\mathbf{V}(I)$. Let $f_1, \ldots, f_s \in k[x_1, \ldots, x_n]$ be a standard basis for I with respect to a local order, and let $g \in k[x_1, \ldots, x_n]$. If the remainder of g on division by $F = (f_1, \ldots, f_s)$ using the Mora algorithm from Corollary (3.13) is zero, then $g \in \mathbf{I}(W)$ (but not necessarily in I).*

PROOF. If the remainder is zero, the Mora algorithm yields an equation

$$u \cdot g = a_1 f_1 + \cdots + a_s f_s,$$

where $u \in k[x_1, \ldots, x_n]$ is a unit in $k[x_1, \ldots, x_n]_{\langle x_1, \ldots, x_n \rangle}$. Since $W \subset V(I)$, $u \cdot g$ is an element of $\mathbf{I}(W)$. But W is irreducible, so $\mathbf{I}(W)$ is a prime ideal, and hence $u \in \mathbf{I}(W)$ or $g \in \mathbf{I}(W)$. The first alternative is not possible since $u(0) \neq 0$. Hence $g \in \mathbf{I}(W)$.   □

It is natural to ask if we can carry out operations on ideals in local rings algorithmically in ways similar to the Gröbner basis methods reviewed in Chapter 1 for ideals in polynomial rings. In the final part of this section, we will show that the answer is yes when $R = k[x_1, \ldots, x_n]_{\langle x_1, \ldots, x_n \rangle}$. Since many of the proofs in the polynomial case use elimination, we first need to study elimination in the local context. The essential point will be to work the new ring $k[x_1, \ldots, x_n]_{\langle x_1, \ldots, x_n \rangle}[t]$, whose elements can be thought of first as polynomials in t whose coefficients are elements of $k[x_1, \ldots, x_n]_{\langle x_1, \ldots, x_n \rangle}$.

In this situation, if we have an ideal $I \subset k[x_1, \ldots, x_n]_{\langle x_1, \ldots, x_n \rangle}[t]$, the basic problem is to find the intersection

$$I_0 = I \cap k[x_1, \ldots, x_n]_{\langle x_1, \ldots, x_n \rangle}.$$

Note that I_0 is analogous to an *elimination ideal* of a polynomial ideal. This elimination problem can be solved using a local order $>$ on the local ring to construct a suitable semigroup order on $S = k[x_1, \ldots, x_n]_{\langle x_1, \ldots, x_n \rangle}[t]$ as follows (see [AMR] and [Grä] for the details).

(5.3) Definition. An *elimination order* on S is any semigroup order $>_{elim}$ on the monomials on S defined in the following way. Let $>$ be a local order in $k[x_1, \ldots, x_n]_{\langle x_1, \ldots, x_n \rangle}$. Then define

$$t^k x^\alpha >_{elim} t^l x^\beta$$

for $k, l \in \mathbb{Z}_{\geq 0}$, and $\alpha, \beta \in \mathbb{Z}_{\geq 0}^n$ if and only if $k > l$, or $k = l$ and $\alpha > \beta$. In other words, an elimination order is a product order combining the degree order on powers of t and the given local order $>$ on x^α in $k[x_1, \ldots, x_n]_{\langle x_1, \ldots, x_n \rangle}$.

Elimination orders on S are *neither* local nor well-orders. Hence, the full strength of the Mora algorithm for general semigroup orders is needed here. We have the following analog of the Elimination Theorem stated in Chapter 2, §1.

(5.4) Theorem (Local Elimination). *Fix an elimination order* $>_{elim}$ *on* $S = k[x_1, \ldots, x_n]_{\langle x_1, \ldots, x_n \rangle}[t]$. *Let* $I \subset S$ *be an ideal, and let* G *be a* *polynomial standard basis for I with respect to $>_{elim}$. Then*

$$G \cap k[x_1, \ldots, x_n] = \{g \in G : \text{LT}(g) \text{ does not contain } t\}$$

and this is a standard basis of $I_0 = I \cap k[x_1, \ldots, x_n]_{\langle x_1, \ldots, x_n \rangle}$.

PROOF. Let $G = \{g_1, \ldots, g_t\}$ be a standard basis of I and $G_0 = \{g \in G : \text{LT}(g) \text{ does not contain } t\}$. By the definition of $>_{elim}$, the condition that $\text{LT}(g)$ does not contain t implies that g does not contain t. Since $G_0 \subset I_0$, we need only show that if $f \in I_0 \cap k[x_1, \ldots, x_n]$, then f can be written as a combination of elements in G_0 with coefficients in $k[x_1, \ldots, x_n]_{\langle x_1, \ldots, x_n \rangle}$. Since $f \in I$ and $\{g_1, \ldots, g_t\}$ is a standard basis of I, the Mora algorithm gives an expression

$$f = a_1 g_1 + \cdots + a_t g_t$$

(see Exercise 2 of §4), where $\text{LT}(f) \geq \text{LT}(a_i g_i)$ for all $a_i \neq 0$. By our choice of order, we have $a_i = 0$ for $g_i \notin G_0$ and $g_i \in k[x_1, \ldots, x_n]_{\langle x_1, \ldots, x_n \rangle}$ otherwise, since t does not appear in $\text{LT}(f)$.   □

With this out of the way, we can immediately prove the following.

(5.5) Theorem. *Let $I, J \subset k[x_1, \ldots, x_n]_{\langle x_1, \ldots, x_n \rangle}$ and $f \in k[x_1, \ldots, x_n]$.*
a. $I \cap J = (t \cdot I + (1 - t) \cdot J) \cap k[x_1, \ldots, x_n]_{\langle x_1, \ldots, x_n \rangle}$.
b. $I : \langle f \rangle = \frac{1}{f} \cdot (I \cap \langle f \rangle)$.
c. $I : f^\infty = (I + \langle 1 - f \cdot t \rangle) \cap k[x_1, \ldots, x_n]_{\langle x_1, \ldots, x_n \rangle}$.
d. $f \in \sqrt{I}$ *if and only if* $1 \in I + \langle 1 - f \cdot t \rangle$ *in* $k[x_1, \ldots, x_n]_{\langle x_1, \ldots, x_n \rangle}[t]$.

PROOF. The proofs are the same as for polynomial ideals. (See Chapter 1 of this book, §2 and §3 of Chapter 4 of [CLO], and [AL] or [BW].)

We remind the reader that the *stable quotient* of I with respect to f, denoted $I : f^\infty$, is defined to be the ideal

$$I : f^\infty = \{g \in R : \text{ there exists } n \geq 1 \text{ for which } f^n g \in I\}.$$

The stable quotient is frequently useful in applications of local algebra. We also remark that the division in part b, where one divides the common factor f out from all generators of $I \cap \langle f \rangle$ in $k[x_1, \ldots, x_n]_{\langle x_1, \ldots, x_n \rangle}$, uses the Mora algorithm.   □

Just as the ability to do computations in polynomial rings extends to allow one to do computations in quotients (i.e., homomorphic images of polynomial rings), so, too, the ability to do computations in local rings extends to allow one to do computations in quotients of local rings. Suppose that $J \subset k[x_1, \ldots, x_n]_{\langle x_1, \ldots, x_n \rangle}$ and let $\overline{R} = k[x_1, \ldots, x_n]_{\langle x_1, \ldots, x_n \rangle}/J$. Then one can do computations algorithmically in $\overline{R}$ due to the following elementary proposition.

(5.6) Proposition. *Let* $\overline{I_1}, \overline{I_2} \subset \overline{R}$ *be ideals, and let* I_1, I_2 *denote their preimages in* $k[x_1, \ldots, x_n]_{\langle x_1, \ldots, x_n \rangle}$. *Let* $f \in k[x_1, \ldots, x_n]_{\langle x_1, \ldots, x_n \rangle}$ *and* $[f] \in \overline{R}$ *be its coset. Then:*

a. $\overline{I_1} \cap \overline{I_2} = (I_1 \cap I_2)/J;$

b. $\overline{I_1} : [f] = (I_1 : f)/J;$

c. $\overline{I_1} : [f]^\infty = (I_1 : f^\infty)/J.$

Using a standard basis of J allows one to determine whether $f, g \in R$ represent the same element in $\overline{R}$ (that is, whether $[f] = [g]$.) One can also compute Hilbert functions and syzygies over $\overline{R}$.

The techniques we have outlined above also extend to rings that are finite algebraic extensions of $k[x_1, \ldots, x_n]_{\langle x_1, \ldots, x_n \rangle}$ in $k[[x_1, \ldots, x_n]]$. This allows us to handle computations involving algebraic power series in $k[[x_1, \ldots, x_n]]$ algorithmically. See [AMR] for details. There are still many open questions in this area, however. Basically, one would hope to handle any operations on ideals whose generators are defined in some suitably algebraic fashion (not just ideals generated by polynomials), but there are many instances where no algorithms are known.

ADDITIONAL EXERCISES FOR §5

Exercise 2.

a. Let $f_1, \ldots, f_n \in k[x_1, \ldots, x_n]$ be homogeneous polynomials of degrees $d_1, \ldots, d_n$, respectively. Assume that $I = \langle f_1, \ldots, f_n \rangle$ is zero-dimensional, and that the origin is the only point in $\mathbf{V}(I)$. Show that the multiplicity is also the dimension of

$$k[x_1, \ldots, x_n]/\langle f_1, \ldots, f_n \rangle,$$

and then prove that the multiplicity of 0 as a solution of $f_1 = \cdots = f_n = 0$ is $d_1 \cdots d_n$. Hint: Regard $f_1, \ldots, f_n$ as homogeneous polynomials in $x_0, x_1, \ldots, x_n$, where x_0 is a new variable. Using $x_0, x_1, \ldots, x_n$ as homogeneous coordinates for $\mathbb{P}^n$, show that $f_1 = \cdots = f_n = 0$ have no nontrivial solutions when $x_0 = 0$, so that there are no solutions at ∞ in the sense of Chapter 3. Then use Bézout's Theorem as stated in Chapter 3.

b. Let $f(x_1, \ldots, x_n)$ be a homogeneous polynomial of degree d with an isolated singularity at the origin. Show that the Milnor number of f at the origin is $(d-1)^n$.

Exercise 3. Determine the multiplicity of the solution at the origin for each of the following systems of polynomial equations.

a. $x^2 + 2xy^4 - y^2 = xy - y^3 = 0.$

b. $x^2 + 2y^2 - y - 2z = x^2 - 8y^2 + 10z = x^2 - 7yz = 0.$

c. $x^2 + y^2 + z^2 - 2x^4 = x^3 - yz - x = x - y + 2z = 0.$

Exercise 4. Compute the Milnor and Tjurina numbers at the origin of the following polynomials (all of which have an isolated singularity at 0).

a. $f(x, y) = (x^2 + y^2)^3 - 4x^2y^2$. The curve $\mathbf{V}(f) \subset \mathbb{R}^2$ is the four-leaved rose—see Exercise 11 of [CLO], Chapter 3, §5.

b. $f(x, y) = y^2 - x^n$, $n \geq 2$. Express the Milnor number as a function of the integer n.

c. $f(x, y, z) = xyz + x^4 + y^4 + z^4$.

Exercise 5. (Tangent Cones) For each $f \in \langle x_1, \ldots, x_n \rangle$, let f_{min} be the homogeneous component of lowest degree in f. Let $V = \mathbf{V}(f_1, \ldots, f_s) \subset k^n$ be a variety containing the origin.

a. Let $G = \{g_1, \ldots, g_t\}$ be a standard basis for

$$I = \langle f_1, \ldots, f_s \rangle k[x_1, \ldots, x_n]_{\langle x_1, \ldots, x_n \rangle}$$

with respect to a degree-anticompatible order $>$. Explain why $\mathrm{LT}_>(g_i)$ is one of the terms in $g_{i,min}$ for each i.

b. Show that $\mathbf{V}(g_{1,min}, \ldots, g_{t,min})$ is the tangent cone of V at the origin.

c. Consider the variety $V = \mathbf{V}(x^3 - yz - x, y^2 + 2z^3)$ in k^3. Using the $>_{alex}$ order on $k[x, y, z]_{\langle x,y,z \rangle}$, with $x > y > z$, show that the two given polynomials in the definition of V are a standard basis for the ideal they generate, and compute the tangent cone of V at the origin using part b.

Exercise 6. For an r-dimensional linear subspace $L \subset \mathbb{C}^n$, a polynomial $f \in \mathbb{C}[x_1, \ldots, x_n]$ restricts to a polynomial function f_L on L.

a. Show that if f has an isolated singularity at the origin in $\mathbb{C}^n$, then for almost all r-dimensional subspaces $L \subset \mathbb{C}^n$, f_L has an isolated singularity at the origin in L.

b. One can show, in fact, that there is an open dense set $\mathcal{N}$ of all r-dimensional subspaces of $\mathbb{C}^n$ such that the Milnor number $\mu(f_L)$ of f_L at the origin does not depend on the choice of L in $\mathcal{N}$. This number is denoted $\mu^r(f)$. Show that $\mu^1(f) = \mathrm{mult}(f) - 1$ where $\mathrm{mult}(f)$ (the multiplicity of f) is the degree of the lowest degree term of f that occurs with nonzero coefficient.

c. Compute $\mu^2(f)$ and $\mu^3(f)$ if
  1. $f = x^5 + y^4 + z^7$;
  2. $f = x^4 + y^5 + z^6 + xyz$;
  3. $f = x^5 + xy^6 + y^7z + z^{15}$;
  4. $f = x^5 + y^7z + z^{15}$.

Note that if n is the number of variables, then $\mu^n(f) = \mu(f)$, so that $\mu^3(f)$ is just the usual Milnor number for these examples. To compute these numbers, use the `milnor` package in **Singular** and note that planes of the form $z = ax + by$ are an open set in the set of all planes in $\mathbb{C}^3$. One could also compute these Milnor numbers by hand. Note that examples 1, 3, and 4 are weighted homogeneous polynomials. For further background, the reader may wish to consult [Dim] or [AGV].

d. A family $\{f_t \in \mathbb{C}[x_1, \ldots, x_n]\}$ of polynomials with an isolated singularity at the origin for t near 0 is μ-*constant* if $\mu(f_0) = \mu(f_t)$ for t near 0. Show that $f_t = x^5 + y^4 + z^7 + tx^8y^2$ and $f_t = x^5 + txy^6 + y^7z + z^{15}$ are μ-constant families but $f_t = x^4 + y^5 + z^6 + txyz$ is not.

e. If $f \in \mathbb{C}[x_1, \ldots, x_n]$ has an isolated singularity at the origin, the n-tuple of integers $(\mu^1(f), \ldots, \mu^n(f))$ is called the *Teissier μ^*-invariant* of f. One says that a family $\{f_t\}$ is μ^*-*constant* if $\mu^*(f_0) = \mu^*(f_t)$. Show that $f_t = x^5 + txy^6 + y^7z + z^{15}$ is μ-constant, but not μ^* constant. This is a famous example due to Briançon and Speder—there are very few known examples of μ-constant families that are not μ^*-constant. At the time of writing, it is not known whether there exist μ-constant families in which μ^1 is not constant. The attempt to find such examples was one of the issues that motivated the development of early versions of Singular.

Chapter 5

Modules

Modules are to rings what vector spaces are to fields: elements of a given module over a ring can be added to one another and multiplied by elements of the ring. Modules arise in algebraic geometry and its applications because a geometric structure on a variety often corresponds algebraically to a module or an element of a module over the coordinate ring of the variety. Examples of geometric structures on a variety that correspond to modules in this way include subvarieties, various sets of functions, and vector fields and differential forms on a variety. In this chapter, we will introduce modules over polynomial rings (and other related rings) and explore some of their algebra, including a generalization of the theory of Gröbner bases for ideals.

§1 Modules over Rings

Formally, if R is a commutative ring with identity, an R-module is defined as follows.

(1.1) Definition. A *module over a ring* R (or R-module) is a set M together with a binary operation, usually written as addition, and an operation of R on M, called (scalar) multiplication, satisfying the following properties.

a. M is an abelian group under addition. That is, addition in M is associative and commutative, there is an additive identity element $0 \in M$, and each element $f \in M$ has an additive inverse $-f$ satisfying $f + (-f) = 0$.
b. For all $a \in R$ and all $f, g \in M$, $a(f + g) = af + ag$.
c. For all $a, b \in R$ and all $f \in M$, $(a + b)f = af + bf$.
d. For all $a, b \in R$ and all $f \in M$, $(ab)f = a(bf)$.
e. If 1 is the multiplicative identity in R, $1f = f$ for all $f \in M$.

The properties in the definition of a module may be summarized by saying that the scalar multiplication by elements of R interacts as nicely as possible with the addition operation in M. The simplest modules are those consisting of all $m \times 1$ columns of elements of R with componentwise addition and scalar multiplication:

$$
\begin{pmatrix} a_1 \\ a_2 \\ \vdots \\ a_m \end{pmatrix} + \begin{pmatrix} b_1 \\ b_2 \\ \vdots \\ b_m \end{pmatrix} = \begin{pmatrix} a_1 + b_1 \\ a_2 + b_2 \\ \vdots \\ a_m + b_m \end{pmatrix}, \qquad c \begin{pmatrix} a_1 \\ a_2 \\ \vdots \\ a_m \end{pmatrix} = \begin{pmatrix} ca_1 \\ ca_2 \\ \vdots \\ ca_m \end{pmatrix},
$$

for any $a_1, \ldots, a_m, b_1, \ldots, b_m, c \in R$. We call any such column a *vector* and the set of all such R^m.

One obtains other examples of R-modules by considering *submodules* of R^m, that is, subsets of R^m which are closed under addition and scalar multiplication by elements of R and which are, therefore, modules in their own right.

We might, for example, choose a finite set of vectors $\mathbf{f_1}, \ldots, \mathbf{f_s}$ and consider the set of all column vectors which can be written as an R-linear combination of these vectors:

$$
\{ a_1 \mathbf{f_1} + \cdots + a_s \mathbf{f_s} \in R^m, \text{where } a_1, \ldots, a_s \in R \}.
$$

We denote this set $\langle \mathbf{f_1}, \ldots, \mathbf{f_s} \rangle$ and leave it to you to show that this is an R-module.

Alternatively, consider an $l \times m$ matrix A with entries in the ring R. If we define matrix multiplication in the usual way, then for any $\mathbf{f} \in R^m$, the product $A\mathbf{f}$ is a vector in R^l. We claim (and leave it to you to show) that the set

$$
\ker A \equiv \{ \mathbf{f} \in R^m : A\mathbf{f} = \mathbf{0} \}
$$

where $\mathbf{0}$ denotes the vector in R^l all of whose entries are 0 is an R-module.

Exercise 1. Let R be any ring, and R^m the $m \times 1$ column vectors with entries in R.

a. Show that the set $\langle \mathbf{f_1}, \ldots, \mathbf{f_s} \rangle$ of R-linear combinations of any finite set $\mathbf{f_1}, \ldots, \mathbf{f_s}$ of elements of R^m is a submodule of R^m.

b. If A is an $l \times m$ matrix with entries in R, show that $\ker A$ is a submodule of R^m.

c. Let A be as above. Show that the set

$$
\operatorname{im} A \equiv \{ \mathbf{g} \in R^l : \mathbf{g} = A\mathbf{f}, \text{ for some } \mathbf{f} \in R^m \}
$$

is a submodule of R^l. In fact, show that it is the submodule consisting of all R-linear combinations of the columns of A considered as elements of R^l.

d. Compare parts a and c, and conclude that $\langle \mathbf{f_1}, \ldots, \mathbf{f_s} \rangle = \operatorname{im} F$ where F is the $m \times s$ matrix whose columns are the vectors $\mathbf{f_1}, \ldots, \mathbf{f_s}$.

The modules R^m are close analogues of vector spaces. In fact, if $R = k$ is a field, then the properties in Definition (1.1) are exactly the same as those defining a vector space over k, and it is a basic fact of linear algebra that the vector spaces k^m exhaust the collection of (finite-dimensional) k-vector spaces. (More precisely, any finite dimensional k-vector space is isomorphic to k^m for some m.) However, submodules of R^m when R is a polynomial ring can exhibit behavior very different from vector spaces, as the following exercise shows.

Exercise 2. Let $R = k[x, y, z]$.
a. Let $M \subset R^3$ be the module $\langle \mathbf{f}_1, \mathbf{f}_2, \mathbf{f}_3 \rangle$ where

$$\mathbf{f}_1 = \begin{pmatrix} y \\ -x \\ 0 \end{pmatrix}, \quad \mathbf{f}_2 = \begin{pmatrix} z \\ 0 \\ -x \end{pmatrix}, \quad \mathbf{f}_3 = \begin{pmatrix} 0 \\ z \\ -y \end{pmatrix}.$$

Show that $M = \ker A$ where A is the 1×3 matrix $(\, x \quad y \quad z \,)$.
b. Show that the set $\{\mathbf{f}_1, \mathbf{f}_2, \mathbf{f}_3\}$ is minimal in the sense that $M \neq \langle \mathbf{f}_i, \mathbf{f}_j \rangle$, $1 \leq i < j \leq 3$.
c. Show that the set $\{\mathbf{f}_1, \mathbf{f}_2, \mathbf{f}_3\}$ is R-linearly dependent. That is, show that there exist $a_1, a_2, a_3 \in R = k[x, y, z]$, not all zero, such that $a_1 \mathbf{f}_1 + a_2 \mathbf{f}_2 + a_3 \mathbf{f}_3 = \mathbf{0}$, where $\mathbf{0}$ is the zero vector in R^3.
d. Note that the preceding two parts give an example of a submodule of $k[x, y, z]^3$ in which there is a minimal generating set which is not linearly independent. This phenomenon cannot occur in any vector space.
e. In fact more is true. Show that there is *no* linearly independent set of vectors which generate the module M. Hint: First show that any linearly independent set could have at most two elements. A fairly brutal computation will then give the result.

On the other hand, some of the familiar properties of vector spaces carry over to the module setting.

Exercise 3. Let M be a module over a ring R.
a. Show that the additive identity $0 \in M$ is unique.
b. Show that each $f \in M$ has a unique additive inverse.
c. Show that $0f = 0$ where $0 \in R$ on the left hand side is the zero element of R and $0 \in M$ on the right hand side is the identity element in M.

Before moving on, we remark that up to this point in this book, we have used the letters f, g, h most often for single polynomials (or elements of the ring $R = k[x_1, \ldots, x_n]$). In discussing modules, however, it will be convenient to reserve the letters e, f, g, h to mean elements of modules over some ring R, most often in fact over $R = k[x_1, \ldots, x_n]$. In addition, we will use boldface letters $\mathbf{e}, \mathbf{f}, \mathbf{g}, \mathbf{h}$ for column vectors (that is, elements of the module R^m). This is not logically necessary, and may strike you as slightly schizophrenic, but we feel that it makes the text easier to read. For single ring elements, we will use letters such as a, b, c. Occasionally, for typographical reasons, we will need to write vectors as rows. In these cases, we use the notation $(a_1, \ldots, a_m)^T$ to indicate the column vector which is the *transpose* of the row vector $(a_1 \quad \ldots \quad a_m)$.

Many of the algebraic structures studied in Chapters 1 through 4 of this text may also be incorporated into the general context of modules as the exercise below shows. Part of what makes the concept of a module so useful is that it simultaneously generalizes the notion of ideal and quotient ring.

Exercise 4.

a. Show that an ideal $I \subset R$ is an R-module, using the sum and product operations from R.

b. Conversely, show that if a subset $M \subset R$ is a module over R, then M is an ideal in R.

c. Let I be an ideal in R. Show that the quotient ring $M = R/I$ is an R-module under the quotient ring sum operation, and the scalar multiplication defined for cosets $[g] \in R/I$, and $f \in R$ by $f[g] = [fg] \in R/I$.

d. (For readers of Chapter 4) Show that the localization $M = R_P$ of R at a prime ideal $P \subset R$ is a module over R, where the sum is the ring sum operation from R_P, and the scalar product of $b/c \in R_P$ by $a \in R$ is defined as $a \cdot b/c = ab/c \in R_P$.

e. Let M, N be two R-modules. The *direct sum* $M \oplus N$ is the set of all ordered pairs (f, g) with $f \in M$, and $g \in N$. Show that $M \oplus N$ is an R-module under the component-wise sum and scalar multiplication operations. Show that we can think of R^m as the direct sum

$$R^m = R \oplus R \oplus \ldots \oplus R$$

of R with itself m times.

We have already encountered examples of submodules of R^m. More generally, a subset of any R-module M which is itself an R-module (that is, which is closed under addition and multiplication by elements of R) is called a *submodule* of M. These are the analogues of vector subspaces of a vector space.

Exercise 5. Let $F \subset M$ be a subset and let $N \subset M$ be the collection of all $f \in M$ which can be written in the form

$$f = a_1 f_1 + \cdots + a_n f_n,$$

with $a_i \in R$ and $f_i \in F$ for all i. Show that N is a submodule of M.

The submodule N constructed in this exercise is called the *submodule of M generated by F*. Since these submodules are natural generalizations of the ideals generated by given subsets of the ring R, we will use the same notation—the submodule generated by a set F is denoted by $\langle F \rangle$. If $\langle F \rangle = M$, we say that F *spans* (or *generates*) M. If there is a finite set that generates M, then we say that M is *finitely generated*.

Exercise 6.
a. Let R be a ring. Show that $M = R^m$ is finitely generated for all m. Hint: Think of the standard basis for the vector space k^m and generalize.
b. Show that $M = k[x, y]$ is a module over $R = k[x]$ using the ring sum operation from $k[x, y]$ and the scalar multiplication given by polynomial multiplication of general elements of M by elements in R. However, show that M is not finitely generated as an R-module.

If N is a submodule of M, then the set of equivalence classes of elements of M where $f \in M$ is deemed equivalent to $g \in M$ if and only if $f - g \in N$ forms an R-module with the operations induced from M (we ask you to check this below). It is called the *quotient* of M by N and denoted by M/N.

Exercise 7. As above, let M, N be R-modules with $N \subset M$, let $[f] = \{g \in M : g - f \in N\}$ be the set of all elements of M equivalent to f, and denote the set of all sets of equivalent elements by M/N. These sets of equivalent elements are called *equivalence classes* or *cosets*. Note that we can write $[f] = f + N$. Show that M/N is an R-module if we define addition by $[f] + [g] = [f + g]$ and the scalar multiplication by $a[f] = [af]$ by $a \in R$. Hint: You need to show that these are well-defined. Also show that the zero element is the set $[0] = N$.

The quotient module construction takes a little getting used to, but is extremely powerful. Several other constructions of modules and operations on modules are studied in the additional exercises.

After defining any algebraic structure, one usually defines maps that preserve that structure. Accordingly, we consider module homomorphisms, the analogues of linear mappings between vector spaces.

(1.2) Definition. An *R-module homomorphism* between two R-modules M and N is an R-linear map between M and N. That is, a map $\varphi : M \rightarrow N$ is an R-module homomorphism if for all $a \in R$ and all $f, g \in M$, we have

$$\varphi(af + g) = a\varphi(f) + \varphi(g).$$

This definition implies, of course, that $\varphi(f + g) = \varphi(f) + \varphi(g)$ and $\varphi(af) = a\varphi(f)$ for all $a \in R$ and all $f, g \in M$.

When M and N are free modules, we can describe module homomorphisms in the same way that we specify linear mappings between vector spaces. For example, letting $M = N = R$, every R-module homomorphism $\varphi : R \to R$ is given by multiplication by a fixed $f \in R$—if $g \in R$, then $\varphi(g) = fg$. To see this, given φ, let $f = \varphi(1)$. Then for any $a \in R$,

$$\varphi(a) = \varphi(a \cdot 1) = a \cdot \varphi(1) = af = fa.$$

Conversely, by the distributive law in R, multiplication by any f defines an R-module homomorphism from R to itself.

More generally φ is a module homomorphism from R^l to R^m if and only if there exist l elements $\mathbf{f}_1, \ldots, \mathbf{f}_l \in R^m$ such that

$$\varphi((a_1, \ldots, a_l)^T) = a_1 \mathbf{f}_1 + \cdots + a_l \mathbf{f}_l$$

for all $(a_1, \ldots, a_l)^T \in R^l$. Given φ, and letting $\mathbf{e}_1, \mathbf{e}_2, \ldots, \mathbf{e}_l$ be the standard basis vectors in R^l (that is $\mathbf{e}_i$ is the column vector with a 1 in the ith row and a 0 in all other rows), we can see this as follows. For each i, let $\mathbf{f}_i = \varphi(\mathbf{e}_i)$. Each $(a_1, \ldots, a_l)^T$ can be written uniquely as $(a_1, \ldots, a_l)^T = a_1 \mathbf{e}_1 + \cdots + a_l \mathbf{e}_l$. But then, since φ is a homomorphism, knowing $\varphi(\mathbf{e}_j) = \mathbf{f}_j$ determines the value of $\varphi((a_1, \ldots, a_l)^T)$ for all $(a_1, \ldots, a_l)^T \in R^l$. Then expanding each $\mathbf{f}_j$ in terms of the standard basis in R^m, we see that φ may be represented as multiplication by a fixed $m \times l$ matrix $A = (a_{ij})$ with coefficients in R. The entries in the jth column give the coefficients in the expansion of $\mathbf{f}_j = \varphi(\mathbf{e}_j)$ in terms of the standard basis in R^m. We record the result of this discussion as follows (the second part of the proposition is a result of Exercise 1).

(1.3) Proposition. *Given any R-module homomorphism $\varphi : R^m \to R^l$, there exists an $l \times m$ matrix A with coefficients in R such that $\varphi(\mathbf{f}) = A\mathbf{f}$ for all $\mathbf{f} \in R^m$. Conversely, multiplication by any $l \times m$ matrix defines an R-module homomorphism from R^m to R^l.*

The discussion above actually shows that an R-module homomorphism $\varphi : M \to N$ between two R-modules is always determined once one knows the action of φ on a set of generators of M. However, unlike the situation in which $M = R^m$, one cannot define a homomorphism φ on M by specifying φ arbitrarily on a set of generators of M. The problem is that there may be relations among the generators, so that one must be careful to choose values of φ on the generators so that φ is well-defined. The following exercise should illustrate this point.

Exercise 8. Let $R = k[x, y]$.
a. Is there any R-module homomorphism φ from $M = \langle x^2, y^3 \rangle \subset R$ to R satisfying $\varphi(x^2) = y$ and $\varphi(y^3) = x$? Why or why not?
b. Describe all $k[x, y]$-homomorphisms of $\langle x^2, y^3 \rangle$ into $k[x, y]$.

As in the case of vector spaces, one can develop a theory of how the same homomorphism can be represented by matrices with respect to different sets of generators. We carry out some of this development in the exercises. We have already defined the kernel and image of matrix multiplication. The same definitions carry over to arbitrary homomorphisms.

(1.4) Definition. If $\varphi : M \to N$ is an R-module homomorphism between two R-modules M and N, define the *kernel* of φ, denoted $\ker(\varphi)$, to be the set

$$\ker(\varphi) = \{f \in M : \varphi(f) = 0\},$$

and the *image* of φ, denoted $\mathrm{im}(\varphi)$, to be the set

$$\mathrm{im}(\varphi) = \{g \in N : \text{there exists } f \in M \text{ with } \varphi(f) = g\}.$$

The homomorphism φ is said to be an *isomorphism* if it is both one-to-one and onto, and two R-modules M, N are called *isomorphic*, written $M \cong N$ if there is some isomorphism $\varphi : M \to N$.

The proofs of the following statements are the same as those of the corresponding statements for linear mappings between vector spaces, and they are left as exercises for the reader.

(1.5) Proposition. *Let $\varphi : M \to N$ be an R-module homomorphism between two R-modules M and N. Then*
a. $\varphi(0) = 0$.
b. $\ker(\varphi)$ *is a submodule of M.*
c. $\mathrm{im}(\varphi)$ *is a submodule of N.*
d. φ *is one-to one (injective) if and only if $\ker(\varphi) = \{0\}$.*

PROOF. See Exercise 16. □

When we introduce the notions of linear combinations and linear independence and R is not a field (for example when $R = k[x_1, \ldots, x_n]$), the theory of modules begins to develop a significantly different flavor from that of vector spaces. As in linear algebra, we say that a subset $F = \{f_1, \ldots, f_n\}$ of a module M is *linearly independent* over R (or R-linearly independent) if the only linear combination $a_1 f_1 + \cdots + a_n f_n$ with $a_i \in R$ and $f_i \in F$ which equals $0 \in M$ is the trivial one in which $a_1 = \cdots = a_n = 0$. A set $F \subset M$ which is R-linearly independent and which spans M is said to be a *basis* for M.

Recall from linear algebra that every vector space over a field has a basis. In Exercise 2, we saw that not every module has a basis. An even simpler example is supplied by the ideal $M = \langle x^2, y^3 \rangle \subset R$ studied in Exercise 8 (which is the same as the R-module generated by x^2 and y^3 in R). The set $\{x^2, y^3\}$ is not a basis for M as a module because x^2 and y^3 are not R-linearly independent. For example, there is a linear dependence relation $y^3 x^2 - x^2 y^3 = 0$, but the coefficients y^3 and $-x^2$ are certainly not 0. On the other hand, because $\{x^2, y^3\}$ spans M, it is a basis for M *as an ideal*.

More generally, any ideal M in $R = k[x_1, \ldots, x_n]$ which requires more than a single polynomial to generate it *cannot* be generated by an R-linearly independent set. This is true because any pair of polynomials $f_1, f_2 \in R$ that might appear in a generating set (an ideal basis) satisfies a non-trivial linear dependence relation $f_2 f_1 - f_1 f_2 = 0$ with coefficients in R. Thus the meaning of the word "basis" depends heavily on the context, and we will strive to make it clear to the reader which meaning is intended by using the phrases "ideal basis" or "module basis" to distinguish between the alternatives when there is a possibility of confusion. The following proposition gives a characterization of module bases.

(1.6) Proposition. *Let M be a module over a ring R. A set $F \subset M$ is a module basis for M if and only if every $f \in M$ can be written in one and only one way as a linear combination*

$$f = a_1 f_1 + \cdots + a_n f_n,$$

where $a_i \in R$, and $f_i \in F$.

PROOF. The proof is the same as the corresponding statement for bases of vector spaces.   □

The examples above show that, unlike vector spaces, modules need not have any generating set which is linearly independent. Those that do are given a special name.

(1.7) Definition. Let M be a module over a ring R. M is said to be a *free module* if M has a module basis (that is, a generating set that is R-linearly independent).

For instance, the R-module $M = R^m$ is a free module. The *standard basis* elements

$$\mathbf{e}_1 = \begin{pmatrix} 1 \\ 0 \\ 0 \\ \vdots \\ 0 \end{pmatrix}, \mathbf{e}_2 = \begin{pmatrix} 0 \\ 1 \\ 0 \\ \vdots \\ 0 \end{pmatrix}, \ldots, \mathbf{e}_m = \begin{pmatrix} 0 \\ 0 \\ 0 \\ \vdots \\ 1 \end{pmatrix},$$

form one basis for M as an R-module. There are many others as well. See Exercise 19 below.

We remark that just because a module has a single generator, it need not be free. As an example, let R be any polynomial ring and $f \in R$ a nonzero polynomial. Then $M = R/\langle f \rangle$ is generated by the set $[1]$ of elements equivalent to 1. But $[1]$ is not a basis because $f \cdot [1] = [f] = [0] = 0 \in M$.

In the following exercise we will consider another very important class of modules whose construction parallels one construction of vector subspaces in k^n, and we will see a rather less trivial example of free modules.

Exercise 9. Let $a_1, \ldots, a_m \in R$, and consider the set M of all solutions $(X_1, \ldots, X_m)^T \in R^m$ of the linear equation

$$a_1 X_1 + \cdots + a_m X_m = 0.$$

a. Show that M is a submodule of R^m. (In fact, this follows from Exercise 1 because $M = \ker A$ where A is the row matrix $A = (\,a_1 \quad \cdots \quad a_m\,)$.)

b. Take $R = k[x, y]$, and consider the following special case of the linear equation above:

$$X_1 + x^2 X_2 + (y - 2)X_3 = 0.$$

Show that $\mathbf{f}_1 = (-x^2, 1, 0)^T$, and $\mathbf{f}_2 = (-y + 2, 0, 1)^T$ form a basis for M as an R-module in this case.

c. Generalizing the previous part, show that if $R = k[x_1, \ldots, x_n]$, and one of the coefficients a_i in $a_1 X_1 + \cdots + a_m X_m = 0$ is a non-zero constant, then the module M of solutions is a free R-module.

It can be difficult to determine whether a submodule of R^m is free. For example, the following, seemingly innocuous, generalization of Exercise 9 follows from the solution in 1976 by Quillen [Qui] and Suslin [Sus] of a famous problem raised by Serre [Ser] in 1954. We will have more to say about this problem in the Exercises 25–27 and later in this chapter.

(1.8) Theorem (Quillen-Suslin). *Let $R = k[x_1, \ldots, x_n]$ and suppose that $a_1, \ldots, a_m \in R$ are polynomials that generate all of R (that is $\langle a_1, \ldots, a_m \rangle = \langle 1 \rangle = R$). Then the module M of all solutions $(X_1, \ldots, X_m)^T \in R^m$ of the linear equation*

$$a_1 X_1 + \cdots + a_m X_m = 0$$

is free.

In 1992, Logar and Sturmfels [LS] gave an algorithmic proof of the Quillen-Suslin result, and in 1994 Park and Woodburn [PW] gave an algorithmic procedure that allows one to compute a basis of $\ker A$ where A is an explicitly given unimodular row. The procedure depends on some algorithms that we will outline later in this chapter (and is quite complicated).

Exercise 10.

a. Let $a_1, \ldots, a_m \in R$. Show that the homomorphism $R^m \to R$ given by matrix multiplication $\mathbf{f} \mapsto A\mathbf{f}$ by the row matrix $A = (\,a_1 \quad \cdots \quad a_m\,)$ is onto if and only if $\langle a_1, \ldots, a_m \rangle = R$. Hint: A is onto if and only if $1 \in \mathrm{im}(A)$. Such a matrix is often called a *unimodular row*.

b. Show that Theorem (1.8) generalizes Exercise 9c.
c. Let $R = k[x, y]$ and consider the equation

$$(1 + x)X_1 + (1 - y)X_2 + (x + xy)X_3 = 0.$$

That is, consider $\ker A$ in the special case $A = (\, a_1 \ \ a_2 \ \ a_3 \,) = (\, 1 + x \ \ 1 - y \ \ x + xy \,)$. Show $1 \in \langle a_1, a_2, a_3 \rangle$.

d. Theorem (1.8) guarantees that one can find a basis for $M = \ker A$ in the special case of part c. Try to find one. Hint: This is hard to do directly—feel free to give up after trying and to look at Exercise 25.

e. In Exercise 25, we will show that the "trivial" relations,

$$\mathbf{h}_1 = \begin{pmatrix} a_2 \\ -a_1 \\ 0 \end{pmatrix}, \mathbf{h}_2 = \begin{pmatrix} a_3 \\ 0 \\ -a_1 \end{pmatrix}, \mathbf{h}_3 = \begin{pmatrix} 0 \\ a_3 \\ -a_2 \end{pmatrix}$$

generate $\ker A$. Assuming this, show that $\{\mathbf{h}_1, \mathbf{h}_2, \mathbf{h}_3\}$ is not linearly independent and no proper subset generates $\ker A$. This gives an example of a minimal set of generators of a free module that does not contain a basis.

The fact that some modules do not have bases (and the fact that even when they do, one may not be able to find them) raises the question of how one explicitly handles computations in modules. The first thing to note is that one not only needs a generating set, but also the set of all relations satisfied by the generators—otherwise, we have no way of knowing in general whether two elements expressed in terms of the generators are equal or not.

For instance, suppose you know that M is a $\mathbb{Q}[x, y]$-module and that f_1, f_2, f_3 is a generating set. If someone asks you whether $4f_1 + 5f_2 + 6f_3$ and $f_1 + 3f_2 + 4f_3$ represent the same element, then you cannot tell unless you know whether the difference, $3f_1 + 2f_2 + 2f_3$, equals zero in M. To continue the example, if you knew that every relation on the f_1, f_2, f_3 was a $\mathbb{Q}[x, y]$-linear combination of the relations $3f_1 + (1 + x)f_2 = 0, f_1 + (2x + 3)f_2 + 4yf_3 = 0$, and $(2 - 2x)f_2 + 4f_3 = 0$, then you could settle the problem provided that you could decide whether $3f_1 + 2f_2 + 2f_3 = 0$ is a $\mathbb{Q}[x, y]$-linear combination of the given relations (which it is).

Exercise 11. Verify that (no matter what f_i are), if every linear relation on the f_1, f_2, f_3 is a $\mathbb{Q}[x, y]$-linear combination of the relations $3f_1 + (1 + x)f_2 = 0, f_1 + (2x + 3)f_2 + 4yf_3 = 0$ and $(2 - 2x)f_2 + 4f_3 = 0$, then $3f_1 + 2f_2 + 2f_3 = 0$ is a $\mathbb{Q}[x, y]$-linear combination of the given relations.

It is worthwhile to say a few more words about relations at this point. Suppose that $F = (f_1, \ldots, f_t)$ is an ordered t-tuple of elements of some R-module M, so that $f_1, \ldots, f_t \in M$. Then a relation on F is an R-linear combination of the f_i which is equal to 0:

$$a_1 f_1 + \cdots + a_t f_t = 0 \in M.$$

We think of a relation on F as a t-tuple $(a_1, \ldots, a_t)$ of elements of R. Equivalently. we think of a relation as an element of R^t. Such relations are also called *syzygies* from the Greek word $\sigma v \zeta v \gamma \iota \alpha$ meaning "yoke" (and "copulation"). In fact we have the following statement.

(1.9) Proposition. *Let* $(f_1, \ldots, f_t)$ *be an ordered t-tuple of elements* $f_i \in M$. *The set of all* $(a_1, \ldots, a_t)^T \in R^t$ *such that* $a_1 f_1 + \cdots + a_t f_t = 0$ *is an R-submodule of* R^t, *called the (first) syzygy module of* $(f_1, \ldots, f_t)$, *and denoted* $\mathrm{Syz}(f_1, \ldots, f_t)$.

PROOF. Let $(a_1, \ldots, a_t)^T, (b_1, \ldots, b_t)^T$ be elements of $\mathrm{Syz}(f_1, \ldots, f_t)$, and let $c \in R$. Then

$$a_1 f_1 + \cdots + a_t f_t = 0$$
$$b_1 f_1 + \cdots + b_t f_t = 0$$

in M. Multiplying the first equation on both sides by $c \in R$, adding to the second equation and using the distributivity properties from the definition of a module, we obtain

$$(ca_1 + b_1)f_1 + \cdots (ca_t + b_t)f_t = 0.$$

This shows $(ca_1 + b_1, \ldots, ca_t + b_t)^T$ is also an element of $\mathrm{Syz}(f_1, \ldots, f_t)$. Hence $\mathrm{Syz}(f_1, \ldots, f_t)$ is a submodule of R^t.   □

This proposition allows us to be precise about what it means to "know" all relations on a fixed set of generators of a module. If there are t generators, then the set of relations is just a submodule of R^t. In Exercise 32 (and in the next section), we will show that any submodule of R^t, and hence any syzygy module, is finitely generated as a module, provided only that every ideal of R is finitely generated (i.e., provided that R is Noetherian). Hence, we "know" all relations on a set of generators of a module if we can find a set of generators for the first syzygy module.

Since we think of elements of R^t as column vectors, we can think of a finite collection of syzygies as columns of a matrix. If M is a module spanned by the t generators $f_1, \ldots, f_t$, then a *presentation matrix* for M is any matrix whose columns generate $\mathrm{Syz}(f_1, \ldots, f_t) \subset R^t$. So, for example, a presentation matrix A for the module of Exercise 11 would be

$$A = \begin{pmatrix} 3 & 1 & 0 \\ 1+x & 2x+3 & 2-2x \\ 0 & 4y & 4 \end{pmatrix}.$$

If A is a presentation matrix for a module M with respect to some generating set of M, then we shall say that A *presents* the module M. Note that the number of rows of A is equal to the number of generators in the generating set of M. The following proposition is easy to prove, but exceedingly useful.

(1.10) Proposition. *Suppose that A is an $l \times m$ matrix with entries in R, and suppose that A is the presentation matrix for two different R-modules M and N. Then*

a. *M and N are isomorphic as R-modules*

b. *M (and, hence, N) is isomorphic to R^l/AR^m where AR^m denotes the image im A of R^m under multiplication by A.*

PROOF. Part b clearly implies part a, but it is more instructive to prove part a directly. Since A is a presentation matrix for M, there is a set of generators $m_1, \ldots, m_l$ such that the columns of A generate the module of syzygies on $m_1, \ldots, m_l$. Similarly, there is a set of generators $n_1, \ldots, n_l$ of N such that the columns of A generate $\mathrm{Syz}(n_1, \ldots, n_l)$. Define a homomorphism $\varphi : M \to N$ by setting $\varphi(m_i) = n_i$ and extending linearly. That is, for any $c_1, \ldots, c_l \in R$, set $\varphi(\sum c_i m_i) = \sum c_i n_i$. We leave it to the reader to show that φ is well-defined (that is, if $\sum c_i m_i = \sum d_i m_i$ in M for $d_1, \ldots, d_l \in R$, then $\varphi(\sum c_i m_i) = \varphi(\sum d_i m_i)$) and one-one. It is clearly onto.

To show part b, note that if A is an $l \times m$ matrix, then AR^m is the submodule of R^l generated by the columns of A. The quotient module R^l/AR^m is generated by the cosets $\mathbf{e}_1 + AR^m, \ldots, \mathbf{e}_l + AR^m$ (where $\mathbf{e}_1, \ldots, \mathbf{e}_l$ denotes the standard basis of unit vectors in R^l), and $(c_1, \ldots, c_l)^T \in \mathrm{Syz}(\mathbf{e}_1 + AR^m, \ldots, \mathbf{e}_l + AR^m)$ if and only if $(c_1, \ldots, c_l)^T \in AR^m$ if and only if $(c_1, \ldots, c_l)^T$ is in the span of the columns of A. This says that A is a presentation matrix for R^l/AR^m. Now apply part a. $\square$

The presentation matrix of a module M is not unique. It depends on the set of generators that one chooses for M, and the set of elements that one chooses to span the module of syzygies on the chosen set of generators of M. We could, for example, append the column $(3, 2, 2)^T$ to the matrix A in the example preceding Proposition (1.10) above to get a 3×4 presentation matrix (see Exercise 11) of the same module. For a rather more dramatic example, see Exercise 30 below. In the exercises, we shall give a characterization of the different matrices that can present the same module. The following exercise gives a few more examples of presentation matrices.

Exercise 12. Let $R = k[x, y]$.

a. Show that the 2×1 matrix $\begin{pmatrix} x \\ 0 \end{pmatrix}$ presents the R module $k[y] \oplus k[x, y]$ where $k[y]$ is viewed as an R-module by defining multiplication by x to be 0.

b. What module does the 1×2 matrix $(\, x \quad 0 \,)$ present? Why does the 1×1 matrix (x) present the same module?

c. Find a matrix which presents the ideal $M = \langle x^2, y^3 \rangle \subset R$ as an R-module.

The importance of presentation matrices is that once we have a presentation matrix A for a module M, we have a concrete set of generators and relations for M (actually for an isomorphic copy of M), and so can work concretely with M. As an example, we characterize the homomorphisms of M into a free module.

(1.11) Proposition. *If A is an $l \times m$ presentation matrix for an R-module M, then any R-module homomorphism $\varphi : M \to R^t$ can be represented by a $t \times l$ matrix B such that $BA = 0$, where 0 denotes the $t \times m$ zero matrix. Conversely, if B is any $t \times l$ matrix with entries in R such that $BA = 0$, then B defines a homomorphism from M to R^t.*

PROOF. To see this, note that for M to have an $l \times m$ presentation matrix means that M can be generated by l elements $f_1, \ldots, f_l$, say. Hence, φ is determined by $\varphi(f_1), \ldots, \varphi(f_l)$, which we think of as columns of the $t \times l$ matrix B. We leave it as an exercise to show that φ is well-defined if and only if $BA = 0$.

Conversely, if A is a presentation matrix of M with respect to a generating set $\{f_1, \ldots, f_l\}$, and if B is any $t \times l$ matrix with entries in R such that $BA = 0$, then B defines a homomorphism from M to R^t by mapping $\sum c_i m_i$ to $B\mathbf{c}$ where $\mathbf{c} = (\, c_1 \quad \cdots \quad c_l \,)^T \in R^l$. Again, we leave the proof that the homomorphism is well-defined if $BA = 0$ as an exercise. $\square$

ADDITIONAL EXERCISES FOR §1

Exercise 13. The ring $k[x, y]$ can be viewed as a k-module, as a $k[x]$-module, as a $k[y]$-module, or as a $k[x, y]$-module. Illustrate the differences between these structures by providing a nontrivial example of a map from $k[x, y]$ to itself which is

a. a k-module homomorphism, but not a $k[x]$-module, $k[y]$-module, or $k[x, y]$-module homomorphism,

b. a $k[x]$-module homomorphism, but not a $k[y]$-module, or $k[x, y]$-module homomorphism,

c. a $k[y]$-module homomorphism, but not a $k[x]$-module, or $k[x, y]$-module homomorphism,

d. a ring homomorphism, but not a $k[x, y]$-module homomorphism.

Exercise 14. Let N_1, N_2 be submodules of an R-module M.

a. Show that $N_1 + N_2 = \{f_1 + f_2 \in M : f_i \in N_i\}$ is also a submodule of M.

b. Show that $N_1 \cap N_2$ is a submodule of M.

c. If N_1 and N_2 are finitely generated, show that $N_1 + N_2$ is also finitely generated.

Exercise 15. Show that every free module with a finite basis is isomorphic to R^m for some m. One can actually show more: namely, that any finitely generated free module is isomorphic to R^m. See Exercise 19.

Exercise 16. Prove Proposition (1.5).

Exercise 17. Let $R = k[x, y, z]$ and let $M \subset R^3$ be the module described in Exercise 2. Explicitly describe all homomorphisms $M \mapsto R^l$. Hint: The set of relations on $\mathbf{f}_1, \mathbf{f}_2, \mathbf{f}_3$ is generated by a single element which you can easily find.

Exercise 18. Complete the proof of Proposition (1.11).

Exercise 19. Let $R = k[x_1, \ldots, x_n]$.

a. Show that if $A = (a_{ij})$ is any invertible $s \times s$ matrix with coefficients in k, then the vectors

$$\mathbf{f}_i = a_{i1}\mathbf{e}_1 + \cdots + a_{is}\mathbf{e}_s,$$

$i = 1, \ldots, s$ also form a basis of the free module R^s.

b. Show that a finitely generated module N over a ring R is free if and only if N is isomorphic to $M = R^s$ as a module, for some s. (In view of Exercise 15, the point is to show that if a module is free and has a finite set of generators, then it has a finite basis.)

c. Show that $A = (a_{ij})$ is an invertible $s \times s$ matrix with coefficients in R if and only if $\det A$ is a non-zero element of k. Repeat part a with A invertible with coefficients in R. Hint: Consider the adjoint matrix of A as defined in linear algebra.

Exercise 20. Let M and N be R-modules.

a. Show that the set $\hom(M, N)$ of all R-module homomorphisms from M to N is an R-module with a suitable definition of addition and scalar multiplication.

b. If M is presented by a matrix A, and N is presented by a matrix B, what conditions must a matrix C representing a homomorphism $\varphi : M \to N$ satisfy? Hint: Compare with Proposition (1.11).

c. Find a matrix D presenting $\hom(M, N)$.

Exercise 21. Suppose that M, N are R-modules and $N \subset M$.

a. Show that the mapping $\nu : M \to M/N$ defined by $\nu(f) = [f] = f + N$ is an R-module homomorphism.

b. Let $\varphi : M \to N$. Show that there is an R-module isomorphism between $M/\ker(\varphi)$ and $\text{im}(\varphi)$.

Exercise 22. Let N_1 and N_2 be submodules of an R-module M, and define

$$(N_1 : N_2) = \{a \in R : af \in N_1 \text{ for all } f \in N_2\}.$$

Show that $(N_1 : N_2)$ is an ideal in R. The ideal $(0 : N)$ is also called the *annihilator* of N, denoted $\text{ann}(N)$.

Exercise 23.
a. Let M be an R-module, and let $I \subset R$ be an ideal. Show that $IM = \{af : a \in I, f \in M\}$ is a submodule of M.
b. We know that M/IM is an R-module. Show that M/IM is also an R/I-module.

Exercise 24.
a. Let L, M, N be R-modules with $L \subset M \subset N$. Describe the homomorphisms which relate the three quotient modules and show that N/M is isomorphic to $(N/L)/(M/L)$.
b. Let M, N be submodules of an R-module P. Show that $(M + N)/N$ is isomorphic to $M/(M \cap N)$.

(Note: The result in part a is often called the *Third Isomorphism Theorem* and that in part b the *Second Isomorphism Theorem*. The *First Isomorphism Theorem* is the result established in part b of Exercise 21.)

Exercise 25. This is a continuation of Exercise 10. We let $R = k[x, y]$ and consider the equation

$$(1 + x)X_1 + (1 - y)X_2 + (x + xy)X_3 = 0.$$

That is, we consider $\ker A$ in the special case $A = (\, a_1 \quad a_2 \quad a_3 \,) = (\, 1 + x \quad 1 - y \quad x + xy \,)$. Since $1 \in \langle a_1, a_2, a_3 \rangle$ (part c of Exercise 10), Theorem (1.8) guarantees that one can find a basis for $M = \ker A$ in the special case of Exercise 10c. We find a basis for M as follows.
a. Find a triple of polynomials $\mathbf{f} = (f_1, f_2, f_3)^T \in R^3$ such that $(1+x)f_1 + (1 - y)f_2 + (x + xy)f_3 = 1$.
b. By multiplying the relation $A\mathbf{f} = 1$ in part a by $1 + x$ and transposing, then by $1 - y$ and transposing, and finally by $x + xy$ and transposing, find three vectors $\mathbf{g}_1, \mathbf{g}_2, \mathbf{g}_3 \in \ker A$ (these vectors are the columns of the 3×3 matrix $I - \mathbf{f} \cdot A$, where I is the 3×3 identity matrix). Show these vectors span $\ker A$. Hint: If $A\mathbf{f} = 0$, then $\mathbf{f} = (I - \mathbf{f} \cdot A)\mathbf{f}$ is a linear combination of the colums of $I - \mathbf{f} \cdot A$.
c. Show that $\{\mathbf{g}_1, \mathbf{g}_2\}$ is a basis.

d. Use part b to show that the "trivial" relations

$$\mathbf{h}_1 = \begin{pmatrix} a_2 \\ -a_1 \\ 0 \end{pmatrix}, \quad \mathbf{h}_2 = \begin{pmatrix} a_3 \\ 0 \\ -a_1 \end{pmatrix}, \quad \mathbf{h}_3 = \begin{pmatrix} 0 \\ a_3 \\ -a_2 \end{pmatrix}$$

generate ker A. As pointed out in Exercise 10, they supply an example of a minimal set of generators of a free module that does not contain a basis.

Exercise 26. The goal of this exercise is to show how Theorem (1.8) follows from the solution of the Serre problem. An R-module M is called *projective* if it is a direct summand of a free module: that is, if there is an R-module N such that $M \oplus N$ is a finitely generated free module. In 1954, Serre asked whether every projective R-module when R is a polynomial ring is free and Quillen and Suslin independently proved that this was the case in 1976.

a. Show that $\mathbb{Z}/6 = \mathbb{Z}/3 \oplus \mathbb{Z}/2$, so that $\mathbb{Z}/3$ is a projective $\mathbb{Z}/6$-module which is clearly not a free $\mathbb{Z}/6$-module. (So, the answer to Serre's question is definitely negative if R is not a polynomial ring $k[x_1, \ldots, x_n]$.)

b. Let $R = k[x_1, \ldots, x_n]$ and let $A = (a_1 \cdots a_l)$ be a $1 \times l$ matrix such that $1 \in \langle a_1, \ldots, a_l \rangle$. Then multiplication by A defines an onto map $R^l \to R$. Show that $(\ker A) \oplus R \cong R^l$, so that $\ker A$ is projective. Hint: Fix $\mathbf{f} \in R^l$ such that $A\mathbf{f} = 1$. Given any $\mathbf{h} \in R^l$, write $\mathbf{h} = \mathbf{h}_1 + \mathbf{h}_2$ (uniquely) with $\mathbf{h}_2 = (A\mathbf{h})\mathbf{f}$ and $\mathbf{h}_1 = \mathbf{h} - (A\mathbf{h})\mathbf{f} \in \ker A$. The Quillen-Suslin result now implies Theorem (1.8).

Exercise 27. The purpose of this exercise is to generalize the methods of Exercise 25 to further investigate the result of Theorem (1.8). Let $R = k[x_1, \ldots, x_n]$ and let $A = (a_1 \cdots a_l)$ be a $1 \times l$ matrix such that $1 \in \langle a_1, \ldots, a_l \rangle$.

a. Choose $\mathbf{f} \in R^l$ such that $A\mathbf{f} = 1$. Generalize the result of Exercise 25b to show that the columns of $I - \mathbf{f} \cdot A$ are elements of Syz $(a_1, \ldots, a_l)$ that generate Syz $(a_1, \ldots, a_l)$.

b. Show that one can extract a basis from the columns of $I - \mathbf{f} \cdot A$ in the case that one of the entries of f is a nonzero element of R.

c. The preceding part shows Theorem (1.8) in the special case that there exists $\mathbf{f} \in R^l$ such that $A\mathbf{f} = 1$ and some entry of $\mathbf{f}$ is a non-zero element of k. Show that this includes the case examined in Exercise 9c. Also show that if $\mathbf{f}$ is as above, then the set $\{\mathbf{h} \in R^l : A\mathbf{h} = 1\} = \mathbf{f} + \text{Syz}(a_1, \ldots, a_l)$.

d. There exist unimodular rows A with the property that no $\mathbf{f} \in R^l$ such that $A\mathbf{f} = 1$ has an entry which is a nonzero element of k. (In the case $R = k[x, y]$, the matrix $A = (1 + xy + x^4 \quad y^2 + x - 1 \quad xy - 1)$ provides such an example.)

Exercise 28. Let $\varphi : M \to N$ be an R-module homomorphism. The *cokernel* of φ is by definition the module $\operatorname{coker}(\varphi) = N/\operatorname{im}(\varphi)$. Show that φ is onto if and only if $\operatorname{coker}(\varphi) = \{0\}$. (Note that in terms of this definition, Proposition (1.10) says that if M is an R-module with an $l \times m$ presentation matrix A, show that M is isomorphic to the cokernel of the homomorphism from R^l to R^m given by multiplication by A.)

Exercise 29. We have just seen that a presentation matrix determines a module up to isomorphism. The purpose of this exercise is to characterize the operations on a matrix which do not change the module it presents.

a. Let A be the $m \times n$ matrix representing a homomorphism $\varphi : R^n \to R^m$ with respect to bases $F = (f_1, \ldots, f_n)$ of R^n and bases $G = (g_1, \ldots, g_m)$ of R^m. Let $F' = (f_1', \ldots, f_n')$ be another basis of R^n and $P = (p_{ij})$ the $n \times n$ invertible matrix with $p_{ij} \in R$ such that $F = F'P$. Similarly, let $G' = (g_1', \ldots, g_m')$ be another basis of R^m and $Q = (q_{ij})$ the $m \times m$ invertible matrix with $q_{ij} \in R$ such that $G = G'Q$. Show that $A' = QAP^{-1}$ represents φ with respect to the bases F' of R^n and G' of R^m. Hint: Adapt the proof of the analogous result for vector spaces.

b. If A is an $m \times n$ presentation matrix for an R-module M, and if $A' = QAP^{-1}$ with P any $n \times n$, and Q any $m \times m$, invertible matrices with coefficients in R, show that A' also presents M.

c. In particular if A' is an $m \times n$ matrix obtained from A by adding c, $c \in R$, times the ith column of A to the jth column of A, or c times the ith row of A to the jth row of A, show that A' and A present the same module. Hint: If A' is obtained from A by adding c times the ith column of A to the jth column of A then $A' = AP$ where P is the $m \times m$ matrix with ones along the diagonal and all other entries zero except the ijth, which equals c.

d. If A' is obtained from A by deleting a column of zeroes (assume that A is not a single column of zeroes), show that A and A' present the same module. Hint: A column of zeroes represents the trivial relation.

e. Suppose that A has at least two columns and that its jth column is e_i (the standard basis vector of R^m with 1 in the ith row and all other entries zero). Let A' be obtained from A by deleting the ith row and jth column of A. Show that A and A' present the same module. Hint: To say that a column of A is e_i is to say that the ith generator of the module being presented is zero.

Exercise 30. Let $R = k[x, y]$ and consider the R-module M presented by the matrix

$$A = \begin{pmatrix} 3 & 1 & 0 \\ 1 + x & 2x + 3 & 2 - 2x \\ 0 & 4y & 4 \end{pmatrix}$$

(compare Exercise 6 and the discussion preceding Exercise 7). Use the 1 in row 1 and column 2 and elementary row operations to make the second column e_2. Use the operation in part e of the preceding exercise to reduce to a 2×2 matrix. Make the entry in row 2 and column 2 a 1 and use row operations to clear the entry in row 2 column 1, and repeat the operation in part e. Conclude that the 1×1 matrix $(-8 - 5x + 6y(x - 1))$ also presents M, whence $M = k[x, y]/\langle -8 - 5x + 6y(x - 1) \rangle$.

Exercise 31. The purpose of this exercise is to show that two matrices present the same module if and only if they can be transformed into one another by the operations of Exercise 29.

a. Let A be a presentation matrix of the R-module M with respect to a generating set $f_1, \ldots, f_m$. Suppose that $g_1, \ldots, g_s \in M$ and write $g_i = \sum b_{ji} f_j$ with $b_{ji} \in R$. Let $B = (b_{ij})$. Show that the block matrix

$$\begin{pmatrix} A & -B \\ 0 & I \end{pmatrix}$$

presents M with respect to the generators $(f_1, \ldots, f_m; g_1, \ldots, g_s)$.

b. Suppose that $g_1, \ldots, g_s$ also generate M and that A' presents M with respect to this set of generators. Write $f_i = \sum c_{ji} g_j$ and let $C = (c_{ij})$. Show that the block matrix

$$D = \begin{pmatrix} A & -B & I & 0 \\ 0 & I & -C & A' \end{pmatrix}$$

presents M with respect to the generators $(f_1, \ldots, f_m; g_1, \ldots, g_s)$.

c. Show that D can be reduced to both A and to A' by repeatedly applying the operations of Exercise 29. Hint: Show that row operations give the block matrix

$$\begin{pmatrix} A & 0 & I - BC & BA' \\ 0 & I & -C & A' \end{pmatrix}$$

which reduces by part d of Exercise 29 to the matrix

$$(A \quad I - BC \quad BA').$$

Show that the columns of $I - BC$ and of BA' are syzygies, hence spanned by the columns of A.

d. Show that any presentation matrix of a module can be transformed into any other by a sequence of operations from Exercise 29 and their inverses.

Exercise 32.

a. Show that if every ideal I of R is finitely generated (that is, if R is Noetherian), then any submodule M of R^t is finitely generated. Hint: Proceed by induction. If $t = 1$, M is an ideal, hence finitely generated. If $t > 1$, show that the set of first components of vectors in M is an ideal in R , hence finitely generated. Suppose that $r_i \in R, 1 \le i \le s$ generate this ideal, and choose column vectors $f_1, \ldots, f_s \in M$ with first components $r_1, \ldots, r_s$ respectively. Show that the submodule M' of M consisting of vectors in M with first component 0 is finitely generated. Show that $f_1, \ldots, f_s$ together with any generating set of M' is a generating set of M.

b. Show that if R is Noetherian, any submodule of a finitely generated R-module M is finitely generated. Hint: If M is generated by $f_1, \ldots, f_s$ it is an image of R^s under a surjective homomorphism.

Exercise 33. There is another way to view Exercise 31 which is frequently useful, and which we outline here. If A and A' are $m \times t$ matrices such that $A' = QAP^{-1}$ for an invertible $m \times m$ matrix Q and an invertible $t \times t$ matrix P, then we say that A' and A are *equivalent*. Equivalent matrices present the same modules (because we can view $P \in \mathrm{GL}(t, R)$ and $Q \in \mathrm{GL}(m, R)$ as a change of basis in R^t and R^m respectively, where for any n, $\mathrm{GL}(n, R)$ is the group of $n \times n$ invertible matrices with entries in R).

a. Let A be an $m \times t$ matrix and A' an $r \times s$ matrix with coefficients in R. Show that A and A' present identical modules if and only if the matrices

$$\begin{pmatrix} A & 0 & 0 & 0 \\ 0 & I_r & 0 & 0 \end{pmatrix} \quad \text{and} \quad \begin{pmatrix} 0 & 0 & I_m & 0 \\ 0 & 0 & 0 & A' \end{pmatrix}$$

are equivalent. Hint: This is equivalent to Exercise 31.

b. In part a above, show that we can take $P = I$.

c. Two matrices A and A' are called *Fitting equivalent* if there exist identity and zero matrices such that

$$\begin{pmatrix} A & 0 & 0 \\ 0 & I & 0 \end{pmatrix} \quad \text{and} \quad \begin{pmatrix} I & 0 & 0 \\ 0 & A' & 0 \end{pmatrix}$$

are equivalent. Show that A and A' present the same module if and only if A and A' are Fitting equivalent.

§2 Monomial Orders and Gröbner Bases for Modules

Throughout this section R will stand for a polynomial ring $k[x_1, \ldots, x_n]$. The goals of this section are to develop a theory of monomial orders in the free modules R^m and to introduce Gröbner bases for submodules $M \subset R^m$, in order to be able to solve the following problems by methods generalizing the ones introduced in Chapter 1 for ideals in R.

(2.1) Problems.

a. *(Submodule Membership) Given a submodule $M \subset R^m$ and $\mathbf{f} \in R^m$, determine if $\mathbf{f} \in M$.*

b. *(Syzygies) Given an ordered s-tuple of elements $(\mathbf{f}_1, \ldots, \mathbf{f}_s)$ of R^m (for example, an ordered set of generators), find a set of generators for the module $\mathrm{Syz}(\mathbf{f}_1, \ldots, \mathbf{f}_s) \subset R^s$ of syzygies.*

One can restate problem 2.1b as that of finding a presentation matrix for a submodule of R^m. It is easy to see why Gröbner bases might be involved in solving the submodule membership problem. When $m = 1$, a submodule of R^m is the same as an ideal in R (see Exercise 4b of §1). Division with respect to a Gröbner basis gives an algorithmic solution of the ideal membership problem, so it is natural to hope that a parallel theory for submodules of R^m might be available for general m. In the next section, we shall see that Gröbner bases are also intimately related to the problem of computing syzygies.

As we will see, one rather pleasant surprise is the way that, once we introduce the terminology needed to extend the notion of monomial orders to the free modules R^m, the module case follows the ideal case almost exactly. (Also see Exercise 6 below for a way to encode a module as a portion of an ideal and apply Gröbner bases for ideals.)

Let us first agree that a *monomial* $\mathbf{m}$ in R^m is an element of the form $x^\alpha \mathbf{e}_i$ for some i. We say $\mathbf{m}$ *contains* the standard basis vector $\mathbf{e}_i$. Every element $\mathbf{f} \in R^m$ can be written in a unique way as a k-linear combination of monomials $\mathbf{m}_i$

$$\mathbf{f} = \sum_{i=1}^{n} c_i \mathbf{m}_i,$$

where $c_i \in k, c_i \neq 0$. Thus for example, in $k[x, y]^3$

$$(2.2) \quad \begin{aligned} \mathbf{f} &= \begin{pmatrix} 5xy^2 - y^{10} + 3 \\ 4x^3 + 2y \\ 16x \end{pmatrix} \\ &= 5\begin{pmatrix} xy^2 \\ 0 \\ 0 \end{pmatrix} - \begin{pmatrix} y^{10} \\ 0 \\ 0 \end{pmatrix} + 3\begin{pmatrix} 1 \\ 0 \\ 0 \end{pmatrix} \\ &\quad + 4\begin{pmatrix} 0 \\ x^3 \\ 0 \end{pmatrix} + 2\begin{pmatrix} 0 \\ y \\ 0 \end{pmatrix} + 16\begin{pmatrix} 0 \\ 0 \\ x \end{pmatrix} \\ &= 5xy^2\mathbf{e}_1 - y^{10}\mathbf{e}_1 + 3\mathbf{e}_1 + 4x^3\mathbf{e}_2 + 2y\mathbf{e}_2 + 16x\mathbf{e}_3, \end{aligned}$$

which is a k-linear combination of monomials. The product $c \cdot \mathbf{m}$ of a monomial $\mathbf{m}$ with an element $c \in k$ is called a *term* and c is called its coefficient. We say that the terms $c_i \mathbf{m}_i, c_i \neq 0$, in the expansion of $\mathbf{f} \in R^m$ and the corresponding monomials $\mathbf{m}_i$ *belong* to $\mathbf{f}$.

If $\mathbf{m}, \mathbf{n}$ are monomials in R^m, $\mathbf{m} = x^\alpha \mathbf{e}_i$, $\mathbf{n} = x^\beta \mathbf{e}_j$, then we say that $\mathbf{n}$ *divides* $\mathbf{m}$ (or $\mathbf{m}$ *is divisible by* $\mathbf{n}$) if and only if $i = j$ and x^β divides

x^α. If $\mathbf{n}$ divides $\mathbf{m}$ we define the *quotient* $\mathbf{m}/\mathbf{n}$ to be $x^\alpha/x^\beta \in R$ (that is, $\mathbf{m}/\mathbf{n} = x^{\alpha-\beta}$). Note that the quotient is an element of the ring R, and if $\mathbf{n}$ divides $\mathbf{m}$, we have $(\mathbf{m}/\mathbf{n}) \cdot \mathbf{n} = \mathbf{m}$, which we certainly want. If $\mathbf{m}$ and $\mathbf{n}$ are monomials containing the same basis element $\mathbf{e}_i$, we define the greatest common divisor, $\mathrm{GCD}(\mathbf{m}, \mathbf{n})$, and least common multiple, $\mathrm{LCM}(\mathbf{m}, \mathbf{n})$ to be the greatest common divisor and least common multiple, respectively, of x^α and x^β, times $\mathbf{e}_i$. On the other hand, if $\mathbf{m}, \mathbf{n}$ contain different standard basis vectors, we define $\mathrm{LCM}(\mathbf{m}, \mathbf{n}) = \mathbf{0}$.

We say that a submodule $M \subset R^m$ is a *monomial submodule* if M can be generated by a collection of monomials. As for monomial ideals, it is easy to see that $\mathbf{f}$ is in a monomial submodule M if and only if every term belonging to $\mathbf{f}$ is in M. Monomial submodules have properties closely paralleling those of monomial ideals.

(2.3) Proposition.

a. *Every monomial submodule of R^m is generated by a finite collection of monomials.*

b. *Every infinite ascending chain $M_1 \subset M_2 \subset \cdots$ of monomial submodules of R^m stabilizes. That is, there exists N such that $M_N = M_{N+1} = \cdots = M_{N+\ell} = \cdots$ for all $\ell \geq 0$.*

c. *Let $\{\mathbf{m}_1, \ldots, \mathbf{m}_t\}$ be a set of monomial generators for a monomial submodule of R^m, and let $\epsilon_1, \ldots, \epsilon_t$ denote the standard basis vectors in R^t. Let $\mathbf{m}_{ij} = \mathrm{LCM}(\mathbf{m}_i, \mathbf{m}_j)$. The syzygy module $\mathrm{Syz}(\mathbf{m}_1, \ldots, \mathbf{m}_t)$ is generated by the syzygies $\sigma_{ij} = (\mathbf{m}_{ij}/\mathbf{m}_i)\epsilon_i - (\mathbf{m}_{ij}/\mathbf{m}_j)\epsilon_j$, for all $1 \leq i < j \leq t$ ($\sigma_{ij} = \mathbf{0}$ unless $\mathbf{m}_i$ and $\mathbf{m}_j$ contain the same standard basis vector in R^m).*

PROOF. For part a, let M be a monomial submodule of R^m. For each i, let $M_i = M \cap R\mathbf{e}_i$ be the subset of M consisting of elements whose jth components are zero for all $j \neq i$. In Exercise 5 below, you will show that M_i is an R-submodule of M. Each element of M_i has the form $f\mathbf{e}_i$ for some $f \in R$. By Exercise 4 of §1 of this chapter, $M_i = I_i\mathbf{e}_i$ for some ideal $I_i \subset R$, and it follows that I_i must be a monomial ideal. By Dickson's Lemma for monomial ideals (see, for instance, [CLO], Theorem 5 of Chapter 2, §4), it follows that I_i has a finite set of generators $x^{\alpha(i1)}, \ldots, x^{\alpha(id_i)}$. But then the

$$x^{\alpha(11)}\mathbf{e}_1, \ldots, x^{\alpha(1d_1)}\mathbf{e}_1$$
$$x^{\alpha(21)}\mathbf{e}_2, \ldots, x^{\alpha(2d_2)}\mathbf{e}_2$$
$$\vdots$$
$$x^{\alpha(m1)}\mathbf{e}_m, \ldots, x^{\alpha(md_m)}\mathbf{e}_m$$

generate M.

Part b follows from part a. See Exercise 5 below.

For part c, first observe that if $(a_1, \ldots, a_t)^T$ is a syzygy on a collection of monomials and we expand in terms of the standard basis in R^m:

$$0 = a_1 \mathbf{m}_1 + \cdots + a_t \mathbf{m}_t = f_1 \mathbf{e}_1 + \cdots + f_n \mathbf{e}_n,$$

then $f_1 = \cdots = f_n = 0$, and the syzygy is a sum of syzygies on subsets of the $\mathbf{m}_j$ containing the same $\mathbf{e}_i$. Hence we can restrict to considering collections of monomials containing the same $\mathbf{e}_i$:

$$\mathbf{m}_1 = x^{\alpha_1} \mathbf{e}_i, \ldots, m_s = x^{\alpha_s} \mathbf{e}_i.$$

Now, if $(a_1, \ldots, a_s)^T$ is a syzygy in $\mathrm{Syz}(\mathbf{m}_1, \ldots, \mathbf{m}_s)$, we can collect terms of the same multidegree in the expansion $a_1 x^{\alpha_1} + \cdots + a_s x^{\alpha_s} = 0$. Each sum of terms of the same multidegree in this expansion must also be zero, and the only way this can happen is if the coefficients (in k) of those terms sum to zero. Hence $(a_1, \ldots, a_s)^T$ can be written as a sum of syzygies of the form

$$(c_1 x^{\alpha - \alpha_1}, \ldots, c_s x^{\alpha - \alpha_s})^T,$$

with $c_1, \ldots, c_s \in k$ satisfying $c_1 + \cdots + c_s = 0$. Such a syzygy is called a *homogeneous syzygy*, and we have just shown that all syzygies are sums of homogeneous syzygies. (Compare with Lemma 7 of Chapter 2, §9 of [CLO].)

When $s = 3$, for instance, we can write a syzygy

$$(c_1 x^{\alpha - \alpha_1}, c_2 x^{\alpha - \alpha_2}, c_3 x^{\alpha - \alpha_3})^T$$

with $c_1 + c_2 + c_3 = 0$ as a sum:

$$(c_1 x^{\alpha - \alpha_1}, -c_1 x^{\alpha - \alpha_2}, 0)^T + (0, (c_1 + c_2) x^{\alpha - \alpha_2}, c_3 x^{\alpha - \alpha_3})^T,$$

where $(c_1 x^{\alpha - \alpha_1}, -c_1 x^{\alpha - \alpha_2})^T = c_1 (x^{\alpha - \alpha_1}, -x^{\alpha - \alpha_2})^T$ is a syzygy on the pair of monomials $x^{\alpha_1}, x^{\alpha_2}$ and $((c_1 + c_2) x^{\alpha - \alpha_2}, c_3 x^{\alpha - \alpha_3})^T = -c_3 (x^{\alpha - \alpha_2}, -x^{\alpha - \alpha_3})^T$ is a syzygy on the pair $x^{\alpha_2}, x^{\alpha_3}$.

In fact, for any s, every homogeneous syzygy can be written as a sum of syzygies between pairs of monomials in a similar way (see Exercise 5 below). Also observe that given two monomials x^α and x^β and some x^γ that is a multiple of each, then the syzygy $(x^{\gamma - \alpha}, -x^{\gamma - \beta})^T$ is a monomial times

$$\sigma = (\mathrm{LCM}(x^\alpha, x^\beta)/x^\alpha, -\mathrm{LCM}(x^\alpha, x^\beta)/x^\beta)^T.$$

From here, part c of the proposition follows easily.   □

If $M = \langle \mathbf{m}_1, \ldots, \mathbf{m}_t \rangle$ and $\mathbf{f}$ is an arbitrary element of R^m, then $\mathbf{f} \in M$ if and only if every term of $\mathbf{f}$ is divisible by some $\mathbf{m}_i$. Thus, it is easy to solve the submodule membership problem for monomial submodules.

Extending the theory of Gröbner bases to modules will involve three things: defining orders on the monomials of R^m, constructing a division algorithm on elements of R^m, and extending the Buchberger algorithm to R^m. Let us consider each in turn.

The definition of a monomial order on R^m is the same as the definition in R (see (2.1) from Chapter 1 of this book). Namely, we say that an ordering relation $>$ on the monomials of R^m is a *monomial ordering* if:

a. $>$ is a total order,
b. for every pair of monomials $\mathbf{m}, \mathbf{n} \in R^m$ with $\mathbf{m} > \mathbf{n}$, we have $x^\alpha \mathbf{m} > x^\alpha \mathbf{n}$ for every monomial $x^\alpha \in R$, and
c. $>$ is a *well-ordering*.

Exercise 1. Show that condition c is equivalent to $x^\alpha \mathbf{m} > \mathbf{m}$ for all monomials $\mathbf{m} \in R^m$ and all monomials $x^\alpha \in R$ such that $x^\alpha \neq 1$.

Some of the most common and useful monomial orders on R^m come by extending monomial orders on R itself. There are two especially natural ways to do this, once we choose an ordering on the standard basis vectors. We will always use the "downward" ordering on the entries in a column:

$$\mathbf{e}_1 > \mathbf{e}_2 > \cdots > \mathbf{e}_m,$$

although any other ordering could be used as well. (Note that this is the reverse of the numerical order on the subscripts.)

(2.4) Definition. Let $>$ be any monomial order on R.
a. (TOP extension of $>$) We say $x^\alpha \mathbf{e}_i >_{TOP} x^\beta \mathbf{e}_j$ if $x^\alpha > x^\beta$, or if $x^\alpha = x^\beta$ and $i < j$.
b. (POT extension of $>$) We say $x^\alpha \mathbf{e}_i >_{POT} x^\beta \mathbf{e}_j$ if $i < j$, or if $i = j$ and $x^\alpha > x^\beta$.

This terminology follows [AL], Definitions 3.5.2 and 3.5.3 (except for the ordering on the $\mathbf{e}_i$). Following Adams and Loustaunau, TOP stands for "term-over-position," which is certainly appropriate since a TOP order sorts monomials first by the term order on R, then breaks ties using the position within the vector in R^m. On the other hand, POT stands for "position-over-term."

Exercise 2. Verify that for any monomial order $>$ on R, both $>_{TOP}$ and $>_{POT}$ define monomial orders on R^m.

As a simple example, if we extend the *lex* order on $k[x, y]$ with $x > y$ to a TOP order on $k[x, y]^3$ we get an order $>_1$ such that the terms in (2.2) are ordered as follows.

$$\begin{pmatrix} 0 \\ x^3 \\ 0 \end{pmatrix} >_1 \begin{pmatrix} xy^2 \\ 0 \\ 0 \end{pmatrix} >_1 \begin{pmatrix} 0 \\ 0 \\ x \end{pmatrix} >_1 \begin{pmatrix} y^{10} \\ 0 \\ 0 \end{pmatrix} >_1 \begin{pmatrix} 0 \\ y \\ 0 \end{pmatrix} >_1 \begin{pmatrix} 1 \\ 0 \\ 0 \end{pmatrix}.$$

If we extend the same *lex* order to a POT order on $k[x, y]^3$ we get an order $>_2$ such that

$$\begin{pmatrix} xy^2 \\ 0 \\ 0 \end{pmatrix} >_2 \begin{pmatrix} y^{10} \\ 0 \\ 0 \end{pmatrix} >_2 \begin{pmatrix} 1 \\ 0 \\ 0 \end{pmatrix} >_2 \begin{pmatrix} 0 \\ x^3 \\ 0 \end{pmatrix} >_2 \begin{pmatrix} 0 \\ y \\ 0 \end{pmatrix} >_2 \begin{pmatrix} 0 \\ 0 \\ x \end{pmatrix}.$$

In either case, we have $e_1 > e_2$.

Once we have an ordering $>$ on monomials, we can write any element $\mathbf{f} \in R^m$ as a sum of terms

$$\mathbf{f} = \sum_{i=1}^{t} c_i \mathbf{m}_i$$

with $c_i \neq 0$ and $\mathbf{m}_1 > \mathbf{m}_2 > \cdots > \mathbf{m}_t$. We define the *leading coefficient, leading monomial,* and *leading term* of f just as in the ring case:

$$\mathrm{LC}_>(\mathbf{f}) = c_1$$
$$\mathrm{LM}_>(\mathbf{f}) = \mathbf{m}_1$$
$$\mathrm{LT}_>(\mathbf{f}) = c_1\mathbf{m}_1.$$

If, for example,

$$\mathbf{f} = \begin{pmatrix} 5xy^2 - y^{10} + 3 \\ 4x^3 + 2y \\ 16x \end{pmatrix} \in k[x, y]^3$$

as in (2.2), and $>_{TOP}$ is the TOP extension of the *lex* order on $k[x, y]$ $(x > y)$, then

$$\mathrm{LC}_{>_{TOP}}(f) = 4, \quad \mathrm{LM}_{>_{TOP}}(f) = x^3\mathbf{e}_2, \quad \mathrm{LT}_{>_{TOP}}(f) = 4x^3\mathbf{e}_2.$$

Similarly, if $>_{POT}$ is the POT extension of the *lex* order, then

$$\mathrm{LC}_{>_{POT}}(f) = 5, \quad \mathrm{LM}_{>_{POT}}(f) = xy^2\mathbf{e}_1, \quad \mathrm{LT}_{>_{POT}}(f) = 5xy^2\mathbf{e}_1.$$

Once we have a monomial ordering in R^m we can divide by a set $F \subset R^m$ in exactly the same way we did in R.

(2.5) Theorem (Division Algorithm in R^m). *Fix any monomial ordering on R^m and let $F = (\mathbf{f}_1, \ldots, \mathbf{f}_s)$ be an ordered s-tuple of elements of R^m. Then every $\mathbf{f} \in R^m$ can be written as*

$$\mathbf{f} = a_1\mathbf{f}_1 + \cdots + a_s\mathbf{f}_s + \mathbf{r},$$

where $a_i \in R$, $\mathbf{r} \in R^m$, $\mathrm{LT}(a_i\mathbf{f}_i) \leq \mathrm{LT}(\mathbf{f})$ for all i, and either $\mathbf{r} = 0$ or $\mathbf{r}$ is a k-linear combination of monomials none of which is divisible by any of $\mathrm{LM}(\mathbf{f}_1), \ldots, \mathrm{LM}(\mathbf{f}_s)$. We call $\mathbf{r}$ the remainder on division by F.

PROOF. To prove the existence of the $a_i \in R$ and $\mathbf{r} \in R^m$ it is sufficient to give an algorithm for their construction. The algorithm is word-for-word the same as that supplied in [CLO], Theorem 3 of Chapter 2, §3, or [AL], Algorithm 1.5.1 in the ring case. (The module version appears in [AL] as Algorithm 3.5.1). The proof of termination is also identical.   □

Instead of reproducing the formal statement of the algorithm, we describe it in words. The key operation in carrying out the division process is the reduction of the partial dividend $\mathbf{p}$ ($\mathbf{p} = \mathbf{f}$ to start) by an $\mathbf{f}_i$ such that $\mathrm{LT}(\mathbf{f}_i)$ divides $\mathrm{LT}(\mathbf{p})$. If $\mathrm{LT}(\mathbf{p}) = t \cdot \mathrm{LT}(\mathbf{f}_i)$ for a term $t \in R$, we define

$$\mathrm{Red}\,(\mathbf{p}, \mathbf{f}_i) = \mathbf{p} - t\mathbf{f}_i$$

and say that we have *reduced* $\mathbf{p}$ by $\mathbf{f}_i$. One divides $\mathbf{f}$ by F by successively reducing by the first $\mathbf{f}_i$ in the list (that is, the element with the smallest index i) for which reduction is possible, and keeping track of the quotients. If at some point, reduction is not possible, then $\mathrm{LT}(\mathbf{p})$ is not divisible by any of the $\mathrm{LT}(\mathbf{f}_i)$. In this case, we subtract the lead term of $\mathbf{p}$, place it into the remainder and again try to reduce with the $\mathbf{f}_i$. The process stops when $\mathbf{p}$ is reduced to $\mathbf{0}$.

The following exercise gives a simple example of division in the module setting. When calculating by hand, it is sometimes convenient to use a format similar to the polynomial long division from [CLO] Chapter 2, but we will not do that here.

Exercise 3. Let

$$\mathbf{f} = (5xy^2 - y^{10} + 3, 4x^3 + 2y, 16x)^T \in k[x, y]^3$$

as in (2.2), and let

$$\mathbf{f}_1 = (xy + 4x, 0, y^2)^T$$
$$\mathbf{f}_2 = (0, y - 1, x - 2)^T.$$

Let $>$ stand for the POT extension of the *lex* order on $k[x, y]$ with $x > y$. Then $\mathrm{LT}(\mathbf{f}) = 5x^2y\mathbf{e}_1$, $\mathrm{LT}(\mathbf{f}_1) = xy\mathbf{e}_1$, and $\mathrm{LT}(\mathbf{f}_2) = y\mathbf{e}_2$. Let $\mathbf{p}$ be the intermediate dividend at each step of the algorithm—set $\mathbf{p} = \mathbf{f}$ to start and $a_1 = a_2 = 0$ and $\mathbf{r} = \mathbf{0}$.

a. Since $\mathrm{LT}(\mathbf{f}_1)$ divides $\mathrm{LT}(\mathbf{f})$, show that the first step in the division will yield intermediate values $a_1 = 5y$, $a_2 = 0$, $\mathbf{r} = \mathbf{0}$, and $\mathbf{p} = \mathrm{Red}\,(\mathbf{f}, \mathbf{f}_1) = (-20xy - y^{10} + 3, 4x^3 + 2y, 16x - 5y^3)^T$.

b. $\mathrm{LT}(\mathbf{p})$ is still divisible by $\mathrm{LT}(\mathbf{f}_1)$, so we can reduce by $\mathbf{f}_1$ again. Show that this step yields intermediate values $a_1 = 5y - 20$, $a_2 = 0$, $\mathbf{r} = \mathbf{0}$, and $\mathbf{p} = (80x - y^{10} + 3, 4x^3 + 2y, 16x - 5y^3 + 20y^2)^T$.

c. Show that in the next three steps in the division, the leading term of $\mathbf{p}$ is in the first component, but is not divisible by the leading term of

either of the divisors. Hence after three steps we obtain intermediate
values $a_1 = 5y - 10$, $a_2 = 0$, $\mathbf{r} = (80x - y^{10} + 3, 0, 0)^T$, and $\mathbf{p} =$
$(0, 4x^3 + 2y, 16x - 5y^3 + 20y^2)^T$.

d. The leading term of $\mathbf{p}$ at this point is $4x^3\mathbf{e}_2$, which is still not divisible
by the leading terms of either of the divisors. Hence the next step will
remove the term $4x^3\mathbf{e}_2$ and place that into $\mathbf{r}$ as well.

e. Complete the division algorithm on this example.

f. Now use the TOP extension of the *lex* order and divide $\mathbf{f}$ by $(\mathbf{f}_1, \mathbf{f}_2)$
using this new order.

The division algorithm behaves best when the set of divisors has the
defining property of a Gröbner basis.

(2.6) Definition. Let M be a submodule of R^m, and let $>$ be a monomial
order.

a. We will denote by $\langle \mathrm{LT}(M) \rangle$ the monomial submodule generated by the
leading terms of all $\mathbf{f} \in M$ with respect to $>$.

b. A finite collection $\mathcal{G} = \{\mathbf{g}_1, \ldots, \mathbf{g}_s\} \subset M$ is called a *Gröbner basis* for
M if $\langle \mathrm{LT}(M) \rangle = \langle \mathrm{LT}(\mathbf{g}_1), \ldots, \mathrm{LT}(\mathbf{g}_s) \rangle$.

The good properties of ideal Gröbner bases with respect to division
extend immediately to this new setting, and with the same proofs.

(2.7) Proposition. *Let $\mathcal{G}$ be a Gröbner basis for a submodule $M \subset R^m$,
and let $\mathbf{f} \in R^m$.*

a. *$\mathbf{f} \in M$ if and only if the remainder on division by $\mathcal{G}$ is zero.*

b. *A Gröbner basis for M generates M as a module: $M = \langle \mathcal{G} \rangle$.*

Part a of this proposition gives a solution of the submodule membership
problem stated at the start of this section, provided that we have a Gröbner
basis for the submodule M in question. For example, the divisors $\mathbf{f}_1, \mathbf{f}_2$ in
Exercise 3 do form a Gröbner basis for the submodule M they generate,
with respect to the POT extension of the *lex* order. (This will follow from
Theorem (2.9) below, for instance.) Since the remainder on division of $\mathbf{f}$ is
not zero, $\mathbf{f} \notin M$.

Some care must be exercised in summarizing part b of the proposition
in words. It is not usually true that a Gröbner basis is a *basis* for M as
an R-module—a Gröbner basis is a set of generators for M, but it need
not be linearly independent over R. However, Gröbner bases do exist for
all submodules of R^m, by essentially the same argument as for ideals.

Exercise 4. By Proposition (2.3), $\langle \mathrm{LT}(M) \rangle = \langle \mathbf{m}_1, \ldots, \mathbf{m}_t \rangle$ for some
finite collection of monomials. Let $\mathbf{f}_i \in M$ be an element with $\mathrm{LT}(\mathbf{f}_i) = \mathbf{m}_i$.

a. Show that $\{\mathbf{f}_1, \ldots, \mathbf{f}_t\}$ is a Gröbner basis for M.

b. Use Proposition (2.7) to show that every submodule of R^m is finitely generated.

Reduced and *monic* Gröbner bases may be defined as for ideals, and there is a unique monic (reduced) Gröbner basis for each submodule in R^m once we choose a monomial order.

Exercise 4 also shows that every submodule of R^m is finitely generated. Using this, it is straightforward to show that submodules of R^m satisfy the *ascending chain condition* (ACC), which asserts that every infinite ascending chain $M_1 \subset M_2 \subset \cdots$ of submodules of R^m stabilizes; that is, there is an N such that $M_N = M_{N+1} = \cdots = M_{M+\ell} = \cdots$ for all $\ell \geq 0$. We proved this for monomial submodules in Proposition (2.3) and you will prove it for arbitrary submodules of R^m in Exercise 13 below.

Now we turn to the extension of Buchberger's Algorithm to the module case.

(2.8) Definition. Fix a monomial order on R^m, and let $\mathbf{f}, \mathbf{g} \in R^m$. The *S-vector* of $\mathbf{f}$ and $\mathbf{g}$, denoted $S(\mathbf{f}, \mathbf{g})$, is the following element of R^m. Let $\mathbf{m} = \mathrm{LCM}(\mathrm{LT}(\mathbf{f}), \mathrm{LT}(\mathbf{g}))$ as defined above. Then

$$S(\mathbf{f}, \mathbf{g}) = \frac{\mathbf{m}}{\mathrm{LT}(\mathbf{f})} \mathbf{f} - \frac{\mathbf{m}}{\mathrm{LT}(\mathbf{g})} \mathbf{g}.$$

For example, if $\mathbf{f} = (xy - x, x^3 + y)^T$ and $\mathbf{g} = (x^2 + 2y^2, x^2 - y^2)^T$ in $k[x, y]^2$ and we use the POT extension of the *lex* order on $R = k[x, y]$ with $x > y$, then

$$S(\mathbf{f}, \mathbf{g}) = xf - yg$$
$$= (-x^2 - 2y^3, x^4 - x^2y + xy + y^3)^T.$$

The foundation for an algorithm for computing Gröbner bases is the following generalization of Buchberger's Criterion.

(2.9) Theorem (Buchberger's Criterion for Submodules). *A set* $\mathcal{G} = \{\mathbf{g}_1, \ldots, \mathbf{g}_s\} \subset R^m$ *is a Gröbner basis for the module it generates if and only if the remainder on division by* $\mathcal{G}$ *of* $S(\mathbf{g}_i, \mathbf{g}_j)$ *is* $\mathbf{0}$ *for all* i, j.

PROOF. The proof is essentially the same as in the ideal case. □

For example, $\mathcal{G} = \{\mathbf{f}_1, \mathbf{f}_2\}$ from Exercise 3 is a Gröbner basis for the submodule it generates in $k[x, y]^3$, with respect to the POT extension of the *lex* order. The reason is that the leading terms of the $\mathbf{f}_i$ contain different standard basis vectors, so their least common multiple is zero. As a result, the S-vector satisfies $S(\mathbf{f}_1, \mathbf{f}_2) = \mathbf{0}$, and Buchberger's Criterion implies that $\mathcal{G}$ is a Gröbner basis.

For a less trivial example, if we define M to be the matrix

$$M = \begin{pmatrix} a^2 + b^2 & a^3 - 2bcd & a - b \\ c^2 - d^2 & b^3 + acd & c + d \end{pmatrix},$$

over $R = k[a, b, c, d]$ then a TOP *grevlex* Gröbner basis $\mathcal{G}$ for the submodule generated by the columns of M has four elements:

$$\mathbf{g}_1 = (b^2, -ac/2 - bc/2 + c^2/2 - ad/2 - bd/2 - d^2/2)^T,$$
$$\mathbf{g}_2 = (a - b, c + d)^T$$
(2.10) $\mathbf{g}_3 = (-2bcd, b^3 - abc/2 + b^2c/2 - ac^2 + bc^2/2 - abd/2 + b^2d/2$
$$+ acd + ad^2 - bd^2/2)^T$$
$$\mathbf{g}_4 = (0, a^2c + b^2c - ac^2 + bc^2 + a^2d + b^2d + ad^2 - bd^2)^T.$$

Note that $\mathrm{LT}(\mathbf{g}_1) = b^2\mathbf{e}_1$ and $\mathrm{LT}(\mathbf{g}_2) = a\mathbf{e}_1$ for the TOP extension of *grevlex* on $k[a, b, c, d]$. Hence

$$\begin{aligned} S(\mathbf{g}_1, \mathbf{g}_2) &= a\mathbf{g}_1 - b^2\mathbf{g}_2 \\ &= (b^3, -a^2c/2 - abc/2 + ac^2/2 - a^2d/2 - abd/2 - ad^2/2 \\ &\quad - b^2c - b^2d)^T \\ &= b\mathbf{g}_1 - (1/2)\mathbf{g}_4, \end{aligned}$$

so that $S(\mathbf{g}_1, \mathbf{g}_2)$ reduces to $\mathbf{0}$ on division by $\mathcal{G}$. It is easy to check that all the other S-vectors reduce to $\mathbf{0}$ modulo $\mathcal{G}$ as well.

To compute Gröbner bases, we need a version of Buchberger's Algorithm. Using Theorem (2.9), this extends immediately to the module case.

(2.11) Theorem (Buchberger's Algorithm for Submodules). *Let $F = (\mathbf{f}_1, \ldots, \mathbf{f}_t)$ where $\mathbf{f}_i \in R^m$, and fix a monomial order on R^m. The following algorithm computes a Gröbner basis $\mathcal{G}$ for $M = \langle F \rangle \subset R^m$, where $\overline{S(\mathbf{f}, \mathbf{g})}^{\mathcal{G}'}$ denotes the remainder on division by $\mathcal{G}'$, using Theorem (2.5):*

> Input: $F = (\mathbf{f}_1, \ldots, \mathbf{f}_t) \subset R^m$, an order $>$
>
> Output: a Gröbner basis $\mathcal{G}$ for $M = \langle F \rangle$, with respect to $>$
>
> $\mathcal{G} := F$
>
> REPEAT
>
> > $\mathcal{G}' := \mathcal{G}$
> >
> > FOR each pair $\mathbf{f} \neq \mathbf{g}$ in $\mathcal{G}'$ DO
> >
> > > $S := \overline{S(\mathbf{f}, \mathbf{g})}^{\mathcal{G}'}$
> > >
> > > IF $S \neq 0$ THEN $\mathcal{G} := \mathcal{G} \cup \{S\}$
> >
> > UNTIL $\mathcal{G} = \mathcal{G}'$.

PROOF. Once again, the proof is the same as in the ideal case, using the fact from Proposition (2.3) that the ascending chain condition holds for monomial submodules to ensure termination.   □

Unfortunately, the Gröbner basis packages in Maple and *Mathematica* do not allow computation of Gröbner bases for submodules of R^m for $m > 1$ by the methods introduced above. The CALI package for REDUCE, CoCoA, Singular and *Macaulay 2* do have this capability however. For instance, the Gröbner basis in (2.10) was computed using the implementation of Buchberger's Algorithm in the computer algebra system *Macaulay 2* (though in this small example, the computations would also be feasible by hand). In Exercise 8 below, you will see how the computation was done. In Exercise 9, we re-do the computation using the computer algebra system Singular. Exercises 10 and 11 explain how to trick Maple into doing module computations, and Exercise 12 presents an additional application of the techniques of this section—computation in the quotient modules R^m/M.

While the exercises below illustrate how to use *Macaulay 2*, Singular, and Maple, we will not cover CoCoA. The interested reader should consult [KR] for information about this capable system.

ADDITIONAL EXERCISES FOR §2

Exercise 5. This exercise will supply some of the details for the proof of Proposition (2.3).
a. Show that if M is a submodule of R^m and $M_i = M \cap Re_i$, then M_i is a submodule of M.
b. Using part a of Proposition (2.3), show that monomial submodules of R^m satisfy the ascending chain condition. That is, for every infinite ascending chain $M_1 \subset M_2 \subset \cdots$ of monomial submodules of R^m, there exists N such that $M_N = M_{N+1} = \cdots = M_{N+\ell}$ for all $\ell \geq 0$. Hint: Consider $\cup_{n=1}^{\infty} M_n$, which is also a monomial submodule.

Exercise 6. In this exercise, we will see how the theory of submodules of R^m can be "emulated" via ideals in a larger polynomial ring obtained by introducing additional variables $X_1, \ldots, X_m$ corresponding to the standard basis vectors in R^m. Write $S = k[x_1, \ldots, x_n, X_1, \ldots, X_m]$, and define a mapping $\varphi : R^m \to S$ as follows. For each $\mathbf{f} \in R^m$, expand $\mathbf{f} = \sum_{j=1}^{m} f_j \mathbf{e}_j$, where $f_j \in R$, and let $F = \varphi(\mathbf{f}) \in S$ be the polynomial $F = \sum_{j=1}^{m} f_j X_j$.
a. Show that S can be viewed as an R-module, where the scalar multiplication by elements of R is the restriction of the ring product in S.
b. Let $S_1 \subset S$ denote the vector subspace of S consisting of polynomials that are homogeneous linear in the X_j (the k-span of the collection of monomials of the form $x^\alpha X_j$). Show that S_1 is an R-submodule of S.

c. For each submodule $M = \langle \mathbf{f}_1, \ldots, \mathbf{f}_s \rangle \subset R^m$, let $F_i = \varphi(\mathbf{f}_i) \in S$. Show that $\varphi(M)$ equals $\langle F_1, \ldots, F_s \rangle S \cap S_1$, where $\langle F_1, \ldots, F_s \rangle S$ denotes the ideal generated by the F_i in S.

d. Show that a Gröbner basis for the module M could be obtained by applying a modified form of Buchberger's Algorithm for ideals to $I = \langle F_1, \ldots, F_s \rangle S$. The modification would be to compute remainders only for the S-polynomials $S(F_i, F_j)$ that are contained in S_1, and ignore all other S-pairs.

Exercise 7. Let $R = k[x]$, the polynomial ring in one variable. Let M be a submodule of R^m for some $m \geq 1$. Describe the form of the unique monic reduced Gröbner basis for M with respect to the POT extension of the degree order on R. In particular, how many Gröbner basis elements are there whose leading terms contain each $\mathbf{e}_i$? What is true of the ith components of the other basis elements if some leading term contains $\mathbf{e}_i$?

Using *Macaulay 2* to Compute in Modules.

Since we have not used *Macaulay 2* before in this book, and since it is rather different in design from general computer algebra systems such as Maple, a few words about its set-up are probably appropriate at this point. For more information on this program, we refer the reader to the *Macaulay 2* website http://www.math.uiuc.edu/Macaulay2/. *Macaulay 2* is a computer algebra system specifically designed for computations in algebraic geometry and commutative algebra. Its basic computational engine is a full implementation of Buchberger's algorithm for modules over polynomial rings. Built-in commands for manipulating ideals and submodules in various ways, performing division as in Theorem (2.5), computing Gröbner bases, syzygies, Hilbert functions, free resolutions (see Chapter 6 of this book), displaying results of computations, etc. Introductions to *Macaulay 2* may be found in [EGSS] and Appendix C of [Vas].

Before working with a submodule of R^m in *Macaulay 2*, the base ring R must be defined. In our examples, R is always a polynomial ring over a field. In *Macaulay 2*, $\mathbb{Q}$ is written QQ, while a finite field such as $\mathbb{Z}/\langle 31991 \rangle$ is written ZZ/31991. Over the latter, polynomials in variables x, y, z are entered via the command

$$R = ZZ/33191[x, y, z]$$

at the *Macaulay 2* prompt "in :", where i stands for "input" and n is the number of the input. The default is the *grevlex* order on the polynomial ring, and a TOP extension to submodules in R^m with the standard basis ordered $\mathbf{e}_1 < \cdots < \mathbf{e}_m$. This "upward" order is the opposite of what we used in Definition (2.4). To change to our "downward" order, enter the command

$$R = ZZ/33191[x, y, z, \mathtt{MonomialOrder} \Rightarrow \{\mathtt{Position} \Rightarrow \mathtt{Down}\}].$$

Exercise 8.

a. In these examples, we will work over $R = k[a, b, c, d]$ with $k = \mathbb{Q}$, so enter

$$\mathtt{R = QQ[a..d]}$$

at the prompt in a *Macaulay 2* session.

b. To define the submodule generated by the columns of

$$M = \begin{pmatrix} a^2 + b^2 & a^3 - 2bcd & a - b \\ c^2 - d^2 & b^3 + acd & c + d \end{pmatrix},$$

enter the command

$$\mathtt{M = matrix\{\{a\char`\^2 + b\char`\^2, a\char`\^3 - 2*b*c*d, a - d\},}$$

$$\mathtt{\{c\char`\^2 - d\char`\^2, b\char`\^3 + a*c*d, c + d\}\}.}$$

c. The **gb** command computes the Gröbner basis for the module generated by the columns of a matrix. The result of this computation has the data type "GroebnerBasis". To get a matrix, one uses the command **gens** to get the generators in matrix form. Thus you should issue the command **gens gb M**. Compare with the Gröbner basis given above in (2.10)—they should be the same.

Using Singular to Compute in Modules.

We used **Singular** in the last chapter for computations in local rings. This very powerful program is also very well-suited for computations in modules over polynomial rings. We demonstrate by redoing Exercise 8 using **Singular**.

Exercise 9.

a. We will work over $R = k[a, b, c, d]$ with $k = \mathbb{Q}$, so enter

$$\mathtt{ring\ R=0,\ (a,b,c,d),\ (dp,C);}$$

at the > prompt in a **Singular** session. The first term "**ring R=0**" asserts that R will be a ring with characteristic 0 and the second that R have indeterminates a, b, c, d. Had we wanted $k = \mathbb{Z}/31991$ and indeterminates x, y, z, we would have entered "**ring R=31991, (x,y,z)**" at the prompt >. The third term specifies the ordering. Examples of possible well-orderings on R are *lex*, *grevlex*, and *grlex*, specified by **lp**, **dp** and **Dp** respectively. In our case, we chose grevlex. The letter **C** indicates the "downward" order $e_1 > e_2 > \cdots > e_m$ on the standard basis elements of the free module R^m. The lower-case **c** indicates the reverse, "upward" order $e_m > \cdots > e_2 > e_1$ on basis elements. The pair (**dp**, **C**) indicates the TOP extension of **dp** to R^m using the downward order

on standard basis elements. This is the ordering we used in Exercise
8. Had we wanted a POT extension, we would have written (C, dp).
A POT extension of a pure lex ordering on R to R^m with the upward
ordering $e_1 < \cdots < e_m$ would be specified by entering (c, lp).

b. To define the submodule M generated by the columns of the matrix M
in part b of Exercise 8 above, enter, for example,

$$> \texttt{vector s1} = [\texttt{a2} + \texttt{b2}, \texttt{c2} - \texttt{d2}];$$

$$> \texttt{vector s2} = [\texttt{a3} - \texttt{2bcd}, \texttt{b3} + \texttt{acd}];$$

$$> \texttt{vector s3} = [\texttt{a} - \texttt{b}, \texttt{c} + \texttt{d}];$$

$$> \texttt{module M} = \texttt{s1}, \texttt{s2}, \texttt{s3};$$

(We have shown the prompt > which you should not re-enter.) Note that
the command $\texttt{vector s1} = [\texttt{a2+b2,c2-d2}]$; defines the vector $s1 =$
$(a^2 + b^2, c^2 - d^2)^T \in R^2$.

c. To define a module N generated by a Gröbner basis of $s1, s2, s3$, enter

$$\texttt{module N} = \texttt{std(M)};$$

after the prompt >.

d. To see the result, type $\texttt{N}$; after the prompt >. Verify that you get the
same result (up to multiplication by 2) as in (2.10).

e. In addition to the TOP, downward extension of graded reverse lex, ex-
periment with the following different extensions of the graded reverse
lex order on R to the free modules R^m: POT and upward; TOP and
upward; POT and downward. For which of these does the Gröbner ba-
sis of $M = \langle s1, s2, s3 \rangle$ have the fewest number of elements? the least
degree? What about different extensions of the lex order on R?

Using Maple to Compute in Modules.

It is also possible to use Maple to find Gröbner bases of submodules
of R^m. The basic idea is the observation from Exercise 6 that submodules
of R^m can be "emulated" in a larger polynomial ring. We first need to
study how this works with respect to monomial orders on modules.

Exercise 10. Let $R = k[x_1, \ldots, x_n]$ and $S = k[x_1, \ldots, x_n, X_1, \ldots, X_m]$.
We also have the map φ which sends $\mathbf{f} = \sum_{j=1}^m f_j \mathbf{e}_j \in R^m$ to $F =$
$\sum_{j=1}^m f_j X_j$. In the notation of part b of Exercise 6, we have an isomorphism
$\varphi : R^m \cong S_1$. Now consider the following monomial orders $>_1$ and $>_2$ on S:

$$x^\alpha X^\beta >_1 x^\gamma X^\delta \iff x^\alpha >_{grevlex} x^\gamma, \text{ or } x^\alpha = x^\gamma \text{ and } X^\beta >_{grevlex} X^\delta$$

$$x^\alpha X^\beta >_2 x^\gamma X^\delta \iff X^\beta >_{grevlex} X^\delta, \text{ or } X^\beta = X^\delta \text{ and } x^\alpha >_{grevlex} x^\gamma.$$

(These are examples of elimination orders.) Also let $>_{TOP}$ and $>_{POT}$
denote the TOP and POT extensions to R^m of the *grevlex* order on R.

a. Prove that the restriction of $>_1$ to S_1 agrees with $>_{TOP}$ via the isomorphism φ.

b. Similarly, prove that the restriction of $>_2$ to S_1 agrees with $>_{POT}$ via φ.

By using the Ore_algebra and Groebner packages of Maple, we can compute Gröbner bases as follows.

Exercise 11. We will compute the Gröbner basis of (2.10) as follows. In Maple, first issue the commands

```
with(Ore_algebra):
S := poly_algebra(a,b,c,d,e1,e2);
F :=[(a^2+b^2)*e1+(c^2-d^2)*e2,
(a^3-2*b*c*d)*e1+(b^3+a*c*d)*e2,(a-b)*e1+(c+d)*e2];
T := termorder(S,lexdeg([d,c,b,a],[e1,e2]),[e1,e2]);
```

to enter the TOP extension of *grevlex*. (The Maple command lexdeg implements the orders $>_1$ and $>_2$ of Exercise 10, depending on which set of variables is put first, though due to a bug in the Groebner package, one needs to reverse the variables in the first argument of lexdeg.) Then show that the command

$$\text{gbasis(F, T);}$$

computes the Gröbner basis (2.10). Hint: Use the command

$$\text{collect(GB, \{e1, e2\}, distributed);}$$

to make the Gröbner basis easier to read.

Exercise 12. In this exercise, we will show how Gröbner bases can be applied to perform calculations in the quotient modules R^m/M for $M \subset R^m$.

a. Let $\mathcal{G}$ be a Gröbner basis for M with respect to any monomial order on R^m. Use Theorem (2.5) to define a one-to-one correspondence between the cosets in R^m/M and the remainders on division of $f \in R^m$ by $\mathcal{G}$.

b. Deduce that the set of monomials in the complement of $\langle \text{LT}(M) \rangle$ forms a vector space basis for R^m/M over k.

c. Let $R = k[a, b, c, d]$. Find a vector space basis for the quotient module R^2/M where M is generated by the columns of the matrix from Exercise 8, using the TOP *grevlex* Gröbner basis from (2.10). (Note: R^2/M is not finite-dimensional in this case.)

d. Explain how to compute the sum and scalar multiplication operations in R^m/M using part a. Hint: see Chapter 2, §2 of this book for a discussion of the ideal case.

e. $R = k[x_1, \ldots, x_n]$. State and prove a criterion for finite-dimensionality of R^m/M as a vector space over k generalizing the Finiteness Theorem from Chapter 2, §2 of this book.

Exercise 13. Prove that submodules of R^m satisfy the ACC. Hint: Given $M_1 \subset M_2 \subset \cdots$, show that $M = M_1 \cup M_2 \cup \cdots$ is a submodule. Then use part b of Exercise 4.

§3 Computing Syzygies

In this section, we begin the study of syzygies on a set of elements of a module, and we shall show how to solve Problem (2.1) b of the last section. Once again R will stand for a polynomial ring $k[x_1, \ldots, x_n]$. Solving this problem will allow us to find a matrix presenting any submodule of R^m for which we know a set of generators.

Gröbner bases play a central role here because of the following key observation. In computing a Gröbner basis $\mathcal{G} = \{g_1, \ldots, g_s\}$ for an ideal $I \subset R$ with respect to some fixed monomial ordering using Buchberger's algorithm, a slight modification of the algorithm would actually compute a set of generators for the module of syzygies $\mathrm{Syz}(g_1, \ldots, g_s)$ as well as the g_i themselves. The main idea is that Buchberger's S-polynomial criterion for ideal Gröbner bases is precisely the statement that a set of generators is a Gröbner basis if and only if every homogeneous syzygy on the leading terms of the generators "lifts" to a syzygy on the generators, in the sense described in the following theorem. The "lifting" is accomplished by the division algorithm.

To prepare for the theorem, let $S(g_i, g_j)$ be the S-polynomial of g_i and g_j:

$$S(g_i, g_j) = \frac{x^{\gamma_{ij}}}{\mathrm{LT}(g_i)} \cdot g_i - \frac{x^{\gamma_{ij}}}{\mathrm{LT}(g_j)} \cdot g_j,$$

where $x^{\gamma_{ij}}$ is the least common multiple of $\mathrm{LM}(g_i)$ and $\mathrm{LM}(g_j)$ (see (2.2) of Chapter 1 of this book). Since $\mathcal{G}$ is a Gröbner basis, by Buchberger's Criterion from §3 of Chapter 1, the remainder of $S(g_i, g_j)$ upon division by $\mathcal{G}$ is 0, and the division algorithm gives an expression

$$S(g_i, g_j) = \sum_{k=1}^{s} a_{ijk} g_k,$$

where $a_{ijk} \in R$, and $\mathrm{LT}(a_{ijk} g_k) \leq \mathrm{LT}(S(g_i, g_j))$ for all i, j, k.

Let $\mathbf{a}_{ij} \in R^s$ denote the column vector

$$\mathbf{a}_{ij} = a_{ij1}\mathbf{e}_1 + a_{ij2}\mathbf{e}_2 + \cdots + a_{ijs}\mathbf{e}_s = \begin{pmatrix} a_{ij1} \\ a_{ij2} \\ \vdots \\ a_{ijs} \end{pmatrix} \in R^s$$

and define $\mathbf{s}_{ij} \in R^s$ by setting

$$(3.1) \qquad \mathbf{s}_{ij} = \frac{x^{\gamma_{ij}}}{\mathrm{LT}(g_i)} \mathbf{e}_i - \frac{x^{\gamma_{ij}}}{\mathrm{LT}(g_j)} \mathbf{e}_j - \mathbf{a}_{ij}$$

in R^s. Then we have the following theorem of Schreyer from [Schre1].

(3.2) Theorem. *Let* $\mathcal{G} = \{g_1, \ldots, g_s\}$ *be a Gröbner basis of an ideal* I *in* R *with respect to some fixed monomial order, and let* $M = \mathrm{Syz}(g_1, \ldots, g_s)$. *The collection* $\{\mathbf{s}_{ij}, 1 \leq i, j \leq s\}$ *from (3.1) generates* M *as an* R-*module.*

Part of this statement is easy to verify from the definition of the $\mathbf{s}_{ij}$.

Exercise 1. Prove that $\mathbf{s}_{ij}$ is an element of the syzygy module M for all i, j.

The first two terms

$$\frac{x^{\gamma_{ij}}}{\mathrm{LT}(g_i)} \mathbf{e}_i - \frac{x^{\gamma_{ij}}}{\mathrm{LT}(g_j)} \mathbf{e}_j$$

in expression (3.1) for $\mathbf{s}_{ij}$ form a column vector which is a syzygy *on the leading terms* of the g_i (that is, an element of $\mathrm{Syz}(\mathrm{LT}(g_1), \ldots, \mathrm{LT}(g_s))$). The "lifting" referred to above consists of adding the additional terms $-\mathbf{a}_{ij}$ in (3.1) to produce a syzygy on the g_i themselves (that is, an element of $\mathrm{Syz}(g_1, \ldots, g_s)$).

A direct proof of this theorem can be obtained by a careful reconsideration of the proof of Buchberger's Criterion (see Theorem 6 of [CLO], Chapter 2, §6). Schreyer's proof, which is actually significantly simpler, comes as a byproduct of the theory of Gröbner bases for modules, and it establishes quite a bit more in the process. So we will present it here.

First, let us note that we can parallel in the module case the observations made above. Let $\mathcal{G} = \{\mathbf{g}_1, \ldots, \mathbf{g}_s\}$ be a Gröbner basis for any submodule $M \subset R^m$ with respect to some fixed monomial order $>$. Since $\mathcal{G}$ is a Gröbner basis, by Theorem (2.9) now, the remainder of $S(\mathbf{g}_i, \mathbf{g}_j)$ on division by $\mathcal{G}$ is $\mathbf{0}$, and the division algorithm gives an expression

$$S(\mathbf{g}_i, \mathbf{g}_j) = \sum_{k=1}^{s} a_{ijk} \mathbf{g}_k,$$

where $a_{ijk} \in R$, and $\mathrm{LT}(a_{ijk} \mathbf{g}_k) \leq \mathrm{LT}(S(\mathbf{g}_i, \mathbf{g}_j))$ for all i, j, k.

Write $\epsilon_1, \ldots, \epsilon_s$ for the standard basis vectors in R^s. Let $\mathbf{m}_{ij} = \mathrm{LCM}(\mathrm{LT}(\mathbf{g}_i), \mathrm{LT}(\mathbf{g}_j))$, and let $\mathbf{a}_{ij} \in R^s$ denote the column vector

$$\mathbf{a}_{ij} = a_{ij1} \epsilon_1 + a_{ij2} \epsilon_2 + \cdots + a_{ijs} \epsilon_s \in R^s.$$

For the pairs (i, j) such that $\mathbf{m}_{ij} \neq \mathbf{0}$, define $\mathbf{s}_{ij} \in R^s$ by setting

$$\mathbf{s}_{ij} = \frac{\mathbf{m}_{ij}}{\mathrm{LT}(\mathbf{g}_i)} \epsilon_i - \frac{\mathbf{m}_{ij}}{\mathrm{LT}(\mathbf{g}_j)} \epsilon_j - \mathbf{a}_{ij}$$

in R^s, and let $\mathbf{s}_{ij}$ be zero otherwise. Since a Gröbner basis for a module generates the module by Proposition (2.7), the following theorem includes Theorem (3.2) as a special case. Hence, by proving this more general result we will establish Theorem (3.2) as well.

(3.3) Theorem (Schreyer's Theorem). *Let $\mathcal{G} \subset R^m$ be a Gröbner basis with respect to any monomial order $>$ on R^m. The $\mathbf{s}_{ij}$ form a Gröbner basis for the syzygy module $M = \mathrm{Syz}(\mathbf{g}_1, \ldots, \mathbf{g}_s)$ with respect to a monomial order $>_{\mathcal{G}}$ on R^s defined as follows: $x^\alpha \epsilon_i >_{\mathcal{G}} x^\beta \epsilon_j$ if $\mathrm{LT}_>(x^\alpha \mathbf{g}_i) > \mathrm{LT}_>(x^\beta \mathbf{g}_j)$ in R^m, or if $\mathrm{LT}_>(x^\alpha \mathbf{g}_i) = \mathrm{LT}_>(x^\beta \mathbf{g}_j)$ and $i < j$.*

PROOF. We leave it to the reader as Exercise 1 below to show that $>_{\mathcal{G}}$ is a monomial order on R^s. Since $S(\mathbf{g}_i, \mathbf{g}_j)$ and $S(\mathbf{g}_j, \mathbf{g}_i)$ differ only by a sign, it suffices to consider the $\mathbf{s}_{ij}$ for $i < j$ only. We claim first that if $i < j$, then

$$(3.4) \qquad\qquad \mathrm{LT}_{>_{\mathcal{G}}}(\mathbf{s}_{ij}) = \frac{\mathbf{m}_{ij}}{\mathrm{LT}(\mathbf{g}_i)} \epsilon_i.$$

Since we take $i < j$, this term is larger than $(\mathbf{m}_{ij}/\mathrm{LT}(\mathbf{g}_j))\epsilon_j$ in the $>_{\mathcal{G}}$ order. It is also larger than any of the terms in $\mathbf{a}_{ij}$, for the following reason. The a_{ijk} are obtained via the division algorithm, dividing $\mathbf{S} = S(\mathbf{g}_i, \mathbf{g}_j)$ with respect to $\mathcal{G}$. Hence $\mathrm{LT}_>(\mathbf{S}) \geq \mathrm{LT}_>(a_{ij\ell}\mathbf{g}_\ell)$ for all $\ell = 1, \ldots, s$ (in R^m). However, by the definition of the S-vector,

$$\mathrm{LT}_> \left(\frac{\mathbf{m}_{ij}}{\mathrm{LT}(\mathbf{g}_i)} \mathbf{g}_i \right) > \mathrm{LT}_>(\mathbf{S}),$$

since the S-vector is guaranteed to produce a cancellation of leading terms. Putting these two inequalities together establishes (3.4).

Now let $\mathbf{f} = \sum_{i=1}^s f_i \epsilon_i$ be any element of the syzygy module M, let $\mathrm{LT}_{>_{\mathcal{G}}}(f_i \epsilon_i) = m_i \epsilon_i$ for some term m_i appearing in f_i. Further, let $\mathrm{LT}_{>_{\mathcal{G}}}(\mathbf{f}) = m_v \epsilon_v$ for some v. With this v fixed, we set

$$\mathbf{s} = \sum_{u \in S} m_u \epsilon_u,$$

where $S = \{u : m_u \mathrm{LT}_>(\mathbf{g}_u) = m_v \mathrm{LT}_>(\mathbf{g}_v)\}$.

One can show without difficulty that $\mathbf{s}$ is an element of $\mathrm{Syz}(\{\mathrm{LT}_>(\mathbf{g}_u) : u \in S\})$. By part c of Proposition (2.3) of this chapter, it follows that $\mathbf{s}$ is an element of the submodule of R^s generated by the

$$\sigma_{uw} = \frac{\mathbf{m}_{uw}}{\mathrm{LT}_>(\mathbf{g}_u)} \epsilon_u - \frac{\mathbf{m}_{uw}}{\mathrm{LT}_>(\mathbf{g}_w)} \epsilon_w,$$

where $u < w$ are elements of S. Then (3.4) implies that $\mathrm{LT}_\mathcal{G}(\mathbf{s})$ is divisible by $\mathrm{LT}_\mathcal{G}(\mathbf{s}_{ij})$ for some $i < j$. So by definition the $\mathbf{s}_{ij}$ form a Gröbner basis for M with respect to the $>_\mathcal{G}$ order.   □

Exercise 2. Verify that the order $>_\mathcal{G}$ introduced in the statement of the theorem is a monomial order on R^m.

Theorem (3.3) gives the outline for an algorithm for computing $\mathrm{Syz}(\mathbf{g}_1, \ldots, \mathbf{g}_s)$ for any Gröbner basis $\mathcal{G} = \{\mathbf{g}_1, \ldots, \mathbf{g}_s\}$ using the division algorithm. Hence we have a solution of Problem (2.1) b in the special case of computing syzygies for Gröbner bases. Using this, we will see how to compute the module of syzygies on any set of generators $\{\mathbf{f}_1, \ldots, \mathbf{f}_t\}$ for a submodule of R^m.

So suppose that we are given $\mathbf{f}_1, \ldots, \mathbf{f}_t \in R^m$, and that we compute a Gröbner basis $\mathcal{G} = \{\mathbf{g}_1, \ldots, \mathbf{g}_s\}$ for $M = \langle \mathbf{f}_1, \ldots, \mathbf{f}_t \rangle = \langle \mathcal{G} \rangle$. Let $F = (\mathbf{f}_1, \ldots, \mathbf{f}_t)$ and $G = (\mathbf{g}_1, \ldots, \mathbf{g}_s)$ be the $m \times t$ and $m \times s$ matrices in which the $\mathbf{f}_i$'s and $\mathbf{g}_i$'s are columns, respectively. Since the columns of F and G generate the same module, there are a $t \times s$ matrix A and an $s \times t$ matrix B, both with entries in R, such that $G = FA$ and $F = GB$. The matrix B can be computed by applying the division algorithm with respect to G, expressing each $\mathbf{f}_i$ in terms of the $\mathbf{g}_j$. The matrix A can be generated as a byproduct of the computation of G. This is because each S-vector remainder that is added to G in the basic Buchberger algorithm, Theorem (2.11), comes with an expression in terms of the $\mathbf{f}_i$, computed by division and substitution. However, the matrix A can also be computed in an ad hoc way as in simple examples such as the following.

Suppose, for example, that $m = 1$, so that M is an ideal, say $M = \langle f_1, f_2 \rangle$ in $R = k[x, y]$, where

$$f_1 = xy + x, \qquad f_2 = y^2 + 1.$$

Using the *lex* monomial order with $x > y$, the reduced Gröbner basis for M consists of

$$g_1 = x, \qquad g_2 = y^2 + 1.$$

Then it is easy to check that

$$f_1 = (y+1)g_1$$
$$g_1 = -(1/2)(y-1)f_1 + (1/2)xf_2,$$

so that

(3.5) $$G = (g_1, g_2) = (f_1, f_2)\begin{pmatrix} -(y-1)/2 & 0 \\ x/2 & 1 \end{pmatrix} = FA$$

and

(3.6) $F = (f_1, f_2) = (g_1, g_2) \begin{pmatrix} y + 1 & 0 \\ 0 & 1 \end{pmatrix} = GB.$

If we express G in terms of F using the equation $G = FA$ then substitute $F = GB$ on the right, we obtain an equation $G = GBA$. Similarly, $F = FAB$. What we have done here is analogous to changing from one basis to another in a vector space, then changing back to the first basis. However, in the general R-module case, it is *not* necessarily true that AB and BA equal identity matrices of the appropriate sizes. This is another manifestation of the fact that a module need not have a basis over R. For instance, in the example from (3.5) and (3.6) above, we have

$$AB = \begin{pmatrix} -(y^2 - 1)/2 & 0 \\ (xy + x)/2 & 1 \end{pmatrix},$$

and

$$BA = \begin{pmatrix} -(y^2 - 1)/2 & 0 \\ x/2 & 1 \end{pmatrix}.$$

In addition to connecting F with G, the matrices A and B also connect the syzygies on F and G in the following ways.

(3.7) Lemma.
a. *Let* $\mathbf{s} \in R^s$ *(a column vector) be an element of* $\mathrm{Syz}(\mathbf{g}_1, \ldots, \mathbf{g}_s)$, *then the matrix product* $A\mathbf{s}$ *is an element of* $\mathrm{Syz}(\mathbf{f}_1, \ldots, \mathbf{f}_t)$.
b. *Similarly, if* $\mathbf{t} \in R^t$ *(also a column vector) is an element of* $\mathrm{Syz}(\mathbf{f}_1, \ldots, \mathbf{f}_t)$, *then* $B\mathbf{t} \in R^s$ *is an element of* $\mathrm{Syz}(\mathbf{g}_1, \ldots, \mathbf{g}_s)$.
c. *Each column of the matrix* $I_t - AB$ *also defines an element of* $\mathrm{Syz}(\mathbf{f}_1, \ldots, \mathbf{f}_t)$.

PROOF. Take the matrix equation $G = FA$ and multiply on the right by the column vector $\mathbf{s} \in \mathrm{Syz}(\mathbf{g}_1, \ldots, \mathbf{g}_s)$. Since matrix multiplication is associative, we get the equation

$$0 = G\mathbf{s} = FA\mathbf{s} = F(A\mathbf{s}).$$

Hence $A\mathbf{s}$ is an element of $\mathrm{Syz}(\mathbf{f}_1, \ldots, \mathbf{f}_t)$. Part b is proved similarly, starting from the equation $F = GB$. Finally, $F = FAB$ implies

$$F(I_t - AB) = F - FAB = F - F = 0,$$

and part c follows immediately.   □

Our next result gives the promised solution to the problem of computing syzygies for a general ordered t-tuple $F = (\mathbf{f}_1, \ldots, \mathbf{f}_t)$ of elements of R^m (not necessarily a Gröbner basis).

(3.8) Proposition. *Let* $F = (\mathbf{f}_1, \dots, \mathbf{f}_t)$ *be an ordered t-tuple of elements of* R^m, *and let* $G = (\mathbf{g}_1, \dots, \mathbf{g}_s)$ *be an ordered Gröbner basis for* $M = \langle F \rangle$ *with respect to any monomial order in* R^m. *Let* A *and* B *be the matrices introduced above, and let* $\mathbf{s}_{ij}$, $1 \le i, j, \le s$ *be the basis for* $\mathrm{Syz}(\mathbf{g}_1, \dots, \mathbf{g}_s)$ *given by Theorem (3.3) or (3.2). Finally, let* $\mathbf{S}_1, \dots, \mathbf{S}_t$ *be the columns of the* $t \times t$ *matrix* $I_t - AB$. *Then*

$$\mathrm{Syz}(\mathbf{f}_1, \dots, \mathbf{f}_t) = \langle A\mathbf{s}_{ij}, \mathbf{S}_1, \dots, \mathbf{S}_t \rangle.$$

PROOF. $\langle A\mathbf{s}_{ij}, \mathbf{S}_1, \dots, \mathbf{S}_t \rangle$ is a submodule of $\mathrm{Syz}(\mathbf{f}_1, \dots, \mathbf{f}_t)$, so to prove the proposition, we must show that every syzygy on F can be expressed as an R-linear combination of the $A\mathbf{s}_{ij}$ and the $\mathbf{S}_k$. To see this, let $\mathbf{t}$ be any element of $\mathrm{Syz}(\mathbf{f}_1, \dots, \mathbf{f}_t)$. By part b of Lemma (3.7), $B\mathbf{t}$ is an element of $\mathrm{Syz}(\mathbf{g}_1, \dots, \mathbf{g}_s)$. Since the $\mathbf{s}_{ij}$ are a Gröbner basis for that module, and hence a generating set for $\mathrm{Syz}(\mathbf{g}_1, \dots, \mathbf{g}_s)$, there are $a_{ij} \in R$ such that

$$B\mathbf{t} = \sum_{ij} a_{ij} \mathbf{s}_{ij}.$$

But multiplying both sides by A on the left, this implies

$$AB\mathbf{t} = \sum_{ij} a_{ij} A\mathbf{s}_{ij},$$

so that

$$
\begin{aligned}
\mathbf{t} &= ((I_t - AB) + AB)\mathbf{t} \\
&= (I_t - AB)\mathbf{t} + \sum_{ij} a_{ij} A\mathbf{s}_{ij}.
\end{aligned}
$$

The first term on the right in the last equation is an R-linear combination of the columns $\mathbf{S}_1, \dots, \mathbf{S}_t$ of $(I_t - AB)$, hence $\mathbf{t} \in \langle A\mathbf{s}_{ij}, \mathbf{S}_1, \dots, \mathbf{S}_t \rangle$. Since this is true for all $\mathbf{t}$, the proposition is proved. $\square$

Note that the hypothesis in Proposition (3.8) above that G is a Gröbner basis is needed only to ensure that the $\mathbf{g}_i$ generate and that the $\mathbf{s}_{ij}$ are a basis for the module of syzygies. More generally, if we have any set of generators for a module M, and a set of generators for the module of syzygies of that set of generators, then we can find a generating set of syzygies on any other generating set of M.

(3.9) Corollary. *Let the notation be as above, but assume only that* $G = (\mathbf{g}_1, \dots, \mathbf{g}_s)$ *is a set of generators for* M *and that* D *is a matrix presenting* M, *so the columns of* D *generate* $\mathrm{Syz}(\mathbf{g}_1, \dots, \mathbf{g}_s)$. *Then the block matrix*

$$(AG \quad I_t - AB)$$

presents M *with respect to the generating set* $\mathbf{f}_1, \dots, \mathbf{f}_t$.

PROOF. This follows immediately from the proof of Proposition (3.8) above. We have also seen this result in part c of Exercise 31 in §1.   □

As an example of Proposition (3.8), we use the F, G, A, B from (3.5) and (3.6) and proceed as follows. Since

$$S(g_1, g_2) = y^2 g_1 - x g_2 = -x = -g_1,$$

by Theorem (3.3) we have that

$$\mathbf{s}_{12} = \begin{pmatrix} y^2 + 1 \\ -x \end{pmatrix}$$

generates $\mathrm{Syz}(g_1, g_2)$. Multiplying by A we get

$$A\mathbf{s}_{12} = \begin{pmatrix} -(y-1)/2 & 0 \\ x/2 & 1 \end{pmatrix} \begin{pmatrix} y^2 + 1 \\ -x \end{pmatrix}$$

$$= \begin{pmatrix} -(y^3 - y^2 + y - 1)/2 \\ (xy^2 - x)/2 \end{pmatrix}.$$

Exercise 3. Verify directly that $A\mathbf{s}_{12}$ gives a syzygy on (f_1, f_2).

Continuing the example, the columns of $I_2 - AB$ are

$$\mathbf{S}_1 = \begin{pmatrix} (y^2 + 1)/2 \\ -(xy + x)/2 \end{pmatrix}, \qquad \mathbf{S}_2 = \begin{pmatrix} 0 \\ 0 \end{pmatrix}.$$

So by the proposition

$$\mathrm{Syz}(f_1, f_2) = \langle A\mathbf{s}_{12}, \mathbf{S}_1 \rangle.$$

This example has another instructive feature as shown in the following exercise.

Exercise 4. Show that $A\mathbf{s}_{12}$ above is actually a multiple of $\mathbf{S}_1$ by a nonconstant element of R. Deduce that $\mathbf{S}_1$ alone generates $\mathrm{Syz}(f_1, f_2)$, yet $A\mathbf{s}_{12}$ does not. Compare to Exercise 12.

Hence, the $A\mathbf{s}_{ij}$ are alone not sufficient to generate the syzygies on F in some cases.

Let us now return to the situation where M is a submodule of R^m and $\mathbf{f}_1, \ldots, \mathbf{f}_t$ and $\mathbf{g}_1, \ldots, \mathbf{g}_s$ are different sets of generators of M. At first glance, Corollary (3.9) seems a little asymmetric in that it privileges the $\mathbf{g}$'s over the $\mathbf{f}$'s (in Proposition (3.8), this issue does not arise, because it seems sensible to privilege a Gröbner basis over the set of generators from which it was computed). Given presentations for $\mathrm{Syz}(\mathbf{f}_1, \ldots, \mathbf{f}_t)$ and $\mathrm{Syz}(\mathbf{g}_1, \ldots, \mathbf{g}_s)$, Exercise 31 of §1 provides a block matrix which presents

M with respect the combined set of generators $\mathbf{f}_1, \ldots, \mathbf{f}_t, \mathbf{g}_1, \ldots, \mathbf{g}_s$ (and which reduces to the matrices F and G separately). It is worth pausing, and phrasing a result that links the set of syzygies on any two generating sets of the same module M.

(3.10) Proposition. *Suppose that* $\mathbf{f}_1, \ldots, \mathbf{f}_t$ *and* $\mathbf{g}_1, \ldots, \mathbf{g}_s$ *are ordered sets of elements of* R^m *which generate the same module* M. *Then, there are free* R-*modules* L *and* L' *such that*

$$\mathrm{Syz}(\mathbf{f}_1, \ldots, \mathbf{f}_t) \oplus L \cong \mathrm{Syz}(\mathbf{g}_1, \ldots, \mathbf{g}_s) \oplus L'.$$

PROOF. We claim that $N = \mathrm{Syz}(\mathbf{f}_1, \ldots, \mathbf{f}_t, \mathbf{g}_1, \ldots, \mathbf{g}_s)$ is a direct sum of a module isomorphic to $\mathrm{Syz}(\mathbf{f}_1, \ldots, \mathbf{f}_t)$ and a free module. In fact, N is the set of vectors $(c_1, \ldots, c_t, d_1, \ldots, d_s)^T \in R^{t+s}$ such that

$$c_1 \mathbf{f}_1 + \cdots + c_t \mathbf{f}_t + d_1 \mathbf{g}_1 + \cdots + d_s \mathbf{g}_s = \mathbf{0}.$$

Now consider the submodule $K \subset N$ obtained by taking those elements with all $d_i = 0$. Note that K is clearly isomorphic to $\mathrm{Syz}(\mathbf{f}_1, \ldots, \mathbf{f}_t)$. Moreover, since the $\mathbf{f}_j$ generate M, we can write $\mathbf{g}_i = \sum a_{ij} \mathbf{f}_j$. Then, each of the t vectors $\mathbf{n}_k = (a_{k1}, \ldots, a_{kt}, 0, \ldots, 0, -1, 0, \ldots, 0)^T$, with all terms in the $(t+j)$th place, $0 < j \le s$, equal to 0 except the $(t+k)$th term which is -1, belongs to N. Moreover, the $\mathbf{n}_k$, $1 \le k \le t$, are clearly linearly independent, so they generate a free submodule of N, which we will call L. Clearly $K \cap L = \emptyset$. To see that $N = K + L$, suppose that we have an element $(c_1, \ldots, c_t, d_1, \ldots, d_s)^T \in N$. Then

$$\begin{aligned}
\mathbf{0} &= c_1 \mathbf{f}_1 + \cdots + c_t \mathbf{f}_t + d_1 \mathbf{g}_1 + \cdots + d_s \mathbf{g}_s \\
&= c_1 \mathbf{f}_1 + \cdots + c_t \mathbf{f}_t + d_1 \sum a_{ij} \mathbf{f}_j + \cdots + d_s \sum a_{sj} \mathbf{f}_j \\
&= \sum_i (c_i + \sum_j d_j a_{ji}) \mathbf{f}_i,
\end{aligned}$$

so that

$$\begin{aligned}
&(c_1, \ldots, c_t, d_1, \ldots, d_s)^T + \sum_j d_j \mathbf{n}_j \\
&= (c_1 + \sum_j d_j a_{j1}, \ldots, c_t + \sum_j d_j a_{jt}, 0, \ldots, 0)^T,
\end{aligned}$$

which belongs to K. This proves the claim. Similarly, we show that N is a direct sum of $\mathrm{Syz}(\mathbf{g}_1, \ldots, \mathbf{g}_s)$ and the result follows. $\square$

Modules which become isomorphic after addition of free modules are often called *equivalent*. The proposition shows that any two modules of syzygies on any different sets of generators of the same module are equivalent. We leave it as an exercise to show that any modules of syzygies on equivalent modules are equivalent.

As an application, we will develop a syzygy-based algorithm for computing ideal and submodule intersections. For ideals, this method is more efficient than the elimination-based algorithm given in Chapter 1, §3, Exercise 11. We start with the ideal case. The following statement gives a connection between the intersection $I \cap J$ and syzygies on a certain collection of elements of R^2.

(3.11) Proposition. *Let* $I = \langle f_1, \ldots, f_t \rangle$ *and* $J = \langle g_1, \ldots, g_s \rangle$ *be ideals in* R. *A polynomial* $h_0 \in R$ *is an element of* $I \cap J$ *if and only if* h_0 *appears as the first component in a syzygy*

$$(h_0, h_1, \ldots, h_t, h_{t+1}, \ldots, h_{t+s})^T \in R^{s+t+1}$$

in the module

$$S = \mathrm{Syz}(\mathbf{v}_0, \mathbf{v}_1, \ldots, \mathbf{v}_t, \mathbf{v}_{t+1}, \ldots, \mathbf{v}_{s+t})$$

where

$$\mathbf{v}_0 = \begin{pmatrix} 1 \\ 1 \end{pmatrix}, \mathbf{v}_1 = \begin{pmatrix} f_1 \\ 0 \end{pmatrix}, \ldots, \mathbf{v}_t = \begin{pmatrix} f_t \\ 0 \end{pmatrix},$$

$$\mathbf{v}_{t+1} = \begin{pmatrix} 0 \\ g_1 \end{pmatrix}, \ldots, \mathbf{v}_{s+t} = \begin{pmatrix} 0 \\ g_s \end{pmatrix}$$

in R^2.

PROOF. Suppose that

$$\mathbf{0} = h_0 \mathbf{v}_0 + h_1 \mathbf{v}_1 + \cdots + h_t \mathbf{v}_t + h_{t+1} \mathbf{v}_{t+1} + \cdots + h_{s+t} \mathbf{v}_{s+t}.$$

From the first components, we obtain an equation

$$0 = h_0 + h_1 f_1 + \cdots + h_t f_t + 0 + \cdots + 0,$$

so $h_0 \in \langle f_1, \ldots, f_t \rangle = I$. Similarly from the second components, $h_0 \in \langle g_1, \ldots, g_s \rangle = J$. Hence $h_0 \in I \cap J$.

On the other hand, in Exercise 7 below, you will show that every $h_0 \in I \cap J$ appears as the first component in some syzygy on the $\mathbf{v}_0, \ldots, \mathbf{v}_{s+t}$. □

Exercise 5. Show that Proposition (3.11) extends to submodules $M, N \subset R^m$ in the following way. Say $M = \langle \mathbf{f}_1, \ldots, \mathbf{f}_t \rangle$, $N = \langle \mathbf{g}_1, \ldots, \mathbf{g}_s \rangle$ where now the $\mathbf{f}_i, \mathbf{g}_j \in R^m$. In R^{2m}, consider the vectors $\mathbf{v}_{01}, \ldots, \mathbf{v}_{0m}$, where $\mathbf{v}_{0i}$ is formed by concatenating two copies of the standard basis vector $\mathbf{e}_i \in R^m$ to make a vector in R^{2m}. Then take $\mathbf{v}_1, \ldots, \mathbf{v}_t$, where $\mathbf{v}_i$ is formed by appending m zeros *after* the components of $\mathbf{f}_i$, and $\mathbf{v}_{t+1}, \ldots, \mathbf{v}_{t+s}$, where $\mathbf{v}_{t+j}$ is formed by appending m zeroes *before* the components of $\mathbf{g}_j$. Show that the statement of Proposition (3.11) goes over to this setting in the following way: $(\mathbf{h}_{01}, \ldots, \mathbf{h}_{0m})^T \in M \cap N$ if and only if the $\mathbf{h}_{01}, \ldots, \mathbf{h}_{0m}$ appear as the first m components in a syzygy in the module

$$\mathrm{Syz}(\mathbf{v}_{01}, \ldots, \mathbf{v}_{0m}, \mathbf{v}_1, \ldots, \mathbf{v}_t, \mathbf{v}_{t+1}, \ldots, \mathbf{v}_{s+t})$$

in R^{n+t+s}.

Exercise 6.
a. Using Propositions (3.8) and (3.11) and the previous exercise, develop an algorithm for computing a set of generators for $M \cap N$.
b. In the ideal case $(m = 1)$, show that if a POT extension of a monomial order $>$ on R is used and $\mathcal{G}$ is a Gröbner basis for the syzygy module S from Proposition (3.11), then the first components of the vectors in $\mathcal{G}$ give a Gröbner basis for $I \cap J$ with respect to $>$.

Macaulay 2 has a built-in command `syz` for computing the module of syzygies on the columns of a matrix using the method developed here. For instance, with the matrix

$$M = \begin{pmatrix} a^2 + b^2 & a^3 - 2bcd & a - b \\ c^2 - d^2 & b^3 + acd & c + d \end{pmatrix},$$

from the examples in §2, we could use: `syz M` to compute syzygies. Note: this produces a set of generators for the syzygies on the columns of the original matrix M, not the Gröbner basis for the module they generate. Try it! Your output should be:

```
i7 : syz M
```

```
o7 = {2} | -ab3+b4+a3c+a3d-a2cd+abcd-2bc2d-2bcd2    |
     {3} | -a2c-b2c+ac2-bc2-a2d-b2d-ad2+bd2          |
     {1} | a2b3+b5-a3c2+a3cd+ab2cd+2bc3d+a3d2-2bcd3  |
```

```
           3     1
o7 : Matrix R <--- R
```

One can also use `Singular` (or CoCoA or CALI) to compute modules of syzygies. To do this in `Singular` define R and M as in Exercise 9 of §2, we enter `syz(M)`; at the prompt $>$. Depending on how wide your screen size is set, your output will be:

```
>syz(M);
_[1]=a2b3*gen(3)+b5*gen(3)+a3c2*gen(3)+a3cd*gen(3)+ab2cd
*gen(3)+2bc3d*gen(3)+a3d2*gen(3)-2bcd3*gen(3)-ab3*gen(1)
+b4*gen(1)+a3c*gen(1)+a3d*gen(1)-a2cd*gen(1)+abcd*gen(1)
-2bc2d*gen(1)-2bcd2*gen(1)-a2c*gen(2)-b2c*gen(2)+ac2*gen
(2)-bc2*gen(2)-a2d*gen(2)-b2d*gen(2)-ad2*gen(2)+bd2*gen(
2)
```

Note that `Singular` uses the notation `gen(1)`, `gen(2)`, ... to refer to the module elements $\mathbf{e}_1, \mathbf{e}_2, \ldots$. There are a range of options for formatting output. To get output in a format closer to that given by *Macaulay 2* above, change the ordering on the module to POT, upward. That is, define the ring R using the command

$$\text{ring } R = 0, (a, b, c, d), (c, dp);$$

Try it.

Additional Exercises for §3

Exercise 7. Complete the proof of Proposition (3.11) by showing that every element of the intersection $I \cap J$ appears as the first component h_0 in some syzygy in

$$S = \text{Syz}(v_0, v_1, \ldots, v_t, v_{t+1}, \ldots, v_{s+t}).$$

Exercise 8. Let $I = \langle F \rangle = \langle xz - y, y^2 + z, yz + 2x \rangle$ in $k[x, y, z]$.
a. Find the monic reduced *lex* Gröbner basis $G = (g_1, \ldots, g_s)$ for I and the "change of basis" matrices A, B such that $G = FA$, $F = GB$.
b. Find a set of generators for $\text{Syz}(G)$ using Theorem (3.3).
c. Compute a set of generators for $\text{Syz}(F)$ using Proposition (3.8).

Exercise 9. Let $(m_1, \ldots, m_t)$ be any ordered t-tuple of elements of R^m, and let $S = \text{Syz}(m_1, \ldots, m_t) \subset R^t$. Show that for any $1 \leq s \leq t$, the projection of S onto the first (that is, the top) s components (that is, the collection N of $(a_1, \ldots, a_s) \in R^s$ such that $a_1, \ldots, a_s$ appear as the first s elements in some element of S) forms a submodule of R^s. Hint: N is *not the same* as $\text{Syz}(m_1, \ldots, m_s)$.

Exercise 10. In this exercise, you use syzygies to compute the ideal quotient $I : J$. Recall from part b of Exercise 13 from Chapter 1, §1 of this book that if $I \cap \langle h \rangle = \langle g_1, \ldots, g_t \rangle$, then $I : \langle h \rangle = \langle g_1/h, \ldots, g_t/h \rangle$.
a. Using Proposition (3.11) (not elimination), give an algorithm for computing $I : \langle h \rangle$. Explain how the g_i/h can be computed *without factoring*.
b. Now generalize part a to compute $I : J$ for any ideals I, J. Hint: If $J = \langle h_1, \ldots, h_s \rangle$, then by [CLO], Chapter 4, §4, Proposition 10,

$$I : J = \bigcap_{j=1}^{s} (I : \langle h_j \rangle).$$

Exercise 11. Show that a homogeneous syzygy $(c_1 x^{\alpha - \alpha_1}, \ldots, c_s x^{\alpha - \alpha_s})^T$ on a collection of monomials $x^{\alpha_1}, \ldots, x^{\alpha_s}$ in R can be written as a sum of homogeneous syzygies between pairs of the x^{α_i}. (See the proof of Proposition (2.3) part c.)

Exercise 12. If $f_1, f_2 \in R$ are nonzero, use unique factorization to show that $\mathrm{Syz}(f_1, f_2)$ is generated by a single element. Compare with Exercise 4.

Exercise 13.
a. Show that the notion of equivalence defined after Proposition (3.10) is an equivalence relation on R-modules. (That is, show that it is reflexive, symmetric and transitive.)
b. Suppose that M and M' are two R-modules which are equivalent in the sense described after Proposition (3.10). That is, there are free modules L and L' such that $M \oplus L$ is isomorphic to $M' \oplus L'$. Show that any two modules of syzygies on M and M' are equivalent.

Exercise 14. Re-do Exercise 27, parts a and b, from §1 using Proposition (3.8). In fact, write out and prove Proposition (3.8) in the special case that $F = (f_1, \ldots, f_t)$ is an ordered set of elements of R such that $1 \in \langle f_1, \ldots, f_t \rangle$ (in which case the Gröbner basis G consists of the single element $\{1\}$).

Exercise 15. This exercise will show that one Gröbner basis computation can accomplish multiple tasks simultaneously. Let R be a polynomial ring with a monomial order $>$, and for any integer $m \geq 1$, let $>_m$ denote the POT extension of $>$ to R^m. Given $f_1, \ldots, f_s \in R$, our goal is to compute the following:
- A Gröbner basis G with respect to $>$ for the ideal $\langle f_1, \ldots, f_s \rangle$,
- For each $g \in G$, polynomials $h_1, \ldots, h_s$ such that $g = \sum_{i=1}^{s} h_i f_i$, and
- A Gröbner basis G' with respect to $>_s$ for the syzygy module $\mathrm{Syz}(f_1, \ldots, f_s)$.

To do this, we will work in the free module of R^{s+1} with standard basis $\mathbf{e}_0, \mathbf{e}_1, \ldots, \mathbf{e}_s$. Then consider the submodule $M \subset R^{s+1}$ generated by

$$\mathbf{m}_i = f_i \mathbf{e}_0 + \mathbf{e}_i = (f_i, 0, \ldots, 0, 1, 0, \ldots, 0), \quad i = 1, \ldots, s.$$

In other words, $\mathbf{m}_i$ has f_i in the 0th component, 1 in the ith component, and 0s elsewhere. Let G'' be a reduced Gröbner basis of M with respect to $>_{s+1}$.

a. Prove that $M \cap (\{0\} \times R^s) = \{0\} \times \mathrm{Syz}(f_1, \ldots, f_s)$.
b. Prove that the set $G = \{g \in R \mid g \neq 0 \text{ and there are } h_1, \ldots, h_s \in R \text{ with } (g, h_1, \ldots, h_s) \in G''\}$ is a reduced Gröbner basis with respect to $>$ for the ideal $\langle f_1, \ldots, f_s \rangle$.
c. If $g \in G$ and $(g, h_1, \ldots, h_s) \in G''$ as in part b, then show that $g = h_1 f_1 + \cdots + h_s f_s$.

d. Prove that the set G' defined by $\{0\} \times G' = G'' \cap (\{0\} \times R^s)$ is a reduced
   Gröbner basis with respect to $>_s$ for the syzygy module $\mathrm{Syz}(f_1, \ldots, f_s)$.
This exercise is based on an observation of T. Sederberg.

§4 Modules over Local Rings

The last two sections have dealt with modules over polynomial rings. In
this section, we consider modules over local rings. It turns out that the
adaptation of Gröbner basis techniques to local rings outlined in Chapter 4,
extends without difficulty to the case of modules over local rings. Moreover,
as we shall see, modules over local rings are simpler in many ways than
modules over a polynomial ring.

As in the preceding sections, R will denote the polynomial ring
$k[x_1, \ldots, x_n]$ and we shall let Q denote any one of the local rings obtained
from R considered in Chapter 4. More precisely, corresponding to any point
$p = (p_1, \ldots, p_n)$ of affine n-space k^n, we obtain the *localization* R_p of R,

$$R_p = \{f/g : f, g \in R \text{ and } g(p) \neq 0\}$$
$$= \{\text{rational functions defined at } p\}.$$

If $k = \mathbb{R}$ or $\mathbb{C}$, we can also consider the ring of convergent power series at
p, denoted $k\{x_1 - p_1, \ldots, x_n - p_n\}$, and for general k, we can study the
ring of formal power series at p, denoted

$$k[[x_1 - p_1, \ldots, x_n - p_n]].$$

The notation Q will refer to any of these. By *the* local ring at the point
p, we will mean R_p. Whenever convenient, we take the point p to be the
origin $0 \in k^n$ in which case $R_p = R_0 = k[x_1, \ldots, x_n]_{\langle x_1, \ldots, x_n \rangle}$.

In Chapter 4, we restricted ourselves to ideals in Q generated by polynomials. We make the analogous restriction in the case of modules. That is,
we shall only consider Q-modules which are either submodules of Q^s which
can be generated by polynomials (that is by vectors all of whose entries
are polynomials) or modules that have a presentation matrix all of whose
entries are in R.

Exercise 1. If $Q = k[x_1, \ldots, x_n]_{\langle x_1, \ldots, x_n \rangle}$, show that any submodule of
Q^m can be generated by generators which are finite k-linear combinations
of monomials.

Given any R-module M and any point $p \in k^n$, there is a natural R_p-
module, denoted M_p and called the *localization* of M at p, obtained by
allowing the elements of M to be multiplied by elements of R_p. If M is an
ideal I in R, then M_p is just the ideal IR_p. If $M \subset R^s$ is generated by
vectors $\mathbf{f}_1, \ldots, \mathbf{f}_t$, then M_p is generated by $\mathbf{f}_1, \ldots, \mathbf{f}_t$, where the entries in

the vectors $\mathbf{f}_i$ are considered as rational functions and one allows multiplication by elements of R_p. If M is presented by the $m \times n$ matrix A, then M_p is also presented by A. We leave the proof as an exercise.

Exercise 2. Let M be an R-module, and A a presentation matrix for M. If $p \in k^n$ is any point, show that A is a presentation matrix for the R_p-module M_p. Hint: The columns of A continue to be syzygies over R_p, so one only needs to observe that any R_p-linear relation on the generators is a multiple of an R-linear relation on the generators.

It is worth noting, however, that even though the presentation matrix A of M is also a presentation matrix for M_p, the matrix A may simplify much more drastically over R_p than over R. For example, let $R = k[x, y]$ and consider the matrix

$$A = \begin{pmatrix} x & x^2 \\ 1+y & y^2 \\ xy & 0 \end{pmatrix}.$$

This does not simplify substantially over R under the rules of Exercise 29 from §1. However, over R_0, we can divide the second row by $1 + y$, which is a unit in R_0, and use the resulting 1 in the first column, second row, to clear out all other elements in the first column. We obtain the matrix on the left which reduces further as shown

$$\begin{pmatrix} 0 & x^2 - \frac{xy^2}{1+y} \\ 1 & \frac{y^2}{1+y} \\ 0 & \frac{-xy^3}{1+y} \end{pmatrix} \longrightarrow \begin{pmatrix} x^2 - \frac{xy^2}{1+y} \\ \frac{-xy^3}{1+y} \end{pmatrix} \longrightarrow \begin{pmatrix} x^2 - \frac{xy^2}{1+y} \\ -xy^3 \end{pmatrix}$$

$$\longrightarrow \begin{pmatrix} x^2 + yx^2 - xy^2 \\ -xy^3 \end{pmatrix}.$$

Thus, the matrix A presents an R_0-module isomorphic to the ideal $\langle y^3, x + xy - y^2 \rangle$.

Exercise 3. Let A be as above.
a. Consider the R-module M presented by the matrix A. Prove that M is isomorphic to the ideal $\langle y^3, yx^2, -y^2 + x + xy \rangle$.
b. Show that in R_0 the ideal $\langle y^3, yx^2, -y^2 + x + xy \rangle$ is equal to the ideal $\langle y^3, x + xy - y^2 \rangle$.

To extend the algorithmic methods outlined in Chapter 4 to submodules of Q^m, one first extends local orders on Q to orders on Q^m. Just as for well-orderings, there are many ways to extend a given local order. In particular, given a local order on Q one has both TOP and POT extensions to Q^m. The local division algorithm (Theorem (3.10) of Chapter 4) extends to elements of $k[x_1, \ldots, x_n]_{\langle x_1, \ldots, x_n \rangle}^n$ in exactly the same way as the ordinary division algorithm extends to $k[x_1, \ldots, x_n]^n$. One has to give up the determinate remainder, and the proof of termination is delicate, but exactly the same as the proof of the local division algorithm. One defines Gröbner (or standard) bases, and S-vectors exactly as in the polynomial case and checks that Buchberger's criterion continues to hold: that is, a set $\{\mathbf{f}_1, \ldots, \mathbf{f}_t\}$ of vectors in Q^m is a standard basis exactly when each S-vector on any pair $\mathbf{f}_i, \mathbf{f}_j \in Q^m$ has remainder $\mathbf{0}$ when divided by $\{\mathbf{f}_1, \ldots, \mathbf{f}_t\}$, the division being done using the extended local division algorithm. This immediately provides an algorithm for extending a set of *polynomial vectors* in Q^m to a standard basis of polynomial vectors (provided only that one can show termination after a finite number of steps, which follows exactly as in the case of Mora's algorithm for elements of Q). This algorithm is often called Mora's algorithm for modules, and is implemented on the computer algebra programs CALI and `Singular`.

Once we have a method for getting standard bases, we immediately get an algorithm for determining whether an element belongs to a submodule of Q^m generated by polynomial vectors. Likewise, everything we said about syzygies in the last section continues to hold for modules over local rings. In particular, a set of generators $\{\mathbf{f}_1, \ldots, \mathbf{f}_s\}$ for a submodule of Q^m is a standard basis precisely when every syzygy on the leading terms of the $\mathbf{f}_i$ lifts to a syzygy on the $\mathbf{f}_i$, Schreyer's Theorem for computation of syzygies given a standard basis carries over word for word, and the analogues of Proposition (3.8) and Corollary (3.9) continue to hold without change. Thus, we can compute syzygies on any set of polynomial vectors in Q^m.

In the rest of this section, we shall detail a number of ways in which modules over local rings are different, and better behaved, than modules over polynomial rings. This is important, because one can often establish facts about modules over polynomial rings, by establishing the corresponding facts for their localizations.

Minimal generating sets

Given a finitely generated module M over a ring, define the *minimal number of generators* of the module M, often denoted $\mu(M)$, to be the smallest number of elements in any generating set of M. If the module M is free, one can show that any basis has $\mu(M)$ elements (in particular, all bases have the same number of elements). However, if M is not free (or if you don't know whether it is free), it can be quite difficult to compute $\mu(M)$. The reason is that an arbitrary set of generators for M will not, in general, contain a subset of $\mu(M)$ elements that generate. In fact, one can easily find examples of sets of generators which are *unshortenable* in the sense that no proper subset of them generates.

Exercise 4. Let R be the ring $k[x, y]$ and let M be the ideal generated by $\{xy(y - 1), xy(x - 1), x(y - 1)(x - 1)\}$.

a. Show that this set is unshortenable. Hint: The least inspired way of doing this is to compute Gröbner bases, which can be done by hand. A more elegant way is to argue geometrically. Each of the generators defines a union of three lines and the variety corresponding to M is the intersection of the three sets of three lines.

b. Show that $M = \langle xy^2 - x^2y, x^2 - x \rangle$.

We should also mention that Exercise 10e of §1 gives an example of a free module with an unshortenable set of $\mu(M) + 1$ generators.

For modules M over a local ring Q, however, this problem does not arise. Unshortenable sets of generators are minimal, and any set of generators contains an unshortenable set.

Exercise 5. Let $R = k[x, y]$ and M be as in Exercise 4. Let M_0 be the ideal in R_0 obtained by localizing at the origin.

a. Since $\{xy(y - 1), xy(x - 1), x(y - 1)(x - 1)\}$ generates M in R, it generates M_0 in R_0. Show that this set of generators is shortenable. What is the shortest unshortenable subset of it that generates M_0?

b. Answer the same questions for the set $\langle xy^2 - x^2y, x^2 - x \rangle$.

c. With the notation of Exercise 10 of §1, let N be the R-module generated by $\{\mathbf{h}_1, \mathbf{h}_2, \mathbf{h}_3\} \subset R^3$ and $N_0 \subset (R_0)^3$ the R_0-module they generate. Find an unshortenable subset of $\{\mathbf{h}_1, \mathbf{h}_2, \mathbf{h}_3\}$ that generates N_0.

Moreover, it turns out to be easy to compute $\mu(M)$ when M is a module over a local ring Q. The reason is the following extremely simple, and extremely useful, result and its corollaries which hold for all finitely-generated modules over a local ring.

(4.1) Lemma (Nakayama's Lemma). *Let Q be a local ring with maximal ideal $\mathfrak{m}$, and let M be a finitely generated Q-module. If $\mathfrak{m}M = M$, then $M = 0$.*

PROOF. Suppose that $M \neq 0$, and let $f_1, \ldots, f_s$ be a minimal set of generators of M. Then $f_s \in \mathfrak{m}M$. Thus, $f_s = a_1 f_1 + \cdots + a_s f_s$ for some $a_1, \ldots, a_s \in \mathfrak{m}$. Hence,

$$(1 - a_s)f_s = a_1 f_1 + \cdots + a_{s-1} f_{s-1}.$$

But $1 - a_s$ is a unit because $a_s \in \mathfrak{m}$, so we have that f_s is a Q-linear combination of $f_1, \ldots, f_{s-1}$. This contradicts the minimality of the generating set. $\qquad\square$

As a corollary, we obtain the following (equivalent) statement.

(4.2) Corollary. *Let Q be a local ring with maximal ideal $\mathfrak{m}$, let M be a finitely generated Q-module, and let N be a submodule of M. If $M = \mathfrak{m}M + N$, then $M = N$.*

PROOF. Note that $\mathfrak{m}(M/N) = (\mathfrak{m}M + N)/N$. Now apply Nakayama's lemma to M/N.   □

Recall from Exercise 23 of §1 of this chapter that if R is any ring, I any ideal of R, and M an R-module, then M/IM is an R/I-module. If, in addition, I is a maximal ideal, then R/I is a field, so M/IM is a vector space over R/I (any module over a field k is a k-vector space). If M is finitely generated, then M/IM is finite-dimensional. In fact, if $f_1, \ldots, f_s$ generate M as an R-module, then the residue classes $[f_1], \ldots, [f_s]$ in M/IM span M/IM as a vector space. If R is a local ring Q with maximal ideal $\mathfrak{m}$, then the converse is true: if $[f_1], \ldots, [f_s]$ span $M/\mathfrak{m}M$, then $M = \langle f_1, \ldots, f_s \rangle + \mathfrak{m}M$ and Corollary (4.2) to Nakayama's lemma implies that $M = \langle f_1, \ldots, f_s \rangle$. In fact, we can say more.

(4.3) Proposition. *Let $Q, \mathfrak{m}$ be a local ring, $k = Q/\mathfrak{m}$ its residue field (the underlying field of constants), and M any finitely generated Q-module.*
a. *$f_1, \ldots, f_s$ is a minimal generating set of M if and only if the cosets $[f_1], \ldots, [f_s]$ form a basis of the k-vector space $M/\mathfrak{m}M$.*
b. *Any generating set of M contains a minimal generating set. Any unshortenable set of generators of M is a minimal set of generators.*
c. *One can extend the set $f_1, \ldots, f_t$ to a minimal generating set of M, if and only if the cosets $[f_1], \ldots, [f_t]$ are linearly independent over k.*

PROOF. The first statement follows from the discussion preceding the proposition. The second two statements follow as in linear algebra and are left as exercises.   □

An example may make this clearer. Suppose that $Q = k[[x, y]]$ and let $M = \langle \mathbf{f}_1, \mathbf{f}_2 \rangle \subset k[[x, y]]^2$ be the Q-module generated by

$$\mathbf{f}_1 = \begin{pmatrix} x^2 + y^2 + xy \\ x^3 \end{pmatrix}, \quad \mathbf{f}_2 = \begin{pmatrix} x \\ y^2 + x^5 \end{pmatrix}.$$

Then, $\mathfrak{m}M = \langle x, y \rangle M$ is generated by $x\mathbf{f}_1, y\mathbf{f}_1, x\mathbf{f}_2, y\mathbf{f}_2$. Anything in M is of the form $p(x, y)\mathbf{f}_1 + q(x, y)\mathbf{f}_2$ where p, q are formal power series. Since we can always write $p(x, y) = p(0, 0) + xp_1(x, y) + yp_2(x, y)$ for some (non-unique) choice of power series p_1, p_2 and, similarly, $q(x, y) = q(0, 0) + xq_1(x, y) + yq_2(x, y)$ we see that $p(x, y)\mathbf{f}_1 + q(x, y)\mathbf{f}_2$ is congruent to $p(0, 0)\mathbf{f}_1 + q(0, 0)\mathbf{f}_2$ modulo $\langle x\mathbf{f}_1, y\mathbf{f}_1, x\mathbf{f}_2, y\mathbf{f}_2 \rangle$. The latter is a k-linear combination of $[\mathbf{f}_1]$ and $[\mathbf{f}_2]$. (The reader can also check that $[\mathbf{f}_1]$ and $[\mathbf{f}_2]$ are k-linearly independent.)

If M is a module over a local ring, then Proposition (4.3) gives a method to determine $\mu(M)$ in principle. One might ask, however, if there is a way to determine $\mu(M)$ from a presentation of M. Can one, perhaps, find a presentation matrix of $M/\mathfrak{m}M$. There is a very satisfying answer to this. First we need a little lemma, which applies to *any* ring.

(4.4) Lemma. *Let P be any ring (e.g. R, Q, R/J, ...), let I be an ideal in P, let M be any finitely generated P-module, and let A be a presentation matrix for M. If we let $\overline{A}$ denote the matrix obtained from A by interpreting each entry in A as its residue class modulo I, then $\overline{A}$ presents M/IM.*

PROOF. To say an $m \times s$ matrix A presents M is to say that there are generators $f_1, \ldots, f_m$ of M and that if $a_1 f_1 + \cdots + a_m f_m = 0$ with $a_1, \ldots, a_m \in P$ is any relation, then the column vector $(a_1, \ldots, a_m)^T$ is a P-linear combination of the columns of M. It is clear that the images $[f_1], \ldots, [f_m]$ generate M/IM. So we need only show that the columns of $\overline{A}$ span the set of all syzygies on the $[f_i]$. So, suppose that $[r_1][f_1] + \cdots + [r_m][f_m] = 0$ in P/I (here $r_i \in P$ and $[r_i]$ is the coset $r_i + I$ it represents). Then $r_1 f_1 + \cdots + r_m f_m \in IM$. Thus,

$$r_1 f_1 + \cdots + r_m f_m = b_1 f_1 + \cdots + b_m f_m$$

for some $b_i \in I$, whence

$$(r_1 - b_1)f_1 + \cdots + (r_m - b_m)f_m = 0.$$

By assumption, $(r_1 - b_1, \ldots, r_m - b_m)^T$ is a P-linear combination of the columns of A. Hence $([r_1 - b_1], \ldots, [r_m - b_m])^T$ is a P/I linear combination of the columns of $\overline{A}$. But $[r_i - b_i] = [r_i]$ because $b_i \in I$, for all $i = 1, \ldots, m$. Thus the columns of $\overline{A}$ generate all syzygies on $[f_1], \ldots, [f_m]$, and this completes the proof. □

And, now, for the result we alluded to above.

(4.5) Proposition. *Let M be an R-module, $R = k[x_1, \ldots, x_n]$, and suppose that A is a matrix presenting M. If $p \in k^n$ is any point in affine n-space, let $A(p)$ be the matrix obtained by evaluating all the entries of A (which are polynomials $a_{ij} \in R$) at p. Then $A(p)$ presents $M_p/\mathfrak{m}_p M_p$, where $\mathfrak{m}_p$ is the unique maximal ideal $\langle x_1 - p_1, \ldots, x_n - p_n \rangle$ in R_p.*

PROOF. Write $A = (a_{ij})$. Since A presents M, it also presents the R_p-module M_p by Exercise 2 above. By Lemma (4.4), $[A] = (a_{ij} \bmod \mathfrak{m}_p)$ presents $M_p/\mathfrak{m}_p M_p$. But $a_{ij} \equiv a_{ij}(p) \bmod \mathfrak{m}_p$ (exercise!). □

(4.6) Corollary. *Let M and A be as above. For any $p \in k^n$, $\mu(M_p) = m - \mathrm{rk}(A(p))$, where $\mathrm{rk}(A(p))$ denotes the usual rank of a matrix over a field k (that is, the number of linearly independent rows or columns).*

PROOF. By Proposition (4.3), $\mu(M_p) = \dim M_p/\mathfrak{m}_p M_p$. Suppose that A is an $m \times s$ matrix. We know that $A(p)$ presents $M_p/\mathfrak{m}_p M_p$. Then, by

Proposition (1.10), $M_p/\mathfrak{m}_p M_p$ is isomorphic to $k^m/A(p)k^s$ (where $A(p)k^s$ is the image of $A(p)$), and the dimension of the latter is $m - \mathrm{rk}(A(p))$.   □

Minimal presentations

As a result of the discussion above, we have have a privileged set of presentation matrices of any finitely generated module M over a local ring Q. Namely, we choose a minimal set of generators of M. The set of syzygies on this set is again a module over the local ring Q, so we choose a minimal generating set for this set of syzygies. As usual, we arrange the syzygies as columns to obtain a matrix, which we call a *minimal presentation matrix* of M. We claim that the dimensions of this matrix do not depend on the choice of minimal generating set of M, and that any minimal presentation matrix can be obtained from another by a change of generators.

(4.7) Proposition.

a. *Minimal presentation matrices for finitely generated modules M over a local ring Q are essentially unique in the following sense. Let $F = (f_1, \ldots, f_m)$ and $G = (g_1, \ldots, g_m)$ be two minimal generating sets for M. Let A be an $m \times s$ minimal presentation matrix for M with respect to F. Similarly, let B be an $m \times t$ minimal presentation matrix for M with respect to G. Then $s = t$ and $B = CAD$, where C is the $m \times m$ change of basis matrix satisfying $F = GC$, and D is an invertible $s \times s$ matrix with entries in Q.*

b. *If a presentation matrix A for M is a minimal presentation matrix then all entries of A belong to the maximal ideal of Q.*

PROOF. To prove part a, first note that we have $F = GC$ and $G = FC'$ for some $m \times m$ matrices with entries in Q. By Proposition (4.3), reducing mod $\mathfrak{m}$, the matrices $\overline{C}$ and $\overline{C'}$ are invertible $m \times m$ matrices over k. By Corollary (3.9) of this chapter, the columns of CA are in $T = \mathrm{Syz}(G)$, and by the preceding remarks, the cosets of those columns in $T/\mathfrak{m}T$ must be linearly independent over k. Hence, we must have $s \leq t$. Similarly, the columns of $C'B$ are in $S = \mathrm{Syz}(F)$, and the cosets of those columns in $S/\mathfrak{m}S$ must be linearly independent over k. Hence, $t \leq s$. It follows that $s = t$, so that by Proposition (4.3) the columns of CA are a minimal generating set for $\mathrm{Syz}(G)$. Hence, $B = CAD$ for some invertible $s \times s$ matrix D.

For part b, we claim that no entry of a minimal presentation matrix A can be a unit of Q. Indeed, if the i,j entry were a unit, then f_i could be expressed in terms of the other f_k, contradicting the assertion that $\{f_1, \ldots, f_m\}$ is minimal.   □

If we are given an explicit set of generators for a submodule M of Q^m, then the last assertion of the lemma provides an algorithm for computing the minimal presentation of M. One prunes the given set of generators so that it is minimal, computes a basis of the module of syzygies on the chosen set, and discards any syzygies which involve units.

For example, the minimal presentation matrix of the ideal $\langle x, y, z \rangle \subset k[[x, y, z]]$ is the matrix with Koszul relations as columns

$$ A = \begin{pmatrix} y & z & 0 \\ -x & 0 & z \\ 0 & -x & -y \end{pmatrix} $$

We have seen that free modules are the simplest modules. However, it is sometimes difficult to actually determine whether a given module is free. Given a presentation matrix of a module over a local ring, there is a criterion which allows one to determine whether or not the module is free. To do this, we introduce a sequence of ideals which are defined in terms of the presentation matrix for a module, but which turn out to be independent of the presentation.

Let M be a finitely generated R-module (R any ring) and let A be a presentation matrix for M. Then the *ideal of ith minors $I_i(A)$* is the ideal generated by the ith minors of A (that is, by the determinants of $i \times i$ submatrices of A). Here, we define the 0th minor of A to be 1 (so that $I_0(A) = R$). More generally, if $i < 0$, we define $I_i(A) = R$. If i exceeds the number of rows or number of columns of A, we define the ith minor to be 0, so that $I_i(A) = 0$ for sufficiently large i. Although defined in terms of a presentation matrix A, the ideals will turn out to yield invariants of the module M.

(4.8) Lemma. *Let M be an R-module, R any ring. If A and B are matrices that both present M, and that have the same number of rows, then $I_i(A) = I_i(B)$ for all i.*

PROOF. We leave the proof as an exercise—see Exercise 10. $\qquad\square$

The restriction that the presentation matrices have the same number of rows is irksome, but necessary. The matrices $A = (0)$ and $B = \begin{pmatrix} 1 \\ -1 \end{pmatrix}$ clearly present the same module (namely, the free module R). Note that $I_0(A) = R, I_1(A) = \langle 0 \rangle$, while $I_0(B) = R, I_1(B) = R$. It turns out to be more convenient to change the indexing of the ideals of minors.

(4.9) Definition. If M is an R-module presented by A, the ith Fitting invariant $F_i(M)$ is defined by setting $F_i(M) = I_{m-i}(A)$ where A has m rows.

Notice that with this shift in index, the Fitting invariants of the free R-module R are $F_i(R) = R$ for $i > 0$ and $F_i(R) = \langle 0 \rangle$ for $i \leq 0$, no matter whether we use the matrix A or B above to compute the F_i.

(4.10) Proposition. *The Fitting invariants of a module depend only on the module, and not on the presentation. That is, isomorphic modules have isomorphic Fitting invariants.*

PROOF. This is an immediate corollary of Lemma (4.8) and the definition of the Fitting invariants. See Exercise 10.   □

For modules over local rings, it is easy to show that necessary and sufficient conditions for a module to be free can be given in terms of the Fitting invariants.

(4.11) Proposition. *Let Q be a local ring, M a finitely generated Q-module. Then M is free of rank r if and only if $F_i(M) = 0$ for $i < r$ and $F_i(M) = R$ for $i \geq r$.*

PROOF. By Proposition (4.10), $F_i(M)$ does not depend on the choice of matrix A presenting M. If M is free of rank r, then we can take A to be the $m \times 1$ matrix all of whose entries are 0. Computing F_i using this presentation gives $F_i(M) = 0$ for $i < r$ and $F_i(M) = R$ for $i \geq r$.

Conversely, suppose that A is some $m \times s$ matrix presenting M, and suppose that $I_0(A) = I_1(A) = \cdots = I_{m-r}(A) = R$ and $I_{m-r+1}(A) = 0$. Since R is local, this means that some $(m-r) \times (m-r)$ minor of A is a unit (an R-linear combination of elements of a local ring which is a unit must be such that one of the elements is a unit). This minor is a sum of terms, each a product of $m - r$ terms of R. Again because R is local, one such summand must be a unit, and, hence, the $m - r$ terms that multiply to give it must be units. By exchanging columns and rows of A, we may assume that $a_{11}, a_{22}, \ldots, a_{m-r,m-r}$ are units. By row and column operations we may arrange that $a_{11} = a_{22} = \cdots = a_{m-r,m-r} = 1$ and that all other entries in the first $m - r$ rows and first $m - r$ columns are zero.

We claim that all other entries of A must be zero. To see this, suppose that some other entry were nonzero, say $f \in A$. We could arrange that $a_{m-r+1,m-r+1} = f$ by leaving the first $m - r$ columns and rows fixed, and exchanging other rows and columns as necessary. But then the $(m - r + 1) \times (m - r + 1)$ minor obtained by taking the determinant of the submatrix consisting of the first $m - r + 1$ rows and columns would equal f and $I_{m-r+1}(A)$ could not equal zero.

Since A is $m \times s$, we conclude that A presents a module with m generators, the first $m - r$ of which are equal to zero and the last r of which only satisfy the trivial relation. This says that M is free of rank r.   □

Projective modules

Besides free modules, there is another class of modules over any ring which are almost as simple to deal with as free modules. These are the so-called projective modules.

(4.12) Definition. If R is any ring, an R-module M is said to be *projective* if there is an R-module N such that $M \oplus N$ is a free module.

That is, a projective module is a summand of a free module. Such a notion arises when dealing with syzygies, as shown by the following exercise (compare Exercise 26 of §1).

Exercise 6.

a. Suppose that a module M has generators $g_1, \ldots, g_s$ so that the module of syzygies $\mathrm{Syz}\,(g_1, \ldots, g_s)$ is free. Then let $f_1, \ldots, f_t$ be another generating set of M. Use Proposition (3.10) of the preceding section to prove that $\mathrm{Syz}\,(f_1, \ldots, f_t)$ is projective.

b. Let $\{f_1, \ldots, f_t\} \subset R$, R any ring, be a set of elements such that $\langle f_1, \ldots, f_t \rangle = R$. Show that $\mathrm{Syz}(f_1, \ldots, f_t)$ is projective. Hint: Use part a.

Every free module is clearly a projective module, but not conversely. In Exercise 26a of §1, we point out that $\mathbb{Z}/6 = \mathbb{Z}/3 \oplus \mathbb{Z}/2$, but $\mathbb{Z}/3$ is clearly not a free $(\mathbb{Z}/6)$-module.

Over a local ring, however, it is easy to show that any projective module is free.

(4.13) Theorem. *If Q is a local ring, and M a projective Q-module, then M is free.*

PROOF. By assumption, there is a module N such that $M \oplus N \cong Q^s$, for some s. We may harmlessly view M as a submodule of Q^s. Choose a minimal generating set $\mathbf{f}_1, \ldots, \mathbf{f}_m$ of M. If we let $\mathfrak{m}$ denote the maximal ideal of Q, then $\mathbf{f}_1 + \mathfrak{m}M, \ldots, \mathbf{f}_m + \mathfrak{m}M$ are a basis of $M/\mathfrak{m}M$ Since $M \cap N = \{0\}$, $\mathbf{f}_1 + \mathfrak{m}M + \mathfrak{m}N, \ldots, \mathbf{f}_m + \mathfrak{m}M + \mathfrak{m}N$ are linearly independent in $M/(\mathfrak{m}M + \mathfrak{m}N) \subset (M + N)/\mathfrak{m}(M + N)$. Therefore, by the second part of Proposition (4.3), $\mathbf{f}_1, \ldots, \mathbf{f}_m$ extend to a minimal generating set of $M \oplus N$, which is a basis, hence linearly independent over Q. But then, $\mathbf{f}_1, \ldots, \mathbf{f}_m$ must be linearly independent over Q, and hence a basis of M. Thus, M is free. $\square$

For a long time, it was an open question as to whether the above result continues to hold for polynomial rings over a field. The assertion that such is the case (that is, that every projective $k[x_1, \ldots, x_n]$-module is free) is known as *Serre's conjecture* and was finally proved by Quillen and Suslin independently in 1976 (see Theorem (1.8), and Exercises 26 and 27 of §1 for more information).

Since modules over local rings are so much simpler than modules over a polynomial ring, one often tries to establish results about modules over polynomial rings by establishing the result for the localizations of the modules at all points. One then hopes that this will be enough to establish the result for modules over the polynomial ring.

We give one example of this here, phrased to make its algorithmic importance clear. We learned it from M. Artin's *Algebra* [Art].

(4.14) Theorem. *Let M be a finitely generated module over a polynomial ring $R = k[x_1, \ldots, x_n]$ with $k = \mathbb{C}$ and let A be an $m \times s$ matrix presenting M. Then M is a free module of rank r if and only if for every $p \in \mathbb{C}^n$ the matrix $A(p)$ has rank $m - r$ (as above, $A(p)$ is the matrix with entries in $\mathbb{C}$ obtained by evaluating the polynomial entries of A at the point p).*

PROOF. We prove the easy direction, and make some comments about the reverse direction. Suppose that A presents M. Choose a free basis $e_1, \ldots, e_r$ and let A' be the $r \times 1$ matrix of zeros presenting M with respect to this basis. It follows from Exercise 33 of §1 that the matrices

$$D = \begin{pmatrix} A & 0 \\ 0 & I_r \end{pmatrix} \quad \text{and} \quad D' = \begin{pmatrix} I_m & 0 \\ 0 & A' \end{pmatrix}$$

are such that $\text{rank}(D(p)) = \text{rank}(D'(p))$ for all $p \in k^n$. (See Exercise 12) However, D' is a constant matrix of rank m. Thus, $D(p)$ has rank m for all p. It follows that $A(p)$ has rank $m - r$ for all p.

To get the converse, in the exercises we ask you to show that if $\text{rank}(A(q)) = m - r$ for all q in some neighborhood of p (we assumed that $k = \mathbb{C}$ to make sense of this), then M_p is free of rank $m - r$. We then ask you to show that if M_p is free of rank $m - r$ for all $p \in \mathbb{C}^n$, then M is projective of rank $m - r$. The Quillen-Suslin theorem then implies that M is free.   □

ADDITIONAL EXERCISES FOR §4

Exercise 7.
a. Let M be a finitely generated free R-module. Show that any two bases of M have the same number of elements.
b. Let M be any finitely generated R-module. Show that the maximal number of R-linearly independent elements in any generating set of M is the same. (This number is called the *rank* of M.)

Exercise 8. Prove the second and third parts of Proposition (4.3).

Exercise 9. Suppose that $f \in R = k[x_1, \ldots, x_n]$ and $p = (p_1, \ldots, p_n) \in k^n$. Show that $\mathfrak{m}_p = \langle x_1 - p_1, \ldots, x_n - p_n \rangle$ is the maximal ideal of R_p. Explain why $f \equiv f(p) \mod \mathfrak{m}_p$. (Compare the proof of Proposition (4.5).)

Exercise 10. Show that the Fitting ideals of M are an ascending sequence of ideals which do not depend on the choice of presentation matrix of M as follows.

a. Given a finite generating set $f_1, \ldots, f_s$ for M, let $A = (a_{ij})$ be the presentation matrix constructed by choosing one set of generators of $\mathrm{Syz}(f_1, \ldots, f_s)$ and let $B = (b_{ij})$ a presentation matrix constructed by choosing another set of syzygies which generate. Show that the Fitting ideals constructed from the matrix A are the same as the Fitting ideals constructed from the matrix B. Hint: The hypotheses imply that the columns of B can be expressed in terms of the columns of A. It is then clear that $I_1(A) \supset I_1(B)$. To see that $I_2(A) \supset I_2(B)$ write out the two by two minors of B in terms of the entries of A. Generalize to show that $I_i(A) \supset I_i(B)$. Expressing the columns of A in terms of those of B gives the reverse containments.

b. Show that the Fitting ideals do not depend on the choice of generators $f_1, \ldots, f_s$ of M. Hint: Compare the ideals generated by the $i \times i$ minors of a presentation matrix with respect to the generators $f_1, \ldots, f_s$ and those generated by the $i \times i$ minors of a presentation matrix with respect to the generators $f_1, \ldots, f_s, f$, where f is any element of M.

c. Show that $0 = F_0(M) \subset F_1(M) \subset \cdots \subset F_{s+1}(M) = R$ where s is as in part a.

Exercise 11. In the ring $\mathbb{Z}[\sqrt{-5}]$, show that the ideal $\langle 2, 1 + \sqrt{-5} \rangle \subset \mathbb{Z}[\sqrt{-5}]$ is a projective $\mathbb{Z}[\sqrt{-5}]$-module which is not free.

Exercise 12. Show directly from Exercise 31 of §1 (that is, do not use Exercise 33) that the matrices

$$D = \begin{pmatrix} A & 0 \\ 0 & I_r \end{pmatrix} \quad \text{and} \quad D' = \begin{pmatrix} I_m & 0 \\ 0 & A' \end{pmatrix}$$

in the proof of Theorem (4.14) are such that $\mathrm{rank}(D(p)) = \mathrm{rank}(D'(p))$ for all $p \in k^n$. Hint: Use the result of Exercise 31, and compare the result of multiplying the matrix therein on the left by

$$\begin{pmatrix} I_m & B \\ 0 & I_r \end{pmatrix} \quad \text{and} \quad \begin{pmatrix} I_m & 0 \\ 0 & A' \end{pmatrix}.$$

Exercise 13. Suppose that $k = \mathbb{C}$, that A presents M. Show that M_p is free of rank r if and only if $\mathrm{rank}(A(q)) = m - r$, for all q in some neighborhood of p.

Exercise 14. Let $R = k[x_1, \ldots, x_n]$. Show that M is a projective R-module if and only if M_p is a projective (hence free) R_p-module for all $p \in k^n$.

Exercise 15. Let R be a ring and let $A = (\, a_1 \; \cdots \; a_m \,)$ be a $1 \times m$ unimodular matrix (in this situation, unimodular means that $R = \langle a_1, \dots, a_m \rangle$). Also note that $\ker A$ is the syzygy module $\mathrm{Syz}(a_1, \dots, a_m)$. Prove that $\ker A$ is a free R-module if and only if there exists an invertible $m \times m$ matrix B with coefficients in R whose first row is A. Thus, the statement that the kernel of any unimodular row is free is equivalent to the statement that any unimodular row with coefficients in $k[x_1, \dots, x_n]$ is the first row of a square matrix with polynomial coefficients and determinant 1.

Chapter 6

Free Resolutions

In Chapter 5, we saw that to work with an R-module M, we needed not just the generators $f_1, \ldots, f_t$ of M, but the relations they satisfy. Yet the set of relations $\mathrm{Syz}\,(f_1, \ldots, f_t)$ is an R-module in a natural way and, hence, to understand it, we need not just its generators $g_1, \ldots, g_s$, but the set of relations $\mathrm{Syz}\,(g_1, \ldots, g_s)$ on these generators, the so-called second syzygies. The second syzygies are again an R-module and to understand it, we again need a set of generators *and* relations, the third syzygies, and so on. We obtain a sequence, called a resolution, of generators and relations of successive syzygy modules of M. In this chapter, we will study resolutions and the information they encode about M. Throughout this chapter, R will denote the polynomial ring $k[x_1, \ldots, x_n]$ or one of its localizations.

§1 Presentations and Resolutions of Modules

Apart from the possible presence of nonzero elements in the module of syzygies on a minimal set of generators, one of the important things that distinguishes the theory of modules from the theory of vector spaces over a field is that many properties of modules are frequently stated in terms of homomorphisms and exact sequences. Although this is primarily cultural, it is very common and very convenient. In this first section, we introduce this language.

To begin with, we recall the definition of exact.

(1.1) Definition. Consider a sequence of R-modules and homomorphisms

$$\cdots \longrightarrow M_{i+1} \xrightarrow{\varphi_{i+1}} M_i \xrightarrow{\varphi_i} M_{i-1} \longrightarrow \cdots$$

a. We say the sequence is *exact at* M_i if $\mathrm{im}(\varphi_{i+1}) = \ker(\varphi_i)$.
b. The entire sequence is said to be *exact* if it is exact at each M_i which is not at the beginning or the end of the sequence.

Many important properties of homomorphisms can be expressed by saying that a certain sequence is exact. For example, we can phrase what it means for an R-module homomorphism $\varphi : M \to N$ to be onto, injective, or an isomorphism:

- $\varphi : M \to N$ is onto (or surjective) if and only if the sequence

$$M \xrightarrow{\varphi} N \to 0$$

is exact, where $N \to 0$ is the homomorphism sending every element of N to 0. To prove this, recall that onto means $\mathrm{im}(\varphi) = N$. Then the sequence is exact at N if and only if $\mathrm{im}(\varphi) = \ker(N \to 0) = N$, as claimed.

- $\varphi : M \to N$ is one-to-one (or injective) if and only if the sequence

$$0 \to M \xrightarrow{\varphi} N$$

is exact, where $0 \to M$ is the homomorphism sending 0 to the additive identity of M. This is equally easy to prove.

- $\varphi : M \to N$ is an isomorphism if and only if the sequence

$$0 \to M \xrightarrow{\varphi} N \to 0$$

is exact. This follows from the above since φ is an isomorphism if and only if it is one-to-one and onto.

Exact sequences are ubiquitous. Given any R-module homomorphism or any pair of modules, one a submodule of the other, we get an associated exact sequence as follows.

(1.2) Proposition.

a. *For any R-module homomorphism $\varphi : M \to N$, we have an exact sequence*

$$0 \to \ker(\varphi) \to M \xrightarrow{\varphi} N \to \mathrm{coker}(\varphi) \to 0,$$

where $\ker(\varphi) \to M$ is the inclusion mapping and $N \to \mathrm{coker}(\varphi) = N/\mathrm{im}(\varphi)$ is the natural homomorphism onto the quotient module, as in Exercise 12 from §1 of Chapter 5.

b. *If $Q \subset P$ is a submodule of an R-module P, then we have an exact sequence*

$$0 \to Q \to P \xrightarrow{\nu} P/Q \to 0,$$

where $Q \to P$ is the inclusion mapping, and ν is the natural homomorphism onto the quotient module.

PROOF. Exactness of the sequence in part a at $\ker(\varphi)$ follows from the above bullets, and exactness at M is the definition of the kernel of a homomorphism. Similarly, exactness at N comes from the definition of the cokernel of a homomorphism (see Exercise 28 of Chapter 5, §1), and exactness at $\text{coker}(\varphi)$ follows from the above bullets. In the exercises, you will show that part b follows from part a.   □

Choosing elements of an R-module M is also conveniently described in terms of homomorphisms.

(1.3) Proposition. *Let M be an R-module.*
a. *Choosing an element of M is equivalent to choosing a homomorphism $R \to M$.*
b. *Choosing t elements of M is equivalent to choosing a homomorphism $R^t \to M$.*
c. *Choosing a set of t generators of M is equivalent to choosing a homomorphism $R^t \to M$ which is onto (i.e., an exact sequence $R^t \to M \to 0$).*
d. *If M is free, choosing a basis with t elements is equivalent to choosing an isomorphism $R^t \to M$.*

PROOF. To see part a, note that the identity 1 is the distinguished element of a ring R. Choosing an element f of a module M is the same as choosing the R-module homomorphism $\varphi : R \to M$ which satisfies $\varphi(1) = f$. This is true since $\varphi(1)$ determines the values of φ on all $g \in R$:

$$\varphi(g) = \varphi(g \cdot 1) = g \cdot \varphi(1) = gf.$$

Thus, choosing t elements in M can be thought of as choosing t R-module homomorphisms from R to M or, equivalently, as choosing an R-module homomorphism from R^t to M. This proves part b. More explicitly, if we think of R^t as the space of column vectors and denote the standard basis in R^t by $e_1, e_2, \ldots, e_t$, then choosing t elements $f_1, \ldots, f_t$ of M corresponds to choosing the R-module homomorphism $\varphi : R^t \to M$ defined by setting $\varphi(e_i) = f_i$, for all $i = 1, \ldots, t$. The image of φ is the submodule $\langle f_1, \ldots, f_t \rangle \subset M$. Hence, choosing a set of t generators for M corresponds to choosing an R-module homomorphism $R^t \to M$ which is onto. By our previous discussion, this is the same as choosing an exact sequence

$$R^t \to M \to 0.$$

This establishes part c, and part d follows immediately.   □

In the exercises, we will see that we can also phrase what it means to be projective in terms of homomorphisms and exact sequences. Even more useful for our purposes, will be the interpretation of presentation matrices in terms of this language. The following terminology will be useful.

(1.4) Definition. Let M be an R-module. A *presentation* for M is a set of generators $f_1, \ldots, f_t$, together with a set of generators for the syzygy module $\text{Syz}(f_1, \ldots, f_t)$ of relations among $f_1, \ldots, f_t$.

One obtains a presentation matrix for a module M by arranging the generators of Syz $(f_1, \ldots, f_t)$ as columns—being given a presentation matrix is essentially equivalent to being given a presentation of M. To reinterpret Definition (1.4) in terms of exact sequences, note that the generators $f_1, \ldots, f_t$ give a surjective homomorphism $\varphi : R^t \to M$ by part c of Proposition (1.3), which means an exact sequence

$$R^t \xrightarrow{\varphi} M \to 0.$$

The map φ sends $(g_1, \ldots, g_t) \in R^t$ to $\sum_{i=1}^{t} g_i f_i \in M$. It follows that a syzygy on $f_1, \ldots, f_t$ is an element of the kernel of φ, i.e.,

$$\text{Syz} (f_1, \ldots, f_t) = \ker(\varphi : R^t \to M).$$

By part c of Proposition (1.3), choosing a set of generators for the syzygy module corresponds to choosing a homomorphism ψ of R^s onto $\ker(\varphi) = $ Syz $(f_1, \ldots, f_t)$. But ψ being onto is equivalent to $\text{im}(\psi) = \ker(\varphi)$, which is just the condition for exactness at R^t in the sequence

$$(1.5) \qquad\qquad R^s \xrightarrow{\psi} R^t \xrightarrow{\varphi} M \to 0.$$

This proves that a presentation of M is equivalent to an exact sequence of the form (1.5). Also note that the matrix of ψ with respect to the standard bases of R^s and R^t is a presentation matrix for M.

We next observe that *every* finitely generated R-module has a presentation.

(1.6) Proposition. *Let M be a finitely generated R-module.*
a. *M has a presentation of the form given by (1.5).*
b. *M is a homomorphic image of a free R-module. In fact, if $f_1, \ldots, f_t$ is a set of generators of M, then $M \cong R^t/S$ where S is the submodule of R^t given by $S = \text{Syz}(f_1, \ldots, f_t)$. Alternatively, if we let the matrix A represent ψ in (1.5), then $AR^s = \text{im}(\psi)$ and $M \cong R^t/AR^s$.*

PROOF. Let $f_1, \ldots, f_t$ be a finite generating set of M. Part a follows from the fact noted in Chapter 5, §2 that every submodule of R^t, in particular Syz $(f_1, \ldots, f_t) \subset R^t$, is finitely generated. Hence we can choose a finite generating set for the syzygy module, which gives the exact sequence (1.5) as above.

Part b follows from part a and Proposition 1.10 of Chapter 5, §1. $\qquad\square$

Here is a simple example. Let $I = \langle x^2 - x, xy, y^2 - y \rangle$ in $R = k[x, y]$. In geometric terms, I is the ideal of the variety $V = \{(0, 0), (1, 0), (0, 1)\}$ in k^2. We claim that I has a presentation given by the following exact sequence:

$$(1.7) \qquad\qquad R^2 \xrightarrow{\psi} R^3 \xrightarrow{\varphi} I \to 0,$$

where φ is the homomorphism defined by the 1×3 matrix
$$A = (\, x^2 - x \quad xy \quad y^2 - y \,)$$
and ψ is defined by the 3×2 matrix
$$B = \begin{pmatrix} y & 0 \\ -x+1 & y-1 \\ 0 & -x \end{pmatrix}.$$

The following exercise gives one proof that (1.7) is a presentation of I.

Exercise 1. Let S denote $\mathrm{Syz}(x^2 - x, xy, y^2 - y)$.

a. Verify that the matrix product AB equals the 1×2 zero matrix, and explain why this shows that $\mathrm{im}(\psi)$ (the module generated by the columns of the matrix B) is contained in S.

b. To show that S is generated by the columns of B, we can use Schreyer's Theorem—Theorem (3.3) from Chapter 5 of this book. Check that the generators for I form a *lex* Gröbner basis for I.

c. Compute the syzygies s_{12}, s_{13}, s_{23} obtained from the S-polynomials on the generators of I. By Schreyer's Theorem, they generate S.

d. Explain how we could obtain a different presentation
$$R^3 \xrightarrow{\psi'} R^3 \xrightarrow{\varphi} I \to 0$$
of I using this computation, and find an explicit 3×3 matrix representation of the homomorphism ψ'.

e. How do the columns of B relate to the generators s_{12}, s_{13}, s_{23} of S? Why does B have only two columns? Hint: Show that $s_{13} \in \langle s_{12}, s_{23} \rangle$ in R^3.

We have seen that specifying any module requires knowing both generators and the relations between the generators. However, in presenting a module M, we insisted only on having a set of generators for the module of syzygies. Shouldn't we have demanded a set of relations on the generators of the syzygy module? These are the so-called *second syzygies*.

For example, in the presentation from part d of Exercise 1, there is a relation between the generators s_{ij} of $\mathrm{Syz}(x^2 - x, xy, y^2 - y)$, namely

(1.8) $$(y-1)s_{12} - s_{13} + xs_{23} = 0,$$

so $(y-1, -1, x)^T \in R^3$ would be a second syzygy.

Likewise, we would like to know not just a generating set for the second syzygies, but the relations among those generators (the third syzygies), and so on. As you might imagine, the connection between a module, its first syzygies, its second syzygies, and so forth can also be phrased in terms of an exact sequence of modules and homomorphisms. The idea is simple—we just iterate the construction of the exact sequence giving a presentation. For instance, starting from the sequence (1.6) corresponding to a presentation for M, if we want to know the second syzygies as well, we need another step in the sequence:

$$R^r \xrightarrow{\lambda} R^s \xrightarrow{\psi} R^t \xrightarrow{\varphi} M \to 0,$$

where now the image of $\lambda : R^r \to R^s$ is equal to the kernel of ψ (the second syzygy module). Continuing in the same way to the third and higher syzygies, we produce longer and longer exact sequences. We wind up with a free resolution of M. The precise definition is as follows.

(1.9) Definition. Let M be an R-module. A *free resolution* of M is an exact sequence of the form

$$\cdots \to F_2 \xrightarrow{\varphi_2} F_1 \xrightarrow{\varphi_1} F_0 \xrightarrow{\varphi_0} M \to 0,$$

where for all i, $F_i \cong R^{r_i}$ is a free R-module. If there is an ℓ such that $F_{\ell+1} = F_{\ell+2} = \cdots = 0$, but $F_\ell \neq 0$, then we say the resolution is *finite*, of *length* ℓ. In a finite resolution of length ℓ, we will usually write the resolution as

$$0 \to F_\ell \to F_{\ell-1} \to \cdots \to F_1 \to F_0 \to M \to 0.$$

For an example, consider the presentation (1.7) for

$$I = \langle x^2 - x, xy, y^2 - y \rangle$$

in $R = k[x, y]$. If

$$a_1 \begin{pmatrix} y \\ -x+1 \\ 0 \end{pmatrix} + a_2 \begin{pmatrix} 0 \\ y-1 \\ -x \end{pmatrix} = \begin{pmatrix} 0 \\ 0 \\ 0 \end{pmatrix},$$

$a_i \in R$, is any syzygy on the columns of B with $a_i \in R$, then looking at the first components, we see that $ya_1 = 0$, so $a_1 = 0$. Similarly from the third components $a_2 = 0$. Hence the kernel of ψ in (1.7) is the zero submodule. An equivalent way to say this is that the columns of B are a basis for $\text{Syz}(x^2 - x, xy, y^2 - y)$, so the first syzygy module is a free module. As a result, (1.7) extends to an exact sequence:

(1.10) $$0 \to R^2 \xrightarrow{\psi} R^3 \xrightarrow{\varphi} I \to 0.$$

According to Definition (1.9), this is a free resolution of length 1 for I.

Exercise 2. Show that I also has a free resolution of length 2 obtained by extending the presentation given in part d of Exercise 1 above:

(1.11) $$0 \to R \xrightarrow{\lambda} R^3 \xrightarrow{\psi} R^3 \xrightarrow{\varphi} I \to 0,$$

where the homomorphism λ comes from the syzygy given in (1.8).

Generalizing the observation about the matrix B above, we have the following characterization of finite resolutions.

(1.12) Proposition. *In a finite free resolution*

$$0 \to F_\ell \xrightarrow{\varphi_\ell} F_{\ell-1} \xrightarrow{\varphi_{\ell-1}} F_{\ell-2} \to \cdots \to F_0 \xrightarrow{\varphi_0} M \to 0,$$

$\ker(\varphi_{\ell-1})$ *is a free module. Conversely, if M has a free resolution in which* $\ker(\varphi_{\ell-1})$ *is a free module for some ℓ, then M has a finite free resolution of length ℓ.*

PROOF. If we have a finite resolution of length ℓ, then φ_ℓ is one-to-one by exactness at F_ℓ, so its image is isomorphic to F_ℓ, a free module. Also, exactness at $F_{\ell-1}$ implies $\ker(\varphi_{\ell-1}) = \text{im}(\varphi_\ell)$, so $\ker(\varphi_{\ell-1})$ is a free module. Conversely, if $\ker(\varphi_{\ell-1})$ is a free module, then the partial resolution

$$F_{\ell-1} \xrightarrow{\varphi_{\ell-1}} F_{\ell-2} \to \cdots \to F_0 \xrightarrow{\varphi_0} M \to 0$$

can be completed to a finite resolution of length ℓ

$$0 \to F_\ell \to F_{\ell-1} \xrightarrow{\varphi_{\ell-1}} F_{\ell-2} \to \cdots \to F_0 \xrightarrow{\varphi_0} M \to 0,$$

by taking F_ℓ to be the free module $\ker(\varphi_{\ell-1})$ and letting the arrow $F_\ell \to F_{\ell-1}$ be the inclusion mapping. $\qquad\square$

Both (1.11) and the more economical resolution (1.10) came from the computation of the syzygies s_{ij} on the Gröbner basis for I. By Schreyer's Theorem again, the same process can be applied to produce a free resolution of any submodule M of a free module over R. If $\mathcal{G} = \{g_1, \ldots, g_s\}$ is a Gröbner basis for M with respect to any monomial order, then the s_{ij} are a Gröbner basis for the first syzygy module (with respect to the $>_{\mathcal{G}}$ order from Theorem (3.3) of Chapter 5). Since this is true, we can iterate the process and produce Gröbner bases for the modules of second, third, and all higher syzygies. In other words, Schreyer's Theorem forms the basis for an *algorithm* for computing any finite number of terms in a free resolution. This algorithm is implemented in **Singular**, in CoCoA, in the CALI package for REDUCE, and in the **resolution** command of *Macaulay 2*.

For example, consider the homogeneous ideal

$$M = \langle yz - xw, y^3 - x^2z, xz^2 - y^2w, z^3 - yw^2 \rangle$$

in $k[x, y, z, w]$. This is the ideal of a rational quartic curve in $\mathbb{P}^3$. Here is a *Macaulay 2* session calculating and displaying a free resolution for M:

```
i1 : R = QQ[x,y,z,w]

o1 = R

o1 : PolynomialRing
```

```
i2 : M = ideal(z^3-y*w^2,y*z-x*w,y^3-x^2*z,x*z^2-y^2*w)
```

$$o2 = ideal (z^3 - y*w^2 , y*z - x*w, y^3 - x^2 z, x*z^2 - y^2 w)$$

```
o2 : Ideal of R

i3 : MR = resolution M
```

$$o3 = R \overset{1}{\longleftarrow} R \overset{4}{\longleftarrow} R \overset{4}{\longleftarrow} R \overset{1}{}$$

```
      0     1     2     3

o3 : ChainComplex

i4 : MR.dd
```

$$o4 = -1 : 0 \overset{1}{\longleftarrow} R : 0$$
```
                     0
```

$$0 : R \overset{1}{\longleftarrow} R : 1$$
```
              {0} | yz-xw y3-x2z xz2-y2w z3-yw2 |
```

$$1 : R \overset{4}{\longleftarrow} R : 2$$
```
              {2} | -y2 -xz -yw -z2 |
              {3} |  z   w   0   0  |
              {3} |  x   y  -z  -w  |
              {3} |  0   0   x   y  |
```

$$2 : R \overset{4}{\longleftarrow} R : 3$$
```
              {4} |  w |
              {4} | -z |
              {4} | -y |
              {4} |  x |
```

```
o4 : ChainComplexMap
```

The output shows the matrices in a finite free resolution of the form

$$(1.13) \qquad 0 \to R \to R^4 \to R^4 \to M \to 0,$$

from the "front" of the resolution "back." In particular, the first matrix
(1×4) gives the generators of M, the columns of the second matrix give generators for the first syzygies, and the third matrix (4×1) gives a generator for the second syzygy module, which is free.

Exercise 3.
a. Verify by hand that at each step in the sequence (1.13), the image of the mapping "coming in" is contained in the kernel of the mapping "going out."
b. Verify that the generators of M form a Gröbner basis of M for the *grevlex* order with $x > y > z > w$, and compute the first syzygy module using Schreyer's theorem. Why is the first syzygy module generated by just 4 elements (the columns of the 4×4 matrix), and not $6 = \binom{4}{2}$ elements s_{ij} as one might expect?

The programs **Singular** and CALI can be used to compute resolutions of ideals whose generators are not homogeneous (and, more generally, modules which are not graded), as well as resolutions of modules over local rings. Here, for example, is a **Singular** session computing a resolution of the ideal

$$(1.14) \qquad I = \langle z^3 - y, yz - x, y^3 - x^2z, xz^2 - y^2 \rangle$$

in $k[x, y, z]$ (note that I is obtained by dehomogenizing the generators of M above).

```
> ring r=0, (x,y,z), dp;
> ideal I=(z3-y,yz-x,y3-x2z,xz2-y2);
> res(I,0);
[1]:
   _[1]=z3-y
   _[2]=yz-x
   _[3]=y3-x2z
   _[4]=xz2-y2
[2]:
   _[1]=x*gen(1)-y*gen(2)-z*gen(4)
   _[2]=z2*gen(2)-y*gen(1)+1*gen(4)
   _[3]=xz*gen(2)-y*gen(4)-1*gen(3)
[3]:
   _[1]=0
```

The first line of the input specifies that the characteristic of the field is 0, the ring variables are x, y, z, and the monomial order is graded reverse lex. The argument "0" in the **res** command says that the resolution should have as many steps as variables (the reason for this choice will become clear in the next section). Here, again, the output is a set of columns that generate (**gen(1)**, **gen(2)**, **gen(3)**, **gen(4)** refer to the standard basis columns e_1, e_2, e_3, e_4 of $k[x, y, z]^4$).

See the exercises below for some additional examples. Of course, this raises the question whether *finite* resolutions always exist. Are we in a situation of potential infinite regress or does this process always stop eventually, as in the examples above? See Exercise 11 below for an example where the answer is no, but where R is not a polynomial ring. We shall return to this question in the next section.

ADDITIONAL EXERCISES FOR §1

Exercise 4.
a. Prove the second bullet, which asserts that $\varphi : M \to N$ is one-to-one if and only if $0 \to M \to N$ is exact.
b. Explain how part b of Proposition (1.2) follows from part a.

Exercise 5. Let M_1, M_2 be R-submodules of an R-module N. Let $M_1 \oplus M_2$ be the direct sum as in Exercise 4 of Chapter 5, §1, and let $M_1 + M_2 \subset N$ be the sum as in Exercise 14 of Chapter 5, §1.
a. Let $\varepsilon : M_1 \cap M_2 \to M_1 \oplus M_2$ be the mapping defined by $\varepsilon(m) = (m, m)$. Show that ε is an R-module homomorphism.
b. Show that $\delta : M_1 \oplus M_2 \to M_1 + M_2$ defined by $\delta(m_1, m_2) = m_1 - m_2$ is an R-module homomorphism.
c. Show that
$$0 \to M_1 \cap M_2 \xrightarrow{\varepsilon} M_1 \oplus M_2 \xrightarrow{\delta} M_1 + M_2 \to 0$$
is an exact sequence.

Exercise 6. Let M_1 and M_2 be submodules of an R-module N.
a. Show that the mappings $\psi_i : M_i \to M_1 + M_2$ $(i = 1, 2)$ defined by $\psi_1(m_1) = m_1 + 0 \in M_1 + M_2$ and $\psi_2(m_2) = 0 + m_2 \in M_1 + M_2$ are one-to-one module homomorphisms. Hence M_1 and M_2 are submodules of $M_1 + M_2$.
b. Consider the homomorphism $\varphi : M_2 \to (M_1 + M_2)/M_1$ obtained by composing the inclusion $M_2 \to M_1 + M_2$ and the natural homomorphism $M_1 + M_2 \to (M_1 + M_2)/M_1$. Identify the kernel of φ, and deduce that there is an isomorphism of R-modules $(M_1 + M_2)/M_1 \cong M_2/(M_1 \cap M_2)$.

Exercise 7.
a. Let
$$0 \to M_n \xrightarrow{\varphi_n} M_{n-1} \xrightarrow{\varphi_{n-1}} M_{n-2} \xrightarrow{\varphi_{n-2}} \cdots \xrightarrow{\varphi_1} M_0 \to 0$$
be a "long" exact sequence of R-modules and homomorphisms. Show that there are "short" exact sequences
$$0 \to \ker(\varphi_i) \to M_i \to \ker(\varphi_{i-1}) \to 0$$

for each $i = 1, \ldots, n$, where the arrow $M_i \to \ker(\varphi_{i-1})$ is given by the homomorphism φ_i.

b. Conversely, given

$$0 \to \ker(\varphi_i) \to M_i \xrightarrow{\varphi_i} N_i \to 0$$

where $N_i = \ker(\varphi_{i-1}) \subset M_{i-1}$, show that these short exact sequences can be spliced together into a long exact sequence

$$0 \to \ker(\varphi_{n-1}) \to M_{n-1} \xrightarrow{\varphi_{n-1}} M_{n-2} \xrightarrow{\varphi_{n-2}} \cdots \xrightarrow{\varphi_2} M_1 \xrightarrow{\varphi_1} \operatorname{im}(\varphi_1) \to 0.$$

c. Explain how a resolution of a module is obtained by splicing together presentations of successive syzygy modules.

Exercise 8. Let V_i, $i = 0, \ldots, n$ be finite dimensional vector spaces over a field k, and let

$$0 \to V_n \xrightarrow{\varphi_n} V_{n-1} \xrightarrow{\varphi_{n-1}} V_{n-2} \xrightarrow{\varphi_{n-2}} \cdots \xrightarrow{\varphi_1} V_0 \to 0$$

be an exact sequence of k-linear mappings. Show that the alternating sum of the dimensions of the V_i satisfies:

$$\sum_{\ell=0}^{n} (-1)^\ell \dim_k(V_\ell) = 0.$$

Hint: Use Exercise 7 and the *dimension theorem* for a linear mapping φ : $V \to W$:

$$\dim_k(V) = \dim_k(\ker(\varphi)) + \dim_k(\operatorname{im}(\varphi)).$$

Exercise 9. Let

$$0 \to F_\ell \to \cdots \to F_2 \to F_1 \to F_0 \to M \to 0$$

be a finite free resolution of a submodule $M \subset R^n$. Show how to obtain a finite free resolution of the quotient module R^n/M from the resolution for M. Hint: There is an exact sequence $0 \to M \to R^n \to R^n/M \to 0$ by Proposition (1.2). Use the idea of Exercise 7 part b to splice together the two sequences.

Exercise 10. For each of the following modules, find a free resolution either by hand or by using a computer algebra system.

a. $M = \langle xy, xz, yz \rangle \subset k[x, y, z]$.

b. $M = \langle xy - uv, xz - uv, yz - uv \rangle \subset k[x, y, z, u, v]$.

c. $M = \langle xy - xv, xz - yv, yz - xu \rangle \subset k[x, y, z, u, v]$.

d. M the module generated by the columns of the matrix

$$M = \begin{pmatrix} a^2 + b^2 & a^3 - 2bcd & a - b \\ c^2 - d^2 & b^3 + acd & c + d \end{pmatrix}$$

in $k[a, b, c, d]^2$.

e. $M = \langle x^2, y^2, z^2, xy, xz, yz \rangle \subset k[x, y, z]$.

f. $M = \langle x^3, y^3, x^2y, xy^2 \rangle \subset k[x, y, z]$.

Exercise 11. If we work over other rings R besides polynomial rings, then it is not difficult to find modules with no finite free resolutions. For example, consider $R = k[x]/\langle x^2 \rangle$, and $M = \langle x \rangle \subset R$.

a. What is the kernel of the mapping $\varphi : R \to M$ given by multiplication by x?

b. Show that

$$\cdots \xrightarrow{x} R \xrightarrow{x} R \xrightarrow{x} M \to 0$$

is an infinite free resolution of M over R, where x denotes multiplication by x.

c. Show that *every* free resolution of M over R is infinite. Hint: One way is to show that any free resolution of M must "contain" the resolution from part b in a suitable sense.

Exercise 12. We say that an exact sequence of R-modules

$$0 \longrightarrow M \xrightarrow{f} N \xrightarrow{g} P \longrightarrow 0$$

splits if there is a homomorphism $\varphi : P \to N$ such that $g \circ \varphi = \mathrm{id}$.

a. Show that the condition that the sequence above splits is equivalent to the condition that $N \cong M \oplus P$ such that f becomes the inclusion $a \mapsto (a, 0)$ and g becomes the projection $(a, b) \mapsto b$.

b. Show that the condition that the sequence splits is equivalent to the existence of a homomorphism $\psi : N \to M$ such that $\psi \circ f = \mathrm{id}$. Hint: use part a.

c. Show that P is a projective module (that is, a direct summand of a free module—see Definition (4.12) of Chapter 5) if and only if every exact sequence of the form above splits.

d. Show that P is projective if and only if given every homomorphism $f : P \to M_1$ and any surjective homomorphism $g : M_2 \to M_1$, there exists a homomorphism $h : P \to M_2$ such that $f = g \circ h$.

§2 Hilbert's Syzygy Theorem

In §1, we raised the question of whether every R-module has a finite free resolution, and we saw in Exercise 11 that the answer is no if R is the finite-dimensional algebra $R = k[x]/\langle x^2 \rangle$. However, when $R = k[x_1, \ldots, x_n]$ the situation is much better, and we will consider only polynomial rings in this section. The main fact we will establish is the following famous result of Hilbert.

(2.1) Theorem (Hilbert Syzygy Theorem). *Let $R = k[x_1, \ldots, x_n]$. Then every finitely generated R-module has a finite free resolution of length at most n.*

A comment is in order. As we saw in the examples in §1, it is not true that all finite free resolutions of a given module have the same length. The Syzygy Theorem only asserts the existence of *some* free resolution of length $\leq n$ for every finitely-generated module over the polynomial ring in n variables. Also, remember from Definition (1.9) that length $\leq n$ implies that an R-module M has a free resolution of the form

$$0 \to F_\ell \to \cdots \to F_1 \to F_0 \to M, \quad \ell \leq n.$$

This has $\ell + 1 \leq n + 1$ free modules, so that the Syzygy Theorem asserts the existence of a free resolution with at most $n + 1$ free modules in it.

The proof we will present is due to Schreyer. It is based on the following observation about resolutions produced by the Gröbner basis method described in §1, using Schreyer's Theorem—Theorem (3.3) of Chapter 5.

(2.2) Lemma. *Let $\mathcal{G}$ be a Gröbner basis for a submodule $M \subset R^t$ with respect to an arbitrary monomial order, and arrange the elements of $\mathcal{G}$ to form an ordered s-tuple $G = (g_1, \ldots, g_s)$ so that whenever $\mathrm{LT}(g_i)$ and $\mathrm{LT}(g_j)$ contain the same standard basis vector e_k and $i < j$, then $\mathrm{LM}(g_i)/e_k >_{lex} \mathrm{LM}(g_j)/e_k$, where $>_{lex}$ is the lex order on R with $x_1 > \cdots > x_n$. If the variables $x_1, \ldots, x_m$ do not appear in the leading terms of $\mathcal{G}$, then $x_1, \ldots, x_{m+1}$ do not appear in the leading terms of the $\mathbf{s}_{ij} \in \mathrm{Syz}(G)$ with respect to the order $>_\mathcal{G}$ used in Theorem (3.3) of Chapter 5.*

PROOF OF THE LEMMA. By the first step in the proof of Theorem (3.3) of Chapter 5,

$$(2.3) \qquad \mathrm{LT}_{>_\mathcal{G}}(\mathbf{s}_{ij}) = (m_{ij}/\mathrm{LT}(g_i))E_i,$$

where $m_{ij} = \mathrm{LCM}(\mathrm{LT}(g_i), \mathrm{LT}(g_j))$, and E_i is the standard basis vector in R^s. As always, it suffices to consider only the $\mathbf{s}_{ij}$ such that $\mathrm{LT}(g_i)$ and $\mathrm{LT}(g_j)$ contain the same standard basis vector e_k in R^t, and such that $i < j$. By the hypothesis on the ordering of the components of G, $\mathrm{LM}(g_i)/e_k >_{lex} \mathrm{LM}(g_j)/e_k$. Since $x_1, \ldots, x_m$ do not appear in the leading terms, this implies that we can write

$$\mathrm{LM}(g_i)/e_k = x_{m+1}^a n_i$$
$$\mathrm{LM}(g_j)/e_k = x_{m+1}^b n_j,$$

where $a \geq b$, and n_i, n_j are monomials in R containing only $x_{m+2}, \ldots, x_n$. But then $\mathrm{lcm}(\mathrm{LT}(g_i), \mathrm{LT}(g_j))$ contains x_{m+1}^a, and by (2.3), $\mathrm{LT}_{>_\mathcal{G}}(\mathbf{s}_{ij})$ does not contain $x_1, \ldots, x_m, x_{m+1}$. $\qquad \square$

We are now ready for the proof of Theorem (2.1).

PROOF OF THE THEOREM.   Since we assume M is finitely generated as
an R-module, by (1.5) of this chapter, there is a presentation for M of the
form

(2.4) $$F_1 \xrightarrow{\varphi_1} F_0 \to M \to 0$$

corresponding to a choice of a generating set $(f_1, \ldots, f_{r_0})$ for M, and a
Gröbner basis $\mathcal{G}_0 = \{g_1, \ldots, g_{r_1}\}$ for $\text{Syz}(f_1, \ldots, f_{r_0}) = \text{im}(\varphi_1) \subset F_0 = R^{r_0}$ with respect to any monomial order on F_0. Order the elements of $\mathcal{G}_0$
as described in Lemma (2.2) to obtain a vector G_0, and apply Schreyer's
Theorem to compute a Gröbner basis $\mathcal{G}_1$ for the module $\text{Syz}(G_0) \subset F_1 = R^{r_1}$ (with respect to the $>_{\mathcal{G}_0}$ order). We may assume that $\mathcal{G}_1$ is reduced.
By the lemma, at least x_1 will be missing from the leading terms of $\mathcal{G}_1$.
Moreover if the Gröbner basis contains r_2 elements, we obtain an exact
sequence

$$F_2 \xrightarrow{\varphi_2} F_1 \xrightarrow{\varphi_1} F_0 \to M \to 0$$

with $F_2 = R^{r_2}$, and $\text{im}(\varphi_2) = \text{Syz}(G_1)$. Now iterate the process to obtain
$\varphi_i : F_i \to F_{i-1}$, where $\text{im}(\varphi_i) = \text{Syz}(G_{i-1})$ and $\mathcal{G}_i \subset R^{r_i}$ is a Gröbner
basis for $\text{Syz}(G_{i-1})$, where each time we order the Gröbner basis $\mathcal{G}_{i-1}$ to
form the vector G_{i-1} so that the hypothesis of Lemma (2.2) is satisfied.

Since the number of variables present in the leading terms of the Gröbner
basis elements decreases by at least one at each step, by an easy induction
argument, after some number $\ell \leq n$ of steps, the leading terms of the
reduced Gröbner basis $\mathcal{G}_\ell$ do not contain any of the variables $x_1, \ldots, x_n$.
At this point, we will have extended (2.4) to an exact sequence

(2.5) $$F_\ell \xrightarrow{\varphi_\ell} F_{\ell-1} \to \cdots \to F_1 \xrightarrow{\varphi_1} F_0 \to M \to 0,$$

and the leading terms in $\mathcal{G}_\ell$ will be non-zero constants times standard
basis vectors from F_ℓ. In Exercise 8 below, you will show that this implies
$\text{Syz}(G_{\ell-1})$ is a free module, and $\mathcal{G}_\ell$ is a module basis as well as a Gröbner
basis. Hence by Proposition (1.12) we can extend (2.5) to another exact
sequence by adding a zero at the left, and as a result we have produced a
free resolution of length $\ell \leq n$ for M.   □

Here are some additional examples illustrating the Syzygy Theorem. In
the examples we saw in the text in §1, we always found resolutions of
length strictly less than the number of variables in R. But in some cases,
the shortest possible resolutions are of length exactly n.

Exercise 1. Consider the ideal $I = \langle x^2 - x, xy, y^2 - y \rangle \subset k[x, y]$ from
(1.7) of this chapter, and let $M = k[x, y]/I$, which is also a module over
$R = k[x, y]$. Using Exercise 9 from §1, show that M has a free resolution
of length 2, of the form

$$0 \to R^2 \to R^3 \to R \to M \to 0.$$

In this case, it is also possible using localization (see Chapter 4) to show that M has no free resolution of length ≤ 1. See Exercise 9 below for a sketch.

On the other hand, we might ask whether having an especially *short* finite free resolution indicates something special about an ideal or a module. For example, if M has a resolution $0 \to R^r \to M \to 0$ of length 0, then M is isomorphic to R^r as an R-module. Hence M is free, and this is certainly a special property! From Chapter 5, §1, we know this happens for ideals only when $M = \langle f \rangle$ is principal. Similarly, we can ask what can be said about free resolutions of length 1. The next examples indicate a special feature of resolutions of length 1 for a certain class of ideals.

Exercise 2. Let $I \subset k[x, y, z, w]$ denote the ideal of the twisted cubic in $\mathbb{P}^3$, with the following generators:

$$I = \langle g_1, g_2, g_3 \rangle = \langle xz - y^2, xw - yz, yw - z^2 \rangle.$$

a. Show that the given generators form a *grevlex* Gröbner basis for I.
b. Apply Schreyer's Theorem to find a Gröbner basis for the module of first syzygies on the given generators for I.
c. Show that s_{12} and s_{23} form a basis for $\mathrm{Syz}(xz - y^2, xw - yz, yw - z^2)$.
d. Use the above calculations to produce a finite free resolution of I, of the form

$$0 \to R^2 \xrightarrow{A} R^3 \to I \to 0.$$

e. Show that the determinants of the 2×2 minors of A are just the g_i (up to signs).

Exercise 3. (For this exercise, you will probably want to use a computer algebra system.) In k^2 consider the points

$$p_1 = (0, 0), \ p_2 = (1, 0), \ p_3 = (0, 1)$$
$$p_4 = (2, 1), \ p_5 = (1, 2), \ p_6 = (3, 3),$$

and let $I_i = \mathbf{I}(\{p_i\})$ for each i, so for instance $I_3 = \langle x, y - 1 \rangle$.
a. Find a *grevlex* Gröbner basis for

$$J = \mathbf{I}(\{p_1, \ldots, p_6\}) = I_1 \cap \cdots \cap I_6.$$

b. Compute a free resolution of J of the form

$$0 \to R^3 \xrightarrow{A} R^4 \to J \to 0,$$

where each entry of A is of total degree at most 1 in x and y.
c. Show that the determinants of the 3×3 minors of A are the generators of J (up to signs).

The examples in Exercises 2 and 3 are instances of the following general result, which is a part of the Hilbert-Burch Theorem.

(2.6) Proposition. *Suppose that an ideal I in $R = k[x_1, \ldots, x_n]$ has a free resolution of the form*

$$0 \to R^{m-1} \xrightarrow{A} R^m \xrightarrow{B} I \to 0$$

for some m. Then there exists a nonzero element $g \in R$ such that $B = (\, g\tilde{f}_1 \; \cdots \; g\tilde{f}_m \,)$, where $\tilde{f}_i$ is the determinant of the $(m-1) \times (m-1)$ submatrix of A obtained by deleting row i. If k is algebraically closed and $\mathbf{V}(I)$ has dimension $n-2$, then we may take $g = 1$.

PROOF. The proof is outlined in Exercise 11 below.   □

The full Hilbert-Burch Theorem also gives a sufficient condition for the existence of a resolution of the form given in the proposition. For example, such a resolution exists when the quotient ring R/I is *Cohen-Macaulay* of codimension 2. This condition is satisfied, for instance, if $I \subset k[x, y, z]$ is the ideal of a finite subset of $\mathbb{P}^2$ (including the case where one or more of the points has multiplicity > 1 as defined in Chapter 4). We will not give the precise definition of the Cohen-Macaulay condition here. Instead we refer the interested reader to [Eis], where this and many of the other known results concerning the shapes of free resolutions for certain classes of ideals in polynomial and local rings are discussed. In particular, the length of the shortest finite free resolution of an R-module M is an important invariant called the *projective dimension* of M.

ADDITIONAL EXERCISES FOR §2

Exercise 4. Let I be the ideal in $k[x, y]$ generated by the *grevlex* Gröbner basis

$$\{g_1, g_2, g_3\} = \{x^2 + 3/2xy + 1/2y^2 - 3/2x - 3/2y, xy^2 - x, y^3 - y\}$$

This ideal was considered in Chapter 2, §2 (with $k = \mathbb{C}$), and we saw there that $\mathbf{V}(I)$ is a finite set containing 5 points in k^2, each with multiplicity 1.

a. Applying Schreyer's Theorem, show that $\mathrm{Syz}(g_1, g_2, g_3)$ is generated by the columns of the matrix

$$A = \begin{pmatrix} y^2 - 1 & 0 \\ -x - 3y/2 + 3/2 & y \\ -y/2 + 3/2 & -x \end{pmatrix}$$

b. Show that the columns of A form a module basis for $\mathrm{Syz}(g_1, g_2, g_3)$, and deduce that I has a finite free resolution of length 1:

$$0 \to R^2 \xrightarrow{A} R^3 \to I \to 0.$$

c. Show that the determinants of the 2×2 minors of A are just the g_i (up to signs).

Exercise 5. Verify that the resolution from (1.8) of §1 has the form given in Proposition (2.6). (In this case too, the module being resolved is the ideal of a finite set of points in k^2, each appearing with multiplicity 1.)

Exercise 6. Let

$$I = \langle z^3 - y, yz - x, y^3 - x^2z, xz^2 - y^2 \rangle$$

be the ideal in $k[x, y, z]$ considered in §1 see (1.16).
a. Show that the generators of I are a Gröbner basis with respect to the *grevlex* order.
b. The **sres** command in **Singular** produces a resolution using Schreyer's algorithm. The **Singular** session is as follows.

```
> ring r=0, (x,y,z), (dp, C);
> ideal I=(z3-y,yz-x,y3-x2z,xz2-y2);
> sres(I,0);
[1]:
  _[1]=yz-x
  _[2]=z3-y
  _[3]=xz2-y2
  _[4]=y3-x2z
[2]:
  _[1]=-z2*gen(1)+y*gen(2)-1*gen(3)
  _[2]=-xz*gen(1)+y*gen(3)+1*gen(4)
  _[3]=-x*gen(2)+y*gen(1)+z*gen(3)
  _[4]=-y2*gen(1)+x*gen(3)+z*gen(4)
[3]:
  _[1]=x*gen(1)+y*gen(3)-z*gen(2)+1*gen(4)
```

Show that the displayed generators are Gröbner bases with respect to the orderings prescribed by Schreyer's Theorem from Chapter 5, §3.
c. Explain why using Schreyer's Theorem produces a longer resolution in this case than that displayed in §1.

Exercise 7. Find a free resolution of length 1 of the form given in Proposition (2.6) for the ideal

$$I = \langle x^4 - x^3y, x^3y - x^2y^2, x^2y^2 - xy^3, xy^3 - y^4 \rangle$$

in $R = k[x, y]$. Identify the matrix A and the element $g \in R$ in this case in Proposition (2.6). Why is $g \neq 1$?

Exercise 8. Let $\mathcal{G}$ be a monic reduced Gröbner basis for a submodule $M \subset R^t$, with respect to some monomial order. Assume that the leading

terms of all the elements of $\mathcal{G}$ are constant multiples of standard basis vectors in R^t.
a. If e_i is the leading term of some element of $\mathcal{G}$, show that it is the leading term of exactly one element of $\mathcal{G}$.
b. Show that $\mathrm{Syz}(\mathcal{G}) = \{0\} \subset R^s$.
c. Deduce that M is a free module.

Exercise 9. In this exercise, we will sketch one way to show that every free resolution of the quotient R/I for

$$I = \langle x^2 - x, xy, y^2 - y \rangle \subset R = k[x, y]$$

has length ≥ 2. In other words, the resolution $0 \to R^2 \to R^3 \to R \to R/I \to 0$ from Exercise 1 is as short as possible. We will need to use some ideas from Chapter 4 of this book.
a. Let M be an R-module, and let P be a maximal ideal in R. Generalizing the construction of the local ring R_P, define the *localization* of M at P, written M_P, to be the set of "fractions" m/f, where $m \in M$, $f \notin P$, subject to the relation that $m/f = m'/f'$ whenever there is some $g \in R$, $g \notin P$ such that $g(f'm - fm') = 0$ in M. Show that M_P has the structure of a module over the local ring R_P. If M is a free R-module, show that M_P is a free R_P-module.
b. Given a homomorphism $\varphi : M \to N$ of R-modules, show that there is an induced homomorphism of the localized modules $\varphi_P : M_P \to N_P$ defined by $\varphi_P(m/f) = \varphi(m)/f$ for all $m/f \in M_P$. Hint: First show that this rule gives a well-defined mapping from M_P to N_P.
c. Let

$$M_1 \xrightarrow{\varphi_1} M_2 \xrightarrow{\varphi_2} M_3$$

be an exact sequence of R-modules. Show that the localized sequence

$$(M_1)_P \xrightarrow{(\varphi_1)_P} (M_2)_P \xrightarrow{(\varphi_2)_P} (M_3)_P$$

is also exact.
d. We want to show that the shortest free resolution of $M = R/I$ for $I = \langle x^2 - x, xy, y^2 - y \rangle$ has length 2. Aiming for a contradiction, suppose that there is some resolution of length 1: $0 \to F_1 \to F_0 \to M \to 0$. Explain why we may assume $F_0 = R$.
e. By part c, after localizing at $P = \langle x, y \rangle \supset I$, we obtain a resolution $0 \to (F_1)_P \to R_P \to M_P \to 0$. Show that M_P is isomorphic to $R_P/\langle x, y \rangle R_P \cong k$ as an R_P-module.
f. But then the image of $(F_1)_P \to R_P$ must be $\langle x, y \rangle$. Show that we obtain a contradiction because this is not a free R_P-module.

Exercise 10. In $R = k[x_1, \ldots, x_n]$, consider the ideals

$$I_m = \langle x_1, x_2, \ldots, x_m \rangle$$

generated by subsets of the variables, for $1 \leq m \leq n$.

a. Find explicit resolutions for the ideals $I_2, \ldots, I_5$ in $k[x_1, \ldots, x_5]$.

b. Show in general that I_m has a free resolution of length $m - 1$ of the form

$$0 \to R^{\binom{m}{m}} \to \cdots \to R^{\binom{m}{3}} \to R^{\binom{m}{2}} \to R^m \to I \to 0,$$

where if we index the basis B_k of $R^{\binom{m}{k}}$ by k-element subsets of $\{1, \ldots, m\}$:

$$B_k = \{e_{i_1 \ldots i_k} : 1 \leq i_1 < i_2 < \cdots < i_k \leq m\},$$

then the mapping $\varphi_k : R^{\binom{m}{k}} \to R^{\binom{m}{k-1}}$ in the resolution is defined by

$$\varphi_k(e_{i_1 \ldots i_k}) = \sum_{j=1}^{k} (-1)^{j+1} x_{i_j} e_{i_1 \ldots i_{j-1} i_{j+1} \ldots i_k},$$

where in the term with index j, i_j is omitted to yield a $(k-1)$-element subset. These resolutions are examples of *Koszul complexes*. See [Eis] for more information about this topic.

Exercise 11. In this exercise, we will sketch a proof of Proposition (2.6). The basic idea is to consider the linear mapping from K^{m-1} to K^m defined by the matrix A in a resolution

$$0 \to R^{m-1} \xrightarrow{A} R^m \xrightarrow{B} I \to 0,$$

where $K = k(x_1, \ldots, x_n)$ is the field of rational functions (the field of fractions of R) and to use some linear algebra over K.

a. Let V be the space of solutions of the the homogeneous system of linear equations $XA = 0$ where $X \in K^m$ is written as a row vector. Show that the dimension over K of V is 1. Hint: The columns $A_1, \ldots, A_{m-1}$ of A are linearly independent over R, hence over K.

b. Let $B = (f_1 \quad \cdots \quad f_m)$ and note that exactness implies that $BA = 0$. Let $\tilde{f}_i = (-1)^{i+1} \det(A_i)$, where A_i is the $(m-1) \times (m-1)$ submatrix of A obtained by deleting row i. Show that $X = (\tilde{f}_1, \ldots, \tilde{f}_m)$ is also an element of the space V of solutions of $XA = 0$. Hint: append any one of the columns of A to A to form an $m \times m$ matrix $\tilde{A}$, and expand $\det(\tilde{A})$ by minors along the new column.

c. Deduce that there is some $r \in K$ such that $r\tilde{f}_i = f_i$ for all $i = 1, \ldots, m$.

d. Write $r = g/h$ where $g, h \in R$ and the fraction is in lowest terms, and consider the equations $g\tilde{f}_i = hf_i$. We want to show that h must be a nonzero constant, arguing by contradiction. If not, then let p be any irreducible factor of h. Show that $A_1, \ldots, A_{m-1}$ are linearly dependent modulo $\langle p \rangle$, or in other words that there exist $r_1, \ldots, r_{m-1}$ not all in $\langle p \rangle$ such that $r_1 A_1 + \cdots + r_{m-1} A_{m-1} = pB$ for some $B \in R^m$.

e. Continuing from part d, show that $B \in \mathrm{Syz}(f_1, \ldots, f_m)$ also, so that $B = s_1 A_1 + \cdots + s_{m-1} A_{m-1}$ for some $s_i \in R$.

f. Continuing from part e, show that $(r_1 - ps_1, \ldots, r_{m-1} - ps_{m-1})^T$ would be a syzygy on the columns of A. Since those columns are linearly independent over R, $r_i - ps_i = 0$ for all i. Deduce a contradiction to the way we chose the r_i.

g. Finally, in the case that $\mathbf{V}(I)$ has dimension $n - 2$, show that g must be a nonzero constant also. Hence by multiplying each f_i by a nonzero constant, we could take $g = 1$ in Proposition (2.6).

§3 Graded Resolutions

In algebraic geometry, free resolutions are often used to study the homogeneous ideals $I = \mathbf{I}(V)$ of projective varieties $V \subset \mathbb{P}^n$ and other modules over $k[x_0, \ldots, x_n]$. The key fact we will use is that these resolutions have an extra structure coming from the *grading* on the ring $R = k[x_0, \ldots, x_n]$, that is the direct sum decomposition

(3.1) $$R = \bigoplus_{s \geq 0} R_s$$

into the additive subgroups (or k-vector subspaces) $R_s = k[x_0, \ldots, x_n]_s$, consisting of the homogeneous polynomials of total degree s, together with 0. To begin this section we will introduce some convenient notation and terminology for describing such resolutions.

(3.2) Definition. A *graded module* over R is a module M with a family of subgroups $\{M_t : t \in \mathbb{Z}\}$ of the additive group of M. The elements of M_t are called the *homogeneous elements* of degree t in the grading, and the M_t must satisfy the following properties.
a. As additive groups,
$$M = \bigoplus_{t \in \mathbb{Z}} M_t.$$

b. The decomposition of M in part a is compatible with the multiplication by elements of R in the sense that $R_s M_t \subset M_{s+t}$ for all $s \geq 0$ and all $t \in \mathbb{Z}$.

It is easy to see from the definition that each M_t is a module over the subring $R_0 = k \subset R$, hence a k-vector subspace of M. If M is finitely-generated, the M_t are finite dimensional over k.

Homogeneous ideals $I \subset R$ are the most basic examples of graded modules. Recall that an ideal is homogeneous if whenever $f \in I$, the homogeneous components of f are all in I as well (see for instance, [CLO], Chapter 8, §3, Definition 1). Some of the other important properties of these ideals are summarized in the following statement.

- (Homogeneous Ideals) Let $I \subset k[x_0, \ldots, x_n]$ be an ideal. Then the following are equivalent:
  a. I is a homogeneous ideal.
  b. $I = \langle f_1, \ldots, f_s \rangle$ where f_i are homogeneous polynomials.
  c. A reduced Gröbner basis for I (with respect to any monomial order) consists of homogeneous polynomials.

(See for instance [CLO], Theorem 2 of Chapter 8, §3.)

To show that a homogeneous ideal I has a graded module structure, set $I_t = I \cap R_t$. For $t \geq 0$, this is the set of all homogeneous elements of total degree t in I (together with 0), and $I_t = \{0\}$ for $t < 0$. By the definition of a homogeneous ideal, we have $I = \oplus_{t \in \mathbb{Z}} I_t$, and $R_s I_t \subset I_{s+t}$ is a direct consequence of the definition of an ideal and the properties of polynomial multiplication.

The free modules R^m are also graded modules over R provided we take $(R^m)_t = (R_t)^m$. We will call this the *standard* graded module structure on R^m. Other examples of graded modules are given by submodules of the free modules R^m with generating sets possessing suitable homogeneity properties, and we have statements analogous to those above for homogeneous ideals.

(3.3) Proposition. *Let $M \subset R^m$ be submodule. Then the following are equivalent.*

a. *The standard grading on R^m induces a graded module structure on M, given by taking $M_t = (R_t)^m \cap M$ —the set of elements in M where each component is a homogeneous polynomial of degree t (or 0).*
b. *$M = \langle f_1, \ldots, f_r \rangle$ in R^m where each f_i is a vector of homogeneous polynomials of the same degree d_i.*
c. *A reduced Gröbner basis (for any monomial order on R^m) consists of vectors of homogeneous polynomials where all the components of each vector have the same degree.*

PROOF. The proof is left to the reader as Exercise 8 below.   □

Submodules, direct sums, and quotient modules extend to graded modules in the following ways. If M is a graded module and N is a submodule of M, then we say N is a *graded submodule* if the additive subgroups $N_t = M_t \cap N$ for $t \in \mathbb{Z}$ define a graded module structure on N. For example, Proposition (3.3) says that the submodules $M = \langle f_1, \ldots, f_r \rangle$ in R^m where each f_i is a vector of homogeneous polynomials of the same degree d_i are graded submodules of R^m.

Exercise 1.
a. Given a collection of graded modules $M_1, \ldots, M_m$, we can produce the direct sum $N = M_1 \oplus \cdots \oplus M_m$ as usual. In N, let

$$N_t = (M_1)_t \oplus \cdots \oplus (M_m)_t.$$

Show that the N_t define the structure of a graded module on N.

b. If $N \subset M$ is a graded submodule of a graded module M, show that the quotient module M/N also has a graded module structure, defined by the collection of additive subgroups

$$(M/N)_t = M_t/N_t = M_t/(M_t \cap N).$$

Given any graded R-module M, we can also produce modules that are isomorphic to M as abstract R-modules, but with different gradings, by the following trick of shifting the indexing of the family of submodules.

(3.4) Proposition. *Let M be a graded R-module, and let d be an integer. Let $M(d)$ be the direct sum*

$$M(d) = \bigoplus_{t \in \mathbb{Z}} M(d)_t,$$

where $M(d)_t = M_{d+t}$. Then $M(d)$ is also a graded R-module.

PROOF. The proof is left to the reader as Exercise 9.   □

For instance, the modules $(R^m)(d) = R(d)^m$ are called shifted or *twisted* graded free modules over R. The standard basis vectors e_i still form a module basis for $R(d)^m$, but they are now homogeneous elements of degree $-d$ in the grading, since $R(d)_{-d} = R_0$. More generally, part a of Exercise 1 shows that we can consider graded free modules of the form

$$R(d_1) \oplus \cdots \oplus R(d_m)$$

for any integers $d_1, \ldots, d_m$, where the basis vector e_i is homogeneous of degree $-d_i$ for each i.

Exercise 2. This exercise will generalize Proposition (3.3). Suppose that we have integers $d_1, \ldots, d_m$ and elements $f_1, \ldots, f_s \in R^m$ such that

$$f_i = (f_{i1}, \ldots, f_{im})^T$$

where the f_{ij} are homogeneous and $\deg f_{i1} - d_1 = \cdots = \deg f_{im} - d_m$ for each i. Then prove that $M = \langle f_1, \ldots, f_s \rangle$ is a graded submodule of $F = R(d_1) \oplus \cdots \oplus R(d_m)$. Also show that every graded submodule of F has a set of generators of this form.

As the examples given later in the section will show, the twisted free modules we deal with are typically of the form

$$R(-d_1) \oplus \cdots \oplus R(-d_m).$$

Here, the standard basis elements $e_1, \ldots, e_m$ have respective degrees $d_1, \ldots, d_m$.

Next we consider how homomorphisms interact with gradings on modules.

(3.5) Definition. Let M, N be graded modules over R. A homomorphism $\varphi : M \to N$ is said to a *graded homomorphism of degree d* if $\varphi(M_t) \subset N_{t+d}$ for all $t \in \mathbb{Z}$.

For instance, suppose that M is a graded R-module generated by homogeneous elements $f_1, \ldots, f_m$ of degrees $d_1, \ldots, d_m$. Then we get a graded homomorphism

$$\varphi : R(-d_1) \oplus \cdots \oplus R(-d_m) \longrightarrow M$$

which sends the standard basis element e_i to $f_i \in M$. Note that φ is onto. Also, since e_i has degree d_i, it follows that φ has degree zero.

Exercise 3. Suppose that M is a finitely generated R-module. As usual, M_t denotes the set of homogeneous elements of M of degree t.
a. Prove that M_t is a finite dimensional vector space over the field k and that $M_l = \{0\}$ for $l \ll 0$. Hint: Use the surjective map φ constructed above.
b. Let $\psi : M \to M$ be a graded homomorphism of degree zero. Prove that ψ is an isomorphism if and only if $\psi : M_t \to M_t$ is onto for every t. Conclude that ψ is an isomorphism if and only if it is onto.

Another example of a graded homomorphism is given by an $m \times p$ matrix A all of whose entries are homogeneous polynomials of degree d in the ring R. Then A defines a graded homomorphism φ of degree d by matrix multiplication

$$\varphi : R^p \to R^m$$

$$f \mapsto Af.$$

If desired, we can also consider A as defining a graded homomorphism of degree zero from the shifted module $R(-d)^p$ to R^m. Similarly, if the entries of the jth column are all homogeneous polynomials of degree d_j, but the degree varies with the column, then A defines a graded homomorphism of degree zero

$$R(-d_1) \oplus \cdots \oplus R(-d_p) \to R^m.$$

Still more generally, a graded homomorphism of degree zero

$$R(-d_1) \oplus \cdots \oplus R(-d_p) \to R(-c_1) \oplus \cdots \oplus R(-c_m)$$

is defined by an $m \times p$ matrix A where the ij entry $a_{ij} \in R$ is homogeneous of degree $d_j - c_i$ for all i, j. We will call a matrix A satisfying this condition for some collection d_j of column degrees and some collection c_i of row degrees a *graded matrix* over R.

The reason for discussing graded matrices in detail is that these matrices appear in free resolutions of graded modules over R. For example, consider the resolution of the homogeneous ideal

$$M = \langle z^3 - yw^2, yz - xw, y^3 - x^2z, xz^2 - y^2w \rangle$$

in $R = k[x, y, z, w]$ from (1.13) of this chapter, computed using *Macaulay 2*. The ideal itself is the image of a graded homomorphism of degree zero

$$R(-3) \oplus R(-2) \oplus R(-3)^2 \to R,$$

where the shifts are just the negatives of the degrees of the generators, ordered as above. The next matrix in the resolution:

$$A = \begin{pmatrix} -y^2 & -xz & -yw & -z^2 \\ z & w & 0 & 0 \\ x & y & -z & -w \\ 0 & 0 & x & y \end{pmatrix}$$

(whose columns generate the module of syzygies on the generators of M) defines a graded homomorphism of degree zero

$$R(-4)^4 \xrightarrow{A} R(-2) \oplus R(-3)^3.$$

In other words, $d_j = 4$ for all j, and $c_2 = c_3 = c_4 = 3, c_1 = 2$ in the notation as above, so all entries on rows 2, 3, 4 of A are homogeneous of degree $4 - 3 = 1$, while those on row 1 have degree $4 - 2 = 2$. The whole resolution can be written in the form

$$(3.6) \qquad 0 \to R(-5) \to R(-4)^4 \to R(-2) \oplus R(-3)^3 \to M \to 0,$$

where all the arrows are graded homomorphisms of degree zero.

Here is the precise definition of a graded resolution.

(3.7) Definition. If M is a graded R-module, then a *graded resolution* of M is a resolution of the form

$$\cdots \to F_2 \xrightarrow{\varphi_2} F_1 \xrightarrow{\varphi_1} F_0 \xrightarrow{\varphi_0} M \to 0,$$

where each F_ℓ is a twisted free graded module $R(-d_1) \oplus \cdots \oplus R(-d_p)$ and each homomorphism φ_ℓ is a graded homomorphism of degree zero (so that the φ_ℓ are given by graded matrices as defined above).

The resolution given in (3.6) is clearly a graded resolution. What's nice is that *every* finitely generated graded R-module has a graded resolution of finite length.

(3.8) Theorem (Graded Hilbert Syzygy Theorem). *Let* $R = k[x_1, \ldots, x_n]$. *Then every finitely generated graded R-module has a finite graded resolution of length at most n.*

PROOF. This follows from the proof of Theorem (2.1) (the Syzygy Theorem in the ungraded case) with minimal changes. The reason is that by Proposition (3.3) and the generalization given in Exercise 2, if we apply Schreyer's theorem to find generators for the module of syzygies on a homogeneous ordered Gröbner basis $(g_1, \ldots, g_s)$ for a graded submodule of $R(-d_1) \oplus \cdots \oplus R(-d_p)$, then the syzygies s_{ij} are also homogeneous and "live" in another graded submodule of the same form. We leave the details of the proof as Exercise 5 below.   □

The `resolution` command in *Macaulay 2* will compute a finite graded resolution using the method outlined in the proof of Theorem (3.8). However, the resolutions produced by *Macaulay 2* are of a very special sort.

(3.9) Definition. Suppose that

$$\cdots \to F_\ell \xrightarrow{\varphi_\ell} F_{\ell-1} \to \cdots \to F_0 \to M \to 0$$

is a graded resolution of M. Then the resolution is *minimal* if for every $\ell \geq 1$, the nonzero entries of the graded matrix of φ_ℓ have positive degree.

For an example, the reader should note that the resolution (3.6) is a minimal resolution. But not all resolutions are minimal, as shown by the following example.

Exercise 4. Show that the resolution from (1.11) can be homogenized to give a graded resolution, and explain why it is not minimal. Also show that the resolution from (1.10) is minimal after we homogenize.

In *Macaulay 2*, `resolution` computes a minimal resolution.

We will soon see that minimal resolutions have many nice properties. But first, let's explain why they are called "minimal". We say that a set of generators of a module is *minimal* if no proper subset generates the module. Now suppose that we have a graded resolution

$$\cdots \to F_\ell \xrightarrow{\varphi_\ell} F_{\ell-1} \to \cdots \to F_0 \to M \to 0.$$

Each φ_ℓ gives a surjective map $F_\ell \to \mathrm{im}(\varphi_\ell)$, so that φ_ℓ takes the standard basis of F_ℓ to a generating set of $\mathrm{im}(\varphi_\ell)$. Then we can characterize minimality as follows.

(3.10) Proposition. *The above resolution is minimal if and only if for each $\ell \geq 0$, φ_ℓ takes the standard basis of F_ℓ to a minimal generating set of* $\operatorname{im}(\varphi_\ell)$.

PROOF. We will prove one direction and leave the other as an exercise. Suppose that for some $\ell \geq 1$ the graded matrix A_ℓ of φ_ℓ has entries of positive degree. We will show that $\varphi_{\ell-1}$ takes the standard basis of $F_{\ell-1}$ to a minimal generating set of $\operatorname{im}(\varphi_{\ell-1})$. Let $e_1, \ldots, e_m$ be the standard basis vectors of $F_{\ell-1}$. If $\varphi_{\ell-1}(e_1), \ldots, \varphi_{\ell-1}(e_m)$ is not a minimal generating set, then some $\varphi_{\ell-1}(e_i)$ can be expressed in terms of the others. Reordering the basis if necessary, we can assume that

$$\varphi_{\ell-1}(e_1) = \sum_{i=2}^{m} a_i \varphi_{\ell-1}(e_i), \quad a_i \in R.$$

Then $\varphi_{\ell-1}(e_1 - a_2 e_2 - \cdots - a_m e_m) = 0$, so $(1, -a_2, \ldots, -a_m) \in \ker(\varphi_{\ell-1})$. By exactness, $(1, -a_2, \ldots, -a_m) \in \operatorname{im}(\varphi_\ell)$. Since A_ℓ is the matrix of φ_ℓ, the columns of A_ℓ generate $\operatorname{im}(\varphi_\ell)$. We are assuming that the nonzero components of these columns have positive degree. Since the first entry of $(1, -a_2, \ldots, -a_m)$ is a nonzero constant, it follows that this vector cannot be an R-linear combination of the columns of A_ℓ. This contradiction proves that the $\varphi_{\ell-1}(e_i)$ give a minimal generating set of $\operatorname{im}(\varphi_{\ell-1})$. □

The above proposition shows that minimal resolutions are very intuitive. For example, suppose that we have built a graded resolution of an R-module M out to stage $\ell - 1$:

$$F_{\ell-1} \xrightarrow{\varphi_{\ell-1}} F_{\ell-2} \to \cdots \to F_0 \to M \to 0.$$

We extend one more step by picking a generating set of $\ker(\varphi_{\ell-1})$ and defining $\varphi_\ell : F_\ell \to \ker(\varphi_{\ell-1}) \subset F_{\ell-1}$ by mapping the standard basis of F_ℓ to the chosen generating set. To be efficient, we should pick a minimal generating set, and if we do this at every step of the construction, then Proposition (3.10) guarantees that we get a minimal resolution.

Exercise 5. Give a careful proof of Theorem (3.8) (the Graded Syzygy Theorem), and then modify the proof to show that every finitely generated graded module over $k[x_1, \ldots, x_n]$ has a *minimal* resolution of length $\leq n$. Hint: Use Proposition (3.10).

We next discuss to what extent a minimal resolution is unique. The first step is to define what it means for two resolutions to be the same.

(3.11) Definition. Two graded resolutions $\cdots \to F_0 \xrightarrow{\varphi_0} M \to 0$ and $\cdots \to G_0 \xrightarrow{\psi_0} M \to 0$ are *isomorphic* if there are graded isomorphisms $\alpha_\ell : F_\ell \to G_\ell$ of degree zero such that $\psi_0 \circ \alpha_0 = \varphi_0$ and, for every $\ell \geq 1$, the diagram

$$(3.12) \qquad \begin{array}{ccc} F_\ell & \xrightarrow{\varphi_\ell} & F_{\ell-1} \\ \alpha_\ell \downarrow & & \downarrow \alpha_{\ell-1} \\ G_\ell & \xrightarrow{\psi_\ell} & G_{\ell-1} \end{array}$$

commutes, meaning $\alpha_{\ell-1} \circ \varphi_\ell = \psi_\ell \circ \alpha_\ell$.

We will now show that a finitely generated graded module M has a unique minimal resolution up to isomorphism.

(3.13) Theorem. *Any two minimal resolutions of M are isomorphic.*

PROOF. We begin by defining $\alpha_0 : F_0 \to G_0$. If $e_1, \ldots, e_m$ is the standard basis of F_0, then we get $\varphi_0(e_i) \in M$, and since $G_0 \to M$ is onto, we can find $g_i \in G_0$ such that $\psi_0(g_i) = \varphi_0(e_i)$. Then setting $\alpha_0(e_i) = g_i$ defines a graded homomorphism $\alpha_0 : F_0 \to G_0$ of degree zero, and it follows easily that $\psi_0 \circ \alpha_0 = \varphi_0$.

A similar argument gives $\beta_0 : G_0 \to F_0$, also a graded homomorphism of degree zero, such that $\varphi_0 \circ \beta_0 = \psi_0$. Thus $\beta_0 \circ \alpha_0 : F_0 \to F_0$, and if $1_{F_0} : F_0 \to F_0$ denotes the identity map, then

$$(3.14) \quad \varphi_0 \circ (1_{F_0} - \beta_0 \circ \alpha_0) = \varphi_0 - (\varphi_0 \circ \beta_0) \circ \alpha_0 = \varphi_0 - \psi_0 \circ \alpha_0 = 0.$$

We claim that (3.14) and minimality imply that $\beta_0 \circ \alpha_0$ is an isomorphism.

To see why, first recall from the proof of Proposition (3.10) that the columns of the matrix representing φ_1 generate $\mathrm{im}(\varphi_1)$. By minimality, the nonzero entries in these columns have positive degree. If we let $\langle x_1, \ldots, x_n \rangle F_0$ denote the submodule of F_0 generated by $x_i e_j$ for all i, j, it follows that $\mathrm{im}(\varphi_1) \subset \langle x_1, \ldots, x_n \rangle F_0$.

However, (3.14) implies that $\mathrm{im}(1_{F_0} - \beta_0 \circ \alpha_0) \subset \ker(\varphi_0) = \mathrm{im}(\varphi_1)$. By the previous paragraph, we see that $v - \beta_0 \circ \alpha_0(v) \in \langle x_1, \ldots, x_n \rangle F_0$ for all $v \in F_0$. In Exercise 11 at the end of the section, you will show that this implies that $\beta_0 \circ \alpha_0$ is an isomorphism. In particular, α_0 is one-to-one.

By a similar argument using the minimality of the graded resolution $\cdots \to G_0 \to M \to 0$, $\alpha_0 \circ \beta_0$ is also an isomorphism, which implies that α_0 is onto. Hence α_0 is an isomorphism as claimed. Then Exercise 12 at the end of the section will show that α_0 induces an isomorphism $\bar{\alpha}_0 : \ker(\varphi_0) \to \ker(\psi_0)$.

Now we can define α_1. Since $\varphi_1 : F_1 \to \mathrm{im}(\varphi_1) = \ker(\varphi_0)$ is onto, we get a minimal resolution

$$\cdots \to F_1 \xrightarrow{\varphi_1} \ker(\varphi_0) \to 0,$$

of $\ker(\varphi_0)$ (see Exercise 7 of §1), and similarly

$$\cdots \to G_1 \xrightarrow{\psi_1} \ker(\psi_0) \to 0$$

is a minimal resolution of $\ker(\psi_0)$. Then, using the isomorphism $\bar{\alpha}_0 : \ker(\varphi_0) \rightarrow \ker(\psi_0)$ just constructed, the above argument easily adapts to give a graded isomorphism $\alpha_1 : F_1 \rightarrow G_1$ of degree zero such that $\bar{\alpha}_0 \circ \varphi_1 = \psi_1 \circ \alpha_1$. Since $\bar{\alpha}_0$ is the restriction of α_0 to $\mathrm{im}(\varphi_1)$, it follows easily that (3.12) commutes (with $\ell = 1$).

If we apply Exercise 12 again, we see that α_1 induces an isomorphism $\bar{\alpha}_1 : \ker(\varphi_1) \rightarrow \ker(\psi_1)$. Repeating the above process, we can now define α_2 with the required properties, and continuing for all ℓ, the theorem now follows easily.   □

Since we know by Exercise 5 that a finitely generated R-module M has a finite minimal resolution, it follows from Theorem (3.13) that *all* minimal resolutions of M are finite. This fact plays a crucial role in the following refinement of the Graded Syzygy Theorem.

(3.15) Theorem. *If*

$$\cdots \rightarrow F_\ell \xrightarrow{\varphi_\ell} F_{\ell-1} \rightarrow \cdots \rightarrow F_0 \rightarrow M \rightarrow 0,$$

is any graded resolution of M over $k[x_1, \ldots, x_n]$, then the kernel $\ker(\varphi_{n-1})$ is free, and

$$0 \rightarrow \ker(\varphi_{n-1}) \rightarrow F_{n-1} \rightarrow \cdots \rightarrow F_0 \rightarrow M \rightarrow 0$$

is a graded resolution of M.

PROOF. We begin by showing how to simplify a given graded resolution $\cdots \rightarrow F_0 \rightarrow M \rightarrow 0$. Suppose that for some $\ell \geq 1$, $\varphi_\ell : F_\ell \rightarrow F_{\ell-1}$ is not minimal, i.e., the matrix A_ℓ of φ_ℓ has a nonzero entry of degree zero. If we order the standard bases $\{e_1, \ldots, e_m\}$ of F_ℓ and $\{u_1, \ldots, u_t\}$ of $F_{\ell-1}$ appropriately, we can assume that

$$(3.16) \qquad \varphi_\ell(e_1) = c_1 u_1 + c_2 u_2 + \cdots + c_t u_t$$

where c_1 is a nonzero constant (note that $(c_1, \ldots, c_t)^T$ is the first column of A_ℓ). Then let $G_\ell \subset F_\ell$ and $G_{\ell-1} \subset F_{\ell-1}$ be the submodules generated by $\{e_2, \ldots, e_m\}$ and $\{u_2, \ldots, u_t\}$ respectively, and define the maps

$$F_{\ell+1} \xrightarrow{\psi_{\ell+1}} G_\ell \xrightarrow{\psi_\ell} G_{\ell-1} \xrightarrow{\psi_{\ell-1}} F_{\ell-2}$$

as follows:

- $\psi_{\ell+1}$ is the projection $F_\ell \rightarrow G_\ell$ (which sends $a_1 e_1 + a_2 e_2 + \cdots + a_m e_m$ to $a_2 e_2 + \cdots + a_m e_m$) composed with $\varphi_{\ell+1}$.
- If the first row of A_ℓ is $(c_1, d_2, \ldots, d_m)$, then ψ_ℓ is defined by $\psi_\ell(e_i) = \varphi_\ell(e_i - \frac{d_i}{c_1} e_1)$ for $i = 2, \ldots, m$. Since $\varphi_\ell(e_i) = d_i u_1 + \cdots$ for $i \geq 2$, it follows easily from (3.16) that $\psi_\ell(e_i) \in G_{\ell-1}$.
- $\psi_{\ell-1}$ is the restriction of $\varphi_{\ell-1}$ to the submodule $G_{\ell-1} \subset F_{\ell-1}$.

We claim that

$$\cdots \to F_{\ell+2} \overset{\varphi_{\ell+1}}{\to} F_{\ell+1} \overset{\psi_{\ell+1}}{\to} G_\ell \overset{\psi_\ell}{\to} G_{\ell-1} \overset{\psi_{\ell-1}}{\to} F_{\ell-2} \overset{\varphi_{\ell-2}}{\to} F_{\ell-3} \to \cdots$$

is still a resolution of M. To prove this, we need to check exactness at $F_{\ell+1}$, G_ℓ, $G_{\ell-1}$ and $F_{\ell-2}$. (If we set $M = F_{-1}$ and $F_k = 0$ for $k < -1$, then the above sequence makes sense for all $\ell \geq 1$.)

We begin with $F_{\ell-2}$. Here, note that applying $\varphi_{\ell-1}$ to (3.16) gives

$$0 = c_1 \varphi_{\ell-1}(u_1) + c_2 \varphi_{\ell-1}(u_2) + \cdots + c_2 \varphi_{\ell-1}(u_m).$$

Since c_1 is a nonzero constant, $\varphi_{\ell-1}(u_1)$ is an R-linear combination of $\varphi_{\ell-1}(u_i)$ for $i = 2, \ldots, m$, and then $\operatorname{im}(\varphi_{\ell-1}) = \operatorname{im}(\psi_{\ell-1})$ follows from the definition of $\psi_{\ell-1}$. The desired exactness $\operatorname{im}(\psi_{\ell-1}) = \ker(\varphi_{\ell-2})$ is now an easy consequence of the exactness of the original resolution.

Next consider $G_{\ell-1}$. First note that for $i \geq 2$, $\psi_{\ell-1} \circ \psi_\ell(e_i) = \psi_{\ell-1} \circ \varphi_\ell(e_i - \frac{d_i}{c_1} e_1) = 0$ since $\psi_{\ell-1}$ is just the restriction of $\varphi_{\ell-1}$. This shows that $\operatorname{im}(\psi_\ell) \subset \ker(\psi_{\ell-1})$. To prove the opposite inclusion, suppose that $\psi_{\ell-1}(v) = 0$ for some $v \in G_{\ell-1}$. Since $\psi_{\ell-1}$ is the restriction of $\varphi_{\ell-1}$, exactness of the original resolution implies that $v = \varphi_\ell(a_1 e_1 + \cdots + a_m e_m)$. However, since u_1 does not appear in $v \in G_{\ell-1}$ and $\varphi_\ell(e_i) = d_i u_1 + \cdots$, one easily obtains

$$(3.17) \qquad a_1 c_1 + a_2 d_2 + \cdots + a_m d_m = 0$$

by looking at the coefficients of u_1. Then

$$\begin{aligned}
\psi_\ell(a_2 e_2 + \cdots + a_m e_m) &= a_2 \psi_\ell(e_2) + \cdots + a_m \psi_\ell(e_m) \\
&= a_2 \varphi_\ell(e_2 - \tfrac{d_2}{c_1} e_1) + \cdots + a_m \varphi_\ell(e_m - \tfrac{d_m}{c_1} e_1) \\
&= \varphi_\ell(a_1 e_1 + \cdots + a_m e_m) = v,
\end{aligned}$$

where the last equality follows by (3.17). This completes the proof of exactness at $G_{\ell-1}$.

The remaining proofs of exactness are straightforward and will be covered in Exercise 13 at the end of the section.

Since the theorem we're trying to prove is concerned with $\ker(\varphi_{n-1})$, we need to understand how the kernels of the various maps change under the above simplification process. If $e_1 \in F_\ell$ has degree d, then we claim that:

$$\begin{aligned}
\ker(\varphi_{\ell-1}) &\cong R(-d) \oplus \ker(\psi_{\ell-1}) \\
(3.18) \qquad \ker(\varphi_\ell) &\cong \ker(\psi_\ell) \\
\ker(\varphi_{\ell+1}) &= \ker(\psi_{\ell+1})
\end{aligned}$$

We will prove the first and leave the others for the reader (see Exercise 13). Since $\psi_{\ell-1}$ is the restriction of $\varphi_{\ell-1}$, we certainly have $\ker(\psi_{\ell-1}) \subset \ker(\varphi_{\ell-1})$. Also, $\varphi_\ell(e_1) \in \ker(\varphi_{\ell-1})$ gives the submodule $R\varphi_\ell(e_1) \subset \ker(\varphi_{\ell-1})$, and the map sending $\varphi_\ell(e_1) \mapsto 1$ induces an isomorphism $R\varphi_\ell(e_1) \cong R(-d)$. To prove that we have a direct sum, note that (3.16) implies $R\varphi_\ell(e_1) \cap G_{\ell-1} = \{0\}$ since $G_{\ell-1}$ is generated by $u_2, \ldots, u_m$ and c_1 is a nonzero constant. From this, we conclude $R\varphi_\ell(e_1) \cap \ker(\psi_{\ell-1}) = \{0\}$, which implies

$$R\varphi_\ell(e_1) + \ker(\psi_{\ell-1}) = R\varphi_\ell(e_1) \oplus \ker(\psi_{\ell-1}).$$

To show that this equals all of $\ker(\varphi_{\ell-1})$, let $w \in \ker(\varphi_{\ell-1})$ be arbitrary. If $w = a_1 u_1 + \cdots + a_t u_t$, then set $\widetilde{w} = w - \frac{a_1}{c_1} \varphi_\ell(e_1)$. By (3.16), we have $\widetilde{w} \in G_{\ell-1}$, and then $\widetilde{w} \in \ker(\psi_{\ell-1})$ follows easily. Thus $w = \frac{a_1}{c_1} \varphi_\ell(e_1) + \widetilde{w} \in R\varphi_\ell(e_1) \oplus \ker(\psi_{\ell-1})$, which gives the desired direct sum decomposition.

Hence, we have proved that whenever we have a φ_ℓ with a nonzero matrix entry of degree zero, we create a resolution with smaller matrices whose kernels satisfy (3.18). It follows that if the theorem holds for the smaller resolution, then it automatically holds for the original resolution.

Now the theorem is easy to prove. By repeatedly applying the above process whenever we find a nonzero matrix entry of degree zero in some ψ_ℓ, we can reduce to a minimal resolution. But minimal resolutions are isomorphic by Theorem (3.13), and hence, by Exercise 5, the minimal resolution we get has length $\leq n$. Then Proposition (1.12) shows that $\ker(\varphi_{n-1})$ is free for the minimal resolution, which, as observed above, implies that $\ker(\varphi_{n-1})$ is free for the original resolution as well.

The final assertion of the theorem, that

$$0 \to \ker(\varphi_{n-1}) \to F_{n-1} \to \cdots \to F_0 \to M \to 0$$

is a free resolution, now follows immediately from Proposition (1.12). □

The simplification process used in the proof of Theorem (3.15) can be used to show that, in a suitable sense, every graded resolution of M is the direct sum of a minimal resolution and a trivial resolution. This gives a structure theorem which describes *all* graded resolutions of a given finitely generated module over $k[x_1, \ldots, x_n]$. Details can be found in Theorem 20.2 of [Eis].

Exercise 6. Show that the simplification process from the proof of Theorem (3.15) transforms the homogenization of (1.11) into the homogenization of (1.10) (see Exercise 4).

There is also a version of the theorem just proved which applies to partial resolutions.

(3.19) Corollary. *If*

$$F_{n-1} \overset{\varphi_{n-1}}{\to} F_{n-2} \to \cdots \to F_0 \to M \to 0$$

is a partial graded resolution over $k[x_1, \ldots, x_n]$, *then* $\ker(\varphi_{n-1})$ *is free, and*

$$0 \to \ker(\varphi_{n-1}) \to F_{n-1} \to \cdots \to F_0 \to M \to 0$$

is a graded resolution of M.

PROOF. Since any partial resolution can be extended to a resolution, this follows immediately from Theorem (3.15).   □

One way to think about Corollary (3.19) is that over $k[x_1, \ldots, x_n]$, the process of taking repeated syzygies leads to a free syzygy module after at most $n - 1$ steps. This is essentially how Hilbert stated the Syzygy Theorem in his classic paper [Hil], and sometimes Theorem (3.15) or Corollary (3.19) are called the Syzygy Theorem. Modern treatments, however, focus on the *existence* of a resolution of length $\leq n$, since Hilbert's version follows from existence (our Theorem (3.8)) together with the properties of minimal resolutions.

As an application of these results, let's study the syzygies of a homogeneous ideal in two variables.

(3.20) Proposition. *Suppose that* $f_1, \ldots, f_s \in k[x, y]$ *are homogeneous polynomials. Then the syzygy module* $\mathrm{Syz}\,(f_1, \ldots, f_s)$ *is a twisted free module over* $k[x, y]$.

PROOF. Let $I = \langle f_1, \ldots, f_s \rangle \subset k[x, y]$. Then we get an exact sequence

$$0 \to I \to R \to R/I \to 0$$

by Proposition (1.2). Also, the definition of the syzygy module gives an exact sequence

$$0 \to \mathrm{Syz}\,(f_1, \ldots, f_s) \to R(-d_1) \oplus \cdots \oplus R(-d_s) \to I \to 0$$

where $d_i = \deg f_i$. Splicing these two sequences together as in Exercise 7 of §1, we get the exact sequence

$$0 \to \mathrm{Syz}\,(f_1, \ldots, f_s) \to R(-d_1) \oplus \cdots \oplus R(-d_s) \xrightarrow{\varphi_1} R \to R/I \to 0.$$

Since $n = 2$, Corollary (3.19) implies that $\ker(\varphi_1) = \mathrm{Syz}\,(f_1, \ldots, f_s)$ is free, and the proposition is proved.   □

In §4, we will use the Hilbert polynomial to describe the degrees of the generators of $\mathrm{Syz}\,(f_1, \ldots, f_s)$ in the special case when all of the f_i have the same degree.

ADDITIONAL EXERCISES FOR §3

Exercise 7. Assume that $f_1, \ldots, f_s \in k[x, y]$ are homogeneous and not all zero. We know that $\mathrm{Syz}\,(f_1, \ldots, f_s)$ is free by Proposition (3.20), so that if we ignore gradings, $\mathrm{Syz}\,(f_1, \ldots, f_s) \cong R^m$ for some m. This gives an exact sequence

$$0 \to R^m \to R^s \to I \to 0.$$

Prove that $m = s - 1$ and conclude that we are in the situation of the Hilbert-Burch Theorem from §2. Hint: As in Exercise 11 of §2, let $K = k(x_1, \ldots, x_n)$ be the field of rational functions coming from $R = k[x_1, \ldots, x_n]$. Explain why the above sequence gives a sequence

$$0 \to K^m \to K^s \to K \to 0$$

and show that this new sequence is also exact. The result will then follow from the dimension theorem of linear algebra (see Exercise 8 of §1). The ideas used in Exercise 11 of §2 may be useful.

Exercise 8. Prove Proposition (3.3). Hint: Show a $\Rightarrow$ c $\Rightarrow$ b $\Rightarrow$ a.

Exercise 9. Prove Proposition (3.4).

Exercise 10. Complete the proof of Proposition (3.10).

Exercise 11. Suppose that M is a module over $k[x_1, \ldots, x_n]$ generated by $f_1, \ldots, f_m$. As in the proof of Theorem (3.13), let $\langle x_1, \ldots, x_n \rangle M$ be the submodule generated by $x_i f_j$ for all i, j. Also assume that $\psi : M \to M$ is a graded homomorphism of degree zero such that $v - \psi(v) \in \langle x_1, \ldots, x_n \rangle M$ for all $v \in M$. Then prove that ψ is an isomorphism. Hint: By part b of Exercise 3, it suffices to show that $\psi : M_t \to M_t$ is onto. Prove this by induction on t, using part a of Exercise 3 to start the induction.

Exercise 12. Suppose that we have a diagram of R-modules and homomorphisms

$$\begin{array}{ccc} A & \xrightarrow{\varphi} & B \\ \alpha \downarrow & & \downarrow \beta \\ C & \xrightarrow{\psi} & D \end{array}$$

which commutes in the sense of Definition (3.11). If in addition φ, ψ are onto and α, β are isomorphisms, then prove that α restricted to $\ker(\varphi)$ induces an isomorphism $\bar{\alpha} : \ker(\varphi) \to \ker(\psi)$.

Exercise 13. This exercise is concerned with the proof of Theorem (3.15). We will use the same notation as in that proof, including the sequence of mappings

$$\cdots \to F_{\ell+1} \xrightarrow{\psi_{\ell+1}} G_\ell \xrightarrow{\psi_\ell} G_{\ell-1} \xrightarrow{\psi_{\ell-1}} F_{\ell-2} \to \cdots.$$

a. Prove that $\varphi_\ell(\sum_{i=1}^m a_i e_i) = 0$ if and only if $\psi_\ell(\sum_{i=2}^m a_i e_i) = 0$ and $a_1 c_1 + \sum_{i=2}^m a_i d_i = 0$.
b. Use part a to prove that the above sequence is exact at G_ℓ.
c. Prove that the above sequence is exact at $F_{\ell+1}$. Hint: Do you see why it suffices to show that $\ker(\varphi_{\ell+1}) = \ker(\psi_{\ell+1})$?

d. Prove the second line of (3.18), i.e., that $\ker(\varphi_\ell) \cong \ker(\psi_\ell)$. Hint: Use part a.

e. Prove the third line of (3.18), i.e., that $\ker(\varphi_{\ell+1}) = \ker(\psi_{\ell+1})$. Hint: You did this in part c!

Exercise 14. In the proof of Theorem (3.15), we constructed a certain homomorphism $\psi : G_\ell \to G_{\ell-1}$. Suppose that A_ℓ is the matrix of $\varphi_\ell :$ $F_\ell \to F_{\ell-1}$ with respect to the bases $e_1, \ldots, e_m$ of F_ℓ and $u_1, \ldots, u_t$ of $F_{\ell-1}$. Write A_ℓ in the form

$$A_\ell = \begin{pmatrix} A_{00} & A_{01} \\ A_{10} & A_{11} \end{pmatrix}$$

where $A_{00} = c_1$ and $A_{01} = (c_2, \ldots, c_t)$ as in (3.16), and $A_{10} = (d_2, \ldots, d_m)^T$, where the d_i are from the definition of ψ_ℓ. If we let B_ℓ be the matrix of ψ_ℓ with respect to the bases $e_2, \ldots, e_m$ of G_ℓ and $u_2, \ldots, u_t$ of $G_{\ell-1}$, then prove that

$$B_\ell = A_{00} - A_{01} A_{00}^{-1} A_{10}.$$

What's remarkable is that this formula is identical to equation (6.5) in Chapter 3. As happens often in mathematics, the same idea can appear in very different contexts.

Exercise 15. In $k[x_0, \ldots, x_n]$, $n \geq 2$, consider the homogeneous ideal I_n defined by the determinants of the $\binom{n}{2}$ 2×2 submatrices of the $2 \times n$ matrix

$$M = \begin{pmatrix} x_0 & x_1 & \cdots & x_{n-1} \\ x_1 & x_2 & \cdots & x_n \end{pmatrix}.$$

For instance, $I_2 = \langle x_0 x_2 - x_1^2 \rangle$ is the ideal of a conic section in $\mathbb{P}^2$. We have already seen I_3 in different notation (where?).

a. Show that I_n is the ideal of the *rational normal curve* of degree n in $\mathbb{P}^n$—the image of the mapping given in homogeneous coordinates by

$$\varphi : \mathbb{P}^1 \to \mathbb{P}^n$$
$$(s, t) \mapsto (s^n, s^{n-1}t, \ldots, st^{n-1}, t^n).$$

b. Do explicit calculations to find the graded resolutions of the ideals I_4, I_5.

c. Show that the first syzygy module of the generators for I_n is generated by the three-term syzygies obtained by appending a copy of the first (resp. second) row of M to M, to make a $3 \times n$ matrix M' (resp. M''), then expanding the determinants of all 3×3 submatrices of M' (resp. M'') along the new row.

d. Conjecture the general form of a graded resolution of I_n. (Proving this conjecture requires advanced techniques like the *Eagon-Northcott complex*. This and other interesting topics are discussed in Appendix A2.6 of [Eis].)

§4 Hilbert Polynomials and Geometric Applications

In this section, we will study Hilbert functions and Hilbert polynomials. These are computed using the graded resolutions introduced in §3 and contain some interesting geometric information. We will then give applications to the ideal of three points in $\mathbb{P}^2$, parametric equations in the plane, and invariants of finite group actions.

Hilbert Functions and Hilbert Polynomials

We begin by defining the Hilbert function of a graded module. Because we will be dealing with projective space $\mathbb{P}^n$, it is convenient to work over the polynomial ring $R = k[x_0, \ldots, x_n]$ in $n + 1$ variables.

If M is a finitely generated graded R-module, recall from Exercise 3 of §3 that for each t, the degree t homogeneous part M_t is a finite dimensional vector space over k. This leads naturally to the definition of the Hilbert function.

(4.1) Definition. If M is a finitely generated graded module over $R = k[x_0, \ldots, x_n]$, then the *Hilbert function* $H_M(t)$ is defined by

$$H_M(t) = \dim_k M_t,$$

where as usual, $\dim_k$ means dimension as a vector space over k.

The most basic example of a graded module is $R = k[x_0, \ldots, x_n]$ itself. Since R_t is the vector space of homogeneous polynomials of degree t in $n + 1$ variables, Exercise 19 of Chapter 3, §4 implies that for $t \geq 0$, we have

$$H_R(t) = \dim_k R_t = \binom{t + n}{n},$$

If we adopt the convention that $\binom{a}{b} = 0$ if $a < b$, then the above formula holds for *all* t. Similarly, the reader should check that the Hilbert function of the twisted module $R(d)$ is given by

$$(4.2) \qquad H_{R(d)}(t) = \binom{t + d + n}{n}, \quad t \in \mathbb{Z}.$$

An important observation is that for $t \geq 0$ and n fixed, the binomial coefficient $\binom{t+n}{n}$ is a polynomial of degree n in t. This is because

$$(4.3) \qquad \binom{t + n}{n} = \frac{(t + n)!}{t!n!} = \frac{(t + n)(t + n - 1) \cdots (t + 1)}{n!}.$$

It follows that $H_R(t)$ is given by a polynomial for t sufficiently large ($t \geq 0$ in this case). This will be important below when we define the Hilbert polynomial.

Here are some exercises which give some simple properties of Hilbert functions.

Exercise 1. If M is a finitely generated graded R-module and $M(d)$ is the twist defined in Proposition (3.4), then show that

$$H_{M(d)}(t) = H_M(t + d)$$

for all t. Note how this generalizes (4.2).

Exercise 2. Suppose that M, N and P are finitely generated graded R-modules.
a. The direct sum $M \oplus N$ was discussed in Exercise 1 of §3. Prove that $H_{M \oplus N} = H_M + H_N$.
b. More generally, if we have an exact sequence

$$0 \to M \xrightarrow{\alpha} P \xrightarrow{\beta} N \to 0$$

where α and β are graded homomorphisms of degree zero, then show that $H_P = H_M + H_N$.
c. Explain how part b generalizes part a. Hint: What exact sequence do we get from $M \oplus N$?

It follows from these exercises that we can compute the Hilbert function of any twisted free module. However, for more complicated modules, computing the Hilbert function can be rather nontrivial. There are several ways to study this problem. For example, if $I \subset R = k[x_0, \ldots, x_n]$ is a homogeneous ideal, then the quotient ring R/I is a graded R-module, and in Chapter 9, §3 of [CLO], it is shown than if $\langle \mathrm{LT}(I) \rangle$ is the ideal of initial terms for a monomial order on R, then the Hilbert functions $H_{R/I}$ and $H_{R/\langle \mathrm{LT}(I) \rangle}$ are equal. Using the techniques of Chapter 9, §2 of [CLO], it is relatively easy to compute the Hilbert function of a monomial ideal. Thus, once we compute a Gröbner basis of I, we can find the Hilbert function of R/I. (Note: The Hilbert function $H_{R/I}$ is denoted HF_I in [CLO].)

A second way to compute Hilbert functions is by means of graded resolutions. Here is the basic result.

(4.4) Theorem. Let $R = k[x_0, \ldots, x_n]$ and let M be a graded R-module. Then, for any graded resolution of M

$$0 \to F_k \to F_{k-1} \to \cdots \to F_0 \to M \to 0,$$

we have

$$H_M(t) = \dim_k M_t = \sum_{j=0}^{k} (-1)^j \dim_k (F_j)_t = \sum_{j=0}^{k} (-1)^j H_{F_j}(t).$$

PROOF. In a graded resolution, all the homomorphisms are homogeneous of degree zero, hence for each t, restricting all the homomorphisms to the degree t homogeneous parts of the graded modules, we also have an exact sequence of finite dimensional k-vector spaces

$$0 \to (F_k)_t \to (F_{k-1})_t \to \cdots \to (F_0)_t \to M_t \to 0.$$

The alternating sum of the dimensions in such an exact sequence is 0, by Exercise 8 of §1. Hence

$$\dim_k M_t = \sum_{j=0}^{k} (-1)^j \dim_k (F_j)_t,$$

and the theorem follows by the definition of Hilbert function.   □

Since we know the Hilbert function of any twisted free module (by (4.2) and Exercise 2), it follows that the Hilbert function of a graded module M can be calculated easily from a graded resolution. For example, let's compute the Hilbert function of the homogeneous ideal I of the twisted cubic in $\mathbb{P}^3$, namely

$$(4.5) \qquad I = \langle xz - y^2, xw - yz, yw - z^2 \rangle \subset R = k[x, y, z, w].$$

In Exercise 2 of §2 of this chapter, we found that I has a graded resolution of the form

$$0 \to R(-3)^2 \to R(-2)^3 \to I \to 0.$$

As in the proof of Theorem (4.4), this resolution implies

$$\dim_k I_t = \dim_k R(-2)_t^3 - \dim_k R(-3)_t^2$$

for all t. Applying Exercise 2 and (4.2), this can be rewritten as

$$H_I(t) = 3\binom{t-2+3}{3} - 2\binom{t-3+3}{3}$$

$$= 3\binom{t+1}{3} - 2\binom{t}{3}.$$

Using the exact sequence $0 \to I \to R \to R/I \to 0$, Exercise 2 implies that

$$H_{R/I}(t) = H_R(t) - H_I(t) = \binom{t+3}{3} - 3\binom{t+1}{3} + 2\binom{t}{3}$$

for all t. For $t = 0, 1, 2$, one (or both) of the binomial coefficients from H_I is zero. However, computing $H_{R/I}(t)$ separately for $t \le 2$ and doing some algebra, one can show that

(4.6) $$H_{R/I}(t) = 3t + 1$$

for all $t \geq 0$.

In this example, the Hilbert function is a polynomial once t is sufficiently large ($t \geq 0$ in this case). This is a special case of the following general result.

(4.7) Proposition. *If M is a finitely generated R-module, then there is a unique polynomial HP_M such that*

$$H_M(t) = HP_M(t)$$

for all t sufficiently large.

PROOF. The key point is that for a twisted free module of the form

$$F = R(-d_1) \oplus \cdots \oplus R(-d_m),$$

Exercise 2 and (4.2) imply that

$$H_F(t) = \sum_{i=1}^{m} \binom{t - d_i + n}{n}.$$

Furthermore, (4.3) shows that this is a polynomial in t provided $t \geq \max(d_1, \ldots, d_m)$.

Now suppose that M is a finitely generated R-module. We can find a finite graded resolution

$$0 \to F_\ell \to \cdots \to F_0 \to M \to 0,$$

and Theorem (4.4) tells us that

$$H_M(t) = \sum_{j=0}^{\ell} (-1)^j H_{F_j}(t).$$

The above computation implies that $H_{F_j}(t)$ is a polynomial in t for t sufficiently large, so that the same is true for $H_M(t)$. □

The polynomial HP_M given in Proposition (4.7) is called the *Hilbert polynomial* of M. For example, if I is the ideal given by (4.5), then (4.6) implies that

(4.8) $$HP_{R/I}(t) = 3t + 1$$

in this case.

The Hilbert polynomial contains some interesting geometric information. For example, a homogeneous ideal $I \subset k[x_0, \ldots, x_n]$ determines the projective variety $V = \mathbf{V}(I) \subset \mathbb{P}^n$, and the Hilbert polynomial tells us the following facts about V:

- The degree of the Hilbert polynomial $HP_{R/I}$ is the *dimension* of the variety V. For example, in Chapter 9 of [CLO], this is the *definition* of the dimension of a projective variety.
- If the Hilbert polynomial $HP_{R/I}$ has degree $d = \dim V$, then one can show that its leading term is $(D/d!)\, t^d$ for some positive integer D. The integer D is defined to be the *degree* of the variety V. One can also prove that D equals the number of points where V meets a generic $(n - d)$-dimensional linear subspace of $\mathbb{P}^n$.

For example, the Hilbert polynomial $HP_{R/I}(t) = 3t + 1$ from (4.8) shows that the twisted cubic has dimension 1 and degree 3. In the exercises at the end of the section, you will compute additional examples of Hilbert functions and Hilbert polynomials.

The Ideal of Three Points

Given a homogeneous ideal $I \subset k[x_0, \ldots, x_n]$, we get the projective variety $V = \mathbf{V}(I)$. We've seen that a graded resolution enables us to compute the Hilbert polynomial, which in turn determines geometric invariants of V such as the dimension and degree. However, the actual terms appearing in a graded resolution of the ideal I encode additional geometric information about the variety V. We will illustrate this by considering the form of the resolution of the ideal of a collection of points in $\mathbb{P}^2$. For example, consider varieties consisting of three distinct points, namely $V = \{p_1, p_2, p_3\} \subset \mathbb{P}^2$. There are two cases here, depending on whether the p_i are collinear or not. We begin with a specific example.

Exercise 3. Suppose that $V = \{p_1, p_2, p_3\} = \{(0, 0, 1), (1, 0, 1), (0, 1, 1)\}$.
a. Show that $I = \mathbf{I}(V)$ is the ideal $\langle x^2 - xz, xy, y^2 - yz \rangle \subset R = k[x, y, z]$.
b. Show that we have a graded resolution

$$0 \to R(-3)^2 \to R(-2)^3 \to I \to 0$$

and explain how this relates to (1.10).
c. Compute that the Hilbert function of R/I is

$$H_{R/I}(t) = \binom{t + 2}{2} - 3\binom{t}{2} + 2\binom{t - 1}{2}$$

$$= \begin{cases} 1 & \text{if } t = 0, \\ 3 & \text{if } t \geq 1. \end{cases}$$

The Hilbert polynomial in Exercise 3 is the constant polynomial 3, so the dimension is 0 and the degree is 3, as expected. There is also some nice intuition lying behind the graded resolution

(4.9) $$0 \to R(-3)^2 \to R(-2)^3 \to I \to 0$$

found in part b of the exercise. First, note that $I_0 = \{0\}$ since 0 is the only constant vanishing on the points, and $I_1 = \{0\}$ since the points of $V = \{(0,0,1), (1,0,1), (0,1,1)\}$ are noncollinear. On the other hand, there are quadratics which vanish on V. One way to see this is to let ℓ_{ij} be the equation of the line vanishing on points p_i and p_j. Then $f_1 = \ell_{12}\ell_{13}, f_2 = \ell_{12}\ell_{23}, f_3 = \ell_{13}\ell_{23}$ are three quadratics vanishing precisely on V. Hence it makes sense that I is generated by three quadratics, which is what the $R(-2)^3$ in (4.9) says. Also, notice that f_1, f_2, f_3 have obvious syzygies of degree 1, for example, $\ell_{23}f_1 - \ell_{13}f_2 = 0$. It is less obvious that two of these syzygies are free generators of the syzygy module, but this is what the $R(-3)^2$ in (4.9) means.

From a more sophisticated point of view, the resolution (4.9) is fairly obvious. This is because of the converse of the Hilbert-Burch Theorem discussed at the end of §2, which applies here since $V \subset \mathbb{P}^2$ is a finite set of points and hence is Cohen-Macaulay of dimension $2 - 2 = 0$.

The example presented in Exercise 3 is more general than one might suspect. This is because for three noncollinear points p_1, p_2, p_3, there is a linear change of coordinates on $\mathbb{P}^2$ taking p_1, p_2, p_3 to $(0,0,1), (1,0,1), (0,1,1)$. Using this, we see that if I is the ideal of *any* set of three noncollinear points, then I has a free resolution of the form (4.9), so that the Hilbert function of I is given by part c of Exercise 3.

The next two exercises will study what happens when the three points *are* collinear.

Exercise 4. Suppose that $V = \{(0,1,0), (0,0,1), (0,\lambda,1)\}$, where $\lambda \neq 0$. These points lie on the line $x = 0$, so that V is a collinear triple of points.
a. Show that $I = \mathbf{I}(V)$ has a graded resolution of the form

$$0 \to R(-4) \to R(-3) \oplus R(-1) \to I \to 0.$$

Hint: Show that $I = \langle x, yz(y - \lambda z) \rangle$.
b. Show that the Hilbert function of R/I is

$$H_{R/I}(t) = \begin{cases} 1 & \text{if } t = 0, \\ 2 & \text{if } t = 1, \\ 3 & \text{if } t \geq 2. \end{cases}$$

Exercise 5. Suppose now that $V = \{p_1, p_2, p_3\}$ is *any* triple of collinear points in $\mathbb{P}^2$. Show that $I = \mathbf{I}(V)$ has a graded resolution of the form

(4.10) $$0 \to R(-4) \to R(-3) \oplus R(-1) \to I \to 0,$$

and conclude that the Hilbert function of R/I is as in part b of Exercise 4. Hint: Use a linear change of coordinates in $\mathbb{P}^2$.

The intuition behind (4.10) is that in the collinear case, V is the intersection of a line and a cubic, and the only syzygy between these is the obvious one. In geometric terms, we say that V is a *complete intersection* in this case since its dimension $(= 0)$ is the dimension of the ambient space $(= 2)$ minus the number of defining equations $(= 2)$. Note that a noncollinear triple isn't a complete intersection since there are three defining equations.

This sequence of exercises shows that for triples of points in $\mathbb{P}^2$, their corresponding ideals I all give the same Hilbert polynomial $HP_{R/I} = 3$. But depending on whether the points are collinear or not, we get different resolutions (4.10) and (4.9) and different Hilbert functions, as in part c of Exercise 3 and part b of Exercise 4. This is quite typical of what happens.

Here is a similar but more challenging example.

Exercise 6. Now consider varieties $V = \{p_1, p_2, p_3, p_4\}$ in $\mathbb{P}^2$, and write $I = \mathbf{I}(V) \subset R = k[x, y, z]$ as above.

a. First assume the points of V are in general position in the sense that no three are collinear. Show that I_2 is 2-dimensional over k, and that I is generated by any two linearly independent elements of I_2. Deduce that a graded resolution of I has the form

$$0 \to R(-4) \to R(-2)^2 \to I \to 0,$$

and use this to compute $H_{R/I}(t)$ for all t. Do you see how the $R(-2)^2$ is consistent with Bézout's Theorem?

b. Now assume that three of the points of V lie on a line $L \subset \mathbb{P}^2$ but the fourth does not. Show that every element of I_2 is reducible, containing as a factor a linear polynomial vanishing on L. Show that I_2 does not generate I in this case, and deduce that a graded resolution of I has the form

$$0 \to R(-3) \oplus R(-4) \to R(-2)^2 \oplus R(-3) \to I \to 0.$$

Use this to compute $H_{R/I}(t)$ for all t.

c. Finally, consider the case where all four of the points are collinear. Show that in this case, the graded resolution has the form

$$0 \to R(-5) \to R(-1) \oplus R(-4) \to I \to 0,$$

and compute the Hilbert function of R/I for all t.

d. In which cases is V a complete intersection?

Understanding the geometric significance of the shape of the graded resolution of $I = \mathbf{I}(V)$ in more involved examples is an area of active research in contemporary algebraic geometry. A conjecture of Mark Green concerning the graded resolutions of the ideals of canonical curves has stimulated many of the developments here. See [Schre2] and [EH] for some earlier work on Green's conjecture. Recent articles of Montserrat Teixidor ([Tei]) and Claire Voisin ([Voi]) have proved Green's conjecture for a large class of curves. [EH] contains articles on other topics concerning resolutions. Section 15.12 of [Eis] has some interesting projects dealing with resolutions, and some of the exercises in Section 15.11 are also relevant.

Parametric Plane Curves

Here, we will begin with a curve in k^2 parametrized by rational functions

$$(4.11) \qquad x = \frac{a(t)}{c(t)}, \ y = \frac{b(t)}{c(t)},$$

where $a, b, c \in k[t]$ are polynomials such that $c \neq 0$ and $\mathrm{GCD}(a, b, c) = 1$. We also set $n = \max(\deg a, \deg b, \deg c)$. Parametrizations of this form play an important role in computer-aided geometric design, and a question of particular interest is the *implicitization problem*, which asks how the equation $f(x, y) = 0$ of the underlying curve is obtained from the parametrization (4.11). An introduction to implicitization can be found in Chapter 3 of [CLO].

A basic object in this theory is the ideal

$$(4.12) \qquad I = \langle c(t)x - a(t), c(t)y - b(t) \rangle \subset k[x, y, t].$$

This ideal has the following interpretation. Let $W \subset k$ be the roots of $c(t)$, i.e., the solutions of $c(t) = 0$. Then we can regard (4.11) as the function $F : k - W \to k^2$ defined by

$$F(t) = \left(\frac{a(t)}{c(t)}, \frac{b(t)}{c(t)} \right).$$

In Exercise 14 at the end of the section, you will show that the graph of F, regarded as a subset of k^3, is precisely the variety $\mathbf{V}(I)$. From here, one can prove that the intersection $I_1 = I \cap k[x, y]$ is an ideal in $k[x, y]$ such that $\mathbf{V}(I_1) \subset k^2$ is the smallest variety containing the image of the parametrization (4.11) (see Exercise 14). In the terminology of Chapter 2, $I_1 = I \cap k[x, y]$ is an elimination ideal, which we can compute using a Gröbner basis with respect to a suitable monomial order.

It follows that the ideal I contains a lot of information about the curve parametrized by (4.11). Recently, it was discovered (see [SSQK] and [SC]) that I provides other parametrizations of the curve, different from (4.11). To see how this works, let $I(1)$ denote the subset of I consisting of all elements of I of total degee at most 1 in x and y. Thus

$$(4.13) \qquad I(1) = \{f \in I : f = A(t)x + B(t)y + C(t)\}.$$

An element in $A(t)x + B(t)y + C(t) \in I(1)$ is called a *moving line* since for t fixed, the equation $A(t)x + B(t)y + C(t) = 0$ describes a line in the plane, and as t moves, so does the line.

Exercise 7. Given a moving line $A(t)x + B(t)y + C(t) \in I(1)$, suppose that $t \in k$ satisfies $c(t) \neq 0$. Then show that the point given by (4.11) lies on the line $A(t)x + B(t)y + C(t) = 0$. Hint: Use $I(1) \subset I$.

Now suppose that we have moving lines $f, g \in I(1)$. Then, for a fixed t, we get a pair of lines, which typically intersect in a point. By Exercise 7, each of these lines contains $(a(t)/c(t), b(t)/c(t))$, so this must be the point of intersection. Hence, as we vary t, the intersection of the moving lines will trace out our curve.

Notice that our original parametrization (4.11) is given by moving lines, since we have the vertical line $x = a(t)/c(t)$ and the horizontal line $y = b(t)/c(t)$. However, by allowing more general moving lines, one can get polynomials of smaller degree in t. The following exercise gives an example of how this can happen.

Exercise 8. Consider the parametrization

$$x = \frac{2t^2 + 4t + 5}{t^2 + 2t + 3}, \quad y = \frac{3t^2 + t + 4}{t^2 + 2t + 3}.$$

a. Prove that $p = (5t + 5)x - y - (10t + 7)$ and $q = (5t - 5)x - (t + 2)y + (-7t + 11)$ are moving lines, i.e., $p, q \in I$, where I is as in (4.12).
b. Prove that p and q generate I, i.e., $I = \langle p, q \rangle$.

In Exercise 8, the original parametrization had maximum degree 2 in t, while the moving lines p and q have maximum degree 1. This is typical of what happens, for we will show below that in general, if n is the maximum degree of a, b, c, then there are moving lines $p, q \in I$ such that p has maximum degree $\mu \leq \lfloor n/2 \rfloor$ in t and q has maximum degree $n - \mu$. Furthermore, p and q are actually a basis of the ideal I. In the terminology of [CSC], this is the *moving line basis* or *μ-basis* of the ideal.

Our goal here is to prove this result—the existence of a μ-basis—and to explain what this has to do with graded resolutions and Hilbert functions. We begin by studying the subset $I(1) \subset I$ defined in (4.13). It is closed under addition, and more importantly, $I(1)$ is closed under multiplication by elements of $k[t]$ (be sure you understand why). Hence $I(1)$ has a natural structure as a $k[t]$-module. In fact, $I(1)$ is a syzygy module, which we will now show.

(4.14) Lemma. *Let $a, b, c \in k[t]$ satisfy $c \neq 0$ and $\mathrm{GCD}(a, b, c) = 1$, and set $I = \langle cx - a, cy - b \rangle$. Then, for $A, B, C \in k[t]$,*

$$A(t)x + B(t)y + C(t) \in I \iff A(t)a(t) + B(t)b(t) + C(t)c(t) = 0.$$

Thus the map $A(t)x + B(t)y + C(t) \mapsto (A, B, C)$ defines an isomorphism of $k[t]$-modules $I(1) \cong \mathrm{Syz}\,(a, b, c)$.

PROOF. To prove $\Rightarrow$, consider the ring homomorphism $k[x, y, t] \rightarrow k(t)$ which sends x, y, t to $\frac{a(t)}{c(t)}, \frac{b(t)}{c(t)}, t$. Since the generators of I map to zero, so does $A(t)x + B(t)y + C(t) \in I$. Thus $A(t)\frac{a(t)}{c(t)} + B(t)\frac{b(t)}{c(t)} + C(t) = 0$ in $k(t)$, and multiplying by $c(t)$ gives the desired equation.

For the other implication, let $S = k[t]$ and consider the sequence

$$(4.15) \qquad\qquad S^3 \xrightarrow{\alpha} S^3 \xrightarrow{\beta} S$$

where $\alpha(h_1, h_2, h_3) = (ch_1 + bh_3, ch_2 - ah_3, -ah_1 - bh_2)$ and $\beta(A, B, C) = Aa + Bb + Cc$. One easily checks that $\beta \circ \alpha = 0$, so that $\text{im}(\alpha) \subset \ker(\beta)$. It is less obvious that (4.15) is exact at the middle term, i.e., $\text{im}(\alpha) = \ker(\beta)$. This will be proved in Exercise 15 below. The sequence (4.15) is the *Koszul complex* determined by a, b, c (see Exercise 10 of §2 for another example of a Koszul complex). A Koszul complex is not always exact, but Exercise 15 will show that (4.15) is exact in our case because $\text{GCD}(a, b, c) = 1$.

Now suppose that $Aa + Bb + Cc = 0$. We need to show that $Ax + By + C \in I$. This is now easy, since our assumption on A, B, C implies $(A, B, C) \in \ker(\beta)$. By the exactness of (4.15), $(A, B, C) \in \text{im}(\alpha)$, which means we can find $h_1, h_2, h_3 \in k[t]$ such that

$$A = ch_1 + bh_3, \quad B = ch_2 - ah_3, \quad C = -ah_1 - bh_2.$$

Hence

$$Ax + By + C = (ch_1 + bh_3)x + (ch_2 - ah_3)y - ah_1 - bh_2$$
$$= (h_1 + yh_3)(cx - a) + (h_2 - xh_3)(cy - b) \in I,$$

as desired. The final assertion of the lemma now follows immediately. $\sqcup$

(4.16) Definition. Given a parametrization (4.11), we get the ideal $I = \langle cx - a, cy - b \rangle$ and the syzygy module $I(1)$ from (4.13). Then we define μ to the minimal degree in t of a nonzero element in $I(1)$.

The following theorem shows the existence of a μ-basis of the ideal I.

(4.17) Theorem. *Given (4.11) where $c \neq 0$ and $\text{GCD}(a, b, c) = 1$, set $n = \max(\deg a, \deg b, \deg c)$ and $I = \langle cx - a, cy - b \rangle$ as usual. If μ is as in Definition (4.16), then*

$$\mu \leq \lfloor n/2 \rfloor,$$

and we can find $p, q \in I$ such that p has degree μ in t, q has degree $n - \mu$ in t, and $I = \langle p, q \rangle$.

PROOF. We will study the syzygy module $\text{Syz}(a, b, c)$ using the methods of §3. For this purpose, we need to homogenize a, b, c. Let t, u be homogeneous variables and consider the ring $R = k[t, u]$. Then $\tilde{a}(t, u)$ will denote the degree n homogenization of $a(t)$, i.e.,

$$\tilde{a}(t, u) = u^n a\left(\tfrac{t}{u}\right) \in R$$

In this way, we get degree n homogeneous polynomials $\tilde{a}, \tilde{b}, \tilde{c} \in R$, and the reader should check that $\mathrm{GCD}(a, b, c) = 1$ and $n = \max(\deg a, \deg b, \deg c)$ imply that $\tilde{a}, \tilde{b}, \tilde{c}$ have no common zeros in $\mathbb{P}^1$. In other words, the only solution of $\tilde{a} = \tilde{b} = \tilde{c} = 0$ is $t = u = 0$.

Now let $J = \langle \tilde{a}, \tilde{b}, \tilde{c} \rangle \subset R = k[t, u]$. We first compute the Hilbert polynomial HP_J of J. The key point is that since $\tilde{a} = \tilde{b} = \tilde{c} = 0$ have only one solution, no matter what the field is, the Finiteness Theorem from §2 of Chapter 2 implies that the quotient ring $R/J = k[t, u]/J$ is a finite dimensional vector space over k. But J is a homogeneous ideal, which means that R/J is a graded ring. In order for S/J to have finite dimension, we must have $\dim_k (R/J)_s = 0$ for all s sufficiently large (we use s instead of t since t is now one of our variables). It follows that $HP_{R/J}$ is the zero polynomial. Then the exact sequence

$$0 \to J \to R \to R/J \to 0$$

and Exercise 2 imply that

(4.18) $$HP_J(s) = HP_R(s) = \binom{s+1}{1} = s + 1$$

since $R = k[t, u]$. For future reference, note also that by (4.2),

$$HP_{R(-d)}(s) = \binom{s - d + 1}{1} = s - d + 1.$$

Now consider the exact sequence

$$0 \to \mathrm{Syz}\,(\tilde{a}, \tilde{b}, \tilde{c}) \to R(-n)^3 \xrightarrow{\alpha} J \to 0,$$

where $\alpha(A, B, C) = A\tilde{a} + B\tilde{b} + C\tilde{c}$. By Proposition (3.20), the syzygy module $\mathrm{Syz}\,(\tilde{a}, \tilde{b}, \tilde{c})$ is free, which means that we get a graded resolution

(4.19) $$0 \to R(-d_1) \oplus \cdots \oplus R(-d_m) \xrightarrow{\beta} R(-n)^3 \xrightarrow{\alpha} J \to 0$$

for some $d_1, \ldots, d_m$. By Exercise 2, the Hilbert polynomial of the middle term is the sum of the other two Hilbert polynomials. Since we know HP_J from (4.18), we obtain

$$3(s - n + 1) = (s - d_1 + 1) + \cdots + (s - d_m + 1) + (s + 1)$$
$$= (m + 1)s + m + 1 - d_1 - \cdots - d_m.$$

It follows that $m = 2$ and $3n = d_1 + d_2$. Thus (4.19) becomes

(4.20) $$0 \to R(-d_1) \oplus R(-d_2) \xrightarrow{\beta} R(-n)^3 \xrightarrow{\alpha} J \to 0.$$

The matrix L representing β is a 3×2 matrix

$$(4.21) \qquad\qquad L = \begin{pmatrix} p_1 & q_1 \\ p_2 & q_2 \\ p_3 & q_3 \end{pmatrix},$$

and since β has degree zero, the first column of L consists of homogeneous polynomials of degree $\mu_1 = d_1 - n$ and the second column has degree $\mu_2 = d_2 - n$. Then $\mu_1 + \mu_2 = n$ follows from $3n = d_1 + d_2$.

We may assume that $\mu_1 \leq \mu_2$. Since the first column (p_1, p_2, p_3) of (4.21) satisfies $p_1 \tilde{a} + p_2 \tilde{b} + p_3 \tilde{c} = 0$, setting $u = 1$ gives

$$p_1(t, 1)a(t) + p_2(t, 1)b(t) + p_3(t, 1)c(t) = 0.$$

Thus $p = p_1(t, 1)x + p_2(t, 1)y + p_3(t, 1) \in I(1)$ by Lemma (4.14). Similarly, the second column of (4.21) gives $q = q_1(t, 1)x + q_2(t, 1)y + q_3(t, 1) \in I(1)$. We will show that p and q satisfy the conditions of the theorem.

First observe that the columns of L generate $\text{Syz}\,(\tilde{a}, \tilde{b}, \tilde{c})$ by exactness. In Exercise 16, you will show this implies that p and q generate $I(1)$. Since $cx - a$ and $cy - b$ are in $I(1)$, we obtain $I = \langle cx - a, cy - b \rangle \subset \langle p, q \rangle$. The other inclusion is immediate from $p, q \in I(1) \subset I$, and $I = \langle p, q \rangle$ follows.

The next step is to prove $\mu_1 = \mu$. We begin by showing that p has degree μ_1 in t. This follows because $p_1(t, u), p_2(t, u), p_3(t, u)$ are homogeneous of degree μ_1. If the degree of all three were to drop when we set $u = 1$, then each p_i would be divisible by u. However, since p_1, p_2, p_3 give a syzygy on $\tilde{a}, \tilde{b}, \tilde{c}$, so would $p_1/u, p_2/u, p_3/u$. Hence we would have a syzygy of degree $< \mu_1$. But the columns of L generate the syzygy module, so this is impossible since $\mu_1 \leq \mu_2$. Hence p has degree μ_1 in t, and then $\mu \leq \mu_1$ follows from the definition of μ. However, if $\mu < \mu_1$, then we would have $Ax + By + C \in I(1)$ of degree $< \mu_1$. This gives a syzygy of a, b, c, and homogenizing, we would get a syzygy of degree $< \mu_1$ among $\tilde{a}, \tilde{b}, \tilde{c}$. As we saw earlier in the paragraph, this is impossible.

We conclude that p has degree μ in t, and then $\mu_1 + \mu_2 = n$ implies that q has degree $\mu_2 = n - \mu$ in t. Finally, $\mu \leq \lfloor n/2 \rfloor$ follows from $\mu = \mu_1 \leq \mu_2$, and the proof of the theorem is complete. $\square$

As already mentioned, the basis p, q constructed in Theorem (4.17) is called a μ-*basis* of I. One property of the μ-basis is that it can be used to find the implicit equation of the parametrization (4.11). Here is an example of how this works.

Exercise 9. The parametrization studied in Exercise 8 gives the ideal

$$I = \langle (t^2 + 2t + 3)x - (2t^2 + 4t + 5), (t^2 + 2t + 3)y - (3t^2 + t + 4) \rangle.$$

a. Use Gröbner basis methods to find the intersection $I \cap k[x, y]$. This gives the implicit equation of the curve.

b. Show that the resultant of the generators of I with respect to t gives the implicit equation.

c. Verify that the polynomials $p = (5t + 5)x - y - (10t + 7)$ and $q = (5t - 5)x - (t + 2)y + (-7t + 11)$ are a μ-basis for I. Thus $\mu = 1$, which is the biggest possible value of μ (since $n = 2$).

d. Show that the resultant of p and q also gives the implicit equation.

Parts b and d of Exercise 9 express the implicit equation as a resultant. However, if we use the Sylvester determinant, then part b uses a 4×4 determinant, while part d uses a 2×2 determinant. So the μ-basis gives a smaller expression for the resultant. In general, one can show (see [CSC]) that for the μ-basis, the resultant can be expressed as an $(n - \mu) \times (n - \mu)$ determinant. Unfortunately, it can also happen that this method gives a power of the actual implicit equation (see Section 4 of [CSC]).

Earlier in this section, we considered the ideal of three points in $\mathbb{P}^2$. We found that although all such ideals have the same Hilbert polynomial, we can distinguish the collinear and noncollinear cases using the Hilbert function. The situation is similar when dealing with μ-bases. Here, we have the ideal $J = \langle \tilde{a}, \tilde{b}, \tilde{c} \rangle \subset R = k[t, u]$ from the proof of Theorem (4.17). In the following exercise you will compute the Hilbert function of R/J.

Exercise 10. Let $J = \langle \tilde{a}, \tilde{b}, \tilde{c} \rangle$ be as in the proof of Theorem (4.17). In the course of the proof, we showed that the Hilbert polynomial of R/J is the zero polynomial. But what about the Hilbert function?

a. Prove that the Hilbert function $H_{R/J}$ is given by

$$H_{R/J}(s) = \begin{cases} s + 1 & \text{if } 0 \le s \le n - 1 \\ 3n - 2s - 2 & \text{if } n \le s \le n + \mu - 1 \\ 2n - s - \mu - 1 & \text{if } n + \mu \le s \le 2n - \mu - 1 \\ 0 & \text{if } 2n - \mu \le s. \end{cases}$$

b. Show that the largest value of s such that $H_{R/J}(s) \ne 0$ is $s = 2n - \mu - 2$, and conclude that knowing μ is equivalent to knowing the Hilbert function of the quotient ring R/J.

c. Compute the dimension of R/J as a vector space over k.

In the case of the ideal of three points, note that the noncollinear case is generic. This is true in the naive sense that one expects three randomly chosen points to be noncollinear, and this can be made more precise using the notion of generic given in Definition (5.6) of Chapter 3. Similarly, for μ-bases, there is a generic case. One can show (see [CSC]) that among parametrizations (4.11) with $n = \max(\deg a, \deg b, \deg c)$, the "generic" parametrization has $\mu = \lfloor n/2 \rfloor$, the biggest possible value. More generally, one can compute the dimension of the set of all parametrizations with a given μ. This dimension decreases as μ decreases, so that the smaller the μ, the more special the parametrization.

We should also mention that the Hilbert-Burch Theorem discussed in §2 has the following nice application to μ-bases.

(4.22) Proposition. *The μ-basis coming from the columns of (4.21) can be chosen such that*

$$\tilde{a} = p_2 q_3 - p_3 q_2, \quad \tilde{b} = -(p_1 q_3 - p_3 q_1), \quad \tilde{c} = p_1 q_2 - p_1 q_2.$$

Dehomogenizing, this means that a, b, c can be computed from the coefficients of the μ-basis

$$(4.23) \qquad \begin{aligned} p &= p_1(t, 1)x + p_2(t, 1)y + p_3(t, 1) \\ q &= q_1(t, 1)x + q_2(t, 1)y + q_3(t, 1). \end{aligned}$$

PROOF. To see why this is true, first note that the exact sequence (4.20) has the form required by Proposition (2.6) of §2. Then the proposition implies that if $\tilde{f}_1, \tilde{f}_2, \tilde{f}_3$ are the 2×2 minors of (4.21) (this is the notation of Proposition (2.6)), then there is a polynomial $g \in k[t, u]$ such that $\tilde{a} = g\tilde{f}_1, \tilde{b} = g\tilde{f}_2, \tilde{c} = g\tilde{f}_3$. However, since $\tilde{a}, \tilde{b}, \tilde{c}$ have no common roots, g must be a nonzero constant. If we replace p_i with gp_i, we get a μ-basis with the desired properties. $\qquad\square$

Exercise 11. Verify that the μ-basis studied in Exercises 8 and 9 satisfies Proposition (4.22) after changing p by a suitable constant.

It is also possible to generalize Theorem (4.17) by considering curves in m-dimensional space k^m given parametrically by

$$(4.24) \qquad x_1 = \frac{a_1(t)}{c(t)}, \quad \ldots, x_m = \frac{a_m(t)}{c(t)},$$

where $c \neq 0$ and $\mathrm{GCD}(a_1, \ldots, a_m) = 1$. In this situation, the syzygy module $\mathrm{Syz}(a_1, \ldots, a_m, c)$ and its homogenization play an important role, and the analog of the μ-basis (4.13) consists of m polynomials

$$(4.25) \quad p_j = p_{1j}(t, 1)x_1 + \cdots + p_{mj}(t, 1)x_m + p_{m+1j}(t, 1), \quad 1 \le j \le m,$$

which form a basis for the ideal $I = \langle cx_1 - a_1, \ldots, cx_m - a_m \rangle$. If we fix t in (4.25), then the equation $p_j = 0$ is a hyperplane in k^m, so that as t varies, we get a *moving hyperplane*. One can prove that the common intersection of the m hyperplanes $p_j = 0$ sweeps out the given curve and that if p_j has degree μ_j in t, then $\mu_1 + \cdots + \mu_m = n$. Thus we have an m-dimensional version of Theorem (4.17). See Exercise 17 for the proof.

We can use the Hilbert-Burch Theorem to generalize Proposition (4.22) to the more general situation of (4.24). The result is that up to sign, the polynomials $a_1, \ldots, a_m, c$ are the $m \times m$ minors of the matrix $(p_{ij}(t, 1))$ coming from (4.25). Note that since p_j has degree μ_j in t, the $m \times m$ minors $(p_{ij}(t, 1))$ have degree at most $\mu_1 + \cdots + \mu_m = n$ in t. So the degrees work out nicely. The details will be covered in Exercise 17 below.

The proof given of Theorem (4.17) makes nice use of the results of §3, especially Proposition (3.20), and the generalization (4.24) to curves in k^m shows just how powerful these methods are. The heart of what we did in Theorem (4.17) was to understand the structure of the syzygy module Syz $(\tilde{a}, \tilde{b}, \tilde{c})$ as a free module, and for the m-dimensional case, one needs to understand Syz $(\tilde{a}_1, \ldots, \tilde{a}_m, \tilde{c})$ for $\tilde{a}_1, \ldots, \tilde{a}_m, \tilde{c} \in k[t, u]$. Actually, in the special case of Theorem (4.17), one can give a proof using elementary methods which don't require the Hilbert Syzygy Theorem. One such proof can be found in [CSC], and another was given by Franz Meyer, all the way back in 1887 [Mey].

Meyer's article is interesting, for it starts with a problem completely different from plane curves, but just as happened to us, he ended up with a syzygy problem. He also considered the more general syzygy module Syz $(\tilde{a}_1, \ldots, \tilde{a}_m, \tilde{c})$, and he conjectured that this was a free module with generators of degrees $\mu_1, \ldots, \mu_m$ satisfying $\mu_1 + \cdots + \mu_m = n$. But in spite of many examples in support of this conjecture, his attempts at a proof "ran into difficulties which I have at this time not been able to overcome" [Mey, p. 73]. However, three years later, Hilbert proved everything in his groundbreaking paper [Hil] on syzygies. For us, it is interesting to note that after proving his Syzygy Theorem, Hilbert's first application is to prove Meyer's conjecture. He does this by computing a Hilbert polynomial (which he calls the *characteristic function*) in a manner remarkably similar to what we did in Theorem (4.17)—see [Hil, p. 516]. Hilbert then concludes with the Hilbert-Burch Theorem in the special case of $k[t, u]$.

One can also consider surfaces in k^3 parametrized by rational functions

$$x = \frac{a(s, t)}{d(s, t)}, \quad y = \frac{b(s, t)}{d(s, t)}, \quad z = \frac{c(s, t)}{d(s, t)},$$

where $a, b, c, d \in k[s, t]$ are polynomials such that $d \neq 0$ and

$$\mathrm{GCD}(a, b, c, d) = 1.$$

As above, the goal is to find the implicit equation of the surface. Surface implicitization is an important problem in geometric modeling.

This case is more complicated because of the possible presence of *base points*, which are points (s, t) at which a, b, c, d all vanish simultaneously. As in the curve case, it is best to work homogeneously, though the commutative algebra is also more complicated—for example, the syzygy module is rarely free. However, there are still many situations where the implicit equation can be computed using syzygies. See [Cox2] and [Cox3] for introductions to this area of research and references to the literature.

Rings of Invariants

The final topic we will explore is the invariant theory of finite groups. In contrast to the previous discussions, our presentation will not be self-contained. Instead, we will assume that the reader is familiar with the material presented in Chapter 7 of [CLO]. Our goal is to explain how graded resolutions can be used when working with polynomials invariant under a finite matrix group.

For simplicity, we will work over the polynomial ring $S = \mathbb{C}[x_1, \ldots, x_m]$. Suppose that $G \subset \mathrm{GL}(m, \mathbb{C})$ is a finite group. If we regard $g \in G$ as giving a change of coordinates on $\mathbb{C}^m$, then substituting this coordinate change into $f \in S = \mathbb{C}[x_1, \ldots, x_m]$ gives another polynomial $g \cdot f \in S$. Then define

$$S^G = \{f \in \mathbb{C}[x_1, \ldots, x_m] : g \cdot f = f \text{ for all } g \in G\}.$$

Intuitively, S^G consists of all polynomials $f \in S$ which are unchanged (i.e., invariant) under all of the coordinate changes coming from elements $g \in G$. The set S^G has the following structure:

- (Graded Subring) The set of invariants $S^G \subset S$ is a subring of S, meaning that S is closed under addition and multiplication by elements of S^G. Also, if $f \in S^G$, then every homogeneous component of f also lies in S^G.

(See Propositions 9 and 10 of Chapter 7, §2 of [CLO].) We say that S^G is a *graded subring* of S. Hence the degree t homogeneous part S_t^G consists of all invariants which are homogeneous polynomials of degree t. Note that S^G is *not* an ideal of S.

In this situation, we define the *Molien series* of S^G to be the formal power series

$$(4.26) \qquad\qquad F_G(u) - \sum_{t=0}^{\infty} \dim_{\mathbb{C}}(S_t^G) \, u^t.$$

Molien series are important objects in the invariant theory of finite groups. We will see that they have a nice relation to Hilbert functions and graded resolutions.

A basic result proved in Chapter 7, §3 of [CLO] is:

- (Finite Generation of Invariants) For a finite group $G \subset \mathrm{GL}(m, \mathbb{C})$, there are $f_1, \ldots, f_s \in S^G$ such that every $f \in S^G$ is a polynomial in $f_1, \ldots, f_s$. Furthermore, we can assume that $f_1, \ldots, f_s$ are homogeneous.

This enables us to regard S^G as a module over a polynomial ring as follows. Let $f_1, \ldots, f_s$ be homogeneous generators of the ring of invariants S^G, and set $d_i = \deg f_i$. Then introduce variables $y_1, \ldots, y_s$ and consider the ring $R = \mathbb{C}[y_1, \ldots, y_s]$. The ring R is useful because the map sending y_i to f_i defines a ring homomorphism

$$\varphi : R = \mathbb{C}[y_1, \ldots, y_s] \longrightarrow S^G$$

which is onto since every invariant is a polynomial in $f_1, \ldots, f_s$. An important observation is that φ becomes a graded homomorphism of degree zero *provided* we regard the variable y_i as having degree $d_i = \deg f_i$. Previously, the variables in a polynomial ring always had degree 1, but here we will see that having $\deg y_i = d_i$ is useful.

The kernel $I = \ker \varphi \subset R$ consists of all polynomial relations among the f_i. Since φ is onto, we get an isomorphism $R/I \cong S^G$. Regarding S^G as an R-module via $y_i \cdot f = f_i f$ for $f \in S^G$, $R/I \cong S^G$ is an isomorphism of R-modules. Elements of I are called *syzygies* among the invariants $f_1, \ldots, f_s$. (Historically, syzygies were first defined in invariant theory, and only later was this term used in module theory, where the meaning is slightly different).

For going any further, let's pause for an example. Consider the group $G = \{e, g, g^2, g^3\} \subset \mathrm{GL}(2, \mathbb{C})$, where

$$(4.27) \qquad\qquad g = \begin{pmatrix} 0 & -1 \\ 1 & 0 \end{pmatrix}.$$

The group G acts on $f \in S = \mathbb{C}[x_1, x_2]$ via $g \cdot f(x_1, x_2) = f(-x_2, x_1)$. Then, as shown in Example 4 of Chapter 7, §3 of [CLO], the ring of invariants S^G is generated by the three polynomials

$$(4.28) \qquad f_1 = x_1^2 + x_2^2, \quad f_2 = x_1^2 x_2^2, \quad f_3 = x_1^3 x_2 - x_1 x_2^3.$$

This gives $\varphi : R = \mathbb{C}[y_1, y_2, y_3] \to S^G$ where $\varphi(y_i) = f_i$. Note that y_1 has degree 2 and y_2, y_3 both have degree 4. One can also show that the kernel of φ is $I = \langle y_3^2 - y_1^2 y_2 + 4y_2^2 \rangle$. This means that all syzygies are generated by the single relation $f_3^2 - f_1^2 f_2 + 4f_2^2 = 0$ among the invariants (4.28).

Returning to our general discussion, the R-module structure on S^G shows that the Molien series (4.26) is built from the Hilbert function of the R-module S^G. This is immediate because

$$\dim_{\mathbb{C}}(S_t^G) = H_{S^G}(t).$$

In Exercises 24 and 25, we will see more generally that any finitely generated R-module has a *Hilbert series*

$$\sum_{t=-\infty}^{\infty} H_M(t) \, u^t.$$

The basic idea is that one can compute any Hilbert series using a graded resolution of M. In the case when all of the variables have degree 1, this is explained in Exercise 24.

However, we are in a situation where the variables have degree $\deg y_i = d_i$ (sometimes called the *weight* of y_i). Formula (4.2) no longer applies, so instead we use the key fact (to be proved in Exercise 25) that the Hilbert series of the weighted polynomial ring $R = \mathbb{C}[y_1, \ldots, y_s]$ is

$$(4.29) \qquad \sum_{t=0}^{\infty} H_R(t) \, u^t = \sum_{t=0}^{\infty} \dim_{\mathbb{C}}(R_t) \, u^t = \frac{1}{(1 - u^{d_1}) \cdots (1 - u^{d_s})}.$$

Furthermore, if we define the twisted free module $R(-d)$ in the usual way, then one easily obtains

$$(4.30) \qquad \sum_{t=0}^{\infty} H_{R(-d)}(t)\, u^t = \frac{u^d}{(1 - u^{d_1}) \cdots (1 - u^{d_s})}$$

(see Exercise 25 for the details).

Let us see how this works in the example begun earlier.

Exercise 12. Consider the group $G \subset \mathrm{GL}(2, \mathbb{C})$ given in (4.27) with invariants (4.28) and syzygy $f_3^2 + f_1^2 f_2 + 4f_2^2 = 0$.

a. Show that a minimal free resolution of S^G as a graded R-module is given by

$$0 \longrightarrow R(-8) \stackrel{\psi}{\longrightarrow} R \stackrel{\varphi}{\longrightarrow} R^G \longrightarrow 0$$

where ψ is the map represented by the 1×1 matrix $(y_3^2 + y_1^2 y_2 + 4y_2^2)$.

b. Use part a together with (4.29) and (4.30) to show that the Molien series of G is given by

$$F_G(u) = \frac{1 - u^8}{(1 - u^2)(1 - u^4)^2} = \frac{1 + u^4}{(1 - u^2)(1 - u^4)}$$

$$= 1 + u^2 + 3u^4 + 3u^6 + 5u^8 + 5u^{10} + \cdots$$

c. The coefficient 1 of u^2 tells us that we have a unique (up to constant) invariant of degree 2, namely f_1. Furthermore, the coefficient 3 of u^4 tells us that besides the obvious degree 4 invariant f_1^2, we must have two others, namely f_2 and f_3. Give similar explanations for the coefficients of u^6 and u^8 and in particular explain how the coefficient of u^8 proves that we must have a nontrivial syzygy of degree 8.

In general, one can show that if the invariant ring of a finite group G is generated by homogeneous invariants $f_1, \ldots, f_s$ of degree $d_1, \ldots, d_s$, then the Molien series of G has the form

$$F_G(u) = \frac{P(u)}{(1 - u^{d_1}) \cdots (1 - u^{d_s})}$$

for some polynomial $P(u)$. See Exercise 25 for the proof. As explained in [Sta2], $P(u)$ has the following intuitive meaning. If there are no nontrivial syzygies between the f_i, then the Molien series would have been

$$\frac{1}{(1 - u^{d_1}) \cdots (1 - u^{d_s})}.$$

Had R^G been generated by homogeneous elements $f_1, \ldots, f_s$ of degrees $d_1, \ldots, d_s$, with homogeneous syzygies $S_1, \ldots, S_w$ of degrees $\beta_1, \ldots, \beta_w$ and no second syzygies, then the Molien series would be corrected to

$$\frac{1 - \sum_j u^{\beta_j}}{\prod_i (1 - u^{d_i})}.$$

In general, by the Syzygy Theorem, we get

$$F_G(u) = (1 - \underbrace{\sum_j u^{\beta_j} + \sum_k u^{\gamma_k} - \cdots}_{\text{at most } s \text{ sums}})/\prod_i (1 - u^{d_i}).$$

One important result not mentioned so far is *Molien's Theorem*, which states that the Molien series (4.26) of a finite group $G \subset \mathrm{GL}(m, \mathbb{C})$ is given by the formula

$$F_G(u) = \frac{1}{|G|} \sum_{g \in G} \frac{1}{\det(I - ug)}$$

where $|G|$ is the number of elements in G and $I \in \mathrm{GL}(m, \mathbb{C})$ is the identity matrix. This theorem is why (4.26) is called a Molien series. The importance of Molien's theorem is that it allows one to compute the Molien series in advance. As shown by part c of Exercise 12, the Molien series can predict the existence of certain invariants and syzygies, which is useful from a computational point of view (see Section 2.2 of [Stu1]). A proof of Molien's Theorem will be given in Exercise 28.

A second crucial aspect we've omitted is that the ring of invariants S^G is Cohen-Macaulay. This has some far-reaching consequences for the invariant theory. For example, being Cohen-Macaulay predicts that there are algebraically independent invariants $\theta_1, \ldots, \theta_r$ such that the invariant ring S^G is a *free* module over the polynomial ring $\mathbb{C}[\theta_1, \ldots, \theta_r]$. For example, in the invariant ring $S^G = \mathbb{C}[f_1, f_2, f_3]$ considered in Exercise 12, one can show that as a module over $\mathbb{C}[f_1, f_2]$,

$$S^G = \mathbb{C}[f_1, f_2] \oplus f_3 \mathbb{C}[f_1, f_2].$$

(Do you see how the syzygy $f_3^2 - f_1 f_2^2 + 4 f_2^3 = 0$ enables us to get rid of terms involving f_3^2, f_3^3, etc?) This has some strong implications for the Molien series, as explained in [Sta2] or [Stu1].

Hence, to really understand the invariant theory of finite groups, one needs to combine the free resolutions discussed here with a variety of other tools, some of which are more sophisticated (such as Cohen-Macaulay rings). Fortunately, some excellent expositions are available in the literature, and we especially recommend [Sta2] and [Stu1]. Additional references are mentioned at the end of Chapter 7, §3 of [CLO].

This brings us to the end of our discussion of resolutions. The examples presented in this section—ideals of three points, μ-bases, and Molien series—are merely the beginning of a wonderful collection of topics related to the geometry of free resolutions. When combined with the elegance of the algebra involved, it becomes clear why the study of free resolutions is one of the richer areas of contemporary algebraic geometry.

To learn more about free resolutions, we suggest the references [Eis], [Schre2] and [EH] mentioned earlier in the section. The reader may also wish to consult [BH, Chapter 4] for a careful study of Hilbert functions.

ADDITIONAL EXERCISES FOR §4

Exercise 13. The Hilbert polynomial has the property that $H_M(t) = HP_M(t)$ for all t sufficiently large. In this exercise, you will derive an explicit bound on how large t has to be in terms of a graded resolution of M.

a. Equation (4.3) shows that the binomial coefficient $\binom{t+n}{n}$ is given by a polynomial of degree n in t. Show that this identity holds for all $t \geq -n$ and also explain why it fails to hold when $t = -n - 1$.

b. For a twisted free module $M = R(-d_1) \oplus \cdots \oplus R(-d_m)$, show that $H_M(t) = HP_M(t)$ holds for $t \geq \max_i(d_i - n)$.

c. Now suppose we have a graded resolution $\cdots \to F_0 \to M$ where $F_j = \oplus_i R(-d_{ij})$. Then show that $H_M(t) = HP_M(t)$ holds for all $t \geq \max_{ij}(d_{ij} - n)$.

d. For the ideal $I \subset k[x, y, z, w]$ from (4.5), we found the graded resolution

$$0 \to R(-3)^2 \to R(-2)^3 \to R \to R/I \to 0.$$

Use this and part c to show that $H_{R/I}(t) = HP_{R/I}(t)$ for all $t \geq 0$. How does this relate to (4.6)?

Exercise 14. Given a parametrization as in (4.11), we get the ideal $I = \langle c(t)x - a(t), c(t)y - b(t) \rangle \subset k[x, y, t]$. We will assume $\mathrm{GCD}(a, b, c) = 1$.

a. Show that $\mathbf{V}(I) \subset k^3$ is the graph of the function $F : k - W \to k^2$ defined by $F(t) = (a(t)/c(t), b(t)/c(t))$, where $W = \{t \in k : c(t) = 0\}$.

b. If $I_1 = I \cap k[x, y]$, prove that $\mathbf{V}(I_1) \subset k^2$ is the smallest variety containing the parametrization (4.11). Hint: This follows by adapting the proof of Theorem 1 of Chapter 3, §3 of [CLO].

Exercise 15. This exercise concerns the Koszul complex used in the proof of Proposition (4.14).

a. Assuming $\mathrm{GCD}(a, b, c) = 1$ in $S = k[t]$, prove that the sequence (4.15) is exact at its middle term. Hint: Our hypothesis implies that there are polynomials $p, q, r \in k[t]$ such that $pa + qb + rc = 1$. Then if $(A, B, C) \in \ker(\beta)$, note that

$$A = paA + qbA + rcA$$

$$= p(-bB - cC) + qbA + rcA$$

$$= c(-pC + rA) + b(-pB + qA).$$

b. Using Exercise 10 of §2 as a model, show how to extend (4.15) to the full Koszul complex

$$0 \to S \to S^3 \xrightarrow{\alpha} S^3 \xrightarrow{\beta} S \to 0$$

of a, b, c. Also, when $\mathrm{GCD}(a, b, c) = 1$, prove that the entire sequence is exact.

c. More generally, show that $a_1, \ldots, a_m \in k[t]$ give a Koszul complex and prove that it is exact when $\mathrm{GCD}(a_1, \ldots, a_m) = 1$. (This is a challenging exercise.)

Exercise 16. In the proof of Theorem (4.17), we noted that the columns of the matrix (4.20) generate the syzygy module $\mathrm{Syz}(\tilde{a}, \tilde{b}, \tilde{c})$. If we define p, q using (4.23), then prove that p, q generate $I(1)$.

Exercise 17. In this exercise, you will study the m-dimensional version of Theorem (4.17). Thus we assume that we have a parametrization (4.24) of a curve in k^m such that $c \neq 0$ and $\mathrm{GCD}(a_1, \ldots, a_m) = 1$. Also let

$$I = \langle cx_1 - a_1, \ldots, cx_m - a_m \rangle \subset k[x_1, \ldots, x_m, t]$$

and define

$$I(1) = \{ f \in I : f = A_1(t)x_1 + \cdots + A_m(t)x_m + C(t) \}.$$

a. Prove the analog of Lemma (4.14), i.e., show that there is a natural isomorphism $I(1) \cong \mathrm{Syz}(a_1, \ldots, a_m, c)$. Hint: You will use part c of Exercise 15.

b. If $n = \max(\deg a_1, \ldots, \deg a_m, c)$ and $\tilde{a}_i, \tilde{c} \in R = k[t, u]$ are the degree n homogenizations of a_i, c, then explain why there is an injective map

$$\beta : R(-d_1) \oplus \cdots \oplus R(-d_s) \to R(-n)^{m+1}$$

whose image is $\mathrm{Syz}(\tilde{a}_1, \ldots, \tilde{a}_m, \tilde{c})$.

c. Use Hilbert polynomials to show that $s = m$ and that $d_1 + \cdots + d_m = (m+1)n$.

d. If L is the matrix representing β, show that the jth column of L consists of homogeneous polynomials of degree $\mu_j = d_j - n$. Also explain why $\mu_1 + \cdots + \mu_s = n$.

e. Finally, by dehomogenizing the entries of the jth column of L, show that we get the polynomial p_j as in (4.25), and prove that $I = \langle p_1, \ldots, p_m \rangle$.

f. Use the Hilbert-Burch Theorem to show that if p_1 is modified by a suitable constant, then up to a constant, $a_1, \ldots, a_m, c$ are the $m \times m$ minors of the matrix $(p_{ij}(t, 1))$ coming from (4.25).

Exercise 18. Compute the Hilbert function and Hilbert polynomial of the ideal of the rational quartic curve in $\mathbb{P}^3$ whose graded resolution is given in (3.6). What does the Hilbert polynomial tell you about the dimension and the degree?

Exercise 19. In $k[x_0, \ldots, x_n]$, $n \geq 2$, consider the homogeneous ideal I_n defined by the determinants of the $\binom{n}{2}$ 2×2 submatrices of the $2 \times n$ matrix

$$M = \begin{pmatrix} x_0 & x_1 & \cdots & x_{n-1} \\ x_1 & x_2 & \cdots & x_n \end{pmatrix}.$$

(We studied this ideal in Exercise 15 of §3.) Compute the Hilbert functions and Hilbert polynomials of I_4 and I_5. Also determine the degrees of the curves $\mathbf{V}(I_4)$ and $\mathbf{V}(I_5)$ and verify that they have dimension 1. Hint: In part b of Exercise 15 of §3, you computed graded resolutions of these two ideals.

Exercise 20. In this exercise, we will show how the construction of the rational normal curves from the previous exercise and Exercise 15 of §3 relates to the moving lines considered in this section.
a. Show that for each $(t, u) \in \mathbb{P}^1$, the intersection of the lines $\mathbf{V}(tx_0 + ux_1)$ and $\mathbf{V}(tx_1 + ux_2)$ lies on the conic section $\mathbf{V}(x_0 x_2 - x_1^2)$ in $\mathbb{P}^2$. Express the equation of the conic as a 2×2 determinant.
b. Generalizing part a, show that for all $n \geq 2$, if we construct n *moving hyperplanes* $H_i(t, u) = \mathbf{V}(tx_{i-1} + ux_i)$ for $i = 1, \ldots, n$, then for each (t, u) in $\mathbb{P}^1$, the intersection $H_1(t, u) \cap \cdots \cap H_n(t, u)$ is a point on the standard rational normal curve in $\mathbb{P}^n$ given as in Exercise 15 of §3, and show how the determinantal equations follow from this observation.

Exercise 21. In $k[x_0, \ldots, x_n]$, $n \geq 3$, consider the homogeneous ideal J_n defined by the determinants of the $\binom{n-1}{2}$ 2×2 submatrices of the $2 \times (n-1)$ matrix

$$N = \begin{pmatrix} x_0 & x_2 & \cdots & x_{n-1} \\ x_1 & x_3 & \cdots & x_n \end{pmatrix}.$$

The varieties $\mathbf{V}(J_n)$ are surfaces called *rational normal scrolls* in $\mathbb{P}^n$. For instance, $J_3 = \langle x_0 x_3 - x_1 x_2 \rangle$ is the ideal of a smooth quadric surface in $\mathbb{P}^3$.
a. Find a graded resolution of J_4 and compute its Hilbert function and Hilbert polynomial. Check that the dimension is 2 and compute the degree of the surface.
b. Do the same for J_5.

Exercise 22. The (degree 2) *Veronese surface* $V \subset \mathbb{P}^5$ is the image of the mapping given in homogeneous coordinates by

$$\varphi : \mathbb{P}^2 \to \mathbb{P}^5$$
$$(x_0, x_1, x_2) \mapsto (x_0^2, x_1^2, x_2^2, x_0 x_1, x_0 x_2, x_1 x_2).$$

a. Compute the homogeneous ideal $I = \mathbf{I}(V) \subset k[x_0, \ldots, x_5]$.

b. Find a graded resolution of I and compute its Hilbert function and Hilbert polynomial. Also check that the dimension is equal to 2 and the degree is equal to 4.

Exercise 23. Let $p_1 = (0, 0, 1), p_2 = (1, 0, 1), p_3 = (0, 1, 1), p_4 = (1, 1, 1)$ in $\mathbb{P}^2$, and let $I = \mathbf{I}(\{p_1, p_2, p_3, p_4\})$ be the homogeneous ideal of the variety $\{p_1, p_2, p_3, p_4\}$ in $R = k[x_0, x_1, x_2]$.
a. Show that I_3 (the degree 3 graded piece of I) has dimension exactly 6.
b. Let $f_0, \ldots, f_5$ be any vector space basis for I_3, and consider the rational mapping $\varphi : \mathbb{P}^2 - - \to \mathbb{P}^5$ given in homogeneous coordinates by

$$\varphi(x_0, x_1, x_2) = (y_0, \ldots, y_5) = (f_0(x_0, x_1, x_2), \ldots, f_5(x_0, x_1, x_2)).$$

Find the homogeneous ideal J of the image variety of φ.
c. Show that J has a graded resolution as an $S = k[y_0, \ldots, y_5]$-module of the form

$$0 \to S(-5) \to S(-3)^5 \xrightarrow{A} S(-2)^5 \to J \to 0.$$

d. Use the resolution above to compute the Hilbert function of J.

The variety $V = \mathbf{V}(J) = \varphi(\mathbb{P}^2)$ is called a *quintic del Pezzo surface*, and the resolution given in part d has some other interesting properties. For instance, if the ideal basis for J is ordered in the right way and signs are adjusted appropriately, then A is skew-symmetric, and the determinants of the 4×4 submatrices obtained by deleting row i and column i ($i = 1, \ldots, 5$) are the squares of the generators of J. This is a reflection of a remarkable structure on the resolutions of *Gorenstein codimension 3* ideals proved by Buchsbaum and Eisenbud. See [BE].

Exercise 24. One convenient way to "package" the Hilbert function H_M for a graded module M is to consider its *generating function*, the formal power series

$$H(M, u) = \sum_{t=-\infty}^{\infty} H_M(t) u^t.$$

We will call $H(M, u)$ the *Hilbert series* for M.
a. Show that for $M = R = k[x_0, \ldots, x_n]$, we have

$$H(R, u) = \sum_{t=0}^{\infty} \binom{n+t}{n} u^t$$

$$= 1/(1 - u)^{n+1},$$

where the second equality comes from the formal geometric series identity $1/(1 - u) = \sum_{t=0}^{\infty} u^t$ and induction on n.
b. Show that if $R = k[x_0, \ldots, x_n]$ and

$$M = R(-d_1) \oplus \cdots \oplus R(-d_m)$$

is one of the twisted graded free modules over R, then

$$H(M, u) = (u^{d_1} + \cdots + u^{d_m})/(1 - u)^{n+1}.$$

c. Let I be the ideal of the twisted cubic in $\mathbb{P}^3$ studied in Exercise 2 of §2, and let $R = k[x, y, z, w]$. Find the Hilbert series $H(R/I, u)$.

d. Using part b and Theorem (4.4) deduce that the Hilbert series of any graded $k[x_0, \ldots, x_n]$-module M can be written in the form

$$H(M, u) = P(u)/(1 - u)^{n+1}$$

where P is a polynomial in u with coefficients in $\mathbb{Z}$.

Exercise 25. Consider the polynomial ring $R = k[y_1, \ldots, y_s]$, where y_i has weight or degree $\deg y_i = d_i > 0$. Then a monomial $y^{a_1} \cdots y_s^{a_s}$ has (weighted) degree $t = d_1 a_1 + \cdots + d_s a_s$. This gives a grading on R such that R_t is the set of k-linear combinations of monomials of degree t.
a. Prove that the Hilbert series of R is given by

$$\sum_{t=0}^{\infty} \dim_k(R_t) \, u^t = \frac{1}{(1 - u^{d_1}) \cdots (1 - u^{d_s})}.$$

Hint: $1/(1 - u^{d_i}) = \sum_{a_i=0}^{\infty} u^{d_i a_i}$. When these series are multiplied together for $i = 1, \ldots, s$, do you see how each monomial of weighted degree t contributes to the coefficient of u^t?
b. Explain how part a relates to part a of Exercise 24.
c. If $R(-d)$ is defined by $R(-d)_t = R_{t-d}$, then prove (1.30).
d. Generalize parts b, c and d of Exercise 24 to $R = k[y_1, \ldots, y_s]$.

Exercise 26. Suppose that $a, b, c \in k[t]$ have maximum degree 6. As usual, we will assume $c \neq 0$ and $\mathrm{GCD}(a, b, c) = 1$.
a. If $a = t^6 + t^3 + t^2$, $b = t^6 - t^4 - t^2$ and $c = t^6 + t^5 + t^4 - t - 1$, show that $\mu = 2$ and find a μ-basis.
b. Find an example where $\mu = 3$ and compute a μ-basis for your example. Hint: This is the generic case.

Exercise 27. Compute the Molien series for the following finite matrix groups in $\mathrm{GL}(2, \mathbb{C})$. In each case, the ring of invariants $\mathbb{C}[x_1, x_2]^G$ can be computed by the methods of Chapter 7, §3 of [CLO].

a. The Klein four-group generated by $\begin{pmatrix} 1 & 0 \\ 0 & -1 \end{pmatrix}$ and $\begin{pmatrix} -1 & 0 \\ 0 & 1 \end{pmatrix}$.

b. The two-element group generated by $g = \begin{pmatrix} -1 & 0 \\ 0 & -1 \end{pmatrix}$.

c. The four-element group generated by $g = \frac{1}{\sqrt{2}} \begin{pmatrix} 1 & -1 \\ 1 & 1 \end{pmatrix}$.

Exercise 28. Let $G \subset \mathrm{GL}(m, \mathbb{C})$ be a finite group and let $S = \mathbb{C}[x_1, \ldots, x_m]$. The goal of this exercise is to prove Molien's Theorem, which asserts that

$$\sum_{t=0}^{\infty} \dim_{\mathbb{C}}(S_t^G)u^t = \frac{1}{|G|} \sum_{g \in G} \frac{1}{\det(I - ug)}.$$

a. By Chapter 7, §3 of [CLO], the Reynolds operator $R_G(f)(\mathbf{x}) = \frac{1}{|G|} \sum_{g \in G} f(g \cdot \mathbf{x})$ defines a projection operator $R_G : S_t \to S_t^G$. Use this to prove that

$$\dim_{\mathbb{C}}(S_t^G) = \frac{1}{|G|} \sum_{g \in G} \text{trace}(g_t),$$

where $g_t : S_t \to S_t$ is defined by $f(\mathbf{x}) \mapsto f(g \cdot \mathbf{x})$. Hint: First explain why the trace of a projection operator is the dimension of its image.

b. Fix $g \in G$ and a basis $y_1, \ldots, y_m$ of $S_1 \cong \mathbb{C}^m$ such that g_1 is upper triangular on S_1 with eigenvalues $\lambda_1, \ldots, \lambda_m$. Prove that y^α for $|\alpha| = t$ give a basis of S_t such that g_t is upper triangular on S_t with eigenvalues λ^α. Conclude that $\text{trace}(g_t) = \sum_{|\alpha|=t} \lambda^\alpha$.

c. Explain why

$$\sum_{\alpha} \lambda^\alpha u^{|\alpha|} = \prod_{i=1}^{m} \frac{1}{1 - \lambda_i u} = \frac{1}{\det(I - ug)}$$

and use this to complete the proof of Molien's Theorem.

Chapter 7

Polytopes, Resultants, and Equations

In this chapter we will examine some interesting recently-discovered connections between polynomials, resultants, and the geometry of the convex polytopes determined by the exponent vectors of the monomials appearing in polynomials.

§1 Geometry of Polytopes

A set C in $\mathbb{R}^n$ is said to be *convex* if it contains the line segment connecting any two points in C. If a set is not itself convex, its *convex hull* is the smallest convex set containing it. We will use the notation $\text{Conv}(S)$ to denote the convex hull of $S \subset \mathbb{R}^n$.

More explicitly, all the points in $\text{Conv}(S)$ may be obtained by forming a particular set of linear combinations of the elements in S. In Exercise 1 below, you will prove the following proposition.

(1.1) Proposition. *Let S be a subset of $\mathbb{R}^n$. Then*

$$\text{Conv}(S) = \{\lambda_1 s_1 + \cdots + \lambda_m s_m : s_i \in S, \ \lambda_i \geq 0, \ \textstyle\sum_{i=1}^{m}\lambda_i = 1\}.$$

Linear combinations of the form $\lambda_1 s_1 + \cdots + \lambda_m s_m$, where $s_i \in S$, $\lambda_i \geq 0$, and $\sum_{i=1}^{m} \lambda_i = 1$ are called *convex combinations*.

Exercise 1.
a. Show that if $S = \{s_1, s_2\}$ then the set of convex combinations is the straight line segment from s_1 to s_2 in $\mathbb{R}^n$. Deduce that Proposition (1.1) holds in this case.
b. Using part a, show that the set of all convex combinations

$$\{\lambda_1 s_1 + \cdots + \lambda_m s_m : s_i \in S, \ \lambda_i \geq 0, \ \textstyle\sum_{i=1}^{m}\lambda_i = 1\}.$$

is a convex subset of $\mathbb{R}^n$ for every S. Also show that this set contains S.

c. Show that if C is any convex set containing S, then C also contains the set of part b. Hint: One way is to use induction on the number of terms in the sum.

d. Deduce Proposition (1.1) from parts b and c.

By definition, a *polytope* is the convex hull of a *finite* set in $\mathbb{R}^n$. If the finite set is $\mathcal{A} = \{m_1, \dots, m_l\} \subset \mathbb{R}^n$, then the corresponding polytope can be expressed as

$$\text{Conv}(\mathcal{A}) = \{\lambda_1 m_1 + \cdots + \lambda_l m_l : \lambda_i \geq 0, \ \textstyle\sum_{i=1}^{l} \lambda_i = 1\}.$$

In low dimensions, polytopes are familiar figures from geometry:

- A polytope in $\mathbb{R}$ is a line segment.
- A polytope in $\mathbb{R}^2$ is a line segment or a convex polygon.
- A polytope in $\mathbb{R}^3$ is a line segment, a convex polygon lying in a plane, or a three-dimensional polyhedron.

As these examples suggest, every polytope Q has a well-defined *dimension*. A careful definition of $\dim Q$ will be given in the exercises at the end of the section. For more background on convex sets and polytopes, the reader should consult [Zie]. Fig. 7.1 below shows a three-dimensional polytope.

For another example, let $\mathcal{A} = \{(0,0),(2,0),(0,5),(1,1)\} \subset \mathbb{R}^2$. Here, $\text{Conv}(\mathcal{A})$ is the triangle with vertices $(0,0),(2,0)$, and $(0,5)$ since

$$(1.2) \qquad\qquad (1,1) = \tfrac{3}{10}(0,0) + \tfrac{1}{2}(2,0) + \tfrac{1}{5}(0,5)$$

is a convex combination of the other three points in $\mathcal{A}$.

For us, the most important polytopes will be convex hulls of sets of points with *integer* coordinates. These are sometimes called *lattice polytopes*. Thus a lattice polytope is a set of the form $\text{Conv}(\mathcal{A})$, where $\mathcal{A} \subset \mathbb{Z}^n$ is finite. An example of special interest to us is when $\mathcal{A}$ consists of all exponent vectors

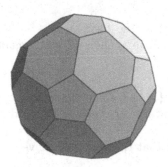

FIGURE 7.1. A three-dimensional polytope

appearing in a collection of monomials. The polytope $Q = \text{Conv}(\mathcal{A})$ will play a *very* important role in this chapter.

Exercise 2. Let $\mathcal{A}_d = \{m \in \mathbb{Z}_{\geq 0}^n : |m| \leq d\}$ be the set of exponent vectors of *all* monomials of total degree at most d.
a. Show that the convex hull of $\mathcal{A}_d$ is the polytope

$$Q_d = \{(a_1, \ldots, a_n) \in \mathbb{R}^n : a_i \geq 0, \ \textstyle\sum_{i=1}^{n} a_i \leq d\}.$$

Draw a picture of $\mathcal{A}_d$ and Q_d when $n = 1, 2, 3$ and $d = 1, 2, 3$.
b. A *simplex* is defined to be the convex hull of $n+1$ points $m_1, \ldots, m_{n+1}$ such that $m_2 - m_1, \ldots, m_{n+1} - m_1$ are a basis of $\mathbb{R}^n$. Show that the polytope Q_d of part a is a simplex.

A polytope $Q \subset \mathbb{R}^n$ has an n-dimensional volume, which is denoted $\text{Vol}_n(Q)$. For example, a polygon Q in $\mathbb{R}^2$ has $\text{Vol}_2(Q) > 0$, but if we regard Q as lying in the xy-plane in $\mathbb{R}^3$, then $\text{Vol}_3(Q) = 0$.
From multivariable calculus, we have

$$\text{Vol}_n(Q) = \int \cdots \int_Q 1 \, dx_1 \cdots dx_n,$$

where $x_1, \ldots, x_n$ are coordinates on $\mathbb{R}^n$. Note that Q has positive volume if and only if it is n-dimensional. A simple example is the unit cube in $\mathbb{R}^n$, which is defined by $0 \leq x_i \leq 1$ for all i and clearly has volume 1.

Exercise 3. Let's compute the volume of the simplex Q_d from Exercise 2.
a. Prove that the map $\phi : \mathbb{R}^n \to \mathbb{R}^n$ defined by

$$\phi(x_1, \ldots, x_n) = (1 - x_1, x_1(1 - x_2), x_1 x_2(1 - x_3), \ldots, x_1 \cdots x_{n-1}(1 - x_n))$$

maps the unit cube $C \subset \mathbb{R}^n$ defined by $0 \leq x_i \leq 1$ to the simplex Q_1. Hint: Use a telescoping sum to show $\phi(C) \subset Q_1$. Be sure to prove the opposite inclusion.
b. Use part a and the change of variables formula for n-dimensional integrals to show that

$$\text{Vol}_n(Q_1) = \int \cdots \int_C x_1^{n-1} x_2^{n-2} \cdots x_{n-1} \, dx_1 \cdots dx_n = \frac{1}{n!}.$$

c. Conclude that $\text{Vol}_n(Q_d) = d^n / n!$.

Polytopes have special subsets called its *faces*. For example, a 3-dimensional polytope in $\mathbb{R}^3$ has:

- faces, which are polygons lying in planes,
- edges, which are line segments connecting certain pairs of vertices, and
- vertices, which are points.

In the general theory, all of these will be called faces. To define a face of an arbitrary polytope $Q \subset \mathbb{R}^n$, let ν be a nonzero vector in $\mathbb{R}^n$. An *affine hyperplane* is defined by an equation of the form $m \cdot \nu = -a$ (the minus sign simplifies certain formulas in §3 and §4—see Exercise 3 of §3 and Proposition (4.6)). If

(1.3) $$a_Q(\nu) = -\min_{m \in Q}(m \cdot \nu),$$

then we call the equation

$$m \cdot \nu = -a_Q(\nu)$$

a *supporting hyperplane* of Q, and we call ν an *inward pointing normal*. Fig. 7.2 below shows a polytope $Q \subset \mathbb{R}^2$ with two supporting hyperplanes (lines in this case) and their inward pointing normals.

In Exercise 13 at the end of the section, you will show that a supporting hyperplane has the property that

$$Q_\nu = Q \cap \{m \in \mathbb{R}^n : m \cdot \nu = -a_Q(\nu)\} \neq \emptyset,$$

and, furthermore, Q lies in the half-space

$$Q \subset \{m \in \mathbb{R}^n : m \cdot \nu \geq -a_Q(\nu)\}.$$

We call $Q_\nu = Q \cap \{m \in \mathbb{R}^n : m \cdot \nu = -a_Q(\nu)\}$ the *face of Q determined by ν*. Fig. 7.2 illustrates two faces, one a vertex and the other an edge.

Exercise 4. Draw a picture of a cube in $\mathbb{R}^3$ with three supporting hyperplanes which define faces of dimensions 0, 1, and 2 respectively. Be sure to include the inward pointing normals in each case.

Every face of Q is a polytope of dimension less than $\dim Q$. *Vertices* are faces of dimension 0 (i.e., points) and *edges* are faces of dimension 1. If Q has dimension n, then *facets* are faces of dimension $n - 1$. Assuming

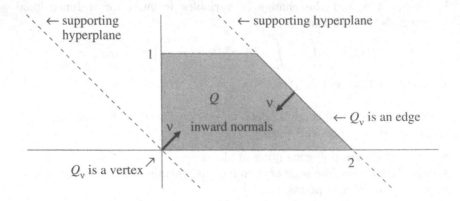

FIGURE 7.2. Supporting hyperplanes, inward normals, and faces

$Q \subset \mathbb{R}^n$, a facet lies on a unique supporting hyperplane and hence has a unique inward pointing normal (up to a positive multiple). In contrast, faces of lower dimension lie in infinitely many supporting hyperplanes. For example, the vertex at the origin in Fig 7.2 is cut out by infinitely many lines through the origin.

We can characterize an n-dimensional polytope $Q \subset \mathbb{R}^n$ in terms of its facets as follows. If $\mathcal{F} \subset Q$ is a facet, we just noted that the inward normal is determined up to a positive constant. Suppose that Q has facets $\mathcal{F}_1, \ldots, \mathcal{F}_N$ with inward pointing normals $\nu_1, \ldots, \nu_N$ respectively. Each facet $\mathcal{F}_j$ has a supporting hyperplane defined by an equation $m \cdot \nu_j = -a_j$ for some a_j. Then one can show that the polytope Q is given by

(1.4) $Q = \{m \in \mathbb{R}^n : m \cdot \nu_j \geq -a_j \text{ for all } j = 1, \ldots, N\}.$

In the notation of (1.3), note that $a_j = a_Q(\nu_j)$.

Exercise 5. How does (1.4) change if we use an *outward* normal for each facet?

When Q is a lattice polytope, we can rescale the inward normal $\nu_{\mathcal{F}}$ of a facet $\mathcal{F}$ so that $\nu_{\mathcal{F}}$ has integer coordinates. We can also assume that the coordinates are relatively prime. In this case, we say the $\nu_{\mathcal{F}}$ is *primitive*. It follows that $\mathcal{F}$ has a *unique* primitive inward pointing normal $\nu_{\mathcal{F}} \in \mathbb{Z}^n$. For lattice polytopes, we will always assume that the inward normals have this property.

Exercise 6. For the lattice polygon Q of Fig. 7.2, find the inward pointing normals. Also, if e_1, e_2 are the standard basis vectors for $\mathbb{R}^2$, then show that the representation (1.4) of Q is given by the inequalities

$$m \cdot e_1 \geq 0, \quad m \cdot e_2 \geq 0, \quad m \cdot (-e_2) \geq -1, \quad m \cdot (-e_1 - e_2) \geq -2.$$

Exercise 7. Let $e_1, \ldots, e_n$ be the standard basis of $\mathbb{R}^n$.
a. Show that the simplex $Q_d \subset \mathbb{R}^n$ of Exercise 2 is given by the inequalities

$$m \cdot \nu_0 \geq -d, \text{ and } m \cdot \nu_j \geq 0, \ j = 1, \ldots, n,$$

where $\nu_0 = -e_1 - \cdots - e_n$ and $\nu_j = e_j$ for $j = 1, \ldots, n$.
b. Show that the square $Q = \text{Conv}(\{(0,0), (1,0), (0,1), (1,1)\}) \subset \mathbb{R}^2$ is given by the inequalities

$$m \cdot \nu_1 \geq 0, \quad m \cdot \nu_2 \geq -1, \quad m \cdot \nu_3 \geq 0, \text{ and } m \cdot \nu_4 \geq -1,$$

where $e_1 = \nu_1 = -\nu_2$ and $e_2 = \nu_3 = -\nu_4$. A picture of this appears in Fig. 7.3 on the next page (with shortened inward normals for legibility).

One of the themes of this chapter is that there is very deep connection between lattice polytopes and polynomials. To describe the connection, we

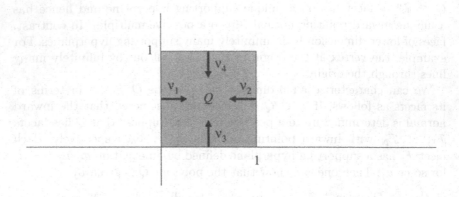

FIGURE 7.3. The unit square

will use the following notation. Let $f \in \mathbb{C}[x_1, \ldots, x_n]$ (or, more generally, in $k[x_1, \ldots, x_n]$ for any field of coefficients), and write

$$f = \sum_{\alpha \in \mathbb{Z}_{\geq 0}^n} c_\alpha x^\alpha.$$

The *Newton polytope* of f, denoted $\mathrm{NP}(f)$, is the lattice polytope

$$\mathrm{NP}(f) = \mathrm{Conv}(\{\alpha \in \mathbb{Z}_{\geq 0}^n : c_\alpha \neq 0\}).$$

In other words, the Newton polytope records the "shape" or "sparsity structure" of a polynomial—it tells us which monomials appear with nonzero coefficients. The actual values of the coefficients do not matter, however, in the definition of $\mathrm{NP}(f)$.

For example, any polynomial of the form

$$f = axy + bx^2 + cy^5 + d$$

with $a, b, c, d \neq 0$ has Newton polytope equal to the triangle

$$Q = \mathrm{Conv}(\{(1,1), (2,0), (0,5), (0,0)\}).$$

In fact, (1.2) shows that polynomials of this form with $a = 0$ would have the same Newton polytope.

Exercise 8. What is the Newton polytope of a 1-variable polynomial $f = \sum_{i=0}^{m} c_i x^i$, assuming that $c_m \neq 0$, so that the degree of f is exactly m? Are there special cases depending on the other coefficients?

Exercise 9. Write down a polynomial whose Newton polytope equals the polytope Q_d from Exercise 2. Which coefficients *must be* non-zero to obtain $\mathrm{NP}(f) = Q_d$? Which can be zero?

We can also go the other way, from exponents to polynomials. Suppose we have a finite set of exponents $\mathcal{A} = \{\alpha_1, \dots, \alpha_l\} \subset \mathbb{Z}_{\geq 0}^n$. Then let $L(\mathcal{A})$ be the set of all polynomials whose terms *all* have exponents in $\mathcal{A}$. Thus

$$L(\mathcal{A}) = \{c_1 x^{\alpha_1} + \cdots + c_l x^{\alpha_l} : c_i \in \mathbb{C}\}.$$

Note that $L(\mathcal{A})$ is a vector space over $\mathbb{C}$ of dimension l (= the number of elements in $\mathcal{A}$).

Exercise 10.
a. If $f \in L(\mathcal{A})$, show that $\mathrm{NP}(f) \subset \mathrm{Conv}(\mathcal{A})$. Give an example to show that equality need not occur.
b. Show that there is a union of proper subspaces $W \subset L(\mathcal{A})$ such that $\mathrm{NP}(f) = \mathrm{Conv}(\mathcal{A})$ for all $f \in L(\mathcal{A}) \setminus W$. This means that $\mathrm{NP}(f) = \mathrm{Conv}(\mathcal{A})$ holds for a *generic* $f \in L(\mathcal{A})$.

Exercise 11. If $\mathcal{A}_d$ is as in Exercise 2, what is $L(\mathcal{A}_d)$?

Finally, we conclude this section with a slight generalization of the notion of monomial and polynomial. Since the vertices of a lattice polytope can have *negative entries*, it will be useful to have the corresponding algebraic objects. This leads to the notion of a polynomial whose terms can have negative exponents.

Let $\alpha = (a_1, \dots, a_n) \in \mathbb{Z}^n$ be an integer vector. The corresponding *Laurent monomial* in variables $x_1, \dots, x_n$ is

$$x^{\alpha} = x_1^{a_1} \cdots x_n^{a_n}.$$

For example, $x^2 y^{-3}$ and $x^{-2} y^3$ are Laurent monomials in x and y whose product is 1. More generally, we have

$$x^{\alpha} \cdot x^{\beta} = x^{\alpha+\beta} \quad \text{and} \quad x^{\alpha} \cdot x^{-\alpha} = 1$$

for all $\alpha, \beta \in \mathbb{Z}^n$. Finite linear combinations

$$f = \sum_{\alpha \in \mathbb{Z}^n} c_\alpha x^\alpha$$

of Laurent monomials are called *Laurent polynomials*, and the collection of all Laurent polynomials forms a commutative ring under the obvious sum and product operations. We denote the ring of Laurent polynomials with coefficients in a field k by $k[x_1^{\pm 1}, \dots, x_n^{\pm 1}]$. See Exercise 15 below for another way to understand this ring.

The definition of the Newton polytope goes over unchanged to Laurent polynomials; we simply allow vertices with negative components. Thus any Laurent polynomial $f \in k[x_1^{\pm 1}, \dots, x_n^{\pm 1}]$ has a Newton polytope $\mathrm{NP}(f)$, which again is a lattice polytope. Similarly, given a finite set $\mathcal{A} \subset \mathbb{Z}^n$, we get the vector space $L(\mathcal{A})$ of Laurent polynomials with exponents in $\mathcal{A}$. Although the introduction of Laurent polynomials might seem unmotivated at this point, they will prove to be very useful in the theory developed in this chapter.

ADDITIONAL EXERCISES FOR §1

Exercise 12. This exercise will develop the theory of affine subspaces. An *affine subspace* $A \subset \mathbb{R}^n$ is a subset with the property that

$$s_1, \ldots, s_m \in A \implies \sum_{i=1}^m \lambda_i s_i \in A \text{ whenever } \sum_{i=1}^m \lambda_i = 1.$$

Note that we do not require that $\lambda_i \geq 0$. We also need the following definition: given a subset $S \subset \mathbb{R}^n$ and a vector $v \in \mathbb{R}^n$, the *translate of S by v* is the set $v + S = \{v + s : s \in S\}$.
a. If $A \subset \mathbb{R}^n$ is an affine subspace and $v \in A$, prove that the translate $-v + A$ is a subspace of $\mathbb{R}^n$. Also show that $A = v + (-v + A)$, so that A is a translate of a subspace.
b. If $v, w \in A$, prove that $-v + A = -w + A$. Conclude that an affine subspace is a translate of a *unique* subspace of $\mathbb{R}^n$.
c. Conversely, if $W \subset \mathbb{R}^n$ is a subspace and $v \in \mathbb{R}^n$, then show that the translate $v + W$ is an affine subspace.
d. Explain how to define the *dimension* of an affine subspace.

Exercise 13. This exercise will define the dimension of a polytope $Q \subset \mathbb{R}^n$. The basic idea is that the dim Q is the dimension of the smallest affine subspace containing Q.
a. Given any subset $S \subset \mathbb{R}^n$, show that

$$\text{Aff}(S) = \{\lambda_1 s_1 + \cdots + \lambda_m s_m : s_i \in S, \ \sum_{i=1}^m \lambda_i = 1\}$$

is the smallest affine subspace containing S. Hint: Use the strategy outlined in parts b, c and d of Exercise 1.
b. Using the previous exercise, explain how to define the dimension of a polytope $Q \subset \mathbb{R}^n$.
c. If $\mathcal{A} = \{m_1, \ldots, m_l\}$ and $Q = \text{Conv}(\mathcal{A})$, prove that $\dim Q = \dim W$, where $W \subset \mathbb{R}^n$ is the subspace spanned by $m_2 - m_1, \ldots, m_l - m_1$.
d. Prove that a simplex in $\mathbb{R}^n$ (as defined in Exercise 2) has dimension n.

Exercise 14. Let $Q \subset \mathbb{R}^n$ be a polytope and $\nu \in \mathbb{R}^n$ be a nonzero vector.
a. Show that $m \cdot \nu = 0$ defines a subspace of $\mathbb{R}^n$ of dimension $n - 1$ and that the affine hyperplane $m \cdot \nu = -a$ is a translate of this subspace. Hint: Use the linear map $\mathbb{R}^n \to \mathbb{R}$ given by dot product with ν.
b. Explain why $\min_{m \in Q}(m \cdot \nu)$ exists. Hint: Q is closed and bounded, and $m \mapsto m \cdot \nu$ is continuous.
c. If $a_Q(\nu)$ is defined as in (1.3), then prove that the intersection

$$Q_\nu = Q \cap \{m \in \mathbb{R}^n : m \cdot \nu = -a_Q(\nu)\}$$

is nonempty and that

$$Q \subset \{m \in \mathbb{R}^n : m \cdot \nu \geq -a_Q(\nu)\}.$$

Exercise 15. There are several ways to represent the ring of Laurent polynomials in $x_1, \ldots, x_n$ as a quotient of a polynomial ring. Prove that

$$k[x_1^{\pm 1}, \ldots, x_n^{\pm 1}] \cong k[x_1, \ldots, x_n, t_1, \ldots, t_n]/\langle x_1 t_1 - 1, \ldots, x_n t_n - 1 \rangle$$

$$\cong k[x_1, \ldots, x_n, t]/\langle x_1 \cdots x_n t - 1 \rangle.$$

Exercise 16. This exercise will study the translates of a polytope. The translate of a set in $\mathbb{R}^n$ is defined in Exercise 12.
a. If $\mathcal{A} \subset \mathbb{R}^n$ is a finite set and $v \in \mathbb{R}^n$, prove that $\mathrm{Conv}(v + \mathcal{A}) = v + \mathrm{Conv}(\mathcal{A})$.
b. Prove that a translate of a polytope is a polytope.
c. If a polytope Q is represented by the inequalites (1.4), what are the inequalities defining $v + Q$?

Exercise 17. If $f \in k[x_1^{\pm 1}, \ldots, x_n^{\pm 1}]$ is a Laurent polynomial and $\alpha \in \mathbb{Z}^n$, how is $\mathrm{NP}(x^\alpha f)$ related to $\mathrm{NP}(f)$? Hint: See the previous exercise.

§2 Sparse Resultants

The multipolynomial resultant $\mathrm{Res}_{d_1,\ldots,d_n}(F_1, \ldots, F_n)$ discussed in Chapter 3 is a *very* large polynomial, partly due to the size of the input polynomials $F_1, \ldots, F_n$. They have *lots* of coefficients, especially as their total degree increases. In practice, when people deal with polynomials of large total degree, they rarely use all of the coefficients. It's much more common to encounter *sparse polynomials*, which involve only exponents lying in a finite set $\mathcal{A} \subset \mathbb{Z}^n$. This suggests that there should be a corresponding notion of *sparse resultant*.

To begin our discussion of sparse resultants, we return to the implicitization problem introduced in §2 of Chapter 3. Consider the surface parametrized by the equations

$$
\begin{aligned}
x &= f(s,t) = a_0 + a_1 s + a_2 t + a_3 st \\
(2.1) \qquad y &= g(s,t) = b_0 + b_1 s + b_2 t + b_3 st \\
z &= h(s,t) = c_0 + c_1 s + c_2 t + c_3 st,
\end{aligned}
$$

where $a_0, \ldots, c_3$ are constants. This is sometimes called a *bilinear surface parametrization*. We will assume

$$(2.2) \qquad \det \begin{pmatrix} a_1 & a_2 & a_3 \\ b_1 & b_2 & b_3 \\ c_1 & c_2 & c_3 \end{pmatrix} \neq 0$$

In Exercise 7 at the end of the section, you will show that this condition rules out the trivial case when (2.1) parametrizes a plane.

Our goal is to find the implicit equation of (2.1). This means finding a polynomial $p(x, y, z)$ such that $p(x, y, z) = 0$ if and only if x, y, z are given by (2.1) for some choice of s, t. In Proposition (2.6) of Chapter 3, we used the resultant

$$(2.3) \qquad p(x, y, z) = \mathrm{Res}_{2,2,2}(F - xu^2, G - yu^2, H - zu^2)$$

to find the implicit equation, where F, G, H are the homogenization of f, g, h with respect to u. Unfortunately, this method fails for the case at hand.

Exercise 1. Show that the resultant (2.3) vanishes identically when F, G, H come from homogenizing the polynomials in (2.1). Hint: You already did a special case of this in Exercise 2 of Chapter 3, §2.

The remarkable fact is that although the multipolynomial resultant from Chapter 3 fails, a *sparse resultant* still exists in this case. In Exercise 2 below, you will show that the implicit equation for (2.1) is given by the determinant

$$(2.4) \qquad p(x, y, z) = \det \begin{pmatrix} a_0 - x & a_1 & a_2 & a_3 & 0 & 0 \\ b_0 - y & b_1 & b_2 & b_3 & 0 & 0 \\ c_0 - z & c_1 & c_2 & c_3 & 0 & 0 \\ 0 & a_0 - x & 0 & a_2 & a_1 & a_3 \\ 0 & b_0 - y & 0 & b_2 & b_1 & b_3 \\ 0 & c_0 - x & 0 & c_2 & c_1 & c_3 \end{pmatrix}.$$

Expanding this 6×6 determinant, we see that $p(x, y, z)$ is a polynomial of total degree 2 in x, y and z.

Exercise 2.
a. If x, y, z are as in (2.1), show that the determinant (2.4) vanishes. Hint: Consider the system of equations obtained by multiplying each equation of (2.1) by 1 and s. You should get 6 equations in the 6 "unknowns" $1, s, t, st, s^2, st^2$. Notice the similarity with Proposition (2.10) of Chapter 3.
b. Next assume (2.4) vanishes. We want to prove the existence of s, t such that (2.1) holds. As a first step, let A be the matrix of (2.4) and explain why we can find a nonzero column vector $v = (\alpha_1, \alpha_2, \alpha_3, \alpha_4, \alpha_5, \alpha_6)^T$ (T denotes transpose) such that $Av = 0$. Then use (2.2) to prove that $\alpha_1 \neq 0$. Hint: Write out $Av = 0$ explicitly and use the first three equations. Then use the final three.
c. If we take the vector v of part b and multiply by $1/\alpha_1$, we can write v in the form $v = (1, s, t, \alpha, \beta, \gamma)$. Explain why it suffices to prove that $\alpha = st$.

d. Use (2.2) to prove $\alpha = st$, $\beta = s^2$ and $\gamma = s\alpha$. This will complete the proof that the implicit equation of (2.1) is given by (2.4). Hint: In the equations $Av = 0$, eliminate $a_0 - x$, $b_0 - y$, $c_0 - z$.

e. Explain why the above proof gives a linear algebra method to find s, t for a given point (x, y, z) on the surface. This solves the *inversion problem* for the parametrized surface. Hint: In the notation of part b, you will show that $s = \alpha_2/\alpha_1$ and $t = \alpha_3/\alpha_1$.

A goal of this section is to explain why a resultant like (2.4) can exist even though the standard multipolynomial resultant (2.3) vanishes identically. The basic reason is that although the equations (2.1) are quadratic in s, t, they do *not* use all monomials of total degree ≤ 2 in s, t. The *sparse resultant* works like the multipolynomial resultant of §2, except that we restrict the exponents occurring in the equations.

For simplicity, we will only treat the special case when all of the equations have exponents lying in the same set, leaving the general case for §6. We will also work exclusively over the field $\mathbb{C}$ of complex numbers. Thus, suppose that the variables are $t_1, \ldots, t_n$, and fix a finite set $\mathcal{A} = \{m_1, \ldots, m_l\} \subset \mathbb{Z}^n$ of exponents. Since negative exponents can occur, we will use the Laurent polynomials

$$f = a_1 t^{m_1} + \cdots + a_l t^{m_l} \in L(\mathcal{A}),$$

as defined in §1. Given $f_0, \ldots, f_n \in L(\mathcal{A})$, we get $n + 1$ equations in n unknowns $t_1, \ldots, t_n$:

(2.5)
$$f_0 = a_{01} t^{m_1} + \cdots + a_{0l} t^{m_l} = 0$$
$$\vdots$$
$$f_n = a_{n1} t^{m_1} + \cdots + a_{nl} t^{m_l} = 0.$$

In seeking solutions of these equations, the presence of negative exponents means that we should consider only *nonzero* solutions of (2.5). We will use the notation

$$\mathbb{C}^* = \mathbb{C} \setminus \{0\}$$

for the set of nonzero complex numbers.

The sparse resultant will be a polynomial in the coefficents a_{ij} which vanishes precisely when we can find a "solution" of (2.5). We put "solution" in quotes because although the previous paragraph suggests that solutions should lie in $(\mathbb{C}^*)^n$, the situation is actually more complicated. For instance, the multipolynomial resultants from Chapter 3 use homogeneous polynomials, which means that the "solutions" lie in projective space. The situation for sparse resultants is similar, though with a twist: a "solution" of (2.5) need not lie in $(\mathbb{C}^*)^n$, but the space where it does lie need not be $\mathbb{P}^n$. For example, we will see in §3 that for equations like (2.1), the "solutions" lie in $\mathbb{P}^1 \times \mathbb{P}^1$ rather than $\mathbb{P}^2$.

To avoid the problem of where the solutions lie, we will take a conservative approach and initially restrict the solutions to lie in $(\mathbb{C}^*)^n$. Then, in (2.5), the coefficients give a point $(a_{ij}) \in \mathbb{C}^{(n+1) \times l}$, and we consider the subset

$$Z_0(\mathcal{A}) = \{(a_{ij}) \in \mathbb{C}^{(n+1) \times l} : (2.5) \text{ has a solution in } (\mathbb{C}^*)^n\}.$$

Since $Z_0(\mathcal{A})$ might not be a variety in $\mathbb{C}^{(n+1)l}$, we use the following fact:

- (Zariski Closure) Given a subset $S \subset \mathbb{C}^m$, there is a smallest affine variety $\overline{S} \subset \mathbb{C}^m$ containing S. We call $\overline{S}$ the *Zariski closure* of S.

(See, for example, [CLO], §4 of Chapter 4.) Then let $Z(\mathcal{A}) = \overline{Z_0(\mathcal{A})}$ be the Zariski closure of $Z_0(\mathcal{A})$.

The sparse resultant will be the equation defining $Z(\mathcal{A}) \subset \mathbb{C}^{(n+1)l}$. To state our result precisely, we introduce a variable u_{ij} for each coefficient a_{ij}. Then, for a polynomial $P \in \mathbb{C}[u_{ij}]$, we let $P(f_0, \ldots, f_n)$ denote the number obtained by replacing each variable u_{ij} with the corresponding coefficient a_{ij} from (2.5). We can now state the basic existence result for the sparse resultant.

(2.6) Theorem. *Let $\mathcal{A} \subset \mathbb{Z}^n$ be a finite set, and assume that $\mathrm{Conv}(\mathcal{A})$ is an n-dimensional polytope. Then there is an irreducible polynomial $\mathrm{Res}_{\mathcal{A}} \in \mathbb{Z}[u_{ij}]$ such that for $(a_{ij}) \in \mathbb{C}^{(n+1)l}$, we have*

$$(a_{ij}) \in Z(\mathcal{A}) \Longleftrightarrow \mathrm{Res}_{\mathcal{A}}(a_{ij}) = 0.$$

In particular, if (2.5) has a solution with $t_1, \ldots, t_n \in \mathbb{C}^*$, then

$$\mathrm{Res}_{\mathcal{A}}(f_0, \ldots, f_n) = 0.$$

PROOF. See [GKZ], Chapter 8.   □

The sparse resultant or $\mathcal{A}$-resultant is the polynomial $\mathrm{Res}_{\mathcal{A}}$. Notice that $\mathrm{Res}_{\mathcal{A}}$ is determined uniquely up to $\pm$ since it is irreducible in $\mathbb{Z}[u_{ij}]$. The condition that the convex hull of $\mathcal{A}$ has dimension n is needed to ensure that we have the right number of equations in (2.5). Here is an example of what can happen when the convex hull has strictly lower dimension.

Exercise 3. Let $\mathcal{A} = \{(1, 0), (0, 1)\} \subset \mathbb{Z}^2$, so that $f_i = a_{i1}t_1 + a_{i2}t_2$ for $i = 0, 1, 2$. Show that rather than one condition for $f_1 = f_2 = f_3 = 0$ to have a solution, there are three. Hint: See part b of Exercise 1 from Chapter 3, §2.

We next show that the multipolynomial resultant from Chapter 3 is a special case of the sparse resultant. For $d > 0$, let

$$\mathcal{A}_d = \{m \in \mathbb{Z}^n_{\geq 0} : |m| \leq d\}.$$

Also consider variables $x_0, \ldots, x_n$, which will be related to $t_1, \ldots, t_n$ by $t_i = x_i / x_0$ for $1 \leq i \leq n$. Then we homogenize the f_i from (2.5) in the usual way, defining

(2.7) $F_i(x_0, \ldots, x_n) = x_0^d f_i(t_1, \ldots, t_n) = x_0^d f_i(x_1/x_0, \ldots, x_n/x_0)$

for $0 \le i \le n$. This gives $n + 1$ homogeneous polynomials F_i in the $n + 1$ variables $x_0, \ldots, x_n$. Note that the F_i all have total degree d.

(2.8) Proposition. *For $\mathcal{A}_d = \{m \in \mathbb{Z}_{\ge 0}^n : |m| \le d\}$, we have*

$$\mathrm{Res}_{\mathcal{A}_d}(f_0, \ldots, f_n) = \pm\mathrm{Res}_{d,\ldots,d}(F_0, \ldots, F_n),$$

where $\mathrm{Res}_{d,\ldots,d}$ is the multipolynomial resultant from Chapter 3.

PROOF. If (2.5) has a solution $(t_1, \ldots, t_n) \in (\mathbb{C}^*)^n$, then $(x_0, \ldots, x_n) = (1, t_1, \ldots, t_n)$ is a nontrivial solution of $F_0 = \cdots = F_n = 0$. This shows that $\mathrm{Res}_{d,\ldots,d}$ vanishes on $Z_0(\mathcal{A}_d)$. By the definition of Zariski closure, it must vanish on $Z(\mathcal{A}_d)$. Since $Z(\mathcal{A}_d)$ is defined by the irreducible equation $\mathrm{Res}_{\mathcal{A}_d} = 0$, the argument of Proposition (2.10) of Chapter 3 shows that $\mathrm{Res}_{d,\ldots,d}$ is a multiple of $\mathrm{Res}_{\mathcal{A}_d}$. But $\mathrm{Res}_{d,\ldots,d}$ is an irreducible polynomial by Theorem (2.3) of Chapter 3, and the desired equality follows.   □

Because $\mathcal{A}_d = \{m \in \mathbb{Z}_{\ge 0}^n : |m| \le d\}$ gives *all* exponents of total degree at most d, the multipolynomial resultant $\mathrm{Res}_{d,\ldots,d}$ is sometimes called the *dense* resultant, in contrast to the *sparse* resultant $\mathrm{Res}_{\mathcal{A}}$.

We next discuss the structure of the polynomial $\mathrm{Res}_{\mathcal{A}}$ in more detail. Our first question concerns its total degree, which is determined by the convex hull $Q = \mathrm{Conv}(\mathcal{A})$. The intuition is that as Q gets larger, so does the sparse resultant. As in §1, we measure the size of Q using its volume $\mathrm{Vol}_n(Q)$. This affects the degree of $\mathrm{Res}_{\mathcal{A}}$ as follows.

(2.9) Theorem. *Let $\mathcal{A} = \{m_1, \ldots, m_l\}$, and assume that every element of $\mathbb{Z}^n$ is an integer linear combination of $m_2 - m_1, \ldots, m_l - m_1$. Then, if we fix i between 0 and n, $\mathrm{Res}_{\mathcal{A}}$ is homogeneous in the coefficients of each f_i of degree $n! \, \mathrm{Vol}_n(Q)$, where $Q = \mathrm{Conv}(\mathcal{A})$. This means that*

$$\mathrm{Res}_{\mathcal{A}}(f_0, \ldots, \lambda f_i, \ldots, f_n) = \lambda^{n! \, \mathrm{Vol}_n(Q)} \mathrm{Res}_{\mathcal{A}}(f_0, \ldots, f_n).$$

Furthermore, the total degree of $\mathrm{Res}_{\mathcal{A}}$ is $(n + 1)! \, \mathrm{Vol}_n(Q)$.

PROOF. The first assertion is proved in [GKZ], Chapter 8. As we observed in Exercise 1 of Chapter 3, §3, the final assertion follows by considering $\mathrm{Res}_{\mathcal{A}}(\lambda f_0, \ldots, \lambda f_n)$.   □

For an example of Theorem (2.9), note that $\mathcal{A}_d = \{m \in \mathbb{Z}_{\ge 0}^n : |m| \le d\}$ satisfies the hypothesis of the theorem, and its convex hull has volume $d^n/n!$ by Exercise 3 of §1. Using Proposition (2.8), we conclude that $\mathrm{Res}_{d,\ldots,d}$ has degree d^n in F_i. This agrees with the prediction of Theorem (3.1) of Chapter 3.

We can also explain how the hypothesis of Theorem (2.9) relates to Theorem (2.6). If the $m_i - m_1$ span over $\mathbb{Z}$, they also span over $\mathbb{R}$, so that the convex hull $Q = \mathrm{Conv}(\mathcal{A})$ has dimension n by Exercise 13 of §1. Thus Theorem (2.9) places a stronger condition on $\mathcal{A} \subset \mathbb{Z}^n$ than Theorem (2.6). The following example shows what can go wrong if the $m_i - m_1$ don't span over $\mathbb{Z}$.

Exercise 4. Let $\mathcal{A} = \{0, 2\} \subset \mathbb{Z}$, so that $\mathrm{Vol}_1(\mathrm{Conv}(\mathcal{A})) = 2$.
a. Let $f_0 = a_{01} + a_{02}t^2$ and $f_1 = a_{11} + a_{12}t^2$. If the equations $f_0 = f_1 = 0$ have a solution in $(\mathbb{C}^*)^2$, show that $a_{01}a_{12} - a_{02}a_{11} = 0$.
b. Use part a to prove $\mathrm{Res}_{\mathcal{A}}(f_0, f_1) = a_{01}a_{12} - a_{02}a_{11}$.
c. Explain why the formula of part b does not contradict Theorem (2.9).

Using Theorem (2.9), we can now determine some sparse resultants using the methods of earlier sections. For example, suppose $\mathcal{A} = \{(0, 0), (1, 0), (0, 1), (1, 1)\} \subset \mathbb{Z}^2$, and consider the equations

$$f(s, t) = a_0 + a_1 s + a_2 t + a_3 st = 0$$
(2.10) $$g(s, t) = b_0 + b_1 s + b_2 t + b_3 st = 0$$
$$h(s, t) = c_0 + c_1 s + c_2 t + c_3 st = 0.$$

The exercise below will show that that in this case, the sparse resultant is given by a determinant:

(2.11) $$\mathrm{Res}_{\mathcal{A}}(f, g, h) = \pm \det \begin{pmatrix} a_0 & a_1 & a_2 & a_3 & 0 & 0 \\ b_0 & b_1 & b_2 & b_3 & 0 & 0 \\ c_0 & c_1 & c_2 & c_3 & 0 & 0 \\ 0 & a_0 & 0 & a_2 & a_1 & a_3 \\ 0 & b_0 & 0 & b_2 & b_1 & b_3 \\ 0 & c_0 & 0 & c_2 & c_1 & c_3 \end{pmatrix}$$

Exercise 5. As above, let $\mathcal{A} = \{(0, 0), (1, 0), (0, 1), (1, 1)\}$.
a. Adapt the argument of Exercise 2 to show that if (2.10) has a solution in $(\mathbb{C}^*)^2$, then the determinant in (2.11) vanishes.
b. Adapt the argument of Proposition (2.10) of Chapter 3 to show that $\mathrm{Res}_{\mathcal{A}}$ divides the determinant in (2.11).
c. By comparing degrees and using Theorem (2.9), show that the determinant is an integer multiple of $\mathrm{Res}_{\mathcal{A}}$.
d. Show that the integer is ± 1 by computing the determinant when $f = 1 + st$, $g = s$ and $h = t$.

It follows that the implicitization problem (2.1) can be solved by setting

(2.12) $$p(x, y, z) = \mathrm{Res}_{\mathcal{A}}(f - x, g - y, h - z),$$

where $\mathcal{A}$ is as above. Comparing this to (2.3), we see from Proposition (2.8) that $\mathrm{Res}_{2,2,2}$ corresponds to $\mathcal{A}_2 = \mathcal{A} \cup \{(2, 0), (0, 2)\}$. The convex hull of

$\mathcal{A}_2$ is strictly larger than the convex hull of $\mathcal{A}$. This explains why our earlier attempt failed—the convex hull was too big!

We also have the following sparse analog of Theorem (3.5) discussed in Chapter 3.

(2.13) Theorem. *When $\mathcal{A}$ satisfies the hypothesis of Theorem (2.9), the resultant $\mathrm{Res}_{\mathcal{A}}$ has the following properties:*

a. *If $g_i = \sum_{i=0}^{n} b_{ij} f_j$, where (b_{ij}) is an invertible matrix, then*

$$\mathrm{Res}_{\mathcal{A}}(g_0, \ldots, g_n) = \det(b_{ij})^{n!\,\mathrm{Vol}(Q)} \mathrm{Res}_{\mathcal{A}}(f_0, \ldots, f_n).$$

b. *Given indices $1 \le k_0 \le \cdots \le k_n \le l$, the **bracket** $[k_0 \ldots k_n]$ is defined to be the determinant*

$$[k_0 \ldots k_n] = \det(u_{i,k_j}) \in \mathbb{Z}[u_{ij}].$$

Then $\mathrm{Res}_{\mathcal{A}}$ is a polynomial in the brackets $[k_0 \ldots k_n]$.

PROOF. See [GKZ], Chapter 8. As explained in the proof of Theorem (3.5) of Chapter 3, the second part follows from the first. In §4, we will prove that $n!\,\mathrm{Vol}(Q)$ is an integer since Q is a lattice polytope.   □

Exercise 6. As in Exercise 5, let $\mathcal{A} = \{(0,0), (1,0), (0,1), (1,1)\}$. Then prove that

$$(2.14) \qquad \mathrm{Res}_{\mathcal{A}}(f, g, h) = [013][023] - [012][123].$$

Hint: Expand the determinant (2.11) three times along certain well-chosen rows and columns.

The answer to Exercise 6 is more interesting than first meets the eye. Label the points in $\mathcal{A} = \{(0,0), (1,0), (0,1), (1,1)\}$ as $0, 1, 2, 3$, corresponding to the subscripts of the coefficients in (2.10). Then the brackets appearing in (2.14) correspond to the two ways of dividing the square $Q = \mathrm{Conv}(\mathcal{A})$ into triangles. This is illustrated in Fig. 7.4 on the next page, where the figure on the left corresponds to $[013][023]$, and the one on the right to $[012][123]$.

The amazing fact is that this is no accident! In general, when we express $\mathrm{Res}_{\mathcal{A}}$ as a polynomial in the brackets $[k_0 \ldots k_n]$, there is a very deep relationship between certain terms in this polynomial and triangulations of the polytope $Q = \mathrm{Conv}(\mathcal{A})$. The details can be found in [KSZ]. See also [Stu4] for some nice examples.

Many of the other properties of multipolynomial resultants mentioned in §3 and §4 have sparse analogs. We refer the reader to [GKZ, Chapter 8] and [PS2] for further details.

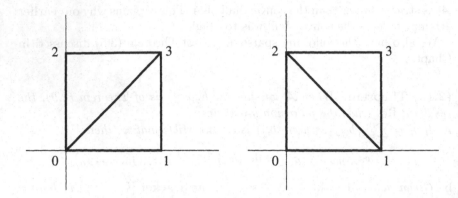

FIGURE 7.4. Triangulations of the unit square

Our account of sparse resultants is by no means complete, and in particular, we have the following questions:

- When $\text{Res}_{\mathcal{A}}(f_0, \ldots, f_n)$ vanishes, the equations (2.5) should have a solution, but where? In §3, we will see that *toric varieties* provide a natural answer to this question.
- What happens when the polynomials in (2.5) have exponents not lying in the same set $\mathcal{A}$? We will explore what happens in §6.
- How do we compute $\text{Res}_{\mathcal{A}}(f_0, \ldots, f_n)$? We will (very briefly) sketch one method in §6 and give references to other methods in the literature.
- What are sparse resultants good for? We've used them for implicitization in (2.12), and applications to solving equations will be covered in §6. A brief discussion of applications to geometric modeling, computational geometry, vision and molecular structure can be found in [Emi2].

We should also mention that besides sparse resultants, some other types of resultants have been studied in recent years. For example:

- The paper [BEM1] defines a notion of resultant which works for any unirational variety. (A projective variety is *unirational* if there is a surjective rational map from $\mathbb{P}^n$ to the variety.)
- When a unirational variety is a blow-up of $\mathbb{P}^n$, the resultant of [BEM1] is called a *residual resultant*. This is studied in [BEM2] when the center of the blow-up is a complete intersection, and [Bus] considers what happens when the center is a local complete intersection in $\mathbb{P}^2$.
- In a different direction, consider polynomials whose Newton polytopes are rectangles with smaller rectangles cut out of each corner. Because we cut out rectangles, we are not using all lattice points in the convex hull. Some interesting formulas for these resultants are given in [ZG] and [Chi].

ADDITIONAL EXERCISES FOR §2

Exercise 7. Let B be the 3×3 matrix in (2.2). In this exercise, we will show that the parametrization (2.1) lies in a plane $\alpha x + \beta y + \gamma z = \delta$ if and only if $\det(B) = 0$.

a. First, if the parametrization lies in the plane $\alpha x + \beta y + \gamma z = \delta$, then show that $Bv = 0$, where $v = (\alpha, \beta, \gamma)^t$. Hint: If a polynomial in s, t equals zero for all values of s and t, then the coefficients of the polynomial must be zero.

b. Conversely, if $\det(B) = 0$, then we can find a nonzero column vector $v = (\alpha, \beta, \gamma)^t$ such that $Bv = 0$. Show that $\alpha x + \beta y + \gamma z = \delta$ for an appropriately chosen δ.

Exercise 8. Given $\mathcal{A} = \{m_1, \ldots, m_l\} \subset \mathbb{Z}^n$ and $v \in \mathbb{Z}^n$, let $v + \mathcal{A} = \{v + m_1, \ldots, v + m_l\}$. Explain why $\mathrm{Res}_{\mathcal{A}} = \mathrm{Res}_{v+\mathcal{A}}$. Hint: Remember that in defining the resultant, we only use solutions of the equations (2.5) with $t_1, \ldots, t_n \in \mathbb{C}^*$.

Exercise 9. For $\mathcal{A} = \{(0,0), (1,0), (0,1), (1,1), (2,0)\}$, compute $\mathrm{Res}_{\mathcal{A}}$ using the methods of Exercise 5. Hint: Let the variables be s, t, and let the equations be $f = g = h = 0$ with coefficients $a_0, \ldots, c_4$. Multiply each of the three equations by $1, s, t$. This will give you a 9×9 determinant. The tricky part is finding polynomials f, g, h such that the determinant is ± 1. See part d of Exercise 5.

Exercise 10. This exercise will explore the *Dixon resultant* introduced by Dixon in 1908. See Section 2.4 of [Stu4] for some nice examples. Let

$$\mathcal{A}_{l,m} = \{(a,b) \in \mathbb{Z}^2 : 0 \le a \le l, \ 0 \le b \le m\}.$$

Note that $\mathcal{A}_{l,m}$ has $(l+1)(m+1)$ elements. Let the variables be s, t. Our goal is to find a determinant formula for $\mathrm{Res}_{\mathcal{A}_{l,m}}$.

a. Given $f, g, h \in L(\mathcal{A}_{l,m})$, we get equations $f = g = h = 0$. Multiplying these equations by $s^a t^b$ for $(a,b) \in \mathcal{A}_{2l-1,m-1}$, show that you get a system of $6lm$ equations in the $6lm$ "unknowns" $s^a t^b$ for $(a,b) \in \mathcal{A}_{3l-1,2m-1}$. Hint: For $l = m = 1$, this is *exactly* what you did in Exercise 1.

b. If A is the matrix of part a, conclude that $\det(A) = 0$ whenever $f = g = h = 0$ has a solution $(s,t) \in (\mathbb{C}^*)^2$. Also show that $\det(A)$ has total degree $2lm$ in the coefficients of f, and similarly for g and h.

c. What is the volume of the convex hull of $\mathcal{A}_{l,m}$?

d. Using Theorems (2.6) and (2.9), show that $\det(A)$ is a constant multiple of $\mathrm{Res}_{\mathcal{A}_{l,m}}$.

e. Show that the constant is ± 1 by considering $f = 1 + s^l t^m$, $g = s^l$ and $h = t^m$. Hint: In this case, A has $4lm$ rows with only one nonzero entry. Use this to reduce to a $2lm \times 2lm$ matrix.

§3 Toric Varieties

Let $\mathcal{A} = \{m_1, \ldots, m_l\} \subset \mathbb{Z}^n$, and suppose that

$$f_i = a_{i1}t^{m_1} + \cdots + a_{il}t^{m_l}, \qquad i = 0, \ldots, n$$

are $n + 1$ Laurent polynomials in $L(\mathcal{A})$. The basic question we want to answer in this section is: *If* $\mathrm{Res}_{\mathcal{A}}(f_0, \ldots, f_n) = 0$, *where do the equations*

$$(3.1) \qquad\qquad f_0 = \cdots = f_n = 0$$

have a solution? In other words, what does it mean for the resultant to vanish?

For $\mathcal{A}_d = \{m \in \mathbb{Z}_{\geq 0}^n : |M| \leq d\}$, we know the answer. Here, we homogenize $f_0, \ldots, f_n$ as in (2.7) to get $F_0, \ldots, F_n$. Proposition (2.8) implies

$$\mathrm{Res}_{\mathcal{A}_d}(f_0, \ldots, f_n) = \mathrm{Res}_{d,\ldots,d}(F_0, \ldots, F_n),$$

and then Theorem (2.3) of Chapter 3 tells us

$$(3.2) \qquad \mathrm{Res}_{d,\ldots,d}(F_0, \ldots, F_n) = 0 \iff \begin{cases} F_0 = \cdots = F_n = 0 \\ \text{has a nontrivial solution.} \end{cases}$$

Recall that a *nontrivial* solution means $(x_0, \ldots, x_n) \neq (0, \ldots, 0)$, i.e., a solution in $\mathbb{P}^n$. Thus, by going from $(\mathbb{C}^*)^n$ to $\mathbb{P}^n$ and changing to homogeneous coordinates in (3.1), we get a space where the vanishing of the resultant means that our equations have a solution.

To understand what happens in the general case, suppose that $\mathcal{A} = \{m_1, \ldots, m_l\} \subset \mathbb{Z}_{\geq 0}^n$, and assume that $Q = \mathrm{Conv}(\mathcal{A})$ has dimension n. Then consider the map

$$\phi_{\mathcal{A}} : (\mathbb{C}^*)^n \longrightarrow \mathbb{P}^{l-1}$$

defined by

$$(3.3) \qquad\qquad \phi_{\mathcal{A}}(t_1, \ldots, t_n) = (t^{m_1}, \ldots, t^{m_l}).$$

Note that $(t^{m_1}, \ldots, t^{m_l})$ is never the zero vector since $t_i \in \mathbb{C}^*$ for all i. Thus $\phi_{\mathcal{A}}$ is defined on all of $(\mathbb{C}^*)^n$, though the image of $\phi_{\mathcal{A}}$ need not be a subvariety of $\mathbb{P}^{l-1}$. Then the *toric variety* $X_{\mathcal{A}}$ is the Zariski closure of the image of $\phi_{\mathcal{A}}$, i.e.,

$$X_{\mathcal{A}} = \overline{\phi_{\mathcal{A}}((\mathbb{C}^*)^n)} \subset \mathbb{P}^{l-1}.$$

Toric varieties are an important area of research in algebraic geometry and feature in many applications. The reader should consult [GKZ] or [Stu2] for an introduction to toric varieties. There is also a more abstract theory of toric varieties, as described in [Ful]. See [Cox4] for an elementary introduction.

For us, the key fact is that the equations $f_i = a_{i1}t^{m_1} + \cdots + a_{il}t^{m_l} = 0$ from (3.1) extend naturally to $X_{\mathcal{A}}$. To see how this works, let $u_1, \ldots, u_l$

be homogeneous coordinates on $\mathbb{P}^{l-1}$. Then consider the linear function $L_i = a_{i1}u_1 + \cdots + a_{il}u_l$, and notice that $f_i = L_i \circ \phi_{\mathcal{A}}$. However, L_i is not a function on $\mathbb{P}^{l-1}$ since $u_1, \ldots, u_l$ are homogeneous coordinates. But the equation $L_i = 0$ still makes sense on $\mathbb{P}^{l-1}$ (be sure you understand why), so in particular, $L_i = 0$ makes sense on $X_{\mathcal{A}}$. Since L_i and f_i have the same coefficients, we can write $\mathrm{Res}_{\mathcal{A}}(L_0, \ldots, L_n)$ instead of $\mathrm{Res}_{\mathcal{A}}(f_0, \ldots, f_n)$. Then we can characterize the vanishing of the resultant as follows.

(3.4) Theorem.

$$\mathrm{Res}_{\mathcal{A}}(L_0, \ldots, L_n) = 0 \Longleftrightarrow \begin{cases} L_0 = \cdots = L_n = 0 \\ \text{has a solution in } X_{\mathcal{A}}. \end{cases}$$

PROOF. See Proposition 2.1 of Chapter 8 of [GKZ]. This result is also discussed in [KSZ]. □

This theorem tells us that the resultant vanishes if and only if (3.1) has a solution in the toric variety $X_{\mathcal{A}}$. From a more sophisticated point of view, Theorem (3.4) says that $\mathrm{Res}_{\mathcal{A}}$ is closely related to the *Chow form* of $X_{\mathcal{A}}$.

To get a better idea of what Theorem (3.4) means, we will work out two examples. First, if $\mathcal{A}_d = \{m \in \mathbb{Z}_{\geq 0}^n : |m| \leq d\}$, let's show that $X_{\mathcal{A}_d} = \mathbb{P}^n$. Let $x_0, \ldots, x_n$ be homogeneous coordinates on $\mathbb{P}^n$, so that by Exercise 19 of Chapter 3, §4, there are $N = \binom{d+n}{n}$ monomials of total degree d in $x_0, \ldots, x_n$. These monomials give a map

$$\Phi_d : \mathbb{P}^n \longrightarrow \mathbb{P}^{N-1}$$

defined by $\Phi_d(x_0, \ldots, x_n) = (\ldots, x^\alpha, \ldots)$, where we use all monomials x^α of total degree d. In Exercise 6 at the end of the section, you will show that Φ_d is well-defined and one-to-one. We call Φ_d the *Veronese map*. The image of Φ_d is a variety by the following basic fact.

• (Projective Images) Let $\Psi : \mathbb{P}^n \to \mathbb{P}^{N-1}$ be defined by $\Psi(x_0, \ldots, x_n) = (h_1, \ldots, h_N)$, where the h_i are homogeneous of the same degree and don't vanish simultaneously on $\mathbb{P}^n$. Then the image $\Psi(\mathbb{P}^n) \subset \mathbb{P}^{N-1}$ is a variety.

(See §5 of Chapter 8 of [CLO].) For $t_1, \ldots, t_n \in \mathbb{C}^*$, observe that

$$(3.5) \qquad \Phi_d(1, t_1, \ldots, t_n) = \phi_{\mathcal{A}_d}(t_1, \ldots, t_n),$$

where $\phi_{\mathcal{A}_d}$ is from (3.3) (see Exercise 6). Thus $\Phi_d(\mathbb{P}^n)$ is a variety containing $\phi_{\mathcal{A}_d}((\mathbb{C}^*)^n)$, so that $X_{\mathcal{A}_d} \subset \Phi_d(\mathbb{P}^n)$. Exercise 6 will show that equality occurs, so that $X_{\mathcal{A}_d} = \Phi_d(\mathbb{P}^n)$. Finally, since Φ_d is one-to-one, $\mathbb{P}^n$ can be identified with its image under Φ_d (we are omitting some details here), and we conclude that $X_{\mathcal{A}_d} = \mathbb{P}^n$. It follows from Theorem (3.4) that for homogeneous polynomials $F_0, \ldots, F_n$ of degree d,

$$\mathrm{Res}_{d,\ldots,d}(F_0, \ldots, F_n) = 0 \Longleftrightarrow \begin{cases} F_0 = \cdots = F_n = 0 \\ \text{has a solution in } \mathbb{P}^n. \end{cases}$$

Thus we recover the characterization of $\text{Res}_{d,\ldots,d}$ given in (3.2).

For a second example, you will show in the next exercise that $\mathbb{P}^1 \times \mathbb{P}^1$ is the toric variety where the equations (2.10) have a solution when the resultant vanishes.

Exercise 1. Let $\mathcal{A} = \{(0,0),(1,0),(0,1),(1,1)\}$. Then $\phi_{\mathcal{A}}(s,t) = (1,s,t,st) \in \mathbb{P}^3$ and $X_{\mathcal{A}}$ is the Zariski closure of the image of $\phi_{\mathcal{A}}$. A formula for $\text{Res}_{\mathcal{A}}$ is given in (2.11).

a. Let the coordinates on $\mathbb{P}^1 \times \mathbb{P}^1$ be (u,s,v,t), so that (u,s) are homogeneous coordinates on the first $\mathbb{P}^1$ and (v,t) are homogeneous coordinates on the second. Show that the *Segre map* $\Phi : \mathbb{P}^1 \times \mathbb{P}^1 \to \mathbb{P}^3$ defined by $\Phi(u,s,v,t) = (uv,sv,ut,st)$ is well-defined and one-to-one.

b. Show that the image of Φ is $X_{\mathcal{A}}$ and explain why this allows us to identify $\mathbb{P}^1 \times \mathbb{P}^1$ with $X_{\mathcal{A}}$.

c. Explain why the "homogenizations" of f, g, h from (2.10) are

$$\begin{aligned}
F(u,s,v,t) &= a_0 uv + a_1 sv + a_2 ut + a_3 st = 0 \\
(3.6) \qquad G(u,s,v,t) &= b_0 uv + b_1 sv + b_2 ut + b_3 st = 0 \\
H(u,s,v,t) &= c_0 uv + c_1 sv + c_2 ut + c_3 st = 0,
\end{aligned}$$

and then prove that $\text{Res}_{\mathcal{A}}(F,G,H) = 0$ if and only if $F = G = H = 0$ has a solution in $\mathbb{P}^1 \times \mathbb{P}^1$. In Exercises 7 and 8 at the end of the section, you will give an elementary proof of this assertion.

Exercise 1 can be restated as saying that $\text{Res}_{\mathcal{A}}(F,G,H) = 0$ if and only if $F = G = H = 0$ has a *nontrivial* solution (u,s,v,t), where nontrivial now means $(u,s) \neq (0,0)$ and $(v,t) \neq (0,0)$. This is similar to (3.2), except that we "homogenized" (3.1) in a different way, and "nontrivial" has a different meaning.

Our next task is to show that there is a systematic procedure for homogenizing the equations (3.1). The key ingredient will again be the polytope $Q = \text{Conv}(\mathcal{A})$. In particular, we will use the facets and inward normals of Q, as defined in §1. If Q has facets $\mathcal{F}_1, \ldots, \mathcal{F}_N$ with inward pointing normals $\nu_1, \ldots, \nu_N$ respectively, each facet $\mathcal{F}_j$ lies in the supporting hyperplane defined by $m \cdot \nu_j = -a_j$, and according to (1.4), the polytope Q is given by

$$(3.7) \qquad Q = \{m \in \mathbb{R}^n : m \cdot \nu_j \geq -a_j \text{ for all } j = 1, \ldots, N\}.$$

As usual, we assume that $\nu_j \in \mathbb{Z}^n$ is the unique primitive inward pointing normal of the facet $\mathcal{F}_j$.

We now explain how to homogenize the equations (3.1) in the general case. Given the representation of Q as in (3.7), we introduce new variables $x_1, \ldots, x_N$. These "facet variables" are related to $t_1, \ldots, t_n$ by the substitution

$$(3.8) \qquad t_i = x_1^{\nu_{1i}} x_2^{\nu_{2i}} \cdots x_N^{\nu_{Ni}}, \qquad i = 1, \ldots, n$$

where ν_{ji} is the ith coordinate of ν_j. Then the "homogenization" of $f(t_1, \ldots, t_n)$ is

$$(3.9) \qquad F(x_1, \ldots, x_n) = \left(\prod_{j=1}^{N} x_j^{a_j} \right) f(t_1, \ldots, t_n),$$

where each t_i is replaced with (3.8). Note the similarity with (2.7). The homogenization of the monomial t^m will be denoted $x^{\alpha(m)}$. An explicit formula for $x^{\alpha(m)}$ will be given below.

Since the inward normals ν_j can have negative coordinates, negative exponents can appear in (3.8). Nevertheless, the following lemma shows that $x^{\alpha(m)}$ has no negative exponents in the case we are interested in.

(3.10) Lemma. *If $m \in Q$, then $x^{\alpha(m)}$ is a monomial in $x_1, \ldots, x_N$ with nonnegative exponents.*

PROOF. Write $m \in \mathbb{Z}^n$ as $m = \sum_{i=1}^{n} a_i e_i$. Since $\nu_{ji} = \nu_j \cdot e_i$, (3.8) implies

$$(3.11) \qquad t^m = x_1^{m \cdot \nu_1} x_2^{m \cdot \nu_2} \cdots x_N^{m \cdot \nu_N},$$

from which it follows that

$$x^{\alpha(m)} = \left(\prod_{j=1}^{N} x_j^{a_j} \right) x_1^{m \cdot \nu_1} x_2^{m \cdot \nu_2} \cdots x_N^{m \cdot \nu_N}$$
$$= x_1^{m \cdot \nu_1 + a_1} x_2^{m \cdot \nu_2 + a_2} \cdots x_N^{m \cdot \nu_N + a_N}.$$

Since $m \in Q$, (3.7) implies that the exponents of the x_j are ≥ 0. $\qquad \square$

Exercise 2. Give a careful proof of (3.11).

Exercise 3. If we used $+a_j$ rather than $-a_j$ in the description of $Q = \mathrm{Conv}(\mathcal{A})$ in (3.7), what effect would this have on (3.9)? This explains the minus signs in (3.7): they give a nicer homogenization formula.

From the equations (3.1), we get the homogenized equations

$$F_0 = a_{01} x^{\alpha(m_1)} + \cdots + a_{0l} x^{\alpha(m_l)} = 0$$

$$\vdots$$

$$F_n = a_{n1} x^{\alpha(m_1)} + \cdots + a_{nl} x^{\alpha(m_l)} = 0,$$

where F_i is the homogenization of f_i. Notice that Lemma (3.10) applies to these equations since $m_i \in \mathcal{A} \subset Q$ for all i. Also note that $F_0, \ldots, F_n$ and $f_0, \ldots, f_n$ have the same coefficients, so we can write the resultant as $\mathrm{Res}_{\mathcal{A}}(F_0, \ldots, F_n)$.

Exercise 4.
a. For $\mathcal{A}_d = \{ m \in \mathbb{Z}_{\geq 0}^n : |m| \leq d \}$, let the facet variables be $x_0, \ldots, x_n$, where we use the labelling of Exercise 3. Show that $t_i = x_i / x_0$ and that the homogenization of $f(t_1, \ldots, t_n)$ is given precisely by (2.7).

b. For $\mathcal{A} = \{(0,0),(1,0),(0,1),(1,1)\}$, the convex hull $Q = \mathrm{Conv}(\mathcal{A})$ in $\mathbb{R}^2$ is given by the inequalities

$$m \cdot \nu_s \geq 0, \quad m \cdot \nu_u \geq -1, \quad m \cdot \nu_t \geq 0, \quad \text{and } m \cdot \nu_v \geq -1,$$

where $e_1 = \nu_s = -\nu_u$ and $e_2 = \nu_t = -\nu_v$. As indicated by the labelling of the facets, the facet variables are u, s, v, t. This is illustrated in Fig. 7.5 on the next page. Show that the homogenization of (2.10) is precisely the system of equations (3.6).

Our final task is to explain what it means for the equations $F_0 = \cdots = F_n = 0$ to have a "nontrivial" solution. We use the *vertices* of polytope Q for this purpose. Since Q is the convex hull of the finite set $\mathcal{A} \subset \mathbb{Z}^n$, it follows that every vertex of Q lies in $\mathcal{A}$, i.e., the vertices are a special subset of $\mathcal{A}$. This in turn gives a special collection of homogenized monomials which will tell us what "nontrivial" means. The precise definitions are as follows.

(3.12) Definition. Let $x_1, \ldots, x_N$ be facet variables for $Q = \mathrm{conv}(\mathcal{A})$.
a. If $m \in \mathcal{A}$ is a vertex of Q, then we say that $x^{\alpha(m)}$ is a *vertex monomial*.
b. A point $(x_1, \ldots, x_N) \in \mathbb{C}^N$ is *nontrivial* if $x^{\alpha(m)} \neq 0$ for at least one vertex monomial.

Exercise 5.
a. Let $\mathcal{A}_d$ and $x_0, \ldots, x_n$ be as in Exercise 4. Show that the vertex monomials are $x_0^d, \ldots, x_n^d$, and conclude that $(x_0, \ldots, x_n)$ is nontrivial if and only if $(x_0, \ldots, x_n) \neq (0, \ldots, 0)$.
b. Let $\mathcal{A}$ and u, s, v, t be as in Exercise 4. Show that the vertex monomials are uv, ut, sv, st, and conclude that (u, s, v, t) is nontrivial if and only if $(u, s) \neq (0, 0)$ and $(v, t) \neq (0, 0)$.

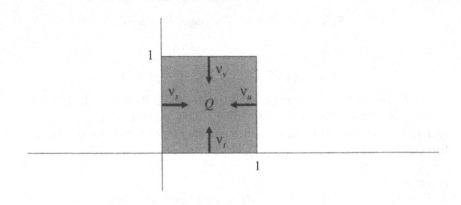

FIGURE 7.5. Facet normals of the unit square

Exercises 4 and 5 show that the homogenizations used in (2.7) and (3.6) are special cases of a theory that works for any set $\mathcal{A}$ of exponents. Once we have the description (3.7) of the convex hull of $\mathcal{A}$, we can read off everything we need, including the facet variables, how to homogenize, and what nontrivial means.

We now come to the main result of this section, which uses the facet variables to give necessary and sufficient conditions for the vanishing of the resultant.

(3.13) Theorem. Let $\mathcal{A} = \{m_1, \ldots, m_l\} \subset \mathbb{Z}_{\geq 0}^n$ be finite, and assume that $Q = \mathrm{Conv}(\mathcal{A})$ is n-dimensional. If $x_1, \ldots, x_N$ are the facet variables, then the homogenized system of equations

$$F_0 = a_{01} x^{\alpha(m_1)} + \cdots + a_{0l} x^{\alpha(m_l)} = 0$$

$$\vdots$$

$$F_n = a_{n1} x^{\alpha(m_1)} + \cdots + a_{nl} x^{\alpha(m_l)} = 0$$

has a nontrivial solution in $\mathbb{C}^N$ if and only if $\mathrm{Res}_{\mathcal{A}}(F_0, \ldots, F_n) = 0$.

PROOF. Let $U \subset \mathbb{C}^N$ consist of all nontrivial points, and notice that $(\mathbb{C}^*)^N \subset U$. Then consider the map Φ defined by

$$\Phi(x_1, \ldots, x_N) = (x^{\alpha(m_1)}, \ldots, x^{\alpha(m_l)}).$$

Since the vertex monomials appear among the $x^{\alpha(m_i)}$, we see that $\Phi(x_1, \ldots, x_N) \neq (0, \ldots, 0)$ when $(x_1, \ldots, x_N) \in U$. Thus Φ can be regarded as a map $\Phi : U \to \mathbb{P}^{l-1}$. By Theorem (3.4), it suffices to prove that the image of Φ is the toric variety $X_{\mathcal{A}}$. To prove this, we will use the following properties of the map Φ:

(i) $\Phi(U)$ is a variety in $\mathbb{P}^{l-1}$.
(ii) $\Phi((\mathbb{C}^*)^N)$ is precisely $\phi_{\mathcal{A}}((\mathbb{C}^*)^n)$.

Assuming (i) and (ii), we see that $\phi_{\mathcal{A}}((\mathbb{C}^*)^n) \subset \Phi(U)$, and since $\Phi(U)$ is a variety, we have $X_{\mathcal{A}} \subset \Phi(U)$. Then the argument of part d of Exercise 6 shows that $X_{\mathcal{A}} = \Phi(U)$, as desired.

The proofs of (i) and (ii) are rather technical and use results from [BC] and [Cox1]. Since Theorem (3.13) has not previously appeared in the literature, we will include the details. What follows is for experts only! For (i), note that [Cox1] implies that Φ factors

$$U \to X_Q \to \mathbb{P}^{l-1},$$

where X_Q is the abstract toric variety determined by Q (see [Ful], §1.5). By Theorem 2.1 of [Cox1], $U \to X_Q$ is a categorical quotient, and in fact, the proof shows that it is a universal categorical quotient (because $\mathbb{C}$ has characteristic 0—see Theorem 1.1 of [FM]). A universal categorical quotient is surjective by §0.2 of [FM], so that $U \to X_Q$ is surjective. This shows that $\Phi(U)$ is the image of $X_Q \to \mathbb{P}^{l-1}$. Since X_Q is a projective variety, a generalization of the Projective Images principle used earlier in this section implies that the image of $X_Q \to \mathbb{P}^{l-1}$ is a variety. We conclude that $\Phi(U)$ is a variety in $\mathbb{P}^{l-1}$.

For (ii), first observe that the restriction of Φ to $(\mathbb{C}^*)^N$ factors

$$(\mathbb{C}^*)^N \xrightarrow{\psi} (\mathbb{C}^*)^n \xrightarrow{\phi_A} \mathbb{P}^{l-1}$$

where ψ is given by (3.8) and ϕ_A is given by (3.3). To prove this, note that by the proof of Lemma (3.11), we can write

$$x^{\alpha(m)} = \left(\prod_{j=1}^N x_j^{a_j} \right) t^m,$$

provided we use ψ to write t^m in terms of $x_0, \ldots, x_N$. It follows that

$$\Phi(x_0, \ldots, x_N) = \left(\prod_{j=1}^N x_j^{a_j} \right) \phi_A(\psi(x_0, \ldots, x_N)).$$

Since we are working in projective space, we conclude that $\Phi = \phi_A \circ \psi$.

Using Remark 8.8 of [BC], we can identify ψ with the restriction of $U \to X_Q$ to $(\mathbb{C}^*)^N$. It follows from [Cox1] (especially the discussion following Theorem 2.1) that ψ is onto, and it follows that

$$\Phi((\mathbb{C}^*)^N) = \phi_A(\psi((\mathbb{C}^*)^N)) = \phi_A((\mathbb{C}^*)^n),$$

which completes the proof of the theorem.   □

The proof of Theorem (3.13) shows that the map $\Phi : U \to X_A$ is surjective, which allows us to think of the facet variables as "homogeneous coordinates" on X_A. However, for this to be useful, we need to understand when two points $P, Q \in U$ correspond to the same point in X_A. In nice cases, there is a simple description of when this happens (see Theorem 2.1 of [Cox1]), but in general, things can be complicated. We should also mention that facet variables and toric varieties have proved to be useful in geometric modeling. See, for example, [CoxKM], [Kra], and [Zub].

There is a *lot* more that one can say about sparse resultants and toric varieties. In Chapter 8, we will discover a different use for toric varieties when we study combinatorial problems arising from magic squares. Toric varieties are also useful in studying solutions of sparse equations, which we will discuss in §5, and the more general sparse resultants defined in §6 also have relations to toric varieties. But before we can get to these topics, we first need to learn more about polytopes.

ADDITIONAL EXERCISES FOR §3

Exercise 6. Consider the Veronese map $\Phi_d : \mathbb{P}^n \to \mathbb{P}^{N-1}$, $N = \binom{n+d}{d}$, as in the text.
a. Show that Φ_d is well-defined. This has two parts: first, you must show that $\Phi_d(x_0, \ldots, x_n)$ doesn't depend on which homogeneous coordinates

you use, and second, you must show that $\Phi_d(x_0, \ldots, x_n)$ never equals the zero vector.

b. Show that Φ_d is one-to-one. Hint: If $\Phi_d(x_0, \ldots, x_n) = \Phi_d(y_0, \ldots, y_n)$, then for some μ, $\mu x^\alpha = y^\alpha$ for all $|\alpha| = d$. Pick i such that $x_i \neq 0$ and let $\lambda = y_i/x_i$. Then show that $\mu = \lambda^d$ and $y_j = \lambda x_j$ for all j.

c. Prove (3.5).

d. Prove that $\Phi_d(\mathbb{P}^n)$ is the Zariski closure of $\phi_{\mathcal{A}_d}((\mathbb{C}^*)^n)$ in $\mathbb{P}^{N-1}$. In concrete terms, this means the following. Let the homogeneous coordinates on $\mathbb{P}^{N-1}$ be $u_1, \ldots, u_N$. If a homogeneous polynomial $H(u_1, \ldots, u_N)$ vanishes on $\phi_{\mathcal{A}_d}((\mathbb{C}^*)^n)$, then prove that H vanishes on $\Phi_d(\mathbb{P}^n)$. Hint: Use (3.5) to show that $x_0 \ldots x_n H \circ \Phi_d$ vanishes identically on $\mathbb{P}^n$. Then argue that $H \circ \Phi_d$ must vanish on $\mathbb{P}^n$.

Exercise 7. Let $\mathcal{A}$ and F, G, H be as in Exercise 1. In this exercise and the next, you will give an elementary proof that $\mathrm{Res}_{\mathcal{A}}(F, G, H) = 0$ if and only if $F = G = H = 0$ has a nontrivial solution (u, s, v, t), meaning $(u, s) \neq (0, 0)$ and $(v, t) \neq (0, 0)$.

a. If $F = G = H = 0$ has a nontrivial solution (u, s, v, t), show that the determinant in (2.11) vanishes. Hint: Multiply the equations by u and s to get 6 equations in the 6 "unknowns" $u^2 v, usv, u^2 t, ust, s^2 v, s^2 t$. Show that the "unknowns" can't all vanish simultaneously.

b. For the remaining parts of the exercise, assume that the determinant (2.11) vanishes. We will find a nontrivial solution of the equations $F = G = H = 0$ by considering 3×3 submatrices (there are four of them) of the matrix

$$\begin{pmatrix} a_0 & a_1 & a_2 & a_3 \\ b_0 & b_1 & b_2 & b_3 \\ c_0 & c_1 & c_2 & c_3 \end{pmatrix}.$$

One of the 3×3 submatrices appears in (2.2), and if its determinant doesn't vanish, show that we can find a solution of the form $(1, s, 1, t)$. Hint: Adapt the argument of Exercise 2 of §2.

c. Now suppose instead that

$$\det \begin{pmatrix} a_0 & a_2 & a_3 \\ b_0 & b_2 & b_3 \\ c_0 & c_2 & c_3 \end{pmatrix} \neq 0.$$

Show that we can find a solution of the form $(u, 1, 1, t)$.

d. The matrix of part b has two other 3×3 submatrices. Show that we can find a nontrivial solution if either of these has nonvanishing determinant.

e. Conclude that we can find a nontrivial solution whenever the matrix of part b has rank 3.

f. If the matrix has rank less than three, explain why it suffices to show that the equations $F = G = 0$ have a nontrivial solution. Hence we are reduced to the case where H is the zero polynomial, which will be considered in the next exercise.

Exercise 8. Continuing the notation of the previous exercise, we will show that the equations $F = G = 0$ always have a nontrivial solution. Write the equations in the form

$$(a_0 u + a_1 s)v + (a_2 u + a_3 s)t = 0$$
$$(b_0 u + b_1 s)v + (b_2 u + b_3 s)t = 0,$$

which is a system of two equations in the unknowns v, t.
a. Explain why we can find $(u_0, s_0) \neq (0, 0)$ such that

$$\det \begin{pmatrix} a_0 u_0 + a_1 s_0 & a_2 u_0 + a_3 s_0 \\ b_0 u_0 + b_1 s_0 & b_2 u_0 + b_3 s_0 \end{pmatrix} = 0.$$

b. Given (u_0, s_0) from part a, explain why we can find $(v_0, t_0) \neq (0, 0)$ such that (u_0, s_0, v_0, t_0) is a nontrivial solution of $F = G = 0$.

Exercise 9. In Exercise 8 of §2, you showed that $\text{Res}_{\mathcal{A}}$ is unchanged if we translate $\mathcal{A}$ by a vector $v \in \mathbb{Z}^n$. You also know that if Q is the convex hull of $\mathcal{A}$, then $v + Q$ is the convex hull of $v + \mathcal{A}$ by Exercise 16 of §1.
a. If Q is represented as in (3.7), show that $v + Q$ is respresented by the inequalities $m \cdot \nu_j \geq -a_j + v \cdot \nu_j$.
b. Explain why $\mathcal{A}$ and $v + \mathcal{A}$ have the same facet variables.
c. Consider $m \in Q$. Show that the homogenization of t^m with respect to $\mathcal{A}$ is equal to the homogenization of t^{v+m} with respect to $v + \mathcal{A}$. This says that the homogenized equations in Theorem (3.13) are unchanged if we replace $\mathcal{A}$ with $v + \mathcal{A}$.

Exercise 10. Let $x_1, \ldots, x_N$ be facet variables for $Q = \text{Conv}(\mathcal{A})$. We say that two monomials x^α and x^β have the same $\mathcal{A}$-*degree* if there is $m \in \mathbb{Z}^n$ such that

$$\beta_j = \alpha_j + m \cdot \nu_j$$

for $j = 1, \ldots, N$.
a. Show that the monomials $x^{\alpha(m)}$, $m \in Q$, have the same $\mathcal{A}$-degree. Thus the polynomials in Theorem (3.13) are $\mathcal{A}$-*homogeneous*, which means that all terms have the same $\mathcal{A}$-degree.
b. If $\mathcal{A}_d$ and $x_0, \ldots, x_n$ are as in part a of Exercise 4, show that two monomials x^α and x^β have the same $\mathcal{A}_d$-degree if and only if they have the same total degree.
c. If $\mathcal{A}$ and u, s, v, t are as in part b of Exercise 4, show that two monomials $u^{a_1} s^{a_2} v^{a_3} t^{a_4}$ and $u^{b_1} s^{b_2} v^{b_3} t^{b_4}$ have the same $\mathcal{A}$-degree if and only if $a_1 + a_2 = b_1 + b_2$ and $a_3 + a_4 = b_3 + b_4$.

Exercise 11. This exercise will explore the notion of "nontrivial" given in Definition (3.12). Let $m \in Q = \text{Conv}(\mathcal{A})$, and let $x_1, \ldots, x_N$ be the facet variables. We define the *reduced monomial* $x_{red}^{\alpha(m)}$ to be the monomial obtained from $x^{\alpha(m)}$ by replacing all nonzero exponents by 1.

a. Prove that

$$x_{red}^{\alpha(m)} = \prod_{m \notin \mathcal{F}_j} x_j.$$

Thus $x_{red}^{\alpha(m)}$ is the product of those facet variables corresponding to the facets *not* containing m. Hint: Look at the proof of Lemma (3.10) and remember that $m \in \mathcal{F}_j$ if and only if $m \cdot \nu_j = -a_j$.

b. Prove that $(x_1, \ldots, x_N)$ is nontrivial if and only if $x_{red}^{\alpha(m)} \neq 0$ for at least one vertex $m \in Q$.

c. Prove that if $m \in Q \cap \mathbb{Z}^n$ is arbitrary, then $x^{\alpha(m)}$ is divisible by some reduced vertex monomial. Hint: The face of Q of smallest dimension containing m is the intersection of those facets $\mathcal{F}_j$ for which $m \cdot \nu_j = -a_j$. Then let m' be a vertex of Q lying in this face.

d. As in the proof of Theorem (3.13), let $U \subset \mathbb{C}^N$ be the set of non-trivial points. If $(x_1, \ldots, x_n) \notin U$, then use parts b and c to show that $(x_1, \ldots, x_n)$ is a solution of the homogenized equations $F_0 = \cdots = F_n = 0$ in the statement of Theorem (3.13). Thus the points in $\mathbb{C}^N - U$ are "trivial" solutions of our equations, which explains the name "nontrivial" for the points of U.

Exercise 12. Let $\mathcal{A} = \{(0,0), (1,0), (0,1), (1,1), (2,0)\}$. In Exercise 9 of §2, you showed that $\text{Res}_{\mathcal{A}}(f, g, h)$ was given by a certain 9×9 determinant. The convex hull of $\mathcal{A}$ is pictured in Fig. 7.2, and you computed the inward normals to be $e_1, e_2, -e_2, -e_1 - e_2$ in Exercise 6 of §1. Let the corresponding facet variables be x_1, x_2, x_3, x_4.

a. What does it mean for (x_1, x_2, x_3, x_4) to be nontrivial? Try to make your answer as nice as possible. Hint: See part b of Exercise 5.

b. Write down explicitly the homogenizations F, G, H of the polynomials f, g, h from Exercise 9 of §2.

c. By combining parts a and b, what is the condition for $\text{Res}_{\mathcal{A}}(F, G, H)$ to vanish?

Exercise 13. In Exercise 10 of §2, you studied the Dixon resultant $\text{Res}_{\mathcal{A}_{l,m}}$, where $\mathcal{A}_{l,m} = \{(a, b) \in \mathbb{Z}^2 : 0 \leq a \leq l, \ 0 \leq b \leq m\}$.

a. Draw a picture of $\text{Conv}(\mathcal{A}_{l,m})$ and label the facets using the variables u, s, v, t (this is similar to what you did in part b of Exercise 4).

b. What is the homogenization of $f \in L(\mathcal{A}_{l,m})$?

c. What does it mean for (u, s, v, t) to be nontrivial?

d. What is the toric variety $X_{\mathcal{A}_{l,m}}$? Hint: It's one you've seen before!

e. Explain how the Dixon resultant can be formulated in terms of *bihomogeneous polynomials*. A polynomial $f \in k[u, s, v, t]$ is bihomogeneous of degree (l, m) if it is homogeneous of degree l as a polynomial in u, s and homogeneous of degree m as a polynomial in v, t.

§4 Minkowski Sums and Mixed Volumes

In this section, we will introduce some important constructions in the theory of convex polytopes. Good general references for this material are [BoF], [BZ], [Ewa] and [Lei]. [Ful] and [GKZ] also contain brief expositions. Throughout, we will illustrate the main ideas using the Newton polytopes (see §1) of the following polynomials:

(4.1)
$$f_1(x, y) = ax^3y^2 + bx + cy^2 + d$$
$$f_2(x, y) = exy^4 + fx^3 + gy.$$

We will assume that the coefficients $a, \ldots, g$ are all non-zero in $\mathbb{C}$.

There are two operations induced by the vector space structure in $\mathbb{R}^n$ that form new polytopes from old ones.

(4.2) Definition. Let P, Q be polytopes in $\mathbb{R}^n$ and let $\lambda \geq 0$ be a real number.
a. The *Minkowski sum* of P and Q, denoted $P + Q$, is

$$P + Q = \{p + q : p \in P \text{ and } q \in Q\},$$

where $p + q$ denotes the usual vector sum in $\mathbb{R}^n$.
b. The polytope λP is defined by

$$\lambda P = \{\lambda p : p \in P\},$$

where λp is the usual scalar multiplication on $\mathbb{R}^n$.

For example, the Minkowski sum of the Newton polytopes $P_1 = \text{NP}(f_1)$ and $P_2 = \text{NP}(f_2)$ from (4.1) is a convex heptagon with vertices $(0, 1), (3, 0), (4, 0), (6, 2), (4, 6), (1, 6)$, and $(0, 3)$. In Fig. 7.6, P_1 is indicated by dashed lines, P_2 by bold lines, and the Minkowski sum $P_1 + P_2$ is shaded.

Exercise 1. In Fig. 7.6, show that the Minkowski sum $P_1 + P_2$ can be obtained by placing a copy of P_1 at every point of P_2. Illustrate your answer with a picture. This works because P_1 contains the origin.

Exercise 2. Let
$$f_1 = a_{20}x^2 + a_{11}xy + a_{02}y^2 + a_{10}x + a_{01}y + a_{00}$$
$$f_2 = b_{30}x^3 + b_{21}x^2y + b_{12}xy^2 + b_{03}y^3 + b_{20}x^2 + \cdots + b_{00}$$

be general ("dense") polynomials of total degrees 2 and 3 respectively. Construct the Newton polytopes $P_i = \text{NP}(f_i)$ for $i = 1, 2$ and find the Minkowski sum $P_1 + P_2$.

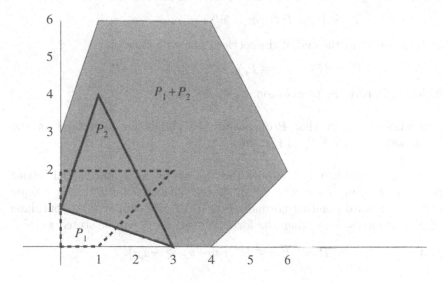

FIGURE 7.6. Minkowski sum of polytopes

Exercise 3.

a. Show that if $f_1, f_2 \in \mathbb{C}[x_1, \ldots, x_n]$ and $P_i = \mathrm{NP}(f_i)$, then $P_1 + P_2 = \mathrm{NP}(f_1 \cdot f_2)$.

b. Show in general that if P_1 and P_2 are polytopes, then their Minkowski sum $P_1 + P_2$ is also convex. Hint: If $P_i = \mathrm{Conv}(\mathcal{A}_i)$, where $\mathcal{A}_i$ is finite, what finite set will give $P_1 + P_2$?

c. Show that a Minkowski sum of lattice polytopes is again a lattice polytope.

d. Show that $P + P = 2P$ for any polytope P. How does this generalize?

Given finitely many polytopes $P_1, \ldots, P_l \subset \mathbb{R}^n$, we can form their Minkowski sum $P_1 + \cdots + P_l$, which is again a polytope in $\mathbb{R}^n$. In §1, we learned about the *faces* of a polytope. A useful fact is that faces of the Minkowski sum $P_1 + \cdots + P_l$ are themselves Minkowski sums. Here is a precise statement.

(4.3) Proposition. *Let $P_1, \ldots, P_r \subset \mathbb{R}^n$ be polytopes in $\mathbb{R}^n$, and let $P = P_1 + \cdots + P_r$ be their Minkowski sum. Then every face P' of P can be expressed as a Minkowski sum*

$$P' = P'_1 + \cdots + P'_r,$$

where each P'_i is a face of P_i.

PROOF. By §1, there is a nonzero vector $\nu \in \mathbb{R}^n$ such that

$$P' = P_\nu = P \cap \{m \in \mathbb{R}^n : m \cdot \nu = -a_P(\nu)\}.$$

In Exercise 12 at the end of the section, you will show that

$$P_\nu = (P_1 + \cdots + P_r)_\nu = (P_1)_\nu + \cdots + (P_r)_\nu,$$

which will prove the proposition.   □

Exercise 4. Verify that Proposition (4.3) holds for each facet of the Minkowski sum $P_1 + P_2$ in Fig. 7.6.

We next show how to compute the volume of an n-dimensional lattice polytope P using its facets. As in §1, each facet $\mathcal{F}$ of P has a unique primitive inward pointing normal $\nu_{\mathcal{F}} \in \mathbb{Z}^n$. If the supporting hyperplane of $\mathcal{F}$ is $m \cdot \nu_{\mathcal{F}} = -a_{\mathcal{F}}$, then the formula (1.4) for P can be stated as

$$(4.4) \qquad P = \bigcap_{\mathcal{F}} \{m \in \mathbb{R}^n : m \cdot \nu_{\mathcal{F}} \geq -a_{\mathcal{F}}\},$$

where the intersection is over all facets $\mathcal{F}$ of P. Recall also that in the notation of (1.3), $a_{\mathcal{F}} = a_P(\nu_{\mathcal{F}})$.

Let $\nu_{\mathcal{F}}^{\perp}$ denote the $(n-1)$-dimensional subspace defined by $m \cdot \nu_{\mathcal{F}} = 0$. Then $\nu_{\mathcal{F}}^{\perp} \cap \mathbb{Z}^n$ is closed under addition and scalar multiplication by integers. One can prove that $\nu_{\mathcal{F}}^{\perp} \cap \mathbb{Z}^n$ is a *lattice of rank* $n-1$, which means there are $n-1$ vectors $w_1, \ldots, w_{n-1} \in \nu_{\mathcal{F}}^{\perp} \cap \mathbb{Z}^n$ such that every element of $\nu_{\mathcal{F}}^{\perp} \cap \mathbb{Z}^n$ is a unique linear combination of $w_1, \ldots, w_{n-1}$ with *integer* coefficients. We call $w_1, \ldots, w_{n-1}$ a *basis* of $\nu_{\mathcal{F}}^{\perp} \cap \mathbb{Z}^n$. The existence of $w_1, \ldots, w_{n-1}$ follows from the fundamental theorem on discrete subgroups of Euclidean spaces. Using $w_1, \ldots, w_{n-1}$, we get the set

$$\mathcal{P} = \{\lambda_1 w_1 + \cdots + \lambda_{n-1} w_{n-1} : 0 \leq \lambda_i \leq 1\},$$

which is the called a *fundamental lattice parallelotope* of the lattice $\nu_{\mathcal{F}}^{\perp} \cap \mathbb{Z}^n$.

If S is subset of $\mathbb{R}^n$ lying in any affine hyperplane, we can define the Euclidean volume $\mathrm{Vol}_{n-1}(S)$. In particular, we can define $\mathrm{Vol}_{n-1}(\mathcal{F})$. However, we also need to take the volume of the fundamental lattice parallelotope $\mathcal{P}$ into account. This leads to the following definition.

(4.5) Definition. The *normalized volume* of the facet $\mathcal{F}$ of the lattice polytope P is given by

$$\mathrm{Vol}'_{n-1}(\mathcal{F}) = \frac{\mathrm{Vol}_{n-1}(\mathcal{F})}{\mathrm{Vol}_{n-1}(\mathcal{P})},$$

where $\mathcal{P}$ is a fundamental lattice parallelotope for $\nu_{\mathcal{F}}^{\perp} \cap \mathbb{Z}^n$.

This definition says that the normalized volume is the usual volume scaled so that the fundamental lattice parallelotope has volume 1. In

Exercise 13, you will show that this definition is independent of which fundamental lattice parallelotope we use. We should also mention the following nice formula:

$$\mathrm{Vol}_{n-1}(\mathcal{P}) = \|\nu_{\mathcal{F}}\|,$$

where $\|\nu_{\mathcal{F}}\|$ is the Euclidean length of the vector $\nu_{\mathcal{F}}$. We omit the proof since we will not use this result.

For example, let $P_2 = \mathrm{NP}(f_2) = \mathrm{Conv}(\{(1,4),(3,0),(0,1)\})$ be the Newton polytope of the polynomial f_2 from (4.1). For the facet

$$\mathcal{F} = \mathrm{Conv}(\{(3,0),(0,1)\}),$$

we have $\nu_{\mathcal{F}} = (1,3)$, and the line containing $\mathcal{F}$ is $x + 3y = 3$. It is easy to check that $(3,0)$ and $(0,1)$ are as close together as any pair of integer points in the line $x + 3y = 3$, so the line segment from $(3,0)$ to $(0,1)$ is a translate of the fundamental lattice parallelotope. It follows that

$$\mathrm{Vol}'_1(\mathcal{F}) = 1.$$

Notice that the usual Euclidean length of $\mathcal{F}$ is $\sqrt{10}$. In general, the normalized volume differs from the Euclidean volume.

Exercise 5. Let $P_2 = \mathrm{NP}(f_2)$ be as above.
a. Show that for the facet $\mathcal{G} = \mathrm{Conv}(\{(3,0),(1,4)\})$, we have $\nu_{\mathcal{G}} = (-2,-1)$ and $\mathrm{Vol}'_1(\mathcal{G}) = 2$.
b. Finally, for the facet $\mathcal{H} = \mathrm{Conv}(\{(0,1),(1,4)\})$, show that $\nu_{\mathcal{H}} = (3,-1)$ and $\mathrm{Vol}'_1(\mathcal{H}) = 1$.

Our main reason for introducing the normalized volume of a facet is the following lovely connection between the n-dimensional volume of a polytope and the $(n-1)$-dimensional normalized volumes of its facets.

(4.6) Proposition. *Let P be a lattice polytope in $\mathbb{R}^n$, and assume that P is represented as in (4.4). Then*

$$\mathrm{Vol}_n(P) = \frac{1}{n} \sum_{\mathcal{F}} a_{\mathcal{F}} \, \mathrm{Vol}'_{n-1}(\mathcal{F}),$$

where the sum is taken over all facets of P.

PROOF. See [BoF], [Lei] or [Ewa], Section IV.3. The formula given in these sources is not specifically adapted to lattice polytopes, but with minor modifications, one gets the desired result. Note also that this proposition explains the minus sign used in the equation $m \cdot \nu_{\mathcal{F}} \geq -a_{\mathcal{F}}$ of a supporting hyperplane. $\square$

For an example of Proposition (4.6), we will compute the area of the polytope $P_2 = \mathrm{NP}(f_2)$ of Exercise 5. First note that if we label the facet normals $\nu_{\mathcal{F}} = (1,3)$, $\nu_{\mathcal{G}} = (-2,-1)$ and $\nu_{\mathcal{H}} = (3,-1)$ as above, then P_2 is defined by

$$m \cdot \nu_{\mathcal{F}} \geq 3, \ m \cdot \nu_{\mathcal{G}} \geq -6, \ \text{and} \ m \cdot \nu_{\mathcal{H}} \geq -1.$$

It follows that $a_{\mathcal{F}} = -3$, $a_{\mathcal{G}} = 6$ and $a_{\mathcal{H}} = 1$. Applying Proposition (4.6), the area of P_2 is given by

$$(4.7) \qquad \mathrm{Vol}_2(P_2) = (1/2)(-3 \cdot 1 + 6 \cdot 2 + 1 \cdot 1) = 5.$$

You should check that this agrees with the result obtained from the elementary area formula for triangles.

Exercise 6. Show that the area of the polytope $P_1 = \mathrm{NP}(f_1)$ for f_1 from (4.1) is equal to 4, by first applying Proposition (4.6), and then checking with an elementary area formula.

Proposition (4.6) enables us to prove results about volumes of lattice polytopes using induction on dimension. Here is a nice example which is relevant to Theorem (2.13).

(4.8) Proposition. *If $P \subset \mathbb{R}^n$ is a lattice polytope, then $n! \, \mathrm{Vol}_n(P)$ is an integer.*

PROOF. The proof is by induction on n. Then case $n = 1$ is obvious, so we may assume inductively that the result is true for lattice polytopes in $\mathbb{R}^{n-1}$. By Proposition (4.6), we get

$$n! \, \mathrm{Vol}_n(P) = \sum_{\mathcal{F}} a_{\mathcal{F}} \cdot (n-1)! \, \mathrm{Vol}'_{n-1}(\mathcal{F}).$$

Note that $a_{\mathcal{F}}$ is an integer. If we can show that $(n-1)! \, \mathrm{Vol}'_{n-1}(\mathcal{F})$ is an integer, the proposition will follow.

A basis $w_1, \ldots, w_{n-1}$ of the lattice $\nu_{\mathcal{F}}^{\perp} \cap \mathbb{Z}^n$ gives $\phi : \nu_{\mathcal{F}}^{\perp} \cong \mathbb{R}^{n-1}$ which carries $\nu_{\mathcal{F}}^{\perp} \cap \mathbb{Z}^n \subset \nu_{\mathcal{F}}^{\perp}$ to the usual lattice $\mathbb{Z}^{n-1} \subset \mathbb{R}^{n-1}$. Since the fundamental lattice polytope $\mathcal{P}$ maps to $\{(a_1, \ldots, a_{n-1}) : 0 \leq a_i \leq 1\}$ under ϕ, it follows easily that

$$\mathrm{Vol}'_{n-1}(S) = \mathrm{Vol}_{n-1}(\phi(S)),$$

where Vol_{n-1} is the usual Euclidean volume in $\mathbb{R}^{n-1}$. By translating $\mathcal{F}$, we get a lattice polytope $\mathcal{F}' \subset \nu_{\mathcal{F}}^{\perp}$, and then $\phi(\mathcal{F}') \subset \mathbb{R}^{n-1}$ is a lattice polytope in $\mathbb{R}^{n-1}$. Since

$$(n-1)! \, \mathrm{Vol}'_{n-1}(\mathcal{F}) = (n-1)! \, \mathrm{Vol}'_{n-1}(\mathcal{F}') = (n-1)! \, \mathrm{Vol}_{n-1}(\phi(\mathcal{F}')),$$

we are done by our inductive assumption. $\qquad\qquad\qquad\qquad\qquad \square$

Our next result concerns the volumes of linear combinations of polytopes formed according to Definition (4.2).

(4.9) Proposition. *Consider any collection* $P_1, \ldots, P_r$ *of polytopes in* $\mathbb{R}^n$, *and let* $\lambda_1, \ldots, \lambda_r \in \mathbb{R}$ *be nonnegative. Then*

$$\mathrm{Vol}_n(\lambda_1 P_1 + \cdots + \lambda_r P_r)$$

is a homogeneous polynomial function of degree n *in the* λ_i.

PROOF. The proof is by induction on n. For $n = 1$, the $P_i = [\ell_i, r_i]$ are all line segments in $\mathbb{R}$ (possibly of length 0 if some $\ell_i = r_i$). The linear combination $\lambda_1 P_1 + \cdots + \lambda_r P_r$ is the line segment $[\sum_i \lambda_i \ell_i, \sum_i \lambda_i r_i]$, whose length is clearly a homogeneous linear function of the λ_i.

Now assume the proposition has been proved for all combinations of polytopes in $\mathbb{R}^{n-1}$, and consider polytopes P_i in $\mathbb{R}^n$ and $\lambda_i \geq 0$. The polytope $Q = \lambda_1 P_1 + \cdots + \lambda_r P_r$ depends on $\lambda_1, \ldots, \lambda_r$, but as long as $\lambda_i > 0$ for all i, the Q's all have the *same* set of inward pointing facet normals (see Exercise 14 at the end of the section). Then, using the notation of (1.3), we can write the formula of Proposition (4.6) as

$$(4.10) \qquad \mathrm{Vol}_n(Q) = \sum_\nu a_Q(\nu) \mathrm{Vol}'_{n-1}(Q_\nu),$$

where the sum is over the set of common inward pointing facet normals ν. In this situation, the proof of Proposition (4.3) tells us that

$$Q_\nu = \lambda_1 (P_1)_\nu + \cdots + \lambda_r (P_r)_\nu.$$

By the induction hypothesis, for each ν, the volume $\mathrm{Vol}'_{n-1}(Q_\nu)$ in (4.10) is a homogeneous polynomial of degree $n - 1$ in $\lambda_1, \ldots, \lambda_r$ (the details of this argument are similar to what we did in Proposition (4.8)).

Turning to $a_Q(\nu)$, we note that by Exercise 12 at the end of the section,

$$a_Q(\nu) = a_{\lambda_1 P_1 + \cdots + \lambda_r P_r}(\nu) = \lambda_1 a_{P_1}(\nu) + \cdots + \lambda_r a_{P_r}(\nu).$$

Since ν is independent of the λ_i, it follows that $a_Q(\nu)$ is a homogeneous linear function of $\lambda_1, \ldots, \lambda_r$. Multiplying $a_Q(\nu)$ and $\mathrm{Vol}'_{n-1}(Q_\nu)$, we see that each term on the right hand side of (4.10) is a homogeneous polynomial function of degree n, and the proposition follows. □

When $r = n$, we can single out one particular term in the polynomial $\mathrm{Vol}_n(\lambda_1 P_1 + \cdots + \lambda_n P_n)$ that has special meaning for the whole collection of polytopes.

(4.11) Definition. The n-dimensional *mixed volume* of a collection of polytopes $P_1, \ldots, P_n$, denoted

$$MV_n(P_1, \ldots, P_n),$$

is the coefficient of the monomial $\lambda_1 \cdot \lambda_2 \cdots \lambda_n$ in $\mathrm{Vol}_n(\lambda_1 P_1 + \cdots + \lambda_n P_n)$.

Exercise 7.

a. If P_1 is the unit square $\mathrm{Conv}(\{(0,0),(1,0),(0,1),(1,1)\})$ and P_2 is the triangle $\mathrm{Conv}(\{(0,0),(1,0),(1,1)\})$, show that

$$\mathrm{Vol}_2(\lambda_1 P_1 + \lambda_2 P_2) = \lambda_1^2 + 2\lambda_1\lambda_2 + \tfrac{1}{2}\lambda_2^2,$$

and conclude that $MV_2(P_1, P_2) = 2$.

b. Show that if $P_i = P$ for all i, then the mixed volume is given by

$$MV_n(P, P, \ldots, P) = n!\,\mathrm{Vol}_n(P).$$

Hint: First generalize part d of Exercise 3 to prove $\lambda_1 P + \cdots + \lambda_n P = (\lambda_1 + \cdots + \lambda_n)\,P$, and then determine the coefficient of $\lambda_1\lambda_2\cdots\lambda_n$ in $(\lambda_1 + \cdots + \lambda_n)^n$.

The basic properties of the n-dimensional mixed volume are given by the following theorem.

(4.12) Theorem.

a. *The mixed volume $MV_n(P_1, \ldots, P_n)$ is invariant if the P_i are replaced by their images under a volume-preserving transformation of $\mathbb{R}^n$ (for example, a translation).*

b. *$MV_n(P_1, \ldots, P_n)$ is symmetric and linear in each variable.*

c. *$MV_n(P_1, \ldots, P_n) \geq 0$. Furthermore, $MV_n(P_1, \ldots, P_n) = 0$ if one of the P_i has dimension zero (i.e., if P_i consists of a single point), and $MV_n(P_1, \ldots, P_n) > 0$ if every P_i has dimension n.*

d. *The mixed volume of any collection of polytopes can be computed as*

$$MV_n(P_1, \ldots, P_n) = \sum_{k=1}^{n} (-1)^{n-k} \sum_{\substack{I \subset \{1,\ldots,n\} \\ |I|=k}} \mathrm{Vol}_n\Big(\sum_{i \in I} P_i\Big),$$

where $\sum_{i \in I} P_i$ is the Minkowski sum of polytopes.

e. *For all collections of lattice polytopes $P_1, \ldots, P_n$,*

$$MV_n(P_1, \ldots, P_n) = \sum_{\nu} a_{P_1}(\nu) MV'_{n-1}((P_2)_\nu, \ldots, (P_n)_\nu),$$

where $a_{P_1}(\nu)$ is defined in (1.3) and the sum is over all primitive vectors $\nu \in \mathbb{Z}^n$ such that $(P_i)_\nu$ has dimension ≥ 1 for $i = 2, \ldots, n$. The notation $MV'_{n-1}((P_2)_\nu, \ldots, (P_n)_\nu)$ on the right stands for the normalized mixed volume analogous to the normalized volume in Definition (4.5):

$$MV'_{n-1}((P_2)_\nu, \ldots, (P_n)_\nu) = \frac{MV_{n-1}((P_2)_\nu, \ldots, (P_n)_\nu)}{\mathrm{Vol}_{n-1}(\mathcal{P})},$$

where $\mathcal{P}$ is a fundamental lattice parallelotope in the hyperplane $\nu^\perp$ orthogonal to ν.

PROOF. Part a follows directly from the definition of mixed volumes, as does part b. We leave the details to the reader as Exercise 15 below.

The nonnegativity assertion of part c is quite deep, and a proof can be found in [Ful], Section 5.4. This reference also proves positivity when the P_i all have dimension n. If P_i has dimension zero, then adding the term $\lambda_i P_i$ merely translates the sum of the other terms in $\lambda_1 P_1 + \cdots + \lambda_n P_n$ by a vector whose length depends on λ_i. The volume of the resulting polytope does not change, so that $\mathrm{Vol}_n(\lambda_1 P_1 + \cdots + \lambda_n P_n)$ is independent of λ_i. Hence the coefficient of $\lambda_1 \cdot \lambda_2 \cdots \lambda_n$ in the expression for the volume must be zero.

For part d, see [Ful], Section 5.4. Part e is a generalization of the volume formula given in Proposition (4.6) and can be deduced from that result. See Exercises 16 and 17 below. Proofs may also be found in [BoF], [Lei] or [Ewa], Section IV.4. Note that by part b of the theorem, only ν with $\dim(P_i)_\nu > 0$ can yield non-zero values for $MV'_{n-1}((P_2)_\nu, \ldots, (P_n)_\nu)$. $\square$

For instance, let's use Theorem (4.12) to compute the mixed volume $MV_2(P_1, P_2)$ for the Newton polytopes of the polynomials from (4.1). In the case of two polytopes in $\mathbb{R}^2$, the formula of part d reduces to:

$$MV_2(P_1, P_2) = -\mathrm{Vol}_2(P_1) - \mathrm{Vol}_2(P_2) + \mathrm{Vol}_2(P_1 + P_2).$$

Using (4.7) and Exercise 5, we have $\mathrm{Vol}_2(P_1) = 4$ and $\mathrm{Vol}_2(P_2) = 5$. The Minkowski sum $P_1 + P_2$ is the heptagon pictured in Fig. 7.6 above. Its area may be found, for example, by subdividing the heptagon into four trapezoids bounded by the horizontal lines $y = 0, 1, 2, 3, 6$. Using that subdivision, we find

$$\mathrm{Vol}_2(P_1 + P_2) = 3 + 11/2 + 23/4 + 51/4 - 27.$$

The mixed volume is therefore

(4.13) $$MV_2(P_1, P_2) = -4 - 5 + 27 = 18.$$

Exercise 8. Check the result of this computation using the formula of part e of Theorem (4.12). Hint: You will need to compute $a_{P_1}(\nu_{\mathcal{F}})$, $a_{P_1}(\nu_{\mathcal{G}})$ and $a_{P_1}(\nu_{\mathcal{H}})$, where $\nu_{\mathcal{F}}, \nu_{\mathcal{G}}, \nu_{\mathcal{H}}$ are the inward normals to the facets $\mathcal{F}, \mathcal{G}, \mathcal{H}$ of P_2.

In practice, computing the mixed volume $MV_n(P_1, \ldots, P_n)$ using the formulas given by parts d and e of Theorem (4.12) can be very time consuming. A better method, due to Sturmfels and Huber [HuS1] and Canny and Emiris [EC], is given by the use of a *mixed subdivision* of the Minkowski sum $P_1 + \cdots + P_n$. A brief description of mixed subdivisions will be given in §6, where we will also give further references and explain how to obtain software for computing mixed volumes.

Exercise 9. Let $P_1, \ldots, P_n$ be lattice polytopes in $\mathbb{R}^n$.
a. Prove that the mixed volume $MV_n(P_1, \ldots, P_n)$ is an integer.

b. Explain how the result of part a generalizes Proposition (4.8). Hint: Use Exercise 7.

We should remark that there are several different conventions in the literature concerning volumes and mixed volumes. Some authors include an extra factor of $1/n!$ in the definition of the mixed volume, so that $MV_n(P, \ldots, P)$ will be exactly equal to $\mathrm{Vol}_n(P)$. When this is done, the right side of the formula from part d of Theorem (4.12) acquires an extra $1/n!$. Other authors include the extra factor of $n!$ in the definition of Vol_n itself (so that the "volume" of the n-dimensional simplex is 1). In other words, care should be taken in comparing the formulas given here with those found elsewhere!

ADDITIONAL EXERCISES FOR §4

Exercise 10. Let $P_1, \ldots, P_r$ be polytopes in $\mathbb{R}^n$. This exercise will show that the dimension of $\lambda_1 P_1 + \cdots + \lambda_r P_r$ is independent of the the λ_i, provided all $\lambda_i > 0$.
a. If $\lambda > 0$ and $p_0 \in P$, show that $(1 - \lambda)p_0 + \mathrm{Aff}(\lambda P + Q) = \mathrm{Aff}(P + Q)$. This uses the affine subspaces discussed in Exercises 12 and 13 of §1. Hint: $(1 - \lambda)p_0 + \lambda p + q = \lambda(p + q) - \lambda(p_0 + q) + p_0 + q$.
b. Conclude that $\dim(\lambda P + Q) = \dim(P + Q)$.
c. Prove that $\dim(\lambda_1 P_1 + \cdots + \lambda_r P_r)$ is independent of the the λ_i, provided all $\lambda_i > 0$.

Exercise 11. Let $m \cdot \nu = -a_P(\nu)$ be a supporting hyperplane of $P = \mathrm{Conv}(\mathcal{A})$, where $\mathcal{A} \subset \mathbb{R}^n$ is finite. Prove that

$$P_\nu = \mathrm{Conv}(\{m \in \mathcal{A} : m \cdot \nu = -a_P(\nu)\}).$$

Exercise 12. Let $a_P(\nu) = -\min_{m \in P}(m \cdot \nu)$ be as in (1.3).
a. Show that $(\lambda P)_\nu = \lambda P_\nu$ and $a_{\lambda P}(\nu) = \lambda a_P(\nu)$.
b. Show that $(P + Q)_\nu = P_\nu + Q_\nu$ and $a_{P+Q}(\nu) = a_P(\nu) + a_Q(\nu)$.
c. Conclude that $(\lambda_1 P_1 + \cdots + \lambda_r P_r)_\nu = \lambda_1 (P_1)_\nu + \cdots + \lambda_r (P_r)_\nu$ and $a_{\lambda_1 P_1 + \cdots + \lambda_r P_r}(\nu) = \lambda_1 a_{P_1}(\nu) + \cdots + \lambda_r a_{P_r}(\nu)$.

Exercise 13. Let $\nu^\perp$ be the hyperplane orthogonal to a nonzero vector $\nu \in \mathbb{Z}^n$, and let $\{w_1, \ldots, w_{n-1}\}$ and $\{w'_1, \ldots, w'_{n-1}\}$ be any two bases for the lattice $\nu^\perp \cap \mathbb{Z}^n$.
a. By expanding the w'_i in terms of the w_j, show that there is an $(n - 1) \times (n - 1)$ integer matrix $A = (a_{ij})$ such that $w'_i = \sum_{i=1}^{n-1} a_{ij} w_j$ for all $i = 1, \ldots, n - 1$.
b. Reversing the roles of the two lattice bases, deduce that A is invertible, and A^{-1} is also an integer matrix.
c. Deduce from part b that $\det(A) = \pm 1$.

d. Show that in the coordinate system defined by $w_1, \ldots, w_{n-1}$, A defines a volume preserving transformation from $\nu^\perp$ to itself. Explain why this shows that any two fundamental lattice parallelotopes in $\nu^\perp$ have the same $(n-1)$-dimensional volume.

Exercise 14. Fix polytopes $P_1, \ldots, P_r$ in $\mathbb{R}^n$ such that $P_1 + \cdots + P_r$ has dimension n. Prove that for any positive reals $\lambda_1, \ldots, \lambda_r$, the polytopes $\lambda_1 P_1 + \cdots + \lambda_r P_r$ all have the *same* inward pointing facet normals. Illustrate your answer with a picture. Hint: If ν is an inward pointing facet normal for $P_1 + \cdots + P_r$, then $(P_1 + \cdots + P_r)_\nu$ has dimension $n-1$. This implies that $(P_1)_\nu + \cdots + (P_r)_\nu$ has dimension $n-1$ by Exercise 12. Now use Exercise 10.

Exercise 15.
a. Using Definition (4.11), show that the mixed volume $MV_n(P_1, \ldots, P_n)$ is invariant under all permutations of the P_i.
b. Show that the mixed volume is linear in each variable:

$$MV_n(P_1, \ldots, \lambda\, P_i + \mu\, P_i', \ldots, P_n)$$
$$= \lambda\, MV_n(P_1, \ldots, P_i, \ldots, P_n) + \mu\, MV_n(P_1, \ldots, P_i', \ldots, P_n)$$

for all $i = 1, \ldots, n$, and all $\lambda, \mu \geq 0$ in $\mathbb{R}$. Hint: When $i = 1$, consider the polynomial representing $\mathrm{Vol}_n(\lambda\, P_1 + \lambda'\, P_1' + \lambda_2 P_2 + \cdots + \lambda_n P_n)$ and look at the coefficients of $\lambda\lambda_2 \cdots \lambda_n$ and $\lambda'\lambda_2 \cdots \lambda_n$.

Exercise 16. In this exercise, we will consider several additional properties of mixed volumes. Let P, Q be polytopes in $\mathbb{R}^n$.
a. If $\lambda, \mu \geq 0$ are in $\mathbb{R}$, show that $\mathrm{Vol}_n(\lambda\, P + \mu\, Q)$ can be expressed in terms of mixed volumes as follows:

$$\frac{1}{n!} \sum_{k=0}^{n} \binom{n}{k} \lambda^k \mu^{n-k} MV_n(P, \ldots, P, Q, \ldots, Q),$$

where in the term corresponding to k, P is repeated k times and Q is repeated $n-k$ times in the mixed volume. Hint: By Exercise 7, $n!\,\mathrm{Vol}_n(\lambda\, P + \mu\, Q) = MV_n(\lambda\, P + \mu\, Q, \ldots, \lambda\, P + \mu\, Q)$.
b. Using part a, show that $MV_n(P, \ldots, P, Q)$ (which appears in the term containing $\lambda^{n-1}\mu$ in the formula of part a) can also be expressed as

$$(n-1)!\, \lim_{\mu \to 0^+} \frac{\mathrm{Vol}_n(P + \mu Q) - \mathrm{Vol}_n(P)}{\mu}.$$

Exercise 17. In this exercise, we will use part b of Exercise 16 to prove part e of Theorem (4.12). Replacing Q by a translate, we may assume that the origin is one of the vertices of Q.
a. Show that the Minkowski sum $P + \mu Q$ can be decomposed into: a sub-polytope congruent to P, prisms over each facet $\mathcal{F}$ of P with height equal

to $\mu \cdot a_Q(\nu) \geq 0$, where $\nu = \nu_{\mathcal{F}}$, and other polyhedra with n-dimensional volume bounded above by a constant times μ^2.

b. From part a, deduce that

$$\mathrm{Vol}_n(P + \mu Q) = \mathrm{Vol}_n(P) + \mu \sum_\nu a_Q(\nu)\mathrm{Vol}'_{n-1}(P_\nu) + O(\mu^2).$$

c. Using part b of Exercise 16, show that

$$MV_n(P, \ldots, P, Q) = (n-1)! \sum_\nu a_Q(\nu)\mathrm{Vol}'_{n-1}(P_\nu),$$

where the sum is over the primitive inward normals ν to the facets of P.

d. Now, to prove part e of Theorem (4.12), substitute

$$P = \lambda_2 P_2 + \cdots + \lambda_n P_n$$

and $Q = P_1$ into the formula of part c and use Exercises 7 and 15.

Exercise 18. Given polytopes $P_1, \ldots, P_r$ in $\mathbb{R}^n$, this exercise will show that *every* coefficient of the polynomial representing

$$\mathrm{Vol}_n(\lambda_1 P_1 + \cdots + \lambda_r P_r)$$

is given by an appropriate mixed volume (up to a constant). We will use the following notation. If $\alpha = (i_1, \ldots, i_r) \in \mathbb{Z}^r_{\geq 0}$ satisfies $|\alpha| = n$, then λ^α is the usual monomial in $\lambda_1, \ldots, \lambda_r$, and let $\alpha! = i_1! i_2! \cdots i_r!$. Also define

$$MV_n(P; \alpha) = MV_n(P_1, \ldots, P_1, P_2, \ldots, P_2, \ldots, P_r, \ldots, P_r),$$

where P_1 appears i_1 times, P_2 appears i_2 times, $\ldots$, P_r appears i_r times. Then prove that

$$\mathrm{Vol}_n(\lambda_1 P_1 + \cdots + \lambda_r P_r) = \sum_{|\alpha|=n} \frac{1}{\alpha!} MV_n(P; \alpha)\lambda^\alpha.$$

Hint: Generalize what you did in part a of Exercise 16.

§5 Bernstein's Theorem

In this section, we will study how the geometry of polytopes can be used to predict the *number* of solutions of a general system of n polynomial (or Laurent polynomial) equations $f_i(x_1, \ldots, x_n) = 0$. We will also indicate how these results are related to a particular class of numerical root-finding methods called *homotopy continuation methods*.

Throughout the section, we will use the following system of equations to illustrate the main ideas:

$$(5.1) \qquad \begin{aligned} 0 &= f_1(x, y) = ax^3y^2 + bx + cy^2 + d \\ 0 &= f_2(x, y) = exy^4 + fx^3 + gy, \end{aligned}$$

where the coefficients $a, \ldots, g$ are in $\mathbb{C}$. These are the same polynomials used in §4. We want to know how many solutions these equations have. We will begin by studying this question using the methods of Chapters 2 and 3, and then we will see that the mixed volume discussed in §4 has an important role to play. This will lead naturally to Bernstein's Theorem, which is the main result of the section.

Let's first proceed as in §1 of Chapter 2 to find the solutions of (5.1). Since different choices of $a, \ldots, g$ could potentially lead to different numbers of solutions, we will initially treat the coefficients $a, \ldots, g$ in (5.1) as symbolic parameters. This means working over the field $\mathbb{C}(a, \ldots, g)$ of rational functions in $a, \ldots, g$. Using a *lex* Gröbner basis to eliminate y, it is easy to check that the reduced Gröbner basis for the ideal $\langle f_1, f_2 \rangle$ in the ring $\mathbb{C}(a, \ldots, g)[x, y]$ has the form

(5.2)
$$0 = y + p_{17}(x)$$
$$0 = p_{18}(x),$$

where $p_{17}(x)$ and $p_{18}(x)$ are polynomials in x alone, of degrees 17 and 18 respectively. The coefficients in p_{17} and p_{18} are rational functions in $a, \ldots, g$. Gröbner basis theory tells us that we can transform (5.2) back into our original equations (5.1), and vice versa. These transformations will also have coefficients in $\mathbb{C}(a, \ldots, g)$.

Now assign numerical values in $\mathbb{C}$ to $a, \ldots, g$. We claim that for "most" choices of $a, \ldots, g \in \mathbb{C}$, (5.1) is still equivalent (5.2). This is because transforming (5.1) into (5.2) and back involves a finite number of elements of $\mathbb{C}(a, \ldots, g)$. If we pick $a, \ldots, g \in \mathbb{C}$ so that none of the denominators appearing in these elements vanish, then our transformations will still work for the chosen numerical values of $a, \ldots, g$. In fact, for most choices, (5.2) remains a Gröbner basis for (5.1)—this is related to the idea of *specialization* of a Gröbner basis, which is discussed in Chapter 6, §3 of [CLO], especially Exercises 7–9.

The equivalence of (5.1) and (5.2) for most choices of $a, \ldots, g \in \mathbb{C}$ can be stated more geometrically as follows. Let $\mathbb{C}^7$ denote the affine space consisting of all possible ways of choosing $a, \ldots, g \in \mathbb{C}$, and let P be the product of all of the denominators appearing in the transformation of (5.1) to (5.2) and back. Note that $P(a, \ldots, g) \neq 0$ implies that all of the denominators are nonvanishing. Thus, (5.1) is equivalent to (5.2) for all coefficients $(a, \ldots, g) \in \mathbb{C}^7$ such that $P(a, \ldots, g) \neq 0$. As defined in §5 of Chapter 3, this means that the two systems of equations are equivalent *generically*. We will make frequent use of the term "generic" in this section.

Exercise 1. Consider the equations (5.1) with symbolic coefficients.

a. Using Maple or another computer algebra system, compute the exact form of the Gröbner basis (5.2) and identify explicitly a polynomial P such that if $P(a, \ldots, g) \neq 0$, then (5.1) is equivalent to a system of the

form (5.2). Hint: One can transform (5.1) into (5.2) using the division
algorithm. Going the other way is more difficult. The Maple package
described in the section on Maple in Appendix D of [CLO] can be used
for this purpose.
b. Show that there is another polynomial P' such that if $P'(a, \ldots, g) \neq 0$,
then the solutions lie in $(\mathbb{C}^*)^2$, where as usual $\mathbb{C}^* = \mathbb{C} \setminus \{0\}$.

Since (5.2) clearly has at most 18 distinct solutions in $\mathbb{C}^2$, the same is true
generically for (5.1). Exercise 8 will show that for generic $(a, \ldots, g)$, p_{18}
has *distinct* solutions, so that (5.1) has precisely 18 solutions in the generic
case. Then, using part b of Exercise 1, we conclude that generically, (5.1)
has 18 solutions, all of which lie in $(\mathbb{C}^*)^2$. This will be useful below.

We next turn to §5 of Chapter 3, where we learned about Bézout's Theorem and solving equations via resultants. Since the polynomials f_1 and f_2
have total degree 5, Bézout's Theorem predicts that (5.1) should have at
most $5 \cdot 5 = 25$ solutions in $\mathbb{P}^2$. If we homogenize these equations using a
third variable z, we get

$$0 = F_1(x, y) = ax^3y^2 + bxz^4 + cy^2z^3 + dz^5$$
$$0 = f_2(x, y) = exy^4 + fx^3z^2 + gyz^4.$$

Here, solutions come in two flavors: affine solutions, which are the solutions
of (5.1), and solutions "at ∞", which have $z = 0$. Assuming $ae \neq 0$ (which
holds generically), it is easy to see that the solutions at ∞ are $(0, 1, 0)$ and
$(1, 0, 0)$. This, combined with Bézout's Theorem, tells us that (5.1) has at
most 23 solutions in $\mathbb{C}^2$.

Why do we get 23 instead of 18, which is the actual number? One way to
resolve this discrepancy is to realize that the solutions $(0, 1, 0)$ and $(1, 0, 0)$
at ∞ have *multiplicities* (in the sense of Chapter 4) bigger than 1. By
computing these multiplicities, one can prove that there are 18 solutions.
However, it is more important to realize that by Bézout's Theorem, *generic*
equations $f_1 = f_2 = 0$ of total degree 5 in x, y have 25 solutions in $\mathbb{C}^2$.
The key point is that the equations in (5.1) are *not* generic in this sense—a
typical polynomial $f(x, y)$ of total degree 5 has 21 terms, while those in
(5.1) have far fewer. In the terminology of §2, we have *sparse* polynomials—
those with fixed Newton polytopes—and what we're looking for is a *sparse*
Bézout's Theorem. As we will see below, this is precisely what Bernstein's
Theorem does for us.

At this point, the reader might be confused about our use of the word
"generic". We just finished saying that the equations (5.1) aren't generic,
yet in our discussion of Gröbner bases, we showed that generically, (5.1)
has 18 solutions. This awkwardness is resolved by observing that *generic is
always relative to a particular set of Newton polytopes*. To state this more
precisely, suppose we fix finite sets $\mathcal{A}_1, \ldots, \mathcal{A}_l \subset \mathbb{Z}^n$. Each $\mathcal{A}_i$ gives the set

$L(\mathcal{A}_i)$ of Laurent polynomials

$$f_i = \sum_{\alpha \in \mathcal{A}_i} c_{i,\alpha} x^\alpha.$$

Note that we can regard each $L(\mathcal{A}_i)$ as an affine space with the coefficients $c_{i,\alpha}$ as coordinates. Then we can define generic as follows.

(5.3) Definition. A property is said to *hold generically* for Laurent polynomials $(f_1, \ldots, f_l) \in L(\mathcal{A}_1) \times \cdots \times L(\mathcal{A}_l)$ if there is a nonzero polynomial in the coefficients of the f_i such that the property holds for all $f_1, \ldots, f_l$ for which the polynomial is nonvanishing.

This definition generalizes Definition (5.6) from Chapter 3. Also observe that by Exercise 10 of §1, the Newton polytope $NP(f_i)$ of a generic $f_i \in L(\mathcal{A}_i)$ satisfies $NP(f_i) = \text{Conv}(\mathcal{A}_i)$. Thus we can speak of *generic polynomials with fixed Newton polytopes*. In particular, for polynomials of total degree 5, Bézout's Theorem deals with generic relative to the Newton polytope determined by *all* monomials $x^i y^j$ with $i + j \leq 5$, while for (5.1), generic means relative to the Newton polytopes of f_1 and f_2. The difference in Newton polytopes explains why there is no conflict between our various uses of the term "generic".

One also could ask if resultants can help solve (5.1). This was discussed in §5 of Chapter 3, where we usually assumed our equations had no solutions at ∞. Since (5.1) does have solutions at ∞, standard procedure suggests making a random change of coordinates in (5.1). With high probability, this would make all of the solutions affine, but it would destroy the sparseness of the equations. In fact, it should be clear that rather than the classical multipolynomial resultants of Chapter 3, we want to use the sparse resultants of §2 of this chapter. Actually, we need something slightly more general, since §2 assumes that the Newton polytopes are all equal, which is not the case for (5.1). In §6 we will learn about more general sparse resultants which can be used to study (5.1).

The above discussion leads to the first main question of the section. Suppose we have Laurent polynomials $f_1, \ldots, f_n \in \mathbb{C}[x_1^{\pm 1}, \ldots, x_n^{\pm 1}]$ such that $f_1 = \cdots = f_n = 0$ have finitely many solutions in $(\mathbb{C}^*)^n$. Then we want to know if there is a way to predict an upper bound on the number of solutions of $f_1 = \cdots = f_n = 0$ in $(\mathbb{C}^*)^n$ that is more refined than the Bézout Theorem bound $\deg(f_1) \cdot \deg(f_2) \cdots \deg(f_n)$. Ideally, we want a bound that uses only information about the forms of the polynomials f_i themselves. In particular, we want to *avoid* computing Gröbner bases and studying the ring $A = \mathbb{C}[x_1, \ldots, x_n]/\langle f_1, \ldots, f_n \rangle$ as in Chapter 2, if possible.

To see how mixed volumes enter the picture, let P_1 and P_2 denote the Newton polytopes of the polynomials f_1, f_2 in (5.1). Referring back to equation (4.13) from the previous section, note that the *mixed volume* of these polytopes satisfies

$$MV_2(P_1, P_2) = 18,$$

which agrees with the number of solutions of the system (5.1) for generic choices of the coefficients. Surely this is no coincidence! As a further test, consider instead two generic polynomials of total degree 5. Here, the Newton polytopes are both the simplex $Q_5 \subset \mathbb{R}^2$ described in Exercise 2 of §1, which has volume $\mathrm{Vol}_2(Q_5) = 25/2$ by Exercise 3 of that section. Using Exercise 7 of §4, we conclude that

$$MV_2(Q_5, Q_5) = 2 \, \mathrm{Vol}_2(Q_5) = 25,$$

so that again, the mixed volume predicts the number of solutions.

Exercise 2. More generally, polynomials of total degrees $d_1, \ldots, d_n$ in $x_1, \ldots, x_n$ have Newton polytopes given by the simplices $Q_{d_1}, \ldots, Q_{d_n}$ respectively. Use the properties of mixed volume from §4 to prove that

$$MV_n(Q_{d_1}, \ldots, Q_{d_n}) = d_1 \cdots d_n,$$

so that the general Bézout bound is the mixed volume of the appropriate Newton polytopes.

The main result of this section is a theorem of Bernstein relating the number of solutions to the mixed volume of the Newton polytopes of the equations. A slightly unexpected fact is that the theorem predicts the numbers of solutions in $(\mathbb{C}^*)^n$ rather than in $\mathbb{C}^n$. We will explain why at the end of the section.

(5.4) Theorem (Bernstein's Theorem). *Given Laurent polynomials $f_1, \ldots, f_n$ over $\mathbb{C}$ with finitely many common zeroes in $(\mathbb{C}^*)^n$, let $P_i = \mathrm{NP}(f_i)$ be the Newton polytope of f_i in $\mathbb{R}^n$. Then the number of common zeroes of the f_i in $(\mathbb{C}^*)^n$ is bounded above by the mixed volume $MV_n(P_1, \ldots, P_n)$. Moreover, for generic choices of the coefficients in the f_i, the number of common solutions is exactly $MV_n(P_1, \ldots, P_n)$.*

PROOF. We will sketch the main ideas in Bernstein's proof, and indicate how $MV_n(P_1, \ldots, P_n)$ solutions of a generic system can be found. However, proving that this construction finds *all* the solutions of a generic system in $(\mathbb{C}^*)^n$ requires some additional machinery. Bernstein uses the theory of Puiseux expansions of algebraic functions for this; a more geometric understanding is obtained via the theory of projective toric varieties. We will state the relevant facts here without proof. For this and other details of the proof, we will refer the reader to [Ber] (references to other proofs will be given below).

The proof is by induction on n. For $n = 1$, we have a single Laurent polynomial $f(x) = 0$ in one variable. After multiplying by a suitable Laurent monomial x^a, we obtain a polynomial equation

$$(5.5) \qquad 0 = \hat{f}(x) = x^a f(x) = c_m x^m + c_{m-1} x^{m-1} + \cdots + c_0,$$

where $m \geq 0$. Multiplying by x^a does not affect the solutions of $f(x) = 0$ in $\mathbb{C}^*$. By the Fundamental Theorem of Algebra, we see that both (5.5) and the original equation $f = 0$ have m roots (counting multiplicity) in $\mathbb{C}^*$ provided $c_m c_0 \neq 0$. Furthermore, as explained in Exercise 8 at the end of the section, $\hat{f}$ has distinct roots when $c_0, \ldots, c_m$ are generic. Thus, generically, $f = 0$ has m distinct roots in $\mathbb{C}^*$. However, the Newton polytope $P = \mathrm{NP}(f)$ is a translate of $\mathrm{NP}(\hat{f})$, which is the interval $[0, m]$ in $\mathbb{R}$. By Exercise 7 of §4, the mixed volume $MV_1(P)$ equals the length of P, which is m. This establishes the base case of the induction.

The induction step will use the geometry of the Minkowski sum $P = P_1 + \cdots + P_n$. The basic idea is that for each primitive inward pointing facet normal $\nu \in \mathbb{Z}^n$ of P, we will deform the equations $f_1 = \cdots = f_n = 0$ by varying the coefficients until some of them are zero. Using the induction hypothesis, we will show that in the limit, the number of solutions of the deformed equations is given by

$$(5.6) \qquad a_{P_1}(\nu) \, MV'_{n-1}((P_2)_\nu, \ldots, (P_n)_\nu),$$

where $a_{P_1}(\nu)$ is defined in (1.3) and $MV'_{n-1}((P_2)_\nu, \ldots, (P_n)_\nu)$ is the normalized $(n-1)$-dimensional mixed volume defined in Theorem (4.12). We will also explain how each of these solutions contributes a solution to our original system. Adding up these solutions over all facet normals ν of P gives the sum

$$(5.7) \qquad \sum_\nu a_{P_1}(\nu) \, MV'_{n-1}((P_2)_\nu, \ldots, (P_n)_\nu) = MV_n(P_1, \ldots, P_n),$$

where the equality follows from Theorem (4.12). To complete the induction step, we would need to show that the total number of solutions of the original system in $(\mathbb{C}^*)^n$ is generically equal to, and in any case no larger than, the sum given by (5.7). The proof is beyond the scope of this book, so we will not do this. Instead, we will content ourselves with showing explicitly how each facet normal ν of P gives a deformation of the equations $f_1 = \cdots = f_n = 0$ which in the limit has (5.6) as its generic number of solutions.

To carry out this strategy, let $\nu \in \mathbb{Z}^n$ be the primitive inward pointing normal to a facet of P. As usual, the facet is denoted P_ν, and we know from §4 that

$$P_\nu = (P_1)_\nu + \cdots + (P_n)_\nu,$$

where $(P_i)_\nu$ is the face (not necessarily a facet) of the Newton polytope $P_i = \mathrm{NP}(f_i)$ determined by ν. By §1, $(P_i)_\nu$ is the convex hull of those α minimizing $\nu \cdot \alpha$ among the monomials x^α from f_i. In other words, if the face $(P_i)_\nu$ lies in the hyperplane $m \cdot \nu = -a_{P_i}(\nu)$, then for all exponents α

of f_i, we have

$$\alpha \cdot \nu \geq -a_{P_i}(\nu),$$

with equality holding if and only if $\alpha \in (P_i)_\nu$. This means that f_i can be written

$$(5.8) \qquad f_i = \sum_{\nu \cdot \alpha = -a_{P_i}(\nu)} c_{i,\alpha} x^\alpha + \sum_{\nu \cdot \alpha > -a_{P_i}(\nu)} c_{i,\alpha} x^\alpha.$$

Before we can deform our equations, we first need to change f_1 slightly. If we multiply f_1 by $x^{-\alpha}$ for some $\alpha \in P_1$, then we may assume that there is a nonzero constant term c_1 in f_1. This means $0 \in P_1$, so that $a_{P_1}(\nu) \geq 0$ by the above inequality. As noted in the base case, changing f_1 in this way affects neither the solutions of the system in $(\mathbb{C}^*)^n$ nor the mixed volume of the Newton polytopes.

We also need to introduce some new coordinates. In Exercise 9 below, you will show that since ν is primitive, there is an invertible $n \times n$ integer matrix B such that ν is its first row and its inverse is also an integer matrix. If we write $B = (b_{ij})$, then consider the coordinate change

$$(5.9) \qquad x_j \mapsto \prod_{i=1}^{n} y_i^{-b_{ij}}.$$

This maps x_j to the Laurent monomial in the new variables $y_1, \ldots, y_n$ whose exponents are the integers appearing in the jth column of the matrix $-B$. (The minus sign is needed because ν is an inward pointing normal.) Under this change of coordinates, it is easy to check that the Laurent monomial x^α maps to the Laurent monomial $y^{-B\alpha}$, where $B\alpha$ is the usual matrix multiplication, regarding α as a column vector. See Exercise 10 below.

If we apply this coordinate change to f_i, note that a monomial x^α appearing in the first sum of (5.8) becomes

$$y^{-B\alpha} = y_1^{a_{P_i}(\nu)} y_2^{\beta_2} \cdots y_n^{\beta_n}$$

(for some integers $\beta_2, \ldots, \beta_n$) since $\nu \cdot \alpha = -a_{P_i}(\nu)$ and ν is the first row of B. Similarly, a monomial x^α in the second sum of (5.8) becomes

$$y^{-B\alpha} = y_1^{\beta_1} y_2^{\beta_2} \cdots y_n^{\beta_n}, \quad \beta_1 < a_{P_i}(\nu).$$

It follows from (5.8) that f_i transforms into a polynomial of the form

$$g_{i\nu}(y_2, \ldots, y_n) y_1^{a_{P_i}(\nu)} + \sum_{j < a_{P_i}(\nu)} g_{ij\nu}(y_2, \ldots, y_n) y_1^j.$$

Note also that the Newton polytope of $g_{i\nu}(y_2, \ldots, y_n)$ is equal to the image under the linear mapping defined by the matrix B of the face $(P_i)_\nu$.

Thus the equations $f_1 = \cdots = f_n = 0$ map to the new system

$$0 = g_{1\nu}(y_2, \ldots, y_n)y_1^{a_{P_1}(\nu)} + \sum_{j < a_{P_1}(\nu)} g_{1j\nu}(y_2, \ldots, y_n)y_1^j$$

$$0 = g_{2\nu}(y_2, \ldots, y_n)y_1^{a_{P_2}(\nu)} + \sum_{j < a_{P_2}(\nu)} g_{2j\nu}(y_2, \ldots, y_n)y_1^j$$

(5.10)

$$\vdots$$

$$0 = g_{n\nu}(y_2, \ldots, y_n)y_1^{a_{P_n}(\nu)} + \sum_{j < a_{P_n}(\nu)} g_{nj\nu}(y_2, \ldots, y_n)y_1^j$$

under the coordinate change $x^\alpha \mapsto y^{-B\alpha}$. As above, the constant term of f_1 is denoted c_1, and we now deform these equations by substituting

$$c_1 \mapsto \frac{c_1}{t^{a_{P_1}(\nu)}}, \quad y_1 \mapsto \frac{y_1}{t}$$

in (5.10), where t is a new variable, and then multiplying the ith equation by $t^{a_{P_i}(\nu)}$. To see what this looks like, first suppose that $a_{P_1}(\nu) > 0$. This means that in the first equation of (5.10), c_1 is the $j = 0$ term in the sum. Then you can check that the deformation has the effect of leaving c_1 and the $g_{i\nu}$ unchanged, and multiplying all other terms by positive powers of t. It follows that the deformed equations can be written in the form

$$0 = y_{1\nu}(y_2, \ldots, y_n)y_1^{a_{P_1}(\nu)} + c_1 + O(t)$$

$$0 = g_{2\nu}(y_2, \ldots, y_n)y_1^{a_{P_2}(\nu)} + O(t)$$

(5.11)

$$\vdots$$

$$0 = g_{n\nu}(y_2, \ldots, y_n)y_1^{a_{P_n}(\nu)} + O(t),$$

where the notation $O(t)$ means a sum of terms each divisible by t.

When $t = 1$, the equations (5.11) coincide with (5.10). Also, from the point of view of our original equations $f_i = 0$, note that (5.11) corresponds to multiplying each term in the second sum of (5.8) by a positive power of t, with the exception of the constant term c_1 of f_1, which is unchanged.

Now, in (5.11), let $t \to 0$ along a general path in $\mathbb{C}$. This gives the equations

$$0 = g_{1\nu}(y_2, \ldots, y_n)y_1^{a_{P_1}(\nu)} + c_1$$

$$0 = g_{2\nu}(y_2, \ldots, y_n)y_1^{a_{P_2}(\nu)}$$

$$\vdots$$

$$0 = g_{n\nu}(y_2, \ldots, y_n)y_1^{a_{P_n}(\nu)},$$

which, in terms of solutions in $(\mathbb{C}^*)^n$, are equivalent to

$$0 = g_{1\nu}(y_2, \ldots, y_n)y_1^{a_{P_1}(\nu)} + c_1$$
$$0 = g_{2\nu}(y_2, \ldots, y_n)$$

(5.12)

$$\vdots$$

$$0 = g_{n\nu}(y_2, \ldots, y_n).$$

It can be shown that for a sufficiently generic original system of equations, the equations $g_{2\nu} = \cdots = g_{n\nu} = 0$ in (5.12) are generic with respect to $B \cdot (P_2)_\nu, \ldots, B \cdot (P_n)_\nu$. Hence, applying the induction hypothesis to the last $n-1$ equations in (5.12), we see that there are

$$MV_{n-1}(B \cdot (P_2)_\nu, \ldots, B \cdot (P_n)_\nu)$$

possible solutions $(y_2, \ldots, y_n) \in (\mathbb{C}^*)^{n-1}$ of these $n-1$ equations. In Exercise 11 below, you will show that

$$MV_{n-1}(B \cdot (P_2)_\nu, \ldots, B \cdot (P_n)_\nu) = MV'_{n-1}((P_2)_\nu, \ldots, (P_n)_\nu),$$

where MV'_{n-1} is the normalized mixed volume from Theorem (4.12).

For each $(y_2, \ldots, y_n)$ solving the last $n-1$ equations in (5.12), there are $a_{P_1}(\nu)$ possible values for $y_1 \in \mathbb{C}^*$ provided $g_{1\nu}(y_2, \ldots, y_n) \neq 0$ and $c_1 \neq 0$. This is true generically (we omit the proof), so that the total number of solutions of (5.12) is

$$a_{P_1}(\nu)\, MV'_{n-1}((P_2)_\nu, \ldots, (P_n)_\nu),$$

which agrees with (5.6).

The next step is to prove that for each solution $(y_1, \ldots, y_n)$ of (5.12), one can find parametrized solutions $(y_1(t), \ldots, y_n(t))$ of the deformed equations (5.11) satisfying $(y_1(0), \ldots, y_n(0)) = (y_1, \ldots, y_n)$. This step involves some concepts we haven't discussed (the functions $y_i(t)$ are *not* polynomials in t), so we will not go into the details here, though the discussion following the proof will shed some light on what is involved.

Once we have the parametrized solutions $(y_1(t), \ldots, y_n(t))$, we can follow them back to $t = 1$ to get solutions $(y_1(1), \ldots, y_n(1))$ of (5.10). Since the inverse of the matrix B has integer entries, each of these solutions $(y_1(1), \ldots, y_n(1))$ can be converted back to a unique $(x_1, \ldots, x_n)$ using the inverse of (5.9) (see Exercise 10 below). It follows that the equations (5.12) give rise to (5.6) many solutions of our original equations.

This takes care of the case when $a_{P_1}(\nu) > 0$. Since we arranged f_1 so that $a_{P_1}(\nu) \geq 0$, we still need to consider what happens when $a_{P_1}(\nu) = 0$. Here, c_1 lies in the first sum of (5.8) for f_1, so that under our coordinate change, it becomes the constant term of $g_{1\nu}$. This means that instead of (5.11), the first deformed equation can be written as

$$0 = g_{1\nu}(y_2, \ldots, y_n) + O(t)$$

since $a_{P_1}(\nu) = 0$ and c_1 appears in $g_{1\nu}$. Combined with the deformed equations from (5.11) for $2 \leq i \leq n$, the limit as $t \to 0$ gives the equations

$$0 = g_{i\nu}(y_2, \ldots, y_n) y_1^{a_{P_i}(\nu)}, \quad 1 \leq i \leq n.$$

As before, the $(\mathbb{C}^*)^n$ solutions are the same as the solutions of the equations

$$0 = g_{i\nu}(y_2, \ldots, y_n), \quad 1 \leq i \leq n.$$

However, one can show that $g_{1\nu}$ is generic and hence *doesn't* vanish at the solutions of $g_{2\nu} = \cdots = g_{n\nu} = 0$. This means that generically, the $t \to 0$ limit of the deformed system has *no* solutions, which agrees with (5.6).

We conclude that each facet contributes (5.6) many solutions to our original equations, and adding these up as in (5.7), we get the mixed volume $MV_n(P_1, \ldots, P_n)$. This completes our sketch of the proof.   □

In addition to Bernstein's original paper [Ber], there are closely related papers by Kushnirenko [Kus] and Khovanskii [Kho]. For this reason, the mixed volume bound $MV_n(P_1, \ldots, P_n)$ on the number of solutions given in Theorem (5.4) is sometimes called the *BKK bound*. A geometric interpretation of the BKK bound in the context of toric varieties is given in [Ful] and [GKZ], and a more refined version can be found in [Roj3]. Also, [HuS1] and [Roj1] study the genericity conditions needed to ensure that exactly $MV_n(P_1, \ldots, P_n)$ different solutions exist in $(\mathbb{C}^*)^n$. These papers use a variety of methods, including sparse elimination theory and toric varieties.

The proof we sketched for the BKK bound uses the formula

$$\sum_{\nu} a_{P_1}(\nu) \cdot MV'_{n-1}((P_2)_\nu, \ldots, (P_n)_\nu) = MV_n(P_1, \ldots, P_n)$$

from Theorem (4.12). If you look back at the statement of this theorem in §4, you'll see that the sum is actually taken over all facet normals ν such that $(P_2)_\nu, \ldots, (P_n)_\nu$ all have dimension *at least one*. This restriction on ν relates nicely to the proof of the BKK bound as follows.

Exercise 3. In the proof of Theorem (5.4), we obtained the system (5.10) of transformed equations. Suppose that for some i between 2 and n, $(P_i)_\nu$ has dimension zero. Then show that in (5.10), the corresponding $g_{i\nu}$ consists of a single term, and conclude that in the limit (5.12) of the deformed equations, the last $n-1$ equations have *no* solutions generically.

Exercise 4. Consider the equations $f_1 = f_2 = 0$ from (5.1). In this exercise, you will explicitly construct the coordinate changes used in the proof of Bernstein's theorem.

a. Use the previous exercise to show that in this case, the vectors ν that must be considered are all among the facet normals of the polytope $P_2 = NP(f_2)$. These normals, denoted $\nu_\mathcal{F}$, $\nu_\mathcal{G}$ and $\nu_\mathcal{H}$, were computed in Exercise 5 of §4 and in the discussion preceeding that exercise. Also, the mixed volume $MV_2(P_1, P_2) = 18$ was computed in (4.13).

b. Show that $a_{P_1}(\nu_{\mathcal{F}}) = 0$. Hence the term from (5.7) with $\nu = \nu_{\mathcal{F}}$ is zero.

c. For $\nu = \nu_{\mathcal{G}}$, show that

$$B = \begin{pmatrix} -2 & -1 \\ 1 & 0 \end{pmatrix}$$

has ν as first row. Also show that B^{-1} has integer entries.

d. Apply the corresponding change of variables

$$x \mapsto z^2 w^{-1}, \qquad y \mapsto z$$

to (5.1). Note that we are calling the "old" variables x, y and the "new" ones z, w rather than using subscripts. In particular, z plays the role of the variable y_1 used in the proof.

e. After substituting $d \mapsto d/t$ and $z \mapsto z/t$, multiply by the appropriate powers of t to obtain

$$0 = aw^{-3}z^8 + d + t^6 \cdot bw^{-1}z^2 + t^6 \cdot cz^2$$
$$0 = (ew^{-1} + fw^{-3})z^6 + t^5 \cdot gz.$$

f. Let $t \to 0$ and count the number of solutions of the deformed system. Show that this number equals $a_{P_1}(\nu_g)MV_1'(\mathcal{G})$.

g. Finally, carry out steps c–f for the facet $\mathcal{H}$ of P_2, and show we obtain 18 solutions.

Exercise 5. Use Bernstein's theorem to deduce a statement about the number of solutions in $(\mathbb{C}^*)^n$ of a generic system of Laurent polynomial equations $f_1 = \cdots = f_n = 0$ when the Newton polytopes of the f_i are all *equal*. (This was the case considered by Khovanskii in [Kho].)

Exercise 6. Use Bernstein's Theorem and Exercise 2 to obtain a version of the usual Bézout theorem. Your version will be slightly different from those discussed in §5 of Chapter 3 because of the $(\mathbb{C}^*)^n$ restriction.

While the BKK bound tells us about the number of solutions in $(\mathbb{C}^*)^n$, one could also ask about the number of solutions in $\mathbb{C}^n$. For example, for (5.1), we checked earlier that generically, these equations have $MV_2(P_1, P_2) = 18$ solutions in either $\mathbb{C}^2$ or $(\mathbb{C}^*)^2$. However, some surprising things happen if we change the equations slightly.

Exercise 7. Suppose that the equations of (5.1) are $f_1 = f_2 = 0$.

a. Show that generically, the equations $f_1 = x\, f_2 = 0$ have 18 solutions in $(\mathbb{C}^*)^2$ and 20 solutions in $\mathbb{C}^2$. Also show that

$$MV_2(\mathrm{NP}(f_1), \mathrm{NP}(x\, f_2)) = 18.$$

Hint: Mixed volume is unaffected by translation.

b. Show that generically, the equations $y\, f_1 = x\, f_2 = 0$ have 18 solutions in $(\mathbb{C}^*)^2$ and 21 solutions in $\mathbb{C}^2$. Also show that

$$MV_2(\mathrm{NP}(y\, f_1), \mathrm{NP}(x\, f_2)) = 18.$$

This exercise illustrates that multiplying f_1 and f_2 by monomials changes neither the solutions in $(\mathbb{C}^*)^2$ nor the mixed volume, while the number of solutions in $\mathbb{C}^2$ *can* change. There are also examples, not obtained by multiplying by monomials, which have more solutions in $\mathbb{C}^n$ than in $(\mathbb{C}^*)^n$ (see Exercise 13 below). The consequence is that the mixed volume is really tied to the solutions in $(\mathbb{C}^*)^n$. In general, finding the generic number of solutions in $\mathbb{C}^n$ is a more subtle problem. For some recent progress in this area, see [HuS2], [LW], [Roj1], [Roj3], and [RW]. An analysis of genericity conditions for solutions in $\mathbb{C}^n$ appears in [Roj3] and an expository account of recent work in this area (including proofs) can be found in [Roj5].

We will conclude this section with some remarks on how the BKK bound can be combined with numerical methods to actually find the solutions of equations like (5.1). First, recall that for (5.1), Bézout's Theorem gives the upper bound of 25 for the number of solutions, while the BKK bound of 18 is smaller (and gives the exact number generically). For the task of computing numerically all complex solutions of (5.1), the better upper bound 18 is useful information to have, since once we have found 18 solutions, there are no others, and whatever method we are using can terminate.

But what sort of numerical method should we use? Earlier, we discussed methods based on Gröbner bases and resultants. Now we will say a few words about numerical *homotopy continuation methods*, which give another approach to practical polynomial equation solving. The method we will sketch is especially useful for systems whose coefficients are known only in some finite precision approximations, or whose coefficients vary widely in size. Our presentation follows [VVC].

We begin with a point we did not address in the proof of Theorem (5.4): exactly *how* do we extend a solution $(y_1, \ldots, y_n)$ of (5.12) to a parametric solution $(y_1(t), \ldots, y_n(t))$ of the deformed equations (5.11)? In general, the problem is to "track" solutions of systems of equations such as (5.11) where the coefficients depend on a parameter t, and the solutions are thought of as functions of t. General methods for doing this were developed by numerical analysts independently, at about the same time as the BKK bound. See [AG] and [Dre] for general discussion of these *homotopy continuation methods*. The idea is the following. For brevity, we will write a system of equations

$$f_1(x_1, \ldots, x_n) = \cdots = f_n(x_1, \ldots, x_n) = 0$$

more compactly as $f(x) = 0$. To solve $f(x) = 0$, we start with a second system $g(x) = 0$ whose solutions are known in advance. In some versions of this approach, $g(x)$ might have a simpler form than $f(x)$. In others, as we will do below, one takes a known system which we expect has the same number of solutions as $f(x) = 0$.

Then we consider the continuous family of systems

$$(5.13) \qquad\qquad 0 = h(x, t) = c(1 - t)g(x) + tf(x),$$

depending on a parameter t, where $c \in \mathbb{C}$ is some constant which is chosen generically to avoid possible bad special behavior.

When $t = 0$, we get the known system $g(x) = 0$ (up to a constant). Indeed, $g(x) = 0$ is often called the *start system* and (5.13) is called a *homotopy* or *continuation system*. As t changes continuously from 0 to 1 along the real line (or more generally along a path in the complex plane), suppose the rank of the Jacobian matrix of $h(x, t)$ with respect to x:

$$J(x, t) = \left(\frac{\partial h_i}{\partial x_j} (x, t) \right)$$

is n for all values of t. Then, by the Implicit Function Theorem, if x_0 is a solution of $g(x) = 0$, we obtain a solution curve $x(t)$ with $x(0) = x_0$ that is parametrized by algebraic functions of t. The goal is to determine the values of $x(t)$ at $t = 1$, since these will yield the solutions of the system $f(x) = 0$ we are interested in.

To find these parametrized solutions, we proceed as follows. Since we want $h(x(t), t)$ to be identically zero as a function of t, its derivative $\frac{d}{dt} h(x(t), t)$ should also vanish identically. By the multivariable chain rule, we see that the solution functions $x(t)$ satisfy

$$0 = \frac{d}{dt} h(x(t), t) = J(x(t), t) \frac{dx(t)}{dt} + \frac{\partial h}{\partial t}(x(t), t),$$

which gives a system of ordinary differential equations (ODEs):

$$J(x(t), t) \frac{dx(t)}{dt} = -\frac{\partial h}{\partial t}(x(t), t)$$

for the solution functions $x(t)$. Since we also know the initial value $x(0) = x_0$, one possible approach is to use the well-developed theory of numerical methods for ODE initial value problems to construct approximate solutions, continuing this process until approximations to the solution $x(1)$ are obtained.

Alternatively, we could apply an iterative numerical root-finding method (such as the Newton-Raphson method) to solve (5.13). The idea is to take a known solution of (5.13) for $t = 0$ and propagate it in steps of size Δt until $t = 1$. Thus, if we start with a solution $x_0 = x(0)$ for $t = 0$, we can use it as the initial guess for solving

$$h(x(\Delta t), \Delta t) = 0$$

using our given numerical method. Then, once we have $x(\Delta t)$, we use it as the initial guess for solving

$$h(x(2\Delta t), 2\Delta t) = 0$$

by our chosen method. We continue in this way until we have solved $h(x(1), 1) = 0$, which will give the desired solution. This method works because $x(t)$ is a continuous function of t, so that at the step with $t = (k + 1)\Delta t$, we will generally have fairly good estimates for initial points from the results of the previous step (i.e., for $t = k\Delta t$), provided Δt is sufficiently small.

When homotopy continuation methods were first developed, the best commonly known bound on the number of expected solutions was the Bézout theorem bound. A common choice for $g(x)$ was a random dense system with equations of the same total degrees as $f(x)$. But many polynomial systems (for instance (5.1)) have fewer solutions than general dense systems of the same total degrees! When this is true, some of the numerically generated approximate solution paths diverge to infinity as $t \to 1$. This is because the start equations $g(x) = 0$ would typically have many more solutions than the sparse system $f(x) = 0$. Much computational effort can be wasted trying to track them accurately.

As a result, the more refined BKK bound is an important tool in applying homotopy continuation methods. Instead of a random *dense* start system $g(x) = 0$, a much better choice in many cases is a randomly chosen start system for which the g_i have the *same Newton polytopes* as the corresponding f_i:

$$\mathrm{NP}(g_i) = \mathrm{NP}(f_i).$$

Of course, the solutions of $g(x) = 0$ must be determined as well. Unless solutions of some specific system with precisely these Newton polytopes is known, some work must be done to solve the start system before the homotopy continuation method can be applied. For this, the authors of [VVC] propose adapting the deformations used in the proof of Bernstein's theorem, and applying a continuation method again to determine the solutions of $g(x) = 0$. A closely related method, described in [HuS1] and [VGC], uses the mixed subdivisions to be defined in §6. Also, some interesting numerical issues are addressed in [HV]. Some other recent papers on this subject include [DKK], [Li], and [Ver2]. The software PHCpack described in [Ver1] solves systems of equations using the polynomial homotopy continuation method described here. This package is available at `http://www2.math.uic.edu/~jan/PHCpack/phcpack.html`. Other software for polynomial homotopies is described in [Li].

The geometry of polytopes provides powerful tools for understanding sparse systems of polynomial equations. The mixed volume is an efficient bound for the number of solutions, and homotopy continuation methods give practical methods for finding the solutions. This is an active area of research, and further progress is likely in the future.

ADDITIONAL EXERCISES FOR §5

Exercise 8. If $f \in \mathbb{C}[x]$ is a polynomial of degree n, its *discriminant* $\mathrm{Disc}(f)$ is defined to be the resultant

$$\mathrm{Disc}(f) = \mathrm{Res}_{n,n-1}(f, f'),$$

where f' is the derivative of f. One can show that $\mathrm{Disc}(f) \neq 0$ if and only if f has no multiple roots (see Exercises 7 and 8 of Chapter 3, §5 of [CLO]).
a. Show that the generic polynomial $f \in \mathbb{C}[x]$ has no multiple roots. Hint: It suffices to show that the discriminant is a nonzero polynomial in the coefficients of f. Prove this by writing down an explicit polynomial of degree n which has distinct roots.
b. Now let $p_{18} \in \mathbb{C}(a, \ldots, g)[x]$ be the polynomial from (5.2). To show that p_{18} has no multiple roots generically, we need to show that $\mathrm{Disc}(p_{18})$ is nonzero as a rational function of $a, \ldots, g$. Computing this discriminant would be unpleasant since the coefficients of p_{18} are so complicated. So instead, take p_{18} and make a random choice of $a, \ldots, g$. This will give a polynomial in $\mathbb{C}[x]$. Show that the discriminant is nonzero and conclude that p_{18} has no multiple roots for generic $a, \ldots, g$.

Exercise 9. Let $\nu \in \mathbb{Z}^n$ be a primitive vector (thus $\nu \neq 0$ and the entries of ν have no common factor > 1). Our goal is to find an integer $n \times n$ matrix with integer inverse and ν as its first row. For the rest of the exercise, we will regard ν as a column vector. Hence it suffices to find an integer $n \times n$ matrix with integer inverse and ν as its first column.
a. Explain why it suffices to find an integer matrix A with integer inverse such that $A\nu = \vec{e}_1$, where $\vec{e}_1 = (1, 0, \ldots, 0)^T$ is the usual standard basis vector. Hint: Multiply by A^{-1}.
b. An *integer row operation* consists of a row operation of the following three types: switching two rows, adding an integer multiple of a row to another row, and multiplying a row by ± 1. Show that the elementary matrices corresponding to integer row operations are integer matrices with integer inverses.
c. Using parts a and b, explain why it suffices to reduce ν to $\vec{e}_1$ using integer row operations.
d. Using integer row operations, show that ν can be transformed to a vector $(b_1, \ldots, b_n)^T$ where $b_1 > 0$ and $b_1 \leq b_i$ for all i with $b_i \neq 0$.
e. With $(b_1, \ldots, b_n)^T$ as in the previous step, use integer row operations to subtract multiples of b_1 from one of the nonzero entries b_i, $i > 1$, until you get either 0 or something positive and smaller than b_1.

f. By repeatedly applying steps d and e, conclude that we can integer row reduce ν to a positive multiple of $\vec{e}_1$.

g. Finally, show that ν being primitive implies that the previous step gives $\vec{e}_1$ exactly. Hint: Using earlier parts of the exercise, show that we have $A\nu = d\vec{e}_1$, where A has an integer inverse. Then use A^{-1} to conclude that d divides every entry of ν.

Exercise 10.

a. Under the coordinate change (5.9), show that the Laurent monomial x^α, $\alpha \in \mathbb{Z}^n$, maps to the Laurent monomial $y^{-B\alpha}$, where $B\alpha$ is the matrix product.

b. Show that (5.9) actually induces a one-to-one correspondence between Laurent monomials in x and Laurent monomials in y.

c. Show that (5.9) defines a one-to-one, onto mapping from $(\mathbb{C}^*)^n$ to itself. Also explain how $-B^{-1}$ gives the inverse mapping.

Exercise 11. Show that

$$MV_{n-1}(B \cdot (P_2)_\nu, \ldots, B \cdot (P_n)_\nu) = MV'_{n-1}((P_2)_\nu, \ldots, (P_n)_\nu),$$

where the notation is as in the proof of Bernstein's Theorem.

Exercise 12. Consider the following system of three equations in three unknowns:

$$0 = a_1 xy^2 z + b_1 x^4 + c_1 y + d_1 z + c_1$$
$$0 = a_2 xyz^2 + b_2 y^3 + c_2$$
$$0 = a_3 x^3 + b_3 y^2 + c_3 z.$$

What is the BKK bound for the generic number of solutions in $(\mathbb{C}^*)^3$?

Exercise 13. Show that generically, the equations (taken from [RW])

$$0 = ax^2 y + bxy^2 + cx + dy$$
$$0 = ex^2 y + fxy^2 + gx + hy$$

have 4 solutions in $(\mathbb{C}^*)^2$ and 5 solutions in $\mathbb{C}^2$.

§6 Computing Resultants and Solving Equations

The sparse resultant $\text{Res}_A(f_1, \ldots, f_n)$ introduced in §2 requires that the Laurent polynomials $f_1, \ldots, f_n$ be built from monomials using the same set A of exponents. In this section, we will discuss what happens when we allow each f_i to involve different monomials. This will lead to the *mixed sparse resultant*. We also have some unfinished business from §2, namely the problem of computing sparse resultants. For this purpose, we will introduce the notion of a *mixed subdivision*. These will enable us not only to compute sparse resultants but also to find mixed volumes and to solve equations using the methods of Chapter 3.

We begin with a discussion of the mixed sparse resultant. Fix $n + 1$ finite sets $\mathcal{A}_0, \ldots, \mathcal{A}_n \subset \mathbb{Z}^n$ and consider $n + 1$ Laurent polynomials $f_i \in L(\mathcal{A}_i)$. The rough idea is that the resultant

$$\mathrm{Res}_{\mathcal{A}_0,\ldots,\mathcal{A}_n}(f_0, \ldots, f_n)$$

will measure whether or not the $n + 1$ equations in n variables

(6.1) $$f_0(x_1, \ldots, x_n) = \cdots = f_n(x_1, \ldots, x_n) = 0$$

have a solution. To make this precise, we proceed as in §2 and let

$$Z(\mathcal{A}_0, \ldots, \mathcal{A}_n) \subset L(\mathcal{A}_0) \times \cdots \times L(\mathcal{A}_n)$$

be the Zariski closure of the set of all $(f_0, \ldots, f_n)$ for which (6.1) has a solution in $(\mathbb{C}^*)^n$.

(6.2) Theorem. *Assume that $Q_i = \mathrm{Conv}(\mathcal{A}_i)$ is an n-dimensional polytope for $i = 0, \ldots, n$. Then there is an irreducible polynomial $\mathrm{Res}_{\mathcal{A}_0,\ldots,\mathcal{A}_n}$ in the coefficients of the f_i such that*

$$(f_0, \ldots, f_n) \in Z(\mathcal{A}_0, \ldots, \mathcal{A}_n) \iff \mathrm{Res}_{\mathcal{A}_0,\ldots,\mathcal{A}_n}(f_0, \ldots, f_n) = 0.$$

In particular, if (6.1) has a solution $(t_1, \ldots, t_n) \in (\mathbb{C}^*)^n$, then

$$\mathrm{Res}_{\mathcal{A}_0,\ldots,\mathcal{A}_n}(f_0, \ldots, f_n) = 0.$$

This theorem is proved in Chapter 8 of [GKZ]. Note that the mixed sparse resultant includes *all* of the resultants considered so far. More precisely, the (unmixed) sparse resultant from §2 is

$$\mathrm{Res}_{\mathcal{A}}(f_0, \ldots, f_n) = \mathrm{Res}_{\mathcal{A},\ldots,\mathcal{A}}(f_0, \ldots, f_n),$$

and the multipolynomial resultant studied in Chapter 3 is

$$\mathrm{Res}_{d_0,\ldots,d_n}(F_0, \ldots, F_n) = \mathrm{Res}_{\mathcal{A}_0,\ldots,\mathcal{A}_n}(f_0, \ldots, f_n),$$

where $\mathcal{A}_i = \{m \in \mathbb{Z}_{\geq 0}^n : |m| \leq d_i\}$ and F_i is the homogenization of f_i.

We can also determine the degree of the mixed sparse resultant. In §2, we saw that the degree of $\mathrm{Res}_{\mathcal{A}}$ involves the volume of the Newton polytope $\mathrm{Conv}(\mathcal{A})$. For the mixed resultant, this role is played by the mixed volume from §4.

(6.3) Theorem. *Assume that $Q_i = \mathrm{Conv}(\mathcal{A}_i)$ is n-dimensional for each $i = 0, \ldots, n$ and that $\mathbb{Z}^n$ is generated by the differences of elements in $\mathcal{A}_0 \cup \cdots \cup \mathcal{A}_n$. Then, if we fix i between 0 and n, $\mathrm{Res}_{\mathcal{A}_0,\ldots,\mathcal{A}_n}$ is homogeneous in the coefficients of f_i of degree $MV_n(Q_0, \ldots, Q_{i-1}, Q_{i+1}, \ldots, Q_n)$. Thus*

$$\mathrm{Res}_{\mathcal{A}_0,\ldots,\mathcal{A}_n}(f_0, \ldots, \lambda f_i, \ldots, f_n) =$$
$$\lambda^{MV_n(Q_0,\ldots,Q_{i-1},Q_{i+1},\ldots,Q_n)} \mathrm{Res}_{\mathcal{A}_0,\ldots,\mathcal{A}_n}(f_0, \ldots, f_n).$$

A proof can be found in Chapter 8 of [GKZ]. Observe that this result generalizes both Theorem (3.1) of Chapter 3 and Theorem (2.9) of this chapter. There are also more general versions of Theorems (6.2) and (6.3) which don't require that the Q_i be n-dimensional. See, for instance, [Stu3]. Exercise 9 at the end of the section gives a simple example of a sparse resultant where all of the Q_i have dimension $< n$.

We next discuss how to compute sparse resultants. Looking back at Chapter 3, recall that there were wonderful formulas for the multipolynomial case, but it general, computing these resultants was not easy. The known formulas for multipolynomial resultants fall into three main classes:

- Special cases where the resultant is given as a determinant. This includes the resultants $\text{Res}_{l,m}$ and $\text{Res}_{2,2,2}$ from §1 and §2 of Chapter 3.
- The general case where the resultant is given as the GCD of $n + 1$ determinants. This is Proposition (4.7) of Chapter 3.
- The general case where the resultant is given as the quotient of two determinants. This is Theorem (4.9) of Chapter 3.

Do sparse resultants behave similarly? In §2 of this chapter, we gave formulas for the Dixon resultant (see (2.12) and Exercise 10 of §2). Other determinantal formulas for sparse resultants can be found in [CK1], [DE], [Khe], [SZ], and [WZ], so that the first bullet definitely has sparse analogs. We will see below that the second and third bullets also have sparse analogs.

We now introduce our main tool for computing sparse resultants. The idea is to subdivide the Minkowski sum $Q = Q_0 + \cdots + Q_n$ in a special way. We begin with what it means to subdivide a polytope.

(6.4) Definition. Let $Q \subset \mathbb{R}^n$ be a polytope of dimension n. Then a *polyhedral subdivision* of Q consists of finitely many n-dimensional polytopes $R_1, \ldots, R_s$ (the *cells* of the subdivision) such that $Q = R_1 \cup \cdots \cup R_s$ and for $i \neq j$, the intersection $R_i \cap R_j$ is a face of both R_i and R_j.

For example, Fig. 7.7 below shows three ways of dividing a square into smaller pieces. The first two are polyedral subdivisions, but the third isn't since $R_1 \cap R_2$ is not a face of R_1 (and $R_1 \cap R_3$ has a similar problem).

We next define what it means for a polyhedral subdivision to be compatible with a Minkowski sum. Suppose that $Q_1, \ldots, Q_m$ are arbitrary polytopes in $\mathbb{R}^n$.

(6.5) Definition. Let $Q = Q_1 + \cdots + Q_m \subset \mathbb{R}^n$ be a Minkowski sum of polytopes, and assume that Q has dimension n. Then a subdivision

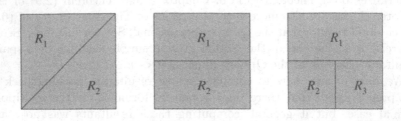

FIGURE 7.7. Subdividing the square

$R_1, \ldots, R_s$ of Q is a *mixed subdivision* if each cell R_i can be written as a Minkowski sum

$$R_i = F_1 + \cdots + F_m$$

where each F_i is a face of Q_i and $n = \dim(F_1) + \cdots + \dim(F_m)$. Furthermore, if $R_j = F_1' + \cdots + F_m'$ is another cell in the subdivision, then $R_i \cap R_j = (F_1 \cap F_1') + \cdots + (F_m \cap F_m')$.

Exercise 1. Consider the polytopes

$$P_1 = \mathrm{Conv}((0,0), (1,0), (3,2), (0,2))$$
$$P_2 = \mathrm{Conv}((0,1), (3,0), (1,4)).$$

The Minkowski sum $P = P_1 + P_2$ was illustrated in Fig. 7.6 of §4.
a. Prove that Fig. 7.8 on the next page gives a mixed subdivision of P.
b. Find a different mixed subdivision of P.

When we have a mixed subdivision, some of the cells making up the subdivision are especially important.

(6.6) Definition. Suppose that $R = F_1 + \cdots + F_m$ is a cell in a mixed subdivision of $Q = Q_1 + \cdots + Q_m$. Then R is called a *mixed cell* if $\dim(F_i) \leq 1$ for all i.

Exercise 2. Show that the mixed subdivision illustrated in Fig. 7.8 has three mixed cells.

As an application of mixed subdivisions, we will give a surprisingly easy formula for mixed volume. Given n polytopes $Q_1, \ldots, Q_n \subset \mathbb{R}^n$, we want to compute the mixed volume $MV_n(Q_1, \ldots, Q_n)$. We begin with a mixed subdivision of $Q = Q_1 + \cdots + Q_n$. In this situation, observe that every mixed cell R is a sum of edges (because the faces $F_i \subset Q_i$ summing to R

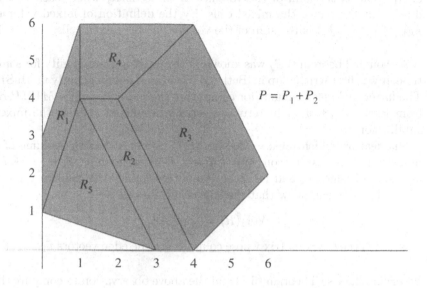

FIGURE 7.8. Mixed subdivision of a Minkowski sum

satisfy $n = \dim(F_1) + \cdots + \dim(F_n)$ and $\dim(F_i) \leq 1$). Then the mixed cells determine the mixed volume in the following simple manner.

(6.7) Theorem. *Given polytopes $Q_1, \ldots, Q_n \subset \mathbb{R}^n$ and a mixed subdivision of $Q = Q_1 + \cdots + Q_n$, the mixed volume $MV_n(Q_1, \ldots, Q_n)$ is computed by the formula*

$$MV_n(Q_1, \ldots, Q_n) = \sum_R \mathrm{Vol}_n(R),$$

where the sum is over all mixed cells R of the mixed subdivision.

PROOF. We will give the main idea of the proof and refer to [HuS1] for the details. The key observation is that mixed subdivisions behave well under scaling. More precisely, let $R_1, \ldots, R_s$ be a mixed subdivision of $Q_1 + \cdots + Q_n$, where each R_i is a Minkowski sum of faces $R_i = F_1 + \cdots + F_n$ as in Definition (6.5). If $\lambda_i > 0$ for $i = 1, \ldots, n$, then one can show that $\lambda_1 Q_1 + \cdots + \lambda_n Q_n$ has a mixed subdivision $R'_1, \ldots, R'_s$ such that

$$R'_i = \lambda_1 F_1 + \cdots + \lambda_n F_n.$$

It follows that

$$\mathrm{Vol}_n(R'_i) = \lambda_1^{\dim(F_1)} \cdots \lambda_n^{\dim(F_n)} \mathrm{Vol}_n(R_i)$$

since $n = \dim(F_1) + \cdots + \dim(F_n)$. Adding these up, we see that $\mathrm{Vol}_n(\lambda_1 Q_1 + \cdots + \lambda_n Q_n)$ is a polynomial in the λ_i and the coefficient of $\lambda_1 \cdots \lambda_n$ is the sum of the volumes of the cells R_i where each F_i has dimension 1, that is, the mixed cells. By the definition of mixed volume, $MV_n(Q_1, \ldots, Q_n)$ is the sum of the volumes of the mixed cells.   □

Although Theorem (6.7) was known in the polytope community for some time, it was first written up in [Bet] and discovered independently in [HuS1]. The latter includes formulas for computing the mixed volumes $MV_n(P, \alpha)$ from Exercise 18 of §4 in terms of certain nonmixed cells in the mixed subdivision.

One feature which makes Theorem (6.7) useful is that the volume of a mixed cell R is easy to compute. Namely, if we write $R = F_1 + \cdots + F_n$ as a sum of edges F_i and let $\vec{v}_i$ be the vector connecting the two vertices of F_i, then one can show that the volume of the cell is

$$\text{Vol}_n(R) = |\det(A)|,$$

where A is the $n \times n$ matrix whose columns are the edge vectors $\vec{v}_1, \ldots, \vec{v}_n$.

Exercise 3. Use Theorem (6.7) and the above observation to compute the mixed volume $MV_2(P_1, P_2)$, where P_1 and P_2 are as in Exercise 1.

Theorem (6.7) has some nice consequences. First, it shows that the mixed volume is nonnegative, which is not obvious from the definition given in §4. Second, since all mixed cells lie inside the Minkowski sum, we can relate mixed volume to the volume of the Minkowski sum as follows:

$$MV_n(Q_1, \ldots, Q_n) \le \text{Vol}_n(Q_1 + \cdots + Q_n).$$

By [Emi1], we have a lower bound for mixed volume as well:

$$MV_n(Q_1, \ldots, Q_n) \ge n! \sqrt[n]{\text{Vol}_n(Q_1) \cdots \text{Vol}_n(Q_n)}.$$

Mixed volume also satisfies the *Alexandrov-Fenchel inequality*, which is discussed in [Ewa] and [Ful].

Exercise 4. Work out the inequalities displayed above for the polytopes P_1 and P_2 from Exercise 1.

All of this is very nice, except for one small detail: how do we find mixed subdivisions? Fortunately, they are fairly easy to compute in practice. We will describe briefly how this is done. The first step is to "lift" the polytopes $Q_1, \ldots, Q_n \subset \mathbb{R}^n$ to $\mathbb{R}^{n+1}$ by picking random vectors $l_1, \ldots, l_n \in \mathbb{Z}^n$ and considering the polytopes

$$\widehat{Q}_i = \{(v, l_i \cdot v) : v \in Q_i\} \subset \mathbb{R}^n \times \mathbb{R} = \mathbb{R}^{n+1}.$$

If we regard l_i as the linear map $\mathbb{R}^n \to \mathbb{R}$ defined by $v \mapsto l_i \cdot v$, then $\widehat{Q}_i$ is the portion of the graph of l_i lying over Q_i.

Now consider the polytope $\hat{Q} = \hat{Q}_1 + \cdots + \hat{Q}_n \subset \mathbb{R}^{n+1}$. We say that a facet $\mathcal{F}$ of $\hat{Q}$ is a *lower facet* if its outward-pointing normal has a *negative* t_{n+1}-coordinate, where t_{n+1} is the last coordinate of $\mathbb{R}^{n+1} = \mathbb{R}^n \times \mathbb{R}$. If the l_i are sufficiently generic, one can show that the projection $\mathbb{R}^{n+1} \to \mathbb{R}^n$ onto the first n coordinates carries the lower facets $\mathcal{F} \subset \hat{Q}$ onto n-dimensional polytopes $R \subset Q = Q_1 + \cdots + Q_n$, and these polytopes form the cells of a mixed subdivision of Q. The theoretical background for this construction is given in [BS] and some nice pictures appear in [CE2] (see also [HuS1], [CE1] and [EC]). Mixed subdivisions arising in this way are said to be *coherent*.

Exercise 5. Let $Q_1 = \mathrm{Conv}((0,0),(1,0),(0,1))$ be the unit simplex in the plane, and consider the vectors $l_1 = (0,4)$ and $l_2 = (2,1)$. This exercise will apply the above strategy to create a coherent mixed subdivision of $Q = Q_1 + Q_2$, where $Q_2 = Q_1$.

a. Write $\hat{Q}_1$ and $\hat{Q}_2$ as convex hulls of sets of three points, and then express $\hat{Q} = \hat{Q}_1 + \hat{Q}_2$ as the convex hull of 9 points in $\mathbb{R}^3$.

b. In $\mathbb{R}^3$, plot the points of $\hat{Q}$ found in part a. Note that such each point lies over a point of Q.

c. Find the lower facets of $\hat{Q}$ (there are 3 of them) and use this to determine the corresponding coherent mixed subdivision of Q. Hint: When one point lies above another, the higher point can't lie in a lower facet.

d. Show that choosing $l_1 = (1,1)$ and $l_2 = (2,3)$ leads to a different coherent mixed subdivision of Q.

It is known that computing mixed volume is #P-complete (see [Ped]). Being #P-complete is similar to being NP complete the difference is that NP-complete refers to a class of hard *decision* problems, while #P-complete refers to certain hard *enumerative problems*. The complexity of computing mixed volume is discussed carefully in [DGH], with some recent developments appearing in [GuS].

There are several known algorithms for computing mixed volumes and mixed subdivisions, some of which have been implemented in publicly available software. In particular, software for computing mixed volumes is available at:

- http://www-sop.inria.fr/galaad/logiciels/emiris/ soft_geo.html, based on [EC] and described in [Emi3];
- http://www2.math.uic.edu/~jan/PHCpack/phcpack.html, described in [Ver1]; and
- http://www.mth.msu.edu/~li/, based on [GL2] and [LL].

Further references for mixed volume are [GL1], [GLW], [VGC], and the references mentioned in Section 6 of [EC].

We now return to our original question of computing the mixed sparse resultant $\mathrm{Res}_{\mathcal{A}_0,\ldots,\mathcal{A}_n}(f_0,\ldots,f_n)$. In this situation, we have $n+1$ polytopes $Q_i = \mathrm{Conv}(\mathcal{A}_i)$. Our goal is to show that a coherent mixed subdivision of the Minkowski sum $Q = Q_0 + \cdots + Q_n$ gives a systematic way to compute the sparse resultant.

To see how this works, first recall what we did in Chapter 3. If we think of the multipolynomial resultant $\text{Res}_{d_0,\ldots,d_n}(F_0,\ldots,F_n)$ in homogeneous terms, then the method presented in §4 of Chapter 3 goes as follows: we fixed the set of monomials of total degree $d_0 + \cdots + d_n - n$ and wrote this set as a disjoint union $S_0 \cup \cdots \cup S_n$. Then, for each monomial $x^\alpha \in S_i$, we multiplied F_i by $x^\alpha/x_i^{d_i}$. This led to the equations (4.1) of Chapter 3:

$$(x^\alpha/x_i^{d_i})F_i = 0, \quad x^\alpha \in S_i, \quad i = 1,\ldots,n.$$

Expressing these polynomials in terms of the monomials in our set gave a system of equations, and the determinant of the coefficient matrix was the polynomial D_n in Definition (4.2) of Chapter 3.

By varying this construction slightly, we got determinants $D_0,\ldots,D_n$ with the following two properties:

- Each D_i is a nonzero multiple of the resultant.
- For i fixed, the degree of D_i as a polynomial in the coefficients of f_i is the same as the degree of the resultant in these coefficients.

(See §4 of Chapter 3, especially Exercise 7 and Proposition (4.6)). From here, we easily proved

$$\text{Res}_{d_0,\ldots,d_n} = \pm\text{GCD}(D_0,\ldots,D_n),$$

which is Proposition (4.7) of Chapter 3.

We will show that this entire framework goes through with little change in the sparse case. Suppose we have exponent sets $\mathcal{A}_0,\ldots,\mathcal{A}_n$, and as above set $Q_i = \text{Conv}(\mathcal{A}_i)$. Also assume that we have a coherent mixed subdivision of $Q = Q_0 + \cdots + Q_n$. The first step in computing the sparse resultant is to fix a set of monomials or, equivalently, a set of exponents. We will call this set $\mathcal{E}$, and we define $\mathcal{E}$ to be

$$\mathcal{E} = \mathbb{Z}^n \cap (Q + \delta),$$

where $\delta \in \mathbb{R}^n$ is a small vector chosen so that for every $\alpha \in \mathcal{E}$, there is a cell R of the mixed subdivision such that α lies in the *interior* of $R + \delta$. Intuitively, we displace the subdivision slightly so that the lattice points lie in the interiors of the cells.

The following exercise illustrates what this looks like in a particularly simple case. We will refer to this exercise several times as we explain how to compute $\text{Res}_{\mathcal{A}_0,\ldots,\mathcal{A}_n}$.

Exercise 6. Consider the equations

$$0 = f_0 = a_1 x + a_2 y + a_3$$
$$0 = f_1 = b_1 x + b_2 y + b_3$$
$$0 = f_2 = c_1 x^2 + c_2 y^2 + c_3 + c_4 xy + c_5 x + c_6 y$$

obtained by setting $z = 1$ in equations (2.9) from Chapter 3. If $\mathcal{A}_i$ is the set of exponents appearing in f_i, then $\mathrm{Res}_{\mathcal{A}_0,\mathcal{A}_1,\mathcal{A}_2}$ is the resultant $\mathrm{Res}_{1,1,2}$ considered in Proposition (2.10) of Chapter 3.

a. If we let $l_0 = (0,4)$, $l_1 = (2,1)$ and $l_2 = (5,7)$, then show that we get the coherent mixed subdivision of Q pictured in Fig. 7.9. This calculation is not easy to do by hand—you should use a program such as qhull (available from the Geometry Center at the University of Minnesota) to compute convex hulls.

b. If $\delta = (\epsilon, \epsilon)$ for small $\epsilon > 0$, show that $\mathcal{E}$ contains the six exponent vectors indicated by dots in Fig. 7.9. We will think of $\mathcal{E}$ as consisting of the monomials

$$x^3y,\ x^2y^2,\ x^2y,\ xy^3,\ xy^2,\ xy.$$

The reason for listing the monomials this way will soon become clear.

c. If $\delta = (-\epsilon, -\epsilon)$ for small $\epsilon > 0$, show that $\mathcal{E}$ consists of 10 exponent vectors. So different δ's can give very different $\mathcal{E}$'s!

Now that we have $\mathcal{E}$, our next task is to break it up into a disjoint union $S_0 \cup \cdots \cup S_n$. This is where the coherent mixed subdivision comes in. Each cell R of the subdivision is a Minkowski sum

$$R = F_0 + \cdots + F_n,$$

where the $F_i \subset Q_i$ are faces such that $n = \dim(F_0) + \cdots + \dim(F_n)$. Note that at least one F_i must have $\dim(F_i) = 0$, i.e., at least one F_i is a vertex. Sometimes R can be written in the above form in several ways (we will see an example below), but using the *coherence* of our mixed subdivision, we get a canonical way of doing this. Namely, R is the projection of a

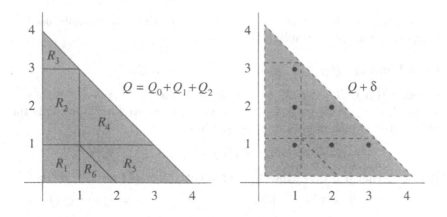

FIGURE 7.9. A coherent mixed subdivision and its shift

lower facet $\mathcal{F} \subset \widehat{Q}$, and one can show that $\mathcal{F}$ can be *uniquely* written as a Minkowski sum

$$\mathcal{F} = \widehat{F}_0 + \cdots + \widehat{F}_n,$$

where $\widehat{F}_i$ is a face of $\widehat{Q}_i$. If $F_i \subset Q_i$ is the projection of $\widehat{F}_i$, then the induced Minkowski sum $R = F_0 + \cdots + F_n$ is called *coherent*. Now, for each i between 0 and n, we define the subset $S_i \subset \mathcal{E}$ as follows:

(6.8)
$$S_i = \{\alpha \in \mathcal{E} : \text{if } \alpha \in R + \delta \text{ and } R = F_0 + \cdots + F_n \text{ is coherent,}$$
$$\text{then } i \text{ is the } smallest \text{ index such that } F_i \text{ is a vertex}\}.$$

This gives a disjoint union $\mathcal{E} = S_0 \cup \cdots \cup S_n$. Furthermore, if $\alpha \in S_i$, we let $v(\alpha)$ denote the vertex F_i in (6.8), i.e., $F_i = \{v(\alpha)\}$. Since $Q_i = \mathrm{Conv}(\mathcal{A}_i)$, it follows that $v(\alpha) \in \mathcal{A}_i$.

Exercise 7. For the coherent subdivision of Exercise 6, show that

$$S_0 = \{x^3 y, x^2 y^2, x^2 y\}, \quad S_1 = \{xy^3, xy^2\}, \quad S_2 = \{xy\},$$

and that

$$x^{v(\alpha)} = \begin{cases} x & \text{for } x^\alpha \in S_0 \\ y & \text{for } x^\alpha \in S_1 \\ 1 & \text{for } x^\alpha \in S_2. \end{cases}$$

(Here, we regard $\mathcal{E}$ and the S_i as consisting of monomials rather than exponent vectors.) Hint: The exponent vector $\alpha = (1,3)$ of xy^3 lies in $R_2 + \delta$, where we are using the labelling of Fig. 7.9. If $\mathcal{F}$ is the lower facet lying over R_2, one computes (using a program such as qhull) that

$$\mathcal{F} = \text{edge of } \widehat{Q}_0 + (0,1,1) + \text{edge of } \widehat{Q}_2$$

which implies that $R_2 = \text{edge of } Q_0 + (0,1) + \text{edge of } Q_2$ is coherent. Thus $xy^3 \in S_1$ and $x^{v(\alpha)} = y$, and the other monomials are handled similarly.

The following lemma will allow us to create the determinants used in computing the sparse resultant.

(6.9) Lemma. *If $\alpha \in S_i$, then $(x^\alpha / x^{v(\alpha)})f_i \in L(\mathcal{E})$.*

PROOF. If $\alpha \in R + \delta = F_0 + \cdots + F_n + \delta$, then $\alpha = \beta_0 + \cdots + \beta_n + \delta$, where $\beta_j \in F_j \subset Q_j$ for $0 \le j \le n$. Since $\alpha \in S_i$, we know that F_i is the vertex $v(\alpha)$, which implies $\beta_i = v(\alpha)$. Thus

$$\alpha = \beta_0 + \cdots + \beta_{i-1} + v(\alpha) + \beta_{i+1} + \cdots + \beta_n + \delta.$$

It follows that if $\beta \in \mathcal{A}_i$, then the exponent vector of $(x^\alpha / x^{v(\alpha)})x^\beta$ is

$$\alpha - v(\alpha) + \beta = \beta_0 + \cdots + \beta_{i-1} + \beta + \beta_{i+1} + \cdots + \beta_n + \delta \subset Q + \delta.$$

This vector is integral and hence lies in $\mathcal{E} = \mathbb{Z}^n \cap (Q + \delta)$. Since f_i is a linear combination of the x^β for $\beta \in \mathcal{A}_i$, the lemma follows. $\square$

Now consider the equations

(6.10) $(x^\alpha / x^{v(\alpha)}) f_i = 0, \quad \alpha \in S_i.$

We get one equation for each α, which means that we have $|\mathcal{E}|$ equations, where $|\mathcal{E}|$ denotes the number of elements in $\mathcal{E}$. By Lemma (6.9), each $(x^\alpha / x^{v(\alpha)}) f_i$ can be written as a linear combination of the monomials x^β for $\beta \in \mathcal{E}$. If we regard these monomials as "unknowns", then (6.10) is a system of $|\mathcal{E}|$ equations in $|\mathcal{E}|$ unknowns.

(6.11) Definition. D_n is the determinant of the coefficient matrix of the $|\mathcal{E}| \times |\mathcal{E}|$ system of linear equations given by (6.10).

Notice the similarity with Definition (4.2) of Chapter 3. Here is a specific example of what this determinant looks like.

Exercise 8. Consider the polynomials f_0, f_1, f_2 from Exercise 6 and the decomposition $\mathcal{E} = S_0 \cup S_1 \cup S_2$ from Exercise 7.
a. Show that the equations (6.10) are *precisely* the equations obtained from (2.11) in Chapter 3 by setting $z = 1$ and multiplying each equation by xy. This explains why we wrote the elements of $\mathcal{E}$ in the order $x^3 y, x^2 y^2, x^2 y, xy^3, xy^2, xy$.
b. Use Proposition (2.10) of Chapter 3 to conclude that the determinant D_2 satisfies

$$D_2 = \pm a_1 \operatorname{Res}_{1,1,2}(f_0, f_1, f_2).$$

This exercise suggests a close relation between D_n and $\operatorname{Res}_{\mathcal{A}_0, \ldots, \mathcal{A}_n}$. In general, we have the following result.

(6.12) Theorem. *The determinant D_n is a nonzero multiple of the mixed sparse resultant $\operatorname{Res}_{\mathcal{A}_0, \ldots, \mathcal{A}_n}$. Furthermore, the degree of D_n as a polynomial in the coefficients of f_n is the mixed volume $MV_n(Q_0, \ldots, Q_{n-1})$.*

PROOF. If the equations $f_0 = \cdots = f_n = 0$ have a solution in $(\mathbb{C}^*)^n$, then the equations (6.10) have a nontrivial solution, and hence the coefficient matrix has zero determinant. It follows that D_n vanishes on the set $Z(\mathcal{A}_0, \ldots, \mathcal{A}_n)$ from Theorem (6.2). Since the resultant is the irreducible defining equation of this set, it must divide D_n. (This argument is similar to one used frequently in Chapter 3.)

To show that D_n is nonzero, we must find $f_0, \ldots, f_n$ for which $D_n \neq 0$. For this purpose, introduce a new variable t and let

(6.13) $f_i = \sum_{\alpha \in \mathcal{A}_i} t^{l_i \cdot \alpha} x^\alpha,$

where the $l_i \in \mathbb{Z}^n$ are the vectors used in the construction of the coherent mixed subdivision of $Q = Q_0 + \cdots + Q_n$. Section 4 of [CE1] shows that

$D_n \neq 0$ for this choice of the f_i. We should also mention that without coherence, it can happen that D_n is identically zero. See Exercise 10 at the end of the section for an example.

Finally, we compute the degree of D_n as a polynomial in the coefficients of f_n. In (6.10), the coefficients of f_n appear in the equations coming from S_n, so that D_n has degree $|S_n|$ in these coefficients. So we need only prove

$$(6.14) \qquad |S_n| = MV_n(Q_0, \ldots, Q_{n-1}).$$

If $\alpha \in S_n$, the word *smallest* in (6.8) means that $\alpha \in R + \delta$, where $R = F_0 + \cdots + F_n$ and $\dim(F_i) > 0$ for $i = 0, \ldots, n-1$. Since the dimensions of the F_i sum to n, we must have $\dim(F_0) = \cdots = \dim(F_{n-1}) = 1$. Thus R is a mixed cell with F_n as the unique vertex in the sum. Conversely, any mixed cell of the subdivision must have exactly one F_i which is a vertex (since the $\dim(F_i) \leq 1$ add up to n). Thus, if R is a mixed cell where F_n is a vertex, then $\mathbb{Z}^n \cap (R+\delta) \subset S_n$ follows from (6.8). This gives the formula

$$|S_n| = \sum_{F_n \text{ is a vertex}} |\mathbb{Z}^n \cap (R + \delta)|,$$

where the sum is over all mixed cells $R = F_0 + \cdots + F_n$ of the subdivision of Q for which F_n is a vertex.

We now use two nice facts. First, the mixed cells R where F_n is a vertex are translates of the mixed cells in a mixed subdivision of $Q_0 + \cdots + Q_{n-1}$. Furthermore, Lemma 5.3 of [Emi1] implies that *all* mixed cells in this subdivision of $Q_0 + \cdots + Q_{n-1}$ appear in this way. Since translation doesn't affect volume, Theorem (6.7) then implies

$$MV_n(Q_0, \ldots, Q_{n-1}) = \sum_{F_n \text{ is a vertex}} \text{Vol}_n(R),$$

where we sum over the same mixed cells as before. The second nice fact is that each of these cells R is a Minkowski sum of edges (up to translation by the vertex F_n), so that by Section 5 of [CE1], the volume of R is the number of lattice points in a generic small translation. This means

$$\text{Vol}_n(R) = |\mathbb{Z}^n \cap (R + \delta)|,$$

and (6.14) now follows immediately. $\qquad \square$

This shows that D_n has the desired properties. Furthermore, we get other determinants $D_0, \ldots, D_{n-1}$ by changing how we choose the subsets $S_i \subset \mathcal{E}$. For instance, if we replace *smallest* by *largest* in (6.8), then we get a determinant D_0 whose degree in the coefficients of f_0 is $MV_n(Q_1, \ldots, Q_n)$. More generally, for each j between 0 and n, we can find a determinant D_j which is a nonzero multiple of the resultant and whose degree in the coefficients of f_j is the mixed volume

$$MV_n(Q_1, \ldots, Q_{j-1}, Q_{j+1}, \ldots, Q_n)$$

(see Exercise 11 below). Using Theorem (6.3) of this section and the argument of Proposition (4.7) of Chapter 3, we conclude that

$$\mathrm{Res}_{\mathcal{A}_0,\ldots,\mathcal{A}_n}(f_0,\ldots,f_n) = \pm \mathrm{GCD}(D_0,\ldots,D_n).$$

As in Chapter 3, the GCD computation needs to be done for $f_0,\ldots,f_n$ with symbolic coefficients.

Recently, D'Andrea showed that there is also a direct formula for the resultant which doesn't involve a GCD computation. Theorem (6.12) tells us that D_n is the product of the resultant times an extraneous factor. The main result of [D'An] states that the extraneous factor is the determinant D'_n of a recursively computable submatrix of the matrix used to compute D_n. This gives the formula

$$\mathrm{Res}_{\mathcal{A}_0,\ldots,\mathcal{A}_n} = \frac{D_n}{D'_n},$$

which generalizes Macaulay's formula for the dense resultant (Theorem (4.9) of Chapter 3).

In practice, this method for computing the sparse resultant is not very useful, mainly because the D_j tend to be enormous polynomials when the f_i have symbolic coefficients. But if we use numerical coefficients for the f_i, the GCD computation doesn't make sense. Two methods for avoiding this difficulty are explained in Section 5 of [CE1]. Fortunately, for many purposes, it suffices to work with just one of the D_j (we will give an example below), and D_j can be computed by the methods discussed at the end of §4 of Chapter 3.

The matrices D_j are sometimes called *Sylvester matrices* since each entry is either 0 or a coefficient, just like Sylvester's formula for the resultant of two univariate polynomials (see (1.2) of Chapter 3). Methods for computing these matrices and their variants are described in [EC], [CE1], and [CE2], and software implementing the resulting algorithms for computing resultants is available from:

- http://www.cs.unc.edu/~geom/MARS, described in [WEM];
- http://www-sop.inria.fr/galaad/logiciels/emiris/
  soft_alg.html, described in [Emi3]; and
- http://www-sop.inria.fr/galaad/logiciels/multires.html,
  described in [Mou2].

In all of these methods, problems arise when the extraneous factor (i.e., the denominator in the resultant formula) vanishes. Methods for avoiding these problems are discussed in [CE2], [D'AE], [Mou1], [Roj2], and [Roj4].

Sparse resultants can be also formulated using *Bézout* or *Dixon* matrices. Here, the entries are more complicated combinations of the coefficients, though the resulting matrices may be smaller. A survey of such matrices appears in [EmM], which includes many references. The paper [BU] has more on Bézout matrices and the `multires` package mentioned above computes

Bézout matrices (this package also computes the matrix $\widetilde{M}$ of Theorem (6.21)—see [Mou2] for examples). The Dixon formulation has been studied extensively, starting with [KSY], [KS1], and [KS2] and more recently in [CK1] and [CK2]. Software packages related to the Dixon resultant formulation are available at:

- `http://www.cs.albany.edu/~artas/dixon/`, related to [CK1] and [CK2]; and
- `http://www.bway.net/~lewis/home.html`, based on the Fermat computer algebra system.

It is also possible to mix Sylvester and Bézout matrices. See [CDS] for some interesting resultant formulas of this type.

We will end this section with a brief discussion (omitting most proofs) of how sparse resultants can be used to solve equations. The basic idea is that given Laurent polynomials $f_i \in L(\mathcal{A}_i)$, we want to solve the equations

$$(6.15) \qquad f_1(x_1, \ldots, x_n) = \cdots = f_n(x_1, \ldots, x_n) = 0.$$

If we assume that the f_i are generic, then by Bernstein's Theorem from §5, the number of solutions in $(\mathbb{C}^*)^n$ is the mixed volume $MV_n(Q_1, \ldots, Q_n)$, where $Q_i = \mathrm{Conv}(\mathcal{A}_i)$.

To solve (6.15), we can use sparse resultants in a variety of ways, similar to what we did in the multipolynomial case studied in Chapter 3. We begin with a sparse version of the u-resultant from §5 of Chapter 3. Let

$$f_0 = u_0 + u_1 x_1 + \cdots + u_n x_n,$$

where $u_0, \ldots, u_n$ are variables. The Newton polytope of f_0 is $Q_0 = \mathrm{Conv}(\mathcal{A}_0)$, where $\mathcal{A}_0 = \{0, \vec{e}_1, \ldots, \vec{e}_n\}$ and $\vec{e}_1, \ldots, \vec{e}_n$ are the usual standard basis vectors. Then the u-resultant of $f_1, \ldots, f_n$ is the resultant $\mathrm{Res}_{\mathcal{A}_0, \ldots, \mathcal{A}_n}(f_0, \ldots, f_n)$, which written out more fully is

$$\mathrm{Res}_{\mathcal{A}_0, \mathcal{A}_1, \ldots, \mathcal{A}_n}(u_0 + u_1 x_1 + \cdots + u_n x_n, f_1, \ldots, f_n).$$

For $f_1, \ldots, f_n$ generic, one can show that there is a nonzero constant C such that

$$(6.16) \qquad \mathrm{Res}_{\mathcal{A}_0, \ldots, \mathcal{A}_n}(f_0, \ldots, f_n) = C \prod_{p \in \mathbf{V}(f_1, \ldots, f_n) \cap (\mathbb{C}^*)^n} f_0(p).$$

This generalizes Theorem (5.8) of Chapter 3 and is proved using a sparse analog (due to Pedersen and Sturmfels [PS2]) of Theorem (3.4) from Chapter 3. If $p = (a_1, \ldots, a_n)$ is a solution of (6.15) in $(\mathbb{C}^*)^n$, then

$$f_0(p) = u_0 + u_1 a_1 + \cdots + u_n a_n,$$

so that factoring the u-resultant gives the solutions of (6.15) in $(\mathbb{C}^*)^n$.

In (6.16), generic means that the solutions all have multiplicity 1. If some of the multiplicities are > 1, the methods of Chapter 4 can be adapted to show that

$$\mathrm{Res}_{A_0,\ldots,A_n}(f_0,\ldots,f_n) = C \prod_{p\in\mathbf{V}(f_1,\ldots,f_n)\cap(\mathbb{C}^*)^n} f_0(p)^{m(p)},$$

where $m(p)$ is the multiplicity of p as defined in §2 of Chapter 4.

Many of the comments about the u-resultant from §5 of Chapter 3 carry over without change to the sparse case. In particular, we saw in Chapter 3 that for many purposes, we can replace the sparse resultant with the determinant D_0. This is true in the sparse case, provided we use D_0 as defined in this section. Thus, (6.16) holds using D_0 in place of the sparse resultant, i.e., there is a constant C' such that

$$D_0 = C' \prod_{p\in\mathbf{V}(f_1,\ldots,f_n)\cap(\mathbb{C}^*)^n} f_0(p).$$

This formula is reasonable since D_0, when regarded as a polynomial in the coefficients $u_0,\ldots,u_n$ of f_0, has degree $MV_n(Q_1,\ldots,Q_n)$, which is the number of solutions of (6.15) in $(\mathbb{C}^*)^n$. There is a similar formula when some of the solutions have multiplicities > 1.

We can also find solutions of (6.15) using the eigenvalue and eigenvector techniques discussed in §6 of Chapter 3. To see how this works, we start with the ring $\mathbb{C}[x_1^{\pm1},\ldots,x_n^{\pm1}]$ of all Laurent polynomials. The Laurent polynomials in our equations (6.15) give the ideal

$$\langle f_1,\ldots,f_n \rangle \subset \mathbb{C}[x_1^{\pm1},\ldots,x_n^{\pm1}].$$

We want to find a basis for the quotient ring $\mathbb{C}[x_1^{\pm1},\ldots,x_n^{\pm1}]/\langle f_1,\ldots,f_n\rangle$.

For this purpose, consider a coherent mixed subdivision of the Minkowski sum $Q_1 + \cdots + Q_n$. If we combine Theorem (6.7) and the proof of Theorem (6.12), we see that if δ is generic, then

$$MV_n(Q_1,\ldots,Q_n) = \sum_R |\mathbb{Z}^n \cap (R + \delta)|,$$

where the sum is over all mixed cells in the mixed subdivision. Thus the set of exponents

$$\widehat{\mathcal{E}} = \{\beta \in \mathbb{Z}^n : \beta \in R + \delta \text{ for some mixed cell } R\}$$

has $MV_n(Q_1,\ldots,Q_n)$ elements. This set gives the desired basis of our quotient ring.

(6.17) Theorem. *For the set $\widehat{\mathcal{E}}$ described above, the cosets $[x^\beta]$ for $\beta \in \widehat{\mathcal{E}}$ form a basis of the quotient ring $\mathbb{C}[x_1^{\pm1},\ldots,x_n^{\pm1}]/\langle f_1,\ldots,f_n\rangle$.*

PROOF. This was proved independently in [ER] and [PS1]. In the terminology of [PS1], the cosets $[x^\beta]$ for $\beta \in \widehat{\mathcal{E}}$ form a *mixed monomial basis* since they come from the mixed cells of a mixed subdivision.

We will prove this in the following special case. Consider $f_0 = u_0 + u_1 x_1 + \cdots + u_n x_n$, and let $\mathcal{A}_0$ and Q_0 be as above. Then pick a coherent mixed subdivision of $Q = Q_0 + Q_1 + \cdots + Q_n$ and let $\mathcal{E} = \mathbb{Z}^n \cap (Q + \delta)$. Also define $S_i \subset \mathcal{E}$ using (6.8) with *smallest* replaced by *largest*. Using the first "nice fact" used in the proof of Theorem (6.12), one can show that the coherent mixed subdivision of Q induces a coherent mixed subdivision of $Q_1 + \cdots + Q_n$. We will show that the theorem holds for the set $\widehat{\mathcal{E}}$ coming from this subdivision.

The first step in the proof is to show that

(6.18) $\alpha \in S_0 \iff \alpha = v(\alpha) + \beta$ for some $v(\alpha) \in \mathcal{A}_0$ and $\beta \in \widehat{\mathcal{E}}$.

This follows from the arguments used in the proof of Theorem (6.12). Now let M_0 be the coefficient matrix of the equations (6.10). These equations begin with

$$(x^\alpha / x^{v(\alpha)}) f_0 = 0, \quad \alpha \in S_0,$$

which, using (6.18), can be rewritten as

(6.19) $x^\beta f_0 = 0, \quad \beta \in \widehat{\mathcal{E}}.$

From here, we will follow the proof of Theorem (6.2) of Chapter 3. We partition M_0 so that the rows and columns of M_0 corresponding to elements of S_0 lie in the upper left hand corner, so that

$$M_0 = \begin{pmatrix} M_{00} & M_{01} \\ M_{10} & M_{11} \end{pmatrix}.$$

By Lemma 4.4 of [Emi1], M_{11} is invertible for generic $f_1, \ldots, f_n$ since we are working with a coherent mixed subdivision—the argument is similar to showing $D_0 \neq 0$ in the proof of Theorem (6.12).

Now let $\widehat{\mathcal{E}} = \{\beta_1, \ldots, \beta_\mu\}$, where $\mu = MV_n(Q_1, \ldots, Q_n)$. Then, for generic $f_1, \ldots, f_n$, we define the $\mu \times \mu$ matrix

(6.20) $$\widetilde{M} = M_{00} - M_{01} M_{11}^{-1} M_{10}.$$

Also, for $p \in \mathbf{V}(f_1, \ldots, f_n) \cap (\mathbb{C}^*)^n$, let $\mathbf{p}^\beta$ denote the column vector

$$\mathbf{p}^\beta = \begin{pmatrix} p^{\beta_1} \\ \vdots \\ p^{\beta_\mu} \end{pmatrix}.$$

Similar to (6.6) in Chapter 3, one can prove

$$\widetilde{M}\, \mathbf{p}^\beta = f_0(p)\, \mathbf{p}^\beta$$

because (6.19) gives the rows of M_0 coming from S_0.

The final step is to show that the cosets $[x^{\beta_1}], \ldots, [x^{\beta_\mu}]$ are linearly independent. The argument is identical to what we did in Theorem (6.2) of Chapter 2. □

Using the mixed monomial basis, the next step is to find the matrix of the multiplication map $m_{f_0} : A \to A$, where

$$A = \mathbb{C}[x_1^{\pm 1}, \ldots, x_n^{\pm 1}]/\langle f_1, \ldots, f_n \rangle$$

and $m_{f_0}([g]) = [f_0 g]$ for $[g] \in A$. As in Chapter 3, this follows immediately from the previous result.

(6.21) Theorem. *Let $f_i \in L(\mathcal{A}_i)$ be generic Laurent polynomials, and let $f_0 = u_0 + u_1 x_1 + \cdots + u_n x_n$. Using the basis from Theorem (6.17), the matrix of the multiplication map $m_{f_0} : A \to A$ defined above is the* **transpose** *of the matrix*

$$\widetilde{M} = M_{00} - M_{01}M_{11}^{-1}M_{10}$$

from (6.20).

If we write $\widetilde{M}$ in the form

$$\widetilde{M} = u_0 I + u_1 \widetilde{M}_1 + \cdots + u_n \widetilde{M}_n,$$

where each $\widetilde{M}_i$ has constant entries, then Theorem (6.21) implies that for all i, $(\widetilde{M}_i)^T$ is the matrix of multiplication by x_i. Thus, as in Chapter 3, $\widetilde{M}$ simultaneously computes the matrices of the multiplication maps by *all* of the variables $x_1, \ldots, x_n$.

Now that we have these multiplication maps, the methods mentioned in Chapters 2 and 3 apply with little change. More detailed discussions of how to solve equations using matrix methods and resultants, including examples, can be found in [Emi1], [Emi2], [Emi3], [EmM], [ER], [Man1], [Mou1], [Mou2], and [Roj4]. It is also possible to apply these methods to study varieties of positive dimension. Here, a typical goal would be to find a point in every irreducible component of the variety. Some references (which employ a variety of approaches) are [ElM3], [KM], [Roj2], and [SVW].

We should mention that other techniques introduced in Chapter 3 can be adapted to the sparse case. For example, the generalized characteristic polynomial (GCP) from §6 of Chapter 3 can be generalized to the *toric GCP* defined in [Roj4]. This is useful for dealing with the types of degeneracies discussed in Chapter 3.

ADDITIONAL EXERCISES FOR §6

Exercise 9. Consider the following system of equations taken from [Stu3]:

$$0 = f_0 = ax + by$$
$$0 = f_1 = cx + dy$$
$$0 = f_2 = ex + fy + g.$$

a. Explain why the hypothesis of Theorem (6.2) is not satisfied. Hint: Look at the Newton polytopes.
b. Show that the sparse resultant exists and is given by $\operatorname{Res}(f_0, f_1, f_2) = ad - bc$.

Exercise 10. In Exercise 7, we defined the decomposition $\mathcal{E} = S_0 \cup S_1 \cup S_2$ using coherent Minkowski sums $R = F_0 + F_1 + F_2$. This exercise will explore what can go wrong if we don't use coherent sums.
a. Exercise 7 gave the coherent Minkowski sum $R_2 = $ edge of $Q_0 + (0, 1) +$ edge of Q_2. Show that $R_2 = (0, 1) + $ edge of $Q_1 + $ edge of Q_2 also holds.
b. If we use coherent Minkowski sums for R_i when $i \neq 2$ and the non-coherent one from part a when $i = 2$, show that (6.8) gives $S_0 = \{x^3 y, x^2 y^2, x^2 y, xy^3, xy^2\}$, $S_1 = \emptyset$ and $S_2 = \{xy\}$.
c. If we compute the determinant D_2 using S_0, S_1, S_2 as in part b, show that D_2 does not involve the coefficients of f_1 and conclude that D_2 is identically zero in this case. Hint: You don't need explicit computations. Argue instead that D_2 is divisible by $\operatorname{Res}_{1,1,2}$.

Exercise 11. This exercise will discuss the determinant D_j for $j < n$. The index j will be fixed throughout the exercise. Given $\mathcal{E}$ as usual, define the subset $S_i \subset \mathcal{E}$ to consist of all $\alpha \in \mathcal{E}$ such that if $\alpha \in R + \delta$, where $R = F_0 + \cdots + F_n$ is coherent, then

$$i = \begin{cases} j & \text{if } \dim(F_k) > 0 \ \forall k \neq j \\ \min(k \neq j : F_k \text{ is a vertex}) & \text{otherwise.} \end{cases}$$

By adapting the proof of Theorem (6.12), explain why this gives a determinant D_j which is a nonzero multiple of the resultant and whose degree as a polynomial in the coefficients of f_j is the mixed volume $MV_n(Q_1, \ldots, Q_{j-1}, Q_{j+1}, \ldots, Q_n)$.

Exercise 12. Prove that as polynomials with integer coefficients, we have

$$\operatorname{Res}_{\mathcal{A}_0, \ldots, \mathcal{A}_n}(f_0, \ldots, f_n) = \pm \operatorname{GCD}(D_0, \ldots, D_n).$$

Hint: Since D_j and $\operatorname{Res}_{\mathcal{A}_0, \ldots, \mathcal{A}_n}$ have the same degrees when regarded as polynomials in the coefficients of f_j, it is relatively easy to prove this over $\mathbb{Q}$. To prove that it is true over $\mathbb{Z}$, it suffices to show that the coefficients of each D_j are relatively prime. To prove this for $j = n$, consider the polynomials f_i defined in (6.13) and use the argument of Section 4 of [CE1] (or, for a more detailed account, Section 5 of [CE2]) to show that D_n has a leading coefficient 1 as a polynomial in t.

Exercise 13. Compute the mixed sparse resultant of the polynomials

$$f_0 = a_1 + a_2 xy + a_3 x^2 y + a_4 x$$
$$f_1 = b_1 y + b_2 x^2 y^2 + b_3 x^2 y + b_4 x$$
$$f_2 = c_1 + c_2 y + c_3 xy + c_4 x.$$

Hint: To obtain a coherent mixed subdivision, let $l_0 = (L, L^2)$, $l_1 = -(L^2, 1)$ and $l_2 = (1, -L)$, where L is a sufficiently large positive integer. Also let $\delta = -(3/8, 1/8)$. The full details of this example, including the explicit matrix giving D_0, can be found in [CE1].

Exercise 14. In Definition (6.5), we require that a mixed subdivision of $Q_1 + \cdots + Q_m$ satisfy the compatibility condition

$$R_i \cap R_j = (F_1 \cap F_1') + \cdots + (F_m \cap F_m'),$$

where $R_i = F_1 + \cdots + F_m$ and $R_j = F_1' + \cdots + F_m'$ are two cells in the subdivision and F_i, F_i' are faces of Q_i. This condition is essential for the scaling used in the proof of Theorem (6.7). To see why, consider the unit square Q in the plane with vertices labeled v_1, v_2, v_3, v_4.

a. Show that $R_i = v_i + Q$, $1 \leq i < 4$, gives a polyhedral subdivision of $Q + Q$ which satisfies Definition (6.5) except for the compatibility condition. Also show that if Theorem (6.7) applied to this subdivision, then the mixed volume $MV_2(Q, Q)$ would be 0.

b. Show that the subdivision of part a does not scale. Hint: Consider $Q + \lambda Q$ and $R_i' = v_i + \lambda Q$.

c. Find a mixed subdivision of $Q + Q$ that satisfies all parts of Definition (6.5) and draw a picture of $Q + \lambda Q$ to illustrate how the subdivision scales.

This example is due to Lyle Ramshaw.

Chapter 8

Polyhedral Regions and Polynomials

In this chapter we will consider a series of interrelated topics concerning polyhedral regions P in $\mathbb{R}^n$ and polynomials. In the first three sections we will see how some of the algebraic methods from earlier chapters give conceptual and computational tools for several classes of problems of intrinsic interest and practical importance. We will begin by considering Gröbner basis methods for integer optimization and combinatorial enumeration problems. We will also use module Gröbner bases to study piecewise polynomial, or *spline*, functions on polyhedral complexes.

The final two sections apply the same polyhedral geometry to furnish some further insight into the foundations of Gröbner basis theory. We will study the *Gröbner fan* of an ideal, a collection of polyhedral cones classifying the ideal's different reduced Gröbner bases, and use the Gröbner fan to develop a general basis conversion algorithm called the *Gröbner Walk*. The walk is applicable even when the ideal is not zero-dimensional, hence is more general than the FGLM algorithm from Chapter 2, §3.

Many of the topics in this chapter are also closely related to the material on polytopes and toric varieties from Chapter 7, but we have tried to make this chapter as independent as possible from Chapter 7 so that it can be read separately.

§1 Integer Programming

This section applies the theory of Gröbner bases to problems in integer programming. Most of the results depend only on the basic algebra of polynomial rings and facts about Gröbner bases for ideals. From Proposition (1.12) on, we will also need to use the language of Laurent polynomials, but the idea should be reasonably clear even if that concept is not familiar. The original reference for this topic is an article by Conti and Traverso, [CT], and another treatment may be found in [AL], Section 2.8. Further developments may be found in the articles [Tho1], [Tho2], [HT], and the book [Stu2]. For a general introduction to linear and integer programming, we recommend [Schri].

To begin, we will consider a very small, but in other ways typical, applied integer programming problem, and we will use this example to illustrate the key features of this class of problems. Suppose that a small local trucking firm has two customers, A and B, that generate shipments to the same location. Each shipment from A is a pallet weighing exactly 400 kilos and taking up 2 cubic meters of volume. Each pallet from B weighs 500 kilos and takes up 3 cubic meters. The shipping firm uses small trucks that can carry any load up to 3700 kilos, and up to 20 cubic meters. B's product is more perishable, though, and they are willing to pay a higher price for on-time delivery: $ 15 per pallet versus $ 11 per pallet from A. The question facing the manager of the trucking company is: How many pallets from each of the two companies should be included in each truckload to maximize the revenues generated?

Using A to represent the number of pallets from company A, and similarly B to represent the number of pallets from company B in a truckload, we want to maximize the revenue function $11A + 15B$ subject to the following constraints:

$$4A + 5B \leq 37 \qquad \text{(the weight limit, in 100's)}$$
$$(1.1) \qquad 2A + 3B \leq 20 \qquad \text{(the volume limit)}$$
$$A, B \in \mathbb{Z}_{\geq 0}.$$

Note that both A, B must be integers. This is, as we will see, an important restriction, and the characteristic feature of *integer programming* problems.

Integer programming problems are generalizations of the mathematical translation of the question above. Namely, in an integer programming problem we seek the *maximum or minimum value* of some *linear* function

$$\ell(A_1, \ldots, A_n) = c_1 A_1 + c_2 A_2 + \cdots + c_n A_n$$

on the set of $(A_1, \ldots, A_n) \in \mathbb{Z}_{\geq 0}^n$ with $A_j \geq 0$ for all $1 \leq j \leq n$ satisfying a set of linear inequalities:

$$a_{11} A_1 + a_{12} A_2 + \cdots + a_{1n} A_n \leq \text{ (or } \geq) \ b_1$$
$$a_{21} A_1 + a_{22} A_2 + \cdots + a_{2n} A_n \leq \text{ (or } \geq) \ b_2$$
$$\vdots$$
$$a_{m1} A_1 + a_{m2} A_2 + \cdots + a_{mn} A_n \leq \text{ (or } \geq) \ b_m.$$

We assume in addition that the a_{ij}, and the b_i are all integers. Some of the coefficients c_j, a_{ij}, b_i may be negative, but we will always assume $A_j \geq 0$ for all j.

Integer programming problems occur in many contexts in engineering, computer science, operations research, and pure mathematics. With large numbers of variables and constraints, they can be *difficult* to solve. It is perhaps instructive to consider our small shipping problem (1.1) in detail. In geometric terms we are seeking a maximum for the function $11A + 15B$ on the integer points in the closed convex polygon P in $\mathbb{R}^2$ bounded above by portions of the lines $4A + 5B = 37$ (slope $-4/5$), $2A + 3B = 20$ (slope $-2/3$), and by the coordinate axes $A = 0$, and $B = 0$. See Fig. 8.1. The set of all points in $\mathbb{R}^2$ satisfying the inequalities from (1.1) is known as the feasible region.

(1.2) Definition. The *feasible region* of an integer programming problem is the set P of all $(A_1, \ldots, A_n) \in \mathbb{R}^n$ satisfying the inequalities in the statement of the problem.

The set of all points in $\mathbb{R}^n$ satisfying a single linear inequality of the form considered here is called a *closed half-space*. A *polyhedral region* or *polyhedron* in $\mathbb{R}^n$ is defined as the intersection of a finite number of closed half-spaces. Equation (1.4) in Chapter 7 shows that polytopes are bounded polyhedral regions. In fact, a polyhedral region is a polytope if and only if it is bounded in $\mathbb{R}^n$ (the other implication is shown for instance in [Ewa], Theorem 1.5). In this chapter we will consider both bounded and unbounded polyhedral regions.

It is possible for the feasible region of an integer programming problem to contain *no* integer points at all. There are no solutions of the optimization

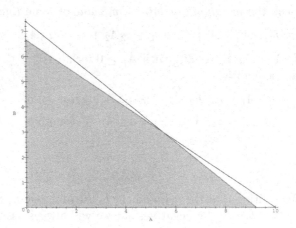

FIGURE 8.1. The feasible region P for (1.1)

problem in that case. For instance in $\mathbb{R}^2$ consider the region defined by

$$A + B \leq 1$$

(1.3) $$3A - B \geq 1$$

$$2A - B \leq 1,$$

and $A, B \geq 0$.

Exercise 1. Verify directly (for example with a picture) that there are no integer points in the region defined by (1.3).

When n is small, it is often possible to analyze the feasible set of an integer programming problem geometrically and determine the integer points in it. However, even this can be complicated since any polyhedral region formed by intersecting half-spaces bounded by affine hyperplanes with equations defined over $\mathbb{Z}$ can occur. For example, consider the set P in $\mathbb{R}^3$ defined by inequalities:

$$
\begin{array}{ll}
2A_1 + 2A_2 + 2A_3 \leq 5 & -2A_1 + 2A_2 + 2A_3 \leq 5 \\
2A_1 + 2A_2 - 2A_3 \leq 5 & -2A_1 + 2A_2 - 2A_3 \leq 5 \\
2A_1 - 2A_2 + 2A_3 \leq 5 & -2A_1 - 2A_2 + 2A_3 \leq 5 \\
2A_1 - 2A_2 - 2A_3 \leq 5 & -2A_1 - 2A_2 - 2A_3 \leq 5.
\end{array}
$$

In Exercise 11, you will show that P is a solid regular octahedron, with 8 triangular faces, 12 edges, and 6 vertices.

Returning to the problem from (1.1), if we did not have the additional constraints $A, B \subset \mathbb{Z}$, (if we were trying to solve a *linear programming* problem rather than an *integer programming* problem), the situation would be somewhat easier to analyze. For instance, to solve (1.1), we could apply the following simple geometric reasoning. The *level curves* of the revenue function $\ell(A, B) = 11A + 15B$ are lines of slope $-11/15$. The values of ℓ increase as we move out into the first quadrant. Since the slopes satisfy $-4/5 < -11/15 < -2/3$, it is clear that the revenue function attains its overall maximum on P at the vertex q in the interior of the first quadrant. Readers of Chapter 7 will recognize q as the face of P in the support line with normal vector $\nu = (-11, -15)$. See Fig. 8.2.

That point has rational, but not *integer* coordinates: $q = (11/2, 3)$. Hence q is *not* the solution of the integer programming problem! Instead, we need to consider only the integer points (A, B) in P. One *ad hoc* method that works here is to fix A, compute the largest B such that (A, B) lies in P, then compute the revenue function at those points and compare values for all possible A values. For instance, with $A = 4$, the largest B giving a point in P is $B = 4$, and we obtain $\ell(4, 4) = 104$. Similarly, with $A = 8$, the largest feasible B is $B = 1$, and we obtain $\ell(8, 1) = 103$. Note incidentally that *both* of these values are larger than the value of ℓ at the integer point closest to q in P—$(A, B) = (5, 3)$, where $\ell(5, 3) = 100$. This shows some

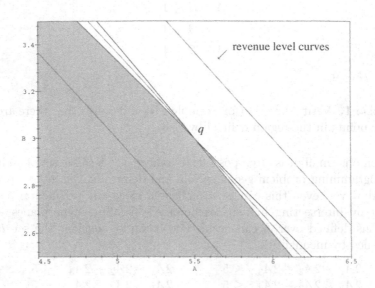

FIGURE 8.2. The linear programming maximum for (1.1)

of the potential subtlety of integer programming problems. Continuing in this way it can be shown that the maximum of ℓ occurs at $(A, B) = (4, 4)$.

Exercise 2. Verify directly (that is, by enumerating integer points as suggested above) that the solution of the shipping problem (1.1) is the point $(A, B) = (4, 4)$.

This sort of approach would be quite impractical for larger problems. Indeed, the general integer programming problem is known to be *NP-complete*, and so as Conti and Traverso remark, "even algorithms with theoretically bad worst case and average complexity can be useful ... , hence deserve investigation."

To discuss integer programming problems in general it will be helpful to standardize their statement to some extent. This can be done using the following observations.

1. We need only consider the problem of *minimizing* the linear function
   $\ell(A_1, \ldots, A_n) = c_1 A_1 + c_2 A_2 + \cdots + c_n A_n$, since maximizing a function
   ℓ on a set of integer n-tuples is the same as minimizing the function $-\ell$.

2. Similarly, by replacing an inequality

$$a_{i1}A_1 + a_{i2}A_2 + \cdots + a_{in}A_n \geq b_i$$

by the equivalent form

$$-a_{i1}A_1 - a_{i2}A_2 - \cdots - a_{in}A_n \leq -b_i,$$

we may consider only inequalities involving $\leq$.

3. Finally, by introducing additional variables, we can rewrite the linear constraint inequalities as *equalities*. The new variables are called "slack variables."

For example, using the idea in point 3 here the inequality

$$3A_1 - A_2 + 2A_3 \leq 9$$

can be replaced by

$$3A_1 - A_2 + 2A_3 + A_4 = 9$$

if $A_4 = 9 - (3A_1 - A_2 + 2A_3) \geq 0$ is introduced as a new variable to "take up the slack" in the original inequality. Slack variables will appear with coefficient zero in the function to be minimized.

Applying 1, 2, and 3 above, any integer programming problem can be put into the *standard form*:

$$\text{Minimize: } c_1 A_1 + \cdots + c_n A_n, \text{ subject to:}$$

(1.4)
$$a_{11}A_1 + a_{12}A_2 + \cdots + a_{1n}A_n = b_1$$
$$a_{21}A_1 + a_{22}A_2 + \cdots + a_{2n}A_n = b_2$$
$$\vdots$$
$$a_{m1}A_1 + a_{m2}A_2 + \cdots + a_{mn}A_n = b_m$$
$$A_j \in \mathbb{Z}_{\geq 0}, \ j = 1, \ldots n,$$

where now n is the total number of variables (including slack variables). As before, we will call the set of all *real* n-tuples satisfying the constraint equations the *feasible region*. Note that this is a polyhedral region in $\mathbb{R}^n$ because the set of all $(A_1, \ldots, A_n) \in \mathbb{R}^n$ satisfying a linear equation $a_{j1}A_1 + \cdots + a_{jn}A_n = b_j$ is the intersection of the two half-spaces defined by $a_{j1}A_1 + \cdots + a_{jn}A_n \geq b_j$ and $a_{j1}A_1 + \cdots + a_{jn}A_n \leq b_j$.

For the rest of this section we will explore an alternative approach to integer programming problems, in which we translate such a problem into a question about polynomials. We will use the standard form (1.4) and first consider the case where all the coefficients are nonnegative: $a_{ij} \geq 0, b_i \geq 0$. The translation proceeds as follows. We introduce an indeterminate z_i for each of the equations in (1.4), and exponentiate to obtain an equality

$$z_i^{a_{i1}A_1 + a_{i2}A_2 + \cdots + a_{in}A_n} = z_i^{b_i}$$

for each $i = 1, \ldots, m$. Multiplying the left and right hand sides of these equations, and rearranging the exponents, we get another equality:

$$(1.5) \qquad \prod_{j=1}^{n} \left(\prod_{i=1}^{m} z_i^{a_{ij}} \right)^{A_j} = \prod_{i=1}^{m} z_i^{b_i}.$$

From (1.5) we get the following direct algebraic characterization of the integer n-tuples in the feasible region of the problem (1.4).

(1.6) Proposition. *Let k be a field, and define $\varphi : k[w_1, \ldots, w_n] \to k[z_1, \ldots, z_m]$ by setting*

$$\varphi(w_j) = \prod_{i=1}^{m} z_i^{a_{ij}}$$

for each $j = 1, \ldots, n$, and $\varphi(g(w_1, \ldots, w_n)) = g(\varphi(w_1), \ldots, \varphi(w_n))$ for a general polynomial $g \in k[w_1, \ldots, w_n]$. Then $(A_1, \ldots, A_n)$ is an integer point in the feasible region if and only if φ maps the monomial $w_1^{A_1} w_2^{A_2} \cdots w_n^{A_n}$ to the monomial $z_1^{b_1} \cdots z_m^{b_m}$.

Exercise 3. Prove Proposition (1.6).

For example, consider the standard form of our shipping problem (1.1), with slack variables C in the first equation and D in the second.

$$\varphi : k[w_1, w_2, w_3, w_4] \to k[z_1, z_2]$$
$$w_1 \mapsto z_1^4 z_2^2$$
$$(1.7) \qquad\qquad w_2 \mapsto z_1^5 z_2^3$$
$$w_3 \mapsto z_1$$
$$w_4 \mapsto z_2.$$

The integer points in the feasible region of this restatement of the problem are the (A, B, C, D) such that

$$\varphi(w_1^A w_2^B w_3^C w_4^D) = z_1^{37} z_2^{20}.$$

Exercise 4. Show that in this case *every* monomial in $k[z_1, \ldots, z_m]$ is the image of some monomial in $k[w_1, \ldots, w_n]$.

In other cases, φ may not be surjective, and the following test for membership in the image of a mapping is an important part of the translation of integer programming problems.

Since the image of φ in Proposition (1.6) is precisely the set of polynomials in $k[z_1, \ldots, z_m]$ that can be expressed as polynomials in the $f_j = \prod_{i=1}^{m} z_i^{a_{ij}}$, we can also write the image as $k[f_1, \ldots, f_n]$, the subring of $k[z_1, \ldots, z_m]$ generated by the f_j. The subring membership test given by parts a and b of the following Proposition is also used in studying rings of invariants for finite matrix groups (see [CLO], Chapter 7, §3).

(1.8) Proposition. *Suppose that $f_1, \ldots, f_n \in k[z_1, \ldots, z_m]$ are given. Fix a monomial order in $k[z_1, \ldots, z_m, w_1, \ldots, w_n]$ with the elimination property: any monomial containing one of the z_i is greater than any monomial containing only the w_j. Let $\mathcal{G}$ be a Gröbner basis for the ideal*

$$I = \langle f_1 - w_1, \ldots, f_n - w_n \rangle \subset k[z_1, \ldots, z_m, w_1, \ldots, w_n]$$

and for each $f \in k[z_1, \ldots, z_m]$, let $\overline{f}^{\mathcal{G}}$ be the remainder on division of f by $\mathcal{G}$. Then

a. *A polynomial f satisfies $f \in k[f_1, \ldots, f_n]$ if and only if $g = \overline{f}^{\mathcal{G}} \in k[w_1, \ldots, w_n]$.*

b. *If $f \in k[f_1, \ldots, f_n]$ and $g = \overline{f}^{\mathcal{G}} \in k[w_1, \ldots, w_n]$ as in part a, then $f = g(f_1, \ldots, f_n)$, giving an expression for f as a polynomial in the f_j.*

c. *If each f_j and f are monomials and $f \in k[f_1, \ldots, f_n]$, then g is also a monomial.*

In other words, part c says that in the situation of Proposition (1.6), if $z_1^{b_1} \cdots z_m^{b_m}$ is in the image of φ, then it is automatically the image of some monomial $w_1^{A_1} \cdots w_n^{A_n}$.

PROOF. Parts a and b are proved in Proposition 7 of Chapter 7, §3 in [CLO], so we will not repeat them here.

To prove c, we note that each generator of I is a difference of two monomials. It follows that in the application of Buchberger's algorithm to compute $\mathcal{G}$, each S-polynomial considered and each nonzero S-polynomial remainder that goes into the Gröbner basis will be a difference of two monomials. This is true since in computing the S-polynomial, we are subtracting one difference of two monomials from another, and the leading terms cancel. Similarly, in the remainder calculation, at each step we subtract one difference of two monomials from another and cancellation occurs. It follows that every element of $\mathcal{G}$ will also be a difference of two monomials. When we divide a monomial by a Gröbner basis of this form, the remainder must be *a monomial*, since at each step we subtract a difference of two monomials from a single monomial and a cancellation occurs. Hence, if we are in the situation of parts a and b and the remainder is $g(w_1, \ldots, w_n) \in k[w_1, \ldots, w_n]$, then g must be a monomial.   □

In the restatement of our example problem in (1.7), we would consider the ideal

$$I = \langle z_1^4 z_2^2 - w_1, z_1^5 z_2^3 - w_2, z_1 - w_3, z_2 - w_4 \rangle.$$

Using the *lex* order with the variables ordered

$$z_1 > z_2 > w_4 > w_3 > w_2 > w_1$$

(chosen to eliminate terms involving slack variables if possible), we obtain a Gröbner basis $\mathcal{G}$:

$$g_1 = z_1 - w_3,$$

$$g_2 = z_2 - w_4,$$

$$g_3 = w_4^2 w_3^4 - w_1$$

(1.9) $$g_4 = w_4 w_3^3 w_2 - w_1^2$$

$$g_5 = w_4 w_3 w_1 - w_2$$

$$g_6 = w_4 w_1^4 - w_3 w_2^3$$

$$g_7 = w_3^2 w_2^2 - w_1^3.$$

(Note: An efficient implementation of Buchberger's algorithm is necessary for working out relatively large explicit examples using this approach, because of the large number of variables involved. We used **Singular** and *Macaulay 2* to compute the examples in this chapter.) So for instance, using g_1 and g_2 the monomial $f = z_1^{37} z_2^{20}$ reduces to $w_3^{37} w_4^{20}$. Hence f is in the image of φ from (1.7). But then further reductions are also possible, and the remainder on division is

$$\overline{f}^{\mathcal{G}} = w_2^4 w_1^4 w_3.$$

This monomial corresponds to the solution of the integer programming problem ($A = 4, B = 4$, and slack $C = 1$) that you verified in Exercise 2. In a sense, this is an accident, since the *lex* order that we used for the Gröbner basis and remainder computations did not take the revenue function ℓ explicitly into account.

To find the solution of an integer programming problem minimizing a given linear function $\ell(A_1, \ldots, A_n)$ we will usually need to use a monomial order specifically tailored to the problem at hand.

(1.10) Definition. A monomial order on $k[z_1, \ldots, z_m, w_1, \ldots, w_n]$ is said to be *adapted* to an integer programming problem (1.4) if it has the following two properties:

a. (Elimination) Any monomial containing one of the z_i is greater than any monomial containing only the w_j.

b. (Compatibility with ℓ) Let $A = (A_1, \ldots, A_n)$ and $A' = (A'_1, \ldots, A'_n)$. If the monomials $w^A, w^{A'}$ satisfy $\varphi(w^A) = \varphi(w^{A'})$ and $\ell(A_1, \ldots, A_n) > \ell(A'_1, \ldots, A'_n)$, then $w^A > w^{A'}$.

(1.11) Theorem. *Consider an integer programming problem in standard form (1.4). Assume all $a_{ij}, b_i \geq 0$ and let $f_j = \prod_{i=1}^{m} z_i^{a_{ij}}$ as before. Let $\mathcal{G}$*

be a Gröbner basis for

$$I = \langle f_1 - w_1, \ldots, f_n - w_n \rangle \subset k[z_1, \ldots, z_m, w_1, \ldots, w_n]$$

with respect to any adapted monomial order. Then if $f = z_1^{b_1} \cdots z_m^{b_m}$ is in $k[f_1, \ldots, f_n]$, the remainder $\overline{f}^{\mathcal{G}} \in k[w_1, \ldots, w_n]$ will give a solution of (1.4) minimizing ℓ. (There are cases where the minimum is not unique and, if so, this method will only find one minimum.)

PROOF. Let $\mathcal{G}$ be a Gröbner basis for I with respect to an adapted monomial order. Suppose that $w^A = \overline{f}^{\mathcal{G}}$ so $\varphi(w^A) = f$, but that $A = (A_1, \ldots, A_n)$ is not a minimum of ℓ. That is, assume that there is some $A' = (A_1', \ldots, A_n') \neq A$ such that $\varphi(w^{A'}) = f$ and $\ell(A_1', \ldots, A_n') < \ell(A_1, \ldots, A_n)$. Consider the difference $h = w^A - w^{A'}$. We have $\varphi(h) = f - f = 0$. In Exercise 5 below, you will show that this implies $h \in I$. But then h must reduce to *zero* under the Gröbner basis $\mathcal{G}$ for I. However, because $>$ is an adapted order, the leading term of h must be w^A, and that monomial is reduced with respect to $\mathcal{G}$ since it is a remainder. This contradiction shows that A must give a minimum of ℓ. $\square$

Exercise 5. Let $f_i \in k[z_1, \ldots, z_m]$, $i = 1, \ldots, n$, as above and define a mapping

$$\varphi : k[w_1, \ldots, w_n] \to k[z_1, \ldots, z_m]$$
$$w_i \mapsto f_i$$

as in (1.6). Let $I = \langle f_1 - w_1, \ldots, f_n - w_n \rangle \subset k[z_1, \ldots, z_m, w_1, \ldots, w_n]$. Show that if $h \in k[w_1, \ldots, w_n]$ satisfies $\varphi(h) = 0$, then $h \in I \cap k[w_1, \ldots, w_n]$. Hint: See the proof of Proposition 3 from Chapter 7, §4 of [CLO].

Exercise 6. Why did the *lex* order used to compute the Gröbner basis in (1.9) correctly find the maximum value of $11A + 15B$ in our example problem (1.1)? Explain, using Theorem (1.11). (Recall, w_4 and w_3 corresponding to the slack variables were taken greater than w_2, w_1.)

Theorem (1.11) yields a Gröbner basis algorithm for solving integer programming problems with all $a_{ij}, b_i \geq 0$:

Input: A, b from (1.4), an adapted monomial order $>$

Output: a solution of (1.4), if one exists

$$f_j := \prod_{i=1}^{m} z_i^{a_{ij}}$$
$$I := \langle f_1 - w_1, \ldots, f_n - w_n \rangle$$

$$\mathcal{G} := \text{Gröbner basis of } I \text{ with respect to } >$$

$$f := \prod_{i=1}^{m} z_i^{b_i}$$

$$g := \overline{f}^{\mathcal{G}}$$

IF $g \in k[w_1, \ldots, w_n]$ THEN

    its exponent vector gives a solution

ELSE

    there is no solution

Monomial orders satisfying both the elimination and compatibility properties from (1.10) can be specified in the following ways.

First, assume that all $c_j \geq 0$. Then it is possible to define a *weight order* $>_\ell$ on the w-variables using the linear function ℓ (see [CLO], Chapter 2, §4, Exercise 12). Namely order monomials in the w-variables alone first by ℓ-values:

$$w_1^{A_1} \cdots w_n^{A_n} >_\ell w_1^{A_1'} \cdots w_n^{A_n'}$$

if $\ell(A_1, \ldots, A_n) > \ell(A_1', \ldots, A_n')$ and break ties using any other fixed monomial order on $k[w_1, \ldots, w_n]$. Then incorporate this order into a product order on $k[z_1, \ldots, z_m, w_1, \ldots, w_n]$ with the z-variables greater than all the w-variables, to ensure that the elimination property from (1.10) holds.

If some $c_j < 0$, then the recipe above produces a total ordering on monomials in $k[z_1, \ldots, z_m, w_1, \ldots, w_n]$ that is compatible with multiplication and that satisfies the elimination property. But it will not be a well-ordering. So in order to apply the theory of Gröbner bases with respect to monomial orders, we will need to be more clever in this case. We begin with the following observation.

In $k[z_1, \ldots, z_m, w_1, \ldots, w_n]$, define a (non-standard) degree for each variable by setting $\deg(z_i) = 1$ for all $i = 1, \ldots, m$, and $\deg(w_j) = d_j = \sum_{i=1}^{m} a_{ij}$ for all $j = 1, \ldots, n$. Each d_j must be strictly positive, since otherwise the constraint equations would not depend on A_j. We say a polynomial $f \in k[z_1, \ldots, z_m, w_1, \ldots, w_n]$ is *homogeneous* with respect to these degrees if all the monomials $z^\alpha w^\beta$ appearing in f have the same (non-standard) total degree $|\alpha| + \sum_j d_j \beta_j$.

(1.12) Lemma. *With respect to the degrees d_j on w_j, the following statements hold.*

a. *The ideal $I = \langle f_1 - w_1, \ldots, f_n - w_n \rangle$ is homogeneous.*

b. *Every reduced Gröbner basis for the ideal I consists of homogeneous polynomials.*

PROOF. Part a follows since the given generators are homogeneous for these degrees—since $f_j = \prod_{i=1}^{m} z_i^{a_{ij}}$, the two terms in $f_j - w_j$ have the same degree.

Part b follows in the same way as for ideals that are homogeneous in the usual sense. The proof of Theorem 2 of Chapter 8, §3 of [CLO] goes over to non-standard assignments of degrees as well. □

For instance, in the *lex* Gröbner basis given in (1.9) above, it is easy to check that all the polynomials are homogeneous with respect to the degrees $\deg(z_i) = 1$, $\deg(w_1) = 6$, $\deg(w_2) = 8$, and $\deg(w_3) = \deg(w_4) = 1$.

Since $d_j > 0$ for all j, given the c_j from ℓ and $\mu > 0$ sufficiently large, all the entries of the vector

$$(c_1, \ldots, c_n) + \mu(d_1, \ldots, d_n)$$

will be positive. Let μ be any fixed number for which this is true. Consider the $(m+n)$-component weight vectors u_1, u_2:

$$u_1 = (1, \ldots, 1, 0, \ldots, 0)$$
$$u_2 = (0, \ldots, 0, c_1, \ldots, c_n) + \mu(0, \ldots, 0, d_1, \ldots, d_n).$$

Then all entries of u_2 are nonnegative, and hence we can define a weight order $>_{u_1, u_2, \sigma}$ by comparing u_1-weights first, then comparing u_2-weights if the u_1-weights are equal, and finally breaking ties with any other monomial order $>_\sigma$.

Exercise 7. Consider an integer programming problem (1.4) in which $a_{ij}, b_i \geq 0$ for all i, j.
a. Show that the order $>_{u_1, u_2, \sigma}$ defined above satisfies the elimination condition from Definition (1.10).
b. Show that if $\varphi(w^A) = \varphi(w^{A'})$, then $w^A - w^{A'}$ is homogeneous with respect to the degrees $d_j = \deg(w_j)$.
c. Deduce that $>_{u_1, u_2, \sigma}$ is an adapted order.

For example, our shipping problem (in standard form) can be solved using the second method here. We take $u_1 = (1, 1, 0, 0, 0, 0)$, and letting $\mu = 2$, we see that

$$u_2 = (0, 0, -11, -15, 0, 0) + 2(0, 0, 6, 8, 1, 1) = (0, 0, 1, 1, 2, 2)$$

has all nonnegative entries. Finally, break ties with $>_\sigma = $ graded reverse lex on all the variables ordered $z_1 > z_2 > w_1 > w_2 > w_3 > w_4$. Here is a `Singular` session performing the Gröbner basis and remainder calculations. Note the definition of the monomial order $>_{u_1, u_2, \sigma}$ by means of weight vectors.

```
> ring R = 0,(z(1..2),w(1..4)),(a(1,1,0,0,0,0),
  a(0,0,1,1,2,2),dp);
```

```
> ideal I = z(1)^4*z(2)^2-w(1), z(1)^5*z(2)^3-w(2), z(1)-w(3),
 z(2)-w(4);
> ideal J = std(I);
> J;
J[1]=w(1)*w(3)*w(4)-1*w(2)
J[2]=w(2)^2*w(3)^2-1*w(1)^3
J[3]=w(1)^4*w(4)-1*w(2)^3*w(3)
J[4]=w(2)*w(3)^3*w(4)-1*w(1)^2
J[5]=w(3)^4*w(4)^2-1*w(1)
J[6]=z(2)-1*w(4)
J[7]=z(1)-1*w(3)
> poly f = z(1)^37*z(2)^20;
> reduce(f,J);
w(1)^4*w(2)^4*w(3)
```

We find

$$\overline{z_1^{37} z_2^{20}}^{\,\mathcal{G}} = w_1^4 w_2^4 w_3$$

as expected, giving the solution $A = 4, B = 4$, and $C = 1, D = 0$.

This computation could also be done using the Maple **Groebner** package or *Mathematica*, since the weight order $>_{u_1,u_1,grevlex}$ can be defined as one of the matrix orders $>_M$ explained in Chapter 1, §2. For example, we could use the 6×6 matrix with first row u_1, second row u_2, and next four rows coming from the matrix defining the *grevlex* order on $k[z_1, z_1, w_1, w_2, w_3, w_4]$ following the patterns from parts b and e of Exercise 6 in Chapter 1, §2:

$$M = \begin{pmatrix} 1 & 1 & 0 & 0 & 0 & 0 \\ 0 & 0 & 1 & 1 & 2 & 2 \\ 1 & 1 & 1 & 1 & 1 & 1 \\ 1 & 1 & 1 & 1 & 1 & 0 \\ 1 & 1 & 1 & 0 & 0 & 0 \\ 1 & 0 & 0 & 0 & 0 & 0 \end{pmatrix}.$$

The other rows from the matrix defining the *grevlex* order are discarded because of linear dependences with previous rows in M.

Finally, we want to discuss general integer programming problems where some of the a_{ij} and b_i may be *negative*. There is no real conceptual difference in that case; the geometric interpretation of the integer programming problem is exactly the same, only the positions of the affine linear spaces bounding the feasible region change. But there is a difference in the algebraic translation. Namely, we cannot view the negative a_{ij} and b_i directly as exponents—that is not legal in an ordinary polynomial. One way to fix this problem is to consider what are called *Laurent polynomials* in the variables z_i instead—polynomial expressions in the z_i and z_i^{-1}, as defined in Chapter 7, §1 of this text. To deal with these more general objects without introducing a whole new set of m variables, we will use the *second* representation of the ring of Laurent polynomials, as presented in Exercise 15 of Chapter 7, §1:

$$k[z_1^{\pm 1}, \ldots, z_m^{\pm 1}] \cong k[z_1, \ldots, z_m, t]/\langle tz_1 \cdots z_m - 1 \rangle.$$

In intuitive terms, this isomorphism works by introducing a single new variable t satisfying $tz_1 \cdots z_m - 1 = 0$, so that formally t is the product of the inverses of the z_i: $t = z_1^{-1} \cdots z_m^{-1}$. Then each of the $\prod_{i=1}^{m} z_i^{a_{ij}}$ involved in the algebraic translation of the integer programming problem can be rewritten in the form $t^{e_j} \prod_{i=1}^{m} z_i^{a'_{ij}}$, where now all $a'_{ij} \geq 0$—we can just take $e_j \geq 0$ to be the negative of the smallest (most negative) a_{ij} that appears, and $a'_{ij} = e_j + a_{ij}$ for each i. Similarly, $\prod_{i=1}^{m} z_i^{b_i}$ can be rewritten in the form $t^e \prod_{i=1}^{m} z_i^{b'_i}$ with $e \geq 0$, and $b_i \geq 0$ for all i. It follows that the equation (1.5) becomes an equation between polynomial expressions in $t, z_1, \ldots, z_n$:

$$\prod_{j=1}^{n} \left(t^{e_j} \prod_{i=1}^{m} z_i^{a'_{ij}} \right)^{A_j} = t^e \prod_{i=1}^{m} z_i^{b'_i},$$

modulo the relation $tz_1 \cdots z_m - 1 = 0$. We have a direct analogue of Proposition (1.6).

(1.13) Proposition. *Define a mapping*

$$\varphi : k[w_1, \ldots, w_n] \to k[z_1^{\pm 1}, \ldots, z_m^{\pm 1}]$$

by setting

$$\varphi(w_j) = t^{e_j} \prod_{i=1}^{m} z_i^{a'_{ij}} \mod \langle tz_1 \cdots z_m - 1 \rangle$$

for each $j = 1, \ldots, n$, and extending to general $g(w_1, \ldots, w_n) \in k[w_1, \ldots, w_n]$ as before. Then $(A_1, \ldots, A_n)$ is an integer point in the feasible region if and only if $\varphi(w_1^{A_1} w_2^{A_2} \cdots w_n^{A_n})$ and $t^e z_1^{b'_1} \cdots z_m^{b'_m}$ represent the same element in $k[z_1^{\pm 1}, \ldots, z_m^{\pm 1}]$ (that is, their difference is divisible by $tz_1 \cdots z_m - 1$).

Similarly, Proposition (1.8) goes over to this more general situation. We will write S for the image of φ in $k[z_1^{\pm 1}, \ldots, z_m^{\pm 1}]$. Then we have the following version of the subring membership test.

(1.14) Proposition. *Suppose that $f_1, \ldots, f_n \in k[z_1, \ldots, z_m, t]$ are given. Fix a monomial order in $k[z_1, \ldots, z_m, t, w_1, \ldots, w_n]$ with the elimination property: any monomial containing one of the z_i or t is greater than any monomial containing only the w_j. Finally, let $\mathcal{G}$ be a Gröbner basis for the ideal*

$$J = \langle tz_1 \cdots z_m - 1, f_1 - w_1, \ldots, f_n - w_n \rangle$$

in $k[z_1, \ldots, z_m, t, w_1, \ldots, w_n]$ *and for each* $f \in k[z_1, \ldots, z_m, t]$, *let* $\overline{f}^{\mathcal{G}}$ *be the remainder on division of* f *by* $\mathcal{G}$. *Then*

a. f *represents an element in* S *if and only if* $g = \overline{f}^{\mathcal{G}} \in k[w_1, \ldots, w_n]$.

b. *If* f *represents an element in* S *and* $g = \overline{f}^{\mathcal{G}} \in k[w_1, \ldots, w_n]$ *as in part a, then* $f = g(f_1, \ldots, f_n)$, *giving an expression for* f *as a polynomial in the* f_j.

c. *If each* f_j *and* f *are monomials and* f *represents an element in* S, *then* g *is also a monomial.*

The proof is essentially the same as the proof for Proposition (1.8) so we omit it.

We should also mention that there is a direct parallel of Theorem (1.11) saying that using monomial orders which have the elimination and compatibility properties will yield minimum solutions for integer programming problems and give an algorithm for their solution. For ℓ with only nonnegative coefficients, adapted orders may be constructed using product orders as above, making t and the z_i greater than any w_j. For a more general discussion of constructing monomial orders compatible with a given ℓ, we refer the reader to [CT].

We will conclude this section with an example illustrating the general case described in the previous paragraph. Consider the following problem in standard form:

(1.15)

Minimize:
$$A + 1000B + C + 100D,$$

Subject to the constraints:
$$3A - 2B + C = -1$$
$$4A + B - C - D = 5$$
$$A, B, C, D \in \mathbb{Z}_{\geq 0}.$$

With the relation $tz_1 z_2 - 1 = 0$, our ideal J in this case is

$$J = \langle tz_1 z_2 - 1, z_1^3 z_2^4 - w_1, t^2 z_2^3 - w_2, tz_1^2 - w_3, tz_1 - w_4 \rangle.$$

If we use an elimination order placing t, z_1, z_2 before the w-variables, and then the use a weight order compatible with ℓ on the w_j (breaking ties with graded reverse lex), then we obtain a Gröbner basis $\mathcal{G}$ for J consisting of the following polynomials:

$$g_1 = w_2 w_3^2 - w_4$$
$$g_2 = w_1 w_4^7 - w_3^3$$
$$g_3 = w_1 w_2 w_4^6 - w_3$$
$$g_4 = w_1 w_2^2 w_3 w_4^5 - 1$$

$$g_5 = z_2 - w_1 w_2^2 w_3 w_4^4$$
$$g_6 = z_1 - w_1 w_2 w_4^5$$
$$g_7 = t - w_2 w_3 w_4.$$

From the right-hand sides of the equations, we consider $f = t z_2^6$. A remainder computation yields

$$\overline{f}^{\mathcal{G}} = w_1 w_2^2 w_4.$$

Since this is still a very small problem, it is easy to check by hand that the corresponding solution $(A = 1, B = 2, C = 0, D = 1)$ really does minimize $\ell(A, B, C, D) = A + 1000B + C + 100D$ subject to the constraints.

Exercise 8. Verify directly that the solution $(A, B, C, D) = (1, 2, 0, 1)$ of the integer programming problem (1.15) is correct. Hint: Show first that $B \geq 2$ in any solution of the constraint equations.

We should also remark that because of the special *binomial* form of the generators of the ideals in (1.11) and (1.13) and the simple polynomial remainder calculations involved here, there are a number of optimizations one could make in special-purpose Gröbner basis integer programming software. See [CT] for some preliminary results and [BLR] for additional developments. Algorithms described in the latter paper have been implemented in the `intprog` package distributed with the current version of CoCoA. The current version of **Singular** also contains an `intprog` library with procedures for integer programming.

Additional Exercises for §1

Exercise 9. What happens if you apply the Gröbner basis algorithm to any optimization problem on the polyhedral region in (1.3)?

Note: For the computational portions of the following problems, you will need to have access to a Gröbner basis package that allows you to specify mixed elimination-weight monomial orders as in the discussion following Theorem (1.11). One way to specify these orders is via suitable weight matrices as explained in Chapter 1, §2. See the example following Exercise 7 above.

Exercise 10. Apply the methods of the text to solve the following integer programming problems:
a.

Minimize: $2A + 3B + C + 5D$, subject to:

$$3A + 2B + C + D = 10$$

$$4A + B + C = 5$$

$$A, B, C, D \in \mathbb{Z}_{\geq 0}.$$

Verify that your solution is correct.

b. Same as a, but with the right-hand sides of the constraint equations changed to $20, 14$ respectively. How much of the computation needs to be redone?

c.

Maximize: $3A + 4B + 2C$, subject to:

$$3A + 2B + C \leq 45$$

$$A + 2B + 3C \leq 21$$

$$2A + B + C \leq 18$$

$$A, B, C \in \mathbb{Z}_{\geq 0}.$$

Also, describe the feasible region for this problem geometrically, and use that information to verify your solution.

Exercise 11. Consider the set P in $\mathbb{R}^3$ defined by inequalities:

$$
\begin{array}{ll}
2A_1 + 2A_2 + 2A_3 \leq 5 \qquad & -2A_1 + 2A_2 + 2A_3 \leq 5 \\
2A_1 + 2A_2 - 2A_3 \leq 5 \qquad & -2A_1 + 2A_2 - 2A_3 \leq 5 \\
2A_1 - 2A_2 + 2A_3 \leq 5 \qquad & -2A_1 - 2A_2 + 2A_3 \leq 5 \\
2A_1 - 2A_2 - 2A_3 \leq 5 \qquad & -2A_1 - 2A_2 - 2A_3 \leq 5.
\end{array}
$$

Verify that P is a solid (regular) octahedron. (What are the vertices?)

Exercise 12.

a. Suppose we want to consider *all* the integer points in a polyhedral region $P \subset \mathbb{R}^n$ as feasible, not just those with non-negative coordinates. How could the methods developed in the text be adapted to this more general situation?

b. Apply your method from part a to find the minimum of $2A_1 - A_2 + A_3$ on the integer points in the solid octahedron from Exercise 11.

§2 Integer Programming and Combinatorics

In this section we will study a beautiful application of commutative algebra and the ideas developed in §1 to combinatorial enumeration problems. For those interested in exploring this rich subject farther, we recommend the marvelous book [Sta1] by Stanley. Our main example is discussed there and far-reaching generalizations are developed using more advanced algebraic tools. There are also connections between the techniques we will develop here, *invariant theory* (see especially [Stu1]), the theory of *toric varieties*

([Ful]), and the *geometry of polyhedra* (see [Stu2]). The prerequisites for this section are the theory of Gröbner bases for polynomial ideals, familiarity with quotient rings, and basic facts about Hilbert functions (see, e.g. Chapter 6, §4 of this book or Chapter 9, §3 of [CLO]).

Most of this section will be devoted to the consideration of the following classical counting problem. Recall that a *magic square* is an $n \times n$ integer matrix $M = (m_{ij})$ with the property that the sum of the entries in each row and each column is the same. A famous 4×4 magic square appears in the well-known engraving *Melancholia* by Albrecht Dürer:

$$
\begin{array}{cccc}
16 & 3 & 2 & 13 \\
5 & 10 & 11 & 8 \\
9 & 6 & 7 & 12 \\
4 & 15 & 14 & 1
\end{array}
$$

The row and column sums in this array all equal 34. Although the extra condition that the m_{ij} are the distinct integers $1, 2, \ldots, n^2$ (as in Dürer's magic square) is often included, we will *not* make that part of the definition. Also, many familiar examples of magic squares have diagonal sums equal to the row and column sum and other interesting properties; we will not require that either. Our problem is this:

(2.1) Problem. *Given positive integers s, n, how many different $n \times n$ magic squares with $m_{ij} \geq 0$ for all i, j and row and column sum s are there?*

There are related questions from statistics and the design of experiments of practical as well as purely mathematical interest. In some small cases, the answer to (2.1) is easily derived.

Exercise 1. Show that the number of 2×2 nonnegative integer magic squares with row and column sum s is precisely $s + 1$, for each $s \geq 0$. How are the squares with sum $s > 1$ related to those with $s = 1$?

Exercise 2. Show that there are exactly six 3×3 magic squares with nonnegative integer entries and $s = 1$, twenty-one with $s = 2$, and fifty-five with $s = 3$. How many are there in each case if we require that the two diagonal sums also equal s?

Our main goal in this section will be to develop a *general* way to attack this and similar counting problems where the objects to be counted can be identified with the integer points in a polyhedral region in $\mathbb{R}^N$ for some N, so that we are in the same setting as in the integer programming problems from §1. We will take a somewhat *ad hoc* approach though, and use only as much general machinery as we need to answer our question (2.1) for small values of n.

To see how (2.1) fits into this context, note that the entire set of $n \times n$ nonnegative integer magic squares M is the set of solutions in $\mathbb{Z}_{\geq 0}^{n \times n}$ of a system of linear equations with integer coefficients. For instance, in the 3×3 case, the conditions that all row and column sums are equal can be expressed as 5 independent equations on the entries of the matrix. Writing

$$\vec{m} = (m_{11}, m_{12}, m_{13}, m_{21}, m_{22}, m_{23}, m_{31}, m_{32}, m_{33})^T,$$

the matrix $M = (m_{ij})$ is a magic square if and only if

$$(2.2) \qquad\qquad\qquad A_3 \vec{m} = 0,$$

where A_3 is the 5×9 integer matrix

$$(2.3) \qquad A_3 = \begin{pmatrix} 1 & 1 & 1 & -1 & -1 & -1 & 0 & 0 & 0 \\ 1 & 1 & 1 & 0 & 0 & 0 & -1 & -1 & -1 \\ 0 & 1 & 1 & -1 & 0 & 0 & -1 & 0 & 0 \\ 1 & -1 & 0 & 1 & -1 & 0 & 1 & -1 & 0 \\ 1 & 0 & -1 & 1 & 0 & -1 & 1 & 0 & -1 \end{pmatrix}$$

and $m_{ij} \geq 0$ for all i, j. Similarly, the $n \times n$ magic squares can be viewed as the solutions of a similar system $A_n \vec{m} = 0$ for an integer matrix A_n with n^2 columns.

Exercise 3.

a. Show that the 3×3 nonnegative integer magic squares are exactly the solutions of the system of linear equations (2.2) with matrix A_3 given in (2.3).

b. What is the minimal number of linear equations needed to define the corresponding space of $n \times n$ magic squares? Describe an explicit way to produce a matrix A_n as above.

As in the discussion following (1.4) of this chapter, the set $\{\vec{m} : A_3 \vec{m} = 0\}$ is a polyhedral region in $\mathbb{R}^{3 \times 3}$. However, there are three important differences between our situation here and the optimization problems considered in §1. First, there is no linear function to be optimized. Instead, we are mainly interested in understanding the structure of the entire set of integer points in a polyhedral region. Second, unlike the regions considered in the examples in §1, the region in this case is *unbounded*, and there are infinitely many integer points. Finally, we have a *homogeneous* system of equations rather than an inhomogeneous system, so the points of interest are elements of the kernel of the matrix A_n. In the following, we will write

$$K_n = \ker(A_n) \cap \mathbb{Z}_{\geq 0}^{n \times n}$$

for the set of all nonnegative integer $n \times n$ magic squares. We begin with a few simple observations.

(2.4) Proposition. *For each n,*

a. K_n *is closed under vector sums in* $\mathbb{Z}^{n \times n}$, *and contains the zero vector.*

b. *The set* $\mathcal{C}_n$ *of solutions of* $A_n \vec{m} = 0$ *satisfying* $\vec{m} \in \mathbb{R}_{\geq 0}^{n \times n}$ *forms a convex polyhedral cone in* $\mathbb{R}^{n \times n}$, *with vertex at the origin.*

PROOF. Part a follows by linearity. For part b, recall that a *convex polyhedral cone with vertex at the origin* is the intersection of finitely many half-spaces containing the origin. Then $\mathcal{C}_n$ is polyhedral since the defining equations are the linear equations $A_n \vec{m} = 0$ and the linear inequalities $m_{ij} \geq 0 \in \mathbb{R}$. It is a cone since any positive real multiple of a point in $\mathcal{C}_n$ is also in $\mathcal{C}_n$. Finally, it is convex since if $\vec{m}$ and $\vec{m}'$ are two points in $\mathcal{C}_n$, any linear combination $x = r\vec{m} + (1 - r)\vec{m}'$ with $r \in [0, 1]$ also satisfies the equations $A_n x = 0$ and has nonnegative entries, hence lies in $\mathcal{C}_n$. □

A set M with a binary operation is said to be a *monoid* if the operation is associative and possesses an identity element in M. For example $\mathbb{Z}_{\geq 0}^{n \times n}$ is a monoid under vector addition. In this language, part a of the proposition says that K_n is a *submonoid* of $\mathbb{Z}_{\geq 0}^{n \times n}$.

To understand the structure of the submonoid K_n, we will seek to find a minimal set of additive generators to serve as building blocks for all the elements of K_n. The appropriate notion is given by the following definition.

(2.5) Definition. Let K be any submonoid of the additive monoid $\mathbb{Z}_{\geq 0}^N$. A finite subset $\mathcal{H} \subset K$ is said to be a *Hilbert basis* for K if it satisfies the following two conditions.

a. For every $k \in K$ there exist $h_i \in \mathcal{H}$ and nonnegative integers c_i such that $k = \sum_{i=1}^{q} c_i h_i$, and

b. $\mathcal{H}$ is minimal with respect to inclusion.

It is a general fact that Hilbert bases exist and are unique for all submonoids $K \subset \mathbb{Z}_{\geq 0}^N$. Instead of giving an existence proof, however, we will present a Gröbner basis *algorithm* for finding the Hilbert basis for the submonoid $K = \ker(A)$ in $\mathbb{Z}_{\geq 0}^N$ for any integer matrix with N columns. (This comes from [Stu1], §1.4.) As in §1, we translate our problem from the context of integer points to Laurent polynomials. Given an integer matrix $A = (a_{ij})$ with N columns and m rows say, we introduce an indeterminate z_i for each row, $i = 1, \ldots, m$, and consider the ring of Laurent polynomials:

$$k[z_1^{\pm 1}, \ldots, z_m^{\pm 1}] \cong k[z_1, \ldots, z_m, t]/\langle t z_1 \cdots z_m - 1 \rangle.$$

(See §1 of this chapter and Exercise 15 of Chapter 7, §1.) Define a mapping

$$(2.6) \qquad \psi : k[v_1, \ldots, v_N, w_1, \ldots, w_N] \rightarrow k[z_1^{\pm 1}, \ldots, z_m^{\pm 1}][w_1, \ldots, w_N]$$

as follows. First take

$$(2.7) \qquad \psi(v_j) = w_j \cdot \prod_{i=1}^{m} z_i^{a_{ij}}$$

and $\psi(w_j) = w_j$ for each $j = 1, \ldots, N$, then extend to polynomials in $k[v_1, \ldots, v_N, w_1, \ldots, w_N]$ so as to make ψ a ring homomorphism.

The purpose of ψ is to detect elements of the kernel of A.

(2.8) Proposition. *A vector $\alpha^T \in \ker(A)$ if and only if $\psi(v^\alpha - w^\alpha) = 0$, that is if and only if $v^\alpha - w^\alpha$ is in the kernel of the homomorphism ψ.*

Exercise 4. Prove Proposition (2.8).

As in Exercise 5 of §1, we can write $J = \ker(\psi)$ as

$$J = I \cap k[v_1, \ldots, v_N, w_1, \ldots, w_N],$$

where

$$I = \left\langle w_j \cdot \prod_{i=1}^{m} z_i^{a_{ij}} - v_j : j = 1, \ldots N \right\rangle$$

in the ring $k[z_1^{\pm 1}, \ldots, z_m^{\pm 1}][v_1, \ldots, v_N, w_1, \ldots, w_N]$. The following theorem of Sturmfels (Algorithm 1.4.5 of [Stu1]) gives a way to find Hilbert bases.

(2.9) Theorem. *Let $\mathcal{G}$ be a Gröbner basis for I with respect to any elimination order $>$ for which all $z_i, t > v_j$, and all $v_j > w_k$. Let S be the subset of $\mathcal{G}$ consisting of elements of the form $v^\alpha - w^\alpha$ for some $\alpha \in \mathbb{Z}_{\geq 0}^N$. Then*

$$\mathcal{H} = \{\alpha : v^\alpha - w^\alpha \in S\}$$

is the Hilbert basis for K.

PROOF. The idea of this proof is similar to that of Theorem (1.11) of this chapter. See [Stu1] for a complete exposition.   □

Here is a first example to illustrate Theorem (2.9). Consider the submonoid of $\mathbb{Z}_{\geq 0}^4$ given as $K = \ker(A) \cap \mathbb{Z}_{\geq 0}^4$, for

$$A = \begin{pmatrix} 1 & 2 & -1 & 0 \\ 1 & 1 & -1 & -2 \end{pmatrix}.$$

To find a Hilbert basis for K, we consider the ideal I generated by

$$w_1 z_1 z_2 - v_1, \quad w_2 z_1^2 z_2 - v_2, \quad w_3 t - v_3, \quad w_4 z_1^2 t^2 - v_4$$

and $z_1 z_2 t - 1$. Computing a Gröbner basis $\mathcal{G}$ with respect to an elimination order as in (2.9), we find only one is of the desired form:

$$v_1 v_3 - w_1 w_3$$

It follows that the Hilbert basis for K consists of a single element: $\mathcal{H} = \{(1, 0, 1, 0)\}$. It is not difficult to verify from the form of the matrix A that every element in K is an integer multiple of this vector. Note that the size of the Hilbert basis is not the same as the dimension of the kernel of the matrix A as a linear mapping on $\mathbb{R}^4$. In general, there is no connection between the size of the Hilbert basis for $K = \ker(A) \cap \mathbb{Z}_{\geq 0}^N$ and $\dim \ker(A)$; the number of elements in the Hilbert basis can be either larger than, equal to, or smaller than the dimension of the kernel, depending on A.

We will now use Theorem (2.9) to continue our work on the magic square enumeration problem. If we apply the method of the theorem to find the Hilbert basis for $\ker(A_3) \cap \mathbb{Z}_{\geq 0}^{3 \times 3}$ (see equation (2.3) above) then we need to compute a Gröbner basis for the ideal I generated by

$$v_1 - w_1 z_1 z_2 z_4 z_5 \qquad v_2 - w_2 z_1^2 z_2^2 z_3^2 z_5 t$$

$$v_3 - w_3 z_1^2 z_2^2 z_3^2 z_4 t \qquad v_4 - w_4 z_2 z_4^2 z_5^2 t$$

$$v_5 - w_5 z_2 z_3 z_5 t \qquad v_6 - w_6 z_2 z_3 z_4 t$$

$$v_7 - w_7 z_1 z_4^2 z_5^2 t \qquad v_8 - w_8 z_1 z_3 z_5 t$$

$$v_9 - w_9 z_1 z_3 z_4 t$$

and $z_1 \cdots z_5 t - 1$ in the ring

$$k[z_1, \ldots, z_5, t, v_1, \ldots, v_9, w_1, \ldots, w_9].$$

Using an elimination order as described in Theorem (2.9) with the computer algebra system *Macaulay 2*, one obtains a very large Gröbner basis. (Because of the simple binomial form of the generators, however, the computation goes extremely quickly.) However, if we identify the subset S as in the theorem, there are only six polynomials corresponding to the Hilbert basis elements:

$$
\begin{array}{lll}
& v_3 v_5 v_7 - w_3 w_5 w_7 & v_3 v_4 v_8 - w_3 w_4 w_8 \\
(2.10) & v_2 v_6 v_7 - w_2 w_6 w_7 & v_2 v_4 v_9 - w_2 w_4 w_9 \\
& v_1 v_6 v_8 - w_1 w_6 w_8 & v_1 v_5 v_9 - w_1 w_5 w_9.
\end{array}
$$

Expressing the corresponding 6-element Hilbert basis in matrix form, we see something quite interesting. The matrices we obtain are precisely the six 3×3 *permutation matrices*—the matrix representations of the permutations of the components of vectors in $\mathbb{R}^3$. (This should also agree with your results in the first part of Exercise 2.) For instance, the Hilbert basis element $(0, 0, 1, 0, 1, 0, 1, 0, 0)$ from the first polynomial in (2.10) corresponds to the matrix

$$
T_{13} = \begin{pmatrix} 0 & 0 & 1 \\ 0 & 1 & 0 \\ 1 & 0 & 0 \end{pmatrix},
$$

which interchanges x_1, x_3, leaving x_2 fixed. Similarly, the other elements of the Gröbner basis give (in the order listed above)

$$S = \begin{pmatrix} 0 & 0 & 1 \\ 1 & 0 & 0 \\ 0 & 1 & 0 \end{pmatrix}, \qquad S^2 = \begin{pmatrix} 0 & 1 & 0 \\ 0 & 0 & 1 \\ 1 & 0 & 0 \end{pmatrix}$$

$$T_{12} = \begin{pmatrix} 0 & 1 & 0 \\ 1 & 0 & 0 \\ 0 & 0 & 1 \end{pmatrix}, \; T_{23} = \begin{pmatrix} 1 & 0 & 0 \\ 0 & 0 & 1 \\ 0 & 1 & 0 \end{pmatrix}, \; I = \begin{pmatrix} 1 & 0 & 0 \\ 0 & 1 & 0 \\ 0 & 0 & 1 \end{pmatrix}.$$

Here S and S^2 are the cyclic permutations, T_{ij} interchanges x_i and x_j, and I is the identity.

Indeed, it is a well-known combinatorial theorem that the $n \times n$ permutation matrices form the Hilbert basis for the monoid K_n for all $n \geq 2$. See Exercise 9 below for a general proof.

This gives us some extremely valuable information to work with. By the definition of a Hilbert basis we have, for instance, that in the 3×3 case every element M of K_3 can be written as a linear combination

$$M = aI + bS + cS^2 + dT_{12} + eT_{13} + fT_{23},$$

where a, b, c, d, e, f are nonnegative integers. This is what we meant before by saying that we were looking for "building blocks" for the elements of our additive monoid of magic squares. The row and column sum of M is then given by

$$s = a + b + c + d + e + f.$$

It might appear at first glance that our problem is solved for 3×3 matrices. Namely for a given sum value s, it might seem that we just need to count the ways to write s as a sum of at most 6 nonnegative integers a, b, c, d, e, f. However, there is an added wrinkle here that makes the problem even more interesting: The 6 permutation matrices are not linearly independent. In fact, there is an obvious relation

$$(2.11) \qquad I + S + S^2 = \begin{pmatrix} 1 & 1 & 1 \\ 1 & 1 & 1 \\ 1 & 1 & 1 \end{pmatrix} = T_{12} + T_{13} + T_{23}.$$

This means that for all $s \geq 3$ there are different combinations of coefficients that produce the same matrix sum. How can we take this (and other possible relations) into account and eliminate multiple counting?

First, we claim that in fact every equality

$$(2.12) \qquad \begin{aligned} aI + bS + cS^2 + dT_{12} + eT_{13} + fT_{23} \\ = a'I + b'S + c'S^2 + d'T_{12} + e'T_{13} + f'T_{23}, \end{aligned}$$

where $a, \ldots, f, a', \ldots, f'$ are nonnegative integers, is a consequence of the relation in (2.11), in the sense that if (2.12) is true, then the difference vector

$$(a, b, c, d, e, f) - (a', b', c', d', e', f')$$

is an integer multiple of the vector of coefficients $(1, 1, 1, -1, -1, -1)$ in the linear dependence relation

$$I + S + S^2 - T_{12} - T_{13} - T_{23} = 0,$$

which follows from (2.11).

This can be verified directly as follows.

Exercise 5.

a. Show that the six 3×3 permutation matrices span a 5-dimensional subspace of the vector space of 3×3 real matrices over $\mathbb{R}$.

b. Using part a, show that in every relation (2.12) with $a, \ldots, f' \in \mathbb{Z}_{\geq 0}$, $(a, b, c, d, e, f) - (a', b', c', d', e', f')$ is an integer multiple of the vector $(1, 1, 1, -1, -1, -1)$.

Given this, we can solve our problem in the 3×3 case by "retranslating" it into algebra. Namely we can identify the 6-tuples of coefficients $(a, b, c, d, e, f) \in \mathbb{Z}_{\geq 0}^6$ with monomials in 6 new indeterminates denoted $x_1, \ldots, x_6$:

$$\alpha = (a, b, c, d, e, f) \leftrightarrow x_1^a x_2^b x_3^c x_4^d x_5^e x_6^f.$$

By (2.11), though, we see that we want to think of $x_1 x_2 x_3$ and $x_4 x_5 x_6$ as being the same. This observation indicates that, in counting, we want to consider the element of the quotient ring

$$R = k[x_1, \ldots, x_6] / \langle x_1 x_2 x_3 - x_4 x_5 x_6 \rangle$$

represented by the monomial x^α. Let $MS_3(s)$ be the number of distinct 3×3 integer magic squares with nonnegative entries, and row and column sum equal to s. Our next goal is to show that $MS_3(s)$ can be reinterpreted as the Hilbert function of the above ring R.

We recall from §4 of Chapter 6 that a homogeneous ideal $I \subset k[x_1, \ldots, x_n]$ gives a quotient ring $R = k[x_1, \ldots, x_n]/I$, and the Hilbert function $H_R(s)$ is defined by

$$(2.13) \quad H_R(s) = \dim_k k[x_1, \ldots, x_n]_s / I_s = \dim_k k[x_1, \ldots, x_n]_s - \dim_k I_s,$$

where $k[x_1, \ldots, x_n]_s$ is the vector space of homogeneous polynomials of total degree s, and I_s is the vector space of homogeneous polynomials of total degree s in I. In the notation of Chapter 9, §3 of [CLO], the Hilbert function of $R = k[x_1, \ldots, x_n]/I$ is written $HF_I(s)$. Since our focus here is on the ideal I, in what follows, we will call both $H_R(s)$ and $HF_I(s)$ the Hilbert function of I. It is a basic result that the Hilbert functions of I and $\langle \mathrm{LT}(I) \rangle$ (for any monomial order) are equal. Hence we can compute the Hilbert function by counting the number of standard monomials with respect to I for each total degree s—that is, monomials of total degree s in the complement of $\langle \mathrm{LT}(I) \rangle$. For this and other information about Hilbert functions, the reader should consult [CLO], Chapter 9, §3 or Chapter 6, §4 of this book.

(2.14) Proposition. *The function $MS_3(s)$ equals the Hilbert function $H_R(s) = HF_I(s)$ of the homogeneous ideal $I = \langle x_1 x_2 x_3 - x_4 x_5 x_6 \rangle$.*

PROOF. The single element set $\{x_1 x_2 x_3 - x_4 x_5 x_6\}$ is a Gröbner basis for the ideal it generates with respect to any monomial order. Fix any order such that the leading term of the generator is $x_1 x_2 x_3$. Then the standard monomials of total degree s in $k[x_1, \ldots, x_6]$ are the monomials of total degree s that are not divisible by $x_1 x_2 x_3$.

Given any monomial $x^\alpha = x_1^a x_2^b x_3^c x_4^d x_5^e x_6^f$, let $A = \min(a, b, c)$, and construct

$$\alpha' = (a - A, b - A, c - A, d + A, e + A, f + A).$$

Since $x^{\alpha'}$ is not divisible by $x_1 x_2 x_3$, it is a standard monomial, and you will show in Exercise 6 below that it is the remainder on division of x^α by $x_1 x_2 x_3 - x_4 x_5 x_6$.

We need to show that the 3×3 magic squares with row and column sum s are in one-to-one correspondence with the standard monomials of degree s. Let M be a magic square, and consider any expression

$$(2.15) \qquad M = aI + bS + cS^2 + dT_{12} + eT_{13} + fT_{23}$$

with $\alpha = (a, \ldots, f) \in \mathbb{Z}_{\geq 0}^6$. We associate to M the standard form in R of the monomial x^α, namely $x^{\alpha'}$ as above. In Exercise 7 you will show that this gives a well-defined mapping from the set of magic squares to the collection of standard monomials with respect to I, since by Exercise 5 any two expressions (2.15) for M yield the same standard monomial $x^{\alpha'}$. Moreover the row and column sum of M is the same as the total degree of the image monomial.

This mapping is clearly onto, since the exponent vector α' of any standard monomial can be used to give the coefficients in an expression (2.15). It is also one-to-one, since if M in (2.15) and

$$M_1 = a_1 I + b_1 S + c_1 S^2 + d_1 T_{12} + e_1 T_{13} + f_1 T_{23}$$

map to the same standard monomial α', then writing $A = \min(a, b, c)$, $A_1 = \min(a_1, b_1, c_1)$, we have

$$(a - A, b - A, c - A, d + A, e + A, f + A)$$
$$= (a_1 - A_1, b_1 - A_1, c_1 - A_1, d_1 + A_1, e_1 + A_1, f_1 + A_1).$$

It follows that $(a, \ldots, f)$ and $(a_1, \ldots, f_1)$ differ by the vector

$$(A - A_1)(1, 1, 1, -1, -1, -1).$$

Hence by (2.11), the magic squares M and M_1 are equal.   □

For readers of Chapter 7, we would like to mention that there is also a much more conceptual way to understand the relationship between the monoid K_3 from our original problem and the ring R and the corresponding variety $\mathbf{V}(x_1 x_2 x_3 - x_4 x_5 x_6)$, using the theory of *toric varieties*. In particular, if $\mathcal{A} = \{\vec{m}_1, \ldots, \vec{m}_6\} \subset \mathbb{Z}^9$ is the set of integer vectors corresponding to the 3×3 permutation matrices as above (the Hilbert basis for K_3), and we define $\phi_{\mathcal{A}} : (\mathbb{C}^*)^9 \to \mathbb{P}^5$ by

$$\phi_{\mathcal{A}}(t) = (t^{\vec{m}_1}, \ldots, t^{\vec{m}_6})$$

as in §3 of Chapter 7, then it follows that the toric variety $X_{\mathcal{A}}$ (the Zariski closure of the image of $\phi_{\mathcal{A}}$) is the projective variety $\mathbf{V}(x_1 x_2 x_3 - x_4 x_5 x_6)$. The ideal $I_{\mathcal{A}} = \langle x_1 x_2 x_3 - x_4 x_5 x_6 \rangle$ is called the *toric ideal* corresponding to $\mathcal{A}$. The defining homogeneous ideal of a toric variety is always generated by differences of monomials, as in this example. See the book [Stu2] for more details.

To conclude, Proposition (2.14) solves the 3×3 magic square counting problem as follows. By the proposition and (2.13), to find $MS_3(s)$, we simply subtract the number of nonstandard monomials of total degree s in 6 variables from the total number of monomials of total degree s in 6 variables. The nonstandard monomials are those divisible by $x_1 x_2 x_3$; removing that factor, we obtain an arbitrary monomial of total degree $s - 3$. Hence one expression is the following:

(2.16)
$$MS_3(s) = \binom{s + 5}{5} - \binom{(s - 3) + 5}{5}$$
$$= \binom{s + 5}{5} - \binom{s + 2}{5}.$$

(Also see Exercise 8 below.) For example, $MS_3(1) = 6$ (binomial coefficients $\binom{m}{\ell}$ with $m < \ell$ are zero), $MS_3(2) = 21$, and $MS_3(3) = 56 - 1 = 55$. This is the first time the relation (2.11) comes into play.

For readers who have studied Chapter 6 of this book, we should also mention how free resolutions can be used to obtain (2.16). The key point is that the ideal $I = \langle x_1 x_2 x_3 - x_4 x_5 x_6 \rangle$ is generated by a polynomial of degree 3, so that $I \cong k[x_1, \ldots, x_6](-3)$ as $k[x_1, \ldots, x_6]$-modules. Hence $R = k[x_1, \ldots, x_6]/I$ gives the exact sequence

$$0 \to k[x_1, \ldots, x_6](-3) \to k[x_1, \ldots, x_6] \to R \to 0.$$

Since $H_R(s) = HF_I(s) = MS_3(s)$ by Proposition (2.14), the formula (2.16) follows immediately by the methods of Chapter 6, §4.

These techniques and more sophisticated ideas from commutative algebra, including the theory of toric varieties, have also been applied to the $n \times n$ magic square problem and other related questions from statistics and the design of experiments. We will consider one aspect of the connection with statistics in Exercises 12 and 13 below. We refer the reader to [Sta1] and [Stu2] for a more complete discussion of this interesting connection between algebra and various other areas of the mathematical sciences.

ADDITIONAL EXERCISES FOR §2

Exercise 6. Let R, α and α' be as in the proof of Proposition (2.14). Show that

$$x^\alpha = q(x_1, \dots, x_6)(x_1 x_2 x_3 - x_4 x_5 x_6) + x^{\alpha'},$$

where

$$q = \left((x_1 x_2 x_3)^{A-1} + (x_1 x_2 x_3)^{A-2}(x_4 x_5 x_6) + \cdots + (x_4 x_5 x_6)^{A-1} \right) \cdot$$
$$\cdot\, x_1^{a-A} x_2^{b-A} x_3^{c-A} x_4^d x_5^e x_6^f.$$

Deduce that $x^{\alpha'}$ is the standard form of x^α in R.

Exercise 7. Use Exercise 5 to show that if we have any two expressions as in (2.15) for a given M with coefficient vectors $\alpha = (a, \dots, f)$ and $\alpha_1 = (a_1, \dots, f_1)$, then the corresponding monomials x^α and x^{α_1} have the same standard form $x^{\alpha'}$ in $R = k[x_1, \dots, x_6]/\langle x_1 x_2 x_3 - x_4 x_5 x_6 \rangle$.

Exercise 8. There is another formula, due to MacMahon, for the number of nonnegative integer magic squares of size 3 with a given sum s:

$$MS_3(s) = \binom{s+4}{4} + \binom{s+3}{4} + \binom{s+2}{4}.$$

Show that this formula and (2.16) are equivalent. Hint: This can be proved in several different ways by applying different binomial coefficient identities.

Exercise 9. Verifying that the Hilbert basis for $K_4 = \ker(A_4) \cap \mathbb{Z}_{\geq 0}^{4 \times 4}$ consists of exactly 24 elements corresponding to the 4×4 permutation matrices is already a *large* calculation if you apply the Gröbner basis method of Theorem (2.9). For larger n, this approach quickly becomes infeasible because of the large number of variables needed to make the polynomial translation. Fortunately, there is also a non-computational proof that every $n \times n$ matrix M with nonnegative integer entries and row and column sums all equal to s is a linear combination of $n \times n$ permutation matrices with nonnegative integer coefficients. The proof is by induction on the number of nonzero entries in the matrix.

a. The base case of the induction is the case where exactly n of the entries are nonzero (why?). Show in this case that M is equal to sP for some permutation matrix P.

b. Now assume that the theorem has been proved for all M with k or fewer nonzero entries and consider an M with equal row and column sums and $k+1$ nonzero entries. Using the transversal form of Hall's "marriage" theorem (see, for instance, [Bry]), show that there is some collection of n nonzero entries in M, one from each row and one from each column.

c. Continuing from b, let $d > 0$ be the smallest element in the collection of nonzero entries found in that part, let P be the permutation matrix corresponding to the locations of those nonzero entries, and apply the induction hypothesis to $M - dP$. Deduce the desired result on M.

d. A *doubly stochastic* matrix is an $n \times n$ matrix with nonnegative real entries, all of whose row and column sums equal 1. Adapt the proof sketched in parts a-c to show that the collection of doubly stochastic matrices is the convex hull of the set of $n \times n$ permutation matrices. (See Chapter 7, §1 for more details about convex hulls.)

Exercise 10.
a. How many 3×3 nonnegative integer magic squares with sum s are there if we add the condition that the two diagonal sums should *also* equal s?

b. What about the corresponding question for 4×4 matrices?

Exercise 11. Study the collections of *symmetric* 3×3 and 4×4 nonnegative integer magic squares. What are the Hilbert bases for the monoids of solutions of the corresponding equations? What relations are there? Find the number of squares with a given row and column sum s in each case.

Exercise 12. In this exercise, we will start to develop some ideas concerning *contingency tables* in statistics and see how they relate to the topics discussed in this section. A "two-way" contingency table is an $m \times n$ matrix C with rows labeled according to the m different possible values of some one characteristic of individuals in a population (e.g., political party affiliation, number of TV sets owned, etc. in a human population) and the columns are similarly labeled according to the n possible values of another different characteristic (e.g., response to an item on a questionnaire, age, etc.). The entries are nonnegative integers recording the numbers of individuals in a sample with each combination of values of the two characteristics. The *marginal distribution* of such a table is the collection of row and column sums, giving the total numbers of individuals having each characteristic. For example, if $m = n = 3$ and

$$C = \begin{pmatrix} 34 & 21 & 17 \\ 23 & 21 & 32 \\ 12 & 13 & 50 \end{pmatrix}$$

we have row sums $72, 76, 75$, and column sums $69, 55, 99$.

a. By following what we did for magic squares in the text, show that the collection of all $m \times n$ contingency tables with a given, *fixed* marginal distribution is the set of nonnegative integer solutions of a system of $m + n$ linear equations in mn variables. Give an explicit form for the matrix of your system.

b. Are your equations from part a independent? Why or why not?

c. Is the set of solutions of your system from part a a monoid in $\mathbb{Z}_{\geq 0}^{mn}$ in this case? Why or why not?

Exercise 13. This application comes originally from the article [DS] by Diaconis and Sturmfels. A typical question that statisticians seek to answer is: can we say two characteristics are correlated on the basis of data from a sample of the population? One way that has been proposed to study this sort of problem is to compare values of some statistical measure of correlation from a given sample contingency table and from the other tables with the *same marginal distribution*. In realistic situations it will usually be too difficult to list all the tables having the given marginal distribution (the number can be huge). So a sort of Monte Carlo approach will usually have to suffice. Some number of *randomly generated* tables having the same marginal distribution can be used instead of the whole set. The problem is then to find some efficient way to generate other elements of the collections of tables studied in Exercise 12, given any one element of that collection. Gröbner bases can be used here as follows.

a. Show that C and C' have the same marginal distribution if and only if the difference $T = C' - C$ is an element of the kernel of the matrix of the system of linear equations you found in Exercise 12, part a.

b. To find appropriate matrices T to generate random walks on the set of tables with a fixed marginal distribution, an idea similar to what we did in Theorem (1.11), Proposition (2.8), and Theorem (2.9) of this chapter can be used. Consider a set of "table entry variables" x_{ij}, $1 \leq i \leq m$, $1 \leq j \leq n$ (one for each entry in our tables), "row variables" y_i, $1 \leq i \leq m$, and "column variables" z_j, $1 \leq j \leq n$. Let I be the elimination ideal

$$I = \langle x_{ij} - y_i z_j : 1 \leq i \leq m, 1 \leq j \leq n \rangle \cap k[x_{ij} : 1 \leq i \leq m, 1 \leq j \leq n].$$

Show that any difference of monomials $x^\alpha - x^\beta$ contained in I gives a matrix T as in part a. Hint: Use the exponents from α as entries with positive signs, and the exponents from β as entries with negative signs.

c. Compute a Gröbner basis for I in the case $m = n = 3$ using a suitable lexicographic order. Interpret the matrices T you get this way.

§3 Multivariate Polynomial Splines

In this section we will discuss a recent application of the theory of Gröbner bases to the problem of constructing and analyzing the *piecewise polynomial* or *spline* functions with a specified degree of smoothness on polyhedral subdivisions of regions in $\mathbb{R}^n$. Two-variable functions of this sort are frequently used in computer-aided design to specify the shapes of curved surfaces, and the degree of smoothness attainable in some specified class of piecewise polynomial functions is an important design consideration. For an introductory treatment, see [Far]. Uni- and multivariate splines are also used to interpolate values or approximate other functions in numerical analysis, most notably in the *finite element method* for deriving approximate solutions to partial differential equations. The application of Gröbner bases to this subject appeared first in papers of L. Billera and L. Rose ([BR1], [BR2], [BR3], [Ros]). For more recent results, we refer the reader to [SS]. We will need to use the results on Gröbner bases for *modules* over polynomial rings from Chapter 5.

To introduce some of the key ideas, we will begin by considering the simplest case of one-variable spline functions. On the real line, consider the subdivision of an interval $[a, b]$ into two subintervals $[a, c] \cup [c, b]$ given by any c satisfying $a < c < b$. In rough terms, a piecewise polynomial function on this subdivided interval is any function of the form

$$(3.1) \qquad f(x) = \begin{cases} f_1(x) & \text{if } x \in [a, c] \\ f_2(x) & \text{if } x \in [c, b], \end{cases}$$

where f_1 and f_2 are polynomials in $\mathbb{R}[x]$. Note that we can always make "trivial" spline functions using *the same* polynomial $f_1 = f_2$ on both subintervals, but those are less interesting because we do not have independent control over the shape of the graph of each piece. Hence we will usually be more interested in finding splines with $f_1 \neq f_2$. Of course, as stated, (3.1) gives us a well-defined function on $[a, b]$ if and only if $f_1(c) = f_2(c)$, and if this is true, then f is *continuous* as a function on $[a, b]$. For instance, taking $a = 0, c = 1, b = 2$, and

$$f(x) = \begin{cases} x + 1 & \text{if } x \in [0, 1] \\ x^2 - x + 2 & \text{if } x \in [1, 2], \end{cases}$$

we get a continuous polynomial spline function. See Fig. 8.3.

Since the polynomial functions f_1, f_2 are C^∞ functions (that is, they have derivatives of all orders) and their derivatives are also polynomials, we can consider the piecewise polynomial derivative functions

$$\begin{cases} f_1^{(r)}(x) & \text{if } x \in [a, c] \\ f_2^{(r)}(x) & \text{if } x \in [c, b] \end{cases}$$

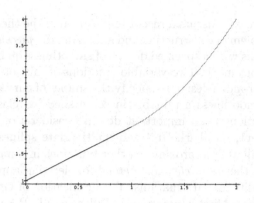

FIGURE 8.3. A continuous spline function

for any $r \geq 0$. As above, we see that f is a C^r function on $[a, b]$ (that is, f is r-times differentiable and its rth derivative, $f^{(r)}$, is continuous) if and only if $f_1^{(s)}(c) = f_2^{(s)}(c)$ for each s, $0 \leq s \leq r$. The following result gives a more algebraic version of this criterion.

(3.2) Proposition. *The piecewise polynomial function f in (3.1) defines a C^r function on $[a, b]$ if and only if the polynomial $f_1 - f_2$ is divisible by $(x - c)^{r+1}$ (that is, $f_1 - f_2 \in \langle (x - c)^{r+1} \rangle$ in $\mathbb{R}[x]$).*

For example, the spline function pictured in Fig. 8.3 is actually a C^1 function since $(x^2 - x + 2) - (x + 1) = (x - 1)^2$. We leave the proof of this proposition to the reader.

Exercise 1. Prove Proposition (3.2).

In practice, it is most common to consider classes of spline functions where the f_i are restricted to be polynomial functions of degree bounded by some fixed integer k. With $k = 2$ we get *quadratic* splines, with $k = 3$ we get *cubic* splines, and so forth.

We will work with two-component splines on a subdivided interval $[a, b] = [a, c] \cup [c, b]$ here. More general subdivisions are considered in Exercise 2 below. We can represent a spline function as in (3.1) by the ordered pair $(f_1, f_2) \in \mathbb{R}[x]^2$. From Proposition (3.2) it follows that the C^r splines form a vector subspace of $\mathbb{R}[x]^2$ under the usual componentwise addition and scalar multiplication. (Also see Proposition (3.10) below, which gives a stronger statement and which includes this one-variable situation as a special case.) Restricting the degree of each component as above, we get elements of the finite-dimensional vector subspace V_k of $\mathbb{R}[x]^2$ spanned by

$$(1, 0), \ (x, 0), \ \ldots, \ (x^k, 0), \ (0, 1), \ (0, x), \ \ldots, \ (0, x^k).$$

The C^r splines in V_k form a vector subspace $V_k^r \subset V_k$. We will focus on the following two questions concerning the V_k^r.

(3.3) Questions.
a. *What is the dimension of V_k^r?*
b. *Given k, what is the biggest r for which there exist C^r spline functions f in V_k^r for which $f_1 \neq f_2$?*

We can answer both of these questions easily in this simple setting. First note that any piecewise polynomial in V_k can be uniquely decomposed as the sum of a spline of the form (f, f), and a spline of the form $(0, g)$:

$$(f_1, f_2) = (f_1, f_1) + (0, f_2 - f_1).$$

Moreover, both terms on the right are again in V_k. Any spline function of the form (f, f) is automatically C^r for every $r \geq 0$. On the other hand, by Proposition (3.2), a spline of the form $(0, g)$ defines a C^r function if and only if $(x - c)^{r+1}$ divides g, and this is possible only if $r + 1 \leq k$. If $r + 1 \leq k$, any linear combination of $(0, (x - c)^{r+1}), \ldots, (0, (x - c)^k)$ gives an element of V_k^r, and these $k - r$ piecewise polynomial functions, together with the $(1, 1), (x, x), \ldots, (x^k, x^k)$ give a basis for V_k^r. These observations yield the following answers to (3.3).

(3.4) Proposition. *For one-variable spline functions on a subdivided interval $[a, b] = [a, c] \cup [c, b]$, The dimension of the space V_k^r is*

$$\dim(V_k^r) = \begin{cases} k + 1 & \text{if } r + 1 > k \\ 2k - r + 1 & \text{if } r + 1 \leq k. \end{cases}$$

The space V_k^r contains spline functions not of the form (f, f) if and only if $r + 1 \leq k$.

For instance, there are C^1 quadratic splines for which $f_1 \neq f_2$, but no C^2 quadratic splines except the ones of the form (f, f). Similarly there are C^2 cubic splines for which $f_1 \neq f_2$, but no C^3 cubic splines of this form. The vector space V_3^2 of C^2 cubic spline functions is 5-dimensional by (3.4). This means, for example, that there is a 2-dimensional space of C^2 cubic splines with any given values $f(a) = A$, $f(c) = C$, $f(b) = B$ at $x = a, b, c$. Because this freedom gives additional control over the shape of the graph of the spline function, one-variable cubic splines are used extensively as interpolating functions in numerical analysis.

The reader should have no difficulty extending all of the above to spline functions on any subdivided interval $[a, b]$, where the subdivision is specified by an arbitrary partition.

Exercise 2. Consider a partition

$$a = x_0 < x_1 < x_2 < \cdots < x_{m-1} < x_m = b$$

of the interval $[a, b]$ into m smaller intervals.

a. Let $(f_1, \ldots, f_m) \in \mathbb{R}[x]^m$ be an m-tuple of polynomials. Define f on $[a, b]$ by setting $f|_{[x_{i-1}, x_i]} = f_i$, Show that f is a C^r function on $[a, b]$ if and only if for each i, $1 \le i \le m - 1$, $f_{i+1} - f_i \in \langle (x - x_i)^{r+1} \rangle$.

b. What is the dimension of the space of C^r splines with $\deg f_i \le k$ for all i? Find a basis. Hint: There exists a nice "triangular" basis generalizing what we did in the text for the case of two subintervals.

c. Show that there is a 2-dimensional space of C^2 cubic spline functions interpolating any specified values at the x_i, $i = 0, \ldots, n$.

We now turn to multivariate splines. Corresponding to subdivisions of intervals in $\mathbb{R}$, we will consider certain subdivisions of *polyhedral* regions in $\mathbb{R}^n$. As in Chapter 7, a *polytope* is the convex hull of a finite set in $\mathbb{R}^n$, and by (1.4) of that chapter, a polytope can be written as the intersection of a collection of affine half-spaces. In constructing partitions of intervals in $\mathbb{R}$, we allowed the subintervals to intersect only at common endpoints. Similarly, in $\mathbb{R}^n$ we will consider subdivisions of polyhedral regions into polytopes that intersect only along common faces.

The major new feature in $\mathbb{R}^n$, $n \ge 2$ is the much greater geometric freedom possible in constructing such subdivisions. We will use the following language to describe them.

(3.5) Definition.
a. A *polyhedral complex* $\Delta \subset \mathbb{R}^n$ is a finite collection of polytopes such that the faces of each element of Δ are elements of Δ, and the intersection of any two elements of Δ is an element of Δ. We will sometimes refer to the k-dimensional elements of a complex Δ as *k-cells*.

b. A polyhedral complex $\Delta \subset \mathbb{R}^n$ is said to be *pure* n-dimensional if every maximal element of Δ (with respect to inclusion) is an n-dimensional polyhedron.

c. Two n-dimensional polytopes in a complex Δ are said to be *adjacent* if they intersect along a common face of dimension $n - 1$.

d. Δ is said to be a *hereditary* complex if for every $\tau \in \Delta$ (including the empty set), any two n-dimensional polytopes σ, σ' of Δ that contain τ can be connected by a sequence $\sigma = \sigma_1, \sigma_2, \ldots, \sigma_m = \sigma'$ in Δ such that each σ_i is n-dimensional, each σ_i contains τ, and σ_i and σ_{i+1} are adjacent for each i.

The cells of a complex give a particularly well-structured subdivision of the polyhedral region $R = \cup_{\sigma \in \Delta} \sigma \subset \mathbb{R}^n$.

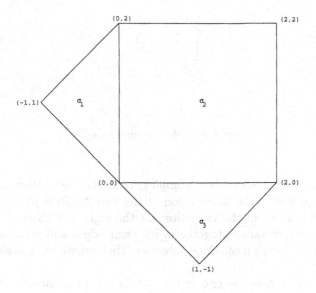

FIGURE 8.4. A polyhedral complex in $\mathbb{R}^2$

Here are some examples to illustrate the meaning of these conditions. For example, Fig. 8.4 is a picture of a polyhedral complex in $\mathbb{R}^2$ consisting of 18 polytopes in all—the three 2-dimensional polygons $\sigma_1, \sigma_2, \sigma_3$, eight 1-cells (the edges), six 0-cells (the vertices at the endpoints of edges), and the empty set, $\emptyset$.

The condition on intersections in the definition of a complex rules out collections of polyhedra such as the ones in Fig. 8.5. In the collection on

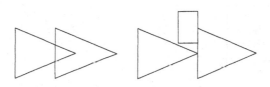

FIGURE 8.5. Collections of polygons that are not complexes

FIGURE 8.6. A non-pure complex

the left (which consists of *two triangles*, their six edges, their six vertices
and the empty set), the intersection of the two 2-cells is not a cell of the
complex. Similarly, in the collection on the right (which consists of two
triangles and a rectangle, together with their edges and vertices, and the
empty set) the 2-cells meet along subsets of their edges, but not along entire
edges.

A complex such as the one in Fig. 8.6 is not pure, since τ is maximal
and only 1-dimensional.

A complex is *not* hereditary if it is not connected, or if it has maximal
elements meeting only along faces of codimension 2 or greater, with no
other connection via n-cells, as is the case for the complex in Fig. 8.7.
(Here, the cells are the two triangles, their edges and vertices, and finally
the empty set.)

Let Δ be any pure n-dimensional polyhedral complex in $\mathbb{R}^n$, let
$\sigma_1, \ldots, \sigma_m$ be a given, fixed, ordering of the n-cells in Δ, and let
$R = \cup_{i=1}^m \sigma_i$. Generalizing our discussion of univariate splines above, we
introduce the following collections of piecewise polynomial functions on R.

FIGURE 8.7. A non-hereditary complex

(3.6) Definition.
a. For each $r \geq 0$ we will denote by $C^r(\Delta)$ the collection of C^r functions f on R (that is, functions such that all rth order partial derivatives exist and are continuous on R) such that for every $\delta \in \Delta$ including those of dimension $< n$, the restriction $f|_\delta$ is a polynomial function $f_\delta \in \mathbb{R}[x_1, \ldots, x_n]$.
b. $C_k^r(\Delta)$ is the subset of $f \in C^r(\Delta)$ such that the restriction of f to each cell in Δ is a polynomial function of degree k or less.

Our goal is to study the analogues of Questions (3.3) for the $C_k^r(\Delta)$. Namely, we wish to compute the dimensions of these spaces over $\mathbb{R}$, and to determine when they contain nontrivial splines.

We will restrict our attention in the remainder of this section to complexes Δ that are both pure and hereditary. If σ_i, σ_j are adjacent n-cells of Δ, then they intersect along an interior $(n-1)$-cell $\sigma_{ij} \in \Delta$, a polyhedral subset of an affine hyperplane $\mathbf{V}(\ell_{ij})$, where $\ell_{ij} \in \mathbb{R}[x_1, \ldots, x_n]$ is a polynomial of total degree 1. Generalizing Proposition (3.2) above, we have the following algebraic characterization of the elements of $C^r(\Delta)$ in the case of a pure, hereditary complex.

(3.7) Proposition. *Let Δ be a pure, hereditary complex with m n-cells σ_i. Let $f \in C^r(\Delta)$, and for each i, $1 \leq i \leq m$, let $f_i = f|_{\sigma_i} \in \mathbb{R}[x_1, \ldots, x_n]$. Then for each adjacent pair σ_i, σ_j in Δ, $f_i - f_j \in \langle \ell_{ij}^{r+1} \rangle$. Conversely, any m-tuple of polynomials $(f_1, \ldots, f_m)$ satisfying $f_i - f_j \in \langle \ell_{ij}^{r+1} \rangle$ for each adjacent pair σ_i, σ_j of n-cells in Δ defines an element $f \in C^r(\Delta)$ when we set $f|_{\sigma_i} = f_i$.*

The meaning of Proposition (3.7) is that for pure n-dimensional complexes $\Delta \subset \mathbb{R}^n$, piecewise polynomial functions are determined by their restrictions to the n-cells $\sigma_1, \ldots, \sigma_m$ in Δ. In addition, for hereditary complexes, the C^r property for piecewise polynomial functions f may be checked by comparing *only* the restrictions $f_i = f|_{\sigma_i}$ and $f_j = f|_{\sigma_j}$ for *adjacent* pairs of n-cells.

PROOF. If f is an element of $C^r(\Delta)$, then for each adjacent pair σ_i, σ_j of n-cells in Δ, $f_i - f_j$ and all its partial derivatives of order up to and including r must vanish on $\sigma_i \cap \sigma_j$. In Exercise 3 below you will show that this implies $f_i - f_j$ is an element of $\langle \ell_{ij}^{r+1} \rangle$.

Conversely, suppose we have $f_1, \ldots, f_m \in \mathbb{R}[x_1, \ldots, x_n]$ such that $f_i - f_j$ is an element of $\langle \ell_{ij}^{r+1} \rangle$ for each adjacent pair of n-cells in Δ. In Exercise 3 below, you will show that this implies that f_i and its partial derivatives of order up to and including r agree with f_j and its corresponding derivatives at each point of $\sigma_i \cap \sigma_j$. But the $f_1, \ldots, f_m$ define a C^r function on R if and only if for *every* $\delta \in \Delta$ and every pair of n-cells σ_p, σ_q containing δ (not only adjacent ones) f_p and its partial derivatives of order up to and including r agree with f_q and its corresponding derivatives at each point of δ. So let p, q be any pair of indices for which $\delta \subset \sigma_p \cap \sigma_q$. Since Δ is hereditary, there is a sequence of n-cells

$$\sigma_p = \sigma_{i_1}, \sigma_{i_2}, \ldots, \sigma_{i_k} = \sigma_q,$$

each containing δ, such that σ_{i_j} and $\sigma_{i_{j+1}}$ are adjacent. By assumption, this implies that for each j, $f_{i_j} - f_{i_{j+1}}$ and all its partial derivatives of orders up to and including r vanish on $\sigma_{i_j} \cap \sigma_{i_{j+1}} \supset \delta$. But

$$f_p - f_q = (f_{i_1} - f_{i_2}) + (f_{i_2} - f_{i_3}) + \cdots + (f_{i_{k-1}} - f_{i_k})$$

and each term on the right and its partials up to and including order r vanish on δ. Hence $f_1, \ldots, f_m$ define an element of $C^r(\Delta)$. $\square$

Exercise 3. Let σ, σ' be two adjacent n-cells in a polyhedral complex Δ, and let $\sigma \cap \sigma' \subset \mathbf{V}(\ell)$ for a linear polynomial $\ell \in \mathbb{R}[x_1, \ldots, x_n]$.
a. Show that if $f, f' \in \mathbb{R}[x_1, \ldots, x_n]$ satisfy $f - f' \in \langle \ell^{r+1} \rangle$, then the partial derivatives of all orders $\leq r$ of f and f' agree at every point in $\sigma \cap \sigma'$.
b. Conversely if the partial derivatives of all orders $\leq r$ of f and f' agree at every point in $\sigma \cap \sigma'$, show that $f - f' \in \langle \ell^{r+1} \rangle$.

Fixing any one ordering on the n-cells σ_i in Δ, we will represent elements f of $C^r(\Delta)$ by ordered m-tuples $(f_1, \ldots, f_m) \in \mathbb{R}[x_1, \ldots, x_n]^m$, where $f_i = f|_{\sigma_i}$.

Consider the polyhedral complex Δ in $\mathbb{R}^2$ from Fig. 8.4, with the numbering of the 2-cells given there. It is easy to check that Δ is hereditary. The interior edges are given by $\sigma_1 \cap \sigma_2 \subset \mathbf{V}(x)$ and $\sigma_2 \cap \sigma_3 \subset \mathbf{V}(y)$. By the preceding proposition, an element $(f_1, f_2, f_3) \in \mathbb{R}[x, y]^3$ gives an element of $C^r(\Delta)$ if and only if

$$f_1 - f_2 \in \langle x^{r+1} \rangle, \quad \text{and}$$
$$f_2 - f_3 \in \langle y^{r+1} \rangle.$$

To prepare for our next result, note that these inclusions can be rewritten in the form

$$f_1 - f_2 + x^{r+1} f_4 = 0$$
$$f_2 - f_3 + y^{r+1} f_5 = 0$$

for some $f_4, f_5 \in \mathbb{R}[x, y]$. These equations can be rewritten again in vector-matrix form as

$$\begin{pmatrix} 1 & -1 & 0 & x^{r+1} & 0 \\ 0 & 1 & -1 & 0 & y^{r+1} \end{pmatrix} \begin{pmatrix} f_1 \\ f_2 \\ f_3 \\ f_4 \\ f_5 \end{pmatrix} = \begin{pmatrix} 0 \\ 0 \end{pmatrix}.$$

Thus, elements of $C^r(\Delta)$ are projections onto the first three components of elements of the kernel of the map $\mathbb{R}[x,y]^5 \to \mathbb{R}[x,y]^2$ defined by

$$(3.8) \qquad M(\Delta, r) = \begin{pmatrix} 1 & -1 & 0 & x^{r+1} & 0 \\ 0 & 1 & -1 & 0 & y^{r+1} \end{pmatrix}.$$

By Proposition (1.10) and Exercise 9 of §3 of Chapter 5, it follows that $C^r(\Delta)$ has the structure of a *module* over the ring $\mathbb{R}[x,y]$. This observation allows us to apply the theory of Gröbner bases to study splines.

Our next result gives a corresponding statement for $C^r(\Delta)$ in general. We begin with some necessary notation. Let Δ be a pure, hereditary polyhedral complex in $\mathbb{R}^n$. Let m be the number of n-cells in Δ, and let e be the number of *interior* $(n-1)$-cells (the intersections $\sigma_i \cap \sigma_j$ for adjacent n-cells). Fix some ordering $\tau_1, \ldots, \tau_e$ for the interior $(n-1)$-cells and let ℓ_s be a linear polynomial defining the affine hyperplane containing τ_s. Consider the $e \times (m+e)$ matrix $M(\Delta, r)$ with the following block decomposition:

$$(3.9) \qquad M(\Delta, r) = (\partial(\Delta) \mid D).$$

(Note: the orderings of the rows and columns are determined by the orderings of the indices of the n-cells and the interior $(n-1)$-cells, but any ordering can be used.) In (3.9), $\partial(\Delta)$ is the $e \times m$ matrix defined by this rule: In the sth row, if $\tau_s = \sigma_i \cap \sigma_j$ with $i < j$, then

$$\partial(\Delta)_{sk} = \begin{cases} +1 & \text{if } k = i \\ -1 & \text{if } k = j \\ 0 & \text{otherwise.} \end{cases}$$

In addition, D is the $e \times e$ diagonal matrix

$$D = \begin{pmatrix} \ell_1^{r+1} & 0 & \cdots & 0 \\ 0 & \ell_2^{r+1} & \cdots & 0 \\ \vdots & \vdots & \ddots & \vdots \\ 0 & 0 & \cdots & \ell_e^{r+1} \end{pmatrix}.$$

Then as in the example above we have the following statement.

(3.10) Proposition. *Let Δ be a pure, hereditary polyhedral complex in $\mathbb{R}^n$, and let $M(\Delta, r)$ be the matrix defined in (3.9) above.*

a. *An m-tuple $(f_1, \ldots, f_m)$ is in $C^r(\Delta)$ if and only if there exist $(f_{m+1}, \ldots, f_{m+e})$ such that $f = (f_1, \ldots, f_m, f_{m+1}, \ldots, f_{m+e})^T$ is an element of the kernel of the map $\mathbb{R}[x_1, \ldots, x_n]^{m+e} \to \mathbb{R}[x_1, \ldots, x_n]^e$ defined by the matrix $M(\Delta, r)$.*

b. *$C^r(\Delta)$ has the structure of a module over the ring $\mathbb{R}[x_1, \ldots, x_n]$. In the language of Chapter 5, it is the image of the projection homomorphism from $\mathbb{R}[x_1, \ldots, x_n]^{m+e}$ onto $\mathbb{R}[x_1, \ldots, x_n]^m$ (in the first m components) of the module of syzygies on the columns of $M(\Delta, r)$.*

c. *$C_k^r(\Delta)$ is a finite-dimensional vector subspace of $C^r(\Delta)$.*

PROOF. Part a is essentially just a restatement of Proposition (3.7). For each interior $(n-1)$-cell $\tau_s = \sigma_i \cap \sigma_j$, $(i < j)$ we have an equation

$$f_i - f_j = -\ell_s^{r+1} f_{m+s}$$

for some $f_{m+s} \in \mathbb{R}[x_1, \ldots, x_n]$. This is the equation obtained by setting the sth component of the product $M(\Delta, r)f$ equal to zero.

Part b follows immediately from part a as in Chapter 5, Proposition (1.10) and Exercise 9 of Chapter 5, §3.

Part c follows by a direct proof, or more succinctly from part b, since $C_k^r(\Delta)$ is closed under sums and products by constant polynomials. □

The Gröbner basis algorithm based on Schreyer's Theorem (Chapter 5, Theorem (3.3)) may be applied to compute a Gröbner basis for the kernel of $M(\Delta, r)$ for each r, and from that information the dimensions of, and bases for, the $C_k^r(\Delta)$ may be determined.

As a first example, let us compute the $C^r(\Delta)$ for the complex $\Delta \subset \mathbb{R}^2$ from (3.8). We consider the matrix as in (3.8) with $r = 1$ first. Using any monomial order in $\mathbb{R}[x, y]^5$ with $e_5 > \cdots > e_1$, we compute a Gröbner basis for $\ker(M(\Delta, 1)$ (that is, the module of syzygies of the columns of $M(\Delta, 1)$) and we find three basis elements, the transposes of

$$g_1 = (1, 1, 1, 0, 0)$$
$$g_2 = (-x^2, 0, 0, 1, 0)$$
$$g_3 = (-y^2, -y^2, 0, 0, 1).$$

(In this simple case, it is easy to write down these syzygies by inspection. They must generate the module of syzygies because of the form of the matrix $M(\Delta, r)$—the last three components of the vector f are arbitrary, and these determine the first two.) The elements of $C^1(\Delta)$ are given by projection on the first three components, so we see that the general element of $C^1(\Delta)$ will have the form

(3.11)
$$\begin{aligned} f(1, 1, 1) + g(-x^2, 0, 0) + h(-y^2, -y^2, 0) \\ = (f - gx^2 - hy^2, f - hy^2, f), \end{aligned}$$

where $f, g, h \in \mathbb{R}[x, y]^2$ are arbitrary polynomials. Note that the triples with $g = h = 0$ are the "trivial" splines where we take the same polynomial on each σ_i, while the other generators contribute terms supported on only one or two of the 2-cells. The algebraic structure of $C^1(\Delta)$ as a module over $\mathbb{R}[x, y]$ is very simple—$C^1(\Delta)$ is a *free* module and the given generators form a module basis. (Billera and Rose show in Lemma 3.3 and Theorem 3.5 of [BR3] that the same is true for $C^r(\Delta)$ for *any* hereditary complex $\Delta \subset \mathbb{R}^2$ and all $r \geq 1$.) Using the decomposition it is also easy to count the dimension of $C_k^1(\Delta)$ for each k. For $k = 0, 1$, we have only the "trivial" splines, so $\dim C_0^1(\Delta) = 1$, and $\dim C_1^1(\Delta) = 3$ (a vector space basis is $\{(1, 1, 1), (x, x, x), (y, y, y)\}$). For $k \geq 2$, there are nontrivial splines as well, and we see by counting monomials of the appropriate degrees in f, g, h that

$$\dim C_k^1(\Delta) = \binom{k+2}{2} + 2\binom{(k-2)+2}{2} = \binom{k+2}{2} + 2\binom{k}{2}.$$

Also see Exercise 9 below for a more succinct way to package the information from the function $\dim C_k^1(\Delta)$.

For larger r, the situation is entirely analogous in this example. A Gröbner basis for the kernel of $M(\Delta, r)$ is given by

$$g_1 = (1, 1, 1, 0, 0)^T$$
$$g_2 = (-x^{r+1}, 0, 0, 1, 0)^T$$
$$g_3 = (-y^{r+1}, -y^{r+1}, 0, 0, 1)^T,$$

and we have that $C^r(\Delta)$ is a free module over $\mathbb{R}[x, y]$ for all $r \geq 0$. Thus

$$\dim C_k^r(\Delta) = \begin{cases} \binom{k+2}{2} & \text{if } k < r + 1 \\ \binom{k+2}{2} + 2\binom{k-r+1}{2} & \text{if } k \geq r + 1. \end{cases}$$

Our next examples, presented as exercises for the reader, indicate some of the subtleties that can occur for more complicated complexes. (Additional examples can be found in the exercises at the end of the section.)

Exercise 4. In $\mathbb{R}^2$, consider the convex quadrilateral

$$R = \mathrm{Conv}(\{(2, 0), (0, 1), (-1, 1), (-1, -2)\})$$

(notation as in §1 of Chapter 7), and subdivide R into triangles by connecting each vertex to the origin by line segments. We obtain in this way a pure, hereditary polyhedral complex Δ containing four 2-cells, eight 1-cells (four interior ones), five 0-cells, and $\emptyset$. Number the 2-cells $\sigma_1, \ldots, \sigma_4$ proceeding counter-clockwise around the origin starting from the triangle $\sigma_1 = \mathrm{Conv}(\{(2, 0), (0, 0), (0, 1)\})$. The interior 1-cells of Δ are then

$$\sigma_1 \cap \sigma_2 \subset \mathbf{V}(x)$$
$$\sigma_2 \cap \sigma_3 \subset \mathbf{V}(x + y)$$
$$\sigma_3 \cap \sigma_4 \subset \mathbf{V}(2x - y)$$
$$\sigma_1 \cap \sigma_4 \subset \mathbf{V}(y).$$

a. Using this ordering on the interior 1-cells, show that we obtain

$$\begin{pmatrix} 1 & -1 & 0 & 0 & x^{r+1} & 0 & 0 & 0 \\ 0 & 1 & -1 & 0 & 0 & (x+y)^{r+1} & 0 & 0 \\ 0 & 0 & 1 & -1 & 0 & 0 & (2x-y)^{r+1} & 0 \\ 1 & 0 & 0 & -1 & 0 & 0 & 0 & y^{r+1} \end{pmatrix}$$

for the matrix $M(\Delta, r)$.

b. With $r = 1$, for instance, show that a Gröbner basis for the $\mathbb{R}[x, y]$-module of syzygies on the columns of $M(\Delta, 1)$ is given by the transposes of the following vectors

$$g_1 = (1, 1, 1, 1, 0, 0, 0, 0)$$
$$g_2 = (1/4)(3y^2, 6x^2 + 3y^2, 4x^2 - 4xy + y^2, 0, 6, -2, -1, -3)$$
$$g_3 = (2xy^2 + y^3, 0, 0, y, -y, 0, -2x - y)$$
$$g_4 = (-3xy^2 - 2y^3, x^3 - 3xy^2 - 2y^3, 0, 0, x, -x + 2y, 0, 3x + 2y)$$
$$g_5 = (x^2y^2, 0, 0, 0, -y^2, 0, 0, -x^2).$$

c. As before, the elements of $C^1(\Delta)$ are obtained by projection onto the first four components. From this, show that there are only "trivial" splines in $C_0^1(\Delta)$ and $C_1^1(\Delta)$, but g_2 and its multiples give nontrivial splines in all degrees $k \geq 2$, while g_3 and g_4 also contribute terms in degrees $k \geq 3$.

d. Show that the g_i form a basis for $C^1(\Delta)$, so it is a free module. Thus

$$\dim C_k^1(\Delta) = \begin{cases} 1 & \text{if } k = 0 \\ 3 & \text{if } k = 1 \\ 7 & \text{if } k = 2 \\ \binom{k+2}{2} + \binom{k}{2} + 2\binom{k-1}{2} & \text{if } k \geq 3. \end{cases}$$

We will next consider a second polyhedral complex Δ' in $\mathbb{R}^2$ which has the same combinatorial data as Δ in Exercise 4 (that is, the numbers of k-cells are the same for all k, the containment relations are the same, and so forth), but which is in special position.

Exercise 5. In $\mathbb{R}^2$, consider the convex quadrilateral

$$R = \text{Conv}(\{(2, 0), (0, 1), (-1, 0), (0, -2)\}).$$

Subdivide R into triangles by connecting each vertex to the origin by line segments. This gives a pure, hereditary polyhedral complex Δ' with four 2-cells, eight 1-cells (four interior ones), five 0-cells, and $\emptyset$. Number the 2-cells $\sigma_1, \ldots, \sigma_4$ proceeding counter-clockwise around the origin starting from the triangle σ_1 with vertices $(2, 0), (0, 0), (0, 1)$. The interior 1-cells of

Δ are then

$$\sigma_1 \cap \sigma_2 \subset \mathbf{V}(x)$$
$$\sigma_2 \cap \sigma_3 \subset \mathbf{V}(y)$$
$$\sigma_3 \cap \sigma_4 \subset \mathbf{V}(x)$$
$$\sigma_1 \cap \sigma_4 \subset \mathbf{V}(y).$$

This is what we meant before by saying that Δ' is in special position—the interior edges lie on only two distinct lines, rather than four of them.

a. Using this ordering on the interior 1-cells, show that we obtain

$$M(\Delta', r) = \begin{pmatrix} 1 & -1 & 0 & 0 & x^{r+1} & 0 & 0 & 0 \\ 0 & 1 & -1 & 0 & 0 & y^{r+1} & 0 & 0 \\ 0 & 0 & 1 & -1 & 0 & 0 & x^{r+1} & 0 \\ 1 & 0 & 0 & -1 & 0 & 0 & 0 & y^{r+1} \end{pmatrix}.$$

b. With $r = 1$, for instance, show that a Gröbner basis for the $\mathbb{R}[x, y]$-module of syzygies on the columns of $M(\Delta', 1)$ is given by the transposes of

$$g_1' = (1, 1, 1, 1, 0, 0, 0, 0)$$
$$g_2' = (0, x^2, x^2, 0, 1, 0, -1, 0)$$
$$g_3' = (y^2, y^2, 0, 0, 0, -1, 0, -1)$$
$$g_4' = (x^2 y^2, 0, 0, 0, -y^2, 0, 0, -x^2).$$

Note that these generators have a different form (in particular, the components have different total degrees) than the generators for the syzygies on the columns of $M(\Delta, 1)$.

c. Check that the g_i' form a basis of $C^1(\Delta')$, and that

$$\dim C_k^1(\Delta') = \begin{cases} 1 & \text{if } k = 0 \\ 3 & \text{if } k = 1 \\ 8 & \text{if } k = 2 \\ 16 & \text{if } k = 3 \\ \binom{k+2}{2} + 2\binom{k}{2} + \binom{k-2}{2} & \text{if } k \geq 3. \end{cases}$$

Comparing Exercises 4 and 5, we see that the dimensions of $C_k^r(\Delta)$ can depend on more than just the combinatorial data of the polyhedral complex Δ—they can vary depending on the positions of the interior $(n-1)$-cells.

The recent paper [Ros] of Lauren Rose sheds some light on examples like these. To describe her results, it will be convenient to use the following notion.

(3.12) Definition. The *dual graph* G_Δ of a pure n-dimensional complex Δ is the graph with vertices corresponding to the n-cells in Δ, and edges corresponding to adjacent pairs of n-cells.

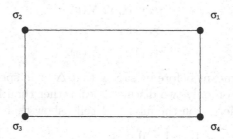

FIGURE 8.8. The dual graph

For instance, the dual graphs for the complexes in Exercises 4 and 5 are both equal to the graph in Fig. 8.8. By an easy translation of the definition in (3.5), the dual graph of a hereditary complex is *connected*.

As before, we will denote by e the number of interior $(n-1)$-cells and let $\delta_1, \ldots, \delta_e$ denote some ordering of them. Choose an ordering on the vertices of G_Δ (or equivalently on the n-cells of Δ), and consider the induced orientations of the edges. If $\delta = jk$ is the oriented edge from vertex j to vertex k in G_Δ, corresponding to the interior $(n-1)$-cell $\delta = \sigma_j \cap \sigma_k$, let ℓ_δ be the equation of the affine hyperplane containing δ. By convention, we take the *negative*, $-\ell_\delta$, as the defining equation for the affine hyperplane containing the edge kj with reversed orientation. For simplicity, we will also write ℓ_i for the linear polynomial ℓ_{δ_i}. Finally, let $\mathcal{C}$ denote the set of cycles in G_Δ. Then, following Rose, we consider a module $B^r(\Delta)$ built out of syzygies on the ℓ_i^{r+1}.

(3.13) Definition. $B^r(\Delta) \subset \mathbb{R}[x_1, \ldots, x_n]^e$ is the submodule defined by

$$B^r(\Delta) = \{(g_1, \ldots, g_e) \in \mathbb{R}[x_1, \ldots, x_n]^e : \text{ for all } c \in \mathcal{C}, \sum_{\delta \in c} g_\delta \ell_\delta^{r+1} = 0\}.$$

The following observation is originally due to Schumaker for the case of bivariate splines (see [Schu]). Our treatment follows Theorem 2.2 of [Ros].

(3.14) Theorem. *If Δ is hereditary, then $C^r(\Delta)$ is isomorphic to $B^r(\Delta) \oplus \mathbb{R}[x_1, \ldots, x_n]$ as an $\mathbb{R}[x_1, \ldots, x_n]$-module.*

PROOF. Consider the mapping

$$\varphi : C^r(\Delta) \to B^r(\Delta) \oplus \mathbb{R}[x_1, \ldots, x_n]$$

defined in the following way. By (3.7), for each $f = (f_1, \ldots, f_m)$ in $C^r(\Delta)$ and each interior $(n-1)$-cell $\delta_i = \sigma_j \cap \sigma_k$, we have $f_j - f_k = g_i \ell_i^{r+1}$ for some $g_i \in \mathbb{R}[x_1, \ldots, x_n]$. Let

$$\varphi(f) = ((g_1, \ldots, g_e), f_1)$$

(the f_1 is the component in the $\mathbb{R}[x_1, \ldots, x_n]$ summand). For each cycle c in the dual graph, $\sum_{\delta \in c} g_\delta \ell_\delta^{r+1}$ equals a sum of the form $\sum (f_j - f_k)$, which cancels completely to 0 since c is a cycle. Hence, the e-tuple $(g_1, \ldots, g_e)$ is an element of $B^r(\Delta)$. It is easy to see that φ is a homomorphism of $\mathbb{R}[x_1, \ldots, x_n]$-modules.

To show that φ is an isomorphism, consider any

$$((g_1, \ldots, g_e), f) \in B^r(\Delta) \oplus \mathbb{R}[x_1, \ldots, x_n].$$

Let $f_1 = f$. For each i, $2 \le i \le m$, since G_Δ is connected, there is some path from vertex σ_1 to σ_i in G_Δ, using the edges in some set E. Let $f_i = f + \sum_{\delta \in E} g_\delta \ell_\delta^{r+1}$, where as above the g_δ are defined by $f_j - f_k = g_\delta \ell_\delta^{r+1}$ if δ is the oriented edge jk. Any two paths between these two vertices differ by a combination of cycles, so since $(g_1, \ldots, g_e) \in B^r(\Delta)$, f_i is a well-defined polynomial function on σ_i, and the m-tuple $(f_1, \ldots, f_m)$ gives a well-defined element of $C^r(\Delta)$ (why?). We obtain in this way a homomorphism

$$\psi : B^r(\Delta) \oplus \mathbb{R}[x_1, \ldots, x_n] \to C^r(\Delta),$$

and it is easy to check that ψ and φ are inverses. $\square$

The algebraic reason for the special form of the generators of the module $C^1(\Delta)$ in Exercise 5 as compared to those in Exercise 4 can be read off easily from the alternate description of $C^1(\Delta)$ given by Theorem (3.14). For the dual graph shown in Fig. 8.8 on the previous page, there is exactly one cycle. In Exercise 4, numbering the edges counterclockwise, we have

$$\ell_1^2 = x^2, \ \ell_2^2 = (x+y)^2, \ \ell_3^2 = (2x-y)^2, \ \ell_4^2 = y^2.$$

It is easy to check that the dimension over $\mathbb{R}$ of the subspace of $B(\Delta)$ with g_i constant for all i is 1, so that applying the mapping ψ from the proof of Theorem (3.14), the quotient of the space $C_2^1(\Delta)$ of quadratic splines modulo the trivial quadratic splines is 1-dimensional. (The spline g_2 from part b of the exercise gives a basis.) On the other hand, in Exercise 5,

$$\ell_1^2 = x^2, \ \ell_2^2 = y^2, \ \ell_3^2 = x^2, \ \ell_4^2 = y^2,$$

so $B^1(\Delta)$ contains both $(1, 0, -1, 0)$ and $(0, 1, 0, -1)$. Under ψ, we obtain that the quotient of $C_2^1(\Delta)$ modulo the trivial quadratic splines is two-dimensional.

As an immediate corollary of Theorem (3.14), we note the following general sufficient condition for $C^r(\Delta)$ to be a free module.

(3.15) Corollary. *If Δ is hereditary and G_Δ is a tree (i.e., a connected graph with no cycles), then $C^r(\Delta)$ is free for all $r \geq 0$.*

PROOF. If there are no cycles, then $B^r(\Delta)$ is equal to the free module $\mathbb{R}[x_1, \ldots, x_n]^e$, and the corollary follows from Theorem (3.14). This result is Theorem 3.1 of [Ros].   □

Returning to bivariate splines, for *generic* pure 2-dimensional hereditary *simplicial* complexes Δ in $\mathbb{R}^2$ (that is, complexes where all 2-cells are *triangles* whose edges are in sufficiently general position) giving triangulations of 2-manifolds with boundary in the plane, there is a simple combinatorial formula for dim $C_k^1(\Delta)$ first conjectured by Strang, and proved by Billera (see [Bil1]). The form of this dimension formula given in [BR1] is the following:

$$(3.16) \qquad \dim C_k^1(\Delta) = \binom{k+2}{2} + (h_1 - h_2)\binom{k}{2} + 2h_2\binom{k-1}{2}.$$

Here h_1 and h_2 are determined by purely combinatorial data from Δ:

$$(3.17) \qquad h_1 = V - 3 \quad \text{and} \quad h_2 = 3 - 2V + E,$$

where V is the number of 0-cells, and E is the number of 1-cells in Δ. (Also see Exercise 12 below for Strang's original dimension formula, and its connection to (3.16).)

For example, the simplicial complex Δ in Exercise 4, in which the interior edges lie on four distinct lines (the generic situation) has $V = 5$ and $E = 8$, so $h_1 = 2$ and $h_2 = 1$. Hence (3.16) agrees with the formula from part d of the exercise. On the other hand, the complex Δ' from Exercise 5 is not generic as noted above, and (3.16) is not valid for Δ'.

Interestingly enough, there is no corresponding statement for $n \geq 3$. Moreover, the modules $C^r(\Delta)$ can fail to be free modules even in very simple cases (see part c of Exercise 10 below for instance). The paper [Sche] gives necessary and sufficient conditions for freeness of $C^r(\Delta)$ and shows that the first three terms of its Hilbert polynomial can be determined from the combinatorics and local geometry of Δ. The case $n = 3$, $r = 1$ is also studied in [ASW]. Nevertheless, this is still an area with many open questions.

ADDITIONAL EXERCISES FOR §3

Exercise 6. Investigate the modules $C^r(\Delta)$ and $C^r(\Delta')$, $r \geq 2$, for the complexes from Exercises 4 and 5. What are dim $C_k^r(\Delta)$ and dim $C_k^r(\Delta')$?

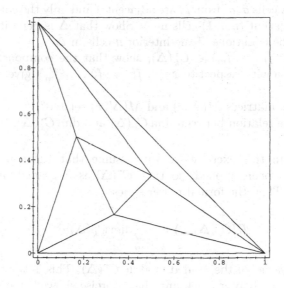

FIGURE 8.9. Figure for Exercise 7

Exercise 7. Let Δ be the simplicial complex in $\mathbb{R}^2$ given in Fig. 8.9. The three interior vertices are at $(1/3, 1/6), (1/2, 1/3)$, and $(1/6, 1/2)$.
a. Find the matrix $M(\Delta, r)$ for each $r \geq 0$.
b. Show that

$$\dim C_k^1(\Delta) = \binom{k+2}{2} + 6\binom{k-1}{2}$$

(where if $k < 3$, by convention, the second term is taken to be zero).
c. Verify that formula (3.16) is valid for this Δ.

Exercise 8. In the examples we presented in the text, the components of our Gröbner basis elements were all homogeneous polynomials. This will not be true in general. In particular, this may fail if some of the interior $(n-1)$-cells of our complex Δ lie on hyperplanes which do not contain the origin in $\mathbb{R}^n$. Nevertheless, there is a variant of *homogeneous coordinates* used to specify points in projective spaces—see [CLO] Chapter 8—that we can use if we want to work with homogeneous polynomials exclusively. Namely, think of a given pure, hereditary complex Δ as a subset of the hyperplane $x_{n+1} = 1$, a copy of $\mathbb{R}^n$ in $\mathbb{R}^{n+1}$. By considering the *cone* $\bar{\sigma}$

over each k-cell $\sigma \in \Delta$ with vertex at $(0, \ldots, 0, 0)$ in $\mathbb{R}^{n+1}$, we get a new polyhedral complex $\overline{\Delta}$ in $\mathbb{R}^{n+1}$.

a. Show that n-cells σ, σ' from Δ are adjacent if and only the corresponding $\overline{\sigma}, \overline{\sigma'}$ are adjacent $(n + 1)$-cells in $\overline{\Delta}$. Show that $\overline{\Delta}$ is hereditary.

b. What are the equations of the interior n-cells in $\overline{\Delta}$?

c. Given $f = (f_1 \ldots, f_m) \in C_k^r(\Delta)$, show that the component-wise homogenization with respect to x_{n+1}, $f^h = (f_1^h \ldots, f_m^h)$, gives an element of $C_k^r(\overline{\Delta})$.

d. How are the matrices $M(\Delta, r)$ and $M(\Delta', r)$ related?

e. Describe the relation between $\dim C_k^r(\Delta)$ and $\dim C_k^r(\overline{\Delta})$.

Exercise 9. In this exercise we will assume that the construction of Exercise 8 has been applied, so that $C^r(\Delta)$ is a graded module over $\mathbb{R}[x_0, \ldots, x_n]$. Then the formal power series

$$H(C^r(\Delta), u) = \sum_{k=0}^{\infty} \dim C_k^r(\Delta) u^k$$

is the *Hilbert series* of the graded module $C^r(\Delta)$. This is the terminology of Exercise 24 of Chapter 6, §4, and that exercise showed that the Hilbert series can be written in the form

(3.18) $$H(C^r(\Delta), u) = P(u)/(1 - u)^{n+1},$$

where $P(u)$ is a polynomial in u with coefficients in $\mathbb{Z}$. We obtain the series from (3.18) by using the formal geometric series expansion

$$1/(1 - u) = \sum_{k=0}^{\infty} u^k.$$

a. Show that the Hilbert series for the module $C^1(\Delta)$ from (3.8) with $r = 1$ is given by

$$(1 + 2u^2)/(1 - u)^3.$$

b. Show that the Hilbert series for the module $C^1(\Delta)$ from Exercise 4 is

$$(1 + u^2 + 2u^3)/(1 - u)^3.$$

c. Show that the Hilbert series for the module $C^1(\Delta')$ from Exercise 5 is

$$(1 + 2u^2 + u^4)/(1 - u)^3.$$

d. What is the Hilbert series for the module $C^1(\Delta)$ from Exercise 7 above?

Exercise 10. Consider the polyhedral complex Δ in $\mathbb{R}^3$ formed by subdividing the octahedron with vertices $\pm e_i$, $i = 1, 2, 3$ into 8 tetrahedra by adding an interior vertex at the origin.

a. Find the matrix $M(\Delta, r)$.
b. Find formulas for the dimensions of $C_k^1(\Delta)$ and $C_k^2(\Delta)$.
c. What happens if we move the vertex of the octahedron at e_3 to $(1, 1, 1)$ to form a new, combinatorially equivalent, subdivided octahedron Δ'? Using *Macaulay 2*'s `hilbertSeries` command, compute the Hilbert series of the graded module $\ker M(\Delta', 1)$ and from the result deduce that $C^1(\Delta')$ cannot be a free module. Hint: In the expression (3.19) for the dimension series of a free module, the coefficients in the numerator $P(t)$ must all be positive; do you see why?

Exercise 11. This exercise uses the language of exact sequences and some facts about graded modules from Chapter 6. The method used in the text to compute dimensions of $C_k^r(\Delta)$ requires the computation of a Gröbner basis for the module of syzygies on the columns of $M(\Delta, r)$, and it yields information leading to explicit bases of the spline spaces $C_k^r(\Delta)$. If bases for these spline spaces are not required, there is another method which can be used to compute the Hilbert series directly from $M(\Delta, r)$ without computing the syzygy module. We will assume that the construction of Exercise 8 has been applied, so that the last e columns of the matrix $M(\Delta, r)$ consist of homogeneous polynomials of degree $r + 1$. Write $R = \mathbb{R}[x_1, \ldots, x_n]$ and consider the exact sequence of graded R-modules

$$0 \to \ker M(\Delta, r) \to R^m \oplus R(-r-1)^e \to \operatorname{im} M(\Delta, r) \to 0.$$

a. Show that the Hilbert series of $R^m \oplus R(-r-1)^e$ is given by

$$(m + eu^{r+1})/(1 - u)^{n+1}.$$

b. Show that the Hilbert series of the graded module $\ker M(\Delta, r)$ is the *difference* of the Hilbert series from part a and the Hilbert series of the image of $M(\Delta, r)$.

The Hilbert series of the image can be computed by applying Buchberger's algorithm to the module M generated by the columns of $M(\Delta, r)$, then applying the fact that M and $\langle \mathrm{LT}(M) \rangle$ have the same Hilbert function.

Exercise 12. Strang's original conjectured formula for the dimension of $C_k^1(\Delta)$ for a simplicial complex in the plane with F triangles, E_0 *interior edges*, and V_0 *interior vertices* was

$$(3.19) \qquad \dim C_k^1(\Delta) = \binom{k+2}{2} F - (2k+1)E_0 + 3V_0,$$

and this is the form proved in [Bil1]. In this exercise, you will show that this form is equivalent to (3.16), under the assumption that Δ gives a triangulation of a topological disk in the plane. Let E and V be the total numbers of edges and vertices respectively.

a. Show that $V - E + F = 1$ and $V_0 - E_0 + F = 1$ for such a triangulation. Hint: One approach is to use induction on the number of triangles. In topological terms, the first equation gives the usual Euler characteristic, and the second gives the Euler characteristic relative to the boundary.

b. Use part a and the edge-counting relation $3F = E + E_0$ to show that $E = 3 + 2E_0 - 3V_0$ and $V = 3 + E_0 - 2V_0$.

c. Show that if F is eliminated using part a, and the expressions for V and E from part b are substituted into (3.16), then (3.19) is obtained. Conversely, show that (3.19) implies (3.16).

Exercise 13. The methods introduced in this section work for some *algebraic*, but non-polyhedral, decompositions of regions in $\mathbb{R}^n$ as well. We will not essay a general development. Instead we will indicate the idea with a simple example. In $\mathbb{R}^2$ suppose we wanted to construct C^r piecewise polynomial functions on the union R of the regions σ_1, σ_2, σ_3 as in Fig. 8.10. The outer boundary is the circle of radius 1 centered at the origin, and the three interior edges are portions of the curves $y = x^2$, $x = -y^2$, and $y = x^3$, respectively.

We can think of this as a non-linear embedding of an abstract 2-dimensional polyhedral complex.

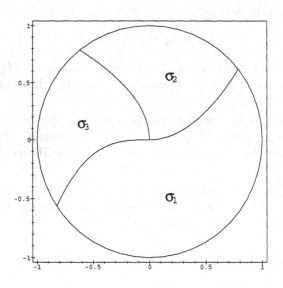

FIGURE 8.10. Figure for Exercise 13

a. Show that a triple $(f_1, f_2, f_3) \in \mathbb{R}[x, y]^3$ defines a C^r spline function on R if and only if

$$f_1 - f_2 \in \langle (y - x^2)^{r+1} \rangle$$
$$f_2 - f_3 \in \langle (x + y^2)^{r+1} \rangle$$
$$f_1 - f_3 \in \langle (y - x^3)^{r+1} \rangle.$$

b. Express the C^1 splines on this subdivided region as the kernel of an appropriate matrix with polynomial entries, and find the Hilbert function for the kernel.

Exercise 14. (The Courant functions and the face ring of a complex, see [Sta1]) Let Δ be a pure n-dimensional, hereditary complex in $\mathbb{R}^n$. Let $v_1, \ldots, v_q$ be the vertices of Δ (the 0-cells).
a. For each i, $1 \leq i \leq q$, show that there is a unique function $X_i \in C_1^0(\Delta)$ (that is, X_i is continuous, and restricts to a linear function on each n-cell) such that

$$X_i(v_j) = \begin{cases} 1 & \text{if } i = j \\ 0 & \text{if } i \neq j. \end{cases}$$

The X_i are called the *Courant functions* of Δ.
b. Show that

$$X_1 + \cdots + X_q = 1,$$

the constant function 1 on Δ.
c. Show that if $\{v_{i_1}, \ldots, v_{i_p}\}$ is any collection of vertices which do *not* form the vertices of any k-cell in Δ, then

$$X_{i_1} \cdot X_{i_2} \cdots X_{i_p} = 0,$$

the constant function 0 on Δ.
d. For a complex Δ with vertices $v_1, \ldots, v_q$, following Stanley and Reisner, we can define the *face ring* of Δ, denoted $\mathbb{R}[\Delta]$, as the quotient ring

$$\mathbb{R}[\Delta] = \mathbb{R}[x_1, \ldots, x_q]/I_\Delta,$$

where I_Δ is the ideal generated by the monomials $x_{i_1} x_{i_2} \cdots x_{i_p}$ corresponding to collections of vertices which are *not* the vertex set of any cell in Δ. Show using part c that there is a ring homomorphism from $\mathbb{R}[\Delta]$ to $\mathbb{R}[X_1, \ldots, X_q]$ (the subalgebra of $C^0(\Delta)$ generated over $\mathbb{R}$ by the Courant functions) obtained by mapping x_i to X_i for each i.

Billera has shown that in fact $C^0(\Delta)$ equals the algebra generated by the Courant functions over $\mathbb{R}$, and that the induced mapping

$$\varphi : \mathbb{R}[\Delta]/\langle x_1 + \cdots + x_q - 1 \rangle \rightarrow C^0(\Delta)$$

(see part b) is an isomorphism of $\mathbb{R}$-algebras. See [Bil2].

§4 The Gröbner Fan of an Ideal

Gröbner bases for the same ideal but with respect to different monomial orders have different properties and can look very different. For example, the ideal

$$I = \langle z^2 - x + y - 1, x^2 - yz + x, y^3 - xz + 2 \rangle \subset \mathbb{Q}[x, y, z]$$

has the following three Gröbner bases.

1. Consider the *grevlex* order with $x > y > z$. Since the leading terms of the generators of I are pairwise relatively prime,

$$\{z^2 - x + y - 1, x^2 - yz + x, y^3 - xz + 2\}$$

   is a monic (reduced) Gröbner basis for I with respect to this monomial order. Note that the basis has three elements.

2. Consider the weight order $>_{\mathbf{w}, grevlex}$ on $\mathbb{Q}[x, y, z]$ with $\mathbf{w} = (2, 1, 5)$. This order compares monomials first according to the weight vector $\mathbf{w}$ and breaks ties with the *grevlex* order. The monic Gröbner basis for I with respect to this monomial order has the form:

$$\{xy^3 + y^2 - xy - y + 2x + y^3 + 2, yz - x^2 - x,$$
$$y^6 + 4y^3 + yx^2 + 4 - y^4 - 2y, x^2y^2 + 2z + xy - x^2 - x + xy^2,$$
$$x^3 - y^4 - 2y + x^2, xz - y^3 - 2, z^2 + y - x - 1\}.$$

   This has seven instead of three elements.

3. Consider the *lex* order with $x > y > z$. The monic Gröbner basis for this ideal is:

$$\{z^{12} - 3z^{10} - 2z^8 + 4z^7 + 6z^6 + 14z^5 - 15z^4 - 17z^3 + z^2 + 9z + 6,$$
$$y + \tfrac{1}{38977}(1055z^{11} + 515z^{10} + 42z^9 - 3674z^8 - 12955z^7 + 5285z^6$$
$$- 1250z^5 + 36881z^4 + 7905z^3 + 42265z^2 - 63841z - 37186),$$
$$x + \tfrac{1}{38977}(1055z^{11} + 515z^{10} + 42z^9 - 3674z^8 - 12955z^7 + 5285z^6$$
$$- 1250z^5 + 36881z^4 + 7905z^3 + 3288z^2 - 63841z + 1791)\}$$

   This basis of three elements has the triangular form described by the Shape Lemma (Exercise 16 of Chapter 2, §4).

Many of the applications discussed in this book make crucial use of the different properties of different Gröbner bases. At this point, it is natural to ask the following questions about the collection of *all* Gröbner bases of a fixed ideal I.

- Is the collection of possible Gröbner bases of I finite or infinite?
- When do two different monomial orders yield the same monic (reduced) Gröbner basis for I?

- Is there some geometric structure underlying the collection of Gröbner bases of I that can help to elucidate properties of I?

Answers to these questions are given by the construction of the *Gröbner fan* of an ideal I. A *fan* consists of finitely many closed convex polyhedral cones with vertex at the origin (as defined in §2) with the following properties.

a. Any face of a cone in the fan is also in the fan. (A *face* of a cone σ is $\sigma \cap \{\ell = 0\}$, where $\ell = 0$ is a nontrivial linear equation such that $\ell \geq 0$ on σ. This is analogous to the definition of a face of a polytope.)
b. The intersection of two cones in the fan is a face of each.

These conditions are similar to the definition of the polyhedral complex given in Definition (3.5). The Gröbner fan encodes information about the different Gröbner bases of I and was first introduced in the paper [MR] of Mora and Robbiano. Our presentation is based on theirs.

The first step in this construction is to show that for each fixed ideal I, as $>$ ranges over all possible monomial orders, the collection of monomial ideals $\langle \mathrm{LT}_>(I) \rangle$ is finite. We use the notation

$$\mathrm{Mon}(I) = \{\langle \mathrm{LT}_>(I) \rangle : > \text{ a monomial order}\}.$$

(4.1) Theorem. *For an ideal* $I \subset k[x_1, \ldots, x_n]$, *the set* $\mathrm{Mon}(I)$ *is finite.*

PROOF. Aiming for a contradiction, suppose that $\mathrm{Mon}(I)$ is an infinite set. For each monomial ideal N in $\mathrm{Mon}(I)$, let $>_N$ be any one particular monomial order such that $N = \langle \mathrm{LT}_{>_N}(I) \rangle$. Let Σ be the collection of monomial orders $\{>_N : N \in \mathrm{Mon}(I)\}$. Our assumption implies that Σ is infinite.

By the Hilbert Basis Theorem we have $I = \langle f_1, \ldots, f_s \rangle$ for polynomials $f_i \in k[x_1, \ldots, x_n]$. Since each f_i contains only a finite number of terms, by a pigeonhole principle argument, there exists an infinite subset $\Sigma_1 \subset \Sigma$ such that the leading terms $\mathrm{LT}_>(f_i)$ agree for all $>$ in Σ_1 and all i, $1 \leq i \leq s$. We write N_1 for the monomial ideal $\langle \mathrm{LT}_>(f_1), \ldots, \mathrm{LT}_>(f_s) \rangle$ (taking any monomial order $>$ in Σ_1).

If $F = \{f_1, \ldots, f_s\}$ were a Gröbner basis for I with respect to some $>_1$ in Σ_1, then we claim that F would be a Gröbner basis for I with respect to *every* $>$ in Σ_1. To see this, let $>$ be any element of Σ_1 other than $>_1$, and let $f \in I$ be arbitrary. Dividing f by F using $>$, we obtain

$$(4.2) \qquad\qquad f = a_1 f_1 + \cdots + a_s f_s + r,$$

where no term in r is divisible by any of the $\mathrm{LT}_>(f_i)$. However, both $>$ and $>_1$ are in Σ_1, so $\mathrm{LT}_>(f_i) = \mathrm{LT}_{>_1}(f_i)$ for all i. Since $r = f - a_1 f_1 - \cdots - a_s f_s \in I$, and F is assumed to be a Gröbner basis for I with respect to $>_1$, this implies that $r = 0$. Since (4.2) was obtained using the division algorithm, $\mathrm{LT}_>(f) = \mathrm{LT}_>(a_i f_i)$ for some i, so $\mathrm{LT}_>(f)$ is divisible by $\mathrm{LT}_>(f_i)$. This shows that F is also a Gröbner basis for I with respect to $>$.

However, this cannot be the case since the original set of monomial orders $\Sigma \supset \Sigma_1$ was chosen so that the monomial ideals $\langle \text{LT}_>(I) \rangle$ for $>$ in Σ were all distinct. Hence, given any $>_1$ in Σ_1, there must be some $f_{s+1} \in I$ such that $\text{LT}_{>_1}(f_{s+1}) \notin \langle \text{LT}_{>_1}(f_1), \ldots, \text{LT}_{>_1}(f_s) \rangle = N_1$. Replacing f_{s+1} by its remainder on division by $f_1, \ldots, f_s$, we may assume in fact that no term in f_{s+1} is divisible by any of the monomial generators for N_1.

Now we apply the pigeonhole principle again to find an infinite subset $\Sigma_2 \subset \Sigma_1$ such that the leading terms of $f_1, \ldots, f_{s+1}$ are the same for all $>$ in Σ_2. Let $N_2 = \langle \text{LT}_>(f_1), \ldots, \text{LT}_>(f_{s+1}) \rangle$ for all $>$ in Σ_2, and note that $N_1 \subset N_2$. The argument given in the preceding paragraph shows that $\{f_1, \ldots, f_{s+1}\}$ cannot be a Gröbner basis with respect to any of the monomial orders in Σ_2, so fixing $>_2 \in \Sigma_2$, we find an $f_{s+2} \in I$ such that no term in f_{s+2} is divisible by any of the monomial generators for $N_2 = \langle \text{LT}_{>_2}(f_1), \ldots, \text{LT}_{>_2}(f_{s+1}) \rangle$.

Continuing in the same way, we produce a descending chain of infinite subsets $\Sigma \supset \Sigma_1 \supset \Sigma_2 \supset \Sigma_3 \supset \cdots$, and an infinite strictly ascending chain of monomial ideals $N_1 \subset N_2 \subset N_3 \subset \cdots$. This contradicts the ascending chain condition in $k[x_1, \ldots, x_n]$, so the proof is complete.   □

We can now answer the first question posed at the start of this section. To obtain a precise result, we introduce some new terminology. It is possible for two monic Gröbner bases of I with respect to different monomial orders to be equal as sets, while the leading terms of some of the basis polynomials are different depending on which order we consider. Examples where I is principal are easy to construct; also see (4.9) below. A *marked Gröbner basis for I* is a set G of polynomials in I, together with an identified leading term in each $g \in G$ such that G is a monic Gröbner basis with respect to some monomial order selecting those leading terms. (More formally, we could define a marked Gröbner basis as a set GM of ordered pairs (g, m) where $\{g : (g, m) \in GM\}$ is a monic Gröbner basis with respect to some order $>$, and $m = \text{LT}_>(g)$ for each (g, m) in GM.) The idea here is that we do not want to build a specific monomial order into the definition of G. It follows from Theorem (4.1) that each ideal in $k[x_1, \ldots, x_n]$ has only finitely many marked Gröbner bases.

(4.3) Corollary. *The set of marked Gröbner bases of I is in one-to-one correspondence with the set* $\text{Mon}(I)$.

PROOF. The key point is that if the leading terms of two marked Gröbner bases generate the same monomial ideal, then the Gröbner bases must be equal. The details of the proof are left to the reader as Exercise 4.   □

Corollary (4.3) also has the following interesting consequence.

Exercise 1. Show that for any ideal $I \subset k[x_1, \ldots, x_n]$, there exists a finite $U \subset I$ such that U is a Gröbner basis simultaneously for all monomial orders on $k[x_1, \ldots, x_n]$.

A set U as in Exercise 1 is called a *universal Gröbner basis* for I. These were first studied by Weispfenning in [Wei], and that article gives an algorithm for constructing universal Gröbner bases. This topic is also discussed in detail in [Stu2].

To answer our other questions we will represent monomial orders using the matrix orders $>_M$ described in Chapter 1, §2. Recall that if M has rows $\mathbf{w}_i$, then $x^\alpha >_M x^\beta$ if there is an ℓ such that $\alpha \cdot \mathbf{w}_i = \beta \cdot \mathbf{w}_i$ for $i = 1, \ldots, \ell - 1$, but $\alpha \cdot \mathbf{w}_\ell > \beta \cdot \mathbf{w}_\ell$.

When $>_M$ is a matrix order, the first row of M plays a special role and will be denoted $\mathbf{w}$ in what follows. We may assume that $\mathbf{w} \neq 0$.

Exercise 2.
a. Let $>_M$ be a matrix order with first row $\mathbf{w}$. Show that

$$\mathbf{w} \in (\mathbb{R}^n)^+ = \{(a_1, \ldots, a_n) : a_i \geq 0, \text{ all } i\}.$$

We call $(\mathbb{R}^n)^+$ the *positive orthant* in $\mathbb{R}^n$. Hint: $x_i >_M 1$ for all i since $>_M$ is a monomial order.
b. Prove that every nonzero $\mathbf{w} \in (\mathbb{R}^n)^+$ is the first row of some matrix M such that $>_M$ is a monomial order.
c. Let M and M' be matrices such that the matrix orders $>_M$ and $>_{M'}$ are equal. Prove that their first rows satisfy $\mathbf{w} = \lambda \mathbf{w}'$ for some $\lambda > 0$.

Exercise 2 implies that each monomial order determines a well-defined ray in the positive orthant $(\mathbb{R}^n)^+$, though different monomial orders may give the same ray. (For example, all graded orders give the ray consisting of positive multiples of $(1, \ldots, 1)$.) Hence it should not be surprising that our questions lead naturally to cones in the positive orthant.

Now we focus on a single ideal I. Let $G = \{g_1, \ldots, g_t\}$ be one of the finitely many marked Gröbner bases of I, with $\mathrm{LT}(g_i) = x^{\alpha(i)}$, and $N = \langle x^{\alpha(1)}, \ldots, x^{\alpha(t)} \rangle$ the corresponding element of $\mathrm{Mon}(I)$. Our next goal is to understand the set of monomial orders for which G is the corresponding marked Gröbner basis of I. This will answer the second question posed at the start of this section. We write

$$g_i = x^{\alpha(i)} + \sum_\beta c_{i,\beta} x^\beta,$$

where $x^{\alpha(i)} > x^\beta$ whenever $c_{i,\beta} \neq 0$. By the above discussion, each such order $>$ comes from a matrix M, so in particular, to find the leading terms we compare monomials first according to the first row $\mathbf{w}$ of the matrix.

If $\alpha(i) \cdot \mathbf{w} > \beta \cdot \mathbf{w}$ for all β with $c_{i,\beta} \neq 0$, the single weight vector $\mathbf{w}$ selects the correct leading term in g_i as the term of highest weight. As we know, however, we may have a tie in the first comparison, in which case we would have to make further comparisons using the other rows of M. This suggests that we should consider the following set of vectors:

$$
\begin{aligned}
(4.4) \quad C_G &= \{\mathbf{w} \in (\mathbb{R}^n)^+ : \alpha(i) \cdot \mathbf{w} \geq \beta \cdot \mathbf{w} \text{ whenever } c_{i,\beta} \neq 0\} \\
&= \{\mathbf{w} \in (\mathbb{R}^n)^+ : (\alpha(i) - \beta) \cdot \mathbf{w} \geq 0 \text{ whenever } c_{i,\beta} \neq 0\}.
\end{aligned}
$$

It is easy to see that C_G is an intersection of closed half-spaces in $\mathbb{R}^n$, hence is a closed convex polyhedral cone contained in the positive orthant. There are many close connections between this discussion and other topics we have considered. For example, we can view the process of finding elements of C_G as finding points in the feasible region of a linear programming problem as in §1 of this chapter. Moreover, given a polynomial, the process of finding its term(s) of maximum weight with respect to a given vector $\mathbf{w}$ is equivalent to an integer programming maximization problem on a feasible region given by the Newton polytope $NP(f)$.

The cone C_G has the property that if $>_M$ is a matrix order such that G is the marked Gröbner basis of I with respect to $>_M$, then the first row $\mathbf{w}$ of M lies in C_G. However, you will see below that the converse can fail, so that the relation between C_G and monomial orders for which G is a marked Gröbner basis is more subtle than meets the eye.

In the following example we determine the cone corresponding to a given marked Gröbner basis for an ideal.

(4.5) Example. Consider the ideal

$$
(4.6) \qquad I = \langle x^2 - y, xz - y^2 + yz \rangle \subset \mathbb{Q}[x, y, z].
$$

The marked Gröbner basis with respect to the *grevlex* order with $x > y > z$ is

$$
G^{(1)} = \{\underline{x^2} - y, \underline{y^2} - xz - yz\},
$$

where the leading terms are underlined. Let $\mathbf{w} = (a, b, c)$ be a vector in the positive orthant of $\mathbb{R}^3$. Then $\mathbf{w}$ is in $C_{G^{(1)}}$ if and only if the following inequalities are satisfied:

$$
\begin{aligned}
(2, 0, 0) \cdot (a, b, c) &\geq (0, 1, 0) \cdot (a, b, c) \quad &\text{or} \quad &2a \geq b \\
(0, 2, 0) \cdot (a, b, c) &\geq (1, 0, 1) \cdot (a, b, c) \quad &\text{or} \quad &2b \geq a + c \\
(0, 2, 0) \cdot (a, b, c) &\geq (0, 1, 1) \cdot (a, b, c) \quad &\text{or} \quad &2b \geq b + c.
\end{aligned}
$$

To visualize $C_{G^{(1)}}$, slice the positive orthant by the plane $a + b + c = 1$ (every nonzero weight vector in the positive orthant can be scaled to make this true). The above inequalities are pictured in Figure 8.11, where the a-axis, b-axis, and c-axis are indicated by dashed lines and you are looking toward the origin from a point on the ray through $(1, 1, 1)$.

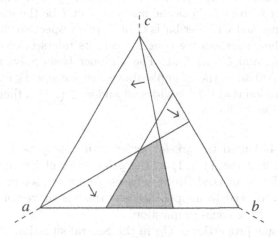

FIGURE 8.11. A slice of the cone $C_{G^{(1)}}$

In this figure, the inequality $2a \geq b$ gives the region in the slice to the left (as indicated by the arrow), the line segment connecting $(0, 0, 1)$ at the top of the triangle to $(\frac{1}{3}, \frac{2}{3}, 0)$ on the base. The other two inequalities are represented similarly, and their intersection in the first orthant gives the shaded quadrilateral in the slice. Then $C_{G^{(1)}}$ consists of all rays emanating from the origin that go through points of the quadrilateral.

Any weight $\mathbf{w}$ corresponding to a point in the interior of $C_{G^{(1)}}$ (where the inequalities above are strict) will select the leading terms of elements of $G^{(1)}$ exactly; a weight vector on one of the boundary planes in the interior of the positive orthant will yield a "tie" between terms in one or more Gröbner basis elements. For instance, $(a, b, c) = (1, 1, 1)$ satisfies $2b = a + c$ and $2b = b + c$, so it is on the boundary of the cone. This weight vector is not sufficient to determine the leading terms of the polynomials.

Now consider a different monomial order, say the *grevlex* order with $z > y > x$. For this order, the monic Gröbner basis for I is

$$G^{(2)} = \{\underline{x^2} - y, \underline{yz} + xz - y^2\},$$

where again the leading terms are underlined. Proceeding as above, the slice of $C_{G^{(2)}}$ in the plane $a + b + c = 1$ is a triangle defined by the inequalities

$$2a \geq b, \ b \geq a, \ c \geq b.$$

You should draw this triangle carefully and verify that $C_{G^{(1)}} \cap C_{G^{(2)}}$ is a common face of both cones (see also Figure 8.12 below).

Exercise 3. Consider the *grlex* order with $x > y > z$. This order comes from a matrix with $(1, 1, 1)$ as the first row. Let I be the ideal from (4.6).

a. Find the marked Gröbner basis G of I with respect to this order.

b. Identify the corresponding cone C_G and its intersections with the two cones $C_{G^{(1)}}$ and $C_{G^{(2)}}$. Hint: The Gröbner basis polynomials contain more terms than in the example above, but some work can be saved by the observation that if $x^{\beta'}$ divides x^β and $\mathbf{w} \in (\mathbb{R}^n)^+$, then $\alpha \cdot \mathbf{w} \geq \beta \cdot \mathbf{w}$ implies $\alpha \cdot \mathbf{w} \geq \beta' \cdot \mathbf{w}$.

Example (4.5) used the *grevlex* order with $z > y > x$, whose matrix has the same first row $(1, 1, 1)$ as the *grlex* order of Exercise 3. Yet they have very different marked Gröbner bases. As we will see in Theorem (4.7) below, this is allowed to happen because the weight vector $(1, 1, 1)$ is on the *boundary* of the cones in question.

Here are some properties of C_G in the general situation.

(4.7) Theorem. *Let I be an ideal in $k[x_1, \dots, x_n]$, and let G be a marked Gröbner basis of I.*

a. *The interior $\mathrm{Int}(C_G)$ of the cone C_G is a nonempty open subset of $\mathbb{R}^n$.*

b. *Let $>_M$ be any matrix order such that the first row of M lies in $\mathrm{Int}(C_G)$. Then G is the marked Gröbner basis of I with respect to $>_M$.*

c. *Let G' be a marked Gröbner basis of I different from G. Then the intersection $C_G \cap C_{G'}$ is contained in a boundary hyperplane of C_G, and similarly for $C_{G'}$.*

d. *The union of all the cones C_G, as G ranges over all marked Gröbner bases of I, is the positive orthant $(\mathbb{R}^n)^+$.*

PROOF. To prove part a, fix a matrix order $>_M$ such that G is a marked Gröbner basis of I with respect to $>_M$ and let $\mathbf{w}_1, \dots, \mathbf{w}_m$ be the rows of M. We will show that $\mathrm{Int}(C_G)$ is nonempty by proving that

$$(4.8) \qquad \mathbf{w} = \mathbf{w}_1 + \epsilon \mathbf{w}_2 + \dots + \epsilon^{m-1} \mathbf{w}_m \in \mathrm{Int}(C_G)$$

provided $\epsilon > 0$ is sufficiently small. In Exercise 5, you will show that given exponent vectors α and β, we have

$$x^\alpha >_M x^\beta \Rightarrow \alpha \cdot \mathbf{w} > \beta \cdot \mathbf{w} \text{ provided } \epsilon > 0 \text{ is sufficiently small,}$$

where "sufficiently small" depends on α, β, and M. It follows that we can arrange this for any finite set of pairs of exponent vectors. In particular, since $x^{\alpha(i)} = \mathrm{LT}_{>_M}(x^{\alpha(i)} + \sum_{i,\beta} c_{i,\beta} x^\beta)$, we can pick ϵ so that

$$\alpha(i) \cdot \mathbf{w} > \beta \cdot \mathbf{w} \text{ whenever } c_{i,\beta} \neq 0$$

in the notation of (4.4). Furthermore, using $x_i >_M 1$ for all i, we can also pick ϵ so that $\mathbf{e}_i \cdot \mathbf{w} > 0$ (where $\mathbf{e}_i$ is the ith standard basis vector). It follows that w is in the interior of the positive orthant. From here, $\mathbf{w} \in \text{Int}(C_G)$ follows immediately.

For part b, let $>_M$ be a matrix order such that the first row of M lies in $\text{Int}(C_G)$. This easily implies that for every $g \in G$, $\text{LT}_{>_M}(g)$ is the marked term of g. From here, it is straightforward to show that G is the marked Gröbner basis of I with respect to $>_M$. See Exercise 6 for the details.

We now prove part c. In Exercise 7, you will show that if $C_G \cap C_{G'}$ contains interior points of either cone, then by part a it contains interior points of both cones. If $\mathbf{w}$ is such a point, we take any monomial order $>_M$ defined by a matrix with first row $\mathbf{w}$. Then by part b, G and G' are both the marked Gröbner bases of I with respect to $>_M$. This contradicts our assumption that $G \neq G'$.

Part d follows immediately from part b of Exercise 2.   □

With more work, one can strengthen part c of Theorem (4.7) to show that $C_G \cap C_{G'}$ is a face of each (see [MR] or [Stu2] for a proof). It follows that as G ranges over all marked Gröbner bases of I, the collection formed by the cones C_G and their faces is a fan, as defined earlier in the section. This is the *Gröbner fan* of the ideal I.

For example, using the start made in Example (4.5) and Exercise 3, we can determine the Gröbner fan of the ideal I from (4.6). In small examples like this one, a reasonable strategy for producing the Gröbner fan is to find the monic (reduced) Gröbner bases for I with respect to "standard" orders (e.g., *grevlex* and *lex* orders with different permutations of the set of variables) first and determine the corresponding cones. Then if the union of the known cones is not all of the positive orthant, select some $\mathbf{w}$ in the complement, compute the monic Gröbner basis for $>_{\mathbf{w},grevlex}$, find the corresponding cone, and repeat this process until the known cones fill the positive orthant.

For the ideal of (4.6), there are seven cones in all, corresponding to the marked Gröbner bases:

$$G^{(1)} = \{\underline{x^2} - y, \underline{y^2} - xz - yz\}$$

$$G^{(2)} = \{\underline{x^2} - y, \underline{yz} + xz - y^2\}$$

$$G^{(3)} = \{\underline{x^4} - x^2z - xz, \underline{y} - x^2\}$$

$$G^{(4)} = \{\underline{x^2} - y, \underline{xz} - y^2 + yz, \underline{y^2z} + xy^2 - y^3 - yz\}$$

(4.9) $$G^{(5)} = \{\underline{y^4} - 2y^3z + y^2z^2 - yz^2, \underline{xz} - y^2 + yz,$$
$$\underline{xy^2} - y^3 + y^2z - yz, \underline{x^2} - y\}$$

$$G^{(6)} = \{\underline{y^2z^2} - 2y^3z + y^4 - yz^2, \underline{xz} - y^2 + yz,$$
$$\underline{xy^2} - y^3 + y^2z - yz, \underline{x^2} - y\}$$

$$G^{(7)} = \{\underline{y} - x^2, \underline{x^2z} - x^4 + xz\}.$$

(Note that $G^{(5)}$ is the Gröbner basis from Exercise 3.)

Figure 8.12 below shows a picture of the slice of the Gröbner fan in the plane $a + b + c = 1$, following the discussion from Example (4.5). The cones are labeled as in (4.9).

For instance, if the Gröbner bases $G^{(1)}, \ldots, G^{(6)}$ in this example are known, the "missing" region of the positive orthant contains (for instance) the vector $\mathbf{w} = (1/10, 2/5, 1/2)$ (see Figure 4.2). Using this weight vector, we find $G^{(7)}$, and the corresponding cone completes the Gröbner fan.

When the number of variables is larger and/or the ideal generators have more terms, this method becomes much less tractable. Mora and Robbiano propose a "parallel Buchberger algorithm" in [MR] which produces the Gröbner fan by considering all potential identifications of leading terms in the computation and reduction of S-polynomials. But their method is certainly not practical on larger examples either. Gröbner fans can be extremely complicated! Fortunately, Gröbner fans are used primarily as conceptual tools—it is rarely necessary to compute large examples.

If we relax our requirement that $\mathbf{w}$ lie in the first orthant and only ask that $\mathbf{w}$ pick out the correct leading terms of a marked Gröbner basis of I, then we can allow weight vectors with negative entries. This leads to a larger "Gröbner fan" denoted $GF(I)$ in [Stu2]. Then the Gröbner fan of Theorem (4.7) (sometimes called the *restricted Gröbner fan*) is obtained by intersecting the cones of $GF(I)$ with the positive orthant. See [MR] and [Stu2] for more about what happens outside the positive orthant.

We close this section with a comment about a closely related topic. In the article [BaM] which appeared at the same time as [MR], Bayer and

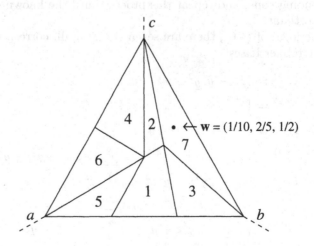

FIGURE 8.12. A slice of the Gröbner fan

Morrison introduced the *state polytope* of a homogeneous ideal. In a sense, this is the dual of the Gröbner fan $GF(I)$ (more precisely, the vertices of the state polytope are in one-to-one correspondence with the elements of $\mathrm{Mon}(I)$, and $GF(I)$ is the normal fan of the state polytope). The state polytope may also be seen as a generalization of the Newton polytope of a single homogeneous polynomial. See [BaM] and [Stu2] for more details.

In the next section, we will see how the Gröbner fan can be used to develop a general Gröbner basis conversion algorithm that, unlike the FGLM algorithm from Chapter 2, does not depend on zero-dimensionality of I.

ADDITIONAL EXERCISES FOR §4

Exercise 4. Using the proof of Proposition (4.1), prove Corollary (4.3).

Exercise 5. Assume that $x^\alpha >_M x^\beta$, where M is an $m \times n$ matrix giving the matrix order $>_M$. Also define $\mathbf{w}$ as in (4.8). Prove that $\alpha \cdot \mathbf{w} > \beta \cdot \mathbf{w}$ provided that $\epsilon > 0$ is sufficiently small.

Exercise 6. Fix a marked Gröbner basis G of an ideal I and let $>$ be a monomial order such that for each $g \in G$, $\mathrm{LT}_>(g)$ is the marked term of the polynomial g. Prove that G is the marked Gröbner basis of I with respect to $>$. Hint: Divide $f \in I$ by G using the monomial order $>$.

Exercise 7. Show that if the intersection of two closed, n-dimensional convex polyhedral cones C, C' in $\mathbb{R}^n$ contains interior points of C, then the intersection also contains interior points of C'.

Exercise 8. Verify the computation of the Gröbner fan of the ideal from (4.6) by finding monomial orders corresponding to each of the seven Gröbner bases given in (4.9) and determining the cones $C_{G^{(k)}}$.

Exercise 9. Determine the Gröbner fan of the ideal of the affine twisted cubic curve: $I = \langle y - x^2, z - x^3 \rangle$. Explain why all of the cones have a common one-dimensional edge in this example.

Exercise 10. This exercise will determine which terms in a polynomial $f = \sum_{i=1}^{k} c_i x^{\alpha(i)}$ can be $\mathrm{LT}(f)$ with respect to some monomial order.
a. Show that $x^{\alpha(1)}$ is $\mathrm{LT}(f)$ for some monomial order if and only if there is some vector $\mathbf{w}$ in the positive orthant such $(\alpha(1) - \alpha(j)) \cdot \mathbf{w} > 0$ for all $j = 2, \ldots, k$.
b. Show that such a $\mathbf{w}$ exists if and only if the origin is *not* in the convex hull of the set of all $(\alpha(1) - \alpha(j))$ for $j = 2, \ldots, k$, together with the standard basis vectors $\mathbf{e}_i$, $i = 1, \ldots, n$ in $\mathbb{R}^n$.

c. Use the result of part b to determine which terms in $f = x^2yz + 2xyw^2$ $+ x^2w - xw + yzw + y^3$ can be LT(f) for some monomial order. Determine an order that selects each of the possible leading terms.

Exercise 11. Determine the Gröbner fan of the following ideals:
a. $I = \langle x^3yz^2 - 2xy^3 - yz^3 + y^2z^2 + xyz \rangle$.
b. $I = \langle x - t^4, y - t^2 - t \rangle$.

§5 The Gröbner Walk

One interesting application of the Gröbner fan is a general Gröbner basis conversion algorithm known as the *Gröbner Walk*. As we saw in the discussion of the FGLM algorithm in Chapter 2, to find a Gröbner basis with respect to an "expensive" monomial order such as a *lex* order or another elimination order, it is often simpler to find some other Gröbner basis first, then convert it to a basis with respect to the desired order. The algorithm described in Chapter 2 does this using linear algebra in the quotient algebra $k[x_1, \ldots, x_n]/I$, so it applies only to zero-dimensional ideals.

In this section, we will present the Gröbner Walk introduced by Collart, Kalkbrener, and Mall in [ColKM]. This method converts a Gröbner basis for any ideal $I \subset k[x_1, \ldots, x_n]$ with respect to any one monomial order into a Gröbner basis with respect to any other monomial order. We will also give examples showing how the walk applies to elimination problems encountered in implicitization.

The basic idea of the Gröbner Walk is pleasingly simple. Namely, we assume that we have a marked Gröbner basis G for I, say the marked Gröbner basis with respect to some monomial order $>_s$. We call $>_s$ the *starting order* for the walk, and we will assume that we have some matrix M_s with first row $\mathbf{w}_s$ representing $>_s$. By the results of the previous section, G corresponds to a cone C_G in the Gröbner fan of I.

The goal is to compute a Gröbner basis for I with respect to some other given *target order* $>_t$. This monomial order can be represented by some matrix M_t with first row $\mathbf{w}_t$. Consider a "nice" (e.g., piecewise linear) path from $\mathbf{w}_s$ to $\mathbf{w}_t$ lying completely in the positive orthant in $\mathbb{R}^n$. For instance, since the positive orthant is convex, we could use the straight line segment between the two points, $(1 - u)\mathbf{w}_s + u\mathbf{w}_t$ for $u \in [0, 1]$, though this is not always the best choice. The Gröbner Walk consists of two basic steps:

- Crossing from one cone to the next;
- Computing the Gröbner basis of I corresponding to the new cone.

These steps are done repeatedly until the end of the path is reached, at which point we have the Gröbner basis with respect to the target order. We will discuss each step separately.

Crossing Cones

Assume we have the marked Gröbner basis G_{old} corresponding to the cone C_{old}, and a matrix M_{old} with first row $\mathbf{w}_{old}$ representing $>_{old}$. As we continue along the path from $\mathbf{w}_{old}$, let $\mathbf{w}_{new}$ be the *last point* on the path that lies in the cone C_{old}.

The new weight vector $\mathbf{w}_{new}$ may be computed as follows. Let $G_{old} = \{x^{\alpha(i)} + \sum_{i,\beta} c_{i,\beta} x^{\beta} : 1 \le i \le t\}$, where $x^{\alpha(i)}$ is the leading term with respect to $>_{M_{old}}$. To simplify notation, let $v_1, \ldots, v_m$ denote the vectors $\alpha(i) - \beta$ where $1 \le i \le t$ and $c_{i,\beta} \ne 0$. By (4.4), C_{old} consists of those points in the positive orthant $(\mathbb{R}^n)^+$ for which

$$\mathbf{w} \cdot v_j \ge 0, \quad 1 \le j \le m.$$

For simplicity say that the remaining portion of the path to be traversed consists of the straight line segment from $\mathbf{w}_{old}$ to $\mathbf{w}_t$. Parametrizing this line as $(1 - u)\mathbf{w}_{old} + u\mathbf{w}_t$ for $u \in [0, 1]$, we see that the point for the parameter value u lies in C_{old} if and only if

$$(5.1) \qquad (1 - u)(\mathbf{w}_{old} \cdot v_j) + u(\mathbf{w}_t \cdot v_j) \ge 0, \quad 1 \le j \le m.$$

Then $\mathbf{w}_{new} = (1 - u_{last})\mathbf{w}_{old} + u_{last}\mathbf{w}_t$, where u_{last} is computed by the following algorithm.

Input: $\mathbf{w}_{old}, \mathbf{w}_t, v_1, \ldots, v_m$

Output: u_{last}

$u_{last} = 1$

(5.2)   FOR $j = 1, \ldots, m$ DO

         IF $\mathbf{w}_t \cdot v_j < 0$ THEN $u_j := \dfrac{\mathbf{w}_{old} \cdot v_j}{\mathbf{w}_{old} \cdot v_j - \mathbf{w}_t \cdot v_j}$

         IF $u_j < u_{last}$ THEN $u_{last} := u_j$

The idea behind (5.2) is that if $\mathbf{w}_t \cdot v_j \ge 0$, then (5.1) holds for all $u \in [0, 1]$ since $\mathbf{w}_{old} \cdot v_j \ge 0$. On the other hand, if $\mathbf{w}_t \cdot v_j < 0$, then the formula for u_j gives the largest value of u such that (5.1) holds for this particular j. Note that $0 \le u_j < 1$ in this case.

Exercise 1. Prove carefully that $\mathbf{w}_{new} = (1 - u_{last})\mathbf{w}_{old} + u_{last}\mathbf{w}_t$ is the last point on the path from $\mathbf{w}_{old}$ to $\mathbf{w}_t$ that lies in C_{old}.

Once we have $\mathbf{w}_{new}$, we need to choose the next cone in the Gröbner fan. Let $>_{new}$ be the weight order where we first compare $\mathbf{w}_{new}$-weights and break ties using the *target* order. Since $>_t$ is represented by M_t, it follows that $>_{new}$ is represented by $\binom{\mathbf{w}_{new}}{M_t}$. This gives the new cone C_{new}.

Furthermore, if we are in the situation where M_t is the bottom of the matrix representing $>_{old}$ (which is what happens in the Gröbner Walk), the following lemma shows that whenever $\mathbf{w}_{old} \neq \mathbf{w}_t$, the above process is guaranteed to move us closer to $\mathbf{w}_t$.

(5.3) Lemma. *Let u_{last} be as in (5.2) and assume that $>_{old}$ is represented by $\binom{\mathbf{w}_{old}}{M_t}$. Then $u_{last} > 0$.*

PROOF. By (5.2), $u_{last} = 0$ implies that $\mathbf{w}_{old} \cdot v_j = 0$ and $\mathbf{w}_t \cdot v_j < 0$ for some j. But recall that $v_j = \alpha(i) - \beta$ for some $g = x^{\alpha(i)} + \sum_{i,\beta} c_{i,\beta} x^\beta \in G$, where $x^{\alpha(i)}$ is the leading term for $>_{old}$ and $c_{i,\beta} \neq 0$. It follows that

$$(5.4) \qquad \mathbf{w}_{old} \cdot \alpha(i) = \mathbf{w}_{old} \cdot \beta \quad \text{and} \quad \mathbf{w}_t \cdot \alpha(i) < \mathbf{w}_t \cdot \beta.$$

Since $>_{old}$ is represented by $\binom{\mathbf{w}_{old}}{M_t}$, the equality in (5.4) tells us that $x^{\alpha(i)}$ and x^β have the same $\mathbf{w}_{old}$-weight, so that we break the tie using M_t. But $\mathbf{w}_t$ is the first row of M_t, so that the inequality in (5.4) implies that $x^{\alpha(i)}$ is not the leading term for $>_{old}$. This contradiction proves the lemma. $\square$

Converting Gröbner Bases

Once we have crossed from C_{old} into C_{new}, we need to convert the marked Gröbner basis G_{old} into a Gröbner basis for I with respect to the monomial order $>_{new}$ represented by $\binom{\mathbf{w}_{new}}{M_t}$. This is done as follows.

The key feature of $\mathbf{w}_{new}$ is that it lies on the boundary of C_{old}, so that some of the inequalities defining C_{old} become equalities. This means that the leading term of some $g \in G_{old}$ has the same $\mathbf{w}_{new}$-weight as some other term in g. In general, given a weight vector $\mathbf{w}$ is the positive orthant $(\mathbb{R}^n)^+$ and a polynomial $f \in k[x_1, \ldots, x_n]$, the *initial form* of f for $\mathbf{w}$, denoted $\text{in}_{\mathbf{w}}(f)$, is the sum of all terms in f of maximum $\mathbf{w}$-weight. Also, given a set S of polynomials, we let $\text{in}_{\mathbf{w}}(S) = \{\text{in}_{\mathbf{w}}(f) : f \in S\}$.

Using this notation, we can form the ideal

$$\langle \text{in}_{\mathbf{w}_{new}}(G_{old}) \rangle$$

of $\mathbf{w}_{new}$-initial forms of elements of G_{old}. Note that $\mathbf{w}_{new} \in C_{old}$ guarantees that the marked term of $g \in G_{old}$ appears in $\text{in}_{\mathbf{w}_{new}}(g)$. The important thing to realize here is that in nice cases, $\text{in}_{\mathbf{w}_{new}}(G_{old})$ consists mostly of monomials, together with a small number of polynomials (in the best case, only one binomial together with a collection of monomials).

It follows that finding a monic Gröbner basis

$$H = \{h_1, \ldots, h_s\}$$

of $\langle \text{in}_{\mathbf{w}_{new}}(G_{old}) \rangle$ with respect to $>_{new}$ may usually be done very quickly. The surprise is that once we have H, it is relatively easy to convert G_{old} into the desired Gröbner basis.

(5.5) Proposition. *Let G_{old} be the marked Gröbner basis for an ideal I with respect to $>_{old}$. Also let $>_{new}$ be represented by $\binom{\mathbf{w}_{new}}{M_t}$, where $\mathbf{w}_{new}$*

is any weight vector in C_{old}, and let H be the monic Gröbner basis of $\langle \text{in}_{\mathbf{w}_{new}}(G_{old})\rangle$ with respect to $>_{new}$ as above. Express each $h_j \in H$ as

$$(5.6) \qquad\qquad h_j = \sum_{g \in G_{old}} p_{j,g}\, \text{in}_{\mathbf{w}_{new}}(g).$$

Then replacing the initial forms by the g themselves, the polynomials

$$(5.7) \qquad\qquad \overline{h}_j = \sum_{g \in G_{old}} p_{j,g}\, g, \quad 1 \le j \le s,$$

form a Gröbner basis of I with respect to $>_{new}$.

Before giving the proof, we need some preliminary observations about weight vectors and monomial orders. A polynomial f is $\mathbf{w}$-*homogeneous* if $f = \text{in}_{\mathbf{w}}(f)$. In other words, all terms of f have the same $\mathbf{w}$-weight. Furthermore, every polynomial can be written uniquely as a sum of $\mathbf{w}$-homogeneous polynomials that are its $\mathbf{w}$-*homogeneous components* (see Exercise 5).

We say that a weight vector $\mathbf{w}$ is *compatible* with a monomial order $>$ if $\text{LT}_>(f)$ appears in $\text{in}_{\mathbf{w}}(f)$ for all nonzero polynomials f. Then we have the following result.

(5.8) Lemma. *Fix $\mathbf{w} \in (\mathbb{R}^n)^+ \setminus \{0\}$ and let G be the marked Gröbner basis of an ideal I for a monomial order $>$.*
a. *If $\mathbf{w}$ is compatible with $>$, then $\text{LT}_>(I) = \text{LT}_>(\text{in}_{\mathbf{w}}(I)) = \text{LT}_>(\langle\text{in}_{\mathbf{w}}(I)\rangle)$.*
b. *If $\mathbf{w} \in C_G$, then $\text{in}_{\mathbf{w}}(G)$ is a Gröbner basis of $\langle\text{in}_{\mathbf{w}}(I)\rangle$ for $>$. In particular,*

$$\langle\text{in}_{\mathbf{w}}(I)\rangle = \langle\text{in}_{\mathbf{w}}(G)\rangle.$$

PROOF. For part a, the first equality $\text{LT}_>(I) = \text{LT}_>(\text{in}_{\mathbf{w}}(I))$ is obvious since the leading term of any $f \in k[x_1, \ldots, x_n]$ appears in $\text{in}_{\mathbf{w}}(f)$. For the second equality, it suffices to show $\text{LT}_>(f) \in \text{LT}_>(\text{in}_{\mathbf{w}}(I))$ whenever $f \in \langle\text{in}_{\mathbf{w}}(I)\rangle$. Given such an f, write it as

$$f = \sum_{i=1}^{t} p_i\, \text{in}_{\mathbf{w}}(f_i), \quad p_i \in k[x_1, \ldots, x_n], \; f_i \in I.$$

Each side is a sum of $\mathbf{w}$-homogeneous components. Since $\text{in}_{\mathbf{w}}(f_i)$ is already $\mathbf{w}$-homogeneous, this implies that

$$\text{in}_{\mathbf{w}}(f) = \sum_{i=1}^{t} q_i\, \text{in}_{\mathbf{w}}(f_i),$$

where we can assume that q_i is $\mathbf{w}$-homogeneous and f and $q_i f_i$ have the same $\mathbf{w}$-weight for all i. It follows that $\text{in}_{\mathbf{w}}(f) = \text{in}_{\mathbf{w}}(\sum_{i=1}^{t} q_i\, f_i) \in \text{in}_{\mathbf{w}}(I)$. Then compatibility implies $\text{LT}_>(f) = \text{LT}_>(\text{in}_{\mathbf{w}}(f)) \in \text{LT}_>(\text{in}_{\mathbf{w}}(I))$.

Turning to part b, first assume that $\mathbf{w}$ is compatible with $>$. Then

$$\langle\text{LT}_>(I)\rangle = \langle\text{LT}_>(G)\rangle = \langle\text{LT}_>(\text{in}_{\mathbf{w}}(G))\rangle,$$

where the first equality follows since G is a Gröbner basis for $>$ and the second follows since $\mathbf{w}$ is compatible with $>$. Combining this with part a, we see that $\langle \mathrm{LT}_> (\langle \mathrm{in}_\mathbf{w}(I) \rangle) \rangle = \langle \mathrm{LT}_> (\mathrm{in}_\mathbf{w}(G)) \rangle$. Hence $\mathrm{in}_\mathbf{w}(G)$ is a Gröbner basis of $\langle \mathrm{in}_\mathbf{w}(I) \rangle$ for $>$, and the final assertion of the lemma follows.

It remains to consider what happens when $\mathbf{w} \in C_G$, which does not necessarily imply that $\mathbf{w}$ is compatible with $>$ (see Exercise 6 for an example). Consider the weight order $>'$ which first compares $\mathbf{w}$-weights and breaks ties using $>$. Note that $\mathbf{w}$ is compatible with $>'$.

The key observation is that since $\mathbf{w} \in C_G$, the leading term of each $g \in G$ with respect to $>'$ is the marked term. By Exercise 6 of §4, it follows that G is the marked Gröbner basis of I for $>'$. Since $\mathbf{w}$ is compatible with $>'$, the earlier part of the argument implies that $\mathrm{in}_\mathbf{w}(G)$ is a Gröbner basis of $\langle \mathrm{in}_\mathbf{w}(I) \rangle$ for $>'$. However, for each $g \in G$, $\mathrm{in}_\mathbf{w}(g)$ has the same leading term with respect to $>$ and $>'$. Using Exercise 6 of §4 again, we conclude that $\mathrm{in}_\mathbf{w}(G)$ is a Gröbner basis of $\langle \mathrm{in}_\mathbf{w}(I) \rangle$ for $>$.   □

We can now prove the proposition.

PROOF OF PROPOSITION (5.5). We will give the proof in three steps. Since $>_{new}$ is represented by $\binom{\mathbf{w}_{new}}{M_t}$, $\mathbf{w}_{new}$ is compatible with $>_{new}$. By part a of Lemma (5.8), we obtain

$$\mathrm{LT}_{>_{new}}(I) = \mathrm{LT}_{>_{new}}(\langle \mathrm{in}_{\mathbf{w}_{new}}(I) \rangle).$$

The second step is to observe that since $\mathbf{w}_{new} \in C_{old}$, the final assertion of part b of Lemma (5.8) implies

$$\langle \mathrm{in}_{\mathbf{w}_{new}}(I) \rangle = \langle \mathrm{in}_{\mathbf{w}_{new}}(G_{old}) \rangle.$$

For the third step, we show that

$$\langle \mathrm{in}_{\mathbf{w}_{new}}(G_{old}) \rangle = \langle \mathrm{LT}_{>_{new}}(H) \rangle = \langle \mathrm{LT}_{>_{new}}(\overline{H}) \rangle,$$

where $H = \{h_1, \ldots, h_t\}$ is the given Gröbner basis of $\langle \mathrm{in}_{\mathbf{w}_{new}}(G_{old}) \rangle$ and $\overline{H} = \{\overline{h}_1, \ldots, \overline{h}_t\}$ is the set of polynomials described in the statement of the proposition. The first equality is obvious, and for the second, it suffices to show that for each j, $\mathrm{LT}_{>_{new}}(h_j) = \mathrm{LT}_{>_{new}}(\overline{h}_j)$. Since the $\mathrm{in}_{\mathbf{w}_{new}}(g)$ are $\mathbf{w}_{new}$-homogeneous, Exercise 7 below shows that the same is true of the h_j and the $q_{j,g}$. Hence for each g, all terms in $q_{j,g}(g - \mathrm{in}_{\mathbf{w}_{new}}(g))$ have smaller $\mathbf{w}_{new}$-weight than those in the initial form. Lifting as in (5.7) to get $\overline{h}_j$ adds only terms with smaller $\mathbf{w}_{new}$-weight. Since $>_{new}$ is compatible with $\mathbf{w}_{new}$, the added terms are also smaller in the new order, so the $>_{new}$-leading term of $\overline{h}_j$ is the same as the leading term of h_j.

Combining the three steps, we obtain

$$\langle \mathrm{LT}_{>_{new}}(I) \rangle = \langle \mathrm{LT}_{>_{new}}(\overline{H}) \rangle.$$

Since $\overline{h}_j \in I$ for all j, we conclude that $\overline{H}$ is a Gröbner basis for I with respect to $>_{new}$, as claimed.   □

The Gröbner basis $\overline{H}$ from Proposition (5.5) is minimal, but not necessarily reduced. Hence a complete interreduction is usually necessary to obtain the marked Gröbner basis G_{new} corresponding to the next cone. In practice, this is a relatively quick process.

In order to use Proposition (5.5), we need to find the polynomials $p_{j,g}$ in (5.6) expressing the Gröbner basis elements h_j in terms of the ideal generators of $\mathrm{in}_{\mathbf{w}_{new}}(G_{old})$. This can be done in two ways:

- First, the $p_{j,g}$ can be computed along with H by an extended Buchberger algorithm (see for instance [BW], Chapter 5, Section 6);
- Second, since $\mathrm{in}_{\mathbf{w}_{new}}(G_{old})$ is a Gröbner basis of $\langle \mathrm{in}_{\mathbf{w}_{new}}(G_{old}) \rangle$ with respect to $>_{old}$ by part b of Lemma (5.8), the $p_{j,g}$ can be obtained by dividing h_j by $\mathrm{in}_{\mathbf{w}_{new}}(G_{old})$ using $>_{old}$.

In practice, the second is often more convenient to implement. The process of replacing the $\mathbf{w}_{new}$-initial forms of the g by the g themselves to go from (5.6) to (5.7) is called *lifting* the initial forms to the new Gröbner basis.

The Algorithm

The following algorithm is a basic Gröbner Walk, following the straight line segment from $\mathbf{w}_s$ to $\mathbf{w}_t$.

(5.9) Theorem. *Let*

1. **NextCone** *be a procedure that computes u_{last} from (5.2). Recall that $\mathbf{w}_{new} = (1 - u_{last})\mathbf{w}_{old} + u_{last}\mathbf{w}_t$ is the last weight vector along the path that lies in the cone C_{old} of the previous Gröbner basis G_{old};*
2. **Lift** *be a procedure that lifts a Gröbner basis for the $\mathbf{w}_{new}$-initial forms of the previous Gröbner basis G_{old} with respect to $>_{new}$ to the Gröbner basis G_{new} following Proposition (5.5); and*
3. **Interreduce** *be a procedure that takes a given set of polynomials and interreduces them with respect to a given monomial order.*

Then the following algorithm correctly computes a Gröbner basis for I with respect to $>_t$ and terminates in finitely many steps on all inputs:

Input: M_s and M_t representing start and target orders with first

rows $\mathbf{w}_s$ and $\mathbf{w}_t$, G_s = Gröbner basis with respect to $>_{M_s}$

Output: last value of G_{new} = Gröbner basis with respect to $>_{M_t}$

$M_{old} := M_s$

$G_{old} := G_s$

$\mathbf{w}_{new} := \mathbf{w}_s$

$$M_{new} := \begin{pmatrix} \mathbf{w}_{new} \\ M_t \end{pmatrix}$$

$done := false$

WHILE $done = false$ DO

$\quad In := \text{in}_{\mathbf{w}_{new}}(G_{old})$

$\quad InG := \text{gbasis}(In, >_{M_{new}})$

$\quad G_{new} := \text{Lift}(InG, G_{old}, In, M_{new}, M_{old})$

$\quad G_{new} := \text{Interreduce}(G_{new}, M_{new})$

$\quad u := \text{NextCone}(G_{new}, \mathbf{w}_{new}, \mathbf{w}_t)$

$\quad$ IF $\mathbf{w}_{new} = \mathbf{w}_t$ THEN

$\quad\quad done := true$

$\quad$ ELSE

$\quad\quad M_{old} := M_{new}$

$\quad\quad G_{old} := G_{new}$

$\quad\quad \mathbf{w}_{new} := (1 - u)\mathbf{w}_{new} + u\mathbf{w}_t$

$$\quad\quad M_{new} := \begin{pmatrix} \mathbf{w}_{new} \\ M_t \end{pmatrix}$$

RETURN(G_{new})

PROOF. We traverse the line segment from $\mathbf{w}_s$ to $\mathbf{w}_t$. To prove termination, observe that by Corollary (4.3), the Gröbner fan of $I = \langle G_s \rangle$ has only finitely many cones, each of which has only finitely many bounding hyperplanes as in (4.4). Discarding those hyperplanes that contain the line segment from $\mathbf{w}_s$ to $\mathbf{w}_t$, the remaining hyperplanes determine a finite set of distinguished points on our line segment.

Now consider $u_{last} = \text{NextCone}(G_{new}, \mathbf{w}_{new}, \mathbf{w}_t)$ as in the algorithm. This uses (5.2) with $\mathbf{w}_{old}$ replaced by the current value of $\mathbf{w}_{new}$. Furthermore, notice that the monomial order always comes from a matrix of the form $\begin{pmatrix} \mathbf{w}_s \\ M_t \end{pmatrix}$. It follows that the hypothesis of Lemma (5.3) is always satisfied. If $u_{last} = 1$, then the next value of $\mathbf{w}_{new}$ is $\mathbf{w}_t$, so that the algorithm terminates after one more pass through the main loop. On the other hand, if $u_{last} = u_j < 1$, then the next value of $\mathbf{w}_{new}$ lies on the hyperplane $\mathbf{w} \cdot v_j = 0$, which is one of our finitely many hyperplanes. However, (5.2) implies that $\mathbf{w}_t \cdot v_j < 0$ and $\mathbf{w}_{new} \cdot v_j \geq 0$, so that the hyperplane meets the line segment in a single point. Hence the next value of $\mathbf{w}_{new}$ is one of our distinguished points. Furthermore, Lemma (5.3) implies that $u_{last} > 0$, so that if the current $\mathbf{w}_{new}$ differs from $\mathbf{w}_t$, then we must move to a distinguished point farther along the line segment. Hence we must eventually reach $\mathbf{w}_t$, at which point the algorithm terminates.

To prove correctness, observe that in each pass through the main loop, the hypotheses of Proposition (5.5) are satisfied. Furthermore, once the value of $\mathbf{w}_{new}$ reaches $\mathbf{w}_t$, the next pass through the loop computes a Gröbner basis of I for the monomial order represented by $\binom{\mathbf{w}_t}{M_t}$. Using Exercise 6 of §4, it follows that the final value of G_{new} is the marked Gröbner basis for $>_t$. $\square$

The complexity of the Gröbner Walk depends most strongly on the number of cones that are visited along the path through the Gröbner fan, and the number of different cones that contain the point $\mathbf{w}_{new}$ at each step. We will say more about this in the examples below.

Examples

We begin with a simple example of the Gröbner Walk in action. Consider the ideal $I = \langle x^2 - y, xz - y^2 + yz \rangle \subset \mathbb{Q}[x, y, z]$ from (4.6). We computed the full Gröbner fan for I in §4 (see Figure 8.12). Say we know

$$G_s = G^{(1)} = \{\underline{x^2} - y, \underline{y^2} - xz - yz\}$$

from (4.9). This is the Gröbner basis of I with respect to $>_{(5,4,1),grevlex}$ (among many others!). Suppose we want to determine the Gröbner basis with respect to $>_{(6,1,3),lex}$ (which is $G^{(6)}$). We could proceed as follows. Let

$$M_s = \begin{pmatrix} 5 & 4 & 1 \\ 1 & 1 & 1 \\ 1 & 1 & 0 \end{pmatrix}$$

so $\mathbf{w}_s = (5, 4, 1)$. Following Exercise 6 from Chapter 1, §2, we have used a square matrix defining the same order instead of the 4×3 matrix with first row $(5, 4, 1)$ and the next three rows from a 3×3 matrix defining the *grevlex* order (as in part b of Exercise 6 of Chapter 1, §2). Similarly,

$$M_t = \begin{pmatrix} 6 & 1 & 3 \\ 1 & 0 & 0 \\ 0 & 1 & 0 \end{pmatrix}$$

and $\mathbf{w}_t = (6, 1, 3)$. We will choose square matrices defining the appropriate monomial orders in all of the following computations by deleting appropriate linearly dependent rows.

We begin by considering the order defined by

$$M_{new} = \begin{pmatrix} 5 & 4 & 1 \\ 6 & 1 & 3 \\ 1 & 0 & 0 \end{pmatrix}$$

(using the weight vector $\mathbf{w}_{new} = (5, 4, 1)$ first, then refining by the target order). The $\mathbf{w}_{new}$-initial forms of the Gröbner basis polynomials with respect to this order are the same as those for G_s, so the basis does not change in the first pass through the main loop.

We then call the NextCone procedure (5.2) with $\mathbf{w}_{new}$ in place of $\mathbf{w}_{old}$. The cone of $>_{M_{new}}$ is defined by the three inequalities obtained by comparing x^2 vs. y and y^2 vs. xz and yz. By (5.2), u_{last} is the largest u such that $(1 - u)(5, 4, 1) + u(6, 1, 3)$ lies in this cone and is computed as follows:

x^2 vs. y :

$$v_1 = (2, -1, 0), \mathbf{w}_t \cdot v_1 = 6 \geq 0 \Rightarrow u_1 = 1$$

y^2 vs. xz :

$$v_2 = (-1, 2, -1), \mathbf{w}_t \cdot v_2 = -7 < 0 \Rightarrow u_2 = \frac{\mathbf{w}_{new} \cdot v_2}{\mathbf{w}_{new} \cdot v_2 - (-7)} = \frac{2}{9}$$

y^2 vs. yz :

$$v_2 = (0, -, -1), \mathbf{w}_t \cdot v_3 = -2 < 0 \Rightarrow u_3 = \frac{\mathbf{w}_{new} \cdot v_3}{\mathbf{w}_{new} \cdot v_3 - (-2)} = \frac{3}{5}.$$

The smallest u value here is $u_{last} = \frac{2}{9}$. Hence the new weight vector is $\mathbf{w}_{new} = (1 - \frac{2}{9})(5, 4, 1) + \frac{2}{9}(6, 1, 3) = (47/9, 10/3, 13/9)$, and M_{old} and

$$M_{new} = \begin{pmatrix} 47/9 & 10/3 & 13/9 \\ 6 & 1 & 3 \\ 1 & 0 & 0 \end{pmatrix}$$

are updated for the next pass through the main loop.

In the second pass, $In = \{y^2 - xz, x^2\}$. We compute the Gröbner basis for $\langle In \rangle$ with respect to $>_{new}$ (with respect to this order, the leading term of the first element is xz), and find

$$H = \{-y^2 + xz, x^2, xy^2, y^4\}.$$

In terms of the generators for $\langle In \rangle$, we have

$$-y^2 + xz = -1 \cdot (y^2 - xz) + 0 \cdot (x^2)$$
$$x^2 = 0 \cdot (y^2 - xz) + 1 \cdot (x^2)$$
$$xy^2 = x \cdot (y^2 - xz) + z \cdot (x^2)$$
$$y^4 = (y^2 + xz) \cdot (y^2 - xz) + z^2 \cdot (x^2).$$

So by Proposition (5.5), to get the next Gröbner basis, we lift to

$$-1 \cdot (y^2 - xz - yz) + 0 \cdot (x^2 - y) = xz + yz - y^2$$
$$0 \cdot (y^2 - xz - yz) + 1 \cdot (x^2 - y) = x^2 - y$$
$$x \cdot (y^2 - xz - yz) + z \cdot (x^2 - y) = xy^2 - xyz - yz$$
$$(y^2 + xz) \cdot (y^2 - xz - yz) + z^2 \cdot (x^2 - y) = y^4 - y^3z - xyz^2 - yz^2.$$

Interreducing with respect to $>_{new}$, we obtain the marked Gröbner basis G_{new} given by

$\{\underline{xz} + yz - y^2, \underline{x^2} - y, \underline{xy^2} - y^3 + y^2z - yz, \underline{y^4} - 2y^3z + y^2z^2 - yz^2\}.$

(This is $G^{(5)}$ in (4.9).) For the call to NextCone in this pass, we use the parametrization $(1 - u)(47/9, 10/3, 13/9) + u(6, 1, 3)$. Using (5.2) as above, we obtain $u_{last} = 17/35$, for which $\mathbf{w}_{new} = (28/5, 11/5, 11/5)$.

In the third pass through the main loop, the Gröbner basis does not change as a set. However, the leading term of the initial form of the last polynomial $y^4 - 2y^3z + y^2z^2 - yz^2$ with respect to $>_{M_{new}}$ is now y^2z^2 since

$$M_{new} = \begin{pmatrix} 28/5 & 11/5 & 11/5 \\ 6 & 1 & 3 \\ 1 & 0 & 0 \end{pmatrix}.$$

Using Proposition (5.5) as usual to compute the new Gröbner basis G_{new}, we obtain

(5.10) $\{\underline{xz} + yz - y^2, \underline{x^2} - y, \underline{xy^2} - y^3 + y^2z - yz, \underline{y^2z^2} - 2y^3z + y^4 - yz^2\},$

which is $G^{(6)}$ in (4.9). The call to NextCone returns $u_{last} = 1$, since there are no pairs of terms that attain equal weight for any point on the line segment parametrized by $(1 - u)(28/5, 11/5, 11/5) + u(6, 1, 3)$. Thus $\mathbf{w}_{new} = \mathbf{w}_t$. After one more pass through the main loop, during which G_{new} doesn't change, the algorithm terminates. Hence the final output is (5.10), which is the marked Gröbner basis of I with respect to the target order.

We note that it is possible to modify the algorithm of Theorem (5.9) so that the final pass in the above example doesn't occur. See Exercise 8.

Exercise 2. Verify the computation of u_{last} in the steps of the above example after the first.

Exercise 3. Apply the Gröbner Walk to convert the basis $G^{(3)}$ for the above ideal to the basis $G^{(4)}$ (see (4.9) and Figure (4.2)). Take $>_s = >_{(2,7,1),grevlex}$ and $>_t = >_{(3,1,6),grevlex}$.

Many advantages of the walk are lost if there are many terms in the $\mathbf{w}_{new}$-initial forms. This tends to happen if a portion of the path lies in a face of some cone, or if the path passes through points where many cones intersect. Hence in [AGK], Amrhein, Gloor, and Küchlin make systematic use of perturbations of weight vectors to keep the path in as general a position as possible with respect to the faces of the cones. For example, one possible variant of the basic algorithm above would be to use (4.8) to obtain a perturbed weight vector in the interior of the corresponding cone each time a new marked Gröbner basis is obtained, and resume the walk to the target monomial order from there. Another variant designed for elimination problems is to take a "sudden-death" approach. If we want a Gröbner basis with respect to a monomial order eliminating the variables $x_1, \ldots, x_n$, leaving $y_1, \ldots, y_m$, and we expect a single generator for the elimination ideal, then we could terminate the walk as soon as some polynomial in $k[y_1, \ldots, y_m]$ appears in the current G_{new}. This is only guaranteed to be a multiple of the generator of the elimination ideal, but even a polynomial satisfying that condition can be useful in some circumstances. We refer the interested reader to [AGK] for a discussion of other implementation issues.

In [Tran], degree bounds on elements of Gröbner bases are used to produce weight vectors in the interior of each cone of the Gröbner fan, which gives a deterministic way to find good path perturbations. A theoretical study of the complexity of the Gröbner Walk and other basis conversion algorithms has been made by Kalkbrener in [Kal].

Our next example is an application of the Gröbner Walk algorithm to an implicitization problem inspired by examples studied in robotics and computer-aided design. Let C_1 and C_2 be two curves in $\mathbb{R}^3$. The *bisector surface* of C_1 and C_2 is the locus of points P equidistant from C_1 and C_2 (that is, P is on the bisector if the closest point(s) to P on C_1 and C_2 are the same distance from P). See, for instance, [EK]. Bisectors are used, for example, in motion planning to find paths avoiding obstacles in an environment. We will consider only the case where C_1 and C_2 are smooth complete intersection algebraic curves $C_1 = \mathbf{V}(f_1, g_1)$ and $C_2 = \mathbf{V}(f_2, g_2)$. (This includes most of the cases of interest in solid modeling, such as lines, circles, and other conics, etc.) $P = (x, y, z)$ is on the bisector of C_1 and C_2 if there exist $Q_1 = (x_1, y_1, z_1) \in C_1$ and $Q_2 = (x_2, y_2, z_2) \in C_2$ such that the distance from P to C_i is a minimum at Q_i, $i = 1, 2$ and the distance from P to Q_1 equals the distance from P to Q_2. Rather than insisting on an absolute minimum of the distance function from P to C_i at Q_i, it is simpler to insist that the distance function simply have a critical point there. It is easy to see that this condition is equivalent to saying that the line segment from P to Q_i is *orthogonal* to the tangent line to C_i at Q_i.

Exercise 4. Show that the distance from C_i to P has a critical point at Q_i if and only if the line segment from P to Q_i is *orthogonal* to the tangent line to C_i at Q_i, and show that this is equivalent to saying that

$$(\nabla f_i(Q_i) \times \nabla g_i(Q_i)) \cdot (P - Q_i) = 0,$$

where $\nabla f_i(Q_i)$ denotes the gradient vector of f_i at Q_i, and $\times$ is the cross product in $\mathbb{R}^3$.

By Exercise 4, we can find the bisector as follows. Let (x_i, y_i, z_i) be a general point Q_i on C_i, and $P = (x, y, z)$. Consider the system of equations

$$0 = f_1(x_1, y_1, z_1)$$
$$0 = g_1(x_1, y_1, z_1)$$
$$0 = f_2(x_2, y_2, z_2)$$

$$0 = g_2(x_2, y_2, z_2)$$
$$0 = (\nabla f_1(x_1, y_1, z_1) \times \nabla g_1(x_1, y_1, z_1)) \cdot (x - x_1, y - y_1, z - z_1)$$
$$(5.11) \quad 0 = (\nabla f_2(x_2, y_2, z_2) \times \nabla g_2(x_2, y_2, z_2)) \cdot (x - x_2, y - y_2, z - z_2)$$
$$0 = (x - x_1)^2 + (y - y_1)^2 + (z - z_1)^2$$
$$- (x - x_2)^2 - (y - y_2)^2 - (z - z_2)^2.$$

Let $J \subset \mathbb{R}[x_1, y_1, z_1, x_2, y_2, z_2, x, y, z]$ be the ideal generated by these seven polynomials. We claim the bisector will be contained in $\mathbf{V}(I)$, where I is the elimination ideal $I = J \cap \mathbb{R}[x, y, z]$. A proof proceeds as follows. $P = (x, y, z)$ is on the bisector of C_1 and C_2 if and only if there exist $Q_i = (x_i, y_i, z_i)$ such that $Q_i \in C_i$, Q_i is a minimum of the distance function to P, restricted to C_i, and $PQ_1 = PQ_2$. Thus P is in the bisector if and only if the equations in (5.11) are satisfied for some $(x_i, y_i, z_i) \in C_i$. Therefore, P is the projection of some point in $\mathbf{V}(J)$, hence in $\mathbf{V}(I)$. Note that (5.11) contains seven equations in nine unknowns, so we expect that $\mathbf{V}(J)$ and its projection $\mathbf{V}(I)$ have dimension 2 in general.

For instance, if C_1 is the twisted cubic $\mathbf{V}(y - x^2, z - x^3)$ and C_2 is the line $\mathbf{V}(x, y - 1)$, then our ideal J is

$$(5.12) \quad J = \langle y_1 - x_1^2, z_1 - x_1^3, x_2, y_2 - 1,$$
$$x - x_1 + 2x_1(y - y_1) + 3x_1^2(z - z_1), z - z_2,$$
$$(x - x_1)^2 + (y - y_1)^2 + (z - z_1)^2$$
$$- (x - x_2)^2 - (y - y_2)^2 - (z - z_2)^2 \rangle.$$

We apply the Gröbner Walk with $>_s$ the *grevlex* order with $x_1 > y_1 > z_1 > x_2 > y_2 > z_2 > x > y > z$, and $>_t$ the $>_{\mathbf{w}, grevlex}$ order, where $\mathbf{w} = (1, 1, 1, 1, 1, 1, 0, 0, 0)$, which has the desired elimination property to compute $J \cap \mathbb{R}[x, y, z]$.

Using our own (somewhat naive) implementation of the Gröbner Walk based on the **Groebner** package in Maple, we computed the $>_{\mathbf{w}, grevlex}$ basis for J as in (5.12). As we expect, the elimination ideal is generated by a single polynomial: $J \cap \mathbb{R}[x, y, z] =$

$$\langle 5832z^6y^3 - 729z^8 - 34992x^2y - 14496yxz - 14328x^2z^2$$
$$+ 24500x^4y^2 - 23300x^4y + 3125x^6 - 5464z^2 - 36356z^4y$$
$$+ 1640xz^3 + 4408z^4 + 63456y^3xz^3 + 28752y^3x^2z^2$$
$$- 201984y^3 - 16524z^6y^2 - 175072y^2z^2 + 42240y^4xz - 92672y^3zx$$
$$+ 99956z^4y^2 + 50016yz^2 + 90368y^2 + 4712x^2 + 3200y^3x^3z$$
$$+ 6912y^4xz^3 + 13824y^5zx + 19440z^5xy^2 + 15660z^3x^3y + 972z^4x^2y^2$$
$$+ 6750z^2x^4y - 61696y^2z^3x + 4644yxz^5 - 37260yz^4x^2$$
$$- 85992y^2x^2z^2 + 5552x^4 - 7134xz^5 + 64464yz^2x^2$$

$$- 5384zyx^3 + 2960zy^2x^3 - 151z^6 + 1936$$
$$+ 29696y^6 + 7074z^6y + 18381z^4x^2 - 2175z^2x^4 + 4374xz^7$$
$$+ 1120zx - 7844x^3z^3 - 139264y^5 - 2048y^7 - 1024y^6z^2$$
$$- 512y^5x^2 - 119104y^3x^2 - 210432y^4z^2 + 48896y^5z^2$$
$$- 104224y^3z^4 + 28944y^4z^4 + 54912y^4x^2 - 20768y + 5832z^5x^3$$
$$- 8748z^6x^2 + 97024y^2x^2 + 58560y^2zx + 240128y^4 + 286912y^3z^2$$
$$+ 10840xyz^3 + 1552x^3z - 3750zx^5\rangle.$$

The computation of the full Gröbner basis (including the initial computation of the *grevlex* Gröbner basis of J) took 43 seconds on a 866 MHz Pentium III using the Gröbner Walk algorithm described in Theorem (5.9). Apparently the cones corresponding to the two monomial orders $>_s, >_t$ are very close together in the Gröbner fan for J, a happy accident. The $\mathbf{w}_{new}$-initial forms in the second step of the walk contained a large number of distinct terms, though. With the "sudden-death" strategy discussed above, the time was reduced to 23 seconds and produced the same polynomial (not a multiple). By way of contrast, a direct computation of the $>_{\mathbf{w},grevlex}$ Gröbner basis using the gbasis command of the Groebner package was terminated after using 20 minutes of CPU time and over 200 Mb of memory. In our experience, in addition to gains in speed, the Gröbner Walk tends also to use much less memory for storing intermediate polynomials than Buchberger's algorithm with an elimination order. This means that even if the walk takes a long time to complete, it will often execute successfully on complicated examples that are not feasible using the Gröbner basis packages of standard computer algebra systems. Similarly encouraging results have been reported from several experimental implementations of the Gröbner Walk.

As of this writing, the Gröbner Walk has not been included in the Gröbner basis packages distributed with general-purpose computer algebra systems such as Maple or *Mathematica*. An implementation is available in Magma, however. The CASA Maple package developed at RISC-Linz (see http://www.risc.uni-linz.ac.at/software/casa/) also contains a Gröbner Walk procedure.

ADDITIONAL EXERCISES FOR §5

Exercise 5. Fix a nonzero weight vector $\mathbf{w} \in (\mathbb{R}^n)^+$. Show that every $f \in k[x_1, \ldots, x_n]$ can be written uniquely as a sum of $\mathbf{w}$-homogeneous polynomials.

Exercise 6. Fix a monomial order $>$ and a nonzero weight vector $\mathbf{w} \in (\mathbb{R}^n)^+$. Also, given an ideal $I \subset k[x_1, \ldots, x_n]$, let $C_>$ be the cone in the Gröbner fan of I corresponding to $\langle \mathrm{LT}_>(I) \rangle \in \mathrm{Mon}(I)$.

a. Prove that $\mathbf{w}$ is compatible with $>$ if and only if $\mathbf{w} \cdot \alpha > \mathbf{w} \cdot \beta$ always implies $x^{\alpha} > x^{\beta}$.

b. Prove that if $\mathbf{w}$ is compatible with $>$, then $\mathbf{w} \in C_>$.

c. Use the example of $>_{lex}$ for $x > y$, $I = \langle x+y \rangle \subset k[x, y]$ and $\mathbf{w} = (1, 1)$ to show that the naive converse to part b is false. (See part d for the real converse.)

d. Prove that $\mathbf{w} \in C_>$ if and only if there is a monomial order $>'$ such that $C_{>'} = C_>$ and $\mathbf{w}$ is compatible with $>'$. Hint: See the proof of part b of Lemma (5.8).

Exercise 7. Suppose that J is an ideal generated by $\mathbf{w}$-homogeneous polynomials. Show that every reduced Gröbner basis of I consists of $\mathbf{w}$-homogeneous polynomials. Hint: This generalizes the corresponding fact for homogeneous ideals. See [CLO], Theorem 2 of Chapter 8, §3.

Exercise 8. It is possible to get a slightly more efficient version of the algorithm described in Theorem (5.9). The idea is to modify (5.2) so that u_{last} is allowed to be greater than 1 if the ray from $\mathbf{w}_{old}$ to $\mathbf{w}_t$ leaves the cone at a point beyond $\mathbf{w}_t$.

a. Modify (5.2) so that it behaves as described above and prove that your modification behaves as claimed.

b. Modify the algorithm described in Theorem (5.9) in two ways: first, $\mathbf{w}_{new}$ is defined using $\min\{1, u_{last}\}$ and second, the IF statement tests whether $u_{last} > 1$ or $\mathbf{w}_{new} = \mathbf{w}_t$. Prove that this modified algorithm correctly converts G_s to G_t.

c. Show that the modified algorithm, when applied to the ideal $I = \langle x^2 - y, y^2 - xz - yz \rangle$ discussed in the text, requires one less pass through the main loop than without the modificiation.

Exercise 9. In a typical polynomial implicitization problem, we are given $f_i \in k[t_1, \ldots, t_m]$, $i = 1, \ldots, n$ (the coordinate functions of a parametrization) and we want to eliminate $t_1, \ldots, t_m$ from the equations $x_i = f_1(t_1, \ldots, t_m)$, $i = 1, \ldots, n$. To do this, consider the ideal

$$J = \langle x_1 - f_1(t_1, \ldots, t_m), \ldots, x_n - f_n(t_1, \ldots, t_m) \rangle$$

and compute $I = J \cap k[x_1, \ldots, x_n]$ to find the implicit equations of the image of the parametrization. Explain how the Gröbner Walk could be applied to the generators of J directly to find I without any preliminary Gröbner basis computation. Hint: They are already a Gröbner basis with respect to a suitable monomial order.

Exercise 10. Apply the Gröbner Walk method suggested in Exercise 9 to compute the implicit equation of the parametric curve

$$\begin{cases} x = t^4 \\ y = t^2 + t. \end{cases}$$

(If you do not have access to an implementation of the walk, you will need to perform the steps "manually" as in the example given in the text.) Also see part b of Exercise 11 in the previous section.

Chapter 9

Algebraic Coding Theory

In this chapter we will discuss some applications of techniques from computational algebra and algebraic geometry to problems in coding theory. After a preliminary section on the arithmetic of finite fields, we will introduce some basic terminology for describing error-correcting codes. We will study two important classes of examples—linear codes and cyclic codes—where the set of codewords possesses additional algebraic structure, and we will use this structure to develop good encoding and decoding algorithms.

§1 Finite Fields

To make our presentation as self-contained as possible, in this section we will develop some of the basic facts about the arithmetic of finite fields. We will do this almost "from scratch," without using the general theory of field extensions. However, we will need to use some elementary facts about finite groups and quotient rings. Readers who have seen this material before may wish to proceed directly to §2. More complete treatments of this classical subject can be found in many texts on abstract algebra or Galois theory.

The most basic examples of finite fields are the *prime fields* $\mathbb{F}_p = \mathbb{Z}/\langle p \rangle$, where p is any prime number, but there are other examples as well. To construct them, we will need to use the following elementary fact.

Exercise 1. Let k be any field, and let $g \in k[x]$ be an *irreducible* polynomial (that is, a non-constant polynomial which is not the product of two nonconstant polynomials in $k[x]$). Show that the ideal $\langle g \rangle \subset k[x]$ is a maximal ideal. Deduce that $k[x]/\langle g \rangle$ is a field if g is irreducible.

For example, let $p = 3$ and consider the polynomial $g = x^2 + x + 2 \in \mathbb{F}_3[x]$. Since g is a quadratic polynomial with no roots in $\mathbb{F}_3$, g is irreducible in $\mathbb{F}_3[x]$. By Exercise 1, the ideal $\langle g \rangle$ is maximal, hence $\mathbb{F} = \mathbb{F}_3[x]/\langle g \rangle$ is a field. As we discussed in Chapter 2, the elements of a quotient ring such as $\mathbb{F}$ are in one-to-one correspondence with the possible remainders on division by g. Hence the elements of $\mathbb{F}$ are the cosets of the polynomials $ax + b$, where a, b are arbitrary in $\mathbb{F}_3$. As a result, $\mathbb{F}$ is a field of $3^2 = 9$ elements.

To distinguish more clearly between polynomials and the elements of our field, we will write α for the element of $\mathbb{F}$ represented by the polynomial x. Thus every element of $\mathbb{F}$ has the form $a\alpha + b$ for $a, b \in \mathbb{F}_3$. Also, note that α satisfies the equation $g(\alpha) = \alpha^2 + \alpha + 2 = 0$.

The addition operation in $\mathbb{F}$ is the obvious one: $(a\alpha + b) + (a'\alpha + b') = (a + a')\alpha + (b + b')$. As in Chapter 2 §2, we can compute products in $\mathbb{F}$ by multiplication of polynomials in α, subject to the relation $g(\alpha) = 0$. For instance, you should verify that in $\mathbb{F}$

$$(\alpha + 1) \cdot (2\alpha + 1) = 2\alpha^2 + 1 = \alpha$$

(recall that the coefficients of these polynomials are elements of the field $\mathbb{F}_3$, so that $1 + 2 = 0$). Using this method, we may compute all the powers of α in $\mathbb{F}$, and we find

$$
\begin{array}{llll}
& \alpha^2 = 2\alpha + 1 & \alpha^3 = 2\alpha + 2 \\
(1.1) & \alpha^4 = 2 & \alpha^5 = 2\alpha \\
& \alpha^6 = \alpha + 2 & \alpha^7 = \alpha + 1,
\end{array}
$$

and $\alpha^8 = 1$. For future reference, we note that this computation also shows that the multiplicative group of nonzero elements of $\mathbb{F}$ is a cyclic group of order 8, generated by α.

The construction of $\mathbb{F}$ in this example may be generalized in the following way. Consider the polynomial ring $\mathbb{F}_p[x]$, and let $g \in \mathbb{F}_p[x]$ be an irreducible polynomial of degree n. The ideal $\langle g \rangle$ is maximal by Exercise 1, so the quotient ring $\mathbb{F} = \mathbb{F}_p[x]/\langle g \rangle$ is a field. The elements of $\mathbb{F}$ may be represented by the cosets modulo $\langle g \rangle$ of the polynomials of degree $n - 1$ or less: $a_{n-1}x^{n-1} + \cdots + a_1 x + a_0, a_i \in \mathbb{F}_p$. Since the a_i are arbitrary, this implies that $\mathbb{F}$ contains p^n distinct elements.

Exercise 2.
a. Show that $g = x^4 + x + 1$ is irreducible in $\mathbb{F}_2[x]$. How many elements are there in the field $\mathbb{F} = \mathbb{F}_2[x]/\langle g \rangle$?

b. Writing α for the element of $\mathbb{F}$ represented by x as above, compute all the distinct powers of α.

c. Show that $\mathbb{K} = \{0, 1, \alpha^5, \alpha^{10}\}$ is a field with four elements contained in $\mathbb{F}$.

d. Is there a field with exactly eight elements contained in $\mathbb{F}$? Are there any other subfields? (For the general pattern, see Exercise 10 below.)

In general we may ask what the possible sizes (numbers of elements) of finite fields are. The following proposition gives a necessary condition.

(1.2) Proposition. *Let $\mathbb{F}$ be a finite field. Then $|\mathbb{F}| = p^n$ where p is some prime number and $n \geq 1$.*

PROOF. Since $\mathbb{F}$ is a field, it contains a multiplicative identity, which we will denote by 1 as usual. Since $\mathbb{F}$ is finite, 1 must have finite additive order: say p is the smallest positive integer such that $p \cdot 1 = 1 + \cdots + 1 = 0$ (p summands). The integer p must be prime. (Otherwise, if $p = mn$ with $m, n > 1$, then we would have $p \cdot 1 = (m \cdot 1)(n \cdot 1) = 0$ in $\mathbb{F}$. But since $\mathbb{F}$ is a field, this would imply $m \cdot 1 = 0$, or $n \cdot 1 = 0$, which is not possible by the minimality of p.) We leave it to the reader to check that the set of elements of the form $m \cdot 1$, $m = 0, 1, \ldots, p - 1$ in $\mathbb{F}$ is a subfield $\mathbb{K}$ isomorphic to $\mathbb{F}_p$. See Exercise 9 below.

The axioms for fields imply that if we consider the addition operation on $\mathbb{F}$ together with scalar multiplication of elements of $\mathbb{F}$ by elements from $\mathbb{K} \subset \mathbb{F}$, then $\mathbb{F}$ has the structure of a vector space over $\mathbb{K}$. Since $\mathbb{F}$ is a finite set, it must be finite-dimensional as a vector space over $\mathbb{K}$. Let n be its dimension (the number of elements in any basis), and let $\{a_1, \ldots, a_n\} \subset \mathbb{F}$ be any basis. Every element of $\mathbb{F}$ can be expressed in exactly one way as a linear combination $c_1 a_1 + \cdots + c_n a_n$, where $c_1, \ldots, c_n \in \mathbb{K}$. There are p^n such linear combinations, which concludes the proof. $\square$

To construct finite fields, we will always consider quotient rings $\mathbb{F}_p[x]/\langle g \rangle$ where g is an irreducible polynomial in $\mathbb{F}_p[x]$. There is no loss of generality in doing this—every finite field can be obtained this way. See Exercise 11 below.

We will show next that for each prime p and each $n \geq 1$, there exist finite fields of every size p^n by counting the irreducible polynomials of fixed degree in $\mathbb{F}_p[x]$. First note that it is enough to consider monic polynomials, since we can always multiply by a constant in $\mathbb{F}_p$ to make the leading coefficient of a polynomial equal 1. There are exactly p^n distinct monic polynomials $x^n + a_{n-1}x^{n-1} + \cdots + a_1 x + a_0$ of degree n in $\mathbb{F}_p[x]$. Consider the *generating function* for this enumeration by degree: the power series in u in which the coefficient of u^n equals the number of monic polynomials of degree n, namely p^n. This is the left hand side in (1.3) below. We treat this as a purely formal series and disregard questions of convergence. The formal geometric series summation formula yields

$$(1.3) \qquad \sum_{n=0}^{\infty} p^n u^n = \frac{1}{1 - pu}.$$

Each monic polynomial factors *uniquely* in $\mathbb{F}_p[x]$ into a product of monic irreducibles. For each n, let N_n be the number of monic irreducibles of degree n in $\mathbb{F}_p[x]$. In factorizations of the form $g = g_1 \cdot g_2 \cdots g_m$ where the g_i are irreducible (but not necessarily distinct) of degrees n_i, we have N_{n_i} choices for the factor g_i for each i. The total degree of g is the sum of the degrees of the factors.

Exercise 3. By counting factorizations as above, show that the number of monic polynomials of degree n (i.e. p^n) can also be expressed as the coefficient of u^n in the formal infinite product

$$(1 + u + u^2 + \cdots)^{N_1} \cdot (1 + u^2 + u^4 + \cdots)^{N_2} \cdots = \prod_{k=1}^{\infty} \frac{1}{(1 - u^k)^{N_k}},$$

where the equality between the left- and right-hand sides comes from the formal geometric series summation formula. Hint: The term in the product with index k accounts for factors of degree k in polynomials.

Hence combining (1.3) with the result of Exercise 3, we obtain the generating function identity

$$(1.4) \qquad \prod_{k=1}^{\infty} \frac{1}{(1 - u^k)^{N_k}} = \frac{1}{1 - pu}.$$

(1.5) Proposition. *We have* $p^n = \sum_{k|n} k N_k$, *where the sum extends over all positive divisors k of the integer n.*

PROOF. Formally taking logarithmic derivatives and multiplying the results by u, from (1.4) we arrive at the identity $\sum_{k=1}^{\infty} k N_k u^k / (1 - u^k) = pu/(1 - pu)$. Using formal geometric series again, this equality can be rewritten as

$$\sum_{k=1}^{\infty} k N_k (u^k + u^{2k} + \cdots) = pu + p^2 u^2 + \cdots.$$

The proposition follows by comparing the coefficients of u^n on both sides of this last equation. $\square$

Exercise 4. (For readers with some background in elementary number theory.) Use Proposition (1.5) and the Möbius inversion formula to derive a general formula for N_n.

We will show that $N_n > 0$ for all $n \geq 1$. For $n = 1$, we have $N_1 = p$ since all $x - \beta$, $\beta \in \mathbb{F}_p$ are irreducible. Then Proposition (1.5) implies that $N_2 = (p^2 - p)/2 > 0$, $N_3 = (p^3 - p)/3 > 0$, and $N_4 = (p^4 - p^2)/4 > 0$.

Arguing by contradiction, suppose that $N_n = 0$ for some n. We may assume $n \geq 5$ by the above. Then from Proposition (1.5),

$$(1.6) \qquad p^n = \sum_{k|n, 0 < k < n} k N_k.$$

We can estimate the size of the right-hand side and derive a contradiction from (1.6) as follows. We write $\lfloor A \rfloor$ for the greatest integer less than or equal to A. Since $N_k \leq p^k$ for all k (the irreducibles are a subset of the whole collection monic polynomials), and any positive proper divisor k of n is at most $\lfloor n/2 \rfloor$, we have

$$p^n \leq \lfloor n/2 \rfloor \sum_{k=0}^{\lfloor n/2 \rfloor} p^k.$$

Applying the finite geometric sum formula, the right-hand side equals

$$\lfloor n/2 \rfloor (p^{\lfloor n/2 \rfloor + 1} - 1)/(p - 1) \leq \lfloor n/2 \rfloor p^{\lfloor n/2 \rfloor + 1}.$$

Hence

$$p^n \leq \lfloor n/2 \rfloor p^{\lfloor n/2 \rfloor + 1}.$$

Dividing each side by $p^{\lfloor n/2 \rfloor}$, we obtain

$$p^{n - \lfloor n/2 \rfloor} \leq \lfloor n/2 \rfloor p.$$

But this is clearly false for all p and all $n \geq 5$. Hence $N_n > 0$ for all n, and as a result we have the following fact.

(1.7) Theorem. *For all primes p and all $n \geq 1$, there exist finite fields $\mathbb{F}$ with $|\mathbb{F}| = p^n$.*

From the examples we have seen and from the proof of Theorem (1.7), it might appear that there are several different finite fields of a given size, since there will usually be more than one irreducible polynomial g of a given degree in $\mathbb{F}_p[x]$ to use in constructing quotients $\mathbb{F}_p[x]/\langle g \rangle$. But consider the following example.

Exercise 5. By Proposition (1.5), there are $(2^3 - 2)/3 = 2$ monic irreducible polynomials of degree 3 in $\mathbb{F}_2[x]$, namely $g_1 = x^3 + x + 1$, and $g_2 = x^3 + x^2 + 1$. Hence $\mathbb{K}_1 = \mathbb{F}_2[x]/\langle g_1 \rangle$ and $\mathbb{K}_2 = \mathbb{F}_2[x]/\langle g_2 \rangle$ are two finite fields with 8 elements. We claim, however, that these fields are *isomorphic*.
a. Writing α for the coset of x in $\mathbb{K}_1$ (so $g_1(\alpha) = 0$ in $\mathbb{K}_1$), show that $g_2(\alpha + 1) = 0$ in $\mathbb{K}_1$.
b. Use this observation to derive an isomorphism between $\mathbb{K}_1$ and $\mathbb{K}_2$ (that is, a one-to-one, onto mapping that preserves sums and products).

The general pattern is the same.

(1.8) Theorem. *Let $\mathbb{K}_1$ and $\mathbb{K}_2$ be two fields with $|\mathbb{K}_1| = |\mathbb{K}_2| = p^n$. Then $\mathbb{K}_1$ and $\mathbb{K}_2$ are isomorphic.*

See Exercise 12 below for one way to prove this. Because of (1.8), it makes sense to adopt the uniform notation $\mathbb{F}_{p^n}$ for any field of order p^n, and we will use this convention for the remainder of the chapter. When we do computations in $\mathbb{F}_{p^n}$, however, we will always use an explicit monic irreducible polynomial $g(x)$ of degree n as in the examples above.

The next general fact we will consider is also visible in (1.1) and in the other examples we have encountered.

(1.9) Theorem. *Let* $\mathbb{F} = \mathbb{F}_{p^n}$ *be a finite field. The multiplicative group of nonzero elements of* $\mathbb{F}$ *is cyclic of order* $p^n - 1$.

PROOF. The statement about the order of the multiplicative group is clear since we are omitting the single element 0. Write $m = p^n - 1$. By Lagrange's Theorem for finite groups ([Her]), every element $\beta \in \mathbb{F} \setminus \{0\}$ is a root of the polynomial equation $x^m = 1$, and the multiplicative order of each is a divisor of m. We must show there is some element of order exactly m to conclude the proof. Consider the prime factorization of m, say $m = q_1^{e_1} \cdots q_k^{e_k}$. Let $m_i = m/q_i$. Since the polynomial $x^{m_i} - 1$ has at most m_i roots in the field $\mathbb{F}$, there is some $\beta_i \in \mathbb{F}$ such that $\beta_i^{m_i} \neq 1$. In Exercise 6 below, you will show that $\gamma_i = \beta_i^{m/q_i^{e_i}}$ has multiplicative order exactly $q_i^{e_i}$ in $\mathbb{F}$. It follows that the product $\gamma_1 \gamma_2 \cdots \gamma_k$ has order m, since the $q_i^{e_i}$ are relatively prime. $\square$

Exercise 6. In this exercise you will supply details for the final two claims in the proof of Theorem (1.9).

a. Using the notation from the proof, show that $\gamma_i = \beta_i^{m/q_i^{e_i}}$ has multiplicative order exactly $q_i^{e_i}$ in $\mathbb{F}$. (That is, show that $\gamma_i^{q_i^{e_i}} = 1$, but that $\gamma_i^k \neq 1$ for all $k = 1, \ldots, q_i^{e_i} - 1$.)

b. Let γ_1, γ_2 be elements of a finite abelian group. Suppose that the orders of γ_1 and γ_2 (n_1 and n_2 respectively) are relatively prime. Show that the order of the product $\gamma_1 \gamma_2$ is equal to $n_1 n_2$.

A generator for the multiplicative group of $\mathbb{F}_{p^n}$ is called a *primitive element*. In the fields studied in (1.1) and in Exercise 2, the polynomials g were chosen so that their roots were primitive elements of the corresponding finite field. This will not be true for *all* choices of irreducible g of a given degree n in $\mathbb{F}_p[x]$.

Exercise 7. For instance, consider the polynomial $g = x^2 + 1$ in $\mathbb{F}_3[x]$. Check that g is irreducible, so that $\mathbb{K} = \mathbb{F}_3[x]/\langle g \rangle$ is a field with 9 elements. However the coset of x in $\mathbb{K}$ is not a primitive element. (Why not? what is its multiplicative order?)

For future reference, we also include the following fact about finite fields.

Exercise 8. Suppose that $\beta \in \mathbb{F}_{p^n}$ is neither 0 nor 1. Then show that $\sum_{j=0}^{p^n-2} \beta^j = 0$. Hint: What is $(x^{p^n-1} - 1)/(x - 1)$?

To conclude this section, we indicate one direct method for performing finite field arithmetic in Maple. Maple provides a built-in facility (via the mod operator) by which polynomial division, row operations on matrices, resultant computations, etc. can be done using coefficients in finite fields. When we construct a quotient ring $\mathbb{F}_p[x]/\langle g \rangle$ the coset of x becomes a root of the equation $g = 0$ in the quotient. In Maple, the elements of a finite field can be represented as (cosets of) polynomials in `RootOf` expressions (see Chapter 2, §1). For example, to declare the field $\mathbb{F}_8 = \mathbb{F}_2[x]/\langle x^3 + x + 1 \rangle$, we could use

$$\texttt{alias(alpha = RootOf(x\^{}3 + x + 1));}$$

Then polynomials in `alpha` represent elements of the field $\mathbb{F}_8$ as before. Arithmetic in the finite field can be performed as follows. For instance, suppose we want to compute $b^3 + b$, where $b = \alpha + 1$. Entering

```
b := alpha + 1;
Normal(b^3 + b) mod 2;
```

yields

$$alpha^2 + alpha + 1.$$

The `Normal` function computes the normal form for the element of the finite field by expanding out $b^3 + b$ as a polynomial in α, then finding the remainder on division by $\alpha^3 + \alpha + 1$, using coefficients in $\mathbb{F}_2$.

A technical point: You may have noticed that the `Normal` function name here is *capitalized*. There is also an uncapitalized `normal` function in Maple which can be used for algebraic simplification of expressions. We *do not* want that function here, however, because we want the function call to be passed *unevaluated* to mod, and all the arithmetic to be performed within the mod environment. Maple uses capitalized names consistently for unevaluated function calls in this situation. Using the command `normal(b^3 + b) mod 2` would instruct Maple to simplify $b^3 + b$, *then reduce mod 2*, which does not yield the correct result in this case. Try it!

ADDITIONAL EXERCISES FOR §1

Exercise 9. Verify the claim made in the proof of Proposition (1.2) that if $\mathbb{F}$ is a field with p^n elements, then $\mathbb{F}$ has a subfield

$$\mathbb{K} = \{0, 1, 2 \cdot 1, \ldots, (p-1) \cdot 1\}$$

isomorphic to $\mathbb{F}_p$.

Exercise 10. Using Theorem (1.9), show that $\mathbb{F}_{p^n}$ contains a subfield $\mathbb{F}_{p^m}$ if and only if m is a divisor of n. Hint: By (1.9), the multiplicative group of the subfield is a subgroup of the multiplicative cyclic group $\mathbb{F}_{p^m} \setminus \{0\}$. What are the orders of subgroups of a cyclic group of order $p^m - 1$?

Exercise 11. In this exercise, you will show that every finite field $\mathbb{F}$ can be obtained (up to isomorphism) as a quotient $\mathbb{F} \cong \mathbb{F}_p[x]/\langle g \rangle$ for some irreducible $g \in \mathbb{F}_p[x]$. For this exercise, we will need the fundamental theorem of ring homomorphisms (see e.g., [CLO] Chapter 5, §2, Exercise 16). Let $\mathbb{F}$ be a finite field, and say $|\mathbb{F}| = p^n$ (see Proposition (1.2)). Let α be a primitive element for $\mathbb{F}$ (see (1.9)). Consider the ring homomorphism defined by

$$\varphi : \mathbb{F}_p[x] \to \mathbb{F}$$
$$x \mapsto \alpha.$$

a. Explain why φ must be onto.
b. Deduce that the kernel of φ must have the form $\ker(\varphi) = \langle g \rangle$ for some irreducible monic polynomial $g \in k[x]$. (The monic generator is called the *minimal polynomial* of α over $\mathbb{F}_p$.)
c. Apply the fundamental theorem to show that

$$\mathbb{F} \cong \mathbb{F}_p[x]/\langle g \rangle.$$

Exercise 12. In this exercise, you will develop one proof of Theorem (1.8), using Theorem (1.9) and the previous exercise. Let $\mathbb{K}$ and $\mathbb{L}$ be two fields with p^n elements. Let β be a primitive element for $\mathbb{L}$, and let $g \in \mathbb{F}_p[x]$ be the minimal polynomial of β over $\mathbb{F}_p$, so that $\mathbb{L} \cong \mathbb{F}_p[x]/\langle g \rangle$ (Exercise 11).
a. Show that g must divide the polynomial $x^{p^n} - x$ in $\mathbb{F}_p[x]$. (Use (1.9).)
b. Show that $x^{p^n} - x$ splits completely into linear factors in $\mathbb{K}[x]$:

$$x^{p^n} - x = \prod_{\alpha \in \mathbb{K}} (x - \alpha).$$

c. Deduce that there is some $\alpha \in \mathbb{K}$ which is a root of $g = 0$.
d. From part c, deduce that $\mathbb{K}$ is *also* isomorphic to $\mathbb{F}_p[x]/\langle g \rangle$. Hence, $\mathbb{K} \cong \mathbb{L}$.

Exercise 13. Find irreducible polynomials g in the appropriate $\mathbb{F}_p[x]$, such that $\mathbb{F}_p[x]/\langle g \rangle \cong \mathbb{F}_{p^n}$, and such that $\alpha = [x]$ is a primitive element in $\mathbb{F}_{p^n}$ for each $p^n \le 64$. (Note: The cases $p^n = 8, 9, 16$ are considered in the text. Extensive tables of such polynomials have been constructed for use in coding theory. See for instance [PH].)

Exercise 14. (The Frobenius Automorphism) Let $\mathbb{F}_q$ be a finite field. By Exercise 10, $\mathbb{F}_q \subset \mathbb{F}_{q^m}$ for each $m \ge 1$. Consider the mapping $F : \mathbb{F}_{q^m} \to \mathbb{F}_{q^m}$ defined by $F(x) = x^q$.
a. Show that F is one-to-one and onto, and that $F(x + y) = F(x) + F(y)$ and $F(xy) = F(x)F(y)$ for all $x, y \in \mathbb{F}_{q^m}$. (In other words, F is an *automorphism* of the field $\mathbb{F}_{q^m}$.)
b. Show that $F(x) = x$ if and only if $x \in \mathbb{F}_q \subset \mathbb{F}_{q^m}$.

For readers familiar with Galois theory, we mention that the Frobenius automorphism F generates the Galois group of $\mathbb{F}_{q^m}$ over $\mathbb{F}_q$—a cyclic group of order m.

§2 Error-Correcting Codes

In this section, we will introduce some of the basic standard notions from algebraic coding theory. For more complete treatments of the subject, we refer the reader to [vLi], [Bla], or [MS].

Communication of information often takes place over "noisy" channels which can introduce errors in the transmitted message. This is the case for instance in satellite transmissions, in the transfer of information within computer systems, and in the process of storing information (numeric data, music, images, etc.) on tape, on compact disks or other media, and retrieving it for use at a later time. In these situations, it is desirable to *encode* the information in such a way that errors can be detected and/or corrected when they occur. The design of coding schemes, together with efficient techniques for encoding and decoding (i.e. recovering the original message from its encoded form) is one of the main goals of coding theory.

In some situations, it might also be desirable to encode information in such a way that unauthorized readers of the received message will not be able to decode it. The construction of codes for secrecy is the domain of *cryptography*, a related but distinct field that will *not* be considered here. Interestingly enough, ideas from number theory and algebraic geometry have assumed a major role there as well. The book [Kob] includes some applications of computational algebraic geometry in modern cryptography.

In this chapter, we will study one specific type of code, in which all information to be encoded consists of strings or *words* of some fixed length k using symbols from a fixed alphabet, and all encoded messages are divided into strings called *codewords* of a fixed block length n, using symbols from the same alphabet. In order to detect and/or correct errors, some *redundancy* must be introduced in the encoding process, hence we will always have $n > k$.

Because of the design of most electronic circuitry, it is natural to consider a binary alphabet consisting of the two symbols $\{0, 1\}$, and to identify the alphabet with the finite field $\mathbb{F}_2$. As in §1, strings of r bits (thought of as the coefficients in a polynomial of degree $r - 1$) can also represent elements of a field $\mathbb{F}_{2^r}$, and it will be advantageous in some cases to think of $\mathbb{F}_{2^r}$ as the alphabet. But the constructions we will present are valid with an arbitrary finite field $\mathbb{F}_q$ as the alphabet.

In mathematical terms, the encoding process for a string from the message will be a one-to-one function $E : \mathbb{F}_q^k \to \mathbb{F}_q^n$. The image $C = E(\mathbb{F}_q^k) \subset \mathbb{F}_q^n$ is referred to as the set of codewords, or more succinctly as *the code*. Mathematically, the decoding operation can be viewed as a function $D : \mathbb{F}_q^n \to \mathbb{F}_q^k$ such that $D \circ E$ is the identity on $\mathbb{F}_q^k$. (This is actually an over-simplification—most real-world decoding functions will also return something like an "error" value in certain situations.)

In principle, the set of codewords can be an arbitrary subset of $\mathbb{F}_q^n$. However, we will almost always restrict our attention to a class of codes with additional structure that is very convenient for encoding and decoding. This is the class of *linear codes*. By definition, a linear code is one where the set of codewords C forms a vector subspace of $\mathbb{F}_q^n$ of dimension k. In this case, as encoding function $E : \mathbb{F}_q^k \to \mathbb{F}_q^n$ we may use a linear mapping whose image is the subspace C. The matrix of E with respect to the standard bases in the domain and target is often called the *generator matrix* G corresponding to E.

It is customary in coding theory to write generator matrices for linear codes as $k \times n$ matrices and to view the strings in $\mathbb{F}_q^k$ as *row vectors* w. Then the encoding operation can be viewed as matrix multiplication of a row vector on the right by the generator matrix, and the rows of G form a basis for C. As always in linear algebra, the subspace C may also be described as the set of solutions of a system of $n - k$ independent linear equations in n variables. The matrix of coefficients of such a system is often called a *parity check matrix* for C. This name comes from the fact that one simple error-detection scheme for binary codes is to require that all codewords have an even (or odd) number of nonzero digits. If one bit error (in fact, any odd number of errors) is introduced in transmission, that fact can be recognized by multiplication of the received word by the parity check matrix $H = (1 \quad 1 \quad \cdots \quad 1)^T$. The parity check matrix for a linear code can be seen as an extension of this idea, in which more sophisticated tests for the validity of the received word are performed by multiplication by the parity check matrix.

Exercise 1. Consider the linear code C with $n = 4$, $k = 2$ given by the generator matrix

$$G = \begin{pmatrix} 1 & 1 & 1 & 1 \\ 1 & 0 & 1 & 0 \end{pmatrix}.$$

a. Show that since we have only the two scalars $0, 1 \in \mathbb{F}_2$ to use in making linear combinations, there are exactly four elements of C:

$$(0,0)G = (0,0,0,0), \qquad (1,0)G = (1,1,1,1),$$
$$(0,1)G = (1,0,1,0), \qquad (1,1)G = (0,1,0,1).$$

b. Show that

$$H = \begin{pmatrix} 1 & 1 \\ 1 & 0 \\ 1 & 1 \\ 1 & 0 \end{pmatrix}$$

   is a parity check matrix for C by verifying that $xH = 0$ for all $x \in C$.

Exercise 2. Let $\mathbb{F}_4 = \mathbb{F}_2[\alpha]/\langle \alpha^2 + \alpha + 1 \rangle$, and consider the linear code C in $\mathbb{F}_4^5$ with generator matrix

$$\begin{pmatrix} \alpha & 0 & \alpha+1 & 1 & 0 \\ 1 & 1 & \alpha & 0 & 1 \end{pmatrix}.$$

How many distinct codewords are there in C? Find them. Also find a parity check matrix for C. Hint: Recall from linear algebra that there is a general procedure using matrix operations for finding a system of linear equations defining a given subspace.

To study the error-correcting capability of codes, we need a measure of how close elements of $\mathbb{F}_q^n$ are, and for this we will use the *Hamming distance*. Let $x, y \in \mathbb{F}_q^n$. Then the Hamming distance between x and y is defined to be

$$d(x, y) = |\{i, 1 \le i \le n : x_i \ne y_i\}|.$$

For instance, if $x = (0, 0, 1, 1, 0)$, and $y = (1, 0, 1, 0, 0)$ in $\mathbb{F}_2^5$, then $d(x, y) = 2$ since only the first and fourth bits in x and y differ.

Let 0 denote the zero vector in $\mathbb{F}_q^n$ and let $x \in \mathbb{F}_q^n$ be arbitrary. Then $d(x, 0)$, the number of nonzero components in x, is called the *weight* of x and denoted $\mathrm{wt}(x)$.

Exercise 3. Verify that the Hamming distance has all the properties of a *metric* or *distance function* on $\mathbb{F}_q^n$. (That is, show $d(x, y) \ge 0$ for all x, y and $d(x, y) = 0$ if and only if $x = y$, the symmetry property $d(x, y) = d(y, x)$ holds for all x, y, and the triangle inequality $d(x, y) \le d(x, z) + d(z, y)$ is valid for all x, y, z.)

Given $x \in \mathbb{F}_q^n$, we will denote by $B_r(x)$ the closed ball of radius r (in the Hamming distance) centered at x:

$$B_r(x) = \{y \in \mathbb{F}_q^n : d(y, x) \le r\}.$$

(In other words, $B_r(x)$ is the set of y differing from x in at most r components.)

The Hamming distance gives a simple but extremely useful way to measure the error-correcting capability of a code. Namely, suppose that every pair of distinct codewords x, y in a code $C \subset \mathbb{F}_q^n$ satisfies $d(x, y) \ge d$ for some $d \ge 1$. If a codeword x is transmitted and errors are introduced, we can view the received word as $z = x + e$ for some nonzero error vector e. If $\mathrm{wt}(e) = d(x, z) \le d - 1$, then under our hypothesis z is not another codeword. Hence any error vector of weight at most $d - 1$ can be *detected*.

Moreover if $d \ge 2t + 1$ for some $t \ge 1$, then for any $z \in \mathbb{F}_q^n$, by the triangle inequality, $d(x, z) + d(z, y) \ge d(x, y) \ge 2t + 1$. It follows immediately that either $d(x, z) > t$ or $d(y, z) > t$, so $B_t(x) \cap B_t(y) = \emptyset$. As a result the only codeword in $B_t(x)$ is x itself. In other words, if an error vector of weight at most t is introduced in transmission of a codeword, those errors can be *corrected* by the *nearest neighbor decoding function*

$$D(x) = E^{-1}(c), \text{ where } c \in C \text{ minimizes } d(x, c)$$

(and an "error" value if there is no unique closest element in C).
From this discussion it is clear that the *minimum distance*

$$d = \min\{d(x, y) : x \neq y \in C\}$$

is an important parameter of codes, and our observations above can be summarized in the following way.

(2.1) Proposition. *Let C be a code with minimum distance d. All error vectors of weight $\leq d - 1$ can be detected. Moreover, if $d \geq 2t + 1$, then all error vectors of weight $\leq t$ can be corrected by nearest neighbor decoding.*

Since the minimum distance of a code contains so much information, it is convenient that for linear codes we need only examine the codewords themselves to determine this parameter.

Exercise 4. Show that for any linear code C the minimum distance d is the same as $\min_{x \in C \setminus \{0\}} |\{i : x_i \neq 0\}|$ (the minimum number of nonzero entries in a nonzero codeword). Hint: Since the set of codewords is closed under vector sums, $x - y \in C$ whenever x and y are.

The *Hamming codes* form a famous family of examples with interesting error-correcting capabilities. One code in the family is a code over $\mathbb{F}_2$ with $n = 7$, $k = 4$. (The others are considered in Exercise 11 below.) For this Hamming code, the generator matrix is

$$(2.2) \qquad G = \begin{pmatrix} 1 & 0 & 0 & 0 & 0 & 1 & 1 \\ 0 & 1 & 0 & 0 & 1 & 0 & 1 \\ 0 & 0 & 1 & 0 & 1 & 1 & 0 \\ 0 & 0 & 0 & 1 & 1 & 1 & 1 \end{pmatrix}.$$

For example $w = (1, 1, 0, 1) \in \mathbb{F}_2^4$ is encoded by multiplication on the right by G, yielding $E(w) = wG = (1, 1, 0, 1, 0, 0, 1)$. From the form of the first four columns of G, the first four components of $E(w)$ will always consist of the four components of w itself.

The reader should check that the 7×3 matrix

$$(2.3) \qquad H = \begin{pmatrix} 0 & 1 & 1 \\ 1 & 0 & 1 \\ 1 & 1 & 0 \\ 1 & 1 & 1 \\ 1 & 0 & 0 \\ 0 & 1 & 0 \\ 0 & 0 & 1 \end{pmatrix}.$$

has rank 3 and satisfies $GH = 0$. Hence H is a parity check matrix for the Hamming code (why?). It is easy to check directly that each of the 15 nonzero codewords of the Hamming code contains at least 3 nonzero components. This implies that $d(x, y)$ is at least 3 when $x \neq y$. Hence the minimum distance of the Hamming code is $d = 3$, since there are exactly three nonzero entries in row 1 of G for example. By Proposition (2.1), any error vector of weight 2 or less can be detected, and any error vector of weight 1 can be corrected by nearest neighbor decoding. The following exercise gives another interesting property of the Hamming code.

Exercise 5. Show that the balls of radius 1 centered at each of the words of the Hamming code are pairwise disjoint, and cover $\mathbb{F}_2^7$ completely. (A code C with minimum distance $d = 2t + 1$ is called a *perfect* code if the union of the balls of radius t centered at the codewords equals $\mathbb{F}_q^n$.)

Generalizing a property of the generator matrix (2.2) noted above, encoding functions with the property that the symbols of the input word appear unchanged in some components of the codeword are known as *systematic encoders*. It is customary to call those components of the codewords the *information positions*. The remaining components of the codewords are called *parity checks*. Systematic encoders are sometimes desirable from a practical point of view because the information positions can be copied directly from the word to be encoded; only the parity checks need to be computed. There are corresponding savings in the decoding operation as well. If information is systematically encoded and no errors occur in transmission, the words in the message can be obtained directly from the received words by simply removing the parity checks. (It is perhaps worthwhile to mention again at this point that the goal of the encoding schemes we are considering here is *reliability* of information transmission, not secrecy!)

Exercise 6. Suppose that the generator matrix for a linear code C has the systematic form $G = (I_k \mid P)$, where I_k is a $k \times k$ identity matrix, and P is some $k \times (n - k)$ matrix. Show that

$$H = \begin{pmatrix} -P \\ I_{n-k} \end{pmatrix}$$

is a parity check matrix for C.

We will refer to a linear code with block length n, dimension k, and minimum distance d as an $[n, k, d]$ code. For instance, the Hamming code given by the generator matrix (2.2) is a $[7,4,3]$ code.

Determining which triples of parameters $[n, k, d]$ can be realized by codes over a given finite field $\mathbb{F}_q$ and constructing such codes are two important problems in coding theory. These questions are directly motivated by the decisions an engineer would need to make in selecting a code for a given application. Since an $[n, k, d]$ code has q^k distinct codewords, the choice of the parameter k will be determined by the size of the collection of words appearing in the messages to be transmitted. Based on the characteristics of the channel over which the transmission takes place (in particular the probability that an error occurs in transmission of a symbol), a value of d would be chosen to ensure that the probability of receiving a word that could not be correctly decoded was acceptably small. The remaining question would be how big to take n to ensure that a code with the desired parameters k and d actually exists. It is easy to see that, fixing k, we can construct codes with d as large as we like by taking n very large. (For instance, our codewords could consist of many concatenated copies of the corresponding words in $\mathbb{F}_q^k$.) However, the resulting codes would usually be too redundant to be practically useful. "Good" codes are ones for which the *information rate $R = k/n$* is not too small, but for which d is relatively large. There is a famous result known as Shannon's Theorem (for the precise statement see, e.g., [vLi]) that ensures the existence of "good" codes in this sense, but the actual construction of "good" codes is one of the main problems in coding theory.

Exercise 7. In the following exercises, we explore some theoretical results giving various bounds on the parameters of codes. One way to try to produce good codes is to fix a block length n and a minimum distance d, then attempt to maximize k by choosing the codewords one by one so as to keep $d(x, y) \geq d$ for all distinct pairs $x \neq y$.

a. Show that $b = |B_{d-1}(c)|$ is given by $b = \sum_{i=0}^{d-1} \binom{n}{i}(q-1)^i$ for each $c \in \mathbb{F}_q^n$.

b. Let d be a given positive integer, and let C be a subset $C \subset \mathbb{F}_q^n$ (not necessarily a linear code) such that $d(x, y) \geq d$ for all pairs $x \neq y$ in C. Assume that for all $z \in \mathbb{F}_q^n \setminus C$, $d(z, c) \leq d - 1$ for some $c \in C$. Then show that $b \cdot |C| \geq q^n$ (b as in part a). This result gives one form of the *Gilbert-Varshamov bound*. Hint: An equivalent statement is that if $b \cdot |C| < q^n$, then there exists some z such that every pair of distinct elements in $C \cup \{z\}$ is still separated by at least d.

c. Show that if k satisfies $b < q^{n-k+1}$, then an $[n, k, d]$ linear code exists. Hint: By induction, we may assume that an $[n, k-1, d]$ linear code C exists. Using part b, consider the linear code C' spanned by C and z, where the distance from z to any word in C is $\geq d$. Show that C' still has minimum distance d.

d. On the other hand, show that for any linear code $d \leq n - k + 1$. This result is known as the *Singleton bound*. Hint: Consider what happens when a subset of $d-1$ components is deleted from each of the codewords.

Many other theoretical results, including both upper and lower bounds on the n, k, d parameters of codes, are also known. See the coding theory texts mentioned at the start of this section.

We now turn to the encoding and decoding operations. Our first observation is that encoding is much simpler to perform for linear codes than for arbitrary codes. For a completely arbitrary C of size q^k there would be little alternative to using some form of table look-up to compute the encoding function. On the other hand, for a linear code all the information about the code necessary for encoding is contained in the generator matrix (only k basis vectors for C rather than the whole set of q^k codewords), and all operations necessary for encoding may be performed using linear algebra.

Decoding a linear code is also correspondingly simpler. A general method, known as *syndrome decoding*, is based on the following observation. If $c = wG$ is a codeword, and some errors $e \in \mathbb{F}_q^n$ are introduced on transmission of c, the received word will be $x = c + e$. Then $cH = 0$ implies that $xH = (c+e)H = cH + eH = 0 + eH = eH$. Hence xH depends only on the error. The possible values for $eH \in \mathbb{F}_q^{n-k}$ are known as *syndromes*, and it is easy to see that the syndromes are in one-to-one correspondence with the cosets of C in $\mathbb{F}_q^n$ (or elements of the quotient space $\mathbb{F}_q^n/C \cong \mathbb{F}_q^{n-k}$), so there are exactly q^{n-k} of them. (See Exercise 12 below.)

Syndrome decoding works as follows. First, a preliminary calculation is performed, before any decoding. We construct a table, indexed by the possible values of the syndrome $s = xH$, of the element(s) in the corresponding coset with the smallest number of nonzero entries. These special elements of the cosets of C are called the *coset leaders*.

Exercise 8. Say $d = 2l + 1$, so we know that any error vector of weight t or less can be corrected. Show that if there are any elements of a coset of C which have t or fewer nonzero entries, then there is only one such element, and as a result the coset leader is unique.

If $x \in \mathbb{F}_q^n$ is received, we first compute the syndrome $s = xH$ and look up the coset leader(s) ℓ corresponding to s in our table. If there is a unique leader, we replace x by $x' = x - \ell$, which is in C (why?). (If $s = 0$, then $\ell = 0$, and $x' = x$ is itself a codeword.) Otherwise, we report an "error" value. By Exercise 8, if no more than t errors occurred in x, then we have found the unique codeword closest to the received word x and we return $E^{-1}(x')$. Note that by this method we have accomplished nearest neighbor decoding *without* computing $d(x, c)$ for all q^k codewords. However, a potentially large collection of information must be maintained to carry out this procedure—the table of coset leader(s) for each of the q^{n-k} cosets of C. In cases of practical interest, $n - k$ and q can be large, so q^{n-k} can be huge.

Exercise 9. Compute the table of coset leaders for the [7,4,3] Hamming code from (2.2). Use syndrome decoding to decode the received word $(1, 1, 0, 1, 1, 1, 0)$.

Here is another example of a linear code, this time over the field $\mathbb{F}_4 = \mathbb{F}_2[\alpha]/\langle \alpha^2 + \alpha + 1 \rangle$. Consider the code C with $n = 8$, $k = 3$ over $\mathbb{F}_4$ defined by the generator matrix:

$$(2.4) \qquad G = \begin{pmatrix} 1 & 1 & 1 & 1 & 1 & 1 & 1 & 1 \\ 0 & 0 & 1 & 1 & \alpha & \alpha & \alpha^2 & \alpha^2 \\ 0 & 1 & \alpha & \alpha^2 & \alpha & \alpha^2 & \alpha & \alpha^2 \end{pmatrix}.$$

Note that G does not have the systematic form we saw above for the Hamming code's generator matrix. Though this is not an impediment to encoding, we can also obtain a systematic generator matrix for C by row-reduction (Gauss-Jordan elimination). This corresponds to changing basis in C; the image of the encoding map $E : \mathbb{F}_4^3 \to \mathbb{F}_4^8$ is not changed. It is a good exercise in finite field arithmetic to perform this computation by hand. It can also be done in Maple as follows. For simplicity, we will write a for α within Maple. To work in $\mathbb{F}_4$ we begin by defining a as a root of the polynomial $x^2 + x + 1$ as above.

```
alias(a=RootOf(x^2+x+1)):
```

The generator matrix G is entered as

```
m :=array(1..3, 1..8, [[1, 1, 1, 1, 1, 1, 1, 1],
     [0, 0, 1, 1, a, a, a^2, a^2], [0, 1, a, a^2, a, a^2, a, a^2]]) :
```

Then the command

```
mr := Gaussjord(m) mod 2;
```

will perform Gauss-Jordan elimination with coefficients treated as elements of $\mathbb{F}_4$. (Recall Maple's capitalization convention for unevaluated function calls, discussed in §1.) The result should be

$$\begin{pmatrix} 1 & 0 & 0 & 1 & a & a+1 & 1 & 0 \\ 0 & 1 & 0 & 1 & 1 & 0 & a+1 & a \\ 0 & 0 & 1 & 1 & a & a & a+1 & a+1 \end{pmatrix}.$$

Note that a^2 is replaced by its reduced form $a + 1$ everywhere here.

In the reduced matrix, the second row has five nonzero entries. Hence the minimum distance d for this code is ≤ 5. By computing all $4^3 - 1 = 63$ nonzero codewords, it can be seen that $d = 5$. It is often quite difficult to determine the exact minimum distance of a code (especially when the number of nonzero codewords, $q^k - 1$, is large).

To conclude this section, we will develop a relationship between the minimum distance of a linear code and the form of parity check matrices for the code.

(2.5) Proposition. *Let C be a linear code with parity check matrix H. If no collection of $\delta - 1$ distinct rows of H is a linearly dependent subset of $\mathbb{F}_q^{n-k}$, then the minimum distance d of C satisfies $d \geq \delta$.*

PROOF. We use the result of Exercise 4. Let $x \in C$ be a nonzero codeword. From the equation $xH = 0$ in $\mathbb{F}_q^{n-k}$, we see that the components of x are the coefficients in a linear combination of the rows of H summing to the zero vector. If no collection of $\delta - 1$ distinct rows is linearly dependent, then x must have at least δ nonzero entries. Hence $d \geq \delta$. $\square$

ADDITIONAL EXERCISES FOR §2

Exercise 10. Consider the formal inner product on $\mathbb{F}_q^n$ defined by

$$\langle x, y \rangle = \sum_{i=1}^n x_i y_i$$

(a bilinear mapping from $\mathbb{F}_q^n \times \mathbb{F}_q^n$ to $\mathbb{F}_q$; there is no notion of positive-definiteness in this context). Given a linear code C, let

$$C^\perp = \{x \in \mathbb{F}_q^n : \langle x, y \rangle = 0 \text{ for all } y \in C\},$$

the subspace of $\mathbb{F}_q^n$ orthogonal to C. If C is k-dimensional, then $C^\perp$ is a linear code of block length n and dimension $n - k$ known as the *dual code* of C.

a. Let $G = (I_k \mid P)$ be a systematic generator matrix for C. Determine a generator matrix for $C^\perp$. How is this related to the parity check matrix for C? (Note on terminology: Many coding theory texts define a parity check matrix for a linear code to be the transpose of what we are calling a parity check matrix. This is done so that the rows of a parity check matrix will form a basis for the dual code.)
b. Find generator matrices and determine the parameters $[n, k, d]$ for the duals of the Hamming code from (2.2), and the code from (2.4).

Exercise 11. (The Hamming codes) Let q be a prime power, and let $m \geq 1$. We will call a set S of vectors in $\mathbb{F}_q^m$ a *maximal pairwise linearly independent subset* of $\mathbb{F}_q^m$ if S has the property that no two distinct elements of S are scalar multiples of each other, and if S is maximal with respect to inclusion. For each pair (q, m) we can construct linear codes C by taking a parity check matrix $H \in M_{n \times m}(\mathbb{F}_q)$ whose rows form a maximal pairwise linearly independent subset of $\mathbb{F}_q^m$, and letting $C \subset \mathbb{F}_q^n$ be the set of solutions of the system of linear equations $xH = 0$. For instance, with $q = 2$, we can take the rows of H to be all the nonzero vectors in $\mathbb{F}_2^k$ (in any order)—see (2.3) for the case $q = 2, k = 3$. The codes with these parity check matrices are called the Hamming codes.
a. Show that if S is a maximal pairwise linearly independent subset of $\mathbb{F}_q^m$, then S has exactly $(q^m - 1)/(q - 1)$ elements. (This is the same as the number of points of the projective space $\mathbb{P}^{m-1}$ over $\mathbb{F}_q$.)
b. What is the dimension k of a Hamming code defined by an $n \times m$ matrix H?

c. Write down a parity check matrix for a Hamming code with $q = 3$, $k = 2$.

d. Show that the minimum distance of a Hamming code is always 3, and discuss the error-detecting and error-correcting capabilities of these codes.

e. Show that all the Hamming codes are *perfect* codes (see Exercise 5 above).

Exercise 12. Let C be an $[n, k, d]$ linear code with parity check matrix H. Show that the possible values for $yH \in \mathbb{F}_q^{n-k}$ (the *syndromes*) are in one-to-one correspondence with the cosets of C in $\mathbb{F}_q^n$ (or elements of the quotient space $\mathbb{F}_q^n/C \cong \mathbb{F}_q^{n-k}$). Deduce that there are q^{n-k} different syndrome values.

§3 Cyclic Codes

In this section, we will consider several classes of linear codes with even more structure, and we will see how some of the algorithmic techniques in symbolic algebra we have developed can be applied to encode them. First we will consider the class of cyclic codes. Cyclic codes may be defined in several ways—the most elementary is certainly the following: A *cyclic code* is a linear code with the property that the set of codewords is closed under cyclic permutations of the components of vectors in $\mathbb{F}_q^n$. Here is a simple example.

In $\mathbb{F}_2^4$, consider the $[4, 2, 2]$ code C with generator matrix

$$(3.1) \qquad\qquad G = \begin{pmatrix} 1 & 1 & 1 & 1 \\ 1 & 0 & 1 & 0 \end{pmatrix}$$

from Exercise 1 in §2. As we saw there, C contains 4 distinct codewords. The codewords $(0, 0, 0, 0)$ and $(1, 1, 1, 1)$ are themselves invariant under all cyclic permutations. The codeword $(1, 0, 1, 0)$ is not itself invariant: shifting one place to the left (or right) we obtain $(0, 1, 0, 1)$. But this *is* another codeword: $(0, 1, 0, 1) = (1, 1)G \in C$. Similarly, shifting $(0, 1, 0, 1)$ one place to the left or right, we obtain the codeword $(1, 0, 1, 0)$ again. It follows that the set C is closed under all cyclic shifts.

The property of invariance under cyclic permutations of the components has an interesting algebraic interpretation. Using the standard isomorphism between $\mathbb{F}_q^n$ and the vector space of polynomials of degree at most $n - 1$ with coefficients in $\mathbb{F}_q$:

$$(a_0, a_1, \ldots, a_{n-1}) \leftrightarrow a_0 + a_1 x + \cdots + a_{n-1} x^{n-1}$$

we may identify a cyclic code C with the corresponding collection of polynomials of degree $n-1$. The right cyclic shift which sends $(a_0, a_1, \ldots, a_{n-1})$ to $(a_{n-1}, a_0, a_1, \ldots, a_{n-2})$ is the same as the result of multiplying the polynomial $a_0 + a_1x + \cdots + a_{n-1}x^{n-1}$ by x, then taking the remainder on division by $x^n - 1$.

Exercise 1. Show that multiplying the polynomial $p(x) = a_0 + a_1x + \cdots + a_{n-1}x^{n-1}$ by x, then taking the remainder on division by $x^n - 1$ yields a polynomial whose coefficients are the same as those of $p(x)$, but cyclically shifted one place to the right.

This suggests that when dealing with cyclic codes we should consider the polynomials of degree at most $n - 1$ as the elements of the quotient ring $R = \mathbb{F}_q[x]/\langle x^n - 1 \rangle$. The reason is that multiplication of $f(x)$ by x followed by division gives the standard representative for the product $xf(x)$ in R. Hence, from now on we will consider cyclic codes as a vector subspaces of the ring R which are closed under multiplication by the coset of x in R. Now we make a key observation.

Exercise 2. Show that if a vector subspace $C \subset R$ is closed under multiplication by $[x]$, then it is closed under multiplication by *every* coset $[h(x)] \in R$.

Exercise 2 shows that cyclic codes have the defining property of ideals in a ring. We record this fact in the following proposition.

(3.2) Proposition. *Let $R = \mathbb{F}_q[x]/\langle x^n - 1 \rangle$. A vector subspace $C \subset R$ is a cyclic code if and only if C is an ideal in the ring R.*

The ring R shares a nice property with its "parent" ring $\mathbb{F}_q[x]$.

(3.3) Proposition. *Each ideal $I \subset R$ is principal, generated by the coset of a single polynomial g of degree $n - 1$ or less. Moreover, g is a divisor of $x^n - 1$ in $\mathbb{F}_q[x]$.*

PROOF. By the standard characterization of ideals in a quotient ring (see e.g. [CLO] Chapter 5, §2, Proposition 10), the ideals in R are in one-to-one correspondence with the ideals in $\mathbb{F}_q[x]$ containing $\langle x^n - 1 \rangle$. Let J be the ideal corresponding to I. Since all ideals in $\mathbb{F}_q[x]$ are principal, J must be generated by some $g(x)$. Since $x^n - 1$ is in J, $g(x)$ is a divisor of $x^n - 1$ in $\mathbb{F}_q[x]$. The ideal $I = J/\langle x^n - 1 \rangle$ is generated by the coset of $g(x)$ in R. $\square$

Naturally enough, the polynomial g in Proposition (3.3) is called a *generator polynomial* for the cyclic code.

Exercise 3. Identifying the 4-tuple $(a, b, c, d) \in \mathbb{F}_2^4$ with $[a + bx + cx^2 + dx^3] \in R = \mathbb{F}_2[x]/\langle x^4 - 1 \rangle$, show that the cyclic code in $\mathbb{F}_2^4$ with generator matrix (3.1) can be viewed as the ideal generated by the coset of $g = 1 + x^2$ in R. Find the codewords of the cyclic code with generator $1 + x$ in R.

The *Reed-Solomon codes* are one particularly interesting class of cyclic codes used extensively in applications. For example, a clever combination of two of these codes is used for error control in playback of sound recordings in the Compact Disc audio system developed by Philips in the early 1980's. They are attractive because they have good *burst error* correcting capabilities (see Exercise 15 below) and also because efficient decoding algorithms are available for them (see the next section). We will begin with a description of these codes via generator matrices, then show that they have the invariance property under cyclic shifts.

Choose a finite field $\mathbb{F}_q$ and consider codes of block length $n = q - 1$ constructed in the following way. Let α be a primitive element for $\mathbb{F}_q$ (see Theorem (1.9) of this chapter), fix $k < q$, and let $L_{k-1} = \{\sum_{i=0}^{k-1} a_i t^i :$ $a_i \in \mathbb{F}_q\}$ be the vector space of polynomials of degree at most $k - 1 < q - 1$ in $\mathbb{F}_q[t]$. We make words in $\mathbb{F}_q^{q-1}$ by *evaluating* polynomials in L_{k-1} at the $q - 1$ nonzero elements of $\mathbb{F}_q$. By definition

$$(3.4) \qquad C = \{(f(1), f(\alpha), \ldots, f(\alpha^{q-2})) \in \mathbb{F}_q^{q-1} : f \in L_{k-1}\}$$

is a Reed-Solomon code, sometimes denoted by $RS(k, q)$. C is a vector subspace of $\mathbb{F}_q^{q-1}$ since it is the image of the vector space L_{k-1} under the linear evaluation mapping

$$f \mapsto (f(1), f(\alpha), \ldots, f(\alpha^{q-2})).$$

Generator matrices for Reed-Solomon codes can be obtained by taking any basis of L_{k-1} and evaluating to form the corresponding codewords. The monomial basis $\{1, t, t^2, \ldots, t^{k-1}\}$ is the simplest. For example, consider the Reed-Solomon code over $\mathbb{F}_9$ with $k = 3$. Using the basis $\{1, t, t^2\}$ for L_3, we obtain the generator matrix

$$(3.5) \qquad G = \begin{pmatrix} 1 & 1 & 1 & 1 & 1 & 1 & 1 & 1 \\ 1 & \alpha & \alpha^2 & \alpha^3 & \alpha^4 & \alpha^5 & \alpha^6 & \alpha^7 \\ 1 & \alpha^2 & \alpha^4 & \alpha^6 & 1 & \alpha^2 & \alpha^4 & \alpha^6 \end{pmatrix},$$

where the first row gives the values of $f(t) = 1$, the second row gives the values of $f(t) = t$, and the third gives the values of $f(t) = t^2$ at the nonzero elements of $\mathbb{F}_9$ (recall, $\alpha^8 = 1$ in $\mathbb{F}_9$). For all $k < q$, the first k columns of the generator matrix corresponding to the monomial basis of L_{k-1} give a submatrix of Vandermonde form with nonzero determinant. It follows that the evaluation mapping is one-to-one, and the corresponding Reed-Solomon code is a linear code with block length $n = q - 1$, and dimension $k = \dim L_{k-1}$.

The generator matrix formed using the monomial basis of L_{k-1} also brings the cyclic nature of Reed-Solomon codes into sharp focus. Observe that each cyclic shift of a row of the matrix G in (3.5) yields a scalar multiple of the same row. For example, cyclically shifting the third row one space to the right, we obtain

$$(\alpha^6, 1, \alpha^2, \alpha^4, \alpha^6, 1, \alpha^2, \alpha^4) = \alpha^6 \cdot (1, \alpha^2, \alpha^4, \alpha^6, 1, \alpha^2, \alpha^4, \alpha^6).$$

Exercise 4. Show that the other rows of (3.5) also have the property that a cyclic shift takes the row to a scalar multiple of the same row. Show that this observation implies this Reed-Solomon code is cyclic. Then generalize your arguments to *all* Reed-Solomon codes. Hint: Use the original definition of cyclic codes—closure under all cyclic shifts. You may wish to begin by showing that the cyclic shifts are linear mappings on $\mathbb{F}_q^n$.

We will give another proof that Reed-Solomon codes are cyclic below, and also indicate how to find the generator polynomial. However, we pause at this point to note one of the other interesting properties of Reed-Solomon codes. Since no polynomial in L_{k-1} can have more than $k-1$ zeroes in $\mathbb{F}_q$, every codeword in C has at least $(q-1) - (k-1) = q-k$ nonzero components (and some have exactly this many). By Exercise 4 of §2, the minimum distance for a Reed-Solomon code is $d = q - k = n - k + 1$. Comparing this with the Singleton bound from part d of Exercise 7 from §2, we see that Reed-Solomon codes have the maximum possible d for the block length $q - 1$ and dimension k. Codes with this property are called *MDS (maximum distance separable) codes* in the literature. So Reed-Solomon codes are good in this sense. However, their fixed, small block length relative to the size of the alphabet is sometimes a disadvantage. There is a larger class of cyclic codes known as *BCH codes* which contain the Reed-Solomon codes as a special case, but which do not have this limitation. Moreover, a reasonably simple lower bound on d is known for all BCH codes. See Exercise 13 below and [MS] or [vLi] for more on BCH codes.

Next, we will see another way to show that Reed-Solomon codes are cyclic that involves somewhat more machinery, but sheds additional light on the structure of cyclic codes of block length $q-1$ in general. Recall from Proposition (3.3) that the generator polynomial of a cyclic code of block length $q - 1$ is a divisor of $x^{q-1} - 1$. By Lagrange's Theorem, each of the $q - 1$ nonzero elements of $\mathbb{F}_q$ is a root of $x^{q-1} - 1 = 0$, hence

$$x^{q-1} - 1 = \prod_{\beta \in \mathbb{F}_q^*} (x - \beta)$$

in $\mathbb{F}_q[x]$, where $\mathbb{F}_q^*$ is the set of nonzero elements of $\mathbb{F}_q$. Consequently, the divisors of $x^{q-1} - 1$ are precisely the polynomials of the form $\prod_{\beta \in S}(x - \beta)$ for subsets $S \subset \mathbb{F}_q^*$. This is the basis for another characterization of cyclic codes.

Exercise 5. Show that a linear code of dimension k in $R = \mathbb{F}_q[x]/\langle x^{q-1} - 1 \rangle$ is cyclic if and only if the codewords, viewed as polynomials of degree at most $q - 2$, have some set S of $q - k - 1$ common zeroes in $\mathbb{F}_q^*$. Hint: If the codewords have the elements in S as common zeroes, then each codeword is divisible by $g(x) = \prod_{\beta \in S} (x - \beta)$.

Using this exercise, we will now determine the generator polynomial of a Reed-Solomon code. Let $f(t) = \sum_{j=0}^{k-1} a_j t^j$ be an element of L_{k-1}. Consider the values $c_i = f(\alpha^i)$ for $i = 0, \ldots, q - 2$. Using the c_i as the coefficients of a polynomial as in the discussion leading up to Proposition (3.2), write the corresponding codeword as $c(x) = \sum_{i=0}^{q-2} c_i x^i$. But then substituting for c_i and interchanging the order of summation, we obtain

$$
\begin{aligned}
c(\alpha^\ell) &= \sum_{i=0}^{q-2} c_i \alpha^{i\ell} \\
&= \sum_{j=0}^{k-1} a_j \left(\sum_{i=0}^{q-2} \alpha^{i(\ell+j)} \right).
\end{aligned}
$$

(3.6)

Assume that $1 \le \ell \le q - k - 1$. Then for all $0 \le j \le k - 1$, we have $1 \le \ell + j \le q - 2$. By the result of Exercise 8 of §1, each of the inner sums on the right is zero so $c(\alpha^\ell) = 0$. Using Exercise 5, we have obtained another proof of the fact that Reed-Solomon codes are cyclic, since the codewords have the set of common zeroes $S = \{\alpha, \alpha^2, \ldots, \alpha^{q-k-1}\}$. Moreover, we have the following result.

(3.7) Proposition. *Let C be the Reed-Solomon code of dimension k and minimum distance $d = q - k$ over $\mathbb{F}_q$. Then the generator polynomial of C has the form*

$$
g = (x - \alpha) \cdots (x - \alpha^{q-k-1}) = (x - \alpha) \cdots (x - \alpha^{d-1}).
$$

For example, the Reed-Solomon codewords corresponding to the three rows of the matrix G in (3.5) above are $c_1 = 1 + x + x^2 + \cdots + x^7$, $c_2 = 1 + \alpha x + \alpha^2 x^2 + \cdots + \alpha^7 x^7$, and $c_3 = 1 + \alpha^2 x + \alpha^4 x^2 + \alpha^6 x^4 + \cdots + \alpha^6 x^7$. Using Exercise 8 of §1, it is not difficult to see that the common roots of $c_1(x) = c_2(x) = c_3(x) = 0$ in $\mathbb{F}_9$ are $x = \alpha, \alpha^2, \ldots, \alpha^5$, so the generator polynomial for this code is

$$
g = (x - \alpha)(x - \alpha^2)(x - \alpha^3)(x - \alpha^4)(x - \alpha^5).
$$

Also see Exercise 11 below for another point of view on Reed-Solomon and related codes.

From the result of Proposition (3.2), it is natural to consider the following generalization of the cyclic codes described above. Let R be a quotient ring of $\mathbb{F}_q[x_1, \ldots, x_m]$ of the form

$$
R = \mathbb{F}_q[x_1, \ldots, x_m] / \langle x_1^{n_1} - 1, \ldots, x_m^{n_m} - 1 \rangle
$$

for some $n_1, \ldots, n_m$. Any ideal I in R will be a linear code closed under products by arbitrary $h(x_1, \ldots, x_n)$ in R. We will call any code obtained in this way an *m-dimensional cyclic code*.

Note first that $\mathcal{H} = \{x_1^{n_1} - 1, \ldots, x_m^{n_m} - 1\}$ is a Gröbner basis for the ideal it generates, with respect to all monomial orders. (This follows for instance from Theorem 3 and Proposition 4 of Chapter 2, §9 of [CLO].) Hence standard representatives for elements of R can be computed by applying the division algorithm in $\mathbb{F}_q[x_1, \ldots, x_m]$ and computing remainders with respect to $\mathcal{H}$. We obtain in this way as representatives of elements of R all polynomials whose degree in x_i is $n_i - 1$ or less for each i.

Exercise 6. Show that as a vector space,

$$R = \mathbb{F}_q[x_1, \ldots, x_m]/\langle x_1^{n_1} - 1, \ldots, x_m^{n_m} - 1 \rangle \cong \mathbb{F}_q^{n_1 \cdot n_2 \cdots n_m}.$$

Multiplication of an element of R by x_1, for example, can be viewed as a sort of cyclic shift in one of the variables. Namely, writing a codeword $c(x_1, \ldots, x_n) \in I$ as a polynomial in x_1, whose coefficients are polynomials in the other variables: $c = \sum_{j=0}^{n_1-1} c_j(x_2, \ldots, x_n)x_1^j$, multiplication by x_1, followed by division by $\mathcal{H}$ yields the standard representative $x_1 c = c_{n_1-1} + c_0 x_1 + c_1 x_1^2 + \cdots + c_{n_1-2}x_1^{n_1-1}$. Since $c \in I$ this shifted polynomial is also a codeword. The same is true for each of the other variables $x_2, \ldots, x_m$.

In the case $m = 2$, for instance, it is customary to think of the codewords of a 2-dimensional cyclic code either as polynomials in two variables, or as $n_1 \times n_2$ matrices of coefficients. In the matrix interpretation, multiplication by x_1 then corresponds to the right cyclic shift on each row, while multiplication by x_2 corresponds to a cyclic shift on each of the columns. Each of these operations leaves the set of codewords invariant.

Exercise 7. Writing $\mathbb{F}_4 = \mathbb{F}_2[\alpha]/\langle \alpha^2 + \alpha + 1 \rangle$, the ideal $I \subset \mathbb{F}_4[x, y]/\langle x^3 - 1, y^3 - 1 \rangle$ generated by $g_1(x, y) = x^2 + \alpha^2 xy + \alpha y$, $g_2(x, y) = y + 1$ gives an example of a 2-dimensional cyclic code with $n = 3^2 = 9$. As an exercise, determine k, the vector space dimension of this 2-dimensional cyclic code, by determining a vector space basis for I over $\mathbb{F}_4$. (Answer: $k = 7$. Also see the discussion following Theorem (3.9) below.) The minimum distance of this code is $d = 2$. Do you see why?

To define an m-dimensional cyclic code, it suffices to give a set of generators $\{[f_1], \ldots, [f_s]\} \subset R$ for the ideal $I \subset R$. The corresponding ideal J in $\mathbb{F}_q[x_1, \ldots, x_m]$ is

$$J = \langle f_1, \ldots, f_s \rangle + \langle x_1^{n_1-1} - 1, \ldots, x_m^{n_m-1} - 1 \rangle.$$

Fix any monomial order on $\mathbb{F}_q[x_1, \ldots, x_m]$. With a Gröbner basis $G = \{g_1, \ldots, g_t\}$ for J with respect to this order we have everything necessary to determine whether a given element of R is in I using the division algorithm in $\mathbb{F}_q[x_1, \ldots, x_m]$.

(3.8) Proposition. *Let R, I, J, G be as above. A polynomial $h(x_1, \ldots, x_n)$ represents an element of I in R if and only if its remainder on division by G is zero.*

PROOF. This follows because $I = J/\langle x_1^{n_1-1} - 1, \ldots, x_m^{n_m-1} - 1 \rangle$ and standard isomorphism theorems (see Theorem 2.6 of [Jac]) give a ring isomorphism

$$R/I \cong \mathbb{F}_q[x_1, \ldots, x_m]/J.$$

See Exercise 14 below for the details.   □

An immediate consequence of Proposition (3.8) is the following systematic encoding algorithm for m-dimensional cyclic codes using division with respect to a Gröbner basis. One of the advantages of m-dimensional cyclic codes over linear codes in general is that their extra structure allows a very compact representation of the encoding function. We only need to know a reduced Gröbner basis for the ideal J corresponding to a cyclic code to perform systematic encoding. A Gröbner basis will generally have fewer elements than a vector space basis of I. This frequently means that much less information needs to be stored. In the following description of a systematic encoder, the *information positions* of a codeword will refer to the k positions in the codeword that duplicate the components of the element of $\mathbb{F}_q^k$ that is being encoded. These will correspond to a certain subset of the coefficients in a polynomial representative for an element of R. Similarly, the parity check positions are the complementary collection of coefficients.

(3.9) Theorem. *Let $I \subset R = \mathbb{F}_q[x_1, \ldots, x_m]/\langle x_1^{n_1} - 1, \ldots, x_m^{n_m} - 1 \rangle$ be an m-dimensional cyclic code, and let G be a Gröbner basis for the corresponding ideal $J \subset \mathbb{F}_q[x_1, \ldots, x_m]$ with respect to some monomial order. Then there is a systematic encoding function for I constructed as follows.*

a. *The information positions are the coefficients of the nonstandard monomials for J in which each x_i appears to a power at most $n_i - 1$. (Non-standard monomials are monomials in $\langle \mathrm{LT}(J) \rangle$.)*

b. *The parity check positions are the coefficients of the standard monomials. (The standard monomials are those not contained in $\langle \mathrm{LT}(J) \rangle$.)*

c. *The following algorithm gives a systematic encoder E for I:*

> Input: the Gröbner basis G for J,
>
> w, a linear combination of nonstandard monomials
>
> Output: $E(w) \in I$

Uses: Division algorithm with respect to given order

$$\overline{w} := \overline{w}^G \qquad \text{(the remainder on division)}$$

$$E(w) := w - \overline{w}$$

PROOF. The dimension of R/I as a vector space over $\mathbb{F}_q$ is equal to the number of standard monomials for J since $R/I \cong \mathbb{F}_q[x_1, \ldots, x_m]/J$. (See for instance Proposition 4 from Chapter 5, §3 of [CLO].) The dimension of I as a vector space over $\mathbb{F}_q$ is equal to the difference $\dim R - \dim R/I$. But this is the same as the number of nonstandard monomials for J, in which each x_i appears to a power at most $n_i - 1$. Hence the span of those monomials is a subspace of R of the same dimension as I. Let w be a linear combination of only these nonstandard monomials. By the properties of the division algorithm, $\overline{w}$ is a linear combination of only standard monomials, so the symbols from w are not changed in the process of computing $E(w) = w - \overline{w}$. By Proposition (3.8), the difference $w - \overline{w}$ is an element of the ideal I, so it represents a codeword. As a result E is a systematic encoding function for I. $\qquad\qquad\qquad\qquad\qquad\qquad\qquad\qquad\qquad\qquad\qquad\qquad\qquad\square$

In the case $m = 1$, the Gröbner basis for J is the generator polynomial g, and the remainder $\overline{w}$ is computed by ordinary 1-variable polynomial division. For example, let $\mathbb{F}_9 = \mathbb{F}_3[\alpha]/\langle \alpha^2 + \alpha + 2 \rangle$ (α is a primitive element by (1.1)) and consider the Reed-Solomon code over $\mathbb{F}_9$ with $n = 8$, $k = 5$. By Proposition (3.7), the generator polynomial for this code is $g = (x - \alpha)(x - \alpha^2)(x - \alpha^3)$, and $\{g\}$ is a Gröbner basis for the ideal J in $\mathbb{F}_9[x]$ corresponding to the Reed-Solomon code. By Theorem (3.9), as information positions for a systematic encoder we can take the coefficients of the nonstandard monomials $x^7, x^6, \ldots, x^3$ in an element of $\mathbb{F}_9[x]/\langle x^8 - 1 \rangle$. The parity check positions are the coefficients of the standard monomials $x^2, x, 1$. To encode a word $w(x) = x^7 + \alpha x^5 + (\alpha + 1)x^3$, for instance, we divide g into w, obtaining the remainder $\overline{w}$. Then $E(w) = w - \overline{w}$. Here is a Maple session performing this computation. We use the method discussed in §§1,2 for dealing with polynomials with coefficients in a finite field. First we find the generator polynomial for the Reed-Solomon code as above, using:

```
alias(alpha = RootOf(t^2 + t + 2));
g := collect(Expand((x-alpha)*(x-alpha^2)*
    (x-alpha^3) mod 3,x);
```

This produces output

$$g := x^3 + alpha\ x^2 + (1 + alpha)x + 2\ alpha + 1.$$

Then

```
w := x^7 + alpha*x^5 + (alpha + 1)*x^3:
(w - Rem(w,g,x)) mod 3;
```

yields output as follows

$$x^7 + alpha\ x^5 + (1 + alpha)x^3 + 2(2 + 2\ alpha)x^2 + x + 2.$$

After simplifying the coefficient of x^2 to $\alpha + 1$, this is the Reed-Solomon codeword.

Next, we consider the 2-dimensional cyclic code in Exercise 7. Recall $I \subset R = \mathbb{F}_4[x, y]/\langle x^3 - 1, y^3 - 1\rangle$ generated by $g_1(x, y) = x^2 + \alpha^2 xy + \alpha y$, $g_2(x, y) = y + 1$. Take $\mathbb{F}_4 = \mathbb{F}_2[\alpha]/\langle \alpha^2 + \alpha + 1\rangle$ and note that $-1 = +1$ in this field. Hence $x^3 - 1$ is the same as $x^3 + 1$, and so forth. As above, we must consider the corresponding ideal

$$J = \langle x^2 + \alpha^2 xy + \alpha y, y + 1, x^3 + 1, y^3 + 1\rangle$$

in $\mathbb{F}_4[x, y]$. Applying Buchberger's algorithm to compute a reduced *lex* Gröbner basis $(x > y)$ for this ideal, we find

$$G = \{x^2 + \alpha^2 x + \alpha, y + 1\}.$$

As an immediate result, the quotient ring $\mathbb{F}_4[x, y]/J \cong R/I$ is 2-dimensional, while R is 9-dimensional over $\mathbb{F}_4$. Hence I has dimension $9 - 2 = 7$. There are also exactly two points in $\mathbf{V}(J)$. According to Theorem (3.9), the information positions for this code are the coefficients of $x^2, y, xy, x^2y, y^2, xy^2, x^2y^2$, and the parity checks are the coefficients of $1, x$. To encode $w = x^2y^2$ for example, we would compute the remainder on division by G, which is $\overline{x^2y^2}^G = \alpha^2 x + \alpha$ then subtract to obtain $E(w) = x^2y^2 + \alpha^2 x + \alpha$.

Gröbner basis computations in polynomial rings over finite fields may be done with Maple's **Groebner** and **Ore_algebra** packages as follows. For example, to compute the example above, we would first load the **Ore_algebra** and **Groebner** packages, then define the polynomial ring using

```
A:= poly_algebra(x,y,a,characteristic=2, alg_relations={a^2+a+1});
```

(This defines a ring A which is isomorphic to $\mathbb{F}_4[x, y]$. Here a is the primitive element for $\mathbb{F}_4$ and the idea is the same as in our earlier computations with a variable aliased as a root of a given irreducible polynomial. However, that method and the **mod** environment are not compatible with the Gröbner basis routines.) Then define the *lex* order as follows.

```
TL:=termorder(A,plex(x,y,a));
```

(Note that a is included.) Finally, if we declare

```
J:=[x^2+a^2*x*y+a*y,y+1,x^3+1,y^3+1];
```

then the command

```
gbasis(J,TL);
```

will do the Gröbner basis computation in the ring A. Other computer algebra systems such as **Singular** and *Macaulay 2* can handle these computations.

ADDITIONAL EXERCISES FOR §3

Exercise 8. Let C be a cyclic code in $R = \mathbb{F}_q[x]/\langle x^n - 1 \rangle$, with monic generator polynomial $g(x)$ of degree $n - k$, so that the dimension of C is k. Write out a generator matrix for C as a linear code, viewing the encoding procedure of Theorem (3.9) as a linear map from the span of $\{x^{n-k}, x^{n-k+1}, \ldots, x^{n-1}\}$ to R. In particular show that every row of the matrix is determined by the first row, i.e. the image $E(x^{n-k})$. This gives another way to understand how the cyclic property reduces the amount of information necessary to describe a code.

Exercise 9. This exercise will study the dual of a cyclic code of block length $q - 1$ or $(q - 1)^m$ more generally. See Exercise 10 from §2 for the definition of the dual of a linear code. Let $R = \mathbb{F}_q[x]/\langle x^{q-1} - 1 \rangle$ as in the discussion of Reed-Solomon codes.

a. Show that if $f(x) = \sum_{i=0}^{q-2} a_i x^i$ and $h(x) = \sum_{i=0}^{q-2} b_i x^i$ represent any two elements of R, then the inner product $\langle a, b \rangle$ of their vectors of coefficients is the same as the constant term in the product $f(x)h(x^{-1}) = f(x)h(x^{q-2})$ in R.

b. Let C be a cyclic code in R. Show that the dual code $C^\perp$ is equal to the collection of polynomials $h(x)$ such that $f(x)h(x^{-1}) = 0$ (product in R) for all $f(x) \in C$.

c. Use part b to describe the generator polynomial for $C^\perp$ in terms of the generator $g(x)$ for C. Hint: recall from the proof of Proposition (3.3) that $g(x)$ is a divisor of $x^{q-1} - 1 = \prod_{\beta \in \mathbb{F}_q^*}(x - \beta)$. The generator polynomial for $C^\perp$ will have the same property.

d. Extend these results to m-dimensional cyclic codes in
$$\mathbb{F}_q[x_1, \ldots, x_m]/\langle x_i^{q-1} - 1 : i = 1, \ldots, m \rangle.$$

Exercise 10. This exercise discusses another approach to the study of cyclic codes of block-length $q - 1$, which recovers the result of Exercise 5 in a different way. Namely, consider the ring $R = \mathbb{F}_q[x]/\langle x^{q-1} - 1 \rangle$. The structure of the ring R and its ideals may be studied as follows.

a. Show that

(3.10)
$$\varphi : R \to \mathbb{F}_q^{q-1}$$
$$c(x) \mapsto (c(1), c(\alpha), \ldots, c(\alpha^{q-2}))$$

defines a bijective mapping, which becomes an *isomorphism of rings* if we introduce the component-wise product

$$(c_0, \ldots, c_{q-2}) \cdot (d_0, \ldots, d_{q-2}) = (c_0 d_0, \ldots, c_{q-2} d_{q-2})$$

as multiplication operation in $\mathbb{F}_q^{q-1}$. (The mapping φ is a discrete analogue of the *Fourier transform* since it takes polynomial products in

R—convolution on the coefficients—to the component-wise products in $\mathbb{F}_q^{q-1}$.)

b. Show that the ideals in the ring $\mathbb{F}_q^{q-1}$ (with the component-wise product) are precisely the subsets of the following form. For each collection of subscripts $S \subset \{0, 1, \ldots, q-2\}$, let

$$I_S = \{(c_0, \ldots, c_{q-2}) : c_i = 0 \text{ for all } i \in S\}.$$

Then each ideal is equal to I_S for some S.

c. Using the mapping φ, deduce from part b and Proposition (3.2) that cyclic codes in R are in one-to-one correspondence with subsets $S \subset \{0, 1, \ldots, q-2\}$, or equivalently subsets of the nonzero elements of the field, $\mathbb{F}_q^*$. Given a cyclic code $C \subset R$, the corresponding subset of $\mathbb{F}_q^*$ is called the set of *zeroes* of C. For Reed-Solomon codes the set of zeroes has the form $\{\alpha, \ldots, \alpha^{q-k-1}\}$ (a "consecutive string" of zeroes starting from α).

Exercise 11.

a. By constructing an appropriate *transform* φ analogous to the map in (3.10), or otherwise, show that the results of Exercise 10 may be modified suitably to cover the case of m-dimensional cyclic codes of block length $n = (q-1)^m$. In particular, an m-dimensional cyclic code I in $\mathbb{F}_q[x_1, \ldots, x_m]/\langle x_1^{q-1} - 1, \ldots, x_m^{q-1} - 1 \rangle$ is uniquely specified by giving a set of zeroes—the points of $\mathbf{V}(J)$—in $(\mathbb{F}_q^*)^m = \mathbf{V}(x_1^{q-1} - 1, \ldots, x_m^{q-1} - 1)$. (Readers of Chapter 2 should compare with the discussion of finite-dimensional algebras in §2 of that chapter.)

b. Consider the 2-dimensional cyclic code I in $\mathbb{F}_9[x, y]/\langle x^8 - 1, y^8 - 1 \rangle$ generated by $g(x, y) = x^7 y^7 + 1$. What is the dimension of I (i.e., the parameter k)? What is the corresponding set of zeroes in $(\mathbb{F}_9^*)^2$?

Exercise 12. In this exercise, we will explore the relation between the zeroes of a cyclic code and its minimum distance. Let α be a primitive element of $\mathbb{F}_q$. Consider a cyclic code C of length $q - 1$ over $\mathbb{F}_q$ and suppose that there exist ℓ and $\delta \geq 2$ such that the $\delta - 1$ consecutive powers of α:

$$\alpha^\ell, \alpha^{\ell+1}, \ldots, \alpha^{\ell+\delta-2}$$

are distinct roots of the generator polynomial of C.

a. By considering the equations $c(\alpha^{\ell+j}) = 0$, $j = 0, \ldots, \delta - 2$, satisfied by the codewords (written as polynomials), show that the vectors

$$(1, \alpha^{\ell+j}, \alpha^{2(\ell+j)}, \ldots, \alpha^{(q-2)(\ell+j)}),$$

can be taken as columns of a parity check matrix H for C.

b. Show that, possibly after removing common factors from the rows, all the determinants of the $(\delta - 1) \times (\delta - 1)$ submatrices of H formed using entries in these columns are Vandermonde determinants.

c. Using Proposition (2.5), show that the minimum distance d of C satisfies $d \geq \delta$.

d. Use the result of part c to rederive the minimum distance of a Reed-Solomon code.

Exercise 13. (The BCH codes) Now consider cyclic codes C of length $q^m - 1$ over $\mathbb{F}_q$ for some $m \geq 1$.

a. Show that the result of Exercise 12 extends in the following way. Let α be a primitive element of $\mathbb{F}_{q^m}$, and suppose that there exist ℓ and $\delta \geq 2$ such that the $\delta - 1$ consecutive powers of α:

$$\alpha^\ell, \alpha^{\ell+1}, \ldots, \alpha^{\ell+\delta-2}$$

are distinct roots of the generator polynomial $g(x) \in \mathbb{F}_q[x]$ of C. Show that C has minimum distance $d \geq \delta$.

b. The "narrow-sense" q-ary BCH code $BCH_q(m, t)$ is the cyclic code over $\mathbb{F}_q$ whose generator polynomial is the *least common multiple* of the minimal polynomials of $\alpha, \alpha^2, \ldots, \alpha^{2t} \in \mathbb{F}_{q^m}$ over $\mathbb{F}_q$. (The minimal polynomial of $\beta \in \mathbb{F}_{q^m}$ over $\mathbb{F}_q$ is the nonzero polynomial of minimal degree in $\mathbb{F}_q[u]$ with β as a root.) Show the the minimum distance of $BCH_q(m, t)$ is at least $2t + 1$. (The integer $2t + 1$ is called the *designed distance* of the BCH code.)

c. Construct the generator polynomial for $BCH_3(2, 2)$ (a code over $\mathbb{F}_3$). What is the dimension of this code?

d. Is it possible for the actual minimum distance of a BCH code to be strictly larger than its designed distance? For example, show using Proposition (2.5) that the actual minimum distance of the binary BCH code $BCH_2(5, 4)$ satisfies $d \geq 11$ even though the designed distance is only 9. Hint: Start by showing that if $\beta \in \mathbb{F}_{2^m}$ is a root of a polynomial $p(u) \in \mathbb{F}_2[u]$, then so are $\beta^2, \beta^4, \ldots, \beta^{2^{m-1}}$. Readers familiar with Galois theory for finite fields will recognize that we are applying the *Frobenius automorphism* of $\mathbb{F}_{2^m}$ over $\mathbb{F}_2$ from Exercise 14 of §1 repeatedly here.

Exercise 14. Prove Proposition (3.8).

Exercise 15. Reed-Solomon codes are now commonly used in situations such as communication to and from deep-space exploration craft, the CD digital audio system, and many others where errors tend to occur in "bursts" rather than randomly. One reason is that Reed-Solomon codes over an alphabet $\mathbb{F}_{2^r}$ with $r > 1$ can correct relatively long bursts of errors on the bit level, even if the minimum distance d is relatively small. Each Reed-Solomon codeword may be represented as a string of $(2^r - 1)r$ bits, since each symbol from $\mathbb{F}_{2^r}$ is represented by r bits. Show that a burst of $r\ell$ consecutive bit errors will change at most $\ell + 1$ of the entries of the codeword, viewed as elements of $\mathbb{F}_{2^r}$. So if $\ell + 1 \leq \lfloor (d-1)/2 \rfloor$, a burst error of length $r\ell$ can be corrected. Compare with Proposition (2.1).

§4 Reed-Solomon Decoding Algorithms

The *syndrome decoding* method that we described in §2 can be applied to decode any linear code. However, as noted there, for codes with large codimension $n - k$, a very large amount of information must be stored to carry it out. In this section, we will see that there are much better methods available for the Reed-Solomon codes introduced in §3—methods which exploit their extra algebraic structure. Several different but related decoding algorithms for these codes have been considered. One well-known method is due to Berlekamp and Massey (see [Bla]). With suitable modifications, it also applies to the larger class of BCH codes mentioned in §3, and it is commonly used in practice. Other algorithms paralleling the Euclidean algorithm for the GCD of two polynomials have also been considered. Our presentation will follow two papers of Fitzpatrick ([Fit1], [Fit2]) which show how Gröbner bases for modules over polynomial rings (see Chapter 5) can be used to give a framework for the computations involved. Decoding algorithms for m-dimensional cyclic codes using similar ideas have been considered by Sakata ([Sak]), Heegard-Saints ([HeS]) and others.

To begin, we introduce some notation. We fix a field $\mathbb{F}_q$ and a primitive element α, and consider the Reed-Solomon code $C \subset \mathbb{F}_q/\langle x^{q-1} - 1 \rangle$ given by a generator polynomial

$$g = (x - \alpha) \cdots (x - \alpha^{d-1})$$

of degree $d - 1$. By Proposition (3.7), we know that the dimension of C is $k = q - d$, and the minimum distance of C is d. For simplicity we will assume that d is odd: $d = 2t + 1$. Then by Proposition (2.1), any error vector of weight t or less should be correctable.

Let $c = \sum_{j=0}^{q-2} c_j x^j$ be a codeword of C. Since C has generator polynomial $g(x)$, this means that in $\mathbb{F}_q[x]$, c is divisible by g. Suppose that c is transmitted, but some errors are introduced, so that the received word is $y = c + e$ for some $e = \sum_{i \in I} e_i x^i$. I is called the set of *error locations* and the coefficients e_i are known as the *error values*. To decode, we must solve the following problem.

(4.1) Problem. *Given a received word y, determine the set of error locations I and the error values e_i. Then the decoding function will return $E^{-1}(y - e)$.*

The set of values $E_j = y(\alpha^j)$, $j = 1, \ldots, d - 1$, serves the same purpose as the syndrome of the received word for a general linear code. (It is not the same thing though—the direct analog of the syndrome would be the remainder on division by the generator. See Exercise 7 below.) First, we can determine whether errors have occurred by computing the values E_j. If $E_j = y(\alpha^j) = 0$ for all $j = 1, \ldots, d - 1$, then y is divisible by g. Assuming the error vector has a weight at most t, y must be the codeword we intended to send. If some $E_j \neq 0$, then there are errors and we can try to use the information included in the E_j to solve Problem (4.1). Note that the E_j are the values of the error polynomial for $j = 1, \ldots, d - 1$:

$$E_j = y(\alpha^j) = c(\alpha^j) + e(\alpha^j) = e(\alpha^j),$$

since c is a multiple of g. (As in Exercise 10 from §3, we could also think of the E_j as a portion of the *transform* of the error polynomial.) The polynomial

$$S(x) = \sum_{j=1}^{d-1} E_j x^{j-1}$$

is called the *syndrome polynomial* for y. Its degree is $d - 2$ or less. By extending the definition of $E_j = e(\alpha^j)$ to *all* exponents j we can also consider the formal power series

(4.2) $$E(x) = \sum_{j=1}^{\infty} E_j x^{j-1}.$$

(Since $\alpha^q = \alpha$, the coefficients in E are periodic, with period at most q, and consequently E is actually the series expansion of a rational function of x; see (4.3) below. One can also solve the decoding problem by finding the recurrence relation of minimal order on the coefficients in E. For the basics of this approach see Exercise 6 below.)

Suppose we knew the error polynomial e. Then

$$E_j = \sum_{i \in I} e_i (\alpha^j)^i = \sum_{i \in I} e_i (\alpha^i)^j.$$

By expanding in formal geometric series, $E(x)$ from (4.2) can be written as

(4.3) $$E(x) = \sum_{i \in I} \frac{e_i \alpha^i}{(1 - \alpha^i x)}$$
$$= \frac{\Omega(x)}{\Lambda(x)},$$

where

$$\Lambda = \prod_{i \in I}(1 - \alpha^i x)$$

and

$$\Omega = \sum_{i \in I} e_i \alpha^i \prod_{\substack{j \neq i \\ j \in I}} (1 - \alpha^j x).$$

The roots of Λ are precisely the α^{-i} for $i \in I$. Since the error locations can be determined easily from these roots, we call Λ the *error locator polynomial*. Turning to the numerator Ω, we see that

$$\deg(\Omega) \leq \deg(\Lambda) - 1.$$

In addition,

$$\Omega(\alpha^{-i}) = e_i \alpha^i \prod_{j \neq i, j \in I} (1 - \alpha^j \alpha^{-i}) \neq 0.$$

Hence Ω has no roots in common with Λ. From this we deduce the important observation that the polynomials Ω and Λ must be *relatively prime*.

Similarly, if we consider the "tail" of the series E,

$$
\begin{aligned}
(4.4) \qquad E(x) - S(x) &= \sum_{j=d}^{\infty} \left(\sum_{i \in I} e_i (\alpha^i)^j \right) x^{j-1} \\
&= x^{d-1} \cdot \frac{\Gamma(x)}{\Lambda(x)},
\end{aligned}
$$

where

$$\Gamma = \sum_{i \in I} e_i \alpha^{id} \prod_{\substack{j \neq i \\ j \in I}} (1 - \alpha^j x).$$

The degree of Γ is also at most $\deg(\Lambda) - 1$.

Combining (4.3) and (4.4), and writing $d - 1 = 2t$ we obtain the relation

$$(4.5) \qquad \Omega = \Lambda S + x^{2t} \Gamma.$$

For some purposes, it will be more convenient to regard (4.5) as a *congruence*. The equation (4.5) implies that

$$(4.6) \qquad \Omega \equiv \Lambda S \bmod x^{2t}.$$

Conversely, if (4.6) holds, there is some polynomial Γ such that (4.5) holds. The congruence (4.6), or sometimes its explicit form (4.5), is called the *key equation* for decoding.

The derivation of the key equation (4.6) assumed e was known. But now consider the situation in an actual decoding problem, assuming an error vector of weight at most t. Given the received word y, S is computed. The key equation (4.6) is now viewed as a relation between the known polynomials S, x^{2t}, and the *unknowns* Ω, Λ. Suppose a solution $(\overline{\Omega}, \overline{\Lambda})$ of the key equation is found, which satisfies the following *degree conditions*:

(4.7)
$$\begin{cases} \deg(\overline{\Lambda}) \leq t \\ \deg(\overline{\Omega}) < \deg(\overline{\Lambda}) \end{cases}$$

and in which $\overline{\Omega}, \overline{\Lambda}$ are relatively prime. We claim that in such a solution $\overline{\Lambda}$ must be a factor of $x^{q-1} - 1$, and its roots give the inverses of the error locations. This is a consequence of the following uniqueness statement.

(4.8) Theorem. *Let S be the syndrome polynomial corresponding to a received word y with an error of weight at most t. Up to a constant multiple, there exists a unique solution (Ω, Λ) of (4.6) that satisfies the degree conditions (4.7), and in which Ω and Λ are relatively prime.*

PROOF. As above, the actual error locator Λ and the corresponding Ω give one such solution. Let $(\overline{\Omega}, \overline{\Lambda})$ be any other. From the congruences

$$\overline{\Omega} \equiv \overline{\Lambda} S \mod x^{2t}$$
$$\Omega \equiv \Lambda S \mod x^{2t},$$

multiplying the second by $\overline{\Lambda}$, the first by Λ and subtracting, we obtain

$$\overline{\Omega}\Lambda \equiv \Omega\overline{\Lambda} \mod x^{2t}.$$

Since the degree conditions (4.7) are satisfied for both solutions, both sides of this congruence are actually polynomials of degree at most $2t - 1$, so it follows that

$$\overline{\Omega}\Lambda = \Omega\overline{\Lambda}.$$

Since Λ and Ω are relatively prime, and similarly for $\overline{\Lambda}$ and $\overline{\Omega}$, Λ must divide $\overline{\Lambda}$ and vice versa. Similarly for Ω and $\overline{\Omega}$. As a result, Λ and $\overline{\Lambda}$ differ at most by a constant multiple. Similarly for Ω and $\overline{\Omega}$, and the constants must agree. □

Given a solution of (4.6) for which the conditions of Theorem (4.8) are satisfied, working backwards, we can determine the roots of $\overline{\Lambda} = 0$ in $\mathbb{F}_q^*$, and hence the error locations—if α^{-i} appears as a root, then $i \in I$ is an error location. Finally, the error values can be determined by the following observation.

Exercise 1. Let (Ω, Λ) be the solution of (4.6) in which the actual error locator polynomial Λ (with constant term 1) appears. If $i \in I$, show that

$$\Omega(\alpha^{-i}) = \alpha^i e_i \chi_i(\alpha^{-i}),$$

where $\chi_i = \prod_{j \neq i}(1 - \alpha^j x)$. Hence we can solve for e_i, knowing the error locations. The resulting expression is called the *Forney formula* for the error value.

Theorem (4.8) and the preceding discussion show that solving the decoding problem (4.1) can be accomplished by solving the key equation (4.6). It is here that the theory of module Gröbner bases can be applied to good effect. Namely, given the integer t and $S \in \mathbb{F}_q[x]$, consider the set of *all* pairs $(\Omega, \Lambda) \in \mathbb{F}_q[x]^2$ satisfying (4.6):

$$K = \{(\Omega, \Lambda) : \Omega \equiv \Lambda S \bmod x^{2t}\}.$$

Exercise 2. Show that K is a $\mathbb{F}_q[x]$-submodule of $\mathbb{F}_q[x]^2$. Also show that every element of K can be written as a combination (with polynomial coefficients) of the two generators

(4.9) $g_1 = (x^{2t}, 0)$ and $g_2 = (S, 1)$.

Hint: For the last part it may help to consider the related module

$$\overline{K} = \{(\Omega, \Lambda, \Gamma) : \Omega = \Lambda S + x^{2t}\Gamma\}$$

and the elements $(\Omega, \Lambda, \Gamma) = (x^{2t}, 0, 1), (S, 1, 0)$ in $\overline{K}$.

The generators for K given in (4.9) involve only the *known polynomials* for the decoding problem with syndrome S. Following Fitzpatrick, we will now show that (4.9) is a *Gröbner basis* for K with respect to one monomial order on $\mathbb{F}_q[x]^2$. Moreover, one of the special solutions $(\Lambda, \Omega) \in K$ given by Theorem (4.8) is guaranteed to occur in a Gröbner basis for K with respect to a second monomial order on $\mathbb{F}_q[x]^2$. These results form the basis for *two different* decoding methods that we will indicate.

To prepare for this, we need to begin by developing some preliminary facts about submodules of $\mathbb{F}_q[x]^2$ and monomial orders. The situation here is very simple compared to the general situation studied in Chapter 5. We will restrict our attention to submodules $M \subset \mathbb{F}_q[x]^2$ such that the quotient $\mathbb{F}_q[x]^2/M$ is *finite-dimensional* as a vector space over $\mathbb{F}_q$. We will see below that this is always the case for the module K with generators as in (4.9). There is a characterization of these submodules that is very similar to the Finiteness Theorem for quotients $k[x_1, \ldots, x_n]/I$ from Chapter 2, §2.

(4.10) Proposition. *Let k be any field, and let M be a submodule of $k[x]^2$. Let $>$ be any monomial order on $k[x]^2$. Then the following conditions are equivalent:*

a. *The k-vector space $k[x]^2/M$ is finite-dimensional.*
b. *$\langle \mathrm{LT}_>(M) \rangle$ contains elements of the form $x^u \mathbf{e}_1 = (x^u, 0)$ and $x^v \mathbf{e}_2 = (0, x^v)$ for some $u, v \geq 0$.*

PROOF. Let $\mathcal{G}$ be a Gröbner basis for M with respect to the monomial order $>$. As in the ideal case, the elements of $k[x]^2/M$ are linear combinations of monomials in the complement of $\langle \mathrm{LT}_>(M) \rangle$. There is a finite number of such monomials if and only if $\langle \mathrm{LT}_>(M) \rangle$ contains multiples of both $\mathbf{e}_1$ and $\mathbf{e}_2$. $\square$

Every submodule we consider from now on in this section will satisfy the equivalent conditions in (4.10), even if no explicit mention is made of that fact.

The monomial orders that come into play in decoding are special cases of *weight* orders on $\mathbb{F}_q[x]^2$. They can also be described very simply "from scratch" as follows.

(4.11) Definition. Let $r \in \mathbb{Z}$, and define an order $>_r$ by the following rules. First, $x^m \mathbf{e}_i >_r x^n \mathbf{e}_i$ if $m > n$ and $i = 1$ or 2. Second, $x^m \mathbf{e}_2 >_r x^n \mathbf{e}_1$ if and only if $m + r \geq n$.

For example, with $r = 2$, the monomials in $k[x]^2$ are ordered by $>_2$ as follows:

$$\mathbf{e}_1 <_2 x \mathbf{e}_1 <_2 x^2 \mathbf{e}_1 <_2 \mathbf{e}_2 <_2 x^3 \mathbf{e}_1 <_2 x \mathbf{e}_2 <_2 x^4 \mathbf{e}_1 <_2 \cdots.$$

Exercise 3.
a. Show that $>_r$ defines a monomial order on $k[x]^2$ for each $r \in \mathbb{Z}$.
b. How are the monomials in $k[x]^2$ ordered under $>_{-2}$?
c. Show that the $>_0$ and $>_{-1}$ orders coincide with *TOP* (*term over position*) orders as introduced in Chapter 5 (for different orderings of the standard basis).
d. Are the *POT* (*position over term*) orders special cases of the $>_r$ orders? Why or why not?

Gröbner bases for submodules with respect to the $>_r$ orders have very special forms.

(4.12) Proposition. *Let M be a submodule of $k[x]^2$, and fix $r \in \mathbb{Z}$. Assume $\langle \mathrm{LT}_{>_r}(M) \rangle$ is generated by $x^u \mathbf{e}_1 = (x^u, 0)$ and $x^v \mathbf{e}_2 = (0, x^v)$ for some $u, v \geq 0$. Then a subset $\mathcal{G} \subset M$ is a reduced Gröbner basis of M with respect to $>_r$ if and only if $\mathcal{G} = \{g_1 = (g_{11}, g_{12}), g_2 = (g_{21}, g_{22})\}$, where the g_i satisfy the following two properties:*
a. $\mathrm{LT}(g_1) = x^u \mathbf{e}_1$ *(in g_{11}), and* $\mathrm{LT}(g_2) = x^v \mathbf{e}_2$ *(in g_{22}) for u, v as above.*
b. $\deg(g_{21}) < u$ *and* $\deg(g_{12}) < v$.

PROOF. Suppose $\mathcal{G}$ is a subset of M satisfying conditions a,b. By a, the leading terms of the elements of $\mathcal{G}$ generate $\langle \mathrm{LT}(M) \rangle$, so by definition $\mathcal{G}$ is a Gröbner basis for M. Condition b implies that no terms in g_1 can be removed by division with respect to g_2 and vice versa, so $\mathcal{G}$ is reduced. Conversely, if $\mathcal{G}$ is a reduced Gröbner basis for M with respect to $>_r$ it must contain exactly two elements. Numbering the generators g_1 and g_2 as above condition a must hold. Finally b must hold if $\mathcal{G}$ is reduced. (Note, fixing the leading terms in g_1 and g_2 implies that the other components satisfy $\deg(g_{12}) + r < u$ and $\deg(g_{21}) \leq v + r$.) $\square$

An immediate, but important, consequence of Proposition (4.12) is the following observation.

(4.13) Corollary. *Let* $\mathcal{G} = \{(S, 1), (x^{2t}, 0)\}$ *be the generators for the module* K *of solutions of the key equation in the decoding problem with syndrome* S. *Then* $\mathcal{G}$ *is a Gröbner basis for* K *with respect to the order* $>_{\deg(S)}$.

Note $\mathrm{LT}_{>_{\deg(S)}}((S, 1)) = (0, 1) = \mathbf{e}_2$, so the module of solutions of the key equation always satisfies the finiteness condition from Proposition (4.10). We leave the proof of Corollary (4.13) as an exercise for the reader.

The final general fact we will need to know is another consequence of the definition of a Gröbner basis. First we introduce some terminology.

(4.14) Definition. Let M be a nonzero submodule of $k[x]^2$. A *minimal element* of M with respect to a monomial order $>$ is a $g \in M \setminus \{0\}$ such that $\mathrm{LT}(g)$ is minimal with respect to $>$.

For instance, from (4.13), $(S, 1)$ is minimal with respect to the order $>_{\deg(S)}$ in $\langle (S, 1), (x^{2t}, 0) \rangle$ since

$$\mathbf{e}_2 = \mathrm{LT}((S, 1)) <_{\deg(S)} \mathrm{LT}((x^{2t}, 0)) = x^{2t} \mathbf{e}_1,$$

and these leading terms generate $\langle \mathrm{LT}(K) \rangle$ for the $>_{\deg(S)}$ order.

Exercise 4. Show that minimal elements of $M \subset k[x]^2$ are *unique*, up to a nonzero constant multiple.

As in the example above, once we fix an order $>_r$, a minimal element for M with respect to that order is *guaranteed* to appear in a Gröbner basis for M with respect to $>_r$.

(4.15) Proposition. *Fix any* $>_r$ *order on* $k[x]^2$, *and let* M *be a submodule. Every Gröbner basis for* M *with respect to* $>_r$ *contains a minimal element of* M *with respect to* $>_r$.

We leave the easy proof to the reader. Now we come to the main point. The special solution of the key equation (4.6) guaranteed by Theorem (4.8) can be characterized as the minimal element of the module K with respect to a suitable order.

(4.16) Proposition. *Let* $g = (\overline{\Omega}, \overline{\Lambda})$ *be a solution of the key equation satisfying the degree conditions (4.7) and with components relatively prime* *(which is unique up to constant multiple by Theorem (4.8)). Then g is a minimal element of K under the $>_{-1}$ order.*

PROOF. An element $\overline{g} = (\overline{\Omega}, \overline{\Lambda}) \in K$ satisfies $\deg(\overline{\Lambda}) > \deg(\overline{\Omega})$ if and only if its leading term with respect to $>_{-1}$ is a multiple of $\mathbf{e}_2$. The elements of K given by Theorem (4.8) have this property and have minimal possible $\deg(\Lambda)$, so their leading term is minimal among leading terms which are multiples of $\mathbf{e}_2$.

Aiming for a contradiction now, suppose that $\overline{g}$ is not minimal, or equivalently that there is some nonzero $h = (A, B)$ in K such that $\mathrm{LT}(h) <_{-1} \mathrm{LT}(\overline{g})$. Then by the remarks above, $\mathrm{LT}(h)$ must be a multiple of $\mathbf{e}_1$, that is, it must appear in A, so

$$(4.17) \qquad \deg(\overline{\Lambda}) > \deg(A) \geq \deg(B).$$

But both h and $\overline{g}$ are solutions of the key equation:

$$A \equiv SB \bmod x^{2t}$$
$$\overline{\Omega} \equiv S\overline{\Lambda} \bmod x^{2t}.$$

Multiplying the second congruence by B, the first by $\overline{\Lambda}$, and subtracting, we obtain

$$(4.18) \qquad \overline{\Lambda}A \equiv B\overline{\Omega} \bmod x^{2t}.$$

We claim this contradicts the inequalities on degrees above. Recall that $\deg(\overline{\Lambda}) \leq t$ and $\deg(\overline{\Omega}) < \deg(\overline{\Lambda})$, hence $\deg(\overline{\Omega}) \leq t - 1$. But from (4.17), it follows that $\deg(A) \leq t - 1$. The product on the left of (4.18) has degree at most $2t - 1$, and the product on the right side has degree strictly less than the product on the left. But that is absurd. □

Combining (4.16) and (4.15), we see that the special solution of the key equation that we seek can be found in a Gröbner basis for K with respect to the $>_{-1}$ order. This gives at least *two* possible ways to proceed in decoding.

1. We could use the generating set

$$\{(S, 1), (x^{2t}, 0)\}$$

for K, apply Buchberger's algorithm (or a suitable variant adapted to the special properties of modules over the one variable polynomial ring $\mathbb{F}_q[x]$), and compute a Gröbner basis for K with respect to $>_{-1}$ directly. Then the minimal element $\overline{g}$ which solves the decoding problem will appear in the Gröbner basis.

2. Alternatively, we could make use of the fact recorded in Corollary (4.13). Since $\mathcal{G} = \{(S, 1), (x^{2t}, 0)\}$ is already a Gröbner basis for K with respect to another order, and $\mathbb{F}_q[x]^2/M$ is finite-dimensional over $\mathbb{F}_q$, we can use an extension of the FGLM Gröbner basis conversion algorithm from §3 of Chapter 2 (see [Fit2]) to convert $\{(S, 1), (x^{2t}, 0)\}$ into a Gröbner basis

$\mathcal{G}'$ for the same module, but with respect to the $>_{-1}$ order. Then as in approach 1, the minimal element in K will be an element of $\mathcal{G}'$.

Yet another possibility would be to build up to the desired solution of the key equation inductively, solving the congruences

$$\Omega \equiv \Lambda S \bmod x^{\ell}$$

for $\ell = 1, 2, \ldots, 2t$ in turn. This approach gives one way to understand the operations from the Berlekamp-Massey algorithm mentioned above. See [Fit1] for a Gröbner basis interpretation of this method.

Of the two approaches detailed above, a deeper analysis shows that the first approach is more efficient for long codes. But both are interesting from the mathematical standpoint. We will discuss the second approach in the text to conclude this section, and indicate how the first might proceed in the exercises. One observation we can make here is that the *full* analog of the FGLM algorithm need not be carried out. Instead, we need only consider the monomials in $\mathbb{F}_q[x]^2$ one by one in increasing $>_{-1}$ order and stop on the *first* instance of a linear dependence among the remainders of those monomials on division by $\mathcal{G}$. Here is the algorithm (see [Fit2], Algorithm 3.5). It uses a subalgorithm called *nextmonom* which takes a monomial u and returns the next monomial after u in $\mathbb{F}_q[x]^2$ in the $>_{-1}$ order. (Since we will stop after one element of the new Gröbner basis is obtained, we do not need to check whether the next monomial is a multiple of the leading terms of the other new basis elements as we did in the full FGLM algorithm in Chapter 2.)

(4.19) Proposition. *The following algorithm computes the minimal element of the module K of solutions of the key equation with respect to the $>_{-1}$ order.*

> *Input:* $\mathcal{G} = \{(S, 1), (x^{2t}, 0)\}$
>
> *Output:* $(\overline{\Omega}, \overline{\Lambda})$ *minimal in* $K = \langle \mathcal{G} \rangle$ *with respect to* $>_{-1}$
>
> *Uses: Division algorithm with respect to* $\mathcal{G}$, *using* $>_{\deg(S)}$ *order ,*
>
>   *nextmonom*
>
> $t_1 := (0, 1); R_1 := \overline{t_1}^{\mathcal{G}}$
>
>   *done := false*
>
>   *WHILE done = false DO*
>
> $t_{j+1} := nextmonom(t_j)$
>
> $R_{j+1} := \overline{t_{j+1}}^{\mathcal{G}}$
>
>   *IF there are $c_i \in \mathbb{F}_q$ with $R_{j+1} = \sum_{i=1}^{j} c_i R_i$ THEN*

$$(\overline{\Omega}, \overline{\Lambda}) := t_{j+1} - \sum_{i=1}^{j} c_i t_i$$

$$done := true$$

$$ELSE$$

$$j := j + 1$$

Exercise 5. Prove that this algorithm always terminates and correctly computes the minimal element of $K = \langle \mathcal{G} \rangle$ with respect to $>_{-1}$. Hint: See the proof of Theorem (3.4) in Chapter 2; this situation is simpler in several ways, though.

We illustrate the decoding method based on this algorithm with an example. Let C be the Reed-Solomon code over $\mathbb{F}_9$, with

$$g = (x - \alpha)(x - \alpha^2)(x - \alpha^3)(x - \alpha^4),$$

and $d = 5$. We expect to be able to correct any error vector of weight 2 or less. We claim that

$$c = x^7 + 2x^5 + x^2 + 2x + 1$$

is a codeword for C. This follows for instance from a Maple computation such as this one. After initializing the field (a below is the primitive element α for $\mathbb{F}_9$), setting c equal to the polynomial above, and g equal to the generator,

$$\texttt{Rem(c,g,x) mod 3;}$$

returns 0, showing that g divides c.

Suppose that errors occur in transmission of c, yielding the received word

$$y = x^7 + \alpha x^5 + (\alpha + 2)x^2 + 2x + 1.$$

(Do you see where the errors occurred?) We begin by computing the syndrome S. Using Maple, we find $y(\alpha) = \alpha + 2$, $y(\alpha^2) = y(\alpha^3) = 2$, and $y(\alpha^4) = 0$. For example, the calculation of $y(\alpha)$ can be done simply by initializing the field, defining y as above, then computing

$$\texttt{Normal(subs(x=a,y)) mod 3;}$$

So we have

$$S = 2x^2 + 2x + \alpha + 2.$$

By Theorem (4.8), we need to consider the module K of solutions of the key equation

$$\Omega \equiv \Lambda S \bmod x^4.$$

By Corollary (4.13), $\mathcal{G} = \{(x^4, 0), (2x^2 + 2x + \alpha + 2, 1)\}$ is the reduced Gröbner basis for K with respect to the order $>_2$. Applying Proposition (4.19), we find

$$
\begin{aligned}
t_1 &= (0, 1) & R_1 &= (x^2 + x + 2\alpha + 1, 0) \\
t_2 &= (1, 0) & R_2 &= (1, 0) \\
t_3 &= (0, x) & R_3 &= (x^3 + x^2 + (2\alpha + 1)x, 0) \\
t_4 &= (x, 0) & R_4 &= (x, 0) \\
t_5 &= (0, x^2) & R_5 &= (x^3 + (2\alpha + 1)x^2, 0).
\end{aligned}
$$

Here for the first time we obtain a linear dependence:

$$R_5 = -(\alpha R_1 + (\alpha + 1)R_2 + 2R_3 + (\alpha + 1)R_4).$$

Hence,

$$\alpha t_1 + (\alpha + 1)t_2 + 2t_3 + (\alpha + 1)t_4 + t_5 = (\alpha + 1 + (\alpha + 1)x, \alpha + 2x + x^2)$$

is the minimal element $(\overline{\Omega}, \overline{\Lambda})$ of K that we are looking for.

The error locations are found by solving

$$\overline{\Lambda(x)} = x^2 + 2x + \alpha = 0.$$

Recall, by definition $\Lambda = \prod_{i \in I}(1 - \alpha^i x)$ has constant term 1, so we need to adjust constants to get the actual error locator polynomial and the correct Ω to use in the determination of the error values, using the Forney formula of Exercise 1. Dividing by α, we obtain $\Lambda = (\alpha + 1)x^2 + (2\alpha + 2)x + 1$. By factoring, or by an exhaustive search for the roots as in

```
for j to 8 do
  Normal(subs(x = a^j,Lambda) mod 3;
end do;
```

we find that the roots are $x = \alpha^3$ and $x = \alpha^6$. Taking the exponents of the *inverses* gives the error locations: $(\alpha^3)^{-1} = \alpha^5$ and $(\alpha^6)^{-1} = \alpha^2$, so the errors occurred in the coefficients of x^2 and x^5. (Check the codeword c and the received word y above to see that this is correct.) Next, we apply Exercise 1 to obtain the error values. We have

$$\Omega = (1/\alpha)((\alpha + 1)x + \alpha + 1) = (\alpha + 2)x + \alpha + 2.$$

For the error location $i = 2$, for instance, we have $\chi_2(x) = 1 - \alpha^5 x$, and

$$e_2 = \frac{\Omega(\alpha^{-2})}{\alpha^2 \chi_2(\alpha^{-2})}$$

$$= \alpha + 1.$$

This also checks. The error value $e_5 = \alpha + 1$ is determined similarly; to decode we subtract $e = (\alpha + 1)x^5 + (\alpha + 1)x^2$ from y, and we recover the correct codeword.

In the Exercises below, we will consider how (part of) a direct calculation of the Gröbner basis for K with respect to $>_{-1}$ can also be used for decoding. Other applications of computational commutative algebra to coding theory are discussed in [dBP1] and [dBP2].

ADDITIONAL EXERCISES FOR §4

Exercise 6. Let $(\overline{\Omega}, \overline{\Lambda})$ be any solution of the congruence (4.6), where S is the syndrome polynomial for some correctable error.

a. Writing $\overline{\Lambda} = \sum_{i=0}^{t} \Lambda_i x^i$ and $S = \sum_{j=1}^{2t} E_j x^{j-1}$ show that (4.6) yields the following system of t homogeneous linear equations for the $t+1$ coefficients in $\overline{\Lambda}$:

$$(4.20) \qquad \sum_{k=0}^{t} \Lambda_k E_{t+\ell-k} = 0$$

for each $\ell = 1, \ldots, t$.

b. Assuming no more than t errors occurred, say in the locations given by a set of indices I, $E_{t+\ell-k} = \sum_{i \in I} e_i \alpha^{i(t+\ell-k)}$ for some polynomial $e(x)$ with t or fewer nonzero terms. Substitute in (4.20) and rearrange to obtain

$$(4.21) \qquad \begin{aligned} 0 &= \sum_{k=0}^{t} \Lambda_k E_{t+\ell-k} \\ &= \sum_{i \in I} e_i \overline{\Lambda}(\alpha^{-i}) \alpha^{i(t+\ell)}. \end{aligned}$$

c. Show that the last equation in (4.21) implies that $\overline{\Lambda}(\alpha^{-i}) = 0$ for all $i \in I$, which gives another proof that Λ divides $\overline{\Lambda}$. Hint: The equations in (4.21) can be viewed as a system of homogeneous linear equations in the unknowns $e_i \overline{\Lambda}(\alpha^{-i})$. The matrix of coefficients has a notable special form. Also, $e_i \neq 0$ for $i \in I$.

Solving the decoding problem can be rephrased as finding the linear recurrence relation (4.20) of minimal order for the E_j sequence. The coefficients Λ_k then give the error locator polynomial.

Exercise 7. A *direct analog* of syndrome decoding for Reed-Solomon codes might begin by computing the remainder on division of a received word y by the generator, giving an expression $y = c + R$, where c is a codeword. How is the remainder R related to the error polynomial e? Is this c necessarily the nearest codeword to y? (There is another decoding method for Reed-Solomon codes, due to Welch and Berlekamp, that uses R rather than the syndrome S. It can also be rephrased as solving a key equation, and Gröbner bases can be applied to solve that equation also.)

Exercise 8. Prove Corollary (4.13).

Exercise 9. Prove Proposition (4.15). Hint: Think about the definition of a Gröbner basis.

Exercise 10. Consider the Reed-Solomon code over $\mathbb{F}_9$ with generator polynomial $g = (x - \alpha)(x - \alpha^2)$ ($d = 3$, so this code is 1 error-correcting). Perform computations using Proposition (4.19) to decode the received words

$$y(x) = x^7 + \alpha x^5 + (\alpha + 2)x^3 + (\alpha + 1)x^2 + x + 2,$$

and

$$y(x) = x^7 + x^6 + \alpha x^5 + (\alpha + 1)x^3 + (\alpha + 1)x^2 + x + 2\alpha.$$

What are the solutions of $\Lambda = 0$ in the second case? How should the decoder handle the situation?

Exercise 11. In this and the following exercise, we will discuss how a portion of a direct calculation of the Gröbner basis for K with respect to $>_{-1}$ starting from the generating set $\{g_1, g_2\} = \{(x^{2t}, 0), (S, 1)\}$ can also be used for decoding. Consider the first steps of Buchberger's algorithm. Recall that S has degree $2t - 1$ or less.
a. Show that the first steps of the algorithm amount to applying the 1-variable division algorithm to divide S into x^{2t}, yielding an equation $x^{2t} = qS + R$, with a quotient q of degree 1 or more, and a remainder R that is either 0 or of degree smaller than $\deg S$. This gives the equation

$$(x^{2t}, 0) = q(S, 1) + (R, -q).$$

b. Deduce that g_2 and $g_3 = (R, -q)$ also generate the module K, so g_1 can actually be discarded for the Gröbner basis computation.
c. Proceed as in the Euclidean algorithm for polynomial GCD's (see e.g. [CLO], Chapter 1, §5), working on the *first* components. For instance, at the next stage we find a relation of the form

$$(S, 1) = q_1(R, -q) + (R_1, q_1 q + 1).$$

In the new module element, $g_4 = (R_1, q_1 q + 1)$, the degree of the first component has decreased, and the degree of the second has increased. Show that after a finite number of steps of this process, we will produce an element (Ω, Λ) of the module K whose second component has degree greater than the degree of the first, so that its $>_{-1}$ leading term is a multiple of e_2.
d. Show that the element obtained in this way is a minimal element K with respect to $>_{-1}$. Hint: It is easy to see that a minimal element could be obtained by removing any factors common to the two components of

this module element; by examining the triple $(\Omega, \Lambda, \Gamma)$ obtained as a solution of the explicit form of the key equation: $\Omega = \Lambda S + x^{2t}\Gamma$, show that in fact Ω and Λ are automatically relatively prime.

Exercise 12. Apply the method from Exercise 11 to the decoding problem from the end of the text of this section. Compare your results with those of the other method. Also compare the amount of calculation needed to carry out each one. Is there a clear "winner"?

Exercise 13. Apply the method from Exercise 11 to the decoding problems from Exercise 10.

Chapter 10

The Berlekamp-Massey-Sakata Decoding Algorithm

Algebraic geometry began to be used extensively in coding theory with the introduction of *geometric Goppa codes*, named after their discoverer, V. D. Goppa. Some of these codes have extremely good parameters and the 1982 paper [TVZ] establishing this fact was a major landmark in the history of coding theory. The original formulation of the geometric Goppa codes required many notions from the classical theory of algebraic curves or function fields of transcendence degree one, as well as topics from number theory. However, there is a class of codes, including the most important geometric Goppa codes, for which a more elementary description is available. We will introduce that treatment here and use it to develop a general version of the Berlekamp-Massey-Sakata decoding algorithm in §2 and §3. For readers with the requisite background, connections with the theory of algebraic curves will be explored in the exercises.

Our presentation follows recent papers of Geil and Pellikaan ([GeP]), O'Sullivan ([O'Su2]), and the synthesis of the earlier work of many other coding theorists made by O'Sullivan in [O'Su1] and Høholdt, van Lint, and Pellikaan in [HvLP].

§1 Codes from Order Domains

We will begin with some motivation for the ideas to be presented in this section. The construction of codes possessing good parameters and efficient decoding methods is the basic problem in coding theory. The Reed-Solomon codes introduced in §3 of Chapter 9 are among the most powerful and successful codes for certain applications. Hence it is natural to try to generalize the construction of Reed-Solomon codes given in (3.4) of Chapter 9 to produce other, potentially even better, codes.

In the Reed-Solomon case, given an

$$f \in L_{k-1} = \{g \in \mathbb{F}_q[t] : \deg(g) \leq k - 1\} \cup \{0\}$$

for some $k < q$, we evaluated f at the nonzero elements of $\mathbb{F}_q$ to form the entries in a codeword of $RS(k, q)$. The set of nonzero elements of $\mathbb{F}_q$ is a collection of points on the affine line and L_{k-1} can be seen as a vector subspace of the ring $R = \mathbb{F}_q[t]$. A possible extension might proceed as follows. Let I be an ideal in $\mathbb{F}_q[x_1, \ldots, x_t]$, let $R = \mathbb{F}_q[x_1, \ldots, x_t]/I$, and let $S = \{P_1, \ldots, P_n\}$ be a set of points with coordinates in $\mathbb{F}_q$ contained in the variety $X = \mathbf{V}(I)$. Such points are called $\mathbb{F}_q$-*rational points* of X and the entire set of such points is sometimes denoted by $X(\mathbb{F}_q)$, so that $S \subseteq X(\mathbb{F}_q)$. We can then follow (3.4) of Chapter 9 to define an evaluation mapping by

$$ev_S : R \to \mathbb{F}_q^n$$
$$f \mapsto (f(P_1), \ldots, f(P_n)).$$

The mapping ev_S is clearly linear, so if L is a finite-dimensional vector subspace of R, the image $E = ev_S(L)$ will be a linear code in $\mathbb{F}_q^n$, called an *evaluation code*. For future reference note that given E, we could also construct the dual code $C = E^{\perp}$ as in Exercise 10 from §2 of Chapter 9.

This gives a very general recipe for constructing codes, but as of yet there is no indication of how the variety X (or the ideal I) and the subspace L might be chosen to yield codes with good parameters and efficient decoding methods. In this chapter, we will see that one way to supply that missing ingredient is based on the notion of an order (or weight) function on a ring, generalizing the degree function on $\mathbb{F}_q[t]$ used in the Reed-Solomon construction.

We first give the formulation in [GeP]. As in Chapter 8, §2, a commutative *monoid* is a set Γ together with an associative, commutative binary operation $+$. We assume that Γ contains an identity element 0 for $+$ and that if $a + b = a + c$ for $a, b, c \in \Gamma$, then $b = c$ (cancellation). We do not assume that Γ contains additive inverses.

(1.1) Definition. Let R be a finitely generated commutative $\mathbb{F}_q$-algebra with identity. Let $(\Gamma, +)$ be a commutative monoid equipped with a total order relation $\succ$ that is compatible with $+$ in the sense that $a \succ b$ implies $a + c \succ b + c$ for all $a, b, c \in \Gamma$. Assume also that Γ is well-ordered under $\succ$. A surjective function $\rho : R \to \{-\infty\} \cup \Gamma$ is said to be an *order function* on R if it satisfies the following properties for all $f, g \in R$, and $\lambda \in \mathbb{F}_q$.
a. $\rho(f) = -\infty$ if and only if $f = 0$.
b. $\rho(\lambda f) = \rho(f)$ for all $\lambda \neq 0$.
c. $\rho(f + g) \preceq \max\{\rho(f), \rho(g)\}$, with equality if $\rho(f) \neq \rho(g)$.
d. If $\rho(f) = \rho(g) \neq -\infty$, then there exists $\lambda \neq 0$ such that $\rho(f + \lambda g) \prec \rho(f)$.
e. $\rho(fg) = \rho(f) + \rho(g)$.

Note that if we let $\Gamma = \mathbb{Z}_{\geq 0}$ with the usual sum operation and ordering, then $\rho(f) = \deg(f)$ for $f \neq 0$ in $R = \mathbb{F}_q[t]$ satisfies the axioms for an order function. The following exercise develops some first general properties of rings with order functions.

Exercise 1. Let ρ be an order function on R.
a. Show using the definition that $\rho(1) = 0$ (the additive identity in Γ), and hence $\rho(f) = 0$ if and only if f is a nonzero constant in $\mathbb{F}_q$.
b. Show that every R having an order function is an integral domain.
c. Show using parts a and c of the definition that any set of elements of R with distinct ρ values must be linearly independent over $\mathbb{F}_q$.
d. Show that there exists an $\mathbb{F}_q$-basis of R consisting of elements with distinct ρ values.

Because of part b of this exercise, we will say that a pair (R, ρ) is an *order domain* if ρ is an order function on R. We will call Γ the *value monoid* of (R, ρ) (the term *value semigroup* is also used). The following order domains will be used as running examples throughout this chapter.

(1.2) Example. Let $R = \mathbb{F}_q[x_1, \ldots, x_t]$, let Γ be the monoid $\mathbb{Z}_{\geq 0}^t$ with the usual vector addition, and let $\succ$ be any monomial order. Given any $f \neq 0 \in R$, we define $\rho(f) = \alpha \in \Gamma$ if $\mathrm{LM}_\succ(f) = x^\alpha$ and $\rho(f) = -\infty$ if $f = 0$. Then it is easy to see from the properties of monomial orders that the axioms in Definition (1.1) are satisfied for ρ.

(1.3) Example. Let $q = 4$, and consider the prime ideal

$$I = \langle x_1^3 + x_2^2 + x_2 \rangle \subset \mathbb{F}_4[x_1, x_2].$$

Let $>$ be the weight order $>_{(2,3),lex}$ (*lex* order with $x_1 > x_2$). Then $\mathrm{LM}_>(x_1^3 + x_2^2 + x_2) = x_1^3$. Since every element of R can be represented uniquely as the class of a remainder f on division by $x_1^3 + x_2^2 + x_2$ using the $>$ order, the classes of the monomials in the complement of $\langle x_1^3 \rangle$, namely the classes of $x_1^i x_2^j$, $0 \leq i \leq 2$ and $j \geq 0$, form a basis for $R = \mathbb{F}_4[x_1, x_2]/I$ as a vector space over $\mathbb{F}_4$. For the basis monomials of R, we define

$$\rho(x_1^i x_2^j) = (2, 3) \cdot (i, j) = 2i + 3j \in \mathbb{Z}_{\geq 0}.$$

It is easy to check that these basis monomials have distinct ρ values since $0 \leq i \leq 2$. We extend ρ to R by setting $\rho(0) = -\infty$ and $\rho(f) = \rho(\mathrm{LM}_>(f))$ if $f \neq 0$ (and f is written as a linear combination of the basis monomials). We claim that this makes R into an order domain with $\Gamma = \langle 2, 3 \rangle = \{2i + 3j : i, j \in \mathbb{Z}_{\geq 0}\} \subset \mathbb{Z}_{\geq 0}$. The easy verification is left to the reader as an exercise.

Exercise 2. Verify that the five properties in the definition of an order function are verified for the function ρ defined in Example (1.3). For property e, note that $G = \{x_1^3 + x_2^2 + x_2\}$ is a Gröbner basis for the ideal it generates with respect to $>_{(2,3),lex}$ (or any other monomial order). In some cases, you will need to consider the remainder on division of the product fg by G in order to find $\rho(fg)$.

In both of our examples above, Γ was a submonoid of $\mathbb{Z}_{\geq 0}^r$ for some $r \geq 1$. By the reasoning used in the proof of Dickson's Lemma (see [CLO], Chapter 2, §4, Theorem 5), any such Γ is finitely generated. We will use the notation

$$\Gamma = \langle m_1, \ldots, m_t \rangle = \{n_1 m_1 + \cdots + n_t m_t : n_i \in \mathbb{Z}_{\geq 0}\}$$

for the monoid generated by $m_1, \ldots, m_t$ in $\mathbb{Z}_{\geq 0}^r$. Conversely, Corollary 5.7 of [GeP], for instance, shows that if the value monoid Γ of an order domain (R, ρ) is finitely generated, then Γ is isomorphic to a submonoid of $\mathbb{Z}_{\geq 0}^r$ and $\succ$ is induced by some monomial order. Although there are interesting examples of order domains whose value monoids are not finitely generated (see [GeP] and [O'Su3]), we will not consider them here. *For the remainder of this chapter, we will make the standing assumption that Γ is finitely generated.*

It will be useful to have a good general description of the class of rings R possessing order functions. We show first that all order domains (R, ρ) with finitely generated Γ are homomorphic images of polynomial rings.

(1.4) Lemma. *Let (R, ρ) be an order domain with value monoid $\Gamma = \langle m_1, \ldots, m_t \rangle$ in $\mathbb{Z}_{\geq 0}^r$. For each i select some $y_i \in R$ such that $\rho(y_i) = m_i$ and let ϕ be the ring homomorphism defined by*

$$\phi : \mathbb{F}_q[x_1, \ldots, x_t] \to R$$

$$x_i \mapsto y_i.$$

Then ϕ is surjective. Moreover if $I = \ker(\phi)$, then $R \cong \mathbb{F}_q[x_1, \ldots, x_t]/I$.

PROOF. Let $f \in R$ be any nonzero element and say $\rho(f) = m = \sum n_i m_i$, where $n_i \in \mathbb{Z}_{\geq 0}$. By part e of Definition (1.1), the element

$$y^n = y_1^{n_1} y_2^{n_2} \cdots y_t^{n_t} = \phi(x_1^{n_1} x_2^{n_2} \cdots x_t^{n_t})$$

satisfies $\rho(y^n) = m$. Hence by part d of Definition (1.1), we can find some $\lambda \neq 0$ in $\mathbb{F}_q$ such that $\rho(f + \lambda y^n) \prec m$. Since $f + \lambda y^n \in R$, we can apply the same argument and reduce the ρ value again. As in the usual division algorithm, the well-ordering property guarantees that this process terminates after a finite number of steps with $f + \sum_k \lambda_k y^k = 0$. It follows that $f = \phi(F)$ for the corresponding linear combination $F = -\sum_k \lambda_k x^k$, so ϕ is surjective. The last claim follows by the First Isomorphism Theorem. $\square$

In the text, we will only consider order domains given explicitly by presentations of the form in Lemma (1.4), but the order domains arising from the original construction of geometric Goppa codes and their generalizations are defined using other standard constructions from the theory of algebraic curves—in particular, using certain vector spaces of rational functions on a curve with bounded pole orders at specified sets of points. Exercises 12 and 13 below study the original description and one way to derive presentations as in Lemma (1.4).

We will next use the structure of Γ to describe the ideals I defining order domains. Let (R, ρ) be as in Lemma (1.4) and let $>_\tau$ be a monomial order on $\mathbb{F}_q[x_1, \ldots, x_t]$. First we define a monomial ordering $>_{(\rho, \succ), \tau}$ on $\mathbb{F}_q[x_1, \ldots, x_t]$ compatible with ρ. We say $x^n >_{(\rho, \succ), \tau} x^m$ if either $\rho(x^n) \succ \rho(x^m)$ in Γ, or $\rho(x^n) = \rho(x^m)$ and $x^n >_\tau x^m$. Exercise 3 asks you to verify that this process does define a monomial order and Exercise 4 gives an alternate description of $>_{(\rho, \succ), \tau}$.

Exercise 3. Write $>$ for the relation $>_{(\rho, \succ), \tau}$. Show that $>$ is a monomial order on $\mathbb{F}_q[x_1, \ldots, x_t]$.

Exercise 4. In this exercise, we develop an alternate description of $>_{(\rho, \succ), \tau}$ when $\Gamma = \langle m_1, \ldots, m_t \rangle \subset \mathbb{Z}_{\geq 0}^r$ with $\rho(x_i) = m_i$, and $\succ$ is a monomial order. Let M be the $r \times t$ matrix with columns $m_1, \ldots, m_t$. Given x^n and x^m, consider n and m as t-component column vectors and let $x^n >_{(M, \succ), \tau} x^m$ if and only if either $Mn \succ Mm$ in $\mathbb{Z}_{\geq 0}^r$, or $Mn = Mm$ and $x^n >_\tau x^m$.
a. Show that $>_{(M, \succ), \tau}$ is the same as the $>_{(\rho, \succ), \tau}$ order in this case.
b. Suppose that the $\succ$ order is given as $>_N$ for some $r \times r$ matrix N as in Chapter 1, §2. How would you determine a matrix representing $>$ from part a?

The standard monomial basis for $R = \mathbb{F}_q[x_1, \ldots, x_t]/I$ has a nice description in this context. Among the set of monomials x^n for which $\rho(x^n)$ equals some fixed element of Γ, there will be one that is minimal under the $>_\tau$ order. Let Δ be the set of all these $>_\tau$-minimal monomials. That is, Δ consists of all monomials x^m in $\mathbb{F}_q[x_1, \ldots, x_t]$ with the property that if x^n and x^m have $\rho(x^n) = \rho(x^m)$ but $x^n \neq x^m$, then $x^n >_\tau x^m$.

For instance, in Example (1.2) above, no $>_\tau$ order is needed since $\succ$ is already a monomial order. The ideal I is $\{0\}$ in this case and Δ is the set of all monomials in $x_1, \ldots, x_t$.

In Example (1.3), $\Gamma = \langle 2, 3 \rangle \subset \mathbb{Z}_{\geq 0}$. There is only one choice for the $\succ$ order on $\mathbb{Z}_{\geq 0}$, namely the usual numerical order. Picking $>_\tau$ to be the *lex* order with $x_1 > x_2$, it is easy to see that $>_{(\rho, \succ), \tau}$ coincides with the $>_{(2,3), lex}$ order used in Example (1.3). Now we consider the set Δ. For instance, $x_1^5 x_2$, $x_1^2 x_2^3$ are the only monomials with $(2, 3)$-weight equal to 13, but $x_1^5 x_2 >_{lex} x_1^2 x_2^3$, so $x_1^2 x_2^3 \in \Delta$. It is easy to see that Δ is the set of monomials $\Delta = \{x_1^i x_2^j : 0 \leq i \leq 2, j \geq 0\}$, since among all monomials with the same $(2, 3)$-weight, the one in Δ has the smallest x_1 exponent, hence is lexicographically minimal. (See Example (1.8) below for a generalization of this reasoning.) Note that Δ is the standard monomial basis for $R = \mathbb{F}_4[x_1, x_2]/I$ from Example (1.3).

The patterns noted in these examples generalize as follows.

(1.5) Proposition. *Let (R, ρ) be an order domain with a presentation as in Lemma (1.4). Let $>$ denote the $>_{(\rho, \succ), \tau}$ order.*

a. *The set Δ is the standard monomial basis for R with respect to $>$ (the set of monomials in the complement of $\langle \mathrm{LT}_>(I) \rangle$).*

b. *Let σ be the minimal monomial generating set of $\langle \mathrm{LT}_>(I) \rangle$. The monic Gröbner basis of I with respect to the $>$ order consists of polynomials of the form $x^s + c_{s,u} x^u + \sum_{x^m \in \Delta, \rho(x^s) \succ \rho(x^m)} c_{s,m} x^m$, where $x^s \in \sigma$, $x^u \in \Delta$ with $\rho(x^s) = \rho(x^u)$, and $c_{s,u} \neq 0$.*

PROOF. For part a, note first that by the definition of Δ, the values $\rho(x^m)$ for $x^m \in \Delta$ are distinct. Hence by Exercise 1, part c, the set Δ is linearly independent in R. By the same argument used in the proof of Lemma (1.4), every $f \in R$ is a linear combination of elements in Δ with ρ values smaller than or equal to $\rho(f)$. Hence Δ also spans R and we see that the classes of the monomials in Δ form a vector space basis for R.

We will show next that the set of standard monomials is contained in Δ. Suppose on the contrary that some standard monomial x^s is not in Δ. By the argument used in the proof of Lemma (1.4), in R we have an equality

$$(1.6) \qquad x^s = - \sum_{x^m \in \Delta} c_m x^m,$$

for some $c_m \in \mathbb{F}_q$, where all the x^m that appear in the sum have $\rho(x^m) \preceq \rho(x^s)$, and there is precisely one term with $\rho(x^m) = \rho(x^s)$. The equation (1.6) implies that $F = x^s + \sum_{x^m \in \Delta} c_m x^m \in I$. By the definition of the $>$ order, $\mathrm{LT}_>(F) = x^s$. But this is a contradiction, since x^s was supposed to be in the complement of $\langle \mathrm{LT}_>(I) \rangle$. It follows that the set of standard monomials is contained in Δ. Since we know the set of standard monomials is also an $\mathbb{F}_q$-basis for R, they must be equal.

We now turn to part b. We have that σ and Δ are disjoint by the result of part a. Hence for each $x^s \in \sigma$, expanding x^s in terms of the basis Δ in R yields an expression $x^s = -c_{s,u} x^u - \sum_{x^m \in \Delta, \rho(x^s) \succ \rho(x^m)} c_{s,m} x^m$, where $x^u \in \Delta$ with $\rho(x^s) = \rho(x^u)$, and $c_{s,u} \neq 0$. Hence we obtain an element $F \in I$ as claimed. By the definition of the $>$ order, $\mathrm{LT}_>(F) = x^s$. By the definition of σ, it follows that the set of all these F is the monic Gröbner basis for I with respect to the $>$ order. $\square$

In what we have said so far, we have only derived consequences of the assumption that an order domain (R, ρ) with a given monoid Γ exists; we have not proved that such order domains do exist for all Γ. However, there is a converse of Proposition (1.5), due to Miura and Matsumoto in the case $r = 1$, and to Geil and Pellikaan in general (see [GeP]), which implies existence. In the statement of the following theorem, M is the matrix from Exercise 4 above, and $>_{(M,\succ),\tau}$ is one of the monomial orders defined in that exercise.

(1.7) Theorem. *Let $\Gamma = \langle m_1, \ldots, m_t \rangle \subseteq \mathbb{Z}_{\geq 0}^r$ be a monoid with ordering $\succ$, and consider the associated monomial order $>_{(M,\succ),\tau}$ (abbreviated $>$ below). Let Δ be the set of $>_\tau$-minimal monomials x^n as Mn runs over Γ. Let σ be the minimal set of monomial generators for the monomial ideal generated by the complement of Δ in $\mathbb{Z}_{\geq 0}^t$. For each $x^s \in \sigma$, let*

$$F_s = x^s + c_{s,u} x^u + \sum_{x^m \in \Delta, Ms \succ Mm} c_{s,m} x^m,$$

where $x^u \in \Delta$ satisfies $Mu = Ms$, and $c_{s,u} \neq 0$ and the $c_{s,n}$ are constants in $\mathbb{F}_q$. Let $I = \langle F_s : x^s \in \sigma \rangle$. If $G = \{F_s : x^s \in \sigma\}$ is a Gröbner basis for I with respect to the $>$ order, then $R = \mathbb{F}_q[x_1, \ldots, x_t]/I$ is an order domain with order function ρ defined by $\rho(f) = Mn$ if $\mathrm{LM}_>(f) = x^n$.

PROOF. The idea of the proof is the same as that of Example (1.3) and Exercise 2 above. We represent elements of R as linear combinations of the elements of Δ. The first four conditions in the definition of an order function follow almost automatically from the construction. The final condition $\rho(fg) = \rho(f) + \rho(g)$ is checked by showing that the remainder of fg on division by G has the same ρ value as fg. This is where the hypothesis $c_{s,u} \neq 0$ is needed.   □

We note that the hypothesis in Theorem (1.7) that G is a Gröbner basis can be checked by Buchberger's Criterion. Exercise 9 below gives a construction of an order domain with a given finitely generated value monoid.

(1.8) Example. To illustrate these results, we consider first the case of order domains (R, ρ) with value monoid $\Gamma = \langle a, b \rangle \subset \mathbb{Z}_{\geq 0}$, where $a < b$ and $\mathrm{GCD}(a, b) = 1$. We will describe the form of the Gröbner basis of I given by Theorem (1.7). Since Γ has two generators, we will have $R = \mathbb{F}_q[x_1, x_2]/I$ for some I, and the order function will be given by $\rho(x_1^i x_2^j) = (a, b) \cdot (i, j) = ai + bj$. We let $>$ be the $>_{(a,b),lex}$ order as in Example (1.3). Among all monomials with the same (a, b)-weight, the one with the smallest x_1 exponent will be lexicographically minimal. By integer division, any ℓ satisfies $\ell = qb + i$ with $0 \leq i \leq b - 1$. So $x_1^\ell x_2^j$ will have the same (a, b)-weight as $x_1^i x_2^{aq+j}$. As a result, the lexicographically minimal monomial of each (a, b)-weight has the form $x_1^i x_2^j$ with $0 \leq i \leq b - 1$, $j \geq 0$, and

$$\Delta = \{x_1^i x_2^j : 0 \leq i \leq b - 1, j \geq 0\}.$$

It follows that the monomial ideal generated by the complement of Δ must be generated by the single monomial x_1^b, and hence that I is the principal ideal generated by a single polynomial of the form:

$$(1.9) \qquad F = x_1^b + cx_2^a + \sum_{n_1 a + n_2 b < ab} c_{n_1, n_2} x_1^{n_1} x_2^{n_2},$$

with $c \neq 0$. Since any such polynomial is a Gröbner basis of the ideal it generates with respect to $>$, Theorem (1.7) implies that *every* $R = \mathbb{F}_q[x_1, x_2]/\langle F \rangle$ with F of the form (1.9) is an order domain (that is, the coefficients $c \neq 0$, c_{n_1, n_2} can be chosen arbitrarily in this case).

One particularly interesting class of examples comes by specializing this general construction as follows. Let m be a prime power, let $a = m$, $b = m + 1$, and consider the polynomial

$$F = x_1^{m+1} - x_2^m - x_2$$

over the field $\mathbb{F}_{m^2}$. (For readers with some background in the theory of algebraic curves, the reason for this will become clearer in Exercise 10 below.) Since F has the form given in (1.9), Theorem (1.7) shows that $R = \mathbb{F}_{m^2}[x_1, x_2]/\langle F \rangle$ is an order domain with $\rho(x_1^i x_2^j) = mi + (m + 1)j$. The order domain considered in Example (1.3) was the case $m = 2$ of this construction. The varieties $\mathbf{V}(F)$ are called (affine) *Hermitian curves*.

In the examples we have seen so far, we have used monomial orders of the form $>_{(\rho, \succ), lex}$. There is no need to make this restriction, and in fact other $>_\tau$ orders can simplify the analysis in some cases. For instance, consider order domains R whose value monoid is

$$(1.10) \qquad \Gamma = \langle 4, 5, 6 \rangle = \{0, 4, 5, 6, j : j \geq 8\} \subset \mathbb{Z}_{\geq 0}.$$

As in Example (1.3), there is only one choice for the $\succ$ order here, so we omit it from the notation. In Exercise 5 below, you will see what happens with the $>_{(4,5,6),lex}$ order. If we use the $>_{(4,5,6),grevlex}$ order instead, then it is easy to check that the *grevlex*-minimal monomials of each $(4, 5, 6)$-weight are the monomials in

$$\Delta = \{x_3^j, x_1 x_3^j, x_2 x_3^j, x_1^2 x_3^j, x_1 x_2 x_3^j, x_1^2 x_2 x_3^j : j \geq 0\}.$$

The simplification here is that the monomial ideal generated by the complement of Δ requires only two generators: $\sigma = \{x_2^2, x_1^3\}$. Hence by Proposition (1.5) any such order domain will be isomorphic to a ring of the form $\mathbb{F}_q[x_1, x_2, x_3]/\langle F, G \rangle$ where

$$F = x_2^2 + ax_1 x_3 + bx_1 x_2 + cx_1^2 + dx_3 + ex_2 + fx_1 + g,$$

and

$$G = x_1^3 + hx_3^2 + ix_2 x_3 + jx_1 x_3 + kx_1 x_2 + lx_1^2 + mx_3 + nx_2 + ox_1 + p.$$

Since the $>_{(4,5,6),grevlex}$ leading terms of F and G are relatively prime, any such polynomials are a Gröbner basis for the ideal they generate. Hence with the order function ρ given in Theorem (1.7) R is an order domain with value monoid Γ. The other coefficients may be chosen arbitrarily (with $a, h \neq 0$) in this case too, but that will not always be true.

Exercise 5. In this exercise, you will consider order domains with the same monoid $\Gamma = \langle 4, 5, 6 \rangle \subset \mathbb{Z}_{\geq 0}$ as in (1.10), but using the $>_{(4,5,6),lex}$ order.

a. Show that every integer $n \geq 20$ is contained in $\Gamma' = \langle 5, 6 \rangle \subset \Gamma$.

b. Use the result of part a to show that using the $>_{(4,5,6),lex}$ order,

$$\Delta = \{x_2^m x_3^n : 0 \leq m \leq 5, n \geq 0\} \cup \Delta',$$

where $\Delta' = \{x_1, x_1^2, x_1 x_2, x_1^2 x_2, x_1 x_2^2, x_1 x_2^4\}$.

c. Find the set σ here. Hint: There are five monomials in σ.

d. What presentation for order domains with $\Gamma = \langle 4, 5, 6 \rangle$ is obtained if the $>_{(4,5,6),lex}$ order is used? Can the coefficients in the polynomials be assigned arbitrarily in this case, or are there relations that must be satisfied?

We will now present the most important examples of evaluation codes as described at the start of this section. Let $X = \mathbf{V}(I)$ where $R = \mathbb{F}_q[x_1, \ldots, x_t]/I$ has an order function ρ. To construct codewords, we will evaluate functions from some vector subspace L in R at the points in $S \subseteq X(\mathbb{F}_q)$. As in the Reed-Solomon case, the most useful vector subspaces L of R will have the form $L = \{f \in R : \rho(f) \preceq m\}$ for some $m \in \Gamma$. For this idea to work in a convenient way, however, we must restrict the orderings we use again at this point. In particular we would like for these L always to be finite-dimensional vector spaces over $\mathbb{F}_q$.

(1.11) Definition. Let (R, ρ) be an order domain with value monoid Γ. We say (R, ρ) is *Archimedean* if there is an order-preserving bijective mapping $\mu : \Gamma \to \mathbb{Z}_{\geq 0}$ (with the usual numerical order).

This terminology is borrowed from the theory of valuations, and is suggested by the following result. Indeed, there is a strong connection between order functions and valuations; see [O'Su3] and [GeP] for more details.

(1.12) Lemma. *Let (R, ρ) be Archimedean. Then for any nonconstant $f \in R$ and any $g \in R$, there exists some $n \geq 1$ such that $\rho(f^n) \succ \rho(g)$.*

PROOF. This follows immediately from the Archimedean property in $\mathbb{Z}_{\geq 0}$. First, by part a of Exercise 1 and the compatibility of $\succ$ and addition in Γ, $0 \prec \rho(f) \prec \rho(f^2) \prec \cdots$ in Γ. Hence, there is some $n \geq 1$ such that $\mu(\rho(f^n)) > \mu(\rho(g))$. Since μ preserves ordering, we have $\rho(f^n) \succ \rho(g)$. $\square$

Any order domain with $\Gamma \subset \mathbb{Z}_{\geq 0}$ is Archimedean, since we can construct the mapping μ in that case simply by listing the elements of Γ in increasing numerical order. Not all order domains constructed from monomial orders on $R = \mathbb{F}_q[x_1, \ldots, x_t]$ are Archimedean, however. For instance, lexicographic orders are not Archimedean since $x_i > x_j$ implies $x_i > x_j^n$ for all $n \geq 1$, so the conclusion of Lemma (1.12) does not hold. The graded lexicographic and graded reverse lexicographic orders, on the other hand, do yield Archimedean order domain structures on R. Since there are only finitely many monomials of each degree, we can construct an order-preserving enumeration $\mu : \Gamma \to \mathbb{Z}_{\geq 0}$ as in the case $\Gamma \subset \mathbb{Z}_{\geq 0}$. There is a simple characterization of monomial orders that do yield Archimedean order domain structures on the polynomial ring $R = \mathbb{F}_q[x_1, \ldots, x_t]$.

Exercise 6.
a. Let $>_N$ be the monomial order on $R = \mathbb{F}_q[x_1, \ldots, x_t]$ defined by a suitable matrix N as in Chapter 1, §2. As in Example (1.2) we have an order domain (R, ρ) with $\Gamma = \mathbb{Z}_{\geq 0}^t$ and $\succ$ given as the matrix order $>_N$. Show that (R, ρ) is Archimedean if and only if all the entries of the first row of N are strictly positive.
b. Let (R, ρ) be an order domain as in Theorem (1.7). Show that (R, ρ) is Archimedean if the matrix M formed from the generators of Γ as in Exercise 4 satisfies the condition given in part a.

From now on in this chapter, we will make the standing assumption that (R, ρ) is Archimedean, and that $\mu : \Gamma \to \mathbb{Z}_{\geq 0}$ is as in Definition (1.11). Following O'Sullivan in [O'Su2], we will write $o = \mu \circ \rho$. The following proposition identifies our good subspaces L_a and studies their properties.

(1.13) Proposition. *Let (R, ρ) be an order domain as above. For $a \in \mathbb{Z}_{\geq 0}$, let $L_a = \{f \in R : o(f) \leq a\}$ and let $L_{-1} = \{0\}$. Then the $\{L_a : a \geq -1\}$ form a nested sequence of $\mathbb{F}_q$-vector subspaces of R whose union is R, and satisfy $\dim_{\mathbb{F}_q} L_a/L_{a-1} = 1$ for all $a \in \mathbb{Z}_{\geq 0}$. In particular L_a is finite-dimensional for all a.*

PROOF. Closure of each L_a under sums and scalar multiples follows from parts b and c of Definition (1.1). The nesting property follows from the definition of the L_a, and the claim that the union of the L_a exhausts R follows from the standing Archimedean assumption. The claim about the quotients L_a/L_{a-1} follows from part d of Definition (1.1).   □

We now present some examples of evaluation codes constructed using the L_a subspaces. We will write $E_a = ev_S(L_a)$, where ev_S is the evaluation mapping as above.

We continue with the order domain structures induced by monomial orders on $\mathbb{F}_q[x_1, \ldots, x_t]$ introduced in Example (1.2). By Exercise 6 above, to obtain an Archimedean order domain structure, we can use monomial orders such as the graded lexicographic order. The o function in this case comes from an enumeration of the monomials in R in increasing *grlex* order. With $t = 2$, $x_1 > x_2$, for instance, we have $o(1) = 0$, $o(x_2) = 1$, $o(x_1) = 2$, $o(x_2^2) = 3$, and so forth. The L_a space is spanned by the monomials $x_1^e x_2^f$ with $o(x_1^e x_2^f) \leq a$. With $a = 5$, for instance, we obtain

$$(1.14) \qquad L_5 = \mathrm{Span}\{1, x_2, x_1, x_2^2, x_1 x_2, x_1^2\}.$$

The ideal I is zero in this case, and the corresponding variety X is the affine plane. The $\mathbb{F}_q$-rational points in X are the ordered pairs $P = (x_1, x_2)$ with $x_i \in \mathbb{F}_q$. Fixing some particular ordering of these points, to form a codeword of the E_5 code we evaluate a polynomial from L_5 at each such P and use the values as the components of a vector of length $n = q^2$. The resulting code is an example of a *Reed-Muller* code. See Exercise 7 below.

We next return to the order domain introduced in Example (1.3). To construct evaluation codes, we need to know the points in $X(\mathbb{F}_4)$, where $X = \mathbf{V}(x_1^3 + x_2^2 + x_2)$. There are exactly eight such points, as you will verify in Exercise 8 below. Writing α for a primitive element of $\mathbb{F}_4$ (a root of $\alpha^2 + \alpha + 1 = 0$), the eight points can be numbered as follows:

$$(1.15) \qquad \begin{array}{rclcrcl} P_1 & = & (0,0) & & P_2 & = & (0,1) \\ P_3 & = & (1,\alpha) & & P_4 & = & (1,\alpha^2) \\ P_5 & = & (\alpha,\alpha) & & P_6 & = & (\alpha,\alpha^2) \\ P_7 & = & (\alpha^2,\alpha) & & P_8 & = & (\alpha^2,\alpha^2). \end{array}$$

The Archimedean condition is satisfied since $\Gamma = \langle 2, 3 \rangle \subset \mathbb{Z}_{\geq 0}$. We use the $>_{(2,3),lex}$ order as above. The code $E_2 = ev(V_2)$ is obtained as follows. The vector space L_2 is spanned by $\{1, x_1, x_2\}$, since $o(1) = 0$, $o(x_1) = 1$, $o(x_2) = 2$, and all other monomials in Δ have o value at least 3. The codewords are obtained by evaluation at the eight points P_i above. This gives the following generator matrix for a code of block length $n = 8$ over $\mathbb{F}_4$:

$$(1.16) \qquad G = \begin{pmatrix} 1 & 1 & 1 & 1 & 1 & 1 & 1 & 1 \\ 0 & 0 & 1 & 1 & \alpha & \alpha & \alpha^2 & \alpha^2 \\ 0 & 1 & \alpha & \alpha^2 & \alpha & \alpha^2 & \alpha & \alpha^2 \end{pmatrix}.$$

This is the same as the code from (2.4) of Chapter 9. The dual code $C_2 = E_2^\perp$ has the transposed matrix G^t as a parity check matrix. These codes are examples of geometric Goppa codes constructed from the Hermitian curve $X = \mathbf{V}(x_1^3 + x_2^2 + x_2)$.

Determining the minimum distance of the evaluation codes can be quite delicate, since it involves the subtle question of how many zeroes a polynomial in L_a can have *at the $\mathbb{F}_q$-rational points on X*. There are both geometric and arithmetic issues involved. In the following simple example the geometry suffices to understand what is going on.

Consider the E_2 code over $\mathbb{F}_4$ studied above. Each codeword is a linear combination of the three rows of the matrix G in (1.16). Hence each codeword is formed by evaluation of some linear function $f = a + bx_1 + cx_2$ at the eight $\mathbb{F}_4$-rational points. We can use *Bézout's Theorem* to give an upper bound for the number of zero entries in a codeword, hence a lower bound for d. Because X is an irreducible cubic curve, it meets each line $\mathbf{V}(a + bx_1 + cx_2)$ in at most three points, and hence $d \geq 5$. Some nonzero words in E_2 have weight exactly 5 since some of these lines intersect X in exactly three affine $\mathbb{F}_4$-rational points. The bound obtained from Bézout's Theorem using the defining equations of X is sharp in this case, but that will not always be true.

We will study the minimum distances of the duals of evaluation codes in the next section, in conjunction with the Berlekamp-Massey-Sakata decoding algorithm.

ADDITIONAL EXERCISES FOR §1

Exercise 7. Fix some numbering $S = \{P_1, P_2, \ldots, P_{q^t}\}$ of the $\mathbb{F}_q$-rational points of t-dimensional affine space. For each $\nu \geq 0$, let $L^{(\nu)}$ be the vector subspace of $\mathbb{F}_q[x_1, \ldots, x_t]$ consisting of all polynomials of total degree $\leq \nu$. Then the Reed-Muller code $RM_q(t, \nu)$ is, by definition, the image $ev_S(L^{(\nu)})$ in $\mathbb{F}_q^{q^t}$.

a. Show that the Reed-Muller code $RM_3(2, 2)$ has a generator matrix

$$
G = \begin{pmatrix}
1 & 1 & 1 & 1 & 1 & 1 & 1 & 1 & 1 \\
0 & 1 & 2 & 0 & 1 & 2 & 0 & 1 & 2 \\
0 & 0 & 0 & 1 & 1 & 1 & 2 & 2 & 2 \\
0 & 1 & 1 & 0 & 1 & 1 & 0 & 1 & 1 \\
0 & 0 & 0 & 0 & 1 & 2 & 0 & 2 & 1 \\
0 & 0 & 0 & 1 & 1 & 1 & 1 & 1 & 1
\end{pmatrix},
$$

where the rows are in one-to-one correspondence with the monomials spanning $L^{(2)}$, listed in the order $\{1, x_2, x_1, x_2^2, x_1x_2, x_1^2\}$ as in (1.14). In what order are the $\mathbb{F}_3$-rational points of the plane listed here?

b. What are the dimension and the minimum distance of the code $RM_3(2, 2)$?

Exercise 8. Verify that the eight points given in (1.15) are all of the $\mathbb{F}_4$-rational points on the variety $\mathbf{V}(x_1^3 + x_2^2 + x_2)$. Hint: How many points are there on each line $x_1 = c \in \mathbb{F}_4$?

Exercise 9. This exercise gives a construction of an order domain with given value monoid Γ and points out a connection between order domains and the *toric varieties* studied in Chapters 7 and 8. We worked over $\mathbb{C}$ there; here we consider toric varieties over finite fields. Let Γ be a submonoid of

$\mathbb{Z}_{\geq 0}^r$, ordered by a monomial order $\succ$. The *monoid ring* associated with Γ is the *subalgebra* $\mathbb{F}_q[\Gamma]$ of $\mathbb{F}_q[T_1, \ldots, T_r]$ generated by the monomials T^m, $m \in \Gamma$.

a. Show that $R = \mathbb{F}_q[\Gamma]$ is an order domain with value monoid Γ by constructing an order function ρ. Hint: This generalizes Example (1.2), where $\Gamma = \mathbb{Z}_{\geq 0}^t$.

b. Let $\Gamma = \langle m_1, \ldots, m_t \rangle$ and consider the polynomial mapping $\phi : \mathbb{F}_q^r \to \mathbb{F}_q^t$ given by $\phi(T_1, \ldots, T_r) = (T^{m_1}, \ldots, T^{m_t})$ as in Chapter 7, §3. Explain how computing the ideal I of the Zariski closure of the image of ϕ by elimination yields a presentation of $\mathbb{F}_q[\Gamma]$ of the form $\mathbb{F}_q[\Gamma] \cong \mathbb{F}_q[x_1, \ldots, x_t]/I$ for some I. The ideal I is called the *toric ideal* associated with Γ and is always generated by differences of monomials. See [Stu2] for more details.

c. Let $\Gamma = \langle 4, 5, 6 \rangle \subset \mathbb{Z}_{\geq 0}$. Find a presentation for $\mathbb{F}_q[\Gamma]$ as in part b, using a Gröbner basis calculation to perform the elimination. Show the presentation has the form given in Theorem (1.7).

The following exercises explore the relation between our evaluation codes and the original formulation of geometric Goppa codes. They require knowledge of the theory of algebraic curves. Let G be an effective divisor on a smooth curve X whose support is some collection of $\mathbb{F}_q$-rational points of X, and let S be a collection of $\mathbb{F}_q$-rational points disjoint from the support of G. The codewords of an evaluation Goppa code are formed by evaluating the rational functions in the vector space $L(G)$ at the points of S. Our evaluation codes come from Goppa's construction in the case where the projective closure of the variety $X = \mathbf{V}(I)$ associated to the order domain R is a curve (i.e., X has dimension 1), X has only one point Q at infinity, and $G = aQ$ for some $a \in \mathbb{Z}_{\geq 0}$ so the rational functions in $L(G)$ are polynomials in the affine coordinates.

Exercise 10. The curve from Example (1.3) is the first of the family of *Hermitian curves*. There is a projective Hermitian curve defined over each field of square order, $\mathbb{F}_{m^2}$: $\overline{H}_m = \mathbf{V}(x_1^{m+1} - x_2^m x_0 - x_2 x_0^m)$. The corresponding affine Hermitian curve is $H_m = \mathbf{V}(x_1^{m+1} - x_2^m - x_2)$. We will develop an interesting property of these curves; there are many others too!

a. Show that the projective Hermitian curve $\overline{H}_m$ is smooth of genus $g = m(m-1)/2$. Use the genus formula for plane curves to compute the genus and verify that this agrees with the cardinality of $\mathbb{Z}_{\geq 0} \setminus \Gamma$ for $\Gamma = \langle m, m+1 \rangle$. This equality is expected by the *Weierstrass Gap theorem* (see [Sti]).

b. Find the set $\mathcal{P}$ of all 27 $\mathbb{F}_9$-rational points on the affine curve H_3 and construct the generator matrix for the code E_4 where the first five elements of Δ are $1, x_1, x_2, x_1^2, x_1 x_2$.

c. How many $\mathbb{F}_{16}$-rational points does the affine curve H_4 have? Find them.

d. Show that for all m, the projective Hermitian curve $\overline{H}_m$ attains the Hasse-Weil bound from [Mor], equation (3.1) or [Sti], V.2.3 over $\mathbb{F}_{m^2}$:

$$|\overline{H}_m(\mathbb{F}_{m^2})| = 1 + m^2 + m(m-1)m = 1 + m^3$$

(do not forget the point at infinity).

Exercise 11. (The Klein Quartic curve) Consider the curve

$$\overline{K} = \mathbf{V}(y_1^3 y_2 + y_2^3 y_0 + y_0^3 y_1)$$

in the projective plane, defined over the field $\mathbb{F}_8 = \mathbb{F}_2[\alpha]/\langle \alpha^3 + \alpha + 1 \rangle$.

a. Show that $\overline{K}$ is smooth of degree 4, hence has genus $g = 3$.

b. In the next parts of the problem, you will show that $\overline{K}$ has 24 points rational over $\mathbb{F}_8$. First show that the three points $Q_0 = (1,0,0)$, $Q_1 = (0,1,0)$, and $Q_2 = (0,0,1)$ are $\mathbb{F}_8$-rational points on $\overline{K}$.

c. Show that the mappings

$$\psi(y_0, y_1, y_2) = (\alpha^4 y_0, \alpha y_1, \alpha^2 y_2) \qquad \tau(y_0, y_1, y_2) = (y_1, y_2, y_0)$$

take the set $\overline{K}(\mathbb{F}_8)$ to itself.

d. Deduce that $\overline{K}(\mathbb{F}_8)$ contains 21 points P_{ij} in addition to the Q_ℓ: $P_{ij} = \tau^i(\psi^j(P_{00}))$, where $P_{00} = (1, \alpha^2, \alpha^2 + \alpha) \in \overline{K}(\mathbb{F}_8)$.

e. Deduce that $|\overline{K}(\mathbb{F}_8)| = 24$ exactly. Hint: Use Serre's improvement of the Hasse-Weil bound from [Sti], V.3.1.

Exercise 12. This exercise relates the original description of the order domains associated with geometric Goppa codes to the order domains studied here.

a. Show that if X is a curve and $Q \in X$ is a smooth point, then we get an order function on the ring $R = \cup_{i=0}^{\infty} L(iQ)$ (the ring of rational functions on X with poles only at Q) by letting $\rho(f)$ be the *order of the pole* of f at Q.

b. We will say an affine curve X defined over $\mathbb{F}_q$ is in *special position* if the following conditions are satisfied.

   1. The projective closure $\overline{X}$ of X is *irreducible and smooth*.
   2. The projective curve $\overline{X}$ has only one point Q in the hyperplane at infinity, $\mathbf{V}(x_0)$, and Q also has coordinates in $\mathbb{F}_q$.
   3. The orders of the poles of the rational functions x_i/x_0 at Q generate the monoid of pole orders at Q of all rational functions on $\overline{X}$ having poles only at Q.

   Show that if the projective closure of $X = \mathbf{V}(F)$ for F as in Example (1.8) is smooth, then it is in special position. Moreover show that the point Q at infinity on $\overline{X}$ is a smooth point of $\overline{X}$ if and only if $b = a + 1$.

c. Let $X = \mathbf{V}(I)$ be a curve in special position. Show that the ring $R = \cup_{i=0}^{\infty} L(iQ)$ is isomorphic to the affine algebra $\mathbb{F}_q[x_1, \ldots, x_t]/I$.

Exercise 13. We will show how to re-embed the Klein Quartic curve K from Exercise 11 to put it in special position, and construct evaluation codes. A similar construction will work for any curve. Consider the point $Q_2 = (y_0, y_1, y_2) = (0, 0, 1)$ on $\overline{K} = \mathbf{V}(y_1^3 y_2 + y_2^3 y_0 + y_0^3 y_1)$ in $\mathbb{P}^2$ over the field $\mathbb{F}_8$.

a. What are the divisors of the rational functions $x = y_1/y_0$ and $y = y_2/y_0$ on $\overline{K}$? (Note x, y are just the usual affine coordinates in the plane.)

b. Show that $\{1, y, xy, y^2, x^2 y\}$ is a basis for the vector space $L(7Q_2)$ of rational functions on $\overline{K}$ with poles of order ≤ 7 at Q_2 and no other poles.

c. Show that the rational mapping defined by the linear system $|7Q_2|$ is an embedding of $\overline{K}$ into $\mathbb{P}^4$. Hint: Show using Riemann-Roch that it separates points and tangent vectors.

d. By part b, in concrete terms, the mapping from c is defined on the curve $\overline{K}$ by $\psi : \overline{K} \to \mathbb{P}^4$, where

$$\psi(y_0, y_1, y_2) = (x_0, x_1, x_2, x_3, x_4) = (y_0^3, y_2 y_0^2, y_1 y_2 y_0, y_2^2 y_0, y_1^2 y_2).$$

Show that the image of ψ is the variety $\overline{K}' =$

$$\mathbf{V}(x_0 x_3 + x_1^2, x_2^2 + x_1 x_4, x_0 x_2 + x_3^2, x_2 x_4, x_1 x_2 x_3 + x_0^2 x_4 + x_0 x_4^2),$$

a curve of degree 7 and genus 3 in $\mathbb{P}^4$ isomorphic to $\overline{K}$.

e. Show that $\overline{K}'$ is in special position, as defined in Exercise 12. Hint: What is $\overline{K}' \cap \mathbf{V}(x_0)$?

§2 The Overall Structure of the BMS Algorithm

We now turn to the Berlekamp-Massey-Sakata (BMS) decoding algorithm, which applies to the duals C_a of the evaluation codes E_a from order domains as in §1. The development of this algorithm, the Feng-Rao bound on the minimum distance of the C_a codes, and the associated majority voting procedure for unknown syndromes form a high point in the recent history of coding theory. We will discuss these topics following [O'Su2], a simplified and generalized synthesis of a body of work carried out by a large group of coding theorists from the mid-1980s through the mid-1990s. The article [HP] surveys the history of decoding algorithms for geometric Goppa codes and the contributions of the many people involved.

We continue all the standing assumptions concerning order domains (R, ρ) from §1. In particular, we will always assume in this section that (R, ρ) is an order domain with finitely generated value monoid Γ. Thus

$$R \cong \mathbb{F}_q[x_1, \ldots, x_t]/I,$$

where Γ is generated by $\rho(x_1), \ldots, \rho(x_t)$. Moreover, we assume (R, ρ) is Archimedean, with an order-preserving bijection

$$\mu : \Gamma \to \mathbb{Z}_{\geq 0},$$

and write $o = \mu \circ \rho : R \setminus \{0\} \to \mathbb{Z}_{\geq 0}$. We also define $o(0) = -\infty$, so that o is defined on all of R.

The $\oplus$ Operation and the $\gg$ Order

We will use the following notions extensively in this section and the next.

(2.1) Definition.
a. Using the monoid operation in Γ and $\mu : \Gamma \to \mathbb{Z}_{\geq 0}$, we define a new binary operation $\oplus$ on $\mathbb{Z}_{\geq 0}$:

$$a \oplus b = c \Leftrightarrow \mu(m+n) = c \text{ when } a = \mu(m), b = \mu(n), \ m, n \in \Gamma.$$

In other words, $(\mathbb{Z}_{\geq 0}, \oplus)$ is a monoid isomorphic to $(\Gamma, +)$ via μ.
b. We write $a = c \ominus b$ if $a \oplus b = c$.
c. Let $a, b \in \mathbb{Z}_{\geq 0}$. We say $a \gg b$ if there exists $c \in \mathbb{Z}_{\geq 0}$ such that $a = b \oplus c$ (including the possibility $c = 0$). In Exercise 1, you will show that $\gg$ is a partial order on $\mathbb{Z}_{\geq 0}$.

Exercise 1.
a. Use $\rho(fg) = \rho(f) + \rho(g)$ to show that

$$a \oplus b = c \Leftrightarrow o(fg) = c$$

for all $f, g \in R$ with $o(f) = a$ and $o(g) = b$.
b. Show that $\oplus$ is cancellative and that $c \ominus b$ is well-defined when it exists.
c. Show that $c \ominus b$ exists if and only if $c \gg b$.
d. Show that $\gg$ is a partial order on $\mathbb{Z}_{\geq 0}$.

The meaning of the $\gg$ order is revealed by the following observation.

(2.2) Lemma. *Let* $a = o(f)$ *and* $b = o(g)$ *for* $f, g \in R \setminus \{0\}$. *Then* $b \gg a$ *if and only if* $\mathrm{LM}_>(f)$ *divides* $\mathrm{LM}_>(g)$, *where* $>$ *is any* $>_{(\rho,\succ),\tau}$ *order associated with* R.

PROOF. We have $b \gg a$ if and only if $b = a \oplus c$ for some $c \geq 0$. By Exercise 1, this holds if and only if $\rho(g) = \rho(fh)$ for $h \in R$ with $\rho(h) = c$. By Definition (1.1) this is equivalent to the existence of some $\lambda \neq 0 \in \mathbb{F}_q$ such that $\rho(g + \lambda fh) \prec \rho(g)$ and $\rho(g + \lambda fh) \prec \rho(fh)$. Such λ and h exist if and only if $\mathrm{LM}_>(f)$ divides $\mathrm{LM}_>(g)$, where $>$ is a $>_{(\rho,\succ),\tau}$ order associated with R. $\square$

This lemma enables us to describe those elements of $\mathbb{Z}_{\geq 0}$ that are $\ll$ a given integer.

(2.3) Corollary. *If* $o(x^n) = b \in \mathbb{Z}_{\geq 0}$, *then*

$$N_b = \{a \in \mathbb{Z}_{\geq 0} : a \ll b\} = \{o(x^m) : x^m \text{ divides } x^n\}.$$

In particular, the set N_b is finite.

PROOF. Note that if $x^n \in \Delta$ (the standard monomial basis for R) and x^m divides x^n, then x^m is in Δ too. Let $g = x^n$ in Lemma (2.2).   □

In the order domain studied in Example (1.3), we have $o(1) = 0$, $o(x_1) = 1$, $o(x_2) = 2$, $o(x_1^2) = 3$, $o(x_1 x_2) = 4$, so $0 \oplus 1 = 1$, $1 \oplus 1 = 3$, $1 \oplus 2 = 4$, and so forth. Thus $4 \gg 2$, but $2 \gg 1$ is false: there is no $a \in \mathbb{Z}_{\geq 0}$ such that $1 \oplus a = 2$, since no multiple $x_1 h$ has $o(x_1 h) = 2$. Finally, $o(x_1 x_2) = 4$ implies that

$$(2.4) \qquad \{a \in \mathbb{Z}_{\geq 0} : a \ll 4\} = \{0, 1, 2, 4\}$$

since the divisors of $x_1 x_2$ are $1, x_1, x_2, x_1 x_2$.

Syndromes and Error Locators

In §1, the evaluation code E_a was defined to be the image of

$$evs : R \to \mathbb{F}_q^n$$
$$f \mapsto (f(P_1), \dots, f(P_n)),$$

where $S = \{P_1, \dots, P_n\}$ consists of the $\mathbb{F}_q$-rational points of $\mathbf{V}(I)$ and

$$L_a = \{f \in R : o(f) \leq a\}.$$

In this section, we will consider the dual code $C_a = E_a^\perp$. By the definition of a dual code, the codewords of E_a furnish parity check equations for the codewords of C_a. It follows that for a word $y \in \mathbb{F}_q^n$,

$$(2.5) \qquad y \in C_a \Leftrightarrow \sum_{j=1}^n y_j f(P_j) = 0 \text{ for all } f \in L_a.$$

We obtain from this an analog of the *syndromes* used in §2 of Chapter 9 for decoding linear codes, and in §4 of Chapter 9 for decoding Reed-Solomon codes. See Exercise 11 below for more on the case of Reed-Solomon codes.

It will be convenient to formulate a *syndrome mapping* associated with $y \in \mathbb{F}_q^n$ as follows:

$$S_y : \mathbb{F}_q[x_1, \dots, x_t] \to \mathbb{F}_q$$
$$(2.6)$$
$$f \mapsto \sum_{j=1}^n y_j f(P_j).$$

Note that $S_y(f) = 0$ if $f \in I$, since then $f(P_j) = 0$ for all j. Hence S_y descends to a mapping from R to $\mathbb{F}_q$. We will use the same notation for the syndrome mapping on R, so that we will write

$$S_y : R \to \mathbb{F}_q.$$

If a codeword $x \in C_a$ is sent over the channel and $y = x + e$ is received, then for the decoding problem, we will use $S_e(f)$ to correct the error. The equation (2.5) implies that $S_y(f) = S_x(f) + S_e(f) = S_e(f)$ for all $f \in L_a$. Hence those error syndrome values, but those only, can be computed from the received word using (2.6). We will call these the *known syndromes*.

As in the Reed-Solomon decoding algorithms considered in §4 of Chapter 9, the first step in BMS decoding is to determine *where* the errors occurred. Then, in a separate second step, the actual components of the error vector are determined. We will focus exclusively on the first step, which is accomplished by determining the ideal defining the error locations. Recall that the entries in codewords $x \in C_a$, and hence also in error vectors, are in one-to-one correspondence with the collection $S = \{P_1, \ldots, P_n\}$ of $\mathbb{F}_q$-rational points on the variety $X = \mathbf{V}(I)$ associated with $R = \mathbb{F}_q[x_1, \ldots, x_t]/I$. We define

$$I_e = \{f \in R : f(P_i) = 0 \text{ whenever } e_i \neq 0\}.$$

It is easy to see that I_e is an ideal of R, called the *error locator ideal* I_e associated with e. The elements of I_e are called *error locators*.

The following proposition describes the quotient ring R/I_e and the connection between error locators $f \in I_e$ and the syndromes.

(2.7) Proposition. *Let e be a vector of weight* $\text{wt}(e) = |\{i : e_i \neq 0\}|$. *Also let*

$$\Delta(e) = \{s \in \mathbb{Z}_{\geq 0} : s \neq o(f) \text{ for all } f \in I_e\}.$$

Then:
a. *Let $f \in R$. We have $f \in I_e$ if and only if $S_e(fg) = 0$ for all $g \in R$.*
b. *R/I_e has dimension $\text{wt}(e)$ as a vector space over $\mathbb{F}_q$.*
c. *For $s \in \Delta(e)$ pick $h_s \in R$ such that $o(h_s) = s$. Then $\{h_s : s \in \Delta(e)\}$ gives a basis of R/I_e. In particular, $|\Delta(e)| = \text{wt}(e)$.*

PROOF. We have $S_e(fg) = \sum_k e_k f(P_k)g(P_k)$ using the definition of the syndrome mapping. If $f \in I_e$, then $f(P_k) = 0$ whenever $e_k \neq 0$. Hence $S_e(fg) = 0$. Conversely, suppose that $S_e(fg) = 0$ for all $g \in R$. By Lemma (2.9) of Chapter 2, for each k, we can find a polynomial $g_k \in R$ such that $g_k(P_k) = 1$, but $g_k(P_j) = 0$ if $j \neq k$. (The ground field was assumed to be $\mathbb{C}$ in that lemma, but the proof actually works over any field.) If $e_k \neq 0$, then $0 = S_e(fg_k) = e_k f(P_k)$ implies that $f(P_k) = 0$. It follows that $f \in I_e$.

Turning to part b, consider the evaluation map $ev : R \to \mathbb{F}_q^{\text{wt}(e)}$ defined by evaluating $f \in R$ at the points P_i such that $e_i \neq 0$. The kernel of ev is I_e, and using the polynomials g_k of part a, one easily sees that ev is onto.

For part c, first observe that any nontrivial linear combination $u = \sum_{s \in \Delta(e)} \lambda_s h_s$ has $o(u) = s$ for some $s \in \Delta(e)$. Then $u \notin I_e$ by the definition of $\Delta(e)$, so that the h_s are linearly independent modulo I_e.

To show that they span modulo I_e, suppose that some $g \in R$ isn't in $\text{Span}(g_s : s \in \Delta(e)) + I_e$. Of all such gs, pick one with $o(g)$ minimal. If $o(g) = o(f)$ for $f \in I_e$, then $o(g - \lambda f) < o(g)$ for some $\lambda \in \mathbb{F}_q \setminus \{0\}$. By minimality, $g - \lambda f \in \text{Span}(h_s : s \in \Delta(e)) + I_e$, which leads to a contradiction. But if $o(g) \neq o(f)$ for all $f \in I_e$, then $o(g) = s \in \Delta(e)$. Hence $o(g) = o(h_s)$. This leads to a contradiction as in the previous case.   □

One of the goals of the BMS algorithm is to compute the set $\Delta(e)$.

We can also study error locators and syndromes using $\mathbb{F}_q[x_1, \ldots, x_t]$ instead of R. Let $\phi : \mathbb{F}_q[x_1, \ldots, x_t] \to R = \mathbb{F}_q[x_1, \ldots, x_t]/I$ be the natural mapping. Corresponding to I_e in R we have the ideal $J_e = \phi^{-1}(I_e)$ in $\mathbb{F}_q[x_1, \ldots, x_t]$ containing I.

Exercise 2. Let $T = \{P_i \in S : e_i \neq 0\}$.
a. Show that $J_e = \mathbf{I}(T)$, the ideal of the set $T \subset \mathbb{F}_q^t$.
b. Prove that there is a natural isomorphism $\mathbb{F}_q[x_1, \ldots, x_t]/J_e \simeq R/I_e$.
c. Let $f \in \mathbb{F}_q[x_1, \ldots, x_t]$. Show that $f \in J_e$ if and only if $S_e(fg) = 0$ for all $g \in \mathbb{F}_q[x_1, \ldots, x_t]$.

Recall that the known syndromes consist of the values $S_e(f)$ for $f \in L_a \subset R$. Suppose that the cosets of elements of L_a span R/I_e as a vector space over $\mathbb{F}_q$. This implies that every $g \in R$ can be written

$$g = f + h,$$

where $f \in L_a$ and $h \in I_e$. Then

$$S_e(g) = S_e(f) + S_e(h) = S_e(f)$$

shows that we can compute $S_e(g)$ for *all* $g \in R$. From this, Proposition (2.7), and Exercise 2, it follows that in principle we have all the information we need to recover the ideals $I_e \subset R$ and $J_e \subset \mathbb{F}_q[x_1, \ldots, x_t]$, which in turn enables us to find the error positions.

Now suppose that the cosets of the elements of L_a do not span R/I_e. (This is a new situation that did not arise in our Reed-Solomon decoding algorithms from §4 of Chapter 9.) A key idea of the version of the BMS decoding algorithm we will present is that when the number of error locations is not too large (this will be made precise using the *Feng-Rao bound*), the known syndromes $S_e(f)$ for $f \in L_a$ determine the $S_e(f)$ for $f \in L_k$, $k > a$. As we will see, the BMS algorithm, including a process called Feng-Rao majority voting for unknown syndromes, *simultaneously* computes the syndromes of $f \in L_k$ for $k > a$ *and* polynomials which, together with a Gröbner basis of I as in Proposition (1.5), form a Gröbner basis of the ideal J_e with respect to a $>_{(\rho, \succ), \tau}$ order associated with R. The structure of the L_k subspaces and the monoid Γ play a key role in the algorithm.

Span, Fail, and Normalized Bases

Proposition (2.7) tells us that we can determine whether $f \in I_e$ by the vanishing of the syndromes $S_e(fg)$. When $f \notin I_e$, we will measure "how close" f is to being in I_e using two integers called the *span* and *fail* of f.

(2.8) Definition. Given $f \in R$, and a vector e, we define the integers $\mathrm{span}_e(f)$ and $\mathrm{fail}_e(f)$ as follows. First,

$$\mathrm{span}_e(f) = \min\{o(g) : S_e(fg) \neq 0 \text{ for some } g \in R\}$$

(minimum in the usual numerical order), or $+\infty$ if $S_e(fg) = 0$ for all $g \in R$. Then

$$\mathrm{fail}_e(f) = o(f) \oplus \mathrm{span}_e(f),$$

where $\oplus$ is the operation from Definition (2.1).

Restating Proposition (2.7), we have $f \in I_e$ if and only if $\mathrm{span}_e(f) = +\infty$, or equivalently if and only if $\mathrm{fail}_e(f) = +\infty$. If $f \notin I_e$, $\mathrm{span}_e(f)$ gives the first (smallest) value of $o(g)$ for which $S_e(fg) = 0$ fails for some g, while $\mathrm{fail}_e(f)$ gives the value of $o(fg)$ for this first failure.

The following exercise will be useful.

Exercise 3.

a. Let $\mathrm{span}_e(f) = c$ and let g be an arbitrary element of R with $o(g) = c$. Show that $S_e(fg) \neq 0$. Hint: Use Definition (1.1).
b. Suppose $o(f) = s$ and $\mathrm{span}_e(f) = c$. Show that if $o(g) = c$, then $\mathrm{span}_e(g) \leq s$ and $\mathrm{fail}_e(g) \leq s \oplus c$.
c. Let $\mathrm{span}_e(f) = \mathrm{span}_e(g) = c$. Show that there exists $\lambda \in \mathbb{F}_q \setminus \{0\}$ such that $\mathrm{span}_e(f + \lambda g) > c$. Hint: Use part a.
d. Show that $\mathrm{fail}_e(f) > k$ if and only if $S_e(fg) = 0$ for all g such that $o(fg) = k$.

The set $\Delta(e)$ is defined in Proposition (2.7) via the o function. It also has a useful characterization using span_e. First, we claim that if $f \notin I_e$ is fixed, the smallest value of $o(g)$ for which $S_e(fg) \neq 0$ must be in $\Delta(e)$. Equivalently,

$$\Delta(e) \supseteq \{\mathrm{span}_e(f) : f \notin I_e\}.$$

This follows since if we had $f \notin I_e$ with $\mathrm{span}_e(f)$ not in $\Delta(e)$, then by the definition of $\Delta(e)$, we would have $\mathrm{span}_e(f) = o(g)$ for some $g \in I_e$. But then Proposition (2.7) would say $S_e(fg) = 0$. This contradicts part a of Exercise 3. In fact, the reverse inclusion is also valid.

(2.9) Proposition. $\Delta(e) = \{\mathrm{span}_e(f) : f \notin I_e\}$.

The proof will be accomplished using Lemma (2.10) below, which involves a new idea. For a fixed $s \in \Delta(e)$, the elements of order s in R do not necessarily all have the same span_e value. The following exercise gives an example of this.

Exercise 4. Let R be the order domain from Example (1.3), and consider the ordering of the eight $\mathbb{F}_4$-rational points on the variety X associated with R from (1.15). Let $e = (0, 1, 0, 0, 0, 0, 0, 1)$.
a. Show using (2.6) that $S_e(1) = 0$, $S_e(x_1) = \alpha^2$, and $S_e(x_2) = \alpha$.
b. Show that $o(x_2) = o(x_2 + \alpha^2 x_1) = 2$, but $\mathrm{span}_e(x_2) = 0$, while $\mathrm{span}_e(x_2 + \alpha^2 x_1) \geq 1$.
c. By computing more syndromes as needed, determine $\mathrm{span}_e(x_2 + \alpha^2 x_1)$.

However, because $\{\mathrm{span}_e(f) : f \notin I_e\}$ is contained in the finite set $\Delta(e)$, for each s, there will be a maximum possible span_e value for f of order s. As is often true, it is productive here to consider the maximum value, denoted

$$\mathrm{sp}_e(s) = \max\{\mathrm{span}_e(f) : o(f) = s\}.$$

By analogy with Definition (2.8), we will also write

$$\mathrm{fl}_e(s) = s \oplus \mathrm{sp}_e(s).$$

The sp_e mapping has the following interesting properties.

(2.10) Lemma.
a. *If $s \in \Delta(e)$, then $\mathrm{sp}_e(s) \in \Delta(e)$ also.*
b. *The mapping $\mathrm{sp}_e : \Delta(e) \to \Delta(e)$ is bijective.*

PROOF. For part a, we argue by contradiction. Suppose $s \in \Delta(e)$, but $\mathrm{sp}_e(s) \notin \Delta(e)$. This says that there are $f \notin I_e$ with $o(f) = s$ and the maximal $\mathrm{span}_e(f) = \mathrm{sp}_e(s)$. Moreover, there is some $g \in I_e$ such that $o(g) = \mathrm{sp}_e(s)$. By Proposition (2.7), $S_e(fg) = 0$. But this contradicts part a of Exercise 3. Hence $\mathrm{sp}_e(s) \in \Delta(e)$.

Turning to part b, since $\Delta(e)$ is a finite set, it suffices to show that sp_e is injective. So consider $s > t$ in $\Delta(e)$ and suppose that $\mathrm{sp}_e(s) = \mathrm{sp}_e(t) = m$. Let f satisfy $o(f) = s$ and $\mathrm{span}_e(f) = m$, and let g satisfy $o(g) = t$ and $\mathrm{span}_e(g) = m$. By part c of Exercise 3, it follows that for some $\lambda \in \mathbb{F}_q \setminus \{0\}$, $\mathrm{span}_e(f + \lambda g) > m$. But we also have $o(f + \lambda g) = o(f) = s$ since we assumed $s > t$. This contradicts the choice of f, and we have proved that sp_e is injective on $\Delta(e)$. $\square$

The proof of Proposition (2.9) follows easily now.

PROOF OF PROPOSITION (2.9). We only need to prove the inclusion $\Delta(e) \subseteq \{\mathrm{span}_e(f) : f \notin I_e\}$. Let $s \in \Delta(e)$. By Lemma (2.10), $s = \mathrm{sp}_e(t)$

for some $t \in \Delta(e)$. By the definition, this means $s = \text{span}_e(f)$ for f with $o(f) = t$ achieving the maximal span_e value.   □

Another tool we will use is a particularly nice basis of the subspaces L_k. Recall from Proposition (1.13) that for all $k \geq -1$, we have $L_k \subset L_{k+1}$ and $\dim(L_k) = k + 1$. We also know from §1 that R has a monomial basis Δ, and since (R, ρ) is Archimedean, we can list the elements of Δ as M_i, $i \in \mathbb{Z}_{\geq 0}$, where $o(M_i) = \mu(\rho(M_i)) = i$ for all $i \in \mathbb{Z}_{\geq 0}$.

Exercise 5. Let $i, j \in \mathbb{Z}_{\geq 0}$.
a. Show that $L_k = \text{Span}(M_0, \ldots, M_k)$ for all $k \geq -1$.
b. Show that $o(M_i M_j) = o(M_{i \oplus j})$ and conclude that there is $\lambda \in \mathbb{F}_q$ such that $o(M_i M_j - \lambda M_{i \oplus j}) < o(M_{i \oplus j})$.

Note that the λ in Exercise 5 depends on i and j. The nicest case is when all of the λs are 1. This leads to the following definition.

(2.11) Definition. We say that $\{z_0, z_1, \ldots\}$ is a *normalized basis* of R if $o(z_i) = i$ for all $i \in \mathbb{Z}_{\geq 0}$ and $o(z_i z_j - z_{i \oplus j}) < o(z_{i \oplus j})$ for all $i, j \in \mathbb{Z}_{\geq 0}$.

Unfortunately, one needs valuation theory to prove the existence of normalized bases. We refer the reader to [GeP], Proposition 6.4, for a proof. In what follows, we will assume that we have a normalized basis $\{z_i\}$ of R.

Exercise 6. For the order domain (R, ρ) of Example (1.3), we know that $\Delta = \{x_1^i x_2^j : 0 \leq i \leq 2, j \geq 0\}$ and $\rho(x_1^i x_2^j) = 2i + 3j$. Show that

$$\{x_1^i x_2^j : 0 \leq i \leq 2, j \geq 0\} = \{1, x_1, x_2, x_1^2, x_1 x_2, x_2^2, \ldots\}$$

is a normalized basis of R. Hint: $x_1^3 + x_2^2 + x_2 = 0$ in R and $\mathbb{F}_4$ has characteristic 2.

In fact, for all of the examples we will consider, the needed elements of the normalized basis will actually be elements of Δ.

The Sets Σ_k and Δ_k

In Proposition (2.7), we defined $\Delta(e) = \{s \in \mathbb{Z}_{\geq 0} : s \neq o(f) \text{ for all } f \in I_e\}$. The complement of $\Delta(e)$ consists of the integers $o(f)$ for all $f \in I_e \setminus \{0\}$ and hence will be important for determining I_e. We call this complement $\Sigma(e)$ and hence:

$$(2.12) \qquad \begin{aligned} \Sigma(e) &= \{o(f) \in \mathbb{Z}_{\geq 0} : f \in I_e \setminus \{0\}\} \\ \Delta(e) &= \mathbb{Z}_{\geq 0} \setminus \Sigma(e). \end{aligned}$$

In an intuitive sense, the larger $\mathrm{fail}_e(f)$ is, the closer f is to being in I_e. The BMS algorithm approximates $\Sigma(e)$ and $\Delta(e)$ using sets Σ_k and Δ_k defined using this idea, as follows.

(2.13) Definition. For $e \in \mathbb{F}_q^n$ and $k \in \mathbb{Z}_{\geq 0}$, let

$$\Sigma_k = \{o(f) \in \mathbb{Z}_{\geq 0} : f \in R \setminus \{0\}, \; \mathrm{fail}_e(f) > k\}$$
$$\Delta_k = \mathbb{Z}_{\geq 0} \setminus \Sigma_k.$$

Exercise 7. Use $f \in I_e \Leftrightarrow \mathrm{fail}_e(f) = +\infty$ to show that we have inclusions

$$\Sigma(e) \subset \cdots \subset \Sigma_k \subset \Sigma_{k-1} \subset \cdots \subset \Sigma_0$$
$$\Delta(e) \supset \cdots \supset \Delta_k \supset \Delta_{k-1} \supset \cdots \supset \Delta_0$$

so that

$$\Sigma(e) = \bigcap_{k=0}^{\infty} \Sigma_k \quad \text{and} \quad \Delta(e) = \bigcup_{k=0}^{\infty} \Delta_k.$$

In the BMS algorithm, the index $k \in \mathbb{Z}_{\geq 0}$ is a counter in a loop such that the kth pass through the loop produces Σ_k and Δ_k from Σ_{k-1} and Δ_{k-1}. Since $\Delta(e)$ is finite, Exercise 7 implies that $\Delta(e) = \Delta_k$ for k sufficiently large. Thus, as the algorithm proceeds, the Δ_k increase and converge to $\Delta(e)$, and Proposition (2.7) tells us that under the o function, $\Delta(e)$ corresponds to the set of standard monomials for the zero-dimensional ideal J_e with respect to a $>_{(\rho, \succ), \tau}$ order associated with R. Similarly, the Σ_k decrease and converge to $\Sigma(e)$. Under o, $\Sigma(e)$ corresponds to the set of monomials in $\langle \mathrm{LT}_>(J_e) \rangle$.

The reasoning used to prove the alternate characterization of $\Delta(e)$ in Proposition (2.9) above gives a similar description of the set Δ_k in terms of the span_e function:

(2.14) $$\Delta_k = \{\mathrm{span}_e(f) : f \in R, \; \mathrm{fail}_e(f) \leq k\}.$$

You will prove this fact in Exercise 12 below.

Probably the most important property of $\Delta(e)$ and Δ_k is that they are *delta-sets*, meaning that if they contain $s \in \mathbb{Z}_{\geq 0}$, then they contain all elements of $\mathbb{Z}_{\geq 0}$ which are $\ll s$. The following exercise makes this precise.

Exercise 8.
a. Prove that $\mathrm{fail}_e(f) = k$ implies that $\mathrm{fail}_e(fg) \geq k$ for all $g \in R$.
b. Show that if $s \in \Delta(e)$ and $a \ll s$, then $a \in \Delta(e)$. This shows that $\Delta(e)$ is a delta-set.
c. Use part a to show that Δ_k is a delta-set.

Because Δ_k is a delta-set, it is natural to describe Δ_k in terms of its set δ_k of maximal elements with respect to $\gg$. We write this as

$$\delta_k = \max_{\gg}(\Delta_k),$$

and then being a delta-set implies that

$$\Delta_k = \{b \in \mathbb{Z}_{\geq 0} : b \ll d \text{ for some } d \in \delta_k\}.$$

Furthermore, since Σ_k is the complement of a delta-set, it has the property that if $s \in \Sigma_k$ and $t \gg s$ then $t \in \Sigma_k$. To describe Σ_k, we will use the set of minimal elements with respect to $\gg$. If we set

$$\sigma_k = \min_{\gg}(\Sigma_k),$$

then

$$\Sigma_k = \{t \in \mathbb{Z}_{\geq 0} : t \gg s \text{ for some } s \in \sigma_k\}.$$

It is clear that δ_k is finite. The finiteness of σ_k is less obvious, but can be proved using an appropriate formulation of Dickson's Lemma. See [BW], Corollary 4.48 for instance. In our case where Σ_k is the complement of the finite delta-set Δ_k, there is also a simpler direct proof leading to an algorithm for producing σ_k from Δ_k. See Exercise 13 below.

The sets δ_k and σ_k are important because they are part of the data structures used by the BMS algorithm to represent Δ_k and Σ_k. Notice that since $\gg$ is different from the usual numerical order in $\mathbb{Z}_{\geq 0}$, the sets δ_k and σ_k can each have more than one element.

The Feng-Rao Bound

The Feng-Rao majority voting procedure used in the BMS algorithm is closely tied to a method for estimating the minimum distances of the C_a codes. Recall from Corollary (2.3) that

$$N_b = \{c \in \mathbb{Z}_{\geq 0} : c \ll b\}$$

is a finite set. We will use the notation

$$n_b = |N_b|.$$

For instance, in our running example started in Example (1.3), $N_5 = \{0, 1, 2, 3, 5\}$, since $0 \oplus 5 = 5$, $1 \oplus 3 = 5$, $2 \oplus 2 = 5$, but $5 \not\gg 4$. Thus we have $n_5 = 5$.

The key fact relating the integers n_b to the minimum distances of the $C_a = evs(L_a)^{\perp}$ codes is the following observation.

(2.15) Lemma.
a. *Suppose* $v \in C_{a-1} \setminus C_a$. *Then* $\mathrm{wt}(v) \geq n_a$.
b. *The minimum distance of* C_a *satisfies* $d(C_a) \geq \min\{n_b : b > a\}$.

PROOF. For part a, we consider span_v, fail_v, and the sets Δ_k, $\Delta(v)$ where the vector v takes the role played by e above. (Although the decoding algorithm focuses on the cases where $v = e$ is an error vector or $v = y = x + e$ is a received word, note that our definitions and results are valid for arbitrary vectors.) Since $v \in C_{a-1}$, $S_v(f) = \sum_{j=1}^n v_j f(P_j) = 0$ for all $f \in L_{a-1}$, but there is some $g \in L_a$ such that $S_v(g) \neq 0$. Hence for $h = 1$ (the constant function in R), $\text{fail}_v(1) = a$, so $a \in \Delta_a$. Since Δ_a is a delta-set, it follows that $N_a \subset \Delta_a$ as well. Hence $n_a = |N_a| \leq |\Delta_a| \leq |\Delta(v)| = \text{wt}(v)$, where the last equality follows from Proposition (2.7).

In Exercise 14 below, you will see that part b follows from part a. $\square$

The lower bound on $d(C_a)$ from part b of the lemma is known as the *Feng-Rao bound* (or *order bound*) on $d(C_a)$. We will write

$$(2.16) \qquad d_{FR}(C_a) = \min\{n_b : b > a\}.$$

The Feng-Rao bound is easy to compute in examples where we have explicit information about the monoid structure of $(\mathbb{Z}_{\geq 0}, \oplus)$.

Exercise 9. Let C_2 be the code from the order domain of Example (1.3).
a. Show that $\mu : (\langle 2, 3 \rangle, +) \to (\mathbb{Z}_{\geq 0}, \oplus)$ is given by

$$\mu(m) = \begin{cases} 0 & \text{if } m = 0 \\ m - 1 & \text{if } m \geq 2 \text{ in } \langle 2, 3 \rangle \subset \mathbb{Z}_{\geq 0}. \end{cases}$$

b. Using the isomorphism μ, show that $n_a = a$ for all $a \geq 2$.
c. Deduce that $d_{FR}(C_2) = 3$.

Exercise 10. Let $R = \mathbb{F}_q[x_1, x_2]$, for fixed $q \geq 3$, with ρ defined as in Example (1.2) using the graded lex order with $x_1 > x_2$. The subspace L_5 is spanned by $\{1, x_2, x_1, x_2^2, x_1 x_2, x_1^2\}$. Consider the $C_5 = E_5^{\perp}$ code.
a. Construct a table giving the values of $a \oplus b$ for $0 \leq a, b \leq 4$ and use it to determine N_j and n_j, $j \leq 5$.
b. Deduce and prove a formula for $d_{FR}(C_a)$ in this case, assuming q is sufficiently large that the basis monomials for L_a are linearly independent at the points of $\mathbb{F}_q^2$.
c. Is the Feng-Rao bound for $d(C_5)$ sharp with $q = 3$?

The Overall Structure

Our final goal in this section is to explain the overall structure of the BMS algorithm. As already mentioned, the algorithm will use the sets δ_k and σ_k. But recall that each $s \in \sigma_k$ is of the form $s = o(f)$ where $\text{fail}_e(f) > k$. We need such elements of R in the data structure, so we will include a function

$$(2.17) \qquad F_k : \sigma_k \to R \setminus \{0\}$$

such that for each $s \in \sigma_k$ we have $o(F_k(s)) = s$ and $\text{fail}_e(F_k(s)) > k$.

We want similar data for δ_k. The precise description is a function

$$G_k : \delta_k \rightarrow R \setminus \{0\}$$

such that for each $c \in \delta_k$ we have $\mathrm{span}_e(G_k(c)) = c$, $\mathrm{fail}_e(G_k(c)) \leq k$, and the normalization $S_e(G_k(c)z_c) = 1$. Existence of such $G_k(c)$ follows from (2.14) above.

The data structure we will use in the BMS algorithm consists of two pieces. The first is

$$\mathrm{Info}(k) = (\sigma_k, F_k, \delta_k, G_k)$$

corresponding to the sets and maps defined above. The second is

$$\mathrm{Syn}(k) = (S_e(z_0), \ldots, S_e(z_k)),$$

where $\{z_i\}$ is our chosen normalized basis of R. This enables us to compute $S_e(f)$ for any $f \in L_k$.

In (2.18), we give a rudimentary outline of the BMS algorithm. We ignore issues such as stopping criteria for the loop in step 2 for the moment. The notation *FRMV* refers to the Feng-Rao majority voting procedure for computing previously unknown syndromes mentioned before, and *InfoUpdate* refers to the procedure that computes Info(k) in terms of Info($k-1$) and Syn(k). Both procedures and the stopping criteria will be described in detail in §3. As above, $\{z_i\}$ is the normalized basis of R.

(2.18) BMS Decoding Algorithm (Overall Structure). Consider the decoding problem for the C_a code from an order domain R.

Input: R, $\{z_i\}$, S_e computing syndromes of elements of L_u, the received word $y = x + e$.

Output: the decoded word x.

1. Initialize

$$\mathrm{Info}(-1) := (\sigma_{-1} = \{0\}, F_{-1}(0) = 1, \delta_{-1} = \emptyset, G_{-1} = \emptyset)$$
$$\mathrm{Syn}(-1) := ()$$

2. FOR $k \in \mathbb{Z}_{\geq 0}$ DO

        IF $k \leq a$ THEN

            compute $S_e(z_k) = S_y(z_k)$ from y

        ELSE

            $S_e(z_k) := FRMV(\mathrm{Info}(k-1), \mathrm{Syn}(k-1))$

            Syn(k) := append $S_e(z_k)$ to Syn($k-1$)

            Info(k) := *InfoUpdate*(Info($k-1$), Syn(k)).

We will see that Info(k) converges to Info $= (\sigma, F, \delta, G)$ after a finite number of steps.

3. Compute $ErrorLocations = \{P \in S : F(s)(P) = 0 \text{ for all } s \in \sigma\}$.
4. Compute error values, e.g., by generalized Forney formulas as in [Leo] or [Lit] to get e, and subtract e from y to recover x.

The precise meaning of termination is that there is k such that

$$\text{Info}(k) = \text{Info}(k + 1) = \text{Info}(k + 2) = \cdots.$$

This gives what we call $\text{Info} = (\sigma, F, \delta, G)$. Since $\Delta_k = \{c \in \mathbb{Z}_{\geq 0} : c \ll d \text{ for some } d \in \delta_k\}$, termination implies that

$$\Delta_k = \Delta_{k+1} = \Delta_{k+2} = \cdots.$$

We showed earlier that $\Delta_\ell = \Delta(e)$ for ℓ sufficiently large. It follows that

$$\Delta(e) = \{c \in \mathbb{Z}_{\geq 0} : c \ll d \text{ for some } d \in \delta\}$$
$$\Sigma(e) = \{t \in \mathbb{Z}_{\geq 0} : t \gg s \text{ for some } s \in \sigma\}.$$

Furthermore, if $s \in \sigma$, then $\sigma = \sigma_\ell$ and $F = F_\ell$ for $\ell \geq k$ imply that $o(F(s)) = s$ and $\text{fail}_e(F(s)) > \ell$ for all $\ell \geq k$. Thus $\text{fail}_e(F(s)) = +\infty$, so that $F(s)$ is an element of the error locator ideal I_e. Moreover, we know from (2.13) that $\Sigma(e) = \{s \in \mathbb{Z}_{\geq 0} : s = o(f) \text{ for some } f \in I_e\}$.

We can also view the $F(s)$ as polynomials in $\mathbb{F}_q[x_1, \ldots, x_t]$ via $\phi : \mathbb{F}_q[x_1, \ldots, x_t] \to R$, and we claim that they, together with the elements of a Gröbner basis for I, form a Gröbner basis for the ideal $J_e = \phi^{-1}(I_e)$ with respect to the $>_{(\rho,\succ),\tau}$ order associated with R. This is a consequence of the following result.

(2.19) Theorem. *Let $\mathcal{G}$ be a Gröbner basis for I with respect to $>_{(\rho,\succ),\tau}$ and let $F : \sigma \to J_e$ satisfy $o(F(s)) = s$ for all $s \in \sigma$. Then*

$$\mathcal{G} \cup \{F(s) : s \in \sigma\}$$

is a Gröbner basis for J_e with respect to $>_{(\rho,\succ),\tau}$.

PROOF. We abbreviate $>_{(\rho,\succ),\tau}$ by $>$ in the following. Let $f \in J_e$ be nonzero. It suffices to show that $\text{LM}_>(f)$ is divisible by either some $\text{LM}_>(g)$, $g \in \mathcal{G}$, or some $\text{LM}_>(F(s))$, $s \in \sigma$.

If $f \in I$, then $\text{LM}_>(f)$ is divisible by some $\text{LM}_>(g)$, $g \in \mathcal{G}$, since $\mathcal{G}$ is a Gröbner basis of I with respect to $>$. On the other hand, if $f \notin I$, then f gives a nonzero element of $R = \mathbb{F}_q[x_1, \ldots, x_t]/I$ which lies in I_e. By definition, this implies $o(f) \in \Sigma(e)$, and since $\sigma = \min_\gg(\Sigma(e))$, we have $o(f) \gg s$ for some $s \in \sigma$. Since $s = o(F(s))$, Lemma (2.2) implies that $\text{LM}_>(f)$ is divisible by $\text{LM}_>(F(s))$. This completes the proof.   □

In the next section we will present the details of the *FRMV* and *InfoUpdate* procedures, and derive the following consequence.

(2.20) Theorem. *The BMS algorithm applied to the C_a code correctly decodes all received words $y = x + e$ for which*

$$\text{wt}(e) \leq \lfloor (d_{FR}(C_a) - 1)/2 \rfloor.$$

Proposition (2.1) of Chapter 9 implies that if a code C has minimum distance d, then all error vectors satisfying $\text{wt}(e) \leq \lfloor \frac{d-1}{2} \rfloor$ can be corrected by nearest neighbor decoding. This theorem says that BMS decoding for the C_a codes will correctly decode errors whose weight is bounded by a similar expression but with d replaced by the lower bound $d_{FR}(C_a) \leq d$. Even though $d_{FR}(C_a)$ often agrees with $d(C_a)$, $d_{FR}(C_a)$ can be strictly less than $d(C_a)$. If that is true the theorem gives no guarantee that BMS takes full advantage of the error-correcting capacity of the code. However, it is often true that BMS will correct errors of weight larger than the bound in Theorem (2.20). See Exercise 6 in §3 below. BMS is an efficient method using the algebraic structure of the C_a codes and it performs very well on many interesting examples, such as the C_a codes from Hermitian curves.

ADDITIONAL EXERCISES FOR §2

Exercise 11. The Reed-Solomon code $RS(k, q)$ is the same as the evaluation code E_{k-1} from the order domain $R = \mathbb{F}_q[t]$, where ρ is the degree function and $S = \mathbb{F}_q \setminus \{0\}$. In this exercise, we will study the dual codes C_{k-1} and their syndromes.
a. Exercises 9 and 10 of §3 of Chapter 9 show that C_{k-1} is also a cyclic code of block length $n = q - 1$. Determine the generator polynomial and the set of zeroes for the dual Reed-Solomon code C_{k-1}.
b. Use the components of a word y in $\mathbb{F}_q^{q-1}$ as coefficients of a polynomial $y(t)$ as in Chapter 9, §3. Show that the values of $y(t)$ at the zeroes of C_{k-1} are the same as certain syndromes defined as in (2.6).

Exercise 12. By adapting the proof of Proposition (2.9), show (2.14) above:

$$\Delta_k = \{\text{span}_e(f) : f \in R, \ \text{fail}_e(f) \leq k\}.$$

Exercise 13. Let $a_1, \ldots, a_t$ be the minimal generators of the monoid $(\mathbb{Z}_{\geq 0}, \oplus)$. Show that

$$\sigma_k = \min_{\gg} \left(\{c \oplus a_i : c \in \Delta_k, 1 \leq i \leq t\} \setminus \Delta_k \right).$$

Since Δ_k is a finite set, this shows σ_k is a finite set and provides an algorithm for computing σ_k, provided we know Δ_k and the minimal generators a_i.

Exercise 14. Complete the proof of part b of Lemma (2.15). Hint: Use the nesting properties of these codes: $\cdots \subseteq C_{a+1} \subseteq C_a \subseteq C_{a-1} \subseteq \cdots$.

§3 The Details of the BMS Algorithm

We continue the development of the BMS decoding algorithm from §2, keeping all notation and standing assumptions from §1 and §2.

From Σ_{k-1} and Δ_{k-1} to Σ_k and Δ_k

Our goal in this section is to study the changes that occur in one *Info Update* step from the main loop in step 2 of (2.18). By Exercise 7 of §2 and the definition of the δ_k and σ_k, the following are possible.

- Some $s \in \Sigma_{k-1}$ can "switch" into Δ_k.
- In the process, the set δ_k can pick up elements that were not in δ_{k-1} from Σ_{k-1}. Elements from δ_{k-1} that are no longer $\gg$-maximal may also not appear in δ_k.
- Similarly, elements from σ_{k-1} can "switch" into Δ_k, and σ_k can pick up new $\gg$-minimal elements of the remaining set Σ_k.

A large part of the "action" of the BMS algorithm occurs in connection with the three sets

$$W_k = \Delta_k \setminus \Delta_{k-1} = \Delta_k \cap \Sigma_{k-1}, \quad \sigma_k \setminus \sigma_{k-1}, \quad \text{and} \quad \sigma_{k-1} \setminus \sigma_k.$$

(W_k is precisely the set of $s \in \Sigma_{k-1}$ that "switch" into Δ_k.) Hence it will be vital to understand them.

We start by pushing the line of reasoning leading to Lemma (2.10) a bit further.

(3.1) Lemma.
a. $\mathrm{sp}_e \circ \mathrm{sp}_e$ is the identity mapping on $\Delta(e)$.
b. $\mathrm{fl}_e(\mathrm{sp}_e(s)) = \mathrm{fl}_e(s)$ for all $s \in \Delta(e)$.

PROOF. By part b of Exercise 3 of §2, it follows that $\mathrm{sp}_e(\mathrm{sp}_e(s)) \le s$ for all $s \in \Delta(e)$. This observation and Lemma (2.10) imply that $\mathrm{sp}_e \circ \mathrm{sp}_e$ is a bijective and nonincreasing mapping from $\Delta(e)$ to itself. Since $\Delta(e)$ is a finite set, it is easy to see that $\mathrm{sp}_e \circ \mathrm{sp}_e$ must be the identity mapping. See Exercise 2 below.

For part b, we use the definition of fl_e and part a to see

$$\mathrm{fl}_e(\mathrm{sp}_e(s)) = \mathrm{sp}_e(s) \oplus \mathrm{sp}_e(\mathrm{sp}_e(s)) = \mathrm{sp}_e(s) \oplus s = \mathrm{fl}_e(s),$$

which is what we wanted to show.   □

Next, we derive an important symmetry property of W_k.

(3.2) Proposition.

a. $W_k = \{s \in \Delta_k : \mathrm{fl}_e(s) = k\}$.

b. *If $s \in W_k$, then $k \gg s$. Moreover, $s \in W_k$ if and only if $k \ominus s \in W_k$.*

c. $\max_{\gg}(W_k) = \{k \ominus s : s \in \sigma_{k-1} \setminus \sigma_k\}$.

PROOF. The definitions of Δ_k and fl_e easily imply that

$$\Delta_k = \{s \in \mathbb{Z}_{\geq 0} : \mathrm{fl}_e(s) \leq k\}.$$

Part a now follows immediately.

We turn to part b. Part a shows $k = s \oplus \mathrm{sp}_e(s)$, so $k \gg s$ and $k \ominus s$ exists. Now, if $s \in W_k$, then $\mathrm{fl}_e(s) = s \oplus \mathrm{sp}_e(s) = k$, so $k \ominus s = \mathrm{sp}_e(s)$. By Lemma (3.1),

$$\mathrm{fl}_e(k \ominus s) = \mathrm{fl}_e(\mathrm{sp}_e(s)) = \mathrm{fl}_e(s) = k,$$

which shows $k \ominus s \in W_k$ by part a. Since it is easy to see $k \ominus (k \ominus s) = s$, the opposite implication here follows as well.

We leave the proof of part c to the reader as Exercise 3. $\qquad\square$

The following corollary will be used at two points later.

(3.3) Corollary. *Let $s \in \sigma_{k-1} \setminus W_k$ and $k \ominus s \in \Sigma_{k-1}$. Then $\mathrm{fail}_e(F_{k-1}(s)) > k$, where $F_{k-1}(s)$ comes from the data structure $\mathrm{Info}(k-1)$ in the BMS algorithm. Moreover, if s is as above, then $\mathrm{fail}_e(F_{k-1}(s)z_a) > k$ for all a.*

PROOF. If the first claim is false, then $\mathrm{fail}_e(F_{k-1}(s)) = k$ exactly. Hence $\mathrm{span}_e(F_{k-1}(s)) = k \ominus s \in \Delta_k$ by (2.14). Our hypotheses then imply that $k \ominus s \in W_k$, and Proposition (3.2) shows $s \in W_k$ as well. But that is a contradiction to $s \in \sigma_{k-1} \setminus W_k$.

For the second claim, we have $\mathrm{fail}_e(F_{k-1}(s)z_a) > k$ by part a of Exercise 8 from §2. $\qquad\square$

Now we turn to the set $\sigma_k \setminus \sigma_{k-1}$ and study the effect of the mapping $t \mapsto k \ominus t$ on this set.

(3.4) Proposition. *Let $t \in \sigma_k \setminus \sigma_{k-1}$. Then either $k \not\gg t$ or $k \ominus t \in \Delta_{k-1}$.*

PROOF. Since $\sigma_k \subset \Sigma_k \subset \Sigma_{k-1}$, there exists $a \geq 0$ and $s \in \sigma_{k-1}$ such that $t = s \oplus a$. Since $t \gg s$, but t is $\gg$-minimal in Σ_k, we must have $s \in \Delta_k$. On the other hand, $s \notin \Delta_{k-1}$ (because it is in σ_{k-1}). So $s \in W_k$ as above. By Proposition (3.2), $k \ominus s \in W_k$ also. Assume $k \gg t$, so $k = t \oplus b$ for some $b \in \mathbb{Z}_{\geq 0}$. Then substituting for t, $k = (s \oplus a) \oplus b$. We have $k \ominus s = a \oplus b \gg b = k \ominus t$. Since Δ_k is a delta-set, this implies $k \ominus t \in \Delta_k$. But $t \notin W_k$ since $t \in \sigma_k \subset \Sigma_k$. By Proposition (3.2) again, $k \ominus t \notin W_k$, which shows $k \ominus t \in \Delta_{k-1}$. $\qquad\square$

We are now ready to develop the *FRMV* and *InfoUpdate* procedures from (2.18).

The Feng-Rao Majority Voting Procedure

We assume that we are in the situation at the start of a pass through the main loop in step 2 of (2.18) for some $k \geq a+1$. That is, we have Info$(k-1)$ and $S_e(z_0), \ldots, S_e(z_{k-1})$ are known. By the linearity of S_e, this means that the syndromes $S_e(f)$ are known for all $f \in L_{k-1} = \mathrm{Span}\{z_j : j < k\}$. The following observation gives the idea for a way to determine $S_e(z_k)$ from known information.

(3.5) Lemma. *Suppose we know $S_e(f)$ for all $f \in L_{k-1}$, and assume $k \gg t$. Let f_t be any element of R with $o(f_t) = t$, so there is $\lambda(t) \in \mathbb{F}_q \setminus \{0\}$ such that $f_t z_{k \ominus t} + \lambda(t) z_k \in L_{k-1}$ and hence*

$$\alpha(t) = S_e(f_t z_{k \ominus t} + \lambda(t) z_k)$$

is known. If in addition

$$\mathrm{fail}_e(f_t) > k,$$

then

$$S_e(z_k) = \lambda(t)^{-1} \alpha(t).$$

PROOF. We have $S_e(f_t z_{k \ominus t}) = 0$ since $\mathrm{fail}_e(f_t) > k$ and $o(f_t z_{k \ominus t}) = k$. Hence

$$\alpha(t) = S_e(f_t z_{k \ominus t}) + \lambda(t) S_e(z_k) = 0 + \lambda(t) S_e(z_k),$$

which gives the desired formula.   □

In order to compute the previously unknown syndrome value $S_e(z_k)$, then, we must find $f_t \in R$ with $o(f_t) = t$, $k \gg t$, and $\mathrm{fail}_e(f_t) > k$. But such functions are provided for us in the data structures used in the BMS algorithm *if we know where to look*. To see this, consider

$$\Gamma_k = \{t \in \Sigma_{k-1} : k \gg t \text{ and } k \ominus t \in \Sigma_{k-1}\}.$$

By Proposition (3.2), $W_k = \Delta_k \setminus \Delta_{k-1} = \Sigma_{k-1} \setminus \Sigma_k$ is contained in Γ_k. But if $t \in W_k$, we saw in part a of Proposition (3.2) that the maximum possible fail_e for f with $o(f) = t$ is exactly k. In fact, it is functions whose orders lie in *the complement $V_k = \Gamma_k \setminus W_k$* that we need to consider. Indeed, Corollary (3.3) above shows that if $s \in V_k \cap \sigma_{k-1}$, then the function $f_s = F_{k-1}(s)$ from Info$(k-1)$ satisfies the hypotheses of Lemma (3.6), and similarly for $f_t = F_{k-1}(s) z_a$ if $t = s \oplus a$, with $s \in V_k \cap \sigma_{k-1}$.

So our basic idea at this point is that we can use any t of the form $t = s \oplus a$ for $s \in V_k \cap \sigma_{k-1}$, and apply Lemma (3.5) to $f_t = F_{k-1}(s) z_a$ to determine $S_e(z_k)$.

However, there is one important point we must address here to turn this line of thought into a procedure for computing $S_e(z_k)$. Even though it is easy in principle to compute Γ_k using Info($k-1$), *we will not yet know* Δ_k and Σ_k, hence we will have no way to determine the sets V_k and W_k at the point in the BMS algorithm when we need to compute the new syndrome value $S_e(z_k)$.

This sounds bad, but it is precisely here that the extremely clever *majority voting* idea comes in! For each $t \in \Gamma_k$, we pick some way of writing $t = s \oplus a$ with $s \in \sigma_{k-1}$ as above. Using $f_t = F_{k-1}(s)z_a$, the value $\lambda(t)^{-1}\alpha(t)$ from Lemma (3.5) can be thought of as *a vote* for the value of $S_e(z_k)$. We have a vote for each $t \in \Gamma_k$.

The following observation shows that if we have several choices for writing $t = s \oplus a$ with $s \in \sigma_{k-1}$, then *any one of them will do*, because if $t \in V_k$, then s will be in V_k.

(3.6) Lemma. *If $t \in V_k$ and $t = s \oplus a$ for some $s \in \sigma_{k-1}$, then $s \in V_k$ also.*

PROOF. The contrapositive is easier to prove here. Let $s \in W_k \cap \sigma_{k-1}$ and $t = s \oplus a$. By Proposition (3.2), $k \ominus s \in W_k$ as well. Moreover, $t \gg s$ implies $k \ominus s \gg k \ominus t$. Since $W_k \subset \Delta_k$ and Δ_k is a delta-set, this shows $k \ominus t \in \Delta_k$. Now by hypothesis, $k \ominus t \in \Gamma_k$ also. It follows that $k \ominus t \in \Sigma_{k-1} \cap \Delta_k = W_k$. But then Proposition (3.2) shows $t \in W_k$.   □

It follows that all the $t \in V_k$ cast votes for the correct value of $S_e(z_k)$. You will show in Exercise 4 below that the elements of $t \in W_k$ cast votes for incorrect values. The majority vote will be the correct value if and only if $|V_k| > |W_k|$. The following theorem shows that this will be the case if $\mathrm{wt}(e)$ is small enough.

(3.7) Theorem (Feng-Rao). *Let e be an error vector satisfying*

$$2\mathrm{wt}(e) < n_k = |N_k|.$$

Then $|V_k| > |W_k|$.

PROOF. We have $|\Delta(e)| = \mathrm{wt}(e)$ by Proposition (2.7). Moreover, Δ_{k-1} and W_k are disjoint subsets of $\Delta(e)$. Hence

(3.8) $$|\Delta_{k-1}| + |W_k| \leq \mathrm{wt}(e).$$

Given s in N_k, we either have $s \in \Sigma_{k-1}$ or $s \in \Delta_{k-1}$, and in the first case either $k \ominus s \in \Sigma_{k-1}$ or $k \ominus s \in \Delta_{k-1}$. By definition Γ_k is the set of $s \in \Sigma_{k-1}$ such that $k \ominus s \in \Sigma_{k-1}$, and we know $\Gamma_k = V_k \cup W_k$. Hence N_k is contained in the union of the four sets

$$V_k, \ W_k, \ \{s \in \Sigma_{k-1} : k \ominus s \in \Delta_{k-1}\}, N_k \cap \Delta_{k-1}.$$

Since the last two sets have cardinality at most $|\Delta_{k-1}|$,

(3.9) $n_k \leq |V_k| + |W_k| + 2|\Delta_{k-1}|$.

From (3.8), $|\Delta_{k-1}| \leq \text{wt}(e) - |W_k|$. Substituting into (3.9), we get

$$n_k - 2\text{wt}(e) \leq |V_k| - |W_k|,$$

which establishes the claim. $\square$

Summarizing our discussions here, we have the following plan for the *FRMV* procedure. For each $t \in \Gamma_k$, write $t = s \oplus a$ for some $s \in \sigma_{k-1}$ and $a \geq 0$. Compute $\lambda(t)^{-1}\alpha(t)$ as in Lemma (3.5), using $f_t = F_{k-1}(s)z_a$. The correct value of $S_e(z_k)$ will be the value obtained for the majority of the t, provided the hypothesis of Theorem (3.7) on the error e is satisfied. In Exercise 5 below, you will develop a pseudocode algorithm based on this informal description.

We also have the following result.

(3.10) Corollary. *Provided that* $\text{wt}(e) \leq \lfloor (d_{FR}(C_a) - 1)/2 \rfloor$, *we have* $|V_k| > |W_k|$ *for all* $k > a$.

PROOF. Recall that $d_{FR}(C_a) = \min\{n_b : b > a\}$. Under the assumption $\text{wt}(e) \leq \lfloor (d_{FR}(C_a) - 1)/2 \rfloor$, the hypothesis of Theorem (3.7) will be satisfied for all $k > a$. $\square$

In intuitive terms, Corollary (3.10) says *all syndromes needed for BMS can be computed by the majority voting process provided that* $\text{wt}(e) \leq \lfloor (d_{FR}(C_a) - 1)/2 \rfloor$. This is part of the proof of Theorem (2.20).

An argument similar to the proof of Theorem (3.7) establishes the same conclusion with the weaker hypothesis $2|N_k \cap \Delta(e)| < n_k$. This shows that BMS decoding can often succeed even for error vectors of weight larger than $\lfloor \frac{d_{FR}(C_a)-1}{2} \rfloor$. See Exercise 6 below.

The InfoUpdate Procedure

The following theorem gives the key *InfoUpdate* procedure of the BMS algorithm, showing how to produce Info(k) from Info($k - 1$). The correctness proof for this algorithm will complete the proof of Theorem (2.20). Note how many of the same facts used to develop the *FRMV* procedure also underlie the different updating steps performed here.

(3.11) Theorem. *Assume* Info($k - 1$) $= (\sigma_{k-1}, F_{k-1}, \delta_{k-1}, G_{k-1})$ *satisfies* $\sigma_{k-1} = \min_{\gg}(\Sigma_{k-1})$, $\delta_{k-1} = \max_{\gg}(\Delta_{k-1})$, *and for each* $s \in \sigma_{k-1}$, $o(F_{k-1}(s)) = s$ *and* $\text{fail}_e(F_{k-1}(s)) > k - 1$, *and for each* $c \in \delta_{k-1}$, $\text{span}_e(G_{k-1}(c)) = c$, $\text{fail}_e(G_{k-1}(c)) \leq k - 1$, *and* $S_e(G_{k-1}(c)z_c) = 1$. *Then the output of the* InfoUpdate *algorithm on the following page is the* *correct* Info(k), *satisfying the above conditions, with $k - 1$ replaced by k everywhere.*

Input: Info($k - 1$), R, a normalized basis $\{z_i\}$, Syn(k)

Output: Info(k) $= (\sigma_k, F_k, \delta_k, G_k)$

$\delta' := \emptyset$

FOR $s \in \sigma_{k-1}$ DO

    IF $k \gg s$ THEN

        $\beta(s) := S_e(F_{k-1}(s)z_{k\ominus s})$

        IF $\beta(s) \neq 0$ and $k \ominus s \in \Sigma_{k-1}$ THEN

            $\delta' := \delta' \cup \{k \ominus s\}$

$\delta_k := \max_{\gg}(\delta' \cup \delta_{k-1})$

$\sigma_k := \min_{\gg}\{t \in \mathbb{Z}_{\geq 0} : \text{there is no } c \in \delta_k \text{ such that } c \gg t\}$

FOR $c \in \delta_k$ DO

    IF $c \in \delta_{k-1}$ THEN

        $G_k(c) := G_{k-1}(c)$

    ELSE

        $G_k(c) := (\beta(k \ominus c))^{-1} F_{k-1}(k \ominus c)$

FOR $t \in \sigma_k$ DO

    find $s \in \sigma_{k-1}, u$ such that $t = s \oplus u$

    IF there exist $c \in \delta_{k-1}, b \in \mathbb{Z}_{\geq 0}$ such that $k \ominus t = c \ominus b$ THEN

        $F_k(t) := F_{k-1}(s)z_u - \beta(s)G_{k-1}(c)z_b$

    ELSE

        $F_k(t) := F_{k-1}(s)z_u$

PROOF. The first loop over $s \in \sigma_{k-1}$ is designed to find the $\gg$-maximal elements of W_k using Proposition (3.2), part c. We claim that $\beta(s) \neq 0$ and $k \ominus s \in \Sigma_{k-1}$ if and only if $s \in \sigma_{k-1} \setminus \sigma_k$. First, if $s \in \sigma_{k-1} \setminus \sigma_k$, then $s \in \Delta_k \setminus \Delta_{k-1} = W_k$, hence $k \ominus s \in W_k \subset \Sigma_{k-1}$ by Proposition (3.2), part b. Moreover, $\beta(s) \neq 0$ follows easily since $\mathrm{fl}_e(s) = k$ by part a of Proposition (3.2). Conversely, if $\beta(s) \neq 0$ and $k \ominus s \in \Sigma_{k-1}$, then $s \in \sigma_{k-1} \setminus \sigma_k$ by Corollary (3.3). The value $k \ominus s = \mathrm{span}_e(F_{k-1}(s))$ is inserted into δ' for those s. After that first loop is complete, the set $\delta_{k-1} \cup \delta'$ contains the $\gg$-maximal elements of Δ_{k-1} and of $W_k = \Delta_k \setminus \Delta_{k-1}$. Hence the assignment for δ_k will yield $\max_{\gg}(\Delta_k)$. The assignment for σ_k is correct since δ_k is the set of $\gg$-maximal elements of the delta-set Δ_k, so $\Sigma_k = \{t \in \mathbb{Z}_{\geq 0} : \text{there is no } c \in \delta_k \text{ such that } c \gg t\}$. The idea for an algorithm for determining σ_k from Δ_k is indicated in Exercise 13 of §2.

The next loop over $c \in \delta_k$ produces $G_k(c)$ satisfying $\text{span}_e(G_k(c)) = c$ and $\text{fail}_e(G_k(c)) \leq k$. If $c \in \delta_{k-1}$, then by hypothesis $G_{k-1}(c)$ has span c and fail at most $k - 1 < k$, so no change is necessary. On the other hand, if $c \in \delta_k \setminus \delta_{k-1}$, then c is one of the $k \ominus s$ from δ' in the first loop and $F_{k-1}(k \ominus c) = F_{k-1}(s)$ has fail exactly k. Multiplying by $(\beta(k \ominus c))^{-1}$ normalizes $G_k(c)$ so that $S_e(G_k(c)z_{k\ominus c}) = 1$.

Finally, the loop over $t \in \sigma_k$ produces $F_k(t)$ satisfying $o(F_k(t)) = t$ and $\text{fail}_e(F_k(t)) > k$. For each such t, there exists $s \in \sigma_{k-1}$ and $u \geq 0$ such that $t = s \oplus u$ since $\sigma_k \subset \Sigma_{k-1}$. There are three cases here.

First, assume $k \not\gg t$. This implies the condition in the if statement cannot be satisfied, so the else block will be executed. We have $o(F_{k-1}(s)z_u) = s \oplus u = t$. Moreover $\text{fail}_e(F_{k-1}(s)z_u) > k - 1$ by hypothesis and Exercise 8 of §2. It cannot equal k since $k \not\gg t$, so there are no $g \in R$ with $o(F_{k-1}(s)z_u g) = k$. Therefore, $\text{fail}_e(F_{k-1}(s)z_u) > k$.

Second, suppose $k \gg t$, but $t = s \oplus u \notin \sigma_{k-1}$. Then Proposition (3.4) implies $k \ominus t \in \Delta_{k-1}$. So $k \ominus t = c \ominus b$ for some $c \in \delta_{k-1}$ and $b \in \mathbb{Z}_{\geq 0}$, and the condition in the if statement will be true. Write $v = k \ominus t = c \ominus b$, so $k = s \oplus u \oplus v$ and $c = b \oplus v$. We must show first that $o(F_k(t)) = t$. We have $o(F_{k-1}(s)z_u) = s \oplus u = t$ as in the first case, so to show $o(F_k(t)) = t$, it suffices to see that $o(G_{k-1}(c)z_b) = o(G_{k-1}(c)) \oplus b < t$. But by hypothesis, $\text{fail}_e(G_{k-1}(c)) = o(G_{k-1}(c)) \oplus c \leq k - 1$. Therefore, $o(G_{k-1}(c)) \oplus c \oplus b < k \oplus b = t \oplus c$. It follows that $o(G_{k-1}(c)) \oplus b < t$ as desired. Next, we must show that $\text{fail}_e(F_k(t)) > k$. If the value of $\beta(s)$ computed in the first loop was zero, then $\text{fail}_e(F_{k-1}(s)) > k$. But then $F_k(t) = F_{k-1}(s)z_u$ has $\text{fail}_e(F_k(t)) > k$ by part a of Exercise 8 from §2. Hence to conclude this case, we can assume $\beta(s) \neq 0$, so $\text{fail}_e(F_{k-1}(s)) = k$. We have $k = t \oplus v$ and $t = s \oplus u$, so $\text{span}_e(F_{k-1}(s)z_u) = v = k \ominus t = c \ominus b$. However, $\text{span}_e(G_{k-1}(c)) = c$, so $\text{span}_e(G_{k-1}(c)z_b) = c \ominus b$ also. By part c of Exercise 3 of §2, some linear combination of $F_{k-1}(s)z_u$ and $G_{k-1}(c)z_b$ will have a larger span_e value, hence fail_e value $> k$. The appropriate linear combination uses the $\beta(s)$ computed in the first loop. Since $k \ominus s = u \oplus v$, and the z_i are assumed to be a normalized basis, we have $o(z_u z_v - z_{k\ominus s}) < k \ominus s$ and $o(z_b z_v - z_c) < c$. So $S_e(F_{k-1}(s)(z_u z_v - z_{k\ominus s})) = 0$ and $S_e(G_{k-1}(c)(z_b z_v - z_c)) = 0$. It follows that $S_e(F_{k-1}(s)z_u z_v) = S_e(F_{k-1}(s)z_{k\ominus s}) = \beta(s)$, and $S_e(G_k(c)z_b z_v) = S_e(G_k(c)z_c) = 1$. Hence $S_e(F_k(t)z_v) = 0$, which shows $\text{fail}_e(F_k(t)) > k$.

Third, we must handle the case $k \gg t$ and $t \in \sigma_{k-1}$. In this case $s = t$ and $u = 0$. If $k \ominus t \in \Delta_{k-1}$, then the previous paragraph applies. If not, then $k \ominus t \in \Sigma_{k-1}$. This is precisely the case covered by Corollary (3.3), and $F_k(t) = F_{k-1}(t)$ has $\text{fail}_e(F_k(t)) > k$. $\square$

Termination of the Main Loop in BMS

The final ingredient needed to make the BMS algorithm explicit is a criterion showing that the loop in step 2 of Algorithm (2.18) can be terminated after a finite number of steps because no further changes in the $\text{Info}(k)$ can occur. At that point, all of the $F_k(s)$ for $s \in \sigma_k$ will be elements of I_e, as we saw in §2.

(3.12) Proposition. *Let $\Delta(e)$ be as above. Let c_{max} be the largest element of $\Delta(e)$ and s_{max} be the largest element of $\sigma = \min_{\gg} \Sigma(e)$.*
a. *Let $K = c_{max} \oplus c_{max}$. Then for all $k \geq K$, $\Delta_k = \Delta(e)$ and $\Sigma_k = \Sigma(e)$.*
b. *The computation in step 2 of (2.18) will be complete after the iteration with $k = M = c_{max} \oplus \max\{c_{max}, s_{max}\}$.*

PROOF. Since the Δ_k are an increasing, nested collection of subsets of $\Delta(e)$, it suffices to show that $\Delta(e) \subseteq \Delta_K$. Take $s \in \Delta(e)$ and pick any f such that $s = o(f)$. The definition of $\Delta(e)$ implies that $f \notin I_e$, and then $\mathrm{span}_e(f) \in \Delta(e)$ by Proposition (2.9). It follows that $\mathrm{fail}_e(f) = o(f) \oplus \mathrm{span}_e(f) \leq c_{max} \oplus c_{max} = K$. Since this is true for all f with $o(f) = s$, we conclude that $s \in \Delta_K$. This establishes part a.

For part b, if $s \in \sigma = \min_{\gg}(\Sigma(e))$ and $k > s \oplus c_{max}$, then $\mathrm{span}_e(F_k(s)) > k \ominus s > c_{max}$. By Proposition (2.9), and the definition of c_{max}, this implies $F_k(s) \in I_e$. Hence once $k \geq M$, no changes will occur in any of the data structures used in the BMS algorithm.   □

When running the BMS algorithm, we don't know $\Delta(e)$ in advance, yet the bound on k given in part b of Proposition (3.12) depends on $\Delta(e)$. We can overcome this seeming difficulty as follows. The key point is the assumption

$$\mathrm{wt}(e) \leq \lfloor (d_{FR}(C_a) - 1)/2 \rfloor$$

from Theorem (2.20). This implies that $\Delta(e)$ is a delta-set of bounded cardinality. In Exercise 8, you will show that there are only finitely many such delta-sets. So by computing the bound given in part b of Proposition (3.12) for each of these delta-sets, we can get a uniform bound for when to terminate the algorithm. An example of how this works in practice will be given in Exercise 9 below. Although this process would be somewhat cumbersome on a larger code, it would only need to be done once. Some other examples of the criterion in Proposition (3.12) are given in [O'Su2].

An Example

We illustrate the BMS decoding algorithm and Feng-Rao majority voting procedure with an extended example.

We will consider the order domain R of Example (1.3). The monomials in the set Δ give a normalized basis $\{z_i\}$ for R in this case by Exercise 6 of §2. Consider the C_4 code (dual of the E_4 code, whose generator matrix is obtained by evaluating $1, x_1, x_2, x_1^2, x_1 x_2$ at the eight $\mathbb{F}_4$-rational points on the Hermitian curve $x_1^3 + x_2^2 + x_2 = 0$ (ordered as in (1.15)). C_4 has parameters $[8, 3, \geq 5]$ since $d_{FR}(C_4) = 5$ by Exercise 9 of §2. The error vector $e = (0, 1, 0, 0, 0, 0, 0, 1)$ has weight 2, hence satisfies the hypotheses of Theorem (3.7) and Corollary (3.10). The known syndromes are $S_e(1) = 0$, $S_e(x_1) = \alpha^2$, $S_e(x_2) = \alpha$, $S_e(x_1^2) = \alpha$, $S_e(x_1x_2) = \alpha$.

Following Algorithm (2.18), we initialize $\sigma_{-1} = \{0\}, F_{-1}(0) = 1$, and $\delta_{-1} = \emptyset$. In the first pass through the loop in step 2 ($k = 0$), the call to *InfoUpdate* from Theorem (3.11) changes nothing since $S_e(F_{-1}(0)) = S_e(1) = 0$. Hence $\delta_0 = \delta_{-1}$, $\sigma_0 = \sigma_{-1}$, and $F_0(0) = F_{-1}(0) = 1$.

In the second pass, we have $k = 1$. Following *InfoUpdate* from Theorem (3.11), with $s = 0$, we have $\beta(0) = S_e(1 \cdot x_1) = \alpha^2$ and $k \ominus s = 1 \ominus 0 = 1 \in \Sigma_0$. Hence $\delta' = \delta_1$ is assigned the value $\{1\}$, $\sigma_1 = \min_{\gg}(\mathbb{Z}_{\geq 0} \setminus \Delta_1) = \{2, 3\}$, and $G_1(1) = 1/\alpha^2$. For both $t \in \sigma_1$, $s = 0$ and $u = t$. The condition in the if statement in this block of the algorithm is false in both cases, so we take the else branch and $F_1(2) = x_2$, $F_1(3) = x_1^2$. (Note that $|\Delta_1| = 2 = \mathrm{wt}(e)$ already in this case.)

In the third pass, we have $k = 2$. With $s = 2 \in \sigma_1$, $\beta(2) = S_e(F_1(2)) = S_e(x_2) = \alpha$, but $k \ominus s = 0 \notin \Sigma_1$. With $s = 3$, $2 \gg 3$ is not true so we do nothing. Hence δ' is empty in this pass, and $\delta_2 = \delta_1$, $\sigma_2 = \sigma_1$ do not change, and $G_2(1) = G_1(1) = 1/\alpha^2$. The $F_1(t)$ are updated as follows. For $t = 2$, $2 \ominus 2 = 0 = 1 \ominus 1$, so the condition in the if statement is satisfied. $F_2(2) = x_2 + \alpha/\alpha^2 x_1 = x_2 + \alpha^2 x_1$. With $t = 3$, $2 \ominus 3$ is not of the form $c \ominus b$ for $c \in \delta_1$. Hence $F_2(3) = F_1(3) = x_1^2$.

In the fourth pass through *InfoUpdate* from Theorem (3.11), $k = 3$. In the loop over $s \in \sigma_2$, $3 \not\gg 2$, but $\beta(3) = S_e(F_2(3)) = \alpha$. However, $3 \ominus 3 \notin \Sigma_2$, so $\delta_3 = \delta_2$ and $\sigma_3 = \sigma_2$ are unchanged again. Hence $G_3(1) = G_2(1) = 1/\alpha^2$. In the loop over $t \in \sigma_3$, with $t = 2$ and $u = 0$, there are no c, b such that $3 \ominus 2 = c \ominus b$. Hence $F_3(2) = F_2(2) = x_2 + \alpha^2 x_1$. But with $t = 3$ (also $u = 0$), $3 \ominus 3 = 1 \ominus 1$, so $F_3(3) = F_2(3) - \alpha/\alpha^2 x_1 = x_1^2 + \alpha^2 x_1$.

In the fifth pass ($k = 4$), in the first loop over $s \in \sigma_3$, we compute $\beta(2) = S_e(F_3(2)x_1) = S_e(x_1x_2) + \alpha^2 S_e(x_1^2) = \alpha + \alpha^2 \cdot \alpha = \alpha^2$. Similarly $\beta(3) = S_e(F_3(3)) = S_e(x_1^2 + \alpha^2 x_1) = \alpha + \alpha \cdot \alpha^2 = \alpha^2$. We have $\delta_4 = \delta_3$, $\sigma_4 = \sigma_3$, and $G_4(1) = G_3(1)$. In the loop over $t \in \sigma_4$, $4 \ominus 2 = 1 \ominus 0$, and $F_4(2) = F_3(2) - \alpha^2/\alpha^2 = x_2 + \alpha^2 x_1 + 1$. There are no c, b such that $4 \ominus 3 = c \ominus b$, so $F_4(3) = F_3(3) = x_1^2 + \alpha^2 x_1$.

By Proposition (3.12), $\Delta(e) = \{0, 1\}$ and $\Sigma(e) = \{j \in \mathbb{Z}_{\geq 0} : j \geq 2\}$. We have $c_{max} \oplus \max\{c_{max}, s_{max}\} = 1 \oplus \max\{1, 3\} = 5$, so one further iteration will suffice. The known syndromes are exhausted, though. So we must begin this iteration by using the Feng-Rao procedure to determine the first unknown syndrome value $S_e(x_2^2)$. We have $\Gamma_5 = \{s \in \Sigma_4 : 5 \ominus s \in \Sigma_4\} = \{2\}$, and $S_e(F_4(2)z_2 - z_5) = S_e((x_2 + \alpha^2 x_1 + 1)x_2 - x_2^2) = \alpha^2$, using the known syndrome values. This is the correct value of $S_e(x_2^2)$. However, in this pass, no further changes are made, and the final output of step 2 of the BMS algorithm is $I_e = \langle x_2 + \alpha^2 x_1 + 1, x_1^2 + \alpha^2 x_1 \rangle$. These two polynomials also give the Gröbner basis of J_e in this case. It is easy to check that $\mathbf{V}(J_e)$ consists of precisely the two points on X corresponding to the nonzero positions in e.

In this example, no changes occurred in our data structures after $k = 4$. In the following exercise, you will see a case where further iterations and computation of initially unknown syndrome values via the *FRMV* procedure are actually necessary.

Exercise 1. Carry out the BMS algorithm for C_4 with the error pattern $e = (0, 0, \alpha, 1, 0, 0, 0, 0)$. How is this case different from the example we worked out above?

Concluding Comments

The theory of order domains that we have presented gives a simplified treatment of the most important geometric Goppa codes and the BMS algorithm. It also creates the possibility of exploiting order domains from varieties of higher dimension for the construction of codes with good parameters and efficient decoding algorithms. This is an area where exploration has just begun as of this writing.

Historically, the BMS algorithm arose from Sakata's work on m-dimensional cyclic codes as in Chapter 9, §3. It was then applied to the geometric Goppa codes from curves mentioned in §1. The original version of BMS did not incorporate the majority voting procedure and hence other (far less efficient) methods for determining the needed unknown syndromes were originally employed. This difficulty was substantially overcome in an efficient way through the ingenious contribution of Feng, Rao, and Duursma (see [FR] and [HP]). Our presentation of *FRMV* follows [O'Su2] and is significantly simpler than the original formulation.

Additional Exercises for §3

Exercise 2. Show that if A is a finite subset of $\mathbb{Z}_{\geq 0}$ and $f : A \to A$ is a bijective mapping such that $f(a) \leq a$ for all $a \in A$, then f is the identity mapping.

Exercise 3. Prove part b of Proposition (3.2). Hint: Use the fact that if $k \gg t, s$, then $t \gg s$ if and only if $k \ominus s \gg k \ominus t$.

Exercise 4. Show that in the situation of the *FRMV* procedure all $t \in W_k$ cast incorrect votes for the value of the unknown syndrome $S_e(z_k)$.

Exercise 5. After Theorem (3.7) in the text, we sketched an informal outline for the *FRMV* procedure. Develop a pseudocode algorithm from this outline. Hint: One way to do this is to keep tallies of how many times each distinct vote occurs.

Exercise 6. In the situation of Theorem (3.7), assume that $2|N_k \cap \Delta(e)| < n_k$. Show that $|V_k| > |W_k|$ and the Feng-Rao majority voting process will produce the correct syndrome value $S_e(z_k)$. Hint: Partition N_k into four subsets by considering whether $c \in N_k$ is contained in $\Sigma(e)$ or $\Delta(e)$ and similarly for $k \ominus c$.

Exercise 7. Let c be a nonzero maximal element (with respect to $\gg$) of a delta-set $\Delta \subset \mathbb{Z}_{\geq 0}$. Also let $a_1, \ldots, a_t$ be the minimal generators of $(\mathbb{Z}_{\geq 0}, \oplus)$.
a. Prove that $\Delta \setminus \{c\}$ is a delta-set.
b. Prove that $c = s \oplus a_i$ for some $s \in \Delta \setminus \{c\}$ and i with $1 \leq i \leq t$.

Exercise 8. Let N be a positive integer. Prove that the number of delta-sets of $\mathbb{Z}_{\geq 0}$ of cardinality N is finite. Hint: Use the previous exercise and induction on N.

Exercise 9. Consider the C_4 code from the order domain R in Example (1.3), used in our example above and in Exercise 1. We have $d_{FR}(C_4) = 5$, so all error vectors of weight 2 or less are correctable by BMS decoding. We will see that it is possible to derive a uniform upper bound on the maximum number of iterations in the main loop in step 2 of (2.18) from Proposition (3.12) needed to correct any of these errors.
a. Show that if $w(e) = 1$, then $\Delta(e) = \{0\}$ and $\sigma = \{1, 2\}$.
b. Show that if $w(e) = 2$, then either $\sigma = \{2, 3\}$ and $\Delta(e) = \{0, 1\}$, or else $\sigma = \{1, 5\}$ and $\Delta(e) = \{0, 2\}$. Hint: What are the possible delta-sets $\Delta(e)$ with exactly two elements for this monoid $(\mathbb{Z}_{\geq 0}, \oplus)$?
c. Deduce that the main loop in step 2 of (2.18) will be complete after the iteration with $k = 8$ for all errors with $\text{wt}(e) \leq 2$.
d. In the same way, determine all possible delta-sets $\Delta(e)$ with $|\Delta(e)| = 3$ in this example.

References

[Act] F. Acton. *Numerical Methods That Work*, MAA, Washington DC, 1990.

[AL] W. Adams and P. Loustaunau. *An Introduction to Gröbner Bases*, AMS, Providence RI, 1994.

[ASW] P. Alfeld, L. Schumaker and W. Whiteley. *The generic dimension of the space of C^1 splines of degree $d \geq 8$ on tetrahedral decompositions*, SIAM J. Numerical Analysis **30** (1993), 889–920.

[AG] E. Allgower and K. Georg. *Numerical Continuation Methods*, Springer-Verlag, New York, 1990.

[AMR] M. Alonso, T. Mora and M. Raimondo. *A computational model for algebraic power series*, J. Pure Appl. Algebra **77** (1992), 1–38.

[AGK] B. Amrhein, O. Gloor and W. Küchlin. *On the walk*, Theoretical Computer Science **187** (1997), 179–202.

[AGV] V. Arnold, S. Guscin-Zade and V. Varchenko. *Singularities of Differential Maps, Volumes 1 and 2*, Birkhäuser, Boston, 1985 and 1988.

[Art] M. Artin. *Algebra*, Prentice-Hall, Englewood Cliffs NJ, 1991.

[AS] W. Auzinger and H. J. Stetter. *An elimination algorithm for the computation of all zeros of a system of multivariate polynomial equations*, in: *Numerical Mathematics (Singapore 1988)*, Internat. Ser. Numer. Math., **86**, Birkhäuser, Basel, 1988, 11–30.

[BGW] C. Bajaj, T. Garrity and J. Warren. *On the applications of multi-equational resultants*, Technical Report CSD-TR-826, Department of Computer Science, Purdue University, 1988.

[BC] V. Batyrev and D. Cox. *On the Hodge structure of projective hypersurfaces*, Duke Math. J. **75** (1994), 293–338.

[BaM] D. Bayer and I. Morrison. *Gröbner bases and geometric invariant theory I*, J. Symbolic Comput. **6** (1988), 209–217.

[BW] T. Becker and V. Weispfenning. *Gröbner Bases*, Springer-Verlag, New York, 1993.

[BKR] M. Ben-Or, D. Kozen and J. Reif. *The complexity of elementary algebra and geometry*, J. of Computation and Systems **32** (1986), 251–264.

[Ber] D. Bernstein. *The number of roots of a system of equations*, Functional Anal. Appl. **9** (1975), 1–4.

[Bet]   U. Betke. *Mixed volumes of polytopes*, Archiv der Mathematik **58** (1992), 388–391.

[BLR]   A. Bigatti, R. LaScala and L. Robbiano. *Computing toric ideals*, J. Symbolic Comput. **27** (1999), 351–365.

[BU]   P. Bikker and A. Yu. Uteshev. *On the Bézout construction of the resultant*, J. Symbolic Comput. **28** (1999), 45–88.

[Bil1] L. Billera. *Homology of smooth splines: Generic triangulations and a conjecture of Strang*, Trans. Amer. Math. Soc. **310** (1988), 325–340.

[Bil2] L. Billera. *The algebra of continuous piecewise polynomials*, Adv. Math. **76** (1989), 170–183.

[BR1]   L. Billera and L. Rose. *Groebner Basis Methods for Multivariate Splines*, in: *Mathematical Methods in Computer Aided Geometric Design* (T. Lyche and L. Schumaker, eds.), Academic Press, Boston, 1989, 93–104.

[BR2]   L. Billera and L. Rose. *A Dimension Series for Multivariate Splines*, Discrete Comp. Geom. **6** (1991), 107–128.

[BR3]   L. Billera and L. Rose. *Modules of piecewise polynomials and their freeness*, Math. Z. **209** (1992), 485–497.

[BS]   L. Billera and B. Sturmfels, *Fiber polytopes*, Ann. of Math. **135** (1992), 527–549.

[Bla]   R. Blahut. *Theory and Practice of Error Control Codes*, Addison Wesley, Reading MA, 1984.

[BMP]   D. Bondyfalat, B. Mourrain and V. Y. Pan. *Solution of polynomial system of equations via the eigenvector computation*, Linear Algebra Appl. **319** (2000), 193–209.

[BoF]   T. Bonnesen and W. Fenchel. *Theorie der konvexen Körper*, Chelsea, New York, 1971 and Springer-Verlag, New York, 1974.

[BH]   W. Bruns and J. Herzog. *Cohen-Macaulay Rings*, Cambridge U. Press, Cambridge, revised edition, 1998.

[Bry]   V. Bryant. *Aspects of Combinatorics*, Cambridge U. Press, Cambridge, 1993.

[BuM]   B. Buchberger and H. M. Möller. *The construction of multivariate polynomials with preassigned zeros*, in: *Computer algebra (Marseille, 1982)*, Lecture Notes in Computer Science **144**, Springer, Berlin, New York, 1982, 24–31.

[BE]   D. Buchsbaum and D. Eisenbud. *Algebra structures for finite free resolutions and some structure theorems for ideals of codimension 3*, Amer. J. Math. **99** (1977), 447–485.

[BZ]   Yu. Burago and V. Zalgaller, *Geometric Inequalities*, Springer-Verlag, New York, 1988.

[BuF]   R. Burden and J. Faires. *Numerical Analysis*, 5th edition, PWS Publishing, Boston, 1993.

[Bus]   L. Busé. *Residual resultant over the projective plane and the implicitization problem*, in: *Proceedings of International Symposium on Symbolic and Algebraic Computation*, ACM Press, New York, 2001, 48–55.

[BEM1] L. Busé, M. Elkadi and B. Mourrain. *Generalized resultants over unirational algebraic varieties*, J. Symbolic Comput. **29** (2000), 515–526.

[BEM2] L. Busé, M. Elkadi and B. Mourrain. *Resultant over the residual of a complete intersection*, J. Pure Appl. Algebra **164** (2001), 35–57.

[Can1] J. Canny. *Generalised characteristic polynomials*, J. Symbolic Comput. **9** (1990), 241–250.

[Can2] J. Canny. *Some algebraic and geometric computations in PSPACE*, in: *Proc. 20th Annual ACM Symposium on the Theory of Computing*, ACM Press, New York, 1988, 460–467.

[CE1] J. Canny and I. Emiris. *An efficient algorithm for the sparse mixed resultant*, in: *Applied Algebra, Algebraic Algorithms and Error-correcting codes (AAECC-10)* (G. Cohen, T. Mora and O. Moreno, eds.), Lecture Notes in Computer Science **673**, Springer-Verlag, New York, 1993, 89–104.

[CE2] J. Canny and I. Emiris. *A subdivision-based algorithm for the sparse resultant*, J. ACM **47** (2000), 417–451.

[CM] J. Canny and D. Manocha. *Multipolynomial resultant algorithms*, J. Symbolic Comput. **15** (1993), 99–122.

[CKL] J. Canny, E. Kaltofen and Y. Lakshman. *Solving systems of nonlinear polynomial equations faster*, in: *Proceedings of International Symposium on Symbolic and Algebraic Computation*, ACM Press, New York, 1989, 121–128.

[CDS] E. Cattani, A. Dickenstein and B. Sturmfels. *Residues and resultants*, J. Math. Sci. Univ. Tokyo **5** (1998), 119–148.

[Chi] E.-W. Chionh. *Rectangular corner cutting and Dixon A-resultants*, J. Symbolic Comput. **31** (2001), 651–669.

[CK1] A. Chtcherba and D. Kapur. *Conditions for exact resultants using Dixon formulation*, in: *Proceedings of International Symposium on Symbolic and Algebraic Computation*, ACM Press, New York, 2000, 62–70.

[CK2] A. Chtcherba and D. Kapur. *Extracting sparse resultant matrices from Dixon resultant formulation*, in: *Computer Algebra: Proceedings of the Seventh Rhine Workshop (RWCA'00)*, Bergenz, Austria, 2000, 167–182.

[CK3] A. Chtcherba and D. Kapur. *On the efficiency and optimality of Dixon-base resultant methods*, in: *Proceedings of International Symposium on Symbolic and Algebraic Computation*, ACM Press, New York, 2002.

[ColKM] S. Collart, M. Kalkbrener and D. Mall. *Converting bases with the Gröbner walk*, J. Symbolic Comput. **24** (1997), 465–469.

[CT] P. Conti and C. Traverso. *Buchberger algorithm and integer programming*, in: *Applied Algebra, Algebraic Algorithms and Error-correcting codes (AAECC-9)* (H. Mattson, T. Mora and T. Rao, eds.), Lecture Notes in Computer Science **539**, Springer-Verlag, New York, 1991, 130–139.

[Cox1] D. Cox. *The homogeneous coordinate ring of a toric variety*, J. Algebraic Geom. **4** (1995), 17–50.

[Cox2] D. Cox. *Equations of parametric curves and surfaces via syzygies*, in: *Symbolic Computation: Solving Equations in Algebra, Geometry and Engineering* (E. L. Green, S. Hoşten, R. C. Laubenbacher and V. A. Powers, eds.), Contemporary Mathematics **286**, AMS, Providence, RI, 2001, 1–20.

[Cox3] D. Cox. *Curves, Surfaces and Syzygies*, in: *Algebraic Geometry and Geometric Modeling* (R. Goldman and R. Krasauskas, eds.), Contemporary Mathematics **334**, AMS, Providence, RI, 2003, 131–150.

[Cox4] D. Cox. *What is a toric variety?*, in: *Algebraic Geometry and Geometric Modeling* (R. Goldman and R. Krasauskas, eds.), Contemporary Mathematics **334**, AMS, Providence, RI, 2003, 203–223.

[CoxKM] D. Cox, R. Krasauskas and M. Muştaţǎ. *Universal rational parametrizations and toric varieties*, in: *Algebraic Geometry and Geometric Modeling* (R. Goldman and R. Krasauskas, eds.), Contemporary Mathematics **334**, AMS, Providence, RI, 2003, 241–265.

[CLO] D. Cox, J. Little and D. O'Shea. *Ideals, Varieties, and Algorithms*, 2nd edition, Springer-Verlag, New York, 1997.

[CSC] D. Cox, T. Sederberg and F. Chen. *The moving line ideal basis of planar rational curves*, Computer Aided Geometric Design **15** (1998), 803–827.

[DKK] Y. Dai, S. Kim and M. Kojima. *Computing all nonsingular solutions of cyclic-n polynomial using polyhedral homotopy continuation methods*, J. Comput. Appl. Math. **152** (2003), 83–97.

[D'An] C. D'Andrea. *Macaulay style formulas for sparse resultants*, Trans. Amer. Math. Soc. **354** (2002), 2595–2629.

[DD] C. D'Andrea and A. Dickenstein. *Explicit formulas for the multivariate resultant*, in: *Effective Methods in Algebraic Geometry (Bath, 2000)*, J. Pure Appl. Algebra **164** (2001), 59–86.

[D'AE] C. D'Andrea and I. Emiris. *Computing sparse projection operators*, in: *Symbolic Computation: Solving Equations in Algebra, Geometry and Engineering* (E. L. Green, S. Hoşten, R. C. Laubenbacher and V. A. Powers, eds.), Contemporary Mathematics **286**, AMS, Providence, RI, 2001, 121–139.

[dBP1] M. de Boer and R. Pellikaan. *Gröbner bases for codes*, in: *Some Tapas of Computer Algebra* (A. M. Cohen, H. Cuypers, H. Sterk and H. Cohen, eds.), Springer-Verlag, Berlin, 1999, 237–259.

[dBP2] M. de Boer and R. Pellikaan. *Gröbner bases for decoding*, in: *Some Tapas of Computer Algebra* (A. M. Cohen, H. Cuypers, H. Sterk and H. Cohen, eds.), Springer-Verlag, Berlin, 1999, 260–275.

[Dev] R. Devaney. *A First Course in Chaotic Dynamical Systems*, Addison-Wesley, Reading MA, 1992.

[DS] P. Diaconis and B. Sturmfels. *Algebraic algorithms for sampling from conditional distributions*, Annals of Statistics **26** (1998), 363–397.

[DE]   A. Dickenstein and I. Emiris. *Multihomogeneous resultant for-mulae by means of complexes*, J. Symbolic Comput. **36** (2003), 317–342.

[Dim]   A. Dimca. *Singularities and Topology of Hypersurfaces*, Springer-Verlag, New York, 1992.

[Dre]   F. Drexler. *A homotopy method for the calculation of zeros of zero-dimensional ideals*, in: *Continuation Methods* (H. Wacker, ed.), Academic Press, New York, 1978.

[DGH]   M. Dyer, P. Gritzmann and A. Hufnagel. *On the complexity of computing mixed volumes*, SIAM J. Comput. **27** (1998), 356–400.

[Eis]   D. Eisenbud. *Commutative Algebra with a View Toward Algebraic Geometry*, Springer-Verlag, New York, 1995.

[EH]   D. Eisenbud and C. Huneke, eds. *Free Resolutions in Commutative Algebra and Algebraic Geometry*, Jones and Bartlett, Boston, 1992.

[EGSS]   D. Eisenbud, D. Grayson, M. Stillman and B. Sturmfels (eds.). *Computations in Algebraic Geometry with Macaulay 2*, Springer-Verlag, Berlin, 2002.

[EK]   G. Elber and M.-S. Kim. *The bisector surface of rational space curves*, ACM Transactions on Graphics **17** (1998), 32–49.

[ElM1]   M. Elkadi and B. Mourrain. *Some applications of bezoutians in effective algebraic geometry*, Rapport de Recherche **3572**, INRIA, 1998.

[ElM2]   M. Elkadi and B. Mourrain. *Approche effective des résidus algébriques*, Rapport de Recherche **2884**, INRIA, 1996.

[ElM3]   M. Elkadi and B. Mourrain. *A new algorithm for the geometric decomposition of a variety*, in: *Proceedings of International Symposium on Symbolic and Algebraic Computation*, ACM Press, New York, 1999, 9–16.

[Emi1]   I. Emiris. *On the complexity of sparse elimination*, Journal of Complexity **14** (1996), 134–166.

[Emi2]   I. Emiris. *A general solver based on sparse resultants*, in: *Proc. PoSSo (Polynomial System Solving) Workshop on Software*, Paris, 1995, 35–54.

[Emi3]   I. Emiris. *Matrix methods for solving algebraic systems*, in: *Symbolic Algebraic Methods and Verification Methods (Dagstuhl 1999)*, Springer-Verlag, Vienna, 2001, 69–78.

[EC]   I. Emiris and J. Canny. *Efficient incremental algorithms for the sparse resultant and the mixed volume*, J. Symbolic Comput. **20** (1995), 117–149.

[EmM]   I. Emiris and B. Mourrain. *Matrices in elimination theory*, J. Symbolic Comput. **28** (1999), 3–44.

[ER]   I. Emiris and A. Rege. *Monomial bases and polynomial system solving*, in: *Proceedings of International Symposium on Symbolic and Algebraic Computation*, ACM Press, New York, 1994, 114–122.

[Ewa]   G. Ewald. *Combinatorial Convexity and Algebraic Geometry*, Springer-Verlag, New York, 1996.

[Far]   G. Farin. *Curves and Surfaces for Computer Aided Geometric Design*, 2nd edition, Academic Press, Boston, 1990.

[FGLM]   J. Faugère, P. Gianni, D. Lazard and T. Mora. *Efficient computation of zero-dimensional Gröbner bases by change of ordering*, J. Symbolic Comput. **16** (1993), 329–344.

[FR]   G.-L. Feng and T. Rao. *Decoding of algebraic geometric codes up to the designed minimum distance*, IEEE Trans. Inform. Theory **39** (1993), 743–751.

[Fit1]   P. Fitzpatrick. *On the key equation*, IEEE Trans. Inform. Theory **41** (1995), 1290–1302.

[Fit2]   P. Fitzpatrick. *Solving a multivariable congruence by change of term order*, J. Symbolic Comput. **24** (1997), 575–589.

[FM]   J. Fogarty and D. Mumford. *Geometric Invariant Theory*, Second Edition, Springer-Verlag, Berlin, 1982

[FIS]   S. Friedberg, A. Insel and L. Spence. *Linear Algebra*, 3rd edition, Prentice-Hall, Englewood Cliffs, NJ, 1997.

[Ful]   W. Fulton. *Introduction to Toric Varieties*, Princeton U. Press, Princeton NJ, 1993.

[GL1]   T. Gao and T. Y. Li. *Mixed volume computation via linear programming*, Taiwanese J. Math. **4** (2000), 599–619.

[GL2]   T. Gao and T. Y. Li. *Mixed volume computation for semi-mixed systems*, Discrete Comput. Geom. **29** (2003), 257–277.

[GLW]   T. Gao, T. Y. Li and X. Wang. *Finding all isolated zeros of polynomial systems in $\mathbb{C}^n$ via stable mixed volumes*, J. Symbolic Comput. **28** (1999), 187–211.

[GS]   A. Garcia and H. Stichtenoth. *A tower of Artin-Schreier extensions of function fields attaining the Drinfeld-Vladut bound*, Invent. Math. **121** (1995), 211–222.

[GeP]   O. Geil and R. Pellikaan. *On the Structure of Order Domains*, Finite Fields Appl. **8** (2002), 369–396.

[GKZ]   I. Gelfand, M. Kapranov and A. Zelevinsky. *Discriminants, Resultants and Multidimensional Determinants*, Birkhäuser, Boston, 1994.

[GRRT]   L. Gonzalez-Vega, F. Rouillier, M.-F. Roy and G. Trujillo. *Symbolic recipes for real solutions*, in: *Some Tapas of Computer Algebra* (A. M. Cohen, H. Cuypers, H. Sterk and H. Cohen, eds.), Springer-Verlag, Berlin, 1999, 211–167.

[Grä]   H.-G. Gräbe. *Algorithms in Local Algebra*, J. Symbolic Comput. **19** (1995), 545–557.

[GrP]   G.-M. Greuel and G. Pfister. *A **Singular** Introduction to Commutative Algebra*, Springer-Verlag, Berlin, 2002.

[GuS]   L. Gurvits and A. Samorodnitsky. *A deterministic algorithm for approximating the mixed determinant and mixed volume, and a combinatorial corollary*, Discrete Comput. Geom. **27** (2002), 531–550.

[HeS]   C. Heegard and K. Saints. *Algebraic-geometric codes and multidimensional cyclic codes: Theory and algorithms for decoding using Gröbner bases*, IEEE Trans. Inform. Theory **41** (1995), 1733–1751.

[HLS] C. Heegard, J. Little and K. Saints. *Systematic Encoding via Gröbner Bases for a Class of Algebraic-Geometric Goppa Codes*, IEEE Trans. Inform. Theory **41** (1995), 1752–1761.

[Her] I. Herstein. *Topics in Algebra*, 2nd edition, Wiley, New York, 1975.

[Hil] D. Hilbert. *Ueber die Theorie der algebraischen Formen*, Math. Annalen **36** (1890), 473–534.

[Hir] H. Hironaka. *Resolution of singularities of an algebraic variety over a field of characteristic zero*, Ann. of Math. **79** (1964), 109–326.

[HP] T. Høholdt and R. Pellikaan. *On the Decoding of Algebraic Geometric Codes*, IEEE Trans. Inform. Theory **41** (1995), 1589–1614.

[HvLP] T. Høholdt, J. van Lint and R. Pellikaan. *Algebraic geometry codes*, Chapter 10, in: *Handbook of Coding Theory* (V. Pless and W. Huffman, eds.), Elsevier, Amsterdam, 1998.

[HT] S. Hoşten and R. Thomas. *Gröbner bases and integer programming*, in: *Gröbner Bases and Applications (Linz, 1998)* (B. Buchberger and F. Winkler, eds.), London Math. Soc. Lecture Notes Ser. **251**, Cambridge U. Press, (1998), 144–158.

[HuS1] B. Huber and B. Sturmfels. *A Polyhedral Method for Solving Sparse Polynomial Systems*, Math. Comp. **64** (1995), 1541–1555.

[HuS2] B. Huber and B. Sturmfels. *Bernstein's theorem in affine space*, Discrete Comput. Geom. **17** (1997), 137–141.

[HV] B. Huber and J. Verschelde. *Polyhedral end games for polynomial continuation*, Numer. Algorithms **18** (1998), 91–108.

[Jac] N. Jacobson. *Basic Algebra I*, W. II. Freeman, San Francisco, 1974.

[Jou1] J.-P. Jouanolou. *Le formalisme du résultant*, Adv. Math. **90** (1991), 117–263.

[Jou2] J.-P. Jouanolou. *Formes d'inertie et résultant: un formulaire*, Adv. Math. **126** (1997), 119–250.

[Kal] M. Kalkbrener. *On the complexity of Gröbner bases conversion*, J. Symbolic Comput. **28** (1999), 265–273.

[KSZ] M. Kapranov, B. Sturmfels and A. Zelevinsky. *Chow polytopes and general resultants*, Duke Math. J. **67** (1992), 189–218.

[KS1] D. Kapur and T. Saxena. *Comparison of various multivariate resultant formulations*, in: *Proceedings of International Symposium on Symbolic and Algebraic Computation*, ACM Press, New York, 1995.

[KS2] D. Kapur and T. Saxena. *Sparsity considerations in Dixon resultants*, in: *Proceedings of the Symposium on the Theory of Computing*, ACM Press, New York, 1996, 184–191.

[KSY] D. Kapur, T. Saxena and L. Yang. *Algebraic and geometric reasoning using Dixon resultants*, in: *Proceedings of International Symposium on Symbolic and Algebraic Computation*, ACM Press, New York, 1994, 99–107.

[Khe] A. Khetan. *Determinantal formula for the Chow form of a toric surface*, in: *Proceedings of International Symposium on Symbolic and Algebraic Computation*, ACM Press, New York, 2002, 145–150.

[Kho] A.G. Khovanskii. *Newton polytopes and toric varieties*, Functional Anal. Appl. **11** (1977), 289–298.

[Kir] F. Kirwan. *Complex Algebraic Curves*, Cambridge U. Press, Cambridge, 1992.

[Kob] N. Koblitz. *Algebraic aspects of cryptography*, with an appendix by A. J. Menezes, Y.-H. Wu and R. J. Zuccherato, Algorithms and Computation in Mathematics **3**, Springer-Verlag, Berlin, 1998.

[Kra] R. Krasauskas. *Toric surface patches*, Adv. Comput. Math. **17** (2002), 89–113.

[KR] M. Kreuzer and L. Robbiano. *Computational Commutative Algebra 1*, Springer-Verlag, Berlin, 2000.

[KM] S. Krishnan and D. Manocha. *Numeric-symbolic algorithms for evaluating one-dimensional algebraic sets*, in: *Proceedings of International Symposium on Symbolic and Algebraic Computation*, ACM Press, New York, 1995, 59–67.

[Kus] A.G. Kushnirenko. *Newton polytopes and the Bézout theorem*, Functional Anal. Appl. **10** (1976), 233–235.

[Laz] D. Lazard. *Résolution des systèmes d'équations algébriques*, Theor. Comp. Sci. **15** (1981), 77–110.

[Lei] K. Leichtweiss. *Konvexe Mengen*, Springer-Verlag, Berlin, 1980.

[Leo] D. Leonard. *A generalized Forney formula for AG codes*, IEEE Trans. Inform. Theory **46** (1996), 1263–1268.

[Lew] R. Lewis. *Computer algebra system* **Fermat**, available at www.bway.net/~lewis/.

[Li] T. Y. Li. *Solving polynomial systems by polyhedral homotopies*, Taiwanese J. Math. **3** (1999), 252–279.

[LL] T. Y. Li and X. Li. *Finding mixed cells in the mixed volume computation*, Found. Comput. Math. **1** (2001), 161–181.

[LW] T. Li and X. Wang. *The BKK root count in* $\mathbb{C}^n$, Math. Comp. **65** (1996), 1477–1484.

[Lit] J. Little. *A key equation and the computation of error values for codes from order domains*, preprint, 2003, arXiv math.AC/0303299.

[LS] A. Logar and B. Sturmfels. *Algorithms for the Quillen-Suslin Theorem*, J. Algebra **145** (1992), 231–239.

[Mac1] F. Macaulay. *The Algebraic Theory of Modular Systems*, Cambridge U. Press, Cambridge, 1916. Reprint with new introduction, Cambridge U. Press, Cambridge, 1994.

[Mac2] F. Macaulay. *On some formulas in elimination*, Proc. London Math. Soc. **3** (1902), 3–27.

[MS] F. MacWilliams and N. Sloane. *The Theory of Error-Correcting Codes*, North Holland, Amsterdam, 1977.

[Man1] D. Manocha. *Algorithms for computing selected solutions of polynomial equations*, in: *Proceedings of International Symposium on Symbolic and Algebraic Computation*, ACM Press, New York, 1994, 1–8.

[Man2] D. Manocha. *Efficient algorithms for multipolynomial resultant*, The Computer Journal **36** (1993), 485–496.

[Man3] D. Manocha. *Solving systems of polynomial equations*, IEEE Computer Graphics and Applications, March 1994, 46–55.

[MMM1] M. Marinari, H. M. Möller and T. Mora. *Gröbner bases of ideals defined by functionals with an application to ideals of projective points*, Appl. Algebra Engrg. Comm. Comput. **4** (1993), 103–145.

[MMM2] M. Marinari, H. M. Möller and T. Mora. *On multiplicities in polynomial system solving*, Trans. Amer. Math. Soc. **348** (1996), 3283–3321.

[Mey] F. Meyer. *Zur Theorie der reducibeln ganzen Functionen von n Variabeln*, Math. Annalen **30** (1887), 30–74.

[Mil] J. Milnor. *Singular Points of Complex Hypersurfaces*, Princeton U. Press, Princeton, 1968.

[Mis] B. Mishra. *Algorithmic Algebra*, Springer-Verlag, New York, 1993.

[Möl] H. Möller. *Systems of Algebraic Equations Solved by Means of Endomorphisms*, in: *Applied Algebra, Algebraic Algorithms and Error-correcting codes (AAECC-10)* (G. Cohen, T. Mora and O. Moreno, eds.), Lecture Notes in Computer Science **673**, Springer-Verlag, New York, 1993, 43–56.

[MöS] H. M. Möller and H. J. Stetter. *Multivariate polynomial systems with multiple zeros solved by matrix eigenproblems*, Numer. Math. **70** (1995), 311–329.

[MT] H. M. Möller and and R. Tenberg. *Multivariate polynomial system solving using intersections of eigenspaces*, J. Symbolic Comput. **30** (2001), 1–19.

[Mon] C. Monico. *Computing the primary decomposition of zero-dimensional ideals*, J. Symbolic Comput. **34** (2002), 451–459.

[MR] T. Mora and L. Robbiano. *The Gröbner fan of an ideal*, J. Symbolic Comput. **6** (1988), 183–208.

[MPT] T. Mora, G. Pfister and C. Traverso. *An introduction to the tangent cone algorithm*, Advances in Computing Research **6** (1992), 199–270.

[Mor] C. Moreno. *Algebraic Curves over Finite Fields*, Cambridge U. Press, Cambridge, 1991.

[Mou1] B. Mourrain. *Computing the isolated roots by matrix methods*, J. Symbolic Comput. **26** (1998), 715–738.

[Mou2] B. Mourrain. *An introduction to algebraic and geometric methods for solving polynomial equations*, Technical Report ECG-TR-122102-01, INRIA Sophia-Antipolis, 2002.

[O'Su1] M. O'Sullivan. *The key equation for one-point codes and efficient error evaluation*, J. Pure Appl. Algebra **169** (2002), 295–320.

[O'Su2] M. O'Sullivan. *A generalization of the Berlekamp-Massey-Sakata algorithm*, preprint, 2001.

[O'Su3] M. O'Sullivan. *New codes for the Berlekamp-Massey-Sakata algorithm*, Finite Fields Appl. **7** (2001), 293–317.

[PW] H. Park and C. Woodburn. *An algorithmic proof of Suslin's stability theorem for polynomial rings*, J. Algebra **178** (1995), 277–298.

[Ped] P. Pedersen. Lecture at AMS-IMS-SIAM Summer Conference on Continuous Algorithms and Complexity, Mount Holyoke College, South Hadley, MA, 1994.

[PS1] P. Pedersen and B. Sturmfels. *Mixed monomial bases*, in: *Algorithms in Algebraic Geometry and Applications* (L. Gonzalez-Vega and T. Recio, eds.), Birkhäuser, Boston, 1996, 307–316.

[PS2] P. Pedersen and B. Sturmfels. *Product formulas for resultants and Chow forms*, Math. Z., **214** (1993), 377–396.

[PRS] P. Pedersen, M.-F. Roy and A. Szpirglas. *Counting real zeros in the multivariate case*, in: *Computational Algebraic Geometry* (F. Eyssette and A. Galligo, eds.), Birkhäuser, Boston, 1993, 203–224.

[PR] H.-O. Peitgen and P. Richter. *The Beauty of Fractals*, Springer-Verlag, Berlin, 1986.

[PH] A. Poli and F. Huguet. *Error correcting codes*. Prentice Hall International, Hemel Hempstead, 1992.

[Qui] D. Quillen. *Projective modules over polynomial rings*, Invent. Math. **36** (1976), 167–171.

[Rob] L. Robbiano. *Term orderings on the polynomial ring*, in: *Proceedings of EUROCAL 1985*, Lecture Notes in Computer Science **204**, Springer-Verlag New York, 513–517.

[Roj1] J. M. Rojas. *A convex geometric approach to counting roots of a polynomial system*, Theoretical Computer Science **133** (1994), 105–140.

[Roj2] J. M. Rojas. *Solving degenerate sparse polynomial systems faster*, J. Symbolic Comput. **28** (1999), 155–186.

[Roj3] J. M. Rojas. *Toric intersection theory for affine root counting*, J. Pure Appl. Algebra **136** (1999), 67–100.

[Roj4] J. M. Rojas. *Toric laminations, sparse generalized characteristic polynomials, and a refinement of Hilbert's tenth problem*, in: *Foundations of Computational Mathematics, Rio de Janeiro, 1997* (F. Cucker and M. Shub, eds.), Springer-Verlag, Berlin, 1997, 369–381.

[Roj5] J. M. Rojas. *Why polyhedra matter in non-linear equation solving*, in: *Algebraic Geometry and Geometric Modeling* (R. Goldman and R. Krasauskas, eds.), Contemporary Mathematics **334**, AMS, Providence, RI, 2003, 293–320.

[RW] J. M. Rojas and X. Wang. *Counting affine roots of polynomial systems via pointed Newton polytopes*, Journal of Complexity **12** (1996), 116–133.

[Ros] L. Rose. *Combinatorial and Topological Invariants of Modules of Piecewise Polynomials*, Adv. Math. **116** (1995), 34–45.

[Rou] F. Rouillier. *Solving zero-dimensional systems through the rational univariate representation*, Appl. Algebra Engrg. Comm. Comput. **9** (1999), 433–461.

[Sak] S. Sakata. *Extension of the Berlekamp-Massey algorithm to n dimensions*, Inform. Comput. **84** (1989), 207–239.

[Sal] G. Salmon. *Lessons Introductory to the Modern Higher Algebra*, Hodges, Foster and Co., Dublin, 1876.

[Sche] H. Schenck. *A spectral sequence for splines*, Adv. in Appl. Math. **19** (1997), 183–199.

[SS] H. Schenck and M. Stillman. *Local cohomology of bivariate splines*, J. Pure Appl. Algebra **117/118** (1997), 535–548.

[Schre1] F.-O. Schreyer. *Die Berechnung von Syzygien mit dem verallgemeinerten Weierstraßschen Divisionssatz*, Diplom Thesis, University of Hamburg, Germany, 1980.

[Schre2] F.-O. Schreyer. *A Standard Basis Approach to syzygies of canonical curves*, J. reine angew. Math. **421** (1991), 83–123.

[Schri] A. Schrijver. *Theory of Linear and Integer Programming*, Wiley-Interscience, Chichester, 1986.

[Schu] L. Schumaker. *On the dimension of spaces of piecewise polynomials in two variables*, in: *Multivariate Approximation Theory (Proc. Conf., Math. Res. Inst., Oberwolfach, 1979)*, Internat. Ser. Numer. Math. **51**, Birkhäuser, Basel-Boston, 1979, 396–412.

[SC] T. Sederberg and F. Chen. *Implicitization using moving curves and surfaces*, in: *Computer Graphics Proceedings, Annual Conference Series*, 1995, 301–308.

[SSQK] T. Sederberg, T. Saito, D. Qi and K. Klimaszewski. *Curve implicitization using moving lines*, Computer Aided Geometric Design **11** (1994), 687–706.

[Ser] J.-P. Serre. *Faisceaux algébriques cohérents*, Ann. of Math. **61** (1955), 191–278.

[Sha] I. Shafarevich. *Basic Algebraic Geometry*, Springer-Verlag, Berlin, 1974.

[SVW] A. Sommese, J. Verschelde and C. Wampler. *Numerical decomposition of the solution sets of polynomial systems into irreducible components*, SIAM J. Numer. Anal. **36** (2001), 2022–2046.

[Sta1] R. Stanley. *Combinatorics and Commutative Algebra*, 2nd edition, Birkhäuser, Boston, 1996.

[Sta2] R. Stanley. *Invariants of finite groups and their applications to combinatorics*, Bull. Amer. Math. Soc. **1** (1979), 475–511.

[Ste] H. J. Stetter. *Multivariate polynomial equations as matrix eigenproblems*, in: *Contributions in Numerical Analysis* (R. P. Agarwal, ed.), Series in Applicable Analysis **2**, World Scientific, Singapore, 1993, 355–371.

[Sti] H. Stichtenoth. *Algebraic Function Fields and Codes*, Springer-Verlag, Berlin, 1993.

[Stu1] B. Sturmfels. *Algorithms in Invariant Theory*, Springer-Verlag, Vienna, 1993.

[Stu2] B. Sturmfels. *Gröbner Bases and Convex Polytopes*, AMS, Providence RI, 1996.

[Stu3] B. Sturmfels. *On the Newton polytope of the resultant*, J. Algebraic Comb. **3** (1994), 207–236.

[Stu4] B. Sturmfels. *Sparse elimination theory*, in: *Computational Algebraic Geometry and Commutative Algebra* (D. Eisenbud and L. Robbiano, eds.), Cambridge U. Press, Cambridge, 1993, 264–298.

[Stu5] B. Sturmfels. *Solving Systems of Polynomial Equations*, AMS, Providence, RI, 2002.

[SZ]   B. Sturmfels and A. Zelevinsky. *Multigraded resultants of Sylvester type*, J. Algebra **163** (1994), 115–127.

[Sus]   A. Suslin. *Projective modules over a polynomial ring are free*, Soviet Math. Dokl. **17** (1976), 1160–1164.

[Tei]   M. Teixidor i Bigas. *Green's conjecture for the generic r-gonal curve of genus $g \geq 3r - 7$*, Duke Math. J. **111** (2002), 195–222.

[Tho1]   R. Thomas. *A geometric Buchberger algorithm for integer programming*, Math. Operations Research **20** (1995), 864–884.

[Tho2]   R. Thomas. *Applications to integer programming*, in: *Applications of Computational Algebraic Geometry* (D. Cox and B. Sturmfels, eds.), Proc. Sympos. Appl. Math. **53**, AMS, Providence, RI, 1998, 119–142.

[Tran]   Q. N. Tran. *A fast algorithm for Gröbner basis conversion and its applications*, J. Symbolic Comput. **30** (2000), 451–467.

[Trav]   C. Traverso. *Hilbert functions and the Buchberger algorithm*, J. Symbolic Comput. **22** (1996), 355–376.

[TVZ]   M. Tsfasman, S. Vladut and T. Zink. *Modular Curves, Shimura Curves, and Goppa Codes Better than the Varshamov-Gilbert Bound*, Math. Nachr. **109** (1982), 21–28.

[vdW]   B. van der Waerden. *Moderne Algebra, Volume II*, Springer-Verlag, Berlin, 1931. English translations: *Modern Algebra, Volume II*, F. Ungar Publishing Co., New York, 1950; *Algebra, Volume 2*, F. Ungar Publishing Co., New York 1970; and *Algebra, Volume II*, Springer-Verlag, New York, 1991. The chapter on Elimination Theory is included in the first three German editions and the 1950 English translation, but all later editions (German and English) omit this chapter.

[vLi]   J. van Lint. *Introduction to Coding Theory*, 2nd edition, Springer-Verlag, Berlin, 1992.

[vLvG]   J. van Lint and G. van der Geer. *Introduction to Coding Theory and Algebraic Geometry*, Birkhäuser, Basel, 1988.

[Vas]   W. Vasconcelos. *Computational Methods in Commutative Algebra and Algebraic Geometry*, Springer-Verlag, Berlin, 1998.

[Ver1]   J. Verschelde. *Algorithm 795: PHCpack: A general-purpose solver for polynomial systems by homotopy continuation*, ACM Trans. Math. Softw. **25** (1999), 251–276.

[Ver2]   J. Verschelde. *Toric Newton method for polynomial homotopies*, J. Symbolic Comput. **29** (2000), 777–793.

[VGC]   J. Verschelde, K. Gatermann and R. Cools. *Mixed volume computation by dynamic lifting applied to polynomial system solving*, Discrete Comput. Geom. **16** (1996), 69–112.

[VVC]   J. Verschelde, P. Verlinden and R. Cools. *Homotopies Exploiting Newton Polytopes for Solving Sparse Polynomial Systems*, SIAM J. Numer. Anal. **31** (1994), 915–930.

[Voi]   C. Voisin. *Green's generic syzygy conjecture for curves of even genus lying on a K3 surface*, J. Eur. Math. Soc. **4** (2002), 363–404.

[WEM]   A. Wallack, I. Emiris and D. Manocha. *MARS: A MAPLE/MATLAB/C resultant-based solver*, in: *Proceedings of International*

Symposium on Symbolic and Algebraic Computation, ACM Press, New York, 1998, 244–251.

[Wei] V. Weispfenning. *Constructing universal Gröbner bases*, in: *Applied Algebra, Algebraic Algorithms and Error-correcting Codes (AAECC-5)* (L. Huguet and A. Poli, eds.), Lecture Notes in Computer Science **356**, Springer-Verlag, Berlin, 1989, 408–417.

[WZ] J. Weyman and A. Zelevinski, *Determinantal formulas for multigraded resultants*, J. Algebraic Geom. **3** (1994), 569–597.

[Wil] J. Wilkinson. *The evaluation of the zeroes of ill-conditioned polynomials, part 1*, Numerische Mathematik **1** (1959), 150–166.

[YNT] K. Yokoyama, M. Noro and T. Takeshima. *Solutions of systems of algebraic equations and linear maps on residue class rings*, J. Symbolic Comput. **14** (1992), 399–417.

[ZG] M. Zhang and R. Goldman. *Rectangular corner cutting and Sylvester A-resultants*, in: *Proceedings of International Symposium on Symbolic and Algebraic Computation*, ACM Press, New York, 2000, 301–308.

[Zie] G. Ziegler, *Lectures on Polytopes*, Springer-Verlag, New York, 1995.

[Zub] S. Zubè. *The n-sided toric patches and A-resultants*, Computer Aided Geometric Design **17** (2000), 695–714.

Index

Adapted order, *see* monomial order,
adapted

$\mathcal{A}$-degree, 320

adjacent cells or polytopes in a
polyhedral complex, 408,
411–413, 417, 422

$\mathcal{A}$-homogeneous, 330

$\mathcal{A}$-resultant, *see* resultant, sparse

Abel, N., 29

Acton, F., 30, 35, 533

Adams, W., vii, 4, 11, 14, 16, 18, 24,
27, 39, 185, 211, 213, 376, 533

affine
half-space, 408
Hilbert Function, *see* Hilbert
function, affine
hyperplane, 308, 312, 379, 411,
413, 418
linear space, *see* affine, subspace
n-dimensional space over $\mathbb{C}$ ($\mathbb{C}^n$),
90, 97, 105, 114, 125, 316,
327, 343, 344, 346, 352, 353,
357
n-dimensional space over $\mathbb{F}_q$ ($\mathbb{F}_q^n$),
459ff, 468ff, 505, 510, 511
n-dimensional space over k (k^n),
19, 293, 294
n-dimensional space over $\mathbb{R}$ ($\mathbb{R}^n$),
69ff, 305ff, 392ff, 426ff, 436ff
subspace, 312, 340, 388
variety, 20, 24, 45, 495, 501

Agacy, R., xi

alex, 159, 161, 162, 171, 175, 177,
183, 187

Alexandrov-Fenchel inequality, 362

Alfeld, P., 420, 533

algebra over a field k, 38, 71, 495
finite-dimensional algebra, 37ff,
56ff, 96, 478

algebraic
coding theory, viii, ix, 451, 459ff,
468ff, 480ff, 494ff, 508ff, 522ff
curve, 494, 498, 501, 506
decomposition, *see* subdivision,
algebraic (non-polyhedral)

algebraically closed field, 23–25, 27,
42, 96, 148, 150–152, 157,
182, 262

Allgower, E., 353, 533

Alonso, M., 179, 184, 186, 533

alphabet (of a code), 459, 471, 479

Amrhein, B., 445, 446, 533

analytic function, *see* convergent
power series ring

annihilator of N ($\mathrm{ann}(N)$), 203

anti-graded
lex order, *see* alex
revlex order, *see* arevlex
order, *see* degree-anticompatible
order

Archimedean order domain, *see*
order domain, Archimedean

Archimedean property (in $\mathbb{Z}_{\geq 0}$), 502

arevlex, 159, 160, 175

547

(continued from page ii)